Applied Spatial Statistics and Econometrics

This textbook is a comprehensive introduction to applied spatial data analysis using R. Each chapter walks the reader through a different method, explaining how to interpret the results and what conclusions can be drawn. The author team showcases key topics, including unsupervised learning, causal inference, spatial weight matrices, spatial econometrics, heterogeneity and bootstrapping. It is accompanied by a suite of data and R code on Github to help readers practise techniques via replication and exercises.

This text will be a valuable resource for advanced students of econometrics, spatial planning and regional science. It will also be suitable for researchers and data scientists working with spatial data.

Katarzyna Kopczewska is an associate professor at University of Warsaw, Faculty of Economic Sciences. As a quantitative economist, she deals with spatial modelling of geolocalised economic processes – location and co-location, agglomeration, concentration, diffusion, spatial interactions in relation to economic phenomena, companies and real estate but also regional policy or public-sector activities. She conducts methodological research on the implementation of data science methods for spatial analysis and combining them with classical spatial statistics and econometrics in R. She combines quantitative solutions with theory and problems of regional science and economic geography. She serves at the European Regional Science Association (ERSA).

Routledge Advanced Texts in Economics and Finance

For more information about this series, please visit www.routledge.com/Routledge-Advanced-Texts-in-Economics-and-Finance/book-series/SE0757

Applied Spatial Statistics and Econometrics

Data Analysis in R

Katarzyna Kopczewska

Routledge
Taylor & Francis Group

LONDON AND NEW YORK

First published 2021
by Routledge
2 Park Square, Milton Park, Abingdon, Oxon OX14 4RN

and by Routledge
52 Vanderbilt Avenue, New York, NY 10017

Routledge is an imprint of the Taylor & Francis Group, an informa business

British Library Cataloguing-in-Publication Data
A catalogue record for this book is available from the British Library

Library of Congress Cataloging-in-Publication Data
A catalog record for this book has been requested

ISBN: 978-0-367-47077-7 (hbk)
ISBN: 978-0-367-47076-0 (pbk)
ISBN: 978-1-003-03321-9 (ebk)

Typeset in Sabon
by Apex CoVantage, LLC

Contents

Figures

Chapter 3

Chapter 4

Chapter 5

Chapter 6

Chapter 7

Chapter 8

Chapter 9

Appendix B

Tables

Contributors

Piotr Ćwiakowski is a researcher at University of Warsaw, Faculty of Economic Sciences. His research concentrates around modelling of housing valuation, using spatial and a-spatial econometric and machine learning methods. Since 2016, he has been cofounder of a company called LabMasters, where he focuses on training and consulting in data science for private and public entities using R and Python.

Alessandro Festi is a data scientist working on applied solutions for business in spatial data analysis. He graduated with honours in statistics from the University of Bologna in 2019. He has studied at Sorbonne University, Bocconi University and University of Warsaw. Since 2016, he has been developing machine learning models for clients mainly in the teleco, energy and institutional sectors. He currently works as a data scientist at Energy Way in Modena (IT).

Mateusz Kopyt is an assistant professor at University of Warsaw, Faculty of Economic Sciences. He is involved in web and API solutions for economics.

Maria Kubara is a researcher at University of Warsaw, Faculty of Economic Sciences. She was awarded the scholarship for the best young researchers given by the Ministry of Science and Higher Education, Poland. Her interests are around spatial econometrics and statistics, combined with data science problems and solutions.

Piotr Wójcik is an associate professor at the University of Warsaw, Faculty of Economic Sciences. His research interests are focused on two areas. The first is regional and local development and in particular the measurement of inequalities and real economic convergence on a regional and local level. The second area is quantitative finance, in particular construction and testing of algorithmic trading strategies. Both areas of interest involve the use of advanced quantitative tools, mainly in R. In regional economics, he conducts research using spatial econometric models and machine learning algorithms, including image recognition tools.

Kateryna Zabarina is a researcher at University of Warsaw, Faculty of Economic Sciences. She works on modelling of business location models and spatial point patterns. In 2018, she obtained the Jean Paelinck prize for young researchers during the World Conference of the Spatial Econometrics Association.

Introduction

Space as the fundamental feature of matter, in addition to time, is one of the basic variables in scientific research. Each physical phenomenon happens in a specific space. Mathematics is a very spatial field, and mathematicians even have the choice of navigating in differently defined spaces: Banach, Euclidean, functional and so on. In biology, determining the position of occurrence of a given species is as important as its description. In psychology, one distinguishes personal space and sociology analyses social space, while political science explores how territorial division influences the results of elections. Economics, usually pushing spatial dependence studies into economic geography, was for a long against the inclusion of space in research. This was partly due to the beliefs of economists in the universality of economic laws that work equally strongly in the tropics and the Arctic Circle, as well as the lack of methods for analysing these phenomena.

Despite the existence of economic models that take into account space as early as the 19th century, like the widely known model of Von Thunen (1826), the silent revolution in economics did not start until the 1980s. The economic geography law proposed by Tobler (1970): "Everything is related to everything else, but near things are more related than distant things", is a trivial statement in many sciences, and in economics, it destroyed the established perception of the economy as an independent entity that does not interact with its neighbours. Recognition of neighbourhood and distance as factors influencing economic development was slow but only a matter of time. Location models within a framework of classic and behavioural theory, new economic geography or (co)-evolutionary economics gave the theoretical basis for new research, taking into account spatial factors. Spatial development models combined the theory of location and trade theory. As part of these models, changes in the spatial distribution of economic activity caused by the relative centrifugal and centripetal forces are analysed, as well as the concentration of activity in agglomerations, economies of scale and costs of transport, models of location of firms and the potential of regions.

In the last 40 years, the progress in supplementing economic theories with spatial aspects has been really impressive. The need to verify theoretical research has resulted in the development of quantitative spatial methods – spatial statistics and econometrics and, more recently, spatial machine learning. Although the beginnings of spatial dependence analysis can already be found in the works of Moran, Geary and Whittle from the 1940s and early 1950s, this field of econometrics developed rather slowly until the 1970s and the publications of Cliff and Ord (1975a, 1975b, 1975c), Paelinck and Klaassen (1979) and later Cressie (1993b), dealing with the issues of spatial autocorrelation both in terms of statistics and econometric modelling. The 1990s brought the development of applications of these methods and the creation of software, mainly thanks to the work of Luc Anselin. From the beginning of the 21st century, intensive development of the R program and its environment for spatial analysis, created by Roger Bivand, has been observed.

Quantitative methods: statistics, econometrics, machine learning or data analysis have been present in the literature for many years – statistics have been around much longer, since about 1820, and the others for a shorter time: bootstrap since about 1920, econometrics since about 1940 and machine learning from around 1960 (based on statistics from https://books.google.com/ngrams). The popularity of these methods is also different – statistics as a term in books is still about 10 times more popular than data analysis and about 20 times more popular than machine learning or econometrics. In the last 30 years, there has been a decline in the popularity of the econometrics term in favour of the ever-growing interest in machine learning – however, bootstrap is still a more popular term than machine learning or econometrics.

Spatial methods are only part of the world of quantitative methods; hence, their popularity is several times lower. Spatial statistics are still mentioned several times more often than spatial econometrics, but interestingly, both terms are more popular than data science. There is also an increase in interest in spatial econometrics. Observing a steady increase in interest in *Geographic Information System* (GIS) and statistics (statistics are still around four times more popular than GIS), one should forecast an increase in interest in quantitative spatial methods.

This book addresses these trends. Popularisation of quantitative methods in an accessible way for non-mathematicians allows for interdisciplinary implementations and thus the development of science and knowledge. This book is written from the perspective of quantitative economists studying economic phenomena in spatial terms on regional or geolocalised data.

The assumption of this book is to present the topics in an applied manner. Based on the available packages and algorithms in R, the purpose and method of conducting analysis in general as well as on sample data are presented. The obtained results are interpreted and commented. For each example, the R code is presented as completely as possible, from data processing through the proper calculation to result presentation or visualisation. Thanks to this, the book is a guide to R and quantitative methods. The authors' intention is to present methods of data analysis in such a way that the reader can repeat them on their data.

All the sample data used in the book and the codes presented in the book were placed on https://github.com/kkopczewska/spatial_book, which allows for easy replication of the book content. The examples use data for Poland – regional data together with contour maps and point data together with geographical coordinates. This is the basic type of data – most readers probably have similar data at their disposal in their area of interest. We hope that such a guide, and what you can do with this data, will become an inspiration for your own spatial analysis.

The authors

Katarzyna Kopczewska with the team:
Maria Kubara, Piotr Ćwiakowski, Mateusz Kopyt,
Piotr Wójcik, Alessandro Festi, Kateryna Zabarina

Statement by the American Statistical Association on statistical significance and *p*-value – use in the book

This book supports a modern approach to quantitative methods. The interpretation of statistical results is one of the elements. The authors of the book try to include the *p*-value interpretation rules announced by the American Statistical Association (ASA) on statistical significance and *p*-value (2016) available in Wasserstein, R. L., & Lazar, N. A. (2016). The ASA's statement on *p*-values: context, process, and purpose. *The American Statistician*, 70(2), 129–133[1] and as a press release.[2]

This statement contains six rules that relate to errors in the application and interpretation of *p*-value. They are as follows:

1. *P*-values can indicate how incompatible the data are with a specified statistical model.
2. *P*-values do not measure the probability that the studied hypothesis is true, or the probability that the data were produced by random chance alone.
3. Scientific conclusions and business or policy decisions should not be based only on whether a *p*-value passes a specific threshold.
4. Proper inference requires full reporting and transparency.
5. A *p*-value, or statistical significance, does not measure the size of an effect or the importance of a result.
6. By itself, a *p*-value does not provide a good measure of evidence regarding a model or hypothesis.

(Wasserstein & Lazar, 2016)

According to the ASA statement, a *p*-value is the probability under a specified statistical model that a statistical summary of the data (e.g., the sample mean difference between two compared groups) would be equal to or more extreme than its observed value.

The authors of the book try to implement the ASA rule – good statistical practice, as an essential component of good scientific practice, emphasises principles of proper study design and conduct, a variety of numerical and graphical summaries of data, understanding of the phenomenon under study, interpretation of results in context, complete reporting and proper logical and quantitative understanding of what data summaries mean. No single index should substitute for scientific reasoning.

The reader interested in modern principles of statistical inference should read the special issue of the scientific journal *The American Statistician* (vol. 73, 2019, Supplement 1[3] pt. *Statistical Inference in the 21st Century: A World Beyond p < 0.05*), which offers more than 40 scientific papers on this topic. It is also worth recommending the articles in *Nature*: *It's time to talk about ditching statistical significance* (editorial, 2019),[4] *Scientists rise up against statistical significance* (Amrhein, Greenland, & McShane, 2019),[5] *'One-size-fits-all' threshold for P values under fire* (Chawla, 2017)[6] or *Stats: P values akin to 'beyond reasonable doubt'* (Zhang, 2019).[7]

Notes

1 https://amstat.tandfonline.com/doi/full/10.1080/00031305.2016.1154108
2 https://www.amstat.org/asa/files/pdfs/P-ValueStatement.pdf
3 https://amstat.tandfonline.com/toc/utas20/73/sup1?nav=tocList
4 https://www.nature.com/articles/d41586-019-00874-8
5 https://www.nature.com/articles/d41586-019-00857-9
6 https://www.nature.com/news/one-size-fits-all-threshold-for-p-values-under-fire-1.22625
7 https://www.nature.com/articles/d41586-019-01530-x

Acknowledgements

This book is a part of a project supported by the Polish National Science Center (NCN) on "Spatial econometric models with fixed and changing neighborhood structure. Application for real estate valuation and business location" (OPUS 12, contract no. UMO-2016/23/B/ HS4/02363).

The book was edited in the Polish-language version as: *Przestrzenne metody ilościowe w R: statystyka, ekonometria, uczenie maszynowe, analiza danych* (CeDeWu, Warszawa, 2020)

All sample data used in the book and the codes presented in the content of the book were placed on https:// github.com/kkopczewska/spatial_book

## Chapter 1

Basic operations in the R software

Mateusz Kopyt

1.1 About the R software

[1] It is difficult to define what R is clearly. On the one hand, it is software for quantitative analyses in various fields. On the other hand, R is also a programming language that is oriented towards data analysis. What's more, it is also a development (variant) of the S language developed in Bell Laboratories, which is also intended for quantitative analyses. The most appropriate description seems to be the statement that R is an environment (with its own programming language) which is focused on the processing of data and their (statistical) analysis, along with wide visualisation (graphic) possibilities. This program belongs to the so-called free and open software and is distributed and developed under the GNU GPL license. Therefore, the development of R is characterised by both all the advantages and drawbacks associated with such a licensing model and development method. R is available for all major platforms: Windows, MacOS and the UNIX family (i.e. Linux, FreeBSD and others). An important feature of the R environment is its flexibility and the possibility of constantly expanding applications. This is best seen by the number of additional packages available in the official repository. At the moment (June 2020), there are nearly 16,000 packages. For comparison, at the end of 2014, there were around 6000 official packages available. R also integrates well with other software languages such as C, C++ or Fortran.[2] Unlike many other programs distributed under a free license, R has very well ordered and rich documentation and a wide literature on its applications. Due to the fact that the first stable version of R appeared in 2000, it can be safely stated that the software is currently at a mature stage; more and more often it replaces other more specialised statistical packages in scientific research or commercial applications. Specialised packages also enable direct preparation of reports or scientific publications in the R environment by integrating the text with the results of calculations. These packages allow one to use Markdown and LaTeX language variations to format such texts.

1.2 The R software interface

The basic version of R offers (even in MS Windows) a rather limited user interface. Basically, in various versions of the Linux system, there is almost no default interface, and operations in the program are carried out from the command line or by running the appropriate text files (scripts). This is because R is in fact a computational engine, while all forms of the user interface are an addition to facilitate communication with the program. This approach provides great flexibility in the preparation of calculations but requires more knowledge or programming skills. Figure 1.1 shows the basic R interface.

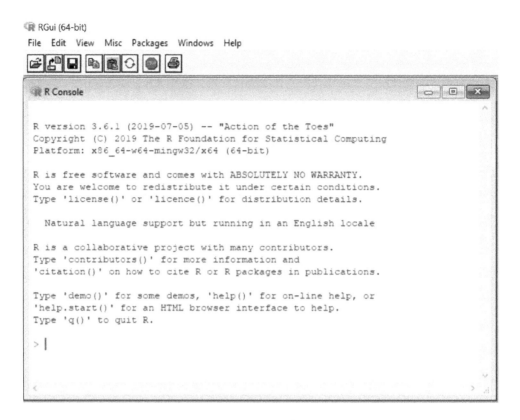

Figure 1.1 Basic interface of the R program (in MS Windows)

Source: Own study

The existence of various interfaces (other than the basic one in the MS Windows system) allows one to choose the form that will be the most convenient for the user. The features of several of the most popular interfaces (apart from the default interface for MS Windows) are briefly presented subsequently. The interface selection belongs to the user. It is worth noting, however, that the chosen interface gives complete freedom in editing the prepared R code and is not limited to providing basic operations on the principle of selecting from the menu.

1.2.1 R Commander

R Commander is one of the longest developed and most stable user interfaces for the R software. It is installed as a classic additional package and is available in standard R repositories. This interface does not require installation of additional external software. As the creators of this overlay indicate, it provides a "clickable" interface and allows for focusing more on calculations and searching for appropriate computational methods (available from the R Commander level) than on the preparation of the script that performs the R language. Even if one "clicks" instead of writing commands, the codes are still visible to the user. It should be emphasised that the use of R Commander – as the authors write themselves – allows analysis only in selected aspects (programmed in this overlay), providing access to only a small portion of the possibilities of the R package.[3] The webpage of this interface is: http://www.rcommander.com, with the latest information and more details available on the website of the author of this project (John Fox): http://socserv.mcmaster.ca/jfox/Misc/Rcmdr/. On this website, one can also find detailed instructions (and other supporting materials, including sample screenshots) for R Commander package. R Commander is an overlay that is quite limiting in the use of the program – at least when it comes to choosing functions from the graphical menu. It should serve, rather, to make the first sketch of the script, which must be supplemented later in the case of more sophisticated calculations.

1.2.2. RStudio

RStudio (Open Source Edition) is an environment that is also available under the free license, as is R itself. It is currently the most advanced and strongly developed development environment for R, so one should consider using it. It is not a typical "clickable" interface and does not relieve the user but provides much better working comfort by facilitating manual programming of scripts in R while giving easier access to data description, the help system, a graphics window, a console with results and so on. RStudio, apart from the standard desktop version, has a server version enabling common work and sharing tasks. The website of the described environment (http://www.rstudio.com/products/rstudio/) contains more information about this software, help and the program itself to download. The creators of RStudio provided – apart from written materials – short videos showing the possibilities of this environment. Before starting to work in this environment, it is worth getting acquainted with the previously mentioned website and the information contained therein. Figure 1.2 presents RStudio when working. More screenshots are available on the program's website.

A few useful remarks about RStudio:

1   The RStudio environment must be additionally installed – installation packages for various systems are available for download from the program's website. One must first install R – otherwise, RStudio will not connect to it and will be useless. In the case of individual use, one should choose the Desktop version; for collaboration between many users, use the server version of RStudio.
2   For Linux-based systems, one needs to download the appropriate package – they are not always in the basic repositories of various distributions (e.g. in Ubuntu, the deb package from the program's website installs without problems, and then one can update it after downloading the new version).
3   By starting RStudio, the R engine starts right away; it is not necessary to do it separately.
4   RStudio has a built-in script editor with syntax highlighting so one can easily build, load and save the work in a quite convenient way.
5   Other useful functions of RStudio are: it allows one to work directly in the console, shows a list and allows one to view already loaded or defined objects/data, facilitates access to the history of executed commands (although in pure R, this is also available) and their selective execution, has a dedicated part of the screen for graphs/graphics, facilitates the management of additional packages and integrates help to R. On the website http://www.rstudio.com/products/rstudio/features/ an interesting wider description of the RStudio function has been provided.

Figure 1.2 RStudio interface

Source: Own study

In addition to these two overlays (three – including the default interface in MS Windows), there are many interfaces that have been created to facilitate work in R, as well as for selected functionalities or applications. Other examples of environments are PMG, JGR or Rattle or even text editors, which are adapted to work with R, such as Tinn-R or Notepad++. Many of the projects that existed before are no longer being developed. Undoubtedly, RStudio is currently the most popular and fastest-growing environment. Attention is also drawn to the well-established position of R Commander, although it seems to be more and more inferior to RStudio. However, it should be noted that the scope of applications of both environments is slightly different and should rather be treated as complementary solutions.

1.3 Using help

The R program has a well-developed help system. All packages, including basic ones loaded by default at program startup, have standardised documentation describing the available functions – description of use, syntax, arguments and examples. Help can be accessed in many ways. Documentation for entire packages in the form of PDF files can be found on the software website (https://cloud.r-project.org/) in the Packages menu (by searching for the appropriate package). It is also worth mentioning here the so-called *Task Views* (appropriate menu on the program page). These are grouped descriptions of possible applications of the R package in selected fields. In addition to information about the capabilities of the software in a given area, *Task Views* indicate the packages that need to be installed to add individual functionalities. Of course, this is also the way to access full package documentation.

Using the help built into R also depends on whether the user uses a "clean" environment or uses selected overlays, that is, a graphical user interface (GUI). Following are some examples of useful commands that allow one to get information about selected R functions regardless of whether one uses the GUI (most examples do not contain results due to their size). One can start searching for a description for a selected program command by entering a command using the question mark or the **help**() function, for example:

```
?cor
help("cor")
```

The basic method of obtaining help indicated previously requires the additional packages to be loaded in advance, as long as the searched commands are in these packages. When you do not know the name of the command you are looking for, you can use a function that searches the help and tries to match the results to the specified text:

```
?? correlation
help.search("correlation")
```

Both previous examples show results not in the program console but in the help window. A similar function, although slightly different, is the **apropos**() command. In this case, R searches only the functions' names and gives them matches with the search text.

```
apropos("cor")
```

```
[1] "cancor"        "cor"           "cor.test"       "cov2cor"
[5] "Harman23.cor" "Harman74.cor" "recordGraphics" "recordPlot"
```

The introduction to the documentation for the selected package, along with the list of functions provided in it, can be obtained in the following way:

```
library(help="stats")
```

Examples of using a given function can be obtained using the **example()** command along with the name of the corresponding function. The way of calling this command and the beginning of the result of calling the example for the function **cor()** available in R are as follows:

```
example("cor")
cor> var(1:10)  # 9.166667
#[1] 9.166667

cor> var(1:5, 1:5)  # 2.5
#[1] 2.5

cor> ## Two simple vectors
cor> cor(1:10, 2:11)  # == 1
#[1] 1

cor> ## Correlation Matrix of Multivariate sample:
cor> (Cl <- cor(longley))
```

```
              GNP.deflator GNP        Unemployed Armed.Forces Population Year      Employed
GNP.deflator  1.0000000    0.9915892  0.6206334  0.4647442    0.9791634  0.9911492 0.9708985
GNP           0.9915892    1.0000000  0.6042609  0.4464368    0.9910901  0.9952735 0.9835516
Unemployed    0.6206334    0.6042609  1.0000000 -0.1774206    0.6865515  0.6682566 0.5024981
Armed.Forces  0.4647442    0.4464368 -0.1774206  1.0000000    0.3644163  0.4172451 0.4573074
Population    0.9791634    0.9910901  0.6865515  0.3644163    1.0000000  0.9939528 0.9603906
Year          0.9911492    0.9952735  0.6682566  0.4172451    0.9939528  1.0000000 0.9713295
Employed      0.9708985    0.9835516  0.5024981  0.4573074    0.9603906  0.9713295 1.0000000`
......
```

As R is available on a free license basis, as indicated earlier, the user has access to the source code of the program, including the source code of the individual functions. It allows anyone to check the method of performing calculations in R and in addition, one can modify such code and expand or adapt it to specific needs. It is possible to show the source code of the function after executing the command name without using any arguments including brackets, for example:

```
cor  # only the beginning and end of the command code will be shown
```

```
1 function (x, y = NULL, use = "everything", method = c("pearson",
2     "kendall", "spearman"))
3 {
4     na.method <- pmatch(use, c("all.obs", "complete.obs",
"pairwise.complete.obs",
5         "everything", "na.or.complete"))
6     if (is.na(na.method))
. . .
108             if (matrix_result)
109                 r
110             else drop(r)
111         }
112     }
113 }
```

In the R software, there is also a rather general function by means of which one can call the "page," which is a collection of supporting materials for the program. There are references to some tutorials. Access to this is obtained using the function:

```
help.start()
```

Finally, we should also mention a set of quite extensive examples present in many packages, both basic and additional ones. Access to the list is obtained by calling the **demo**() command without arguments, while the indication of a specific name will run the desired script in the R language, showing examples of applications of the given function. They often go beyond those present in the documentation for individual packages. The **demo**() command for the **colors**() function is presented subsequently, which gives the names of all defined colours in R. The result of the **demo**(), of course, contains both the code executed and the graphic presentation of colours in R. Because the result is very long, the following example is limited only to the beginning of the results.

```
demo(colors)

Type <Return> to start :

> ### ---------- Show (almost) all named colors --------------------
>
> ## 1) with traditional 'graphics' package:
> showCols1 <- function(bg = "gray", cex = 0.75, srt = 30) {
+     m <- ceiling(sqrt(n <- length(cl <- colors())))
+     length(cl) <- m*m; cm <- matrix(cl, m)
+     ##
+     require("graphics")
+     op <- par(mar=rep(0,4), ann=FALSE, bg = bg); on.exit(par(op))
+     plot(1:m,1:m, type="n", axes=FALSE)
+     text(col(cm), rev(row(cm)), cm, col = cl, cex=cex, srt=srt)
+ }
> showCols1()
Press <Enter> to view the next chart:
```

Some inconvenience for non English-speaking users may be the fact that a large part of the documentation is only available in English. It should also be noted that access to help on functions loaded by additional modules is generally only possible when such a package is loaded. Therefore, it requires the user to search and load the required packages in advance.

If one uses the RStudio graphic environment, start the search with the R-Help item in the Help menu. A "portal" will be opened in the appropriate RStudio panel with the option of switching to online help regarding both R and RStudio.

If you use additional packages (discussed in Section 1.4), invaluable help may be the so-called Vignettes available in the documentation of such packages. They are some kind of examples, tutorials or even fragments of scientific publications showing the application of the functions of a given package to solve examples of problems. Examples of Vignettes can be found at https://cran.r-project.org/web/packages/spdep/ (item Vignettes) in the description of the spdep:: package frequently used in the following chapters of this book.

We should also mention the possibility of finding many educational materials on the website of the R package itself: https://www.r-project.org/. It is worth paying attention to the Documentation menu. One can find there, in addition to quite extensive textbooks, a section on the answers to the most important questions (FAQ) and a link to the *R Journal* website, where articles on the development and the latest uses of the R package are published.

Of course, the resources of the Internet can be an invaluable help. Two webpages specifically specialised in searching for information about R are: http://www.rseek.org/ and http://search.r-project.org/. Also in the Google search engine itself, one can limit the search area to issues related to R.

1.4 Additional packages

After starting the R program, whether with or without a graphic overlay, some basic packages containing the most useful functions and commands of the program are automatically loaded. The following packages are automatically loaded: base::, methods::, datasets::, utils::, graphics::, grDevices:: and stats::. In many cases, these packages are sufficient for basic applications. However, many specific problems/methods/models require additional functions included in the additional packages. They are necessary to perform more advanced calculations, including, for example, operations on spatial data. The number of packages is growing exponentially (which shows continuous and rapid development of R). As it was already mentioned currently (June 2020) there are nearly 16,000 officially supported additional packages available. They are all described on the Contributed Packages website (https://cloud.r-project.org/web/packages/index.html). Checking which packages are currently loaded can be done with the **search**() command. In addition to packages, this command will give information about some other objects that have been included in the current session.

```
search()
```

```
[1] ".GlobalEnv"        "package:stats"      "package:graphics"
[4] "package:grDevices"  "package:utils"      "package:datasets"
[7] "package:methods"    "Autoloads"          "package:base"
```

Preparation for use of the selected additional package in R takes place in two stages. First, one has to install the package on the computer (in the local directory). This operation is performed once, the package can be updated later if new versions are released. Any package can be installed in several ways:

- In the basic R interface from Internet resources using the command *Packages -> Install package(s). . .*;
- Using the command: **install.packages ("package name")**;
- In RStudio, from the menu *Tools -> Install package (s) . . .*;
- In the basic R interface from a local zip file using the command *Packages -> Install package (s) from local zip files. . .* (only used when installing packages from outside the official repository).

Figure 1.3 shows RStudio at the time of installing the spdep:: additional package.

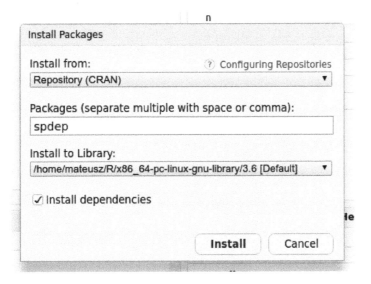

Figure 1.3 **Installation of additional packages in RStudio**

Source: Own study

From time to time, it is worth checking if the previously installed packages are up to date. One can do it from the menu *Packages -> Update packages. . .* in the R interface or in the RStudio menu: *Tools -> Check for package updates. . . .* The same can be obtained by entering the command **update.packages**() in the console:

```
update.packages()
```

After installing additional packages (as described previously), when creating specific projects, each time, it is necessary to attach the desired packages to the program session. Without this step, installed packages stay inactive and their functions cannot be used. In the basic R interface, packages can be loaded using the *Packages -> Load Package* menu. In situations where more packages are used, loading them in such a way can be cumbersome. A better solution is to create startup code and load the packages with the **library**() command. An example of loading the spdep:: package into the R session memory is:

```
library("spdep")
```

Of course, one must use as many **library**() commands as the number of packages being loaded. One can also create a slightly more advanced code that contains a list of packages that must be loaded in a given script while checking that they are already installed. If not, the following code will install them from the default repository R. Using the presented solution is beneficial when running the code on different computers without being sure if all the required packages have been installed. The code shown subsequently obviously contains examples of the package names that the user wants to use.

```
requiredPackages<-c("sp", "spdep", "RColorBrewer") # a list of packages
for(i in requiredPackages){
  if(!require(i,character.only = TRUE)) install.packages(i)
    library(i,character.only = TRUE)}
```

When using RStudio, packages can be loaded into the session by clicking the checkbox next to their name. This is done (with the standard RStudio setting) in the lower-right panel on the *Packages* tab. Figure 1.4 shows the status of the packages window with selected packages that have been included in the current program session.

The less frequently used option is to detach (delete from memory) packages during the session. This is necessary when the loaded package modifies the original version of the given command in R, while at some stage of the work, it is necessary to use the original command. In RStudio, you can do this simply by unchecking the package in the Packages window. The console function used to delete the selected package from memory is **detach**(). An example of disconnecting the spdep:: package is as follows – please pay attention to the required two arguments for this action:

```
detach("package:spdep", unload=TRUE)
```

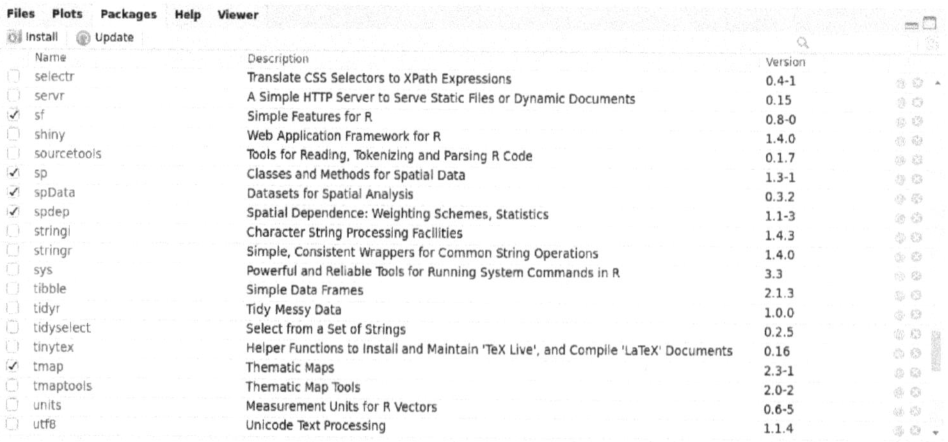

Figure 1.4 Loading additional packages in RStudio

Source: Own study

1.5 R language – basic features

The R environment operates on data organised into specific structures. The basic unit in R is the object. Each object in R must be specified in a given class. The class is a description of the features of the object. This data structure may be known to readers who have already used some programming languages and indicates that the R is close to the so-called object programming languages. It should be mentioned that R also has the features of a functional language, that is, a programming method in which the basic way to solve problems is to create and operate functions.[4] The object in R is every dataset, result, result of the operation and so on. All data that R uses must be of a certain type. The basic data types are: *character, complex, double, expression, integer, list, logical, numeric, single, raw*. These types of data in fact allow the creation of the basic class of objects, which is *vector* (vector), and then enable building further classes using both these types and existing data classes. All this creates a hierarchical structure that organises all data. It should also be added that the *vector* class is a virtual entity, because in reality, depending on the nature of the data assigned to the court of the vector, one of the basic types indicated previously is obtained, for example text, integers, logical values and so on. In addition to the *vector* class, the basic classes for organising data objects are: *matrix, data.frame, list, factor* and *tibble* (simple, but useful for statistical applications, class similar to *data.frame*). In the context of spatial data, these classes allow one to organise such data, although many functions that allow calculation on spatial data require the use of objects with special, more complex classes, which will be discussed later in the book. It is also worth mentioning that the results of calculations may also be stored in objects, and they have their respective classes, that is, structure (e.g. objects of the *htest* class, which is the result of many statistical functions).

In an R environment, one can use any number of objects at once. What are the advantages? First of all, one can load and manipulate simultaneously a lot of different datasets with a different structure into the program and does not need to combine them earlier. Second, the results of calculations or functions can be further used as data for subsequent operations. More information about the data structure, types and classes in R can be found in the program help.[5] The basic syntax that creates the object has the form:

```
object.name <- function (data, options)
```

In further calculations, such new object can be used. Names of objects cannot be only digits or begin with them; they must contain letters. They can be almost arbitrary, though, of course, not all special characters are allowed.

1.6 Defining and loading data

Data used for further analysis can be imported into the R program from many available formats. The basic commands for loading data are included in the default installation R; additional ones are available after installing the foreign:: package. The most commonly used functions that load data have the general syntax of the form: **function.name('file location or address', options)**. The most useful options, among others, are:

- data separator character, for example, (*sep=";"*) for a semicolon, (*sep="\ t"*) for a tab character; (*sep=""*) for spaces;
- a character that points to a decimal separator, for example, (*dec="."*) for dots notation, (*dec=","*) for comma notation;
- indication of whether the column contains the column headings in the first row, which will become the names of individual variables, for example, (*header=T*) if there is a header row – this is the default option – and (*header=F*) otherwise;
- encoding a data file, for example, (*encoding="UTF-8"*) – other values that can be used can be found in the help for R.

It is worth mentioning that the previously mentioned options, as well as all other available ones, can be given in any order chosen by the user. When data is loaded, they are usually assigned to the new object immediately. The entered data can generally be an object, usually in one of a few classes: *vector, matrix, data.frame, list* or *tribble*. For basic statistical and econometric calculations, one needs data in the basic layout, that is, column variables with headers – most often, these are *data.frame* or *tribble* objects.

Examples of basic commands with their syntax for loading data are presented subsequently. Note the options that may be specific to each source data format – their detailed description can be found in the command help.

Particular attention should be paid to the ability to load data from a comma-delimited format (*.csv files) and text (*.txt files). These are the most popular formats for storing source data.

The codes presented use the current working directory *"Working Directory"*, so it is not necessary to enter the full reference path. Using the **getwd()** command, one gets the current working folder, and one can set a new working directory with **setwd()**.

```
getwd() # checking the current working directory
setwd("C:/R/Data/") # setting the required working directory - example
```

In the RStudio environment, one can change the working directory in the console window with the commands given previously or through the use of the graphical menu. Figure 1.5 presents the location of the appropriate menu commands that are used to change the default directory.

The following code shows the commands and basic syntax for loading the most popular formats. These are only examples, and of course one should adapt them to one's needs. It is worth noting that the file being loaded into R may contain national characters (e.g. Polish letters) and spaces in the name – many programs do not do well in such situations – while using national characters in file names may make it difficult to use them in the case of work on systems not adapted to specific language.

```
library(foreign) # library supporting additional file formats
getwd() # checks the location of the default directory

# loading from a text file
data1<-read.table("data.txt", header=TRUE, sep="\t", encoding="UTF-8")

# loading from a CSV file
data2<-read.csv("data.csv",header=TRUE, sep=";",dec=".", encoding="UTF-8")

# loading from tab delimited file
data3<-read.delim("data.dat", header=FALSE, sep="\t")

# loading from a dbf file
data4<-read.dbf("data.dbf")

# loading from the SPSS format
data5<-read.spss("data.sav", use.value.labels=FALSE, to.data.frame=TRUE)

# loading from the STATA format
data6<-read.dta("data.dta", convert.factors=FALSE)
```

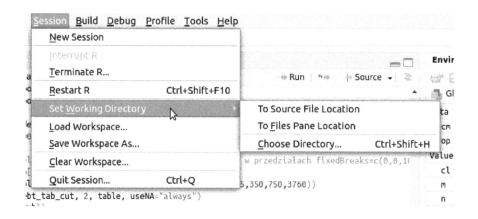

Figure 1.5 **Selecting the working directory in RStudio**
Source: Own study

Importing data from native MS Excel files is basically not (directly) available. Of course, one can fully use the *.csv files obtained from Excel (which is a good practice). When *.xls or *.xlsx files are needed, one can use R Commander, which has a dedicated mechanism for loading data from Excel format sheets. This package also supports the import of data from other popular formats such as MS Access, dBase, SPSS, Minitab, STATA and selected other text formats.

RStudio has recently added support for importing data from MS Excel spreadsheets based on an additional readxl:: package. It is a fairly light package providing basic functions of loading data in this format. Access to the graphical interface is obtained from the menu *File -> Import Dataset -> From Excel*. If the readxl:: package is not installed, RStudio will download it automatically. Of course, one can also use the readxl :: package outside of RStudio by entering the appropriate commands into the console.

Recently, a lot more advanced packages have been created to complement the previous lack of effective data loading from *.xls or *.xlsx files. One of them is the gdata:: package; in addition to readxl:: indicated earlier, it allows for loading a selected sheet from Excel into a *data.frame* object using the **read.xls**() command. The following is an example of how to load an MS Excel file from a working directory. It is worth paying attention to the command syntax for loading files in MS Excel format. In the package help, one can find a lot of options for the **read.xls**() command. It is worth noting that this feature supports files with multiple sheets.

```
library(gdata)
data7=read.xls("myfile.xlsx", sheet=1, header=TRUE)
```

Loading "advanced" data such as maps requires prior installation and loading of appropriate packages for handling spatial objects and will be discussed later in the book. After loading the data, it is worth checking immediately if the data has been loaded correctly. The easiest way to do this is through the **names**() function, which will check the recognition of the variable headers (names), or the function **summary**(), which will calculate the basic statistics. Examples of the use of these commands will be discussed subsequently.

1.7 Basic operations on objects

Before starting operations on the R program objects, one can try the basic math operations. In the following, R will be used as a calculator. Entering the arithmetic operation in the command line, after executing it (by pressing ENTER), the result of the operation is obtained in the result line:

```
3 + 5
#[1]  8
```

R, of course, provides basic mathematical operators (actions) similar to those available in other programs. These are addition (+), subtraction (-), multiplication (*), division (/) and exponentiation (^). In addition, all basic comparison operators can be used in R: less than (<), greater than (>), less than or equal to (<=), greater than or equal to (> =), equal to (==), different (!=). It is worth noting that "equality" is marked by the double character "==." The next group are logical operators AND and OR, represented respectively by (&) and (|), and also denial: NOT (!). One can add to this group the so-called disjoint alternative operator XOR, represented in R by the **xor**() function. More on the subject of logical operators and their special extended characters can be found in the program help – the easiest way is to call it by running the **help** ("&") command.

It is worth remembering that an essential complement to the basic operators are the functions available in the R package. Of course, their number depends on the additional packages loaded. The basic functions – mainly mathematical – should be known from other programs as well. For example, there are several such functions available in standard loaded packages: **abs**() – absolute value; **exp**() – exponential function e^x ($e\hat{}x$); **log**() – natural logarithm; **log10**() – decimal logarithm; **round**() – rounding the number to the indicated order of magnitude; **sign**() – sign of the number: 1 if > 0, −1 if < 0; **sqrt**() – square root and so on.

The basic operation in manipulating objects is to assign the value of the selected name of such an object. This is done by the assignment operator <-. This operator was already presented in the previous subchapter in describing how to load data in different formats. The individual names (data1, data 2, . . .) were assigned to data from files.

The most basic way (although the least effective) is to enter data manually. The following example shows how to create the simplest vector with data – numbers from 1 to 5. This will be a *numeric* class object. By the way, a fairly quick way to define a sequence of integers or a sequence of numbers from . . . to . . . with a set step using the function **seq**() will be presented:

```
vector1<-c(1,2,3,4,5) # vector of numbers 1,2,3,4,5
vector2<-10:20 # vector of integers from 10 to 20
vector3<-seq(5,10,0.2) # a sequence of numbers from 5 to 10 with a step of 0.2
```

As a reminder, below the sample data from the attached *.csv file have been loaded and attached to the *data* object. The *data_nts4_2019* file was used, which is a truncated set of panel data for poviats. A detailed description of the file and the data it contains is available in the attachment to the book. The object data of *data.frame* class was created in the following way.

```
data<-read.csv("data_nts4_2019", header=TRUE, sep=";", dec=",",
encoding="UTF-8")
```

To view the headers of the imported data, use the **names**() command. This is important because in the case of well-imported data, each of the variable names is shown separately (in separate quotation marks), while in the case when data is not seen as numbers, names of numeric variables can be combined with a header. In the case of incorrectly imported data, the data loading options should be changed.

```
names(data)
```

```
 [1]  "ID_MAP"           "code_GUS"        "poviat_name1"    "poviat_name2"
 [5]  "subreg72_name"    "subreg72_nr"     "region_name"     "region_nr"
 [9]  "year"             "core_city"       "dist"            "XA01"
[13]  "XA02"             "XA03"            "XA04"            "XA05"
[17]  "XA06"             "XA07"            "XA08"            "XA09"
[21]  "XA10"             "XA11"            "XA12"            "XA13"
[25]  "XA14"             "XA15"            "XA16"            "XA17"
[29]  "XA18"             "XA19"            "XA20"            "XA21"
[33]  "XA22"
```

It may also be important to check the object class with the **class**() function. For example, one can then make sure that the loaded or created object is actually a *data.frame* class. If this is not the case and one need an object of this class, one can try to convert it using the **as.data.frame**() function. There are similar commands for converting an object to other classes. It may also be helpful to display the structure of the object or its attributes using the **str**() or **attributes**() function, respectively. The functions mentioned here are used subsequently for the previously loaded object. Due to the fact that a broader description of this issue goes beyond the scope of this book, the reader should carry out independent research and experiments in this area.

```
class(data)
#[1] "data.frame"

# converting data to the data.frame class object if possible
# and create a new object if needed
data.df<-as.data.frame(data)
str(data) # structure of the object
```

```
'data.frame': 4560 obs. of 33 variables:
$ ID_MAP : int 1 2 3 4 5 6 7 8 9 10  . . .
$ code_GUS : int 1812000 1813000 1814000 1815000 1816000  . . .
$ poviat_name1 : Factor w/ 380 levels "Powiat st. Warszawa",..: 202 259 260
275 282  . . .
$ poviat_name2 : Factor w/ 370 levels "Powiat \u009credzki",..: 227 272 273
287 291  . . .
```

```
$ subreg72_name      : Factor w/ 72 levels "Podregion 01 -
jeleniog\xf3rski",..: 36 34 34 35 . . .
$ subreg72_nr        : int 36 34 34 35 35 33 36 35 36 33 . . .
$ region_name        : Factor w/ 16 levels "\u008cl¹skie",..: 11 11 11 11 . . .
$ region_nr          : int 18 18 18 18 18 18 18 18 18 18 . . .
$ year               : int 2006 2006 2006 2006 2006 2006 2006 2006 2006 . . .
$ core_city          : int 0 0 0 0 0 0 0 0 0 0 . . .
$ dist               : num 53.84 55.25 41.18 26.23 3.85 . . .
$ XA01               : num 32358040 41340641 43945543 46232049 97537294 . . .
$ XA02               : num 12433922 10668739 15183808 13534635 38065940 . . .
```

```
attributes(dane) # object attributes
```

```
$names
 [1] "ID_MAP"          "code_GUS"       "poviat_name1"    "poviat_name2"
 [5] "subreg72_name"   "subreg72_nr"    "region_name"     "region_nr"
 [9] "year"            "core_city"      "dist"            "XA01"
[13] "XA02"            "XA03"           "XA04"            "XA05"
[17] "XA06"            "XA07"           "XA08"            "XA09"
[21] "XA10"            "XA11"           "XA12"            "XA13"
[25] "XA14"            "XA15"           "XA16"            "XA17"
[29] "XA18"            "XA19"           "XA20"            "XA21"
[33] "XA22"

$class
 [1] "data.frame"

$row.names
 [1]  1   2   3   4   5   6   7   8   9  10  11  12  13
[14] 14  15  16  17  18  19  20  21  22  23  24  25  26
[27] 27  28  29  30  31  32  33  34  35  36  37  38  39
[40] 40  41  42  43  44  45  46  47  48  49  50  51  52
[53] 53  54  55  56  57  58  59  60  61  62  63  64  65
```

The display of object content can be achieved in four ways:

- entering only the name of the object in the command line, for example, *data*;
- using functions such as or **head**(), **tail**() or **str**();
- in the default R interface (in Windows) by selecting the *Edit/Data Editor* menu and entering the object name or after executing the **fix**() command with the object name as an argument;
- in the RStudio interface by clicking on the icon next to the name of the given object or by calling the **View**() command with the object name as an argument.

It is worth noting, however, that in the default R interface, there are limited editing possibilities (change/correction of individual data values).

The data previously loaded and assigned to the object does not necessarily have the desired name. There is the possibility of changing the name of the whole object (indirectly) and the names of particular variables in the object at any time (in particular the *data.frame* class). The simplest way to change the name of an object is to assign it entirely to the new name. After this operation, one can delete the original object, for example, using the command **rm**(). The following example shows this situation based on the previously created *vector3* object.

```
vector3<-seq(5,10,0.2) # vector creation reminder
new_vector3<-vector3 # assignment to a new name
new_vector3# display of the new object
```

```
 [1]]  5.0  5.2  5.4  5.6  5.8  6.0  6.2  6.4  6.6  6.8  7.0  7.2  7.4  7.6
 [15]  7.8  8.0  8.2  8.4  8.6  8.8  9.0  9.2  9.4  9.6  9.8 10.0
```

```
rm(vector3) # deleting vector3 object from the memory
```

It may be much more useful to change the names of particular variables or names of rows in the object. To display current names, use the **colnames**() and **rownames**() functions. In the subsequent example, the *matrix* class object will be created, and then the names of the columns (variables) and lines of the object will be displayed and changed with the conversion to the class *data.frame*.

```
x<-seq(1,20,0.4) # generating a variable x
y<-seq(2,21,0.4) # generating a variable y
z<-seq(3,22,0.4) # generating a variable z
xyz<-cbind(x,y,z) # combining variables into one xyz object

head(xyz) # display the beginning of the xyz object
```

```
       x    y    z
[1,]  1.0  2.0  3.0
[2,]  1.4  2.4  3.4
[3,]  1.8  2.8  3.8
[4,]  2.2  3.2  4.2
[5,]  2.6  3.6  4.6
[6,]  3.0  4.0  5.0
```

```
names(xyz) # request to display headers
#NULL

# adding new names of columns and rows
colnames(xyz)<-c("variable x", "variable y", "variable from")
rownames(xyz)<-rownames(xyz, do.NULL=FALSE, prefix="Obs.")

is.data.frame(xyz) # checking if xyz is a data.frame object
#[1] FALSE
xyz.df<-as.data.frame(xyz) # xyz conversion to data frame

head(xyz.df) # displaying the xyz object after changes
```

```
        variable x    variable y    variable z
Obs.1       1.0          2.0           3.0
Obs.2       1.4          2.4           3.4
Obs.3       1.8          2.8           3.8
Obs.4       2.2          3.2           4.2
Obs.5       2.6          3.6           4.6
Obs.6       3.0          4.0           5.0
```

```
names(xyz.df) # request to display column headers
#[1] " variable x" " variable y" " variable z"

colnames(xyz.df) # request to display column headers
#[1] "variable x" "variable y" "variable z"

head(rownames(xyz.df)) # request to display row headers
#[1] "Obs.1" "Obs.2" "Obs.3" "Obs.4" "Obs.5" "Obs.6"
```

It is worth paying attention to the function **cbind**() (and its equivalent **rbind**() not present in the previous example). The **cbind**() command is used to combine a sequence of *vectors*, *matrices* or *data.frame* objects into one object. The length of the objects should be equal. The result is the creation of a new object, which is a combination of all variables from the output objects – the resulting object has the same length (number of rows) as the output objects. The **rbind**() function works similarly to bind rows from source objects. As a result, one gets the object with the same number of columns as the source objects, but the number of rows is the sum of the number of rows from each object. The latter function is very well suited for combining data obtained from, for example, several sources or in several moments in time from the same source in order to create a complete database. More information and examples of the use of these functions can be found in the R help.

In addition to displaying entire objects (in particular the *data.frame* class), it is also possible to display individual elements (e.g. variables) of these objects. By the way, such a separated variable can be assigned to a new object (in the following example: *unempl*). A reference to specific variables is obtained using the following syntax: object$variable. To display the result, the **head**() function was used, which by default displays the first six rows from the object (one can use the parameter to enter desired number of displayed rows). A full example follows.

```
unempl<-data$XA21
head(unempl)
[1] 21.4 21.1 16.4 18.6 16.5 16.0
```

Another important issue that can help in the preparation of data for further analysis is the conditional selection of data and the creation of new objects on this basis. To create a dataset (object) that meets certain conditions, a square bracket [] after the object name is used. In brackets, the condition for rows and columns is given in sequence, separated by a comma. Conditions may apply to the tested or other variables. The general structure of the condition is in the form: *object [rows, columns]*, where *[rows, columns]* should be understood as *[selected rows, selected columns]* or *[condition for rows, condition for columns]*. It is worth noting that by leaving an empty (unspecified) element in one of the conditions, the entire range for this dimension is selected. The following example selects the first four rows and columns 7, 9, 11 from the given data.

```
# observations from 1 to 4 with variables from columns 3 to 5
section<-data[1:4, c(7,9,11)]

section
```

```
   region_name year dist
1 Podkarpackie 2006 53.84
2 Podkarpackie 2006 55.25
3 Podkarpackie 2006 41.18
4 Podkarpackie 2006 26.23
```

Other examples of conditional choice:

```
# only rows 2,7,10,15 and all columns
section 2<-data[c(2,7,10,15),]

# rows 1 to 10, but without a selected column - here is the fifth
# character (-) deletes the given row or column from the result
section3<-data[1:10,-5]
```

Slightly more complex operations allow one to select data based on a logical condition. In the example, a new object will be created, which will contain all variables but only for the selected year (year = 2015).

```
data15<-data[data$year==2015,]
head(data15) # subset for year 2015
```

Conditions can be combined with each other using appropriate logical operators. Conjunctions (and) or alternatives (or) are used for this purpose. The first one requires the & symbol to be used between the combined conditions, and the second requires the | character (Shift + \). The syntax of such conditional expressions does not differ from the general scheme and, for example, the condition imposed on a row looks as follows.

```
new-object <-source_object[var1 == A & var2 <B,] # for conjunction
new-object <-source_object[var1 == A | var2> = B,] # for an alternative
```

The following example selected data for two different provinces in Poland: Opolskie and Podkarpackie (OR operator) – it is worth noting that data previously limited to the selected year was used:

```
podkarp.or.opol<-data15[data15$voivodeship=="Opolskie" |
                        data15$voivodeship=="Podkarpackie",]
```

There are still many more advanced functions and methods for conditional data selection or manipulation of the object. It is worth paying attention to functions such as **subset**() or **recode**() from the car:: package and others.

Useful functions for manipulating data, especially that obtained from external databases, can be **stack**() and **unstack**(). The first one combines several columns (variables) of data into one column, that is, a variable, assigning observations to the index indicating the original column from which they come. The **unstack**() function has the opposite effect – it separates new variables, that is, columns based on indicated tags/terms, from one column (variable). Depending on the source, the data is usually in one of the previous arrangements. One can then easily try to convert it to a different layout if it is preferred by the user. The following example will allow one to observe the results of these commands.

```
# combining several variables into one object
employment<-cbind(data15$XA03, data15$XA04, data15$XA05)

# limiting the collection for increased transparency
employment<-as.data.frame(employment[1:4,])

# assigning new headers
colnames(employment)<-c("agriculture", "industry", "services")

# displaying the object's content
employment
```

```
    "agriculture", "industry", "services"
1     8253        3029         4300
2    14393        1123         4918
3    13814        4408         6145
4     8230        5539         6132
5    26078       11030        14182
6    10781       10534        10254
```

```
# a new object containing a marker (ind) and values
# displaying the content of the object after applying the stack() function
zat.stack<-stack(employment)
zat.stack
```

```
    values          ind
1     8253      agriculture
2    14393      agriculture
```

```
 3   13814      agriculture
 4    8230      agriculture
 5    3029        industry
 6    1123        industry
 7    4408        industry
 8    5539      industry
 9    4300         services
10    4918         services
11    6145         services
12    6132         services
```

```
unstack(zat.stack) # return to the original data shape
```

```
   "agriculture",  "industry",  "services"
1        8253          3029          4300
2       14393          1123          4918
3       13814          4408          6145
4        8230          5539          6132
5       26078         11030         14182
6       10781         10534         10254
```

An important element in working with data is the correct recognition of missing values. In the case of statistical or econometric calculations, a lack of value is important information and requires an appropriate approach. The R program is well prepared to handle such situations. At the same time, it should be remembered that the missing data should not be encoded as the selected numerical value or any other non-numeric character or left blank. This situation leads to errors in further calculations. Most of the features available in R and its packages have implemented support for missing values or appropriate options that indicate how to have such values treated. The standard indicator for the missing values in R is the NA abbreviation. If the missing values were not recognised correctly in the loaded dataset or are marked with any other string of characters, it is worth recoding them to the standard designation, that is, NA. The function **is.na**() should be used for this purpose. The syntax of this function is unusual (like **colnames**()). The general way of using it is as follows:

```
is.na(object to code NAs)<- condition determining the missing observation
```

In the example code subsequently, a copy of variable XA17 was created (investment expenditures in enterprises), and in a few cases, the values were changed to 99999. Then these values were considered as missing observations and changed to the correct designation (NA). In other words, it was assumed that data was originally missing in the dataset or that such values were invalid and thus cannot have a numeric value.

```
data15$XA17a<-data15$XA17 # coping a variable XA17 with a name XA17a

# creating missing observations coded as 99999
data15$XA17a[c(3,5,9,10)]<-99999

head(data15$XA17a,12) # display of the first 12 data of variable XA17a
```

```
[1]    36.0     125.2    99999.0    323.1    99999.0    208.0    731.9    23.4
[9] 99999.0   99999.0     139.1   1086.5
```

```
# assigning NA to missing observations
is.na(data15$XA17a)<-data15$XA17a>=99998
```

```
head(data15$XA17a,12) # display of the first 12 data of variable XA17a
```

```
 [1]    36.0    125.2     NA   323.1     NA   208.0   731.9   23.4    NA    NA
[11]   139.1   1086.5
```

This chapter presents only very basic methods of operations on objects so that the non-advanced reader will be able to start working in R. The R environment (including additional packages) allows very diverse and often effective ways to manipulate objects and data. Interested readers are referred to their own search for the most convenient methods.

1.8 Basic statistics of the dataset

Counting and presentation of basic statistics is one of the first stages of analysis after loading data, as well as one of the basic and most frequently performed operations. The most frequently used command displaying basic data characteristics, including data contained in the *data.frame* object, is **summary()**. The result of this function will be presented subsequently for sample data (for 2015) previously loaded.

```
summary(data15)
```

```
 GUS code      poviat                voivodeship
 Min.   :1.102e+09   Poiat st. Warszawa  : 1    Mazowieckie : 42
 1st Qu.:2.245e+09   Poviat aleksandrowski: 1   Śląskie      : 36
 Median :3.235e+09   Poviat augustowski   : 1   Wielkopolskie: 35
 Mean   :3.584e+09   Poviat bartoszycki   : 1   Dolnośląskie : 30
 3rd Qu. :5.020e+09  Poviat bełchatowski  : 1   Podkarpackie : 25
 Max.   :6.286e+09   Poviat będziński     : 1   Lubelskie    : 24
                     (Other)            :374  (Other)      :188
 cities with poviat rights municipalities in general urban municipalities
 rural municipalities
 City : 67    Min.   : 0.000   Min.   :0.0000   Min.   : 0.000
 County:313   1st Qu.: 5.000   1st Qu.:0.0000   1st Qu.: 2.000
              Median : 6.000   Median :1.0000   Median : 4.000
              Mean   : 6.529   Mean   :0.8105   Mean   : 4.163
              3rd Qu.: 9.000   3rd Qu.:1.0000   3rd Qu.: 6.000
              Max.   :19.000   Max.   :5.0000   Max.   :17.000

 urban-rural communes    year           XA02             XA03
 Min.   :0.000    Min.   :2015   Min.   :  4359   Min.   :   44
 1st Qu.:0.000    1st Qu.:2015   1st Qu.: 13258   1st Qu.: 2398
 Median :1.000    Median :2015   Median : 20794   Median : 4749
 Mean   :1.555    Mean   :2015   Mean   : 29686   Mean   : 6236
 3rd Qu.:2.000    3rd Qu.:2015   3rd Qu.: 29975   3rd Qu.: 8322
 Max.   :9.000    Max.   :2015   Max.   :868103   Max.   :31718

      XA04             XA05             XA06            XA17
 Min.   :   252   Min.   :  1579   Min.   : 2.40   Min.   :    9.1
 1st Qu.:  3019   1st Qu.:  4881   1st Qu.: 8.30   1st Qu.:  102.0
 Median :  5629   Median :  7396   Median :11.65   Median :  192.1
 Mean   :  7998   Mean   : 15452   Mean   :12.27   Mean   :  439.9
 3rd Qu.:  9727   3rd Qu.: 12705   3rd Qu.:15.45   3rd Qu.:  387.1
 Max.   :107001   Max.   :755385   Max.   :30.80   Max.   :22974.5
```

```
      XA19               XA20               XA21               XA28
Min.   :  3540    Min.   :  12838    Min.   :  3730    Min.   :  1262
1st Qu.:  9929    1st Qu.:  34624    1st Qu.: 10174    1st Qu.:  4300
Median : 14138    Median :  47884    Median : 14170    Median :  6502
Mean   : 18163    Mean   :  63164    Mean   : 19824    Mean   : 11010
3rd Qu.: 20282    3rd Qu.:  69990    3rd Qu.: 20916    3rd Qu.: 10581
Max.   :291837    Max.   :1046632    Max.   :405882    Max.   :401339


      XA30               XA31               XA33               XA40
Min.   :2569      Min.   : 61.90     Min.   : NA       Min.   :     0
1st Qu.:3242      1st Qu.: 78.10     1st Qu.: NA       1st Qu.:  2752
Median :3414      Median : 82.25     Median : NA       Median :  7396
Mean   :3532      Mean   : 85.10     Mean   :NaN       Mean   : 19761
3rd Qu.:3679      3rd Qu.: 88.62     3rd Qu.: NA       3rd Qu.: 18070
Max.   :6956      Max.   :167.60     Max.   : NA       Max.   :593148
                                     NA's   :380


      XA41              XA17a
Min.   :    0.0   Min.   :    9.1
1st Qu.:    0.0   1st Qu.:  102.0
Median :  130.0   Median :  192.1
Mean   :  738.1   Mean   :  441.8
3rd Qu.:  603.0   3rd Qu.:  387.1
Max.   :16403.0   Max.   :22974.5
                  NA's   :4
```

Of course, the previous example also includes variables in the *data.frame* object, which are not numeric values; therefore, the summaries for these variables have not been calculated (except for the number of missing cases – if any).

As one can see, the command used displays the highest and lowest values (minimum and maximum), quantiles (first, second-median, third) and mean. Each of these measures (and other basic ones) can be obtained using separate commands, for example, **mean()** for mean, **sd()** for standard deviation, **median()** for median, **mad()** for median absolute deviation, **var()** for variance, **max()** for maximum, **min()** for minimum. These functions can be used for variables as well as objects of selected classes, with **mean()** and **sd()** not working directly on the entire *data.frame* object. Hence, one needs to use more general commands that allow for applying the selected function to each variable/column from an object of this type: **lapply()** or **sapply()**. Because there are missing values for some observations in the set, it is safe to use *na.rm=TRUE* parameter, which will eliminate any missing data – that is, "NA" values.

In addition, some commands/functions simply do not work in case of missing values occurrence – and one needs to determine how to deal with the missing observations. The way of control over the treatment of NA values in R is very useful. In the example, the average for all variables from the employment object will be determined, as well as for the selected service variable (for certainty, the employment in the appropriate class, that is, *data.frame*, will be created earlier). It is worth noting the differences in results due to the **lapply()** and **sapply()** commands use and the fact that the **mean()** function works fine on a single variable.

```
employment<-as.data.frame(cbind(data15$XA03, data15$XA04, data15$XA05))
colnames(employment)<-c("agriculture", "industry", "services")

# average calculated from all data.frame object variables using lapply()
lapply(employment,mean,na.rm=T)
```

```
$agriculture
#[1] 6236.137

$industry
#[1] 7998.326

$services
#[1] 15451.87
```

```
# average calculated from all data.frame object variables using sapply()
sapply(employment,mean,na.rm=T)
```

```
agriculture industry services
  6236.137    7998.326 15451.866
```

```
# average calculated for one variable (column) with data.frame objects
mean(employment$service, na.rm = TRUE)
#[1] 15451.87
```

Counting quantile for variables can also be performed using the command **quantile**(). By default, without additional parameters, this function returns the min and max as well as the first, second and third quartile. One can also independently determine the probability values from the range [0,1] using the *probs* option. In this case, the values of the desired quantiles are obtained. The following example is for a variable containing 1000 numbers generated randomly from a normal distribution.

```
var1<-rnorm(1000) # 1000 numbers drawn from the normal distribution
quantile(var1)
```

```
        0%          25%          50%          75%         100%
-3.75582395  -0.61900349  0.02644261  0.74143911  2.92377684
```

```
# setting the probability thresholds by the user
quantile(var1, probs=c(1,2,5,95,98,99)/100)
```

```
       1%          2%          5%         95%         98%         99%
-2.206663  -1.986881  -1.609737  1.682391  2.098486  2.362374
```

The values of basic statistics can also be calculated using the **fivenum**() function. According to Tukey's scheme, it counts a five-figure summary: the minimum, the first quartile, the median, the third quartile, the maximum (and in this order, the values appear in the result object). The same statistics can be obtained with **summary**(). Tukey's statistics are extremely useful, e.g. as the basis for constructing the so-called boxplot.

```
fivenum(data15$XA21) # basic statistics for the unemployment in 2015
#[1] 2.40 8.30 11.65 15.50 30.80
```

Text variables that were recognised and imported as a factor during loading can be easily analysed with the **levels**() and **nlevels**() function. The first one in the result gives a list of unique levels of a given factor, and the second indicates the number of such levels. A simple example is presented subsequently – the names and number of provinces in the object were checked, which – for the record – was created from the *data_nts4_2019.csv* file.

```
class(data$voivodeship) # checking if the variable is factor
#[1] "factor"
```

```
levels(data$voivodeship)  # names of levels
```

```
 [1] "Dolnośląskie"          "Kujawsko-Pomorskie"     "Lubelskie"
 [4] "Lubuskie"              "Łódzkie"                "Małopolskie"
 [7] "Mazowieckie"           "Opolskie"               "Podkarpackie"
[10] "Podlaskie"             "Pomorskie"              "Śląskie"
[13] "Świętokrzyskie"        "Warmińsko-Mazurskie"    "Wielkopolskie"
[16] "Zachodniopomorskie"
```

```
nlevels(data$voivodeship)  # number of levels
#[1] 16
```

It is also worth mentioning the correlation measures, which are the basis for analysis of the interdependence of phenomena – links between variables. One usually uses three basic measures of correlation: the Pearson, Kendall or Spearman coefficient. The selection of the type of correlation coefficient is determined by the nature of the analysed data. Values of correlation coefficients are in the range <–1; 1>, which allows one to interpret the strength and direction of the interdependence. A positive value of the calculated correlation coefficient indicates a positive correlation between variables – both variables are essentially either increasing or decreasing. A negative value shows the reverse (negative) relation – the increase of one is essentially accompanied by the decrease of the other. The strength of this relationship is stated by the absolute value of the correlation index; however, there are no fully unambiguous thresholds for classifying dependence power. It is assumed that the weak correlation is in the range (0, 0.3), moderate (0.3, 0.6) and strong (0.6,1). Pearson's coefficient is a correct measure basically when the distribution of the tested variables is close to normal. In other cases, it is worth using Kendall or Spearman rank tests. For the last two correlation coefficients, the relationships between ranks (order) of observations are analysed, not between the observation values as in the Pearson coefficient. Rank tests are more resistant to outliers and sample distributions other than normal.

The basic function calculating the correlation coefficient between two variables is **cor**(). In the simplest case, as input data, one gives two numeric variables of the same length as arguments. In addition, one can still specify which correlation measure to use by specifying it in the method option: (*method="pearson"* or *method="kendall"* or *method="spearman"*). By default, the Pearson coefficient is returned. The example presents calculations of the Pearson correlation coefficient for two selected variables (unemployment rate XA21 and average salaries XA14 in 2015).

```
cor(data15$XA21,data15$XA14)
#[1] -0.3757007
```

The result of correlation coefficient is –0.375 and indicates a moderate negative correlation of these two variables, which means that an increase in one variable (unemployment rate) is accompanied by a decrease in the second variable (average salary).

A slightly wider range of information can be obtained using the **cor.test**() command. In addition to the value of the linear correlation coefficient, the statistical significance test of this statistic will also be shown. In the case of **cor.test**(), one can also specify an alternative hypothesis in the options (alternative parameter); two-sided tests are available – *two.sided* (the default hypothesis that the correlation coefficient is different from 0) and one-sided tests – less/greater (the hypothesis that the correlation coefficient is smaller/greater than 0). The confidence level of the test is determined in the *conf.level* option (0.95 by default). After executing **cor.test**(), a confidence interval is also obtained for the calculated correlation coefficient. For comparison, the results of this function are shown subsequently with the default parameters for the same variables as previously.

```
cor.test(data15$XA21,data15$XA14)
```

```
    Pearson's product-moment correlation
data: data15$XA21 and data15$XA14
t = -7.8819, df = 378, p-value = 3.481e-14
alternative hypothesis: true correlation is not equal to 0
```

```
95 percent confidence interval:
   -0.4589557 -0.2859050
sample estimates:
       cor
-0.3757007
```

The example also shows that the correlation is significant because *p-value=3.481e-14* is very small and below the usual level of significance, that is, 5%, hence the basis for rejecting the null hypothesis H_0 that the correlation coefficient is 0; therefore, one should adopt an alternative hypothesis that this coefficient is not 0, is significantly different from 0. In practice, this means that poviats with a lower rate of unemployment have higher wages.

Another approach to the analysis of the interdependencies of variables can be a graphical correlation study using a plot created, for example, with the **pairs**() function from the graphics:: package, also loaded by default when running R. As an argument, it is best to prepare a *data.frame* object containing only the data used in the analysis. It is also worth changing the headings of the separated variables into readable ones with the **colnames**() command. On the basis of such an object, a matrix of correlation coefficients can be created between all variables using the previously described function **cor**(). The addition of the *use="pairwise"* option is necessary for calculations: it omits missing observations, and only pairs of numbers are taken. In the example (see Figure 1.6), the correlation for several previously selected variables will be examined. In each case (for each pair of variables), the correlation is moderately strong, which confirms the numerical results. Reading the graph is analogous to the covariance/correlation matrix or even the multiplication table. The individual scatterplots refer to the variables indicated by their names from the rows and columns.

```
# employed per capita - number of employees divided by total population
# new variable empl.pc, added to the data15 object
data15$empl.pc<-data15$XA02/(data15$XA19+data15$XA20+data15$XA21)
```

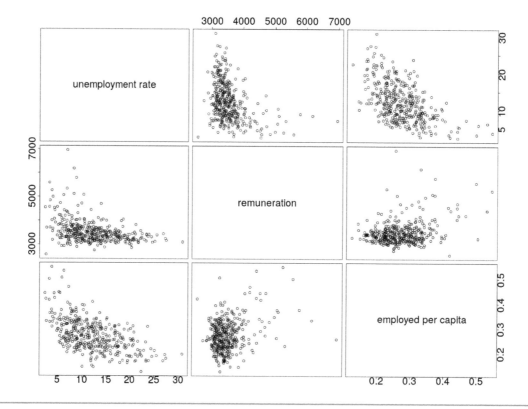

Figure 1.6 Graphical representation of correlation matrix

Source: Own study

```
# creating a new object containing only variables:
# unemployment rate, remuneration, employed per capita
variables.df<-data.frame(cbind(data15$XA21, data15$XA30, data15$empl.pc))
colnames(variables.df)<-c("unemployment rate", "salary ", "employed per
capita")

# correlation matrix, the rounding up to two decimal digits
# was used to increase readability
round(cor(variables.df, use="pairwise"),2)
```

	Unemployment rate	salary	employed per capita
unemployment rate	1.00	-0.38	-0.55
remuneration	-0.38	1.00	0.29
employed per capita	-0.55	0.29	1.00

```
# correlation scatterplot
pairs(variables.df)
```

The last operation presented in this subsection will be an example of variable standardisation. It is based on rescaling the variable by deducting the mean value of this variable from each observation and dividing by the standard deviation of the variable. After transformation, the variable has an average of 0 and a variance/standard deviation equal to 1. In many statistical applications, such an operation is required before further calculations. The general formula for variable scaling is as follows:

$$X_{st} = \frac{X - mean(X)}{sd(X)}$$

where X_{st} stands for a standardised variable.

The simple way (step by step) of standardisation of the variable is presented subsequently – the variable XA14 (salary) for 2015 was selected for this operation. To verify correctness of the transformation, the mean and variance were calculated.

```
variable<-data15$XA14

# standardising variable expression - the new variable is Z_variable
Z_variable<-(variable-mean(variable))/sd(variable)

# the average of the standardized variable is very close to zero
mean(Z_variable)
#[1] 3.302146e-16
# the variance is equal to 1
var(Z_variable)
#[1] 1
```

One can also use the **scale()** function for standardisation. The operation consists of executing this function on the appropriate variable. The results obtained are identical to those achieved previously.

```
variable1<-data15$XA14
Z_variable1<-scale(variable1) # standardized variable
mean(Z_ variable1)
#[1] 3.302146e-16
var(Z_ variable1)
#        [,1]
#[1,]    1
```

Only very simple functions that allow one to obtain basic statistics and to perform a preliminary analysis of the collected data are presented above. This preliminary analysis shows how to become familiar with the collected data and is always a starting point for further research, regardless of whether we use spatial analysis methods or other statistical procedures.

1.9 Basic visualisations

The aim of this subsection is to introduce the visualisation of data in R in the basic form based on standard packages, while more advanced graphics based on ggplot2:: will be presented in the following chapters of this book. The basic forms of graphical commands are usually quite simple to use. Many examples presented in the help for R make the user able to apply them quickly. It is worth knowing the richness of their additional options, which allows one to obtain quite sophisticated results. It is an encouragement to self-search and at the same time a challenge when the task is to create an appropriate graphic form of the analysis.

Graphical presentation of data and results has several advantages over their presentation in numerical form. For recipients of these results, this second form can be an overwhelming with the amount of information or simply unreadable. On the other hand, the graphic form often makes it easier to identify and extract connections between analysed variables. It is worth noting that there are plenty of additional packages in R that facilitate the use of graphic elements. Some of them only make it easy to insert charts that can be created using basic commands; others allow one to insert very complicated graphical elements.

The standard creation of graphics in R consists of applying subsequent layers to the created drawing. Creating a chart is usually started with the **plot**() command. This is the basic command that simultaneously initiates the opening of the graphic window, that is, the "area" in which the desired graph is created. Its argument can be source objects of different classes; that is, the data used in this command can generate specific graphs relevant to their character and structure – the source object class. Appropriate variants of the **plot**() command, dedicated to individual object classes – apart from those supported by default – are included in additional packages as, for example, **plot.lm**() or **plot.nb**(). Usually it is not necessary to use such extended names – just the basic form of the command **plot**() – R will try to choose its appropriate version. In many cases, R automatically recognises the class of source data and adjusts the way it works. The **plot**() command can create a line chart, a point chart, a forecast chart and even a map with spatial objects and others. Subsequent layers of graphs or their elements can be imposed with separate commands such as **lines**(), **points**(), **legend**(), **text**() and so on or by using the basic command, adding the *add=TRUE* option to it.

It should be emphasised once again that creating a graph (or wider graphics) in R consists of adding separate elements on the top of existing graphics in the display window – it graphically corresponds to overlapping successive transparent foils with the desired elements. In the situation when the prepared layer is not correct, the graph should be created from the beginning – it is impossible to remove the layer already applied. So, writing code in the editor makes it possible to correct mistakes and execute it many times.[6]

In summary, one obtains graphs and basic types of visualisation using the **plot**(), **lines**(), **points**(), **barplot**(), **boxplot**(), **pie**(), **hist**() and **curve**() commands. For placing additional elements on graphics such as guides, legends, titles, axes, any text and so on, one can use the **abline**(), **legend**(), **title**(), **axis**() and **text**() commands. Each of these commands has a lot of additional options and parameters that control the appearance of the result. In addition, the drawing area is also flexible and can be modified. A description of the appropriate options can be found in the **par**() command help, which can be used to modify certain chart attributes before they are drawn. The aforementioned multitude of options – and thus the flexibility of creating graphics in R – does not allow to include the description of all of them in this chapter. Subsequently, for example, selected types of charts will be presented along with sample parameters. This will allow the reader to become familiar with the idea of creating graphics in R and should be a good starting point for one's own research, as well as for generating maps and using spatial data, discussed separately later in the book.

1.9.1 Scatterplot and line chart

The basic and most popular type of chart is point or line. Often such charts are prepared for time data, where time is marked on the X axis (abscissa, axis of arguments). Figure 1.7 presents a linear graph of changes in the unemployment rate (variable XA21 in the dataset used previously) in Poland. Average unemployment in subsequent years was determined by the **aggregate**() command. The obtained object is of the *data.frame* class and has headers, so it can be easily used for further calculations.

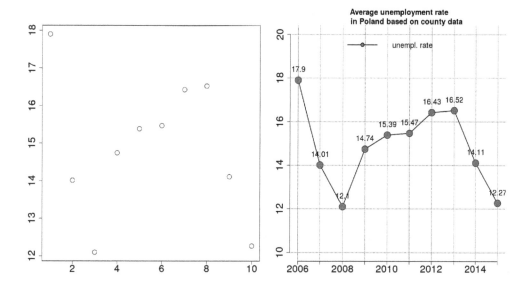

Figure 1.7 Visualisation of time series: a) simple point plot, b) point plot with line and additional options

Source: Own study

```
data<-read.csv("data_nts4_2019.csv", header=TRUE, sep=";", dec=",",
encoding="UTF-8")

# aggregation of data with respect to the year
unemployment <-aggregate(data$XA21, by=list(data$year), mean, na.rm=TRUE)
unemployment
```

```
   Group.1        x
1    2006   17.90211
2    2007   14.01342
3    2008   12.09974
4    2009   14.74289
5    2010   15.38737
6    2011   15.47026
7    2012   16.42658
8    2013   16.52079
9    2014   14.11132
10   2015   12.26868
```

```
par(mfrow=c(1,2), mar=c(2,2,2,1))
plot(unemployment$x) # the simplest point plot

# line chart with additional options
plot(unemployment$x, type= "l", lwd=2, axes=FALSE, xlab="year",
ylim=c(10,20))
axis(1, at=1:10, labels=unemploymet$Group.1, cex=0.8)
axis(2)
abline(h=(5:10)*2, lty=3)
abline(v=(1:10)*1, lty=3)
points(unemployment$x, pch=21, bg="red", cex=1.5)
```

```
title(main=" Average unemployment rate in Poland based on poviat data",
cex.main=0.8)
legend(3,20, pch=21, pt.bg="red", col="black", lwd=2, c("unemployment
rate"), bty="n", cex = 0.6)
text(1:10, unemployment$x+0.5, round(unemployment$x,2), cex=0.6)
```

In the example, the **plot**() command was used. Figure 1.7a is the result of the **plot**() command without any additional options. Figure 1.7b introduces modifications, such as a line graph, a layer of coloured dots, a value label, an *X*-axis scale in years, grid lines and so on. Point connections were obtained by setting the *type="l"* option. The line type can also be set with the parameter *lty=1, 2, 3, 4, 5* or *6*. It is also possible to specify this parameter value by using a word specification (more in the help for the **par**() command in the section describing this parameter). This parameter was also used in the **abline**() command to obtain a dotted grid of auxiliary lines. The thickness of the lines is specified using the *lwd* option, whose values can range from 1 to 5 (from the thinnest to the thickest). Signatures (any text) on the charts are placed using the **text**() command. For this command it is necessary to provide the *x* and *y* coordinates and the content of the signature. The font size of the text or legend can be changed with the *cex* option. The default font size is 1. To double the font size, enter *cex=2*. To reduce the font by half, *cex=0.5* should be specified. Legend is usually applied using the **legend**() function, where the first two arguments are the (*x,y*) coordinates of the legend, and the remaining are the names of lines, line types and thickness. The order of giving these arguments is arbitrary; only the order in each option must be kept. The title was added with a separate **title**() command, although it can be immediately specified as a *title* option of the **plot**() function. Guidelines were created on the graph with the **abline**() command. The first argument determines the position (vertical *v* or horizontal *h*) and the second is the line style. One can also choose the shape of the plot frame. The full frame is achieved by default or by typing in the options *bty="o"*. A graph without a frame is given by the option *bty="n"*. The lower and left part of the frame are coded with the option *bty="l"*, and the upper and right part with the option *bty="7"*. In the case of multilayer charts (e.g. three variables plotted on one graph), several options can be used. The first layer is usually created with the **plot**() command, and additional lines are added with the **lines**() command. It is worth noting that the *add=TRUE* option is not required for this last command, because by default, this function is used for the already created chart – it overlaps lines on an existing chart. The **points**() command works similarly, which places points on the graph, creating a scatterplot.

There are several ways to connect points on a chart, which is determined by the *type = ". . ."* option in the **plot**() and **lines**() command. In the example in Figure 1.8, the most important of available types of such a chart have been presented. The data for the chart has been generated using the **seq**() command as a sequence of natural numbers from 1 to 25, with a step of 1. Parameter selection **par**(*mfrow = . . .*) divides the graph area into several smaller ones, enabling simultaneous display of several plots.

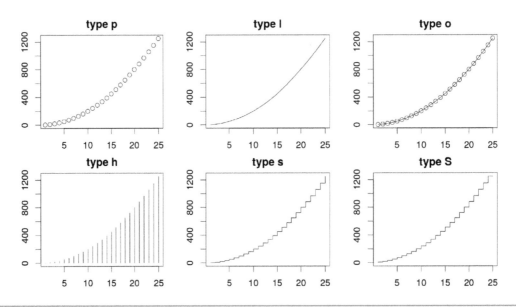

Figure 1.8 Types of point connections on the basic plot produced by the **plot()** command

Source: Own study

```
x<-seq(1,25,1) # generating the x variable
y<-2*x^2 # generating the y variable
# dividing the chart area into parts, and determining the margins
par(mfrow=c(2,3), mar=c(2,2,2,2))
plot(y, type="p", main="typ p")
plot(y, type="l", main="typ l")
plot(y, type="o", main="typ o")
plot(y, type="h", main="typ h")
plot(y, type="s", main="typ s")
plot(y, type="S", main="typ S")
```

1.9.2 Column chart

Quite popular types of graphics are column charts. They look similar to histograms (discussed later). In contrast to the histogram, the column chart is a graphical presentation of a table and has predefined groups. The histogram is a presentation of the distribution of the analysed variable divided according to specific intervals. The column chart is created with the **barplot()** command. Like any other graphic command, it also has many additional options to modify the final effect. One can get a graph in the form of vertical or horizontal bars, with nominal or cumulative values. The data source is usually the values contained in one- or two-dimensional tables. In Figure 1.9, two such charts are presented. The first one is quite simple – it only shows the pre-productive age population (variable XA07) in particular years in the previously used dataset. The second example will present a way to create a cumulative column graph, where the data on populations in pre-working age (XA07), working age (XA08) and post-working age (XA09) will be presented, also according to the years.

```
# loading of data
data<-read.csv("data_nts4_2019.csv", header=TRUE, sep=";", dec=",",
encoding="UTF-8")

# Figure 1.9a
pprod<-aggregate(data$XA19, by=list(data$year), sum, na.rm=TRUE)
colnames(pprod)<-c("year","l_pprod")

# the names.arg option gives labels under the bars
barplot(pprod$l_pprod/1000, col="grey", ylim=c(0,8000), ylab="in thous.",
        names.arg=pprod$year, cex.axis=0.8, main="Pre-working age population",
        cex.names=0.8)
abline(h=c(6000,7000,8000), lty=3, col="red") # horizontal lines

# Figure 1.9b
# cumulated column graph
preprod<-aggregate(data$XA19, by=list(data$year), sum, na.rm=TRUE)
prod<-aggregate(data$XA20, by=list(data$year), sum, na.rm=TRUE)
postprod<-aggregate(data$XA21, by=list(data$year), sum, na.rm=TRUE)
population<-cbind(preprod, prod$x, postprod$x)
colnames(population)<-c("year", "pre-working", "working", "post-working")

# converting the data.frame object to the matrix for the barplot command
population.m<-as.matrix(population)
barplot(t(population.m[,2:4]/1000), ylim=c(0,40000), cex.axis=0.8, names.
arg=population.m[,1], legend.text=TRUE, args.legend=list(x="right",
bg="white"), cex.names=0.8, main="Population structure by years")
abline(h= population.m[1,2]/1000, lty=3, lwd=2, col="blue")
abline(h=sum(population.m[1,2:3])/1000, lty=3, lwd=2, col="blue")
abline(h=sum(population.m[1,2:4])/1000, lty=3, lwd=2, col="blue")
```

The **barplot()** command typically requires input data in the matrix class. Each bar is represented by the next column of the matrix, while the number of series of data in each bar is the number of components of the column.

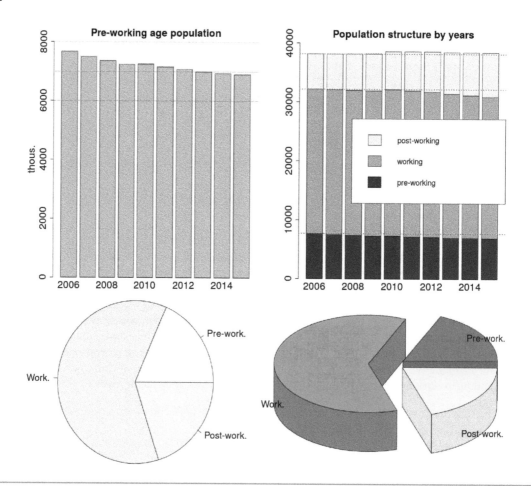

Figure 1.9 **Structural figures: a) simple barplot, b) cumulative barplot, c) simple pie chart, d) 3D pie charts**

Source: Own study

Therefore, when calling a command that creates a bar chart, the function **t()** is used to transpose the matrix, replacing columns with rows. The result of the **t()** command on the previous matrix can be seen subsequently.

```
t(population.m[,2:4])
```

	[,1]	[,2]	[,3]	[,4]	[,5]	[,6]	[,7]
preprod	7671336	7498495	7360049	7241502	7253648	7156800	7076869
prod	24516236	24580056	24625509	24659627	24866837	24774428	24641420
postprod	5991008	6090341	6203758	6319818	6463963	6661963	6869835

	[,8]	[,9]	[,10]
predprod	6995362	6942996	6901795
prod	24422146	24230162	24002168
postprod	7078151 7305444 7533276		

Another unusual element of the **barplot()** command is *args.legend=list (x="right", bg="white")*. It allows one to format the legend created automatically by **boxplot()** using the **legend()** command arguments presented previously. In other words, the option here is used to pass to the function that creates a column chart user-specified options for creating a legend.

It is worth mentioning that apart from the basic form of column/bar charts, one can achieve a bit more aesthetically sophisticated results in R. In this case there is a need to use additional packages. In the case of the type of charts discussed, the plotrix:: package may be indicated. There are several functions in it that allow one to construct column charts (barplot) with more options and a different look. Very aesthetic graphics can also be achieved as part of the ggplot2:: package.

1.9.3 Pie chart

One of the most frequently used charts is also a pie chart. Using such charts, the structure of the phenomenon is illustrated, the percentage share in total. It is considered a fairly attractive form of presenting information, but it should be noted that it is limited in terms of the scope of information that can be presented and can be less readable, as comparing fields and angles in a circle can be deceptive. Also be careful with the number of categories presented in one chart, because if it exceeds 5–6, the colours and parts of the pie chart may become indistinguishable. R makes it possible to create a two-dimensional pie chart by using the **pie()** command, but with the use of additional packages, one can also create three-dimensional pie charts – for example, the **pie3D()** command from the plotrix:: package.

In the example in Figure 1.9c, a table of the age structure of the population in 2015 in the Mazowieckie voivodeship was created, and a pie chart was prepared. Poviat data was aggregated for the selected year, by voivodeships, and data was selected only for the Mazowieckie voivodeship. The figure presents a comparison of a simple two-dimensional graph with a slightly more advanced three-dimensional graph. The code subsequently presents the preparation of data and then the creation of discussed charts. Note the slightly different syntax of the **aggregate()** function, which allows one to group multiple variables according to one criterion. More examples of slightly more sophisticated aggregation can be found in the built-in help for this function.

```
# simple pie chart
data15<-data[data$year==2015,]
structure<-aggregate(cbind(XA19, XA20, XA21) ~ voivodeship, data= data15,
sum)
colnames(structure)<-c("voivodeship", "Pre-working" "Work.", "Post-work")
structure_m<-structure[structure$voivodeship=="Mazowieckie",]
structure_m<-structure_m[2:4] # clearing the name of the voivodeship
```

The next step will be to prepare a comparison between the 2D and 3D charts. To make a 3D chart, the plotrix:: package is necessary.

```
# 3D pie chart
# dividing the chart area into 2 parts, and determining the margins
par(mfrow=c(1,2), mar=c(2,2,2,2))
# converting the data.frame object into a numeric vector
pie(as.numeric(structure_m), labels=names(structure_m), cex=0.6)

library(plotrix) # loading the required package
pie3D(as.numeric(structure_m),edges=1000,radius=1.3,height=0.2,theta=pi/4,
start=0, explode=0.20, col=heat.colors(3),labels=names(structure_m),
labelcex=0.7)
```

Among the options, one can specify the colours of individual parts – three colours from the built-in **heat. colors()** palette and the direction of fields, appearing, for example, *clockwise (clockwise=TRUE)*. Additional options: *edges=1000, radius=1.3, height=0.2, theta=pi/4, start=0, explode=0.20* control the size, height, angle of inclination, spread of the chart and so on. They allow for flexible preparation of the desired graph. A detailed description can be found in the **pie3D()** help.

1.9.4 Boxplot

One of the most popular statistical charts is a boxplot. This graph is based on basic statistics – quartiles and median and an interquartile range. Despite its simplicity, it enables presentation of many aspects of the analysed dataset, including the identification of outliers. A graph of this type is generated by the **boxplot()** command. It has many additional options that can enhance its appearance. The graph can be prepared for one variable, but its value is revealed only in a comparative analysis, where a grouping variable is added. In the basic version, the command syntax is as follows:

```
boxplot(variable_analysed, options)
```

while for a comparative analysis with grouping variable, the syntax looks like this:

```
boxplot(analysed_variable ~ grouping_variable, options)
```

The simplest boxplot for the unemployment rate in poviats in the last available year is the first to be made based on dataset that was used earlier in this chapter. This data will be uploaded again (to be sure that they are properly loaded), and then a subset of data will be extracted only for 2015.

```
data<-read.csv("data_nts4_2019.csv", header=TRUE, sep=";", dec=",")
data15<-data[data$year==2015,]
boxplot(data15$XA06, col="grey78", ylab="unemployment rate",
        main=" Unemployment rate in poviats in 2015")
```

Another example will be presented for the unemployment rate, also for 2015 but with grouping according to voivodeships (see Figure 1.10). The chart will be completed with a legend, and its axes and colours will be changed. The previously loaded data was used. An auxiliary object will also be created with the names of provinces and average values of the variable under investigation. The **for** loop is used to create a legend.

```
a<-aggregate(data15$XA21, by=list(data15$voivodship), mean, na.rm=TRUE)
a$ID<-c(1:16)
colnames(a)<-c("woj", "x", "ID")
boxplot(data15$XA21~data15$voivodship, axes=FALSE, ylim=c(0,65),
col="bisque", border="bisque4")

# average unemployment rate line
abline(h=mean(data15$XA21, na.rm=TRUE), lty=3, col="red")

# axis added
axis(1, at=1:16, labels=a$ID, cex.axis=0.8)
axis(2, at=(0:6)*10, cex.axis=0.8, ylab="%")

# legend text
for(i in 1:16){
text(1,60-3.5*i, a$ID[i], cex=0.6)
```

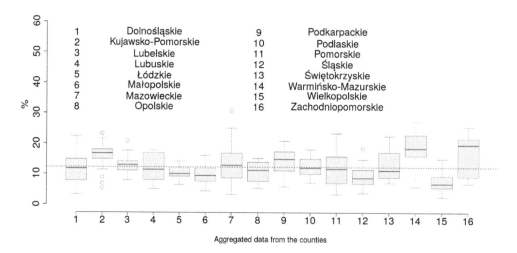

Figure 1.10 **A boxplot with example of grouping according to the selected variable**
Source: Own study

```
text(4,60-3.5*i, a$woj[i], cex=0.6)}
for(i in 9:16){
text(8,60-3.5*(i-8), a$ID[i], cex=0.6)
text(11,60-3.5*(i-8), a$woj[i], cex=0.6)}

title(main="The unemployment rate in 2015 by voivodships", cex.main=0.8,
      sub="Data aggregated from the level of poviats", cex.sub=0.7)
```

In the previous example, colours of "bars" have been changed using *col* option. In this option, one can use the selected colour from the 657 default colour names palette in R. The **colors**() command displays their names. The first 12 names are presented subsequently.

```
head(colors(),12)
# [1] "white"          "aliceblue"     "antiquewhite"   "antiquewhite1"
# [5] "antiquewhite2"  "antiquewhite3" "antiquewhite4"  "aquamarine"
# [9] "aquamarine1"    "aquamarine2"   "aquamarine3"    "aquamarine4"
```

In R, there are also built-in sample colour palettes, such as **heat.colors**() and **terrain.colors**(). An extension of the built-in palettes is the RColorBrewer:: package, which is used to build one's own colour palettes.

1.10 Regression in examples

The construction and analysis of econometric models is one of the most important research method in economics, as well as in many other fields of science. The following are examples of simple linear regression models that provide a good introduction to more advanced models that take into account the spatial factor. In many cases, such a model can be a starting point and a base for assessing whether the use of the spatial factor significantly improves the properties and information value of the created model. In this subchapter, in accordance with the character of this chapter (as an introduction for less advanced users of R), a general scheme of conduct for the analysis of simple regression will be presented without focusing on the substantive role of the model.

The basic least-squares regression procedure is available in the stats:: package, which is installed by default with the basic version of R. The package is also loaded by default when the program starts, so it is immediately available to users. The basic command for linear regression is the function **lm**() (abbreviation for linear model). The general syntax of the regression function is *lm(y ~ x1 + x2 + . . . + xn, options)*, where *y* is an dependent variable, whereas *x1, x2, . . ., xn* are the following independent variables that are to influence *y*. Of course, the number of used independent variables is decided by the user. Variable names do not indicate which object (set) they come from. Such information is specified in the options, for example, *data=GDP.data* (when e.g. using the *GDP.data* object). One can also regress on variables from different objects. In this case, however, in the equation formula, one needs to specify the variable with the object it comes from, for example: *data$x1*.

Before attempting to build a regression formula, it is a good idea to make a graph of a dependent variable relative to individual independent variables, for example, using the **plot**() function discussed earlier. Two variables were chosen to build a simple regression model: official unemployment in % (XA06) and logarithm of average salary in PLN (XA30).

```
plot(log(data15$XA14),data15$XA21, xlab="Logarithm of average salary in
PLN", ylab="Registered unemployment rate (in%)") # Figure 1.11
```

From Figure 1.11, it can be seen that both variables should be negatively correlated with each other. This assumption also seems to be reasonable from the point of view of economic theory. Of course, it cannot be assumed that the only factor affecting unemployment is remuneration – here, for presentation purposes, is a regression model with only one independent variable proposed.

An example of a simple regression on previously selected variables is presented subsequently. It is worth noting that at the beginning, the result of the function **lm**() was assigned to the object. This is the right procedure, because later one can use repeatedly selected elements of this result. The regression result can be shown in several ways: by entering only the name of the regression object, using the **print**() function on the object containing the result or using the **summary**() command. The first and second methods give a rather poor result, displaying only the basic information. The last method displays all components of the result. Of course, one can also

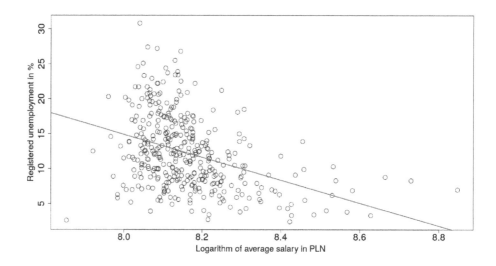

Figure 1.11 **Dependency between the explanatory and dependent variables selected for the model with fitted regression line (discussed later)**

Source: Own study

perform the function **lm**() without assigning the result to the object, but such a result will be only informative (without the possibility of "reusing" it). All the details available in the result object can be checked with the **attributes**() command.

```
model1<-lm(XA21~log(XA14), data=data15)
model1 # regression result - basic version
```

```
Call:
lm(formula = XA21 ~ log(XA14), data = data15)

Coefficients:
(Intercept) log(XA14)
     146.00    -16.39
```

```
summary(model1) # regression result - full version
```

```
Call:
lm(formula    =   XA21    ~   log(XA14),   data   =   data15)
Residuals:
    Min          1Q        Median         3Q          Max
-14.7547    -3.5485      -0.7369      3.4146      16.5429

Coefficients:
        Estimate      Std.      Error      t    value     Pr(>|t|)
(Intercept)   146.00           16.49       8.856    <    2e-16  ***
log(XA14)        -16.39            2.02     -8.113  7.01e-15  ***
---
Signif.  codes:   0 '***' 0.001 '**' 0.01 '*' 0.05 '.' 0.1 ' ' 1

Residual    standard   error:   4.868   on   378   degrees  of  freedom
Multiple   R-squared:   0.1483,          Adjusted R-squared:     0.146
F-statistic: 65.82 on 1 and 378 DF, p-value: 7.009e-15
```

The previous printout shows that the explanatory variable is significant – the *p*-value of the *t* statistic is very small for the beta coefficient. The model fit (R^2, *R*-squared) is unfortunately quite low ($R^2 = 0.1483$), which results from the use of only one explanatory variable. The parameter *Residual Standard Error* indicates that forecasting the explained variable (unemployment rate) with this model may generate an error of on average 4.87 p.p. (in terms of *y* variable).

It may be interesting to compare the theoretical values with those on the basis of which the model was estimated. The previously presented figure was completed with a line showing theoretical values. Information about such values can be obtained from the object to which the result of the model has been assigned. To do this, one needs to refer to the variable fitted: *model1$fitted*. To apply the theoretical values to the previously created graph of empirical values, the **abline()** command was used with the previously mentioned argument from the regression result.

```
plot(log(data15$XA14),data15$XA21, xlab="Logarithm of average salary in
PLN", ylab="Registered unemployment rate (in%)")
abline(model1) # Figure 1.11
```

It is worth pointing out that from the *model1* object, one can identify the slope and intercept of the plotted line separately. To do this, please refer to the variable coefficients in the *model1* object, for example:

```
intercept<-model1$coefficients[1] # extracting of the constant term
slope<-model1$coefficients[2] # extracting the slope of the regression line

intercept
#(Intercept)
# 145.9966
slope
#log(XA14)
#-16.38523
```

In Figure 1.11, the real observations were marked with points, and the regression function was drawn on the basis of model estimates. The chart can be used to identify which observations should be considered outliers and what the overall fit of the model looks like. Figure 1.11 confirms a large spread of observations from the simple regression and thus not perfect fit of the model described by the R^2 value.

It may also be useful to analyse residuals from the estimated model. Their values can be obtained by the **residuals()** command, where the argument is the regression result object. Figure 1.12 presents a scatterplot of

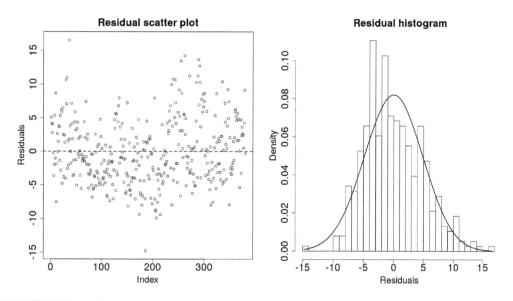

Figure 1.12 Model residuals analysis: a) scatterplot, b) histogram

Source: Own study

residuals with an added horizontal line (*y=0*) using the **abline**() function, which allows a visual evaluation of the distribution of regression residuals. In order to plot a horizontal line, the argument of the **abline**() function is the expression *h=0* (*h*, abbreviation for horizontal) – with the vertical line one would use the argument *v=0* (*v*, short for vertical). In order to more accurately assess the distribution of residuals, it is also worth plotting the histogram and comparing it with the normal distribution. The normal distribution in this case should have parameters (mean and standard deviation) in accordance with the distribution parameters of the analysed residuals.

```
par(mfrow=c(1,2)) # option to show two charts side by side
res<-residuals(model1)
plot(res, ylab="Residual")
abline(h=0, lty=2)
title(main=" Residual scatter plot")

# histogram of residuals
hist(res, breaks=30, main="Residual histogram", freq = FALSE,
xlab="Residual")

# creation of the normal distribution density function
# with parameters consistent with the empirical distribution of the rest
mean.res<-mean(res)
sd.res<-sd(res)
x<-seq(min(res), max(res),length=100)
y<-dnorm(x, mean.res, sd.res)
lines(x,y) # line of theoretical normal distribution
```

It can be seen from the residual graph that the large number of them are in the range of approximately (–5; 5), while the remaining ones (except for two observations) belong to the interval (–10; 15). However, the distribution of residuals is far from normal, as shown by the residual histogram with the density function of the normal distribution superimposed.

The residuals from the model can also be obtained as the difference between the explained variable and the theoretical values created using the **predict**() command. The obtained result is, of course, the same as using the **residuals**() command shown previously. In the following example only residuals are determined without being redrawn.

```
model1.fit<-predict(model1) # theoretical values obtained from the model
# error as a difference between empirical and theoretical values
error<-data15$XA21-model1.fit
```

Testing statistical hypotheses is a separate and very well-programmed issue in R. At the same time, it goes beyond the scope of this introduction. For example, the following is the testing of normality of distribution using the Shapiro-Wilk test available in the stats:: package in the default R installation. The null hypothesis (H_0) of this test assumes that the distribution is normal and the alternative hypothesis (H_1) that the distribution is significantly different from the normal one. The Shapiro-Wilk test is called by the **shapiro.test**() command. The only argument to this function is the data object/vector. The test works on data with the number of observation from 3 to 5000. In the example, the normality of the variable used and residuals from the distribution was tested. This second test may be especially important for the evaluations of regression results (model).

```
shapiro.test(data15$XA14) # normality test for the distribution of XA14
variable
```

```
  Shapiro-Wilk normality test
data: data15$XA14
W = 0.79253, p-value < 2.2e-16
```

```
shapiro.test(res) # normality test of residual distribution
```

```
  Shapiro-Wilk normality test
data: res
W = 0.98196, p-value = 0.0001099
```

The obtained results indicate that it is difficult to recognise the distribution of both variables used in the example as normal. In both cases, the *p*-value (*p-value*) clearly indicates the need to reject the null hypothesis in favour of an alternative hypothesis. In particular, such an conclusion in relation to residuals may cause problems in the context of the constructed (proposed) model and leads to the need to conduct a detailed diagnosis of whether the specification is correct.

It should also be mentioned that dedicated packages were created for a wide range of model diagnostics, e.g. lmtest::, rms:: and nortest:: are worth mentioning.

* * *

The presented issues are only a small part of the basic possibilities of the R program. The purpose of this chapter is to introduce readers to R – a practical compendium of knowledge for beginner R users to get started with this software. Some of the issues discussed here will be useful in the following examples referring to the main plot of the book i.e. spatial analysis. Schemes and syntax of presented simple commands and codes can be helpful to solve more advanced examples from the later chapters of the book.

Notes

1   R logo, https://www.r-project.org/logo/, CC-BY-SA 4.0 0
2   What is R?, https://www.r-project.org/about.html
3   John Fox and Milan Bouchet-Valat, Getting Started With the R Commander, Version 2.4–0, 2017
4   More about functional programming in R in Wickham Hadley (2014).
5   https://stat.ethz.ch/R-manual/R-devel/library/methods/html/BasicClasses.html and **https://stat.ethz.ch/R-manual/R-devel/library/methods/html/Classes_Details.html**
6   In general, the graphs in R are static, and after their construction, one cannot change their parameters, for example, the number of intervals for the histogram. It is worth paying attention to the manipulate:: package, which allows for using dialog boxes and manipulating the results – RStudio fully supports this package. For more sophisticated applications and ideas, an interesting way may be to use the shiny:: package (created by the authors of RStudio), which also allows for creating web applications based on the R code. More about both RStudio and support for the manipulate:: and shiny:: packages are available directly on the website: http://www.rstudio.com.

Chapter 2

Data, spatial classes and basic graphics

Katarzyna Kopczewska

Spatial data is one of the most advanced types of data. It contains information about the location, spatial shapes and values of a studied phenomena in a static or dynamic approach. It has been accepted to divide spatial data into three types (see Cressie, 1993b): a) point data (geostatistical), where the value of the feature at the point is of interest and the locations themselves are treated as the place of measurement; b) point patterns in which the location of the point is of interest – it can be assigned a feature value; then it is a marked point pattern; c) area data (lattice data, areal data), most often for territorial units, resulting from the aggregation of unit data recorded in a given area – the territorial units are polygons. Apart from this data, there is data presented in a grid or raster format – that is, aggregated data for artificial regular division of the area, geographic data for rivers and roads expressed as lines and so on.

Spatial data is most often visualised in a two-dimensional (x, y) system. The X axis gives longitude (east-west directions), and the Y axis gives latitude (north-south directions). For short distances, a typical Cartesian system (planar, flat) without correction for the curvature of the earth does not give a distortion. For larger areas, spherical correction and projection is necessary. Spatial data is determined in these two formats. Differences and the importance of the selection of projections appear in the visualisation (graphics) in relation to the pace of calculations, in particular the mutual distances between many points, as well as when using some commands that require an appropriate projection as input data.

This wealth of spatial data types is an opportunity as well as a threat. Each scientific discipline has its own typical data, the nature of which results from the specifics of research and analysed units. The consequence is a very dynamic development of spatial methods, the effective use of which requires a domain "filter". The same applies to the development of software. R has the richest libraries for spatial methods, dedicated to different disciplines and fields of science. Historically, the chaos of object classes for methods from different disciplines has been avoided – the R community was lucky that a strong leader, Prof. Roger Bivand, has created object classes that integrate all of the individual packages. Thanks to his work, packages can "see" each other, and results from one package or data type can be converted to other packages and data types.

This chapter presents the method of spatial data import (2.1), creation and conversion of spatial classes (2.2), colour palettes (2.3), basic contour maps with a colour layer (2.4), basic graphs for point data (2.5), basic operations on rasters (2.6) and grids (2.7) and spatial geometries (2.8). This chapter primarily uses the *sp* class (*SpatialPoints*, *SpatialPolygons*) and the *RasterLayer* class. In this chapter, the threads of a relatively new *sf* class are not raised – all of Chapter 10 is devoted to this and dedicated to large datasets (big data).

2.1 Loading and basic operations on spatial vector data

In spatial analyses, several types of spatial information are used, among others: static maps (underline), GIS (e.g. from Google or OpenStreetMap), contour maps from shapefile files with values and characteristics assigned to areas or geolocalised points with assigned values in points. The most popular types of spatial data include vector data – points, lines, polygons and pixels. Raster data is often alongside them (see 2.6).

There are many packages in R that process spatial information. Their gofrom description is in TaskViews Spatial at www.r-project.org. The most important of them are:

- sp:: – creates spatial object classes and allows for drawing the spatial classes, data operations, coordinate and coordinate system management methods and so on;
- rgdal:: – imports and exports vector data (OGR) and raster data (GDAL);
- maptools:: – allows one to manipulate geographic data; contains commands that convert object classes for use in PBSmapping::, spatstat::, maps:: or RArcInfo::.

Loading vector data into the *sp* class created within the sp:: package, that is, to *SpatialPoints*, *SpatialLines*, *SpatialPolygons* or *SpatialPixels*, the most universal spatial object standard in R is obtained, which allows switching to other classes and use of the wealth of different packages in R. One can also create the classes of objects extended by a layer of features (values of observations) defined within DataFrame, that is, *SpatialPointsDataFrame*, *SpatialLinesDataFrame*, *SpatialPolygonsDataFrame* and *SpatialPixelsDataFrame*.

The **readOGR**() command from the rgdal:: package reads the shapefile files into the *sp* class. Each spatial layer (division scale into subareas) is loaded separately – for example, maps for provinces (NTS2) and poviats (NTS4) are separate files.

Each shapefile file has its own internal settings, in particular the type of projection, which should be understood as a way of defining geographic coordinates. Several systems operate in parallel with GIS systems. One of the most commonly used and compatible with that used in Google maps is NAD83 (North American Datum 1983) or WGS84 (World Gefrometic System 1984). Differences resulting from different ellipsoids exist, but in the approaches presented in the book, they do not have a key role. The **spTransform**() command from the sp:: package changes the data or map projection to the desired one (e.g. compatible with data from Google). The conversion to these projections requires the second argument to be CRS("*+proj=longlat +datum=NAD83*") or CRS("*+proj=longlat +datum=WGS84*"), respectively. Note that putting extra blank spaces in projection code may stop operations.

The following example loads three contour layers from shapefile files: for the national contour of Poland (file *Panstwo* to object *pl*), contour of the voivodeship (NTS2 in Poland) (file *wojewodztwo* to the *voi* object) and poviat borders (NTS4 in Poland) (*powiaty* file to *pov* object). The loaded objects are in the *sp* class, especially in *SpatialPolygonsDataFrame*.

```
library(rgdal)
library(sp)

# loading shapefile files
pl<-readOGR(".", "Panstwo") # 1 unit
voi<-readOGR(".", "wojewodztwa") # 16 units
pov<-readOGR(".", "powiaty") # 380 units

# change of projection
pov<-spTransform(pov, CRS("+proj=longlat +datum=NAD83"))
voi<-spTransform(voi, CRS("+proj=longlat +datum=NAD83"))
pl<-spTransform(pl, CRS("+proj=longlat +datum=NAD83"))

class(pov)
```

```
#[1] "SpatialPolygonsDataFrame"
attr(,"package")
#[1] "sp"
```

In shapefile files, there are often no specific characteristics of these areas – there is no social, economic or social data. Therefore, for effective work, it is necessary to collect data in a separate file. The most important thing is that the order of units in a given set must be the same as in the shapefile. The order of data in the shapefile can be examined by converting a part of the shapefile database into a *data.frame* object – this is possible with the command **as.data.frame()**. One can also use the **as()** function by specifying the target class of the object in the options. Column *jpt_nazwa_* contains the names of poviats – in this order, an external dataset with data should be determined. Similar access can be obtained by entering the details of the object by @.

```
pov.df<-as.data.frame(pov)
pov.df<-as(pov, "data.frame") # the same operation to the above
class(pov.df)
#[1] "data.frame"

head(pov.df)
head(pov@data) # below only 4 rows instead of 6
```

	iip_przest	iip_identy	iip_wersja	jpt_sjr_ko
0	PL.PZGIK.200	f87cae01-4621-41f4-b768-c777cf94d1d8	2012-09-27T08:01:01+02:00	POW
1	PL.PZGIK.200	f322275e-1c48-431f-889e-8bb0ccb46e81	2012-09-27T08:01:01+02:00	POW
2	PL.PZGIK.200	59e7c5a7-8597-4c55-812a-b30932c08dd2	2012-09-27T08:01:01+02:00	POW
3	PL.PZGIK.200	e86b1e71-8958-42ee-bec5-ca3c87907bc8	2012-09-27T08:01:01+02:00	POW

	jpt_kod_je	jpt_nazwa_	jpt_nazw01	jpt_organ_	jpt_orga01	jpt_jor_id
0	1812	powiat niżański	<NA>	<NA>	NZN	NA
1	1813	powiat przemyski	<NA>	<NA>	NZN	NA
2	1814	powiat przeworski	<NA>	<NA>	NZN	NA
3	1815	powiat ropczycko-sędziszowski	<NA>	<NA>	NZN	NA

	wazny_od	wazny_do	jpt_wazna_	wersja_od	wersja_do	jpt_powier	jpt_kj_iip
0	0000/13/41	5201/20/92 6	NZN	<NA>	0020/12/09	26	0000000000EGIB
1	0000/13/41	5201/20/92 6	NZN	<NA>	0020/12/09	26	0000000000EGIB
2	0000/13/41	5201/20/92 6	NZN	<NA>	0020/12/09	26	0000000000EGIB
3	0000/13/41	5201/20/92 6	NZN	<NA>	0020/12/09	26	0000000000EGIB

	jpt_kj_i01	jpt_kj_i02	jpt_kod_01	id_bufora_	id_bufor01	id_technic	jpt_opis	jpt_sps_ko
0	1812	<NA>	<NA>	NA	13878	NA	829081	UZG
1	1813	<NA>	<NA>	NA	13878	NA	829082	UZG
2	1814	<NA>	<NA>	NA	13878	NA	829083	UZG
3	1815	<NA>	<NA>	NA	13878	NA	829084	UZG

	gra_ids	status_obi	opis_bledu	typ_bledu
0	<NA>	AKTUALNY	<NA>	<NA>
1	<NA>	AKTUALNY	<NA>	<NA>
2	<NA>	AKTUALNY	<NA>	<NA>
3	<NA>	AKTUALNY	<NA>	<NA>

To visualise economic (or other) data, it is necessary to load a set of data with these characteristics. The set must be ordered by the order of objects in the shapefile file. As one can see, the order of the poviats has been preserved in both sets.

```
# loading poviat (NTS4) data
data<-read.csv("data_nts4_2019.csv", header=TRUE, dec=",", sep=";")
summary(data)
head(data[,1:6])
```

```
   ID_MAP code_GUS                 poviat_name1                    poviat_name2
1       1  1812000           Powiat niżański                  Powiat niżański
2       2  1813000          Powiat przemyski                 Powiat przemyski
3       3  1814000         Powiat przeworski                Powiat przeworski
4       4  1815000 Powiat ropczycko-sędziszowski Powiat ropczycko-sędziszowski
5       5  1816000         Powiat rzeszowski                Powiat rzeszowski
6       6  1817000            Powiat sanocki                  Powiat sanocki
                    subreg72_name subreg72_nr
1 Podregion 36 - tarnobrzeski          36
2    Podregion 34 - przemyski          34
3    Podregion 34 - przemyski          34
4    Podregion 35 - rzeszowski          35
5    Podregion 35 - rzeszowski          35
6  Podregion 33 - krośnieński          33
```

The downloaded maps from the shapefile file are drawn with the **plot**() command. These are basic graphs.

```
# graphics - contour maps plotted separately
plot(pl) # map graphics as in appendix
plot(voi)
plot(pov)
```

Basic contour maps can be enriched with additional graphical elements (see Figure 2.1). These are, for example: map scale – drawn with the **map.scale**() command from the maps:: package, arrow pointing north – drawn by **north.arrow**() from the GISTools:: package, map scale applied with the **degAxis(1)** and **degAxis(2)** from the sp:: package, a rose showing compass directions – drawn by **compassRose**() from the sp:: package, geographic grid – drawn by **plot(gridlines())** from the sp:: package.

```
library(maps)
library(GISTools)

# contour map with an arrow and scale
plot(pov) #Figure 2.1a
maps::map.scale(x=14.5, y=49.5, ratio=FALSE, relwidth=0.2, cex=0.8) # from
the maps::
# from the GISTools:: package
north.arrow(xb=15.9, yb=50, len=0.1, lab="N", cex.lab=0.8, col='gray10')

# contour map with a rose of directions and a geographic grid
plot(voi) #Figure 2.1b
degAxis(1) # from the sp:: package
degAxis(2)
compassRose(15, 49.7,rot=0,cex=1) # from the sp:: package
plot(gridlines(voi), add = TRUE) # from the sp:: package
```

The GISTools:: package uses the sp:: environment. The objects are in classes defined by sp::. Among the commands from GISTools::, the following tools are found:

- commands for converting distance units – **ft2miles**(), **miles2ft**(), **ft2km**(), **km2ft**();
- commands operating on spatial geometries: **generalize.polys**(), which generalises the boundaries, reducing the number of details; **poly.areas**(), calculating the area of geometry; **poly.counts**(), counting points within the geometry; **poly.outer**(), masking the area outside the analysed;
- **jitter.points**() and **bstrap.points**() commands that allow bootstrapping points and are used in the bootstrap kernel;
- **kde.points**() and **level.plot**() commands that allow for kernel density estimates from points and graphics.

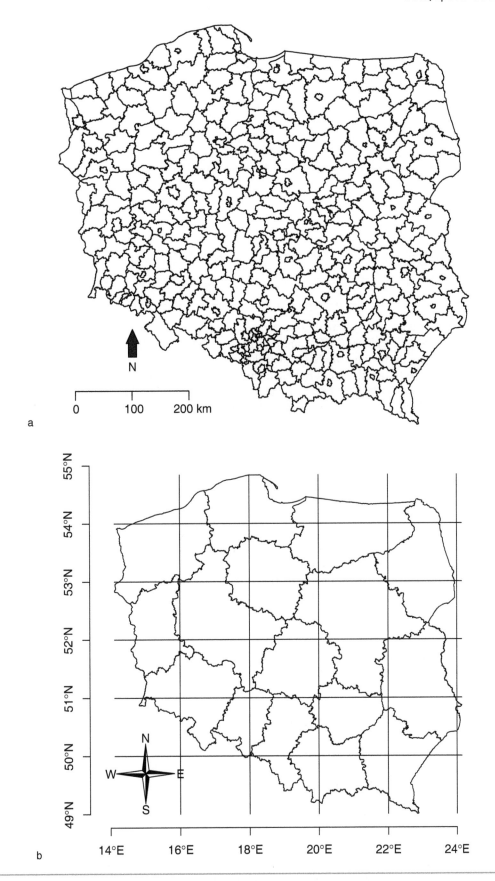

a

b

Figure 2.1 Contour maps from shapefile files – plots with extra supplements

Source: Own study using R and sp::, maptools:: and GISTools:: packages

Spatial geometries, such as contour maps, can be used for calculations. One of the typical calculations is to determine the centres of gravity of these geometries – centroids. The **coordinates**() command from the sp:: package calculates the coordinates of the area centres. The result of this command is the matrix – coordinates of subsequent areas are in the appropriate rows.

```
# coordinates of poviat centroids
crds<-coordinates(pov)
head(crds)
```

```
     [,1]     [,2]
0 22.28163 50.47045
1 22.63489 49.76831
2 22.53846 50.09396
3 21.64403 50.04855
4 22.05174 50.04032
5 22.12398 49.45768
```

Centroids can be drawn as a point layer on the contour map (see Figure 2.2a). Subsequently, three contour layers were imposed: national (NTS0), regional (NTS2) and provincial (NTS4). The following layers are drawn with the **plot**() command by including the option *add=TRUE*. The thickness of the line is determined by the *lwd* option (the higher the natural number, the thicker the line; 1 is the standard thickness). The point layer is added with the **points**() command, which does not require the *add=TRUE* option, since it acts only as another command on the previously prepared graph. One can check more graphical options of the **plot**() command in the **par**() command help.

```
# Figure 2.2 - administrative map and centroids of poviats
plot(pl, lwd=3)
plot(pov, add=TRUE)
plot(voi, add=TRUE, lwd=2)
points(crds, pch=21, bg="red", cex=0.8)
```

Knowing the centres of gravity (centroids) of areas, one can apply text labels to the contour map (see Figure 2.2b). Labels are applied consecutively – the first label is assigned to the first coordinate pair, while the first coordinate pair is calculated for the first area (by order in shapefile). The simplest **text**() command applies the text automatically to the graphic. The more advanced command, **pointLabel**() from the maptools:: package, optimises the location of the labels so that the text overlaps are limited. The graph's margins can be reduced by the **par**() command option. By default, they are set as *mar=c(5, 4, 4, 2)* (successively lower, left, upper, right).

```
# label preparation and coordinate for the provincial map
crds.voi<-coordinates(voi)
voi.df<-as.data.frame(voi)
par(mar=c(1,1,1,1))

# signing regions in an optimized way
plot(voi) #Figure 2.2b
pointLabel(crds.voi, as.character(voi.df$jpt_nazwa_), cex=0.6)

# signing regions in a simple way
plot(voi)
text(crds.voi, as.character(voi.df$jpt_nazwa_), cex=0.6)
par(mar=c(5,4,4,2))
```

Using centroids, it is also possible to identify regions in an interactive way. The **identify**() command from the basic graphics:: package works by comparing the click point with the coordinate set to which the labels are assigned. As a result, the action displays a label at the click location that is assigned to the coordinate pair nearest the selected point. The result of the command can be saved to the object – it enables the dataset to be conditioned and the data to be displayed for the interactive regions indicated.

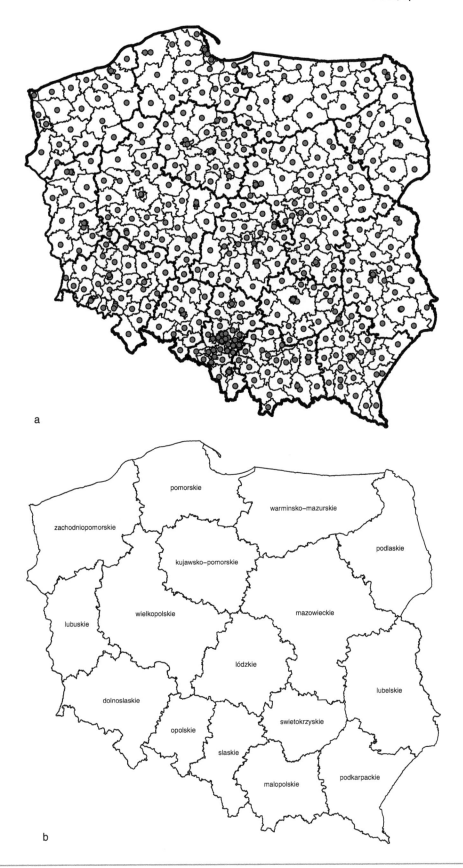

a

b

Figure 2.2 Contour map with centroids: a) as points, b) as text

Source: Own study using R and the sp:: package

```
# interactive edition of the names of regions
plot(pov, border="grey80")
a<-identify(crds, labels=as.character(pov.df$jpt_nazwa_))
data[a,c(3,13:15)]
```

```
          poviat_name1      XA02      XA03      XA04
23   Powiat pruszkowski 132200767  8722309 356314963
90    Powiat tatrzański  16897760  1010412 140907305
305         Powiat Gdynia 213616296 19943782 770562782
```

One can also perform other basic operations on contour maps. The **bbox**() command from the sp:: package specifies the bounding box or coordinates of the vertices of the rectangle described on the contour map. The **proj4string**() command from the sp:: package shows the projection system assigned to the map.

```
bb<-bbox(pov)
bb
```

```
    min max
x 14.12289 24.14578
y 49.00205 54.83579
```

```
proj4string(pov)
#[1] "+proj=longlat +datum=NAD83 +ellps=GRS80 +towgs84=0,0,0"
```

Loaded vector maps can be divided, that is, by selecting one or few spatial units for further investigation. Trimming the map in *SpatialPolygons* class, which consists of the selection of a single or several regions, is carried out by a conditional operation. Choosing regions is equivalent to selecting the appropriate rows of the map. Depending on the data structure, this can be done in several ways. In the dataset connected to the shapefile (part *.dbf of the shapefile), there are often corresponding identifiers that can be used to select a region. This collection can easily be converted to the *data.frame* class using the command **as.data.frame**() to the *SpatialPolygonsDataFrame* object. The map is limited by the condition imposed on the object's rows ([]). However, if the dataset does not contain the appropriate identifiers, for example, due to a different level of aggregation, another conditioning vector should be created, the length of which is the same as the number of regions on the map. The conditioning vector can come from another object – for example, a dataset associated with the map.

In the example subsequently, one province (Lubelskie Province) was selected from the voivfromeship map (NTS2 regions) (see Figure 2.3a). The operation was performed in two ways: 1) by converting the dbf part to *data.frame* and 2) entering the data attribute by @data. In both cases, conditioning is the variable *jpt_nazwa_* from the assigned set of maps. For poviat data, it was necessary to use an external dataset (data), in which the variable *voivodship* determines the poviat's affiliation to voivodeships (see Figure 2.3b).

```
# regional map - for the Lubelskie Voivodeship
# version A
voi.df<-as.data.frame(voi)
lub.voi<-voi[voi.df$jpt_nazwa_=="lubelskie",]
plot(lub.voi, main="Lubelskie NTS2")

# version B
plot(voi[voi@data$jpt_nazwa_=="lubelskie",])

# map of poviats within the Lubelskie Voivodeship
# pov.df<-as.data.frame(pov) # lack of voivodeship identifier
data15<-data[data$year==2015,]
lub.pov<-pov[data15$region_name=="Lubelskie",]
plot(lub.pov, main="Lubelskie NTS4")
plot(lub.voi, add=TRUE, lwd=2)
```

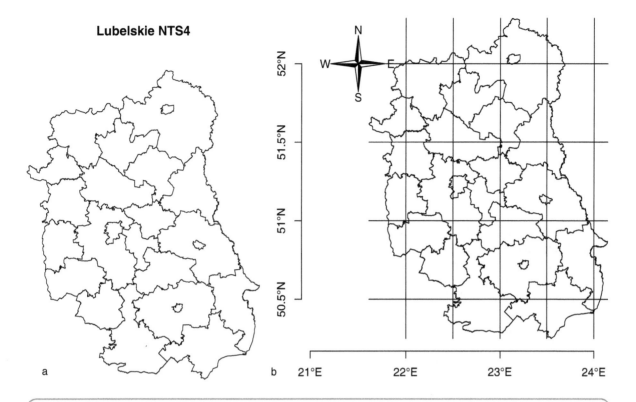

Figure 2.3 Subset of the map: a) many regions selected – poviats in a given voivodeship, b) map subset with grid and axes

Source: Own study with the use of R and the sp:: package

One can also examine the overlapping two contour maps with the **over**() command from the sp:: package. This is particularly useful when there is no external set defining membership. The first argument is data of smaller aggregation (test data whose location is examined), and as the second argument, a base (reference) map is given, from which the ID assigned to the observations from the first argument will be derived. The effect of the **over**() command is the *data.frame* class object, the length as the first argument and content and the second argument dataset. In the following situation, the poviats (NTS4, smaller units) were affiliated with regions (NTS2, voivodships, larger units). The *pov.over* object is of the *data.frame* class and contains subsequent lines from the *data.frame* part of the *SpatialPolygons* object for NTS2 voivodships. The result is shown in Figure 2.3b.

```
# affiliation of poviats (NTS4) to voivodship (NTS2)
pov.over<-over(pov, voi)
head(pov.over)
```

```
  jpt_sjr_ko jpt_kod_je jpt_nazwa_
1    WOJ       18 podkarpackie
2    WOJ       18 podkarpackie
3    WOJ       18 podkarpackie
4    WOJ       18 podkarpackie
5    WOJ       18 podkarpackie
6    WOJ       18 podkarpackie
```

```
lub.pov<-pov[pov.over$jpt_nazwa_=="lubelskie",] # choice of poviats
plot(lub.pov)
```

One can add additional graphic elements to map fragments that are in the parent map class, as shown earlier: grid, compass rose, geographic axes.

```
plot(lub.pov)
degAxis(1)
degAxis(2)
compassRose(21.5, 52,rot=0,cex=1)

plot(gridlines(lub.voi), add = TRUE) # Figure 2.3b
```

The opposite action, combining two spatial objects into one, is possible using the command **spRbind**() from the sp:: package. The command has a structure similar to the usual **rbind**() – by gluing two separate objects in lines. In the example, two Kuyavian-Pomeranian (*Kujawsko-Pomorskie*) and Greater Poland (*Wielkopolskie*) voivodeships were distinguished, after which they were glued into one object. They were further connected by the **spRbind**() command to one object for which typical operations were performed: extracting the database and determining the centroid coordinate with the **coordinates**() command. In the **text**() command, applying labels of voivodeships on the graph, the **jitter**() function was used, which randomly distributes given coordinates in a small radius, thanks to which the labels are slightly above the designated centroid.

```
# separation of two regions
voi.df<-as.data.frame(voi)
voi1<-voi[voi.df$jpt_nazwa_=="kujawsko-pomorskie",]
voi2<-voi[voi.df$jpt_nazwa_=="wielkopolskie",]
plot(voi2) # joint plot
plot(voi1, lwd=2, add=TRUE)

# connection of two regions
voi3<-spRbind(voi1, voi2)
voi3.df<-as.data.frame(voi3)
crds.voi3<-coordinates(voi3)
plot(voi3)
points(crds.voi3)
text(jitter(crds.voi3), labels=voi3.df$jpt_nazwa_)
```

One can also aggregate smaller territorial units to larger ones, for example, to connect poviats in voivodeships. This joining of regions is done with the **unionSpatialPolygons**() command from the maptools:: package. As arguments, the map with the regions to be connected and the connecting vector of the length equal to the number of regions are given. Labels in the connecting vector (text or numbers) indicate which regions should be joined. For example, if the poviat map is to be transformed into a voivodeship, then the connecting vector should contain the name of the voivodeship to which each of the poviats belong (here is the variable *voivodship* from the *sub15* subset for 2015). In the extreme case, when all regions within the map are to be combined into one, enter the same value for all elements of the connecting vector, and then one can use the **rep**() function. In the example, all voivodeships have been merged into the national contour. The **unionSpatialPolygons**() command organises the order of newly created regions according to the values of the connecting vector, that is, alphabetically or ascending.

```
data15<-data[data$year==2015,]
reg1<-unionSpatialPolygons(pov, IDs=data15$region_name) #maptools::
plot(pov) # plot of merged regions
plot(reg1, add=TRUE, border="red", lwd=2)

reg2<-unionSpatialPolygons(voi, IDs=rep(1, times=16)) #maptools::
plot(voi) # plot of all regions merged to one unit
plot(reg2, add=TRUE, border="red")
```

Aggregating areas with the **unionSpatialPolygons**() command results in the loss of the database pinned to the starting object. The *reg1* object is in the *SpatialPolygons* class, although the *pov* object was in the

SpatialPolygonsDataFrame class. New identifiers of territorial units are generated, consistent with the names/ values used in the aggregation. The regions are sorted in ascending/alphabetical order according to this new identifier. One can check this by displaying the object's object summary – using the **lapply**() command. Exit from the mailing list to *data.frame* is possible with the **unlist**() command. A good confirmation is to draw the first and last area and verify their names.

```
plot(reg1)
plot(reg1[1,], border="blue", add=TRUE) # Dolnośląskie on the map
plot(reg1[16,], border="green", add=TRUE) # Zachodniopomorskie on the map
lapply(reg1@polygons, slot, "ID")
```

```
[[1]]
[1] "Dolnośląskie"

[[2]]
[1] "Kujawsko-Pomorskie"

[[3]]
[1] "Lubelskie"

[[4]]
[1] "Lubuskie"
```

```
unlist(lapply(reg1@polygons, slot, "ID"))
```

```
 [1] "Dolnośląskie"  "Kujawsko-Pomorskie"  "Lubelskie" "Lubuskie"
 [5] "Łódzkie"  "Małopolskie"  "Mazowieckie"  "Opolskie"
 [9] "Podkarpackie"  "Podlaskie"  "Pomorskie"  "Śląskie"
[13] "Świętokrzyskie"  "Warmińsko-Mazurskie" "Wielkopolskie"
[16] "Zachodniopomorskie"
```

Aggregation of areas often requires the aggregation of a dataset as well as its sorting to achieve order compatibility with a new map. In the example, selected variables from the poviat dataset were aggregated (average, sum) to the provincial level and sorted. The **aggregate**() command processes data according to the indicated function and sorts it alphabetically according to the grouping variable. One can also add line labels to the **rownames**() command so as not to store the region names as separate variables and change the column headers with the **colnames**() command.

```
# the following variables were selected:
# column 7 - region_name
# column 26 - XA15 - working in total
# column 32 - XA21 - unemployment rate

data15.lim<-data[data$year==2015, c(7,26,32)]
XA15.agg<-aggregate(data15.lim$XA15, by=list(data15.lim$region_name), sum,
na.rm=TRUE)
XA21.agg<-aggregate(data15.lim$XA21, by=list(data15.lim$region_name), mean,
na.rm=TRUE)
data.agg<-cbind(XA15.agg, XA21.agg$x)
rownames(data.agg)<-as.character(XA15.agg$Group.1)
colnames(data.agg)<-c("voi", "XA15", "XA21")
head(data.agg)
```

```
                                  voi   XA15      XA21
Dolnośląskie              Dolnośląskie 825572 12.000000
Kujawsko-Pomorskie  Kujawsko-Pomorskie 556931 15.869565
Lubelskie                    Lubelskie 680432 12.845833
Lubuskie                      Lubuskie 254095 11.835714
Łódzkie                        Łódzkie 752432 10.379167
Małopolskie                Małopolskie 1025463 9.881818
```

2.2 Creating, checking and converting spatial classes

Objects of the *SpatialPolygons* and *SpatialPoints* classes can be created based on transformations of *data.frame* objects.

Having a point dataset, as in *dataset2*, one should specify the variables containing the location of points (*x, y*) and the variable concerning the characteristics (*z*). Following are the columns selected from the company's collection that contain geographical coordinates and company size. Due to the very large number of firms in which the variable poses the minimum value, a decision was made to selectively crop the collection, that is, to select 500 observations with the value of *empl=5* and all observations when *empl=30, 150, 600, 1500*. This was done through subsets – the entire set was divided by rows into subsets according to the value of the *empl* variable, and the subset for *empl=5* was cut and merged by rows into one coherent set. The column names have also been changed to *x, y* and *z*.

```
firms<-read.csv("geoloc_data_firms.csv", header=TRUE, dec=",", sep=";")
firms.lim<-firms[, c(12,13,20)]
head(firms.lim)
```

```
coords.x1 coords.x2 empl
1 22.00263 51.40935   5
2 22.34528 51.35417   5
3 22.96182 50.31523   5
4 21.83905 50.95910   5
5 23.37778 51.96064   5
6 22.25986 51.74347   5
```

```
# division of the dataset into subsets according to the variable empl
firms.5<-firms.lim[firms.lim$empl==5,]
firms.30<-firms.lim[firms.lim$empl==30,]
firms.150<-firms.lim[firms.lim$empl==150,]
firms.600<-firms.lim[firms.lim$empl==600,]
firms.1500<-firms.lim[firms.lim$empl==1500,]
firms.5lim<-firms.5[1:500,] # limiting the subset

# combining subsets by rows into one set
firms.f<-rbind(firms.5lim, firms.30, firms.150, firms.600, firms.1500)
names(firms.f)
colnames(firms.f)<-c("x","y","z") # change of column names
```

Changing the object class from *data.frame* to *SpatialPointsDataFrame* is done by declaring variables – the coordinate in the set using the **coordinates()** command used on the left (resultant) and not the right (command typical) page. In the absence of a fixed projection (which is typical for data in the *data.frame* class), use the **proj4string()** command similarly on the resulting page, declaring the selected projection. If one wants to convert a projection into a planar coordinate system, one should use the **spTransform()** command with the term *merc*, and for spherical coordinates, use the term *longlat*. In the visualisation (see Figure 2.4), for spherical data, it was possible to apply a fragment of the contour map due to coordinate compatibility.

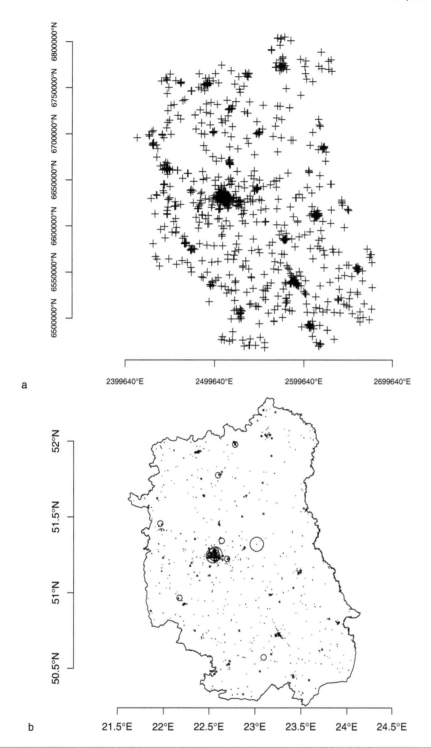

Figure 2.4 Point data in the *SpatialPointsDataFrame* class: a) planar projection, b) spherical projection

Source: Own study using R and the sp:: package

```
class(firms.f)
#[1] "data.frame"

coordinates(firms.f)<-c("x","y") # changing the class of object
class(firms.f)
```

```
#[1] "SpatialPointsDataFrame"
#attr(,"package")
#[1] "sp"

proj4string(firms.f)
#[1] NA

proj4string(firms.f)<-"+proj=longlat +datum=WGS84 +ellps=WGS84"

# planar coordinates
firms.f.merc<-spTransform(firms.f, CRS("+proj=merc +datum=WGS84
+ellps=WGS84"))

# spherical coordinates
firms.f.longlat<-spTransform(firms.f, CRS("+proj=longlat +datum=WGS84
+ellps=WGS84"))

# chart of planar points
plot(firms.f.merc) # Figure 2.4a
degAxis(1)
degAxis(2)

# graph of spherical points # Figure 2.4b
plot(firms.f.longlat, cex=firms.f.longlat$z/500, pch=1)
plot(voi[voi@data$jpt_nazwa_=="lubelskie",], add=TRUE)
degAxis(1)
degAxis(2)
```

It is also possible to create and draw an object of the *SpatialPointsDataFrame* class on the basis of vector coordinates defined, complemented by the created values of the variable (see Figure 2.5a). The *x* and *y* coordinates were created as separate objects – vectors – and further connected and converted using the **SpatialPointsDataFrame()** command. The command's first argument is the coordinates matrix, which is converted on the fly by specifying the shape of the matrix by defining two columns (*ncol=2*). The second argument is the characteristics, which are created on the fly as a *data.frame* object in which there is an *ID* variable with values from 1 to *n*, where *n* is the ordinal number of the last point. The third argument was given a spherical projection.

Further, on the basis of the same points, the *SpatialLines* class object was created. The **sp2sl()** command from the quickPlot:: package generates the *SpatialLines* object (see Figure 2.5b) and requires two arguments: a matrix of "target" points and a matrix of starting points. For this reason, the *xy.from* and *xy.to* objects in *matrix* class were created. In the *xy.to* object, the rows were moved by 1; that is, the last row became the first, the first second and so on.

```
# creating SpatialPointsDataFrame
x<-c(5,6,7,8,8,7,6,5,4,3,2,2,3,4)
y<-c(2,3,4,5,6,7,7,6,7,7,6,5,4,3)
xy<-SpatialPointsDataFrame(matrix(c(x,y), ncol=2),
    data.frame(ID=seq(1:length(x))),
    proj4string=CRS("+proj=longlat +ellps=WGS84 +datum=WGS84"))
plot(xy)
axis(1)
axis(2)

library(quickPlot)
xy.m<-matrix(c(x,y), ncol=2)
xy.from<-xy.m
xy.m<-rbind(xy.m[dim(xy.m)[1],],xy.m) # adding an extra line
xy.to<-xy.m[-dim(xy.m)[1],] # deleting the last row
xy.lines<-sp2sl(xy.to, xy.from) # the SpatialLines class
class(xy.lines)
```

```
#[1] "SpatialLines"
#attr(,"package")
#[1] "sp"

plot(xy.lines, lwd=2)
plot(gridlines(xy), lty=3, add = TRUE) # from the sp package
axis(1)
axis(2)
```

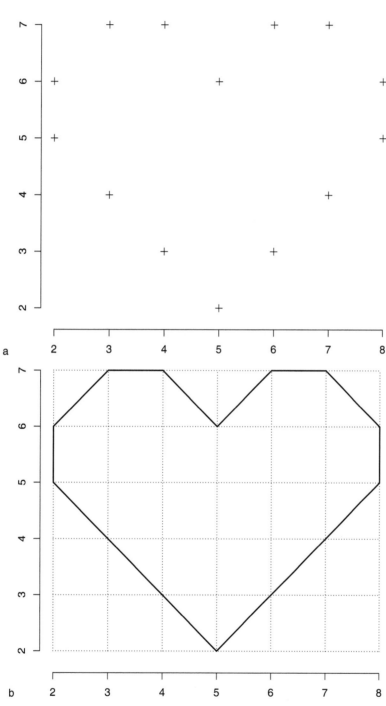

Figure 2.5 Generating point data and merging the points with the line

Source: Own study using R and sp:: and quickPlot:: packages

On the basis of the point matrix, one can also create a polygon in the *SpatialPolygons* class in addition to the *SpatialLines* object. The easy **spPolygons()** command from the raster:: package requires a matrix of points (or a list of points' matrices) and possibly characteristics and projections of the object.

```
library(raster)
xy.poly<-spPolygons(xy.m)
class(xy.poly)

#[1] "SpatialPolygons"
attr(,"package")
#[1] "sp"
```

One can add a dataset to the *SpatialPolygons* class object that one created and convert it to the *SpatialPolygonsDataFrame* class. This is done with the command **SpatialPolygonsDataFrame()**, in which the arguments are given, as in the **SpatialPointsDataFrame()** command, a spatial object – a map in the *SpatialPolygons* class, a dataset and possibly a projection.

```
xy.polydf<-SpatialPolygonsDataFrame(xy.poly, data.frame(ID=1, type="A"))
xy.polydf@data
# ID type
#1 1 A

class(xy.polydf)
#[1] "SpatialPolygonsDataFrame"
#attr(,"package")
#[1] "sp"
```

Screening of the available information in spatial objects should start by displaying the available slots and sockets with the command **str()**. The slot names can be displayed with the **slotNames()** command. By inserting the name of the object as its argument, the names of the sockets are obtained, and by entering a reference to a specific socket (e.g. *Polygons [[1]]*), the names of slots in the socket are obtained.

```
str(xy.polydf)
```

```
Formal class 'SpatialPolygonsDataFrame' [package "sp"] with 5 slots
..@ data :'data.frame': 1 obs. of 2 variables:
.... $ ID : num 1
.... $ type: Factor w/ 1 level "A": 1
..@ polygons :List of 1
.... $ :Formal class 'Polygons' [package "sp"] with 5 slots
.. .. .... @ Polygons :List of 1
.. .. .. .. ..$ :Formal class 'Polygon' [package "sp"] with 5 slots
.. .. .. .. .. .. ..@ labpt : num [1:2] 5 4.98
.. .. .. .. .. .. ..@ area : num 19
.. .. .. .. .. .. ..@ hole : logi FALSE
.. .. .. .. .. .. ..@ ringDir: int 1
.. .. .. .. .. .. ..@ coords : num [1:15, 1:2] 4 3 2 2 3 4 5 6 7 8 . . .
.. .. .... @ plotOrder: int 1
.. .. .... @ labpt : num [1:2] 5 4.98
.. .. .... @ ID : chr "1"
.. .. .... @ area : num 19
..@ plotOrder : int 1
..@ bbox : num [1:2, 1:2] 2 2 8 7
.... - attr(*, "dimnames")=List of 2
.. .. ..$ : chr [1:2] "x" "y"
.. .. ..$ : chr [1:2] "min" "max"
..@ proj4string:Formal class 'CRS' [package "sp"] with 1 slot
.. .. ..@ projargs: chr NA
```

```
slotNames(xy.polydf) # names of slots
#[1] "data" "polygons" "plotOrder" "bbox" "proj4string"

slotNames(xy.polydf@polygons[[1]]) # slot names within the slots
#[1] "Polygons" "plotOrder" "labpt" "ID" "area"

xy.polydf@bbox
# min max
#x 2 8
#y 2 7
```

It is not possible to enter the inside of the slot defined in the slot by $ or @. The **lapply**() command is used in which both the socket and the slot are defined. The result is a list that can be converted to *data.frame* with the **unlist**() command.

```
lapply(xy.polydf@polygons, slot, "ID")
#[[1]]
#[1] "1"
unlist(lapply(reg1@polygons, slot, "ID"))
```

```
 [1] "Dolnośląskie"  "Kujawsko-Pomorskie"  "Lubelskie" "Lubuskie"
 [5] "Łódzkie"  "Małopolskie"  "Mazowieckie"  "Opolskie"
 [9] "Podkarpackie"  "Podlaskie"  "Pomorskie"  "Śląskie"
[13] "Świętokrzyskie"  "Warmińsko-Mazurskie"  "Wielkopolskie"
[16] "Zachodniopomorskie"
```

2.3 Selected colour palettes

In R, a large range of colours and predefined colour palettes are available in various packages. Colour palettes are a very important element of aesthetic graphics. Selected ones will be discussed.

Regardless of the colour generation method, there is a need to visualise them. In some packages, one can easily display selected colours. In other cases, it is worth using the **image**() command, which for a given number of elements and the colours assigned to them creates a neat graph of vertical colour stripes. Its syntax is given successively as x and y locations, z values to be drawn and colours assigned to them. In the following example, for n colours, x is given as a natural number from 1 to n, y is given a value of 1, a matrix of values from 1 to n is given and a vector of created colours is given as colours. Colours can also be displayed as part of a raster. The **raster**() command from the raster:: package creates boxes – specify the beginning (*xmn*) and end (*xmx*) of the x axis and similarly the beginning (*ymn*) and end (*ymx*) of the y axis, and define the number of rows (*nrows*) and columns (*ncols*). It is further necessary to assign any values to bars – in the example, natural numbers from 1 to 660 are assigned. Raster drawn with the **plot**() command requires the definition of colours in the *col* option. The **show_col**() command from the scales:: package is also available. As an argument, a colour vector is given, and the command displays it in boxes.

The most common colour notation is RGB (red/green/blue), which contains six characters preceded by a # sign. Most often, the commands generate shades methodically within a given palette, displaying their code in RGB. RGB colours can also be generated by submitting the **rgb**() command from the grDevices:: basic package of RGB component values from 0 to 1. One must specify arguments for red, green and blue, each from 0 to 1, although there is a parallel notation from 0 to 255. Subsequently, random RGB component vectors were generated and colours were created on this basis.

One can also call individual colours by their names in English – there are 657. One can call them directly with the **colors**() command from the grDevices:: package. The **demo**("**colors**") command displays the colour groups.

```
# generating random colors
a<-rgb(runif(21,0,1), runif(21,0,1), runif(21,0,1))
a
```

```
 [1]  "#9579E0" "#CBA57A" "#50512A" "#30334A" "#633886" "#DE92E4" "#E3FE03"
 [8]  "#818BC8" "#5DB214" "#053E8E" "#381D62" "#C5B4B5" "#FBFFA7" "#BB169A"
 [15] "#3DB2F0" "#ED05AB" "#39AAC6" "#DE97C9" "#837A85" "#ED027A" "#8CB7EE"
```

```
a.dim1<-length(a)
image(1:a.dim1, 1, as.matrix(1:a.dim1), col=a, xlab="grDevices::rgb() /
random colors ") # vertical stripes with colours

colors() # the first few lines of color names
```

```
 [1]  "white"        "aliceblue"     "antiquewhite"
 [4]  "antiquewhite1" "antiquewhite2" "antiquewhite3"
 [7]  "antiquewhite4" "aquamarine"    "aquamarine1"
[10]  "aquamarine2"   "aquamarine3"   "aquamarine4"
[13]  "azure"         "azure1"        "azure2"
[16]  "azure3"        "azure4"        "beige"
[19]  "bisque"        "bisque1"       "bisque2"
```

```
library(raster)
r<-raster(xmn=0, xmx=22, ymn=0, ymx=30, nrows=30, ncols=22)
r[]<-1:660
plot(r, col=colors()) #rasters (cells) with colours
```

There are commands that allow one to navigate between different methods of notation. With the **col2rgb()** command from the grDevices:: package, one can change the English names of colours, as well as colours expressed in hexadecimal, into RGB components:

```
col2rgb(c("azure", "azure1", "azure2"), alpha=FALSE)
```

```
       [,1] [,2] [,3]
red     240  240  224
green   255  255  238
blue    255  255  238
```

```
col2rgb(c("#4B424F", "#BFD15C", "#A44845"), alpha=FALSE)
```

```
       [,1] [,2] [,3]
red      75  191  164
green    66  209   72
blue     79   92   69
```

One of the most popular packages creating attractive colour palettes is the RcolorBrewer:: package. The **brewer.pal()** command from the RcolorBrewer:: package selects colours from different keys. There are several palettes to choose from: sequential – Blues, BuGn (blue-green), BuPu (blue-violet), GnBu (green-blue), Greens, Grays, Oranges, OrRd (orange-red), PuBu (violet-blue), PuBuGn (violet-blue-green), PuRd (violet-red), Purples (violet), RdPu (red-violet), Reds, YlGn (yellow) green), YlGnBu (yellow-green-blue), YlOrBr (yellow-orange-brown), YlOrRd (yellow-orange-red) – and divergent – BrBG (brown-green), PiYG (pink-green), PRGn (violet – green), PuOr (violet-orange), RdBu (red-blue), RdGy (red-gray), RdYlBu (red-yellow-blue), RdYlGn (red-yellow-green) and Spectral (rainbow). The palettes have 9 or 11 colours. As arguments to the **brewer.pal()** command, the number of colours and the name of the palette are specified. The display of the selected palette is possible with the command **display.brewer.pal()**, and the graphics of all palettes are available in the **display.brewer.all()** command.

```
library(RColorBrewer)
display.brewer.all() # all palettes from the package
display.brewer.pal(11,'Spectral') # displaying the palette
display.brewer.pal(9,'OrRd') # displaying the palette
cols<-brewer.pal(n=5, name="RdBu") # saving selected colors
#[1] "#CA0020" "#F4A582" "#F7F7F7" "#92C5DE" "#0571B0"
```

The wesanderson:: colour package, inspired by the colouring of Wes Anderson's films, is becoming more and more popular. One can choose from the following palettes: *BottleRocket1, BottleRocket2, Rushmore1, Royal1, Royal2, Zissou1, Darjeeling1, Darjeeling2, Chevalier1, FantasticFox1, Moonrise1, Moonrise2, Moonrise3, Cavalcanti1, GrandBudapest1, GrandBudapest2, IsleofDogs1, IsleofDogs2*. Most of the palettes have four to five colours available in the discrete version, which in the continuous version are interpolated on a differentiated gradient. A chart appears by calling a colour object.

```
library(wesanderson)
cols1<-wes_palette("GrandBudapest1", 4, type="discrete")
cols1
cols2<-wes_palette("GrandBudapest1", 21, type="continuous")
cols2
```

In the quickPlot:: package, command **divergentColors()** generates 21 divergent colours with the possibility of varying intensity. The output colour, target colour, minimum value, maximum centre value and centre colour are specified as arguments. As a result, successive colours are recorded as RGB in the hexadecimal system.

```
library(quickPlot)
# palette from the command example
a<-divergentColors("darkred", "darkblue", -10, 10, 0, "white")
a
```

```
 [1] "#8B0000" "#961919" "#A23333" "#AD4C4C" "#B96666" "#C47F7F" "#D09999"
 [8] "#DCB2B2" "#E7CCCC" "#F3E5E5" "#FFFFFF" "#E5E5F3" "#CCCCE7" "#B2B2DC"
[15] "#9999D0" "#7F7FC4" "#6666B9" "#4C4CAD" "#3232A2" "#191996" "#00008B"
```

```
a.dim1<-length(a)
image(1:a.dim1, 1, as.matrix(1:a.dim1), col=a,
xlab="quickPlot::divergentColors() / darkred-darkblue")

# another palette
a<-divergentColors("chocolate4", "peachpuff4", -10, 10, 0, "white")
a
```

```
 [1] "#8B4513" "#96572A" "#A26A42" "#AD7C59" "#B98F71" "#C4A289" "#D0B4A0"
 [8] "#DCC7B8" "#E7D9CF" "#F3ECE7" "#FFFFFF" "#F3F1EF" "#E7E3E0" "#DCD6D0"
[15] "#D0C8C1" "#C4BBB2" "#B9ADA2" "#AD9F93" "#A29283" "#968474" "#8B7765"
```

```
a.dim1<-length(a)
image(1:a.dim1, 1, as.matrix(1:a.dim1), col=a,
xlab="quickPlot::divergentColors() / chocolate-peachpuff")
```

An important set of palettes is generated by the viridis:: package. They are palettes with colours fully distinguishable by colour-blind people. The authors in the package vignettes show how different palettes are seen by people who do not see green, red and blue and also after conversion to gray (e.g. in print). By calling the command / object **viridis.map**, one gets a set of 1280 colours, with RGB components (from 0 to 1), divided into five palettes (*opt*): magma (A), inferno (B), plasma (C), viridis (D) and cividis (E). The colours in the discrete version can be obtained with the command **viridis()**, where the number of colours is given, and in the palette type option (e.g. *option="B"*).

```
library(viridis)
library(scales)
viridis.map
```

```
            R            G           B opt
1 0.001461591 0.0004661278 0.01386552   A
2 0.002257640 0.0012949543 0.01833115   A
3 0.003279432 0.0023045299 0.02370833   A
4 0.004512302 0.0034903767 0.02996471   A
5 0.005949770 0.0048428500 0.03712967   A
6 0.007587985 0.0063561362 0.04497308   A
```

```
col1<-viridis(15, option="D") # the default viridis colors
show_col(col1)

col2<-viridis(15, option="B") # inferno palette
show_col(col2)
```

The viridis:: palettes interact with the graphics generated by ggplot2:: much more interestingly. The **scale_fill_viridis()** command automatically adjusts colours to the scale of the phenomenon. Using the code in vignettes for the viridis:: package, it looks like the following. Colours can be selected continuously (in the **scale_color_viridis_c()** command) or in a discrete manner (in the **scale_color_viridis_d()** command).

```
library(ggplot2)
# coloured hexagons circled
ggplot(data.frame(x=rnorm(10000),y =rnorm(10000)), aes(x=x, y=y)) +
  geom_hex() + coord_fixed() + scale_fill_viridis() + theme_bw()

# creating a data set in the data.frame class
data<-data.frame(x=1:100, y=rnorm(100), z=sample(1:100, 100))
head(data)
```

```
  x          y  z
1 1  0.4689631 41
2 2  0.7698945 99
3 3  0.4445094 19
4 4  0.2022598 46
5 5 -0.5204029 18
6 6  1.3330075 65
```

```
# scatterplot in colours
ggplot(data) +aes(x, y, color=z) +geom_point() +scale_color_viridis_c()
```

Improvement of graphics in the range of colour selection and division of variables into ranges is possible with the scales:: package. It contains dozens of commands dedicated to data transformations, data management and division into intervals, as well as advanced colour management from various palettes. Similarly, the advanced package is colorspace::, which allows advanced colour conversions and advanced creation of one's own palettes. One can also use the **smoothScatter()** function from the graphics:: package and the **colorRamp()**

and **colorRampPalette**() functions from the grDevices:: basic package, which return functions that smooth colours and match colours to the next variable values. Another interesting package is RImagePalette::, which takes colours from the analysed photo/graph and creates a palette which allows one to replicate colours. Colour management for raster graphics is available in the quickPlot:: package with the **getColors**() command. Due to the limited space in the book, they will not be discussed in detail here.

2.4 Basic contour maps with a colour layer

There are several schemes for preparing a map with a colour layer. Schemes use different ways to recode the variable value into a colour vector. They differ in the degree of automation of variable division into ranges – from manual to fully automatic. The legend describing variable levels is also different. The schemes also use different colour palettes. The common element is the use of a contour map as a drawing base and the extraction of a vector of values to draw in the same order as the regions on the map.

All colour maps were made on the poviat dataset (NTS4) – *dataset1*. Various variables were selected from the set, among others *XA21* – unemployment rate, *region_nr* – voivodeship number.

```
# loading poviat (NTS4) data
data<-read.csv("data_nts4_2019.csv", header=TRUE, dec=",", sep=";")
summary(data)
```

Scheme 1 – with colorRampPalette() from the grDevices:: package

For a defined number of intervals (*nclr* – number of colours), the palette of shades is chosen (here *Reds*). The **brewer.pal**() command creates a colour vector. Next, the **colorRampPalette**() command from the grDevices:: package "pulls" the colour vector and creates an interpolating function (*fillRed*) that continuously transcodes the variable values into a colour vector. The *colcode* object contains the names of colours that are drawn with the ordinary **plot**() command. The first county layer (NTS4) contains information on the colours of the regions according to the *colcode* object, while the county contours were omitted by the option *lty=0* (see Figure 2.6). The second regional layer (NTS2) has been superimposed on the previous graph thanks to the *add=TRUE* option, while voivodeship contours are drawn in light gray (*border = "gray60"*). The **colorlegend**() command from the shape:: package creates a vertical bar that specifies the colour scale. The method of applying the scale of the map and the direction of the rose is explained earlier.

```
# Figure 2.6a
library(shape)
variable<-data$XA21[data$year==2015]
maxy<-40
breaks<-c(0, 5, 10, 15, 20, 25, 30, 35, 40) # used in the legend
nclr<-8
plotclr<-brewer.pal(nclr, "Reds") # from the RColorBrewer package
fillRed<-colorRampPalette(plotclr) # from the grDevices package
colcode<-fillRed(maxy)[round(variable) + 1] # fillRed is a function

plot(pov, col=colcode, lty=0)
plot(voi, add=TRUE, lwd=1, border="gray60")

maps::map.scale(x=18.0, y=49.3, ratio=FALSE, relwidth=0.2, metric=TRUE)
compassRose(16, 49.8,rot=0,cex=1) # from the sp package

colorlegend(posy=c(0.05,0.9), posx=c(0.9,0.92), col=fillRed(maxy),
zlim=c(0, maxy), zval=breaks, main.cex=0.9) # from the shape:: package

title(main="Unemployment rate in 2015 r.", sub="At the NTS4 level,
according to the Central \n Statistical Office data")
```

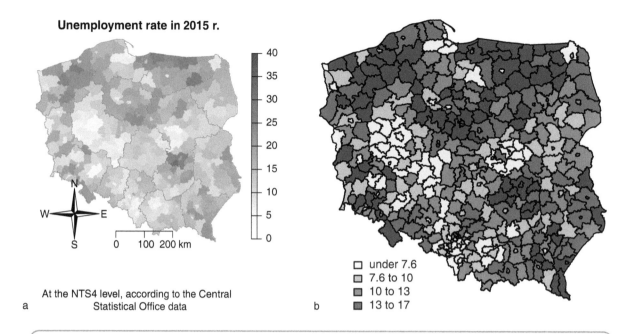

Figure 2.6 Regional map with a colour layer: a) using **colorRampPalette()**, b) using **choropleth()**

Source: Own study using R with sp::, shape:: and GISTools:: packages

Scheme 2 – with choropleth() from the GISTools:: package

The **choropleth**() command from the GISTools:: package draws the specified variable on the specified contour map. The colours and intervals are selected automatically. With the **auto.shading**() command, one can generate a shade vector that has been or will be used in the **choropleth**() command (an automatic / controlled method). With the **choro.legend**() command, one can enter a legend in which colours are indicated.

In the first version, the map was created completely automatically in the first line with the **choropleth**() command. Next, the **auto.shading**() command has saved the colours used previously. They were used in the legend **choro.legend**() (see Figure 2.6b).

In the second version in the **auto.shading**() command, the user can specify the colours himself – here purple (*Purples*) from the **brewer.pal**() palette. The contour map with the colour layer is drawn with the **choropleth**() command.

```
variable<-data$XA21[data$year==2015]
library(GISTools)

choropleth(pov, variable) # from the GISTools:: package, Figure 2.6b
shades<-auto.shading(variable)
choro.legend(15, 50, shades, cex=0.65, bty="n")

shades<-auto.shading(variable, n=6, cols=brewer.pal(6, "Purples"))
choropleth(pov, variable, shading=shades)
```

In the graphics made by **choropleth**(), one can also manage the transparency of colours. The *add.alpha* option on the **auto.shading**() command allows one to get slightly transparent colours (by setting the alpha factor from 0 to 1). The *cutter* option in the **auto.shading**() command allows one to manage how the variable is divided into ranges. Divisions by quantiles in *quantileCuts* or mean and standard deviation in *sdCuts* are possible. In the legend created by the **choro.legend**() command, the variable interval labels are automatically assigned in English, but they can be changed in the options (*under, over, between*).

```
# division into automatic intervals,
# lower transparency
shades<-auto.shading(variable, n=6, cols=add.alpha(brewer.pal(6, "Greens"), 0.5))
```

```
choropleth(pov, variable, shading=shades)
choro.legend(14.3, 50.2, shades, under="below", over="above", between="to",
cex=0.6, bty="n")

# division into intervals by mean and std.dev,
# higher transparency
shades<-auto.shading(variable,     n=6,        cols=add.alpha(brewer.pal(6,
"Greens"),0.35), cutter=sdCuts)
choropleth(pov, variable, shading=shades)
choro.legend(14.3, 50.2, shades, under="below", over="above", between="to",
cex=0.6, bty="n")
```

Scheme 3 – with findInterval() from the base:: package

The **findInterval**() command was one of the first commands in R, so the following method is historically the oldest. The **findInterval**() command checks which range the specified value of the tested variable falls into and further assigns a specific colour, for example, the third colour when the variable is in the third range. This is a fairly effective procedure for determining which colour to use to fill another region.

In the example, the correctness of assigning poviats to voivodeships is checked. Among poviat data, there is a variable *region_nr* (it takes even values from 2 to 32), which determines which province (NTS2) belongs to a given poviat (NTS4). The application of the colour layer according to these codes and the provincial contour allows one to check whether each poviat was properly assigned (see Figure 2.7).

As part of this approach, one needs to prepare the intervals (*brks* object) to which variable values will be assigned (*variable*) and to which subsequent colours will be assigned (*cols*). Labels of voivodeship names were plotted in provincial centroids.

```
variable<-data$region_nr[data$year==2015]
brks<-(0:16)*2
brks
#[1]  0 2 4 6 8 10 12 14 16 18 20 22 24 26 28 30 32

cols<-c("blue3", "cornflowerblue", "seagreen1", "yellow", "chocolate1",
"orangered1", "brown3", "coral4", "salmon4", "aquamarine3", "darkgreen",
"chartreuse3", "cyan4", "darkred", "darkviolet", "cadetblue3", "blue")

par(mar=c(1,1,1,1)) # contour plot & each region with own colour layer
plot(pov, col=cols[findInterval(variable, brks)], border="grey80")
plot(voi, add=TRUE, lwd=1)
par(mar=c(5,4,4,2))

# labels of provincial names
crds.voi<-coordinates(voi)
voi.df<-as.data.frame(voi)
text(crds.voi, label=voi.df$jpt_nazwa_, cex=0.7, font=2)
```

The same applies to the shading of areas instead of the colour layer. In the *dens* option of the **plot**() command, the density of the shading is determined using natural numbers. The **findInterval**() command combines the value of a variable with the appropriate shading.

```
variable<-data$region_nr[data$year==2015]
brks<-(0:16)*2
dens<-(2:length(brks))*3

par(mar=c(1,1,1,1)) # Figure 2.7a
plot(pov, density=dens[findInterval(variable, brks, all.inside=TRUE)],
border="grey80")
plot(voi, add=TRUE, lwd=1)
par(mar=c(5,4,4,2))
```

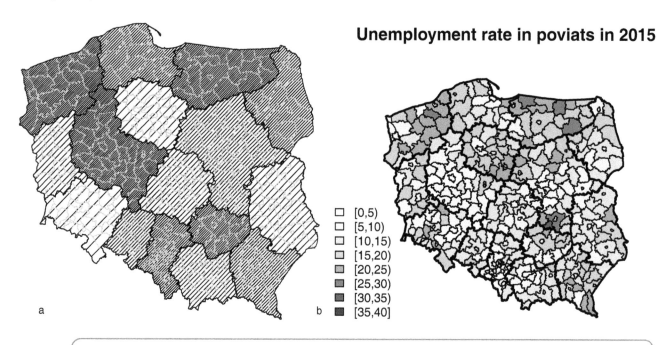

Unemployment rate in poviats in 2015

□ [0,5)
□ [5,10)
□ [10,15)
□ [15,20)
▨ [20,25)
▨ [25,30)
■ [30,35)
■ [35,40]

a   b

Figure 2.7 Regional maps: a) made using **findInterval()** with a shading layer, b) **classIntervals()** and **findColours()**

Source: Own study using R and sp:: and spdep:: packages

Scheme 4 – with findColours() from the classInt:: package

The following diagram uses the **classIntervals()** and **findColours()** commands from the classInt:: package. The **classIntervals()** command cuts the variable into the specified intervals – one can specify the division in the option *style*: by fixed intervals, mean and standard deviation (*sd*), equal ranges (*equal*), ranges by round values (*pretty*), quantile (*quantile*), *k*-means (*kmeans*), hierarchical (*hclust*), so-called bagged clustering (*bclust*), following the Fisher procedure (*fisher*) or Jenks (*jenks*). Next, the **findColours()** command assigns colours to intervals, checking into which interval the next values of the variable fall. The effect of the **findColours()** command is an object that contains an RGB colour vector of a length such as a variable; the attribute *attr(, palette)*, the next colours used and the attribute *attr(, table)*, a table showing the intervals and their number – the same as a class object resulting from the **classIntervals()** command. In the **plot()** command, the colours are taken from the table-colour object. In the legend, the legend labels are taken from the table headings with numbers (these are nicely formatted boundaries of the ranges), and the colours are taken from the colour vector – all from the *color.table* object (see Figure 2.7b).

```
# Figure 2.7b
library(classInt)
variable<-data$XA21[data$year==2015]
summary(variable)

intervals<-8
colors<-brewer.pal(intervals, "BuPu") # choice of colors
classes<-classIntervals(variable, intervals, style="fixed",
    fixedBreaks=c(0, 5, 10, 15, 20, 25, 30, 35, 40))
color.table<-findColours(classes, colors)

plot(pov, col=color.table)
plot(voi, lwd=2, add=TRUE)
legend("bottomleft", legend=names(attr(color.table, "table")),
fill=attr(color.table, "palette"), cex=0.8, bty="n")
title(main="Unemployment rate in poviats in 2015")
```

Scheme 5 – with spplot() from the sp:: package

Another way to draw maps with a colour layer is to use the **spplot()** command from the sp:: package. Unlike the previous schemes, this command requires that the data used for mapping be part of the *SpatialPolygonsData-Frame* object. In the **spplot()** command, the map is given as the first argument, and the second argument is the variable (taken in quotation marks). In the *col.regions* option, one can specify a colour palette, and in the *cuts* option the number of ranges on the colour scale. The **spplot()** command works not only on polygons but also on points, rasters, lines and so on.

The following example shows how to combine data outside the drawing scheme. The colour layer – the value of a variable – comes from the aggregated poviat data. Due to the different spelling of the names of voivodeships (from uppercase and lowercase), it was necessary to reconcile the name and conversion to lowercase letters with the **tolower()** command. The order of territorial units was different – random in the map and alphabetical in the aggregated set of data. The **merge()** command has been attached to the names of provinces in the order from the map (export from *SpatialPolygonsDataFrame* to data.frame). In merging of datasets, sorting has been blocked (*sort=FALSE*) to preserve the original order as in the map. One can also add the identifiers matching the id in *SpatialPolygonsDataFrame* (here from 0 to 15) as row headers (**rownames()** command).

A gradual change was made to the **spplot()** drawing command. The first call is based on automatic settings. In the second approach, the colour palette has been changed to the inverted (**rev()** command) colour vector from the viridis:: **inferno()** shade. In the third approach, the **brewer.pal()** palette was used, and the number of intervals was changed (the *cuts* option). In the fourth approach, the cut points were prepared with the **classIntervals()** command from the classInt:: package, and the divergent colours palette was obtained with the **divergentColors()** command from quickPlot::.

```
# data preparation - aggregation and merging
data15.lim<-data[data$year==2015, c(12,21,25)]
XA15.agg<-aggregate(data15.lim$XA15, by=list(data15.lim$region_name), sum,
na.rm=TRUE)
XA21.agg<-aggregate(data15.lim$XA21, by=list(data15.lim$region_name), mean,
na.rm=TRUE)
data.agg<-cbind(XA15.agg, XA21.agg$x)
colnames(data.agg)<-c("voi", "XA15", "XA21")
data.agg$voi<-tolower(data.agg$voi) # changing the size of letters to small


# combining data with identifiers from the map
order<-data.frame(order=voi@data[,6]) # order of regions from the map
voi.set<-merge(order, data.agg, by.x="order", by.y="voi", sort=FALSE, all.x=TRUE)
rownames(voi.set)<-0:15 # renaming the rows


# attach a dataset to SpatialPolygonsDataFrame
voi<-SpatialPolygonsDataFrame(voi, voi.set)


# map by default settings
spplot(voi, "XA15", main="Employment by voivodships")


# map with changed palette
library(viridis)
spplot(voi,        "XA15",        main="Employment        by        voivodships",
col.regions=rev(viridis(17, option="B")))


# a map with a changed palette and a fixed number of intervals
library(RColorBrewer)
spplot(voi, "XA15", main="Employment by voivodships",
col.regions=brewer.pal(9, "BuPu"), cuts=7)


# map with changed palette and set interval values
library(classInt) # to set intervals
library(quickPlot) # to set colors
```

```
brks<-classIntervals(voi@data$XA15, style="sd")
cols<-divergentColors("darkred", "darkblue", -10, 10, 0, "white")
spplot(voi["XA15"], main="Employment by voivodships", at=brks$brks,
col.regions=cols)
```

2.5 Basic operations and graphs for point data

Visualisation of point data requires the drawing of a contour map in a specific projection and applying points to this layer. The dataset must contain at least the geographical coordinates of the points (x, y) and may contain additional characteristics of these points (z). The projection of points must be consistent with the projection of the map so that the layers "see each other". The basic graphical mechanisms are similar to the area data visualisation.

Scheme 1 – with points() from the graphics:: package – locations only

In the most basic version of graphic geolocalised points, one can be interested in drawing locations only, without values or characteristics at a point. Points, on the prepared contour map, can be drawn on the basis of data from the *data.frame* object. The example uses data from *dataset2* – geolocations of firms. The company's dataset includes geographical coordinates – these are the variables *coords.x1* and *coords.x2*. The contours are drawn with the **plot()** command, and the points are marked with the **points()** command, where the coordinates are given as the first two arguments and then the symbol (*pch*) (see Figure 2.8a).

```
# scatterplot of empirical data, Figure 2.8a
plot(voi[voi@data$jpt_nazwa_=="Lubelskie",])
points(firms$coords.x1, firms$coords.x2, pch=".")
```

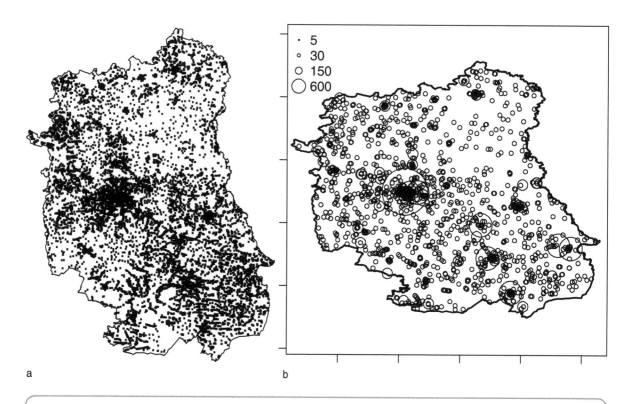

a   b

Figure 2.8 Point data map: a) Scheme 1 – locations only, b) Scheme 2 – locations and values
Source: Own study using R and the sp:: package

Scheme 2 – with spplot() from the sp:: package – locations and values

Points, that is, locations and values, can be automatically drawn with the **spplot()** command from the sp:: package (see Figure 2.9). As with graphics for regions, the **spplot()** command requires the dataset to be saved as a *DataFrame* attribute in the *SpatialPoints* object.

In the example, a separate *SpatialPointsDataFrame* object was created for each drawing, in which there was a variable to draw and geographic coordinates. By drawing graphs for two variables, the scale is automatically shared by both variables – quantitative or qualitative. When drawing from a *SpatialPoints* object, one does not need to specify variables that are coordinates, as this is already defined within the object class itself.

```
firms.lim1<-firms[,c("GR_EMPL", "subreg", "coords.x1", "coords.x2")] #
selection of variables
coordinates(firms.lim1)<-c("coords.x1","coords.x2") # class change
spplot(firms.lim1) # Figure 2.9a

firms.lim1<-firms[,c("SEC_PKD7", "coords.x1", "coords.x2")]
coordinates(firms.lim1)<-c("coords.x1","coords.x2") # class change
spplot(firms.lim1, key.space="left") # Figure 2.9b
```

Data in the *SpatialPoints* or *SpatialPointsDataFrame* class can also be drawn using the **plot()** command. By providing even a full data object, the command will recognise coordinates or variables that determine the location of points. In the *cex* option, one can specify a variable that will vary the size of the symbols drawn. Due to the fact that *cex=1* is the standard size and the cex values should oscillate around this value, it is necessary to rescale the variable.

In the example, 2000 firms were drawn using circles, and the radius of the circle (*cex* option) indicates the size of employment. The legend was based on the employment levels and was scaled more strongly than the

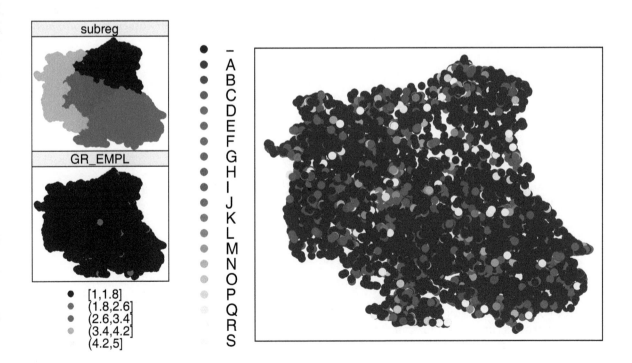

Figure 2.9 **Point data map with values – spplot() command**

Source: Own study using R and the sp:: package

values in the chart. The contour of the region was added for the readability of the point plot. In the option *mar* of **par**() command, the chart margins have been changed.

```
# selected variables: x and y coordinates, employment, sector
firms.sel<-firms[1:2000,c(12,13,20,18)] # selected columns
colnames(firms.sel)<-c("x","y","empl","sector") # names change
coordinates(firms.sel)<-c("x","y") # class change


# Figure 2.8b
par(mar=c(1,1,1,1)) # chart margins
plot(firms.sel, pch=1, cex=sqrt(firms.sel$empl)/3, axes=TRUE)
v<-c(5,30,150,600) # the scale of the legend
legend("topleft", legend=v, pch=1, pt.cex=sqrt(v)/10, bty="n")
plot(voi[voi.df$jpt_nazwa_=="lubelskie",], add=TRUE, lwd=2)
par(mar=c(5,4,4,2))
```

Scheme 3 – with findInterval() from the base:: package – locations, values, different size of symbols

As in the case of regional maps, the colour layer for the point value can be applied using the **findInterval**() command. The **plot**() command, used together with the **findInterval**() command, checks into which interval (*brks* object) the variable value for the next region falls and assigns the appropriate colour; that is, values from the fifth range are assigned by the fifth colour from the colour vector (*cols* object). The colours can be defined manually by name (as in the following) or by using existing colour palettes. The variable division intervals (*brks*) are created manually. In the **points**() command, the *bg* option defines the fill of the point, while the *col* option applies to the point's border. In the legend by command **legend**(), squares are generated automatically. The addition of squares is defined (*fill* option), as are lower boundaries of the ranges as labels (*legend* option) and font size (*cex* option).

In the first case, points were marked on the voivodeship map, whereby, according to the expectations, the locations were revealed only in one province (Lubelskie). In the second case, the points were marked on the regional map. In both cases, the legend was placed automatically in the bottom left corner (the *bottomleft* option). The second graph was saved to WorkingDirectory with the **savePlot**() command.

```
variable<-firms$subreg
locs<-firms[,12:13]
summary(variable)
brks<-c(1, 2, 3, 4)
cols<-c("blue3", "cornflowerblue", "seagreen1", "green2")

plot(voi) #full contour, limited points
points(locs, col=cols[findInterval(variable, brks)], pch=21, cex=0.7,
       bg=cols[findInterval(variable, brks)])
legend("bottomleft", legend=brks, fill=cols, cex=0.8, bty="n")
title(main="Points - colors by values")

# limited contour, limited points
lub.voi<-voi[voi.df$jpt_nazwa_=="lubelskie",]
plot(lub.voi) #Figure 2.10a
points(locs, col=cols[findInterval(variable, brks)], pch=21, cex=0.7,
       bg=cols[findInterval(variable, brks)])
legend("bottomleft", legend=brks, fill=cols, cex=0.8, bty="n")
title(main="Points - colors by values")

savePlot(filename="locations and random values", type="jpg")
```

The previous code will be modified in two ways. First, the area data will be converted into point data – the values assigned to the poviats to date will be assigned to the centroids of these poviats. Second, the drawn points

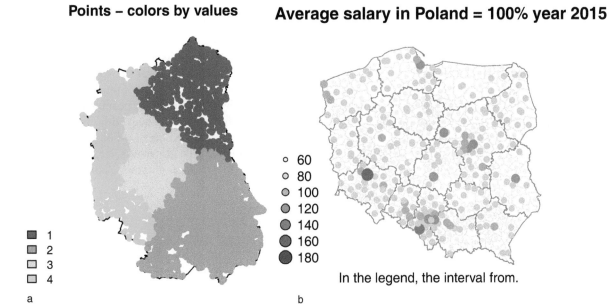

Figure 2.10 Regional map with points in colour

Source: Own study using R and the sp:: package

will be of different sizes, depending on the value of the variable. The colours and sizes of the elements use the same **findInterval**() function, which compares the next element with the assumed intervals, classifies it into a range and assigns the colour/size from the colour/size vector according to the interval number; that is, when the observation falls into the second interval, it is assigned the second colour/size. In the example, a map of average wages in poviats (Poland = 100) in 2015 will be drawn (this is variable *XA14*) (see Figure 2.10b).

```
variable<-data$XA14[data$year==2015]
summary(variable)
crds<-coordinates(pov)
brks<-c(60, 80, 100, 120, 140, 160, 180)
size<-(brks/100)*1.2
cols=brewer.pal(7, "Reds")


plot(pov, border="grey90") #Figure 2.10b
plot(voi, border="grey50", add=TRUE,)
points(crds, col=cols[findInterval(variable, brks)],
       cex=size[findInterval(variable, brks)], pch=21,
       bg=cols[findInterval(variable, brks)])
legend("bottomleft", legend=brks, pt.bg=cols, pt.cex=size, bty="n", pch=21)
title(main="Average salary in Poland = 100% year 2015",
      sub="In the legend, the interval from . . .")
savePlot(filename="Average salary", type="jpg")
```

In the case of point data, their geographical location (*x*, *y*) is often known, but the affiliation with administrative areas (regions) is unknown. This can be easily checked and later used in selective graphics. One can use the **over**() command from the sp:: package to check whether a point belongs to an area. The function's arguments are the points and area. The result is a set of data for the length of the set of points, which contains empty lines if the point is outside the tested area and the characteristics of the region from the set of data from the map if the point belongs to the studied area. The **over()** command can be replaced by the **gIntersects()** command from the rgeos:: package. The command performs the same check of the point's affiliation to the area, where the area and points are given in sequence as the arguments, and the resulting output is a TRUE/FALSE vector.

The following example checks which firms are in the selected county. In the first step, it was checked which poviats are in the Lubelskie voivodeship – a subset for the year was created, a subset for the voivodeship created separately and poviats displayed. In the second step, the Lublin district was selected, a separate map was created for it, and the point set was converted to the *SpatialPoints* class by the **coordinates()** command and was given a projection with the **proj4string()** command. The projections of the dataset and map with the command **spTransform()** were subsequently agreed, and the points belonging to the selected area (Lublin county) were finally checked with the **over()** command from the sp:: package and, alternatively, **gIntersects()** from the rgeos:: package. In the third step, all four layers were drawn: the contour of the province, all points, points from the selected poviat and the contour of the selected poviat.

```
data15<-data[data$year==2015,]
data.lub<-data[data$region_name=="Lubelskie",]
data.lub$poviat_name1
```

```
 [1] Powiat bialski Powiat biłgorajski Powiat chełmski
 [4] Powiat hrubieszowski Powiat janowski Powiat krasnostawski
 [7] Powiat kraśnicki Powiat lubartowski Powiat lubelski
[10] Powiat łęczyński Powiat łukowski Powiat opolski-lubelskie
[13] Powiat parczewski Powiat puławski Powiat radzyński
[16] Powiat rycki Powiat świdnicki-lubelskie Powiat tomaszowski-lubelskie
[19] Powiat Biała Podlaska Powiat włodawski Powiat zamojski
[22] Powiat Lublin Powiat Chełm Powiat Zamość
```

```
pov.sel<-pov[data15$poviat_name1=="Powiat Lublin",]
firms.sp<-firms
coordinates(firms.sp)<-c("coords.x1","coords.x2") # class change
proj4string(firms.sp)<-CRS("+proj=longlat +datum=NAD83")
pov.sel<-spTransform(pov.sel, CRS("+proj=longlat +datum=NAD83"))
firms.sp<-spTransform(firms.sp, CRS("+proj=longlat +datum=NAD83"))

# method with over()
locs.lim<-over(firms.sp, pov.sel)

par(mar=c(1,1,1,1))
plot(voi[voi.df$jpt_nazwa_=="lubelskie",])# plot of whole voivodeship
locs<-firms[,12:13] # all points of the class data.frame
points(locs, col="grey80", pch=".", cex=0.7) # plot of all points
points(locs[locs.lim$jpt_nazwa_=="powiat Lublin",], pch=".", cex=1.1)
plot(pov.sel, add=TRUE, border="red") # plot of selected poviat
par(mar=c(5,4,4,2))

# method with gIntersects()
library(rgeos)
locs.lim<-gIntersects(pov.sel, firms.sp, byid=T)

par(mar=c(1,1,1,1))
plot(voi[voi.df$jpt_nazwa_=="lubelskie",])# plot of whole voivodeship
locs<-firms[,12:13] # all points of the class data.frame
points(locs, col="grey80", pch=".", cex=0.7) # plot of all points
points(locs[locs.lim==TRUE,], pch=".", cex=1.1)
plot(pov.sel, add=TRUE, border="red") # plot of selected poviat
par(mar=c(5,4,4,2))
```

2.6 Basic operations on rasters

Raster data is the second basic spatial data format, along with vector data. In R, it is supported by the raster:: package. In the basic view, the *RasterLayer* class is a lattice that divides the surface into equal cells in which phenomena are aggregated.

To create a raster, one needs geolocalised point data, raster borders (*de facto* bounding box, bbox) and a variable that will be counted (typical statistics are the number of observations and the sum or the average of these observations). With the **raster**() command from the raster:: package, a box with a fixed number of rows and columns is created (in the example, both dimensions are 50, giving 2500 grids) defining the geographical boundaries of the raster range – defined by the territory bbox. The **rasterize**() command divides the point data into grids and determines the indicated statistics for the variable under consideration. The **rasterize**() command has four arguments: coordinate object, raster (defined grid), variable and statistics. Coordinates and variables can be ordered within one object of the class *data.frame* (see Figure 2.11a). It is also necessary to agree on the projection of the contour map and raster. The **crs**() command from the raster:: package is used both to examine the current projection as well as to convert the projection to the desired one. Rasters are drawn with the **plot**() command.

```
# preparation of data based on the previous example
firms.Lublin<-firms[firms$CITY=="Lublin",]    # firms in Lublin
pov.Lublin<-pov.sel    # contour map of Lublin
par(mar=c(1,1,1,1))
plot(pov.Lublin)
points(firms.Lublin[,12:13], pch=".", cex=1.5)
par(mar=c(5,4,4,2))

bbox(pov.Lublin)
```

```
   min max
x 22.45428 22.67354
y 51.13981 51.29656
```

```
# creating a raster

library(raster)

x1<-firms.Lublin[,12]    # coordinates x
y1<-firms.Lublin[,13]    # coordinates y
xy1<-cbind(x1,y1)
p1<-data.frame(xy1, z=firms.Lublin[,20]) # coordinates and population
r<-raster(nrows=50, ncols=50, ymn=51.14, ymx=51.30, xmn=22.45, xmx=22.67)
r1<-rasterize(xy1, r, field=p1$z, fun=sum)

proj.map<-crs(pov.Lublin) # explore the map projection
proj.map
#CRS arguments:
# +proj=longlat +datum=NAD83 +ellps=GRS80 +towgs84=0,0,0

proj.raster<-crs(r1) # examination of the raster projection
proj.raster
#CRS arguments:
# +proj=longlat +datum=WGS84 +ellps=WGS84 +towgs84=0,0,0

crs(r1)<-proj.map # change of raster projection to map projection
crs(r1)
```

```
plot(r1, main="Employment in firms") # Figure 2.11a
plot(pov.Lublin, add=TRUE)

class(r)
class(r1)
#[1] "RasterLayer"
#attr(,"package")
#[1] "raster"
+
```

For testing purposes, one can also create a grid for artificially generated data. Subsequently, a grid for 10,000 grids (100 × 100) was created, centred at (0,0) – hence the boundaries *x* and *y* were set between (–50, +50). Cards can be filled with random values. To specify the number of values to generate, one can count the number of pages with the **ncell()** command from the raster:: package. It is then the first argument of the function generating random numbers **runif()** – here from the uniform distribution on the interval from 1 to 8. It is worth noting that the generated numbers are assigned to the raster, leaving the empty parenthesis as a reference to the whole range. To obtain natural numbers, the results were rounded to 0 decimal places (see Figure 2.11a).

```
# creating a raster on 10,000 grids
r<-raster(ncols=100, nrows=100, ymn=-50, ymx=50, xmn=-50, xmx=50)

ncell(r)
#[1] 10000

# integers, the assignment of values is a raster
r[]<-round(runif(ncell(r), 1,8), digits=0)
plot(r) # Figure 2.11b
```

Raster values can also be easily transformed using basic algebra, for example, the squaring of a raster object. The same values in the raster can be recovered with the **extract()** command from the raster:: package, which specifies the cells from which the observation values should be selected. There is also a **getValues()** function that

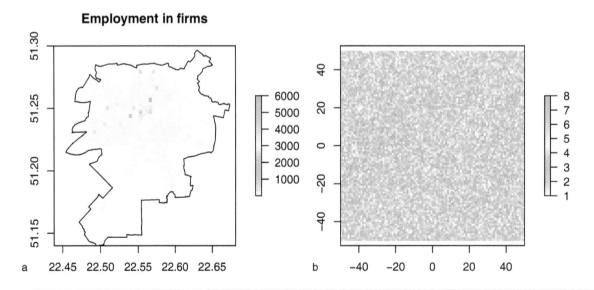

Figure 2.11 Rasterised data: a) employment in companies in Lublin, b) randomly generated data

Source: Own study using R and sp:: and raster:: packages

displays the values for the selected row. In the following case, the **extract()** command retrieved the values for all cells.

```
r1<-r^2
plot(r1)

vals<-extract(r, 1:ncell(r))
head(vals)
#[1] 4 2 7 6 2 3
```

It is possible to create a multilayer raster in the *RasterStack* and *RasterBrick* classes and operations on them using several commands: **stack()** and **brick()** to create a multilayer object; **unstack()** to invert this operation or **subset()**, **addLayer()** and **dropLayer()** to manage layers: select, add and delete.

On the basis of rasters, it is possible to compare similarities of distributions of various point data within the same boxes. Raster spatial distributions of the number of firms were created with the **rasterize()** command from the raster:: package, which requires entering as input: 1) the *data.frame* class object, which contains only coordinates of points; 2) fixed raster division, obtained by the **raster()** command; 3) reference to the value of the characteristic in point and 4) the function that will be applied to the examined feature – the sum (*fun=sum*) is given here. For the **raster()** command, the raster boundaries were determined based on the coordinates obtained from the bbox, and it was decided that the grid would have 50 rows and 50 columns, a total of 2500 cells. In the example subsequently, the number of firms in two sectors will be compared. On the basis of the size distribution of firms in the sectors obtained by the **table()** command, the two most numerous sectors were selected – sectors G and A. Sector subsets were created and an artificial variable with a value of 1 was added, which makes it easier to count the observations in subsets. Rasters are drawn with the **plot()** command, applying the administrative contour (see Figure 2.12).

```
table(firms.Lublin$SEC_PKD7)
```

```
- A   BC  DE  F  G   H    I   J  K
1 689 0 295 21 9 361 1134 298 94 173 163
  L   M   N  O  P   Q   R  S
188 533 110 10 159 369 68 360
```

```
firms.Lublin$ones<-rep(1, times=dim(firms.Lublin)[1])
firms.LublinG<-firms.Lublin[firms.Lublin$SEC_PKD7=="G",]
firms.LublinA<-firms.Lublin[firms.Lublin$SEC_PKD7=="A",]

r<-raster(nrows=50, ncols=50, ymn=51.14, ymx=51.30, xmn=22.45, xmx=22.67)

x1<-firms.LublinG[,12] # company location - x coordinates
y1<-firms.LublinG[,13] # company location - y coordinates
xy1<-cbind(x1,y1) # x and y coordinates as data.frame
p1<-data.frame(xy1, z=firms.LublinG$ones)
r1<-rasterize(xy1, r, field=p1$z, fun=sum) # counting sec.G firms

x2<-firms.LublinA[,12]
y2<-firms.LublinA[,13]
xy2<-cbind(x2,y2)
p2<-data.frame(xy2, z=firms.LublinA$ones) # counting sec.A firms
r2<-rasterize(xy2, r, field=p2$z, fun=sum)

plot(r1, col=colorRampPalette(c("cornsilk2", "indianred1", "brown3"))(255),
main="Spatial distribution firms from the sector G") # Figure 2.12a
plot(pov.Lublin, add=TRUE)

plot(r2, col=colorRampPalette(c("cornsilk2", "indianred1", "brown3"))(255),
main="Spatial distribution of firms from the sector A")
plot(pov.Lublin, add=TRUE)
```

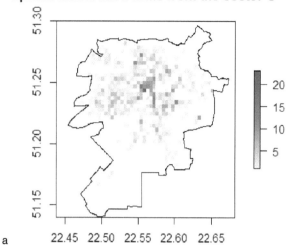
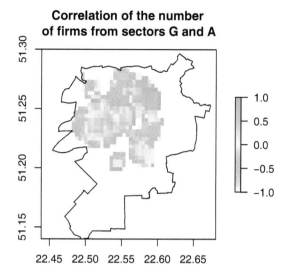

Figure 2.12 Rasterised spatial distributions: a) the number of firms in Lublin, b) Pearson correlation between number of companies in sectors G & A

Source: Own study using R and raster:: package

For the prepared raster distributions, local Pearson correlations were calculated between the two frequencies. The **corLocal()** command from the raster:: package was used. The command uses the so-called focal neighbourhood or neighbourhood in a moving window. The **focal()** command from the raster:: package determines by default the average of the values around a given cell in the square of the indicated size, where the analysed cell is inside. The size of the moving window is determined in the *ngb* option, where *ngb=5* means the neighbourhood $5 \times 5 = 25$ cells. Local correlations are statistics calculated for $n \times n$ cells in both rasters, and the scalar result is assigned to the "central" cell of the moving window. In the **corLocal()** function, it is possible to determine, in addition to the Pearson coefficient, the Kendall and Spearman statistics. Figure 2.12b presents Pearson's statistics (also, their *p*-value can be plotted).

```
# Pearson's correlation
cor.r<-corLocal(r1, r2, ngb=5, method="pearson", test=TRUE)
summary(cor.r)


# Pearson statistics chart
plot(cor.r$pearson, main="Correlation of the number \nof firms from sectors
G and A") # Figure 2.12b
plot(pov.Lublin, add=TRUE)


# p-value graph
plot(cor.r$p.value, main="p-value of correlations between the number \nof
firms from sectors G and A ")
plot(pov.Lublin, add=TRUE)
```

The result of the co-location correlation for firms from various industries should be assessed as follows: the Pearson correlation graph shows that locally, the highest similarities are in the northeastern area (dark-green colour), and the *p*-value graph confirms that these correlations are significant (pale pink colour). On the contrary, there are no associations in the southeastern part (yellow colour), which is confirmed by the high *p*-value (dark-green colour). Statistics are determined for areas in which the appropriate boxes of both distributions and their neighbours have non-zero values.

An interesting way to compare is the **crosstab()** command from the raster:: package. The function lists all pairs of values of the examined features in both rasters and gives the frequency of occurrence of such pairs. In an example comparing the number of firms in sectors A and G (natural numbers), the function lists all available values of the sector G feature in the first side (written as r1), combining them with the first available value of sector A (stored as r2) and further through a replicated block of all values of the G sector characteristic for each successive value of the sector A feature.

```
ct<-crosstab(r1, r2)
head(ct)
```

```
      layer.2
layer.1  1 2 3 4 5 6 7 8 9 12
      1 25 9 9 5 3 0 1 0 1  1
      2 12 8 4 6 1 1 0 1 0  0
      3 14 8 5 3 1 1 0 0 0  0
      4  6 6 4 0 3 2 0 0 0  0
      5  6 5 0 1 4 1 0 0 0  0
      6  5 6 4 0 0 0 0 0 0  0
```

In the SDMTools:: package, there is a **SigDiff()** function that checks the significance of differences between two raster distributions. It refers to the mean and variance of all differences measured in pairs for the corresponding raster cells. The result of the **SigDiff()** command is the raster with the p-values of the test on the equality of two values. The p-value vector for these differences for subsequent cells was saved to the object using the **extract()** command from the raster:: package. There is also a dedicated graphic function, **ImageDiff()**, from the SDMTools:: package – it automatically divides p-value into three groups, internal (insignificant) and two on the edge of the interval [0,1] to indicate significantly lower (p-value < 0.025) and higher (p-value > 0.975) values in cells. The given three colours successively define the values as "significantly lower", "no significant difference" and "significantly higher" (see Figure 2.13a). One can also prepare the figure using **plot()** commands. The raster value colouring is automatically selected from the **terrain.colors()** palette. Colours can be added manually to the descriptive legend.

```
# raster and graph of significant differences
library(SDMTools)
out<-SigDiff(r1,r2, pattern=FALSE)
out.val<-extract(out, 1:ncell(out))

proj.map<-crs(pov.Lublin) # explore the map projection
proj.raster<-crs(out) # examination of the raster projection
crs(out)<-proj.map # change of raster projection to map projection

# figure based on ImageDiff()
plot(pov.Lublin) # Figure 2.13a
ImageDiff(out, main="Pattern Differences", axes=FALSE,
      tcol=c("indianred1","cornsilk2","palegreen1"), add=TRUE)

legend('bottomright', title='significance',
      legend=c('lower','greater','not significant'),
      fill=c("indianred1","palegreen1","cornsilk2"),bg='white', bty='n')

# figure based on plot()
plot(pov.Lublin)
plot(out, add=TRUE)
cols<- rev(terrain.colors(12))
legend('bottomright', title='significance',
      legend=c('lower', 'not significant', 'greater'), fill=c(cols[1],
      cols[6], cols[12]), bg='white', bty='n', cex=0.8)
```

The raster:: package is fully compatible with the sp:: package. It is possible to convert a *raster* class object to the *SpatialPointsDataFrame* class – the **rasterToPoints()** command from the raster:: package is used for this

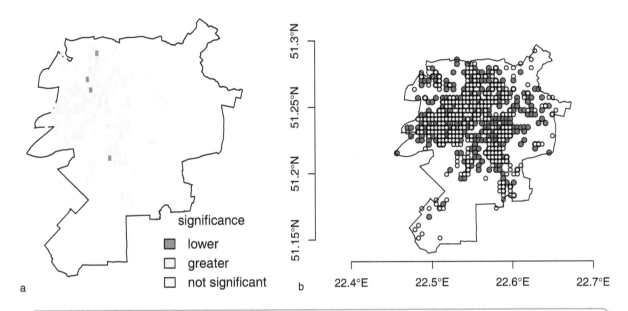

Figure 2.13 Raster plots: a) significant differences between two variables using **ImageDiff()**, b) raster conversion to *SpatialPoints* class

Source: Own study using the raster:: package

purpose. With option *spatial=TRUE*, the result is in the *sp* class, while omitting this option saves the result in the *matrix* class.

```
# raster converted to the SpatialPointsDataFrame class
r1p<-rasterToPoints(r1, spatial=TRUE)
r2p<-rasterToPoints(r2, spatial=TRUE)
plot(r1p, pch=21, bg="red", ylim=c(51.14,51.3))
points(r2p, pch=21, bg="yellow", cex=0.8) #Figure 2.13b
plot(pov.Lublin, add=TRUE)
degAxis(1)
degAxis(2)


# converted raster is a matrix class
r1p<-rasterToPoints(r1)
r2p<-rasterToPoints(r2)
plot(r1p, pch=21, bg="red", ylim=c(51.14,51.3), xlim=c(22.45, 22.67))
points(r2p, pch=21, bg="yellow", cex=0.8)
plot(pov.Lublin, add=TRUE)
```

It is also possible to convert a raster to a polygon and then to create a spatial weights matrix W. The **rasterToPolygons()** command from the raster:: package converts the *raster* class into a *SpatialPolygonsDataFrame* class. With the option *na.rm=TRUE*, only cells with values are included, and NAs are skipped. The option *na.rm=FALSE* allows one to include all cells. The spatial weights matrix is created by default with the **poly2nb()** command to form the neighbourhood matrix and then with the command **nb2listw()** to form the spatial weights matrix W. At the same time, one can convert data to the *data.frame* class. The *xy=TRUE* option stores *xy* coordinates as centroids of areas in this object, and the *xy=FALSE* option skips them.

```
# conversion raster to polygons and W matrix
library(spdep)
r1.poly<-rasterToPolygons(r1, na.rm=FALSE)
r1p.nb<-poly2nb(r1.poly)
r1p.listw<-nb2listw(r1p.nb)
```

```
r1.df<-as.data.frame(r1, na.rm=FALSE, xy=FALSE)
summary(r1.df)
```

```
layer
Min.   :   5.00
1st Qu.:  10.00
Median :  25.00
Mean   :  65.46
3rd Qu.:  70.00
Max.   :6000.00
NA's   :1833
```

```
r1.df<-as.data.frame(r1, na.rm=FALSE, xy=TRUE)
summary(r1.df)
```

```
      x              y            layer
Min.   :22.45  Min.   :51.14  Min.   :   5.00
1st Qu.:22.50  1st Qu.:51.18  1st Qu.:  10.00
Median :22.56  Median :51.22  Median :  25.00
Mean   :22.56  Mean   :51.22  Mean   :  65.46
3rd Qu.:22.61  3rd Qu.:51.26  3rd Qu.:  70.00
Max.   :22.67  Max.   :51.30  Max.   :6000.00
                NA's   :1833
```

2.7 Basic operations on grids

Yet another spatial class is the *grid*. It is possible to create a grid of points regularly spaced inside the rectangle described in the region. The created bounding box then constitutes the area for drawing / generating points.

A grid can be created based on the administrative contour. One uses the **makegrid**() command from the sp:: package, which is "hooked" to the **spsample**() command. The command options usually specify the number of points to be created and a spatial object on which the bounding box will be described. In the mapping, due to the planarity of the coordinates, the order of drawing is of great importance, as confirmed in Figure 2.14.

```
#loading fresh objects
voi<-readOGR(".", "wojewodztwa") # 16 units.
voi<- spTransform(voi, CRS("+proj=longlat +datum=NAD83"))

# creation of a map section - voivodeship (NTS2) Lubelskie
voi.df<-as.data.frame(voi)
lub.voi<-voi[voi.df$jpt_nazwa_=="lubelskie",]

grid.lub<-makegrid(lub.voi, n=100) # grid as part of a bounding box

plot(grid.lub)# planar result
plot(lub.voi, add=TRUE) # spherical layer

plot(lub.voi)    # spherical layer, Figure 2.14a
points(grid.lub, pch=".", cex=1.5) # planar result
```

It is possible to convert a grid from a planar to a spherical layout using the **SpatialPoints**() command from the sp:: package and defining a coordinate system as in the region for which the grid was determined. In drawing a spherical grid together with the contour region of the region, the order does not matter. Narrowing the grid

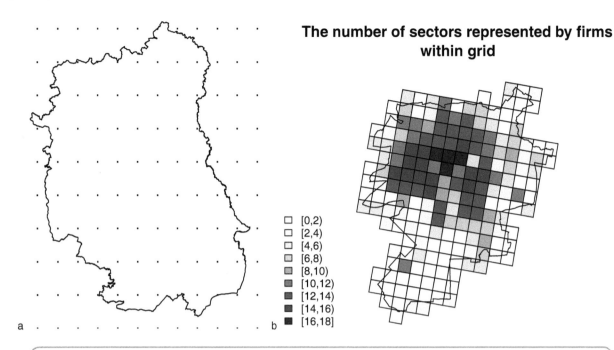

The number of sectors represented by firms within grid

Legend:
- ☐ [0,2)
- ☐ [2,4)
- ☐ [4,6)
- ☐ [6,8)
- ▨ [8,10)
- ▨ [10,12)
- ▨ [12,14)
- ▨ [14,16)
- ■ [16,18)

a   b

Figure 2.14 Grid for a selected bounding box: a) Lubelskie region in a spherical system, grid with the number of sectors represented by companies within each cell

Source: Own study with the use of R and the sp:: and GISTools:: package

to contours of the region is possible by conditional operation – the contour is a condition for grid rows. This operation works only for a spherical grid.

```
# spherical grid
grid.lub.sp<-SpatialPoints(grid.lub, proj4string=CRS(proj4string(lub.voi)))

plot(grid.lub.sp) # layer 1: grid, layer 2: contour
plot(lub.voi, add=TRUE)

plot(lub.voi) # layer 1: contour, layer 2: grid
plot(grid.lub.sp, add=TRUE)

grid.limited<-grid.lub.sp[lub.voi,]
plot(lub.voi) # grid is narrowed to the region border
plot(grid.limited, add=TRUE)
```

Increasingly, statistical data is made available in a grid format. Most often, this is demographic and population data.[1] Data in this format can be loaded into R with the **readOGR()** command from the rgdal:: package. The result is an object of the *SpatialPolygonsDataFrame* class, in which one can change the projection with the **spTransform()** command. In the example grid dataset, there are 315,857 cells, each with an area of 1 km². The grid object data can be accessed via the *@data* slot, or one can convert to the *data.frame* class to a dedicated object. From the loaded object, one can also separate the grid itself, getting rid of data – this is possible with the command **as()**, where the second argument is the type of the target class *SpatialPolygons*. The structure of slots and sockets can typically be checked with the command **str()**, and access to the selected variable is, for example, as *pop@data$FEM_RATIO*. If data is not numeric, it is possible to convert to numeric using the **for()** loop and **as.numeric()** command.

In the following example, two objects were separated from the loaded grid for the population: *pop.df* containing data and *pop.grid* containing only the grid.

```
# reading grid for population and conversion of projection
pop<-readOGR(".", "PD_STAT_GRID_CELL_2011")
pop<-spTransform(pop, CRS("+proj=longlat +datum=NAD83"))
```

```
pop.df<-as.data.frame(pop) # extracting the data to the data.frame object
head(pop.df)
```

```
 TOT TOT_0_14 TOT_15_64 TOT_65__ TOT_MALE TOT_FEM MALE_0_14 MALE_15_64
0 0   0      0         0        0        0       0         0
1 97  15     71        11       51       46      9         38
2 0   0      0         0        0        0       0         0
3 0   0      0         0        0        0       0         0
4 0   0      0         0        0        0       0         0
5 0   0      0         0        0        0       0         0
  MALE_65__ FEM_0_14 FEM_15_64 FEM_65__ FEM_RATIO SHAPE_Leng SHAPE_Area
0 0   0      0 0      0.00000  2070.834  238411.4
1 4   6      33 7     90.19608 4002.039  1001019.4
2 0   0      0 0      0.00000  4002.403  1001201.8
3 0   0      0 0      0.00000  4002.396  1001198.0
4 0   0      0 0      0.00000  2800.241  310494.5
5 0   0      0 0      0.00000  2009.130  128915.9
```

```
pop.grid<-as(pop, "SpatialPolygons") # separation of grid
str(pop) # checking the structure of slots and sockets
summary(pop@data$FEM_RATIO) # summary of the selected variable

# conversion to the numerical data of subsequent columns of the data set
for(i in 1:15) {# does not work for all variables
pop.df[,i]<-as.numeric(levels(pop.df[,i]))[pop.df[,1]]}
```

The example grid can be limited to the selected territory. A good way is to use the **over**() command, in which the first argument is the *SpatialPolygons* class object (only grid without data), and the second argument is the administrative contour in the *SpatialPolygons* or *SpatialPolygonsDataFrame* class. The result is a *data.frame* class object, a length like a grid object and the contents of the NA rows when the cell is not part of the surveyed territory or as in the contour *data.frame* attribute (here *lub.voi*) when the cell belongs to the tested area. Based on the information from the obtained *data.frame*, one can select the cells that lie within the administrative contour. It is also necessary to conditionally truncate data. The **which**() command was used to check which rows are non-empty, and these rows were selected from the dataset. The grid itself can be drawn with the **plot**() command, where the argument is a limited grid. The **choropleth**() command from the GISTools:: package was used to draw the variable value.

In the following example, a contour map is cut to obtain a *SpatialPolygonsDataFrame* class object, a grid is cut to obtain a *pop.grid* object of *SpatialPolygons* class and grid-related data is cut to obtain a *pop.df.lub* data.frame.

```
# cutting the contour map
pov.lub<-pov[pov@data$jpt_nazwa_=="powiat Lublin",]

lim<-over(pop.grid, pov.lub) # limited wider grid with narrow contour map
head(lim[,1:7]) # data.frame class object
```

```
 iip_przest iip_identy iip_version jpt_sjr_ko jpt_code_je jpt_nazwa_ jpt_nazw01
1 <NA>       <NA>       <NA>        <NA>       <NA>        <NA>       <NA>
2 <NA>       <NA>       <NA>        <NA>       <NA>        <NA>       <NA>
3 <NA>       <NA>       <NA>        <NA>       <NA>        <NA>       <NA>
4 <NA>       <NA>       <NA>        <NA>       <NA>        <NA>       <NA>
5 <NA>       <NA>       <NA>        <NA>       <NA>        <NA>       <NA>
6 <NA>       <NA>       <NA>        <NA>       <NA>        <NA>       <NA>
```

```
a<-which(lim$jpt_nazwa_=="powiat Lublin") # rows fulfilling the condition
head(lim[a,1:7])
```

```
       iip_spat        iip_ident
277676 PL.PZGIK.200 5aa84bcb-1088-4c93-ba91-f09d1e8e086e
277729 PL.PZGIK.200 5aa84bcb-1088-4c93-ba91-f09d1e8e086e
277742 PL.PZGIK.200 5aa84bcb-1088-4c93-ba91-f09d1e8e086e
277780 PL.PZGIK.200 5aa84bcb-1088-4c93-ba91-f09d1e8e086e
277795 PL.PZGIK.200 5aa84bcb-1088-4c93-ba91-f09d1e8e086e
278033 PL.PZGIK.200 5aa84bcb-1088-4c93-ba91-f09d1e8e086e
       iip_version jpt_sjr_ko jpt_code_je jpt_nazwa_ jpt_name01
277676 2012-09-27T07:39:43+02:00 POV 0663 poviat Lublin <NA>
277729 2012-09-27T07:39:43+02:00 POV 0663 poviat Lublin <NA>
277742 2012-09-27T07:39:43+02:00 POV 0663 poviat Lublin <NA>
277780 2012-09-27T07:39:43+02:00 POV 0663 poviat Lublin <NA>
277795 2012-09-27T07:39:43+02:00 POV 0663 poviat Lublin <NA>
278033 2012-09-27T07:39:43+02:00 POV 0663 poviat Lublin <NA>
```

```
# conditional grid and data.frame restriction on the selected site
pop.grid.lub<-pop.grid[lim$jpt_nazwa_=="powiat Lublin",]
pop.df.lub<-pop.df[a,]

# administrative contour and grid
plot(pop.grid.lub)
plot(pov.lub, add=TRUE)

library(GISTools)
choropleth(pop.grid.lub, pop.df.lub$TOT) # grid with colour layer
plot(pov.lub, add=TRUE)
```

As in the case of raster, the existing grid with data can be integrated with external point data – that is, limit point data territorially to the grid area and determine statistics from point data within grid cells, which will be done subsequently.

As a reminder, in the previous part of the code, the restriction of the full grid went two ways: in the pop object or the *SpatialPolygons* class, a selected grid was saved, and in the object *pop.df.lub data.frame*, all variables from the full dataset with limited rows were saved. The Lublin district consists of 190 grid cells. The full grid with data for the whole of Poland is in the *pop* object, which has also been divided into two parts: the grid in the *pop.grid* object and the data in the *pop.df* object.

In the next step, the membership of point data for firms to grid cells will be checked with the command **over**(). First of all, the **over**() command needs to add information – preferably *ID* – from the area data to the point data. For this reason, a variable *ID* has been added to the grid, which contains the original (from the base file) row identifiers acquired by the **rownames**() command. Second, the point data for firms (*company.sel* object) was converted with the **coordinates**() command to the *SpatialPointsDataFrame* class, and grid and point data projections were agreed using the **crs**() command from the raster:: package (this command both recovers and gives projections to objects). Finally, as a result of the **over**() command, a new ID variable has been created in the company's data *set.sel*, taking the following values: <NA> if the point lies outside the Lublin poviat and *ID* when the company is in this *poviat*. *ID* indicates membership of a specific grid cell.

```
library(raster)
names(firms)
```

```
 [1] "ID"         "ADDRESS"      "STREET"       "STREET.NO"
 [5] "ZIP"        "CITY_POST"    "CITY"         "region2"
 [9] "poviat"     "region3"      "subreg"       "coords.x1"
[13] "coords.x2"  "LEGAL_FORM1"  "LEGAL_FORM2"  "OWNERSHIP"
[17] "PKD7"       "SEC_PKD7"     "GR_EMPL"      "empl"
```

```
# selected variables: x and y coordinates, employment, sector
firms.sel<-firms[,c(12,13,20,18)]
colnames(firms.sel)<-c("x","y","empl","sector") # change of names
coordinates(firms.sel)<-c("x","y") # changing object class

pop.grid.lub$ID<-rownames(pop.df.lub) # identifier of units in the grid
pop.df.lub$ID<-rownames(pop.df.lub) # identifier of units in data.frame
crs(firms.sel)<-crs(pop.grid.lub) # agreeing the projections

# assigning grid ID to firms
firms.sel$ID<-over(firms.sel, pop.grid.lub)
head(firms.sel)
```

```
  empl Sector ID
1    5 G   <NA>
2    5 K   <NA>
3    5 A   <NA>
4    5 A   <NA>
5    5 A   <NA>
6    5 A   <NA>
```

```
summary(firms.sel)
```

```
Object of class SpatialPointsDataFrame
Coordinates:
    min      max
x 21.64982 24.12993
y 50.26393 52.27482
Is projected: FALSE
proj4string :
[+proj=longlat +datum=NAD83 +ellps=GRS80 +towgs84=0,0,0]
Number of points: 37378
Data attributes:
    empl          sector        ID.ID
Min.  :   5.000   A :20864   Length:37378
1st Qu.:   5.000   G :  4772   Class :character
Median :   5.000   F :  2031   Mode :character
Mean  :   6.336   C :  1374
3rd Qu.:   5.000   M :  1341
Max.  :1500.000   S :  1268
                (Other): 5728
```

Assigning identification numbers to grid cells can be checked in the graph. To write the numbers on the graph, one needs to set the centroids of grid cells using the **coordinates**() command and apply texting – with the **text**() command (in a simple way) or with the **pointLabel**() command from the maptools:: package (in an optimised way).

```
crds<-coordinates(pop.grid.lub)      # grid cell centroids
plot(pop.grid.lub)                   # grid chart
#text(crds, labels=pop.grid.lub$ID, cex=0.4)
library(maptools)
pointLabel(crds, labels=pop.grid.lub$ID, cex=0.4)
```

Point data, enriched with the grid cell ID to which the point belongs, should be aggregated according to these IDs. It's a good idea to use the **aggregate()** command, which works on objects in the *data.frame* class. One can use the **length()** command in options, which counts the observations, or the **sum()** command, which sums up the values of the indicated variable. Subsequently, aggregate results should be added to the grid object. It should be remembered that the ID in the grid file may or may not be ordered. As a rule, the result of the **aggregate()** command is ordered in ascending/alphabetical order by ID. Due to the uncertain order of the ID in the grid set (the grid should not be sorted) and the existence of cells without aggregated data to be assigned, a safer solution is to collect files by using the **merge()** command rather than simply gluing columns with the **cbind()** command. As a result of the **merge()** command, a new variable (*x.x*) was added to the *data.frame* object at the end, indicating the number of firms located within the grid cell and a variable (*x.y*) which sums up the employment in these firms.

```
# change of the object's class
firms.sel.df<-as.data.frame(firms.sel)
head(firms.sel.df)
```

```
   x   y empl sector ID
1 22.00263 51.40935 5 G <NA>
2 22.34528 51.35417 5 K <NA>
3 22.96182 50.31523 5 A <NA>
4 21.83905 50.95910 5 A <NA>
5 23.37778 51.96064 5 A <NA>
6 22.25986 51.74347 5 A <NA>
```

```
# aggregation of point data according to the ID of the grid set
# statistics - the number of observations and the total employment
firms.agg.no<-aggregate(firms.sel.df$ID, by=list(firms.sel.df$ID), length)
firms.agg.sum<-aggregate(firms.sel.df$empl, by=list(firms.sel.df$ID), sum)
head(firms.agg.no)
```

```
Group.1 x
1 277675 1
2 277728 6
3 277741 1
4 277779 2
5 277794 1
6 278032 6
```

```
# combining files by merge()
pop.df.lub.m<-merge(pop.df.lub, firms.agg.no, by.x="ID", by.y="Group.1",
all.x=TRUE)
pop.df.lub.m<-merge(pop.df.lub.m, firms.agg.sum, by.x="ID", by.y="Group.1",
all.x=TRUE)
names(pop.df.lub.m)
```

```
 [1] "ID"          "TOT"         "TOT_0_14"    "TOT_15_64"   "TOT_65__"
 [6] "TOT_MALE"    "TOT_FEM"     "MALE_0_14"   "MALE_15_64"  "MALE_65__"
[11] "FEM_0_14"    "FEM_15_64"   "FEM_65__"    "FEM_RATIO"   "SHAPE_Leng"
[16] "SHAPE_Area"  "CODE"        "x.x"         "x.y"
```

A simple figure with the **choropleth**() command from the GISTools:: package requires that all cells have numeric values. <NA> values are not allowed; therefore, they were converted to 0. To find rows containing <NA>, **which**() and **is.na**() were used.

```
m1<-which(is.na(pop.df.lub.m$x.x))
pop.df.lub.m$x.x[m1]<-0
choropleth(pop.grid.lub, pop.df.lub.m$x.x) # number of firms
title(main="The number of firms in the grid")
shades<-auto.shading(pop.df.lub.m$x.x)
choro.legend(22.65, 51.2, shades, cex=0.65, bty="n")

m2<-which(is.na(pop.df.lub.m$x.y))
pop.df.lub.m$x.y[m2]<-0
choropleth(pop.grid.lub, pop.df.lub.m$x.y) # total employment
title(main="Total employment in firms in the grid")
shades<-auto.shading(pop.df.lub.m$x.y)
choro.legend(22.65, 51.2, shades, cex=0.65, bty="n")
```

It is also possible to count how many sectors are firms within the grid cell. For this purpose, variables for each sector were added to the object with grid data (*pop.df.lub.m*) – their value is the number of firms from a given sector located in a given grid cell. The operation was performed in a loop, iterating after successive names of sectors expressed in letters from A to S. The sector names can be recovered from the set with the command **levels**() and also generated with the **LETTERS**() command. In the loop, each time a sectoral subset was generated, it was aggregated by ID and the aggregation result pasted to the base dataset. The new variables for the sectors have been given the appropriate column names using the **colnames**() command. The number of different sectors was determined by the result as the difference between the maximum value of 19 and the number NA – the missing <NA> was checked in rows with the command **rowSums**(), whose argument is the **is.na**() function for the studied data range. The figure was made using a scheme based on the **classIntervals**() function from the classInt:: package.

```
levels(firms.sel$sector)
#[1] "-" "A" "B" "C" "D" "E" "F" "G" "H" "I" "J" "K" "L" "M" "N" "O" "P" "Q" "R" "S"

head(firms.sel)
```

```
  empl sector   ID
1    5      G <NA>
2    5      K <NA>
3    5      A <NA>
4    5      A <NA>
5    5      A <NA>
6    5      A <NA>
```

```
sectors<-LETTERS[seq(from=1, to=19)]
sectors
#[1] "A" "B" "C" "D" "E" "F" "G" "H" "I" "J" "K" "L" "M" "N" "O" "P" "Q" "R" "S"

for(i in 1:19){
sub<-firms.sel.df[firms.sel.df$sector==sectors[i],]
sub.agg<-aggregate(sub$ID, by=list(sub$ID), length)
pop.df.lub.m<-merge(pop.df.lub.m, sub.agg, by.x="ID", by.y="Group.1", all.x=TRUE)
colnames(pop.df.lub.m)[19+i]<-paste("Sector", sectors[i], sep="_")}

pop.df.lub.m$count<-(19-rowSums(is.na(pop.df.lub.m[,20:38])))
head(pop.df.lub.m)[,20:39]
```

```
  Sector_A Sector_B Sector_C Sector_D Sector_E Sector_F Sector_G Sector_H Sector_I Sector_J
1    NA       NA       NA       NA       NA       NA       NA       NA        1       NA
2     1        1       NA       NA       NA       NA        1        1        1       NA
3    NA       NA       NA       NA       NA       NA       NA       NA       NA       NA
4     1        1       NA       NA       NA       NA       NA       NA       NA       NA
5    NA       NA       NA       NA       NA       NA       NA       NA       NA       NA
6     2        2       NA        2       NA       NA       NA        1       NA       NA
  Sector_K Sector_L Sector_M Sector_N Sector_O Sector_P Sector_Q Sector_R Sector_S x count
1    NA       NA       NA       NA       NA       NA       NA       NA       NA    1
2    NA       NA       NA       NA       NA       NA       NA        2       NA    6
3    NA       NA       NA        1       NA       NA       NA       NA       NA    1
4    NA       NA       NA       NA       NA       NA       NA        1       NA    3
5    NA       NA       NA       NA       NA        1       NA       NA       NA    1
6    NA       NA       NA       NA       NA       NA       NA       NA        1    4
```

```
library(classInt)
variable<-pop.df.lub.m$count
intervals<-9
colors<-brewer.pal(9, "BuPu") # choice of colors
classes<-classIntervals(variable, intervals, style="fixed",
    fixedBreaks=c(0, 2, 4, 6, 8, 10, 12, 14, 16, 18))
table.colors<-findColours(classes, colors)

plot(pop.grid.lub, col= table.colors) #Figure 2.14b
plot(pov.lub, add=TRUE)

legend("bottomleft", legend=names(attr(table.colors, "table")),
fill=attr(table.colors, "palette"), cex=0.8, bty="n")
title(main="The number of sectors represented by firms
within grid")
```

As shown in Figure 2.14b, the most diverse in terms of sectors in which firms operate is the centre of the region – almost all sectors are represented there. In the southern part of the region, there are almost no firms – except for one cell. On the peripheries of the region, industry diversification is weak.

2.8 Spatial geometries

One of R's interesting features are so-called spatial geometries available in the rgeos:: package. The package allows one to study the properties of shapes and spatial objects and carry out manipulations on them. The *sp* class object is used as the input. The package creates an additional class, *SpatialRing* and *SpatialRingData-Frame*. As part of the package, it is possible to determine the area of the geometry with the **gArea**() command, its length with the **gLength**() command, its centroid with the **gCentroid**() command, its boundaries with the **gBoundary**() command, its envelope with the **gEnvelope**() command and its simplification with the **gSimplify**() command. One can specify its properties (commands **gIsEmpty**(), **gIsRing**(), **gIsSimple**(), **gIsValid**()). By examining the relations between geometries, one can specify intersections and overlays with **gCrosses**(), **gIntersects**(), **gIntersection**(), **gRelate**() and **overGeomGeom**(); properties of including with **gContains**(); equality with the **gEquals**() command, **gDifference**() and **gSymdifference**(); the distance between them with the **gDistance**() command and points of touch with the **gTouches**() command. One can create a buffer around the geometry with **gBuffer**() (for example, to make a circle from a dot), indicate the location of the nearest point from two geometries with the command **gNearestPoints**(), assemble two geometries into one by erasing internal boundaries – so-called union – with the **gUnion**() command, perform Delaunay triangulation with the **gDelaunayTriangulation**() command, indicate the location of the point on the geometry surface with the **gPointOnSurface**() command or determine the best location of the area labels with the **polygonsLabel**() command.

The following are the selected commands applied to the data and problems analysed in the book.

One of the possibilities is triangulation, that is, describing triangles on point data. This allows the division of polygon space in a continuous manner. The **gDelaunayTriangulation**() command requires unique spatial coordinates, and input data – points – must be in the *sp* class. In a situation of overlapping points, duplicates can be deleted or all points scattered randomly by a small amount. It should be remembered that points located in the same way may have different values assigned; therefore, a safer solution is the correction of the location with epsilon, which can be executed with the **jitter**() command.

```
# selected variables: x and y coordinates, employment, sector
firms.sel<-firms[,c(12,13,20,18)]
colnames(firms.sel)<-c("x","y","empl","sector") # change of names

# spreading coordinates by epsilon
firms.sel$x<-jitter(firms.sel$x) # coordinate x
firms.sel$y<-jitter(firms.sel$y) # coordinate y
coordinates(firms.sel)<-c("x","y") # change of class to SpatialPoints
library(rgeos)
tri.full<-gDelaunayTriangulation(firms.sel) # triangulation
tri.lim<-gDelaunayTriangulation(firms.sel[1:1000,])

plot(tri.full)

plot(tri.lim) #Figure 2.15a
points(firms.sel[1:1000,], col="red", pch=".", cex=3)
```

Another problem of spatial statistics for point data is the study of the similarity of neighbouring points. Typically, neighbouring points within a given radius from the selected location are analysed. The most commonly used for this purpose is the full distance matrix, of size $n \times n$ for n points, and from this matrix, one selects only the points lying in the given radius. This method is computationally effective only for small datasets, due to the size of the distance matrix.

An alternative method of performing the same task is to create spatial geometries – circles with a given radius – and to examine whether the tested points are included in this geometry. An obtained zero-one vector determines the inclusion within the radius, which is identical to the testing of the distance matrix.

The following example creates a loop that examines consecutive dataset points. In each iteration of the loop, one circle is created and all dataset points in this circle are checked. Points that are contained receive marker 1. Circles are created with the command **gBuffer**() from the rgeos:: package, where as arguments one gives the planar coordinates of the point being examined, the smoothness of the edge of the circle (*quadsegs* option) and the radius (*width* option). The affiliation (belonging) of points to geometry is tested with the command **over**() from the sp:: package – the first argument is given to all tested points and the second geometry to which the points belong. As a result, a length vector of the first argument with values equal to 0 is obtained when the point is out of geometry and values equal to 1 when the point belongs to the geometry. The result of the **over**() command was saved in the dataset as successive columns for subsequent observations. n new columns were created for the n-element set.

The commands subsequently use different coordinate formats. The **gBuffer**() command from the rgeos:: package requires a planar (flat) coordinate; hence, the *SpatialPoints* input object was created with the **SpatialPoints**() command from the sp:: package without giving the projection. In parallel, a *SpatialPoints* object was created with the **coordinates**() command for spherical coordinates using **proj4string**() and **spTransform**(). The circles created with **gBuffer**() are also planar. It is possible (and necessary in this situation) to convert them to the spherical system by assigning any reference system with the **proj4string**() command and then transforming the **spTransform**(). Spatial objects of coordinates of points and circles with a planar projection are input data to the command **over**() from the sp:: package.

In the example, statistics were calculated – the average value of observations (points) inside the circle (geometry). The zero-one vector has been multiplied (belongs / is outside) by the vector of the tested variable, and the mean of the resulting vector was determined. One can also count neighbours within a circle by counting the sum of the zero-one vector.

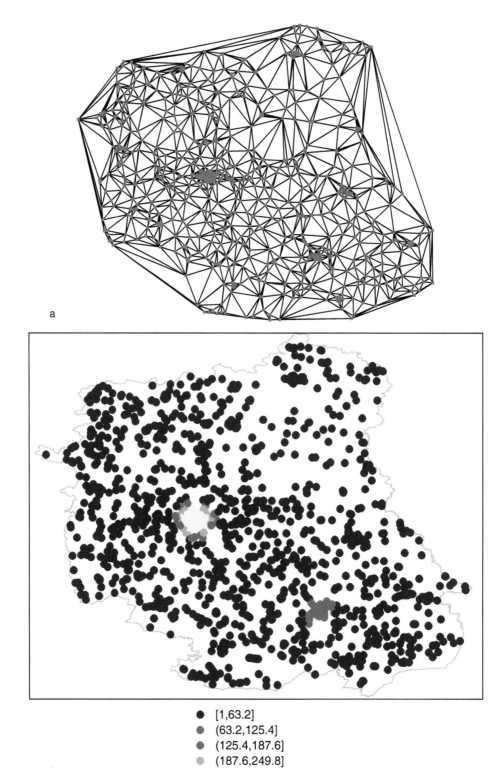

a

b

● [1,63.2]
● (63.2,125.4]
● (125.4,187.6]
● (187.6,249.8]
  (249.8,312]

Figure 2.15 Visualisation of point pattern: a) Delauney's triangulation for data limited to 1000 observations, b) the number of neighbours by location

Source: Own study using R and sp:: and rgeos:: packages

```
# selected variables: x and y coordinates, employment, sector
firms.sel<-firms[1:2000,c(12,13,20,18)]
colnames(firms.sel)<-c("x","y","empl","sector") # change of names
# spreading coordinates by epsilon
firms.sel$x<-jitter(firms.sel$x) # coordinate x
firms.sel$y<-jitter(firms.sel$y) # coordinate y

# preparation of point data in different classes
# required for gBuffer() - planar coordinates
xy<-cbind(firms.sel$x, firms.sel$y)
xy.sp<-SpatialPoints(xy)

# required for over() - spherical coordinates
firms.sp<-firms.sel
coordinates(firms.sp)<-c("x","y")
projection<-"+proj=longlat +datum=WGS84"
proj4string(firms.sp)<-projection
firms.sp<-spTransform(firms.sp, CRS("+proj=longlat +datum=NAD83"))

# creating circles and checking the affiliation of points to a circle
imax<-2000 # for 2000 observations
firms.sel[,5]<-NA # creating an empty column
library(rgeos)
for(i in 1:imax){
ring<-gBuffer(xy.sp[i,], quadsegs=50, byid=TRUE, width=0.1)
proj4string(ring)<-CRS("+proj=longlat +datum=NAD83")
ring<-spTransform(ring, CRS("+proj=longlat +datum=NAD83"))
firms.sel[,i+5]<-over(firms.sp, ring) # checking the affiliation
firms.sel[i,5]<-sum(firms.sel$empl*firms.sel[,i+5], na.rm=TRUE)
/ sum(firms.sel[,i+5], na.rm=TRUE) # average population of neighbors
}

# chart of the number of people populated in point vs. at neighbourhood
plot(firms.sel$empl[1:imax], firms.sel[1:imax,5])

# how many neighbors around a given point
# counted as the sum of subsequent columns
firms.sel$firms.around<-apply(firms.sel[,6:2005], 2, sum, na.rm=TRUE)

# a graph of the number of neighbors nearby - Figure 2.15b
firms.sel1<-firms.sel[,c("x", "y", "firms.around")]
coordinates(firms.sel1)<-c("x","y") # change of class

spplot(firms.sel1, key.space="bottom",
sp.layout=list(voi[voi@data$jpt_nazwa_=="lubelskie",], col='grey'))
```

One can detect that firms with the lowest employment have different neighbours, though they are mostly similar to them. Firms employing 150 people have statistically larger neighbours than smaller firms. Figure 2.15b confirms that the high number of neighbours is typical for central centres (the city of Lublin in the middle of the region and the city of Zamość located in the southeast). In the remaining areas, in the selected radius, there were usually one or two firms.

Another interesting functionality of the rgeos:: package is the integration of several overlapping geometries into one geometry by using **gUnaryUnion()** command and calculating its area with the **gArea()** command. In the example subsequently, the Lublin district was selected as a constraining area, and 100 points were drawn in this area with the **spsample()** command from the sp:: package. Radii values were randomly drawn using the normal **rnorm()** distribution. The **gBuffer()** command converts points to circles with given radii. Overlapping circles were integrated into separated objects. Finally, their fields were counted, and the sum of fields was referred to the total area of the area.

```
pov<-readOGR(".", "powiaty") # 16 units.
pov<- spTransform(pov, CRS("+proj=longlat +datum=NAD83"))

pov.lub<-pov[pov@data$jpt_nazwa_=="powiat Lublin",]
point.lub<-spsample(pov.lub, 100, type="random")

radius<-rnorm(100, mean=0.007, sd=0.003)
xy<-cbind(point.lub@coords[,1], point.lub@coords[,2])
xy.sp<-SpatialPoints(xy) # required for gBuffer()
circles<-gBuffer(xy.sp, quadsegs=50, byid=TRUE, width=radius)

plot(pov.lub, lwd=2) # Figure 2.16a
points(point.lub, pch=".", cex=2)
plot(circles, add=TRUE)

# union (common part) of overlapping circles
circles.union<-gUnaryUnion(circles) # from the package rgeos::
plot(pov.lub, lwd=2) # Figure 2.16b
plot(circles.union, add=TRUE)

gArea(circles)
#[1] 0.01776395

gArea(circles.union)
#[1] 0.01246122

gArea(pov.lub)
#[1] 0.01897292
```

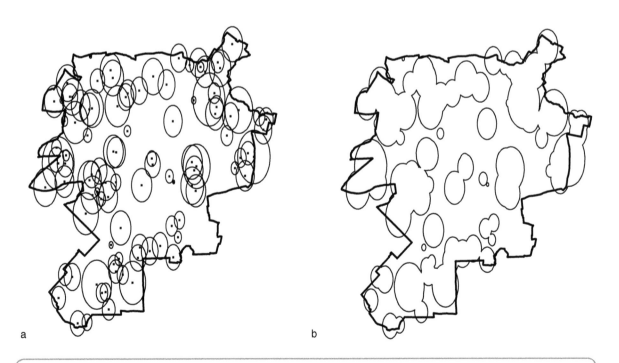

a   b

Figure 2.16 Circles in the plane: a) individual circles, b) circles integrated with the command **gUnaryUnion()**

Source: Own study using R and sp:: and rgeos:: packages

The previous example shows that spatial geometries allow transformations and calculations to be made on points or regions. They can be used in spatial statistics. An example of such statistics is the SPAG: index of the spatial agglomeration (Kopczewska, 2017; Kopczewska, Churski, Ochojski, & Polko, 2019), which is based on the concept of representation of points by poviats and assessment of the land cover with circles, including overlap.

Note

1 For Poland, such data is provided by the Central Statistical Office at: https://geo.stat.gov.pl/inspire

## Chapter 3

Spatial data with Web APIs

Mateusz Kopyt and Katarzyna Kopczewska

This chapter will describe methods for dynamically accessing and downloading various types of external data using the Web application programming interface (Web API) mechanism. This mechanism in the context of spatial analysis allows the use of contextual raster background from existing resources on the Internet, which increases the possibility of presenting localised data. Equally important, vector spatial datasets are beginning to appear in web resources. The use of the Web API gives one the chance to acquire and work on a constantly updated data resources ready for spatial analysis. One should also mention non-spatial resources and databases, which can also be downloaded using APIs. Such data can complement the existing localised database. One of the problems when downloading traditional (non-geocoded) data for spatial analysis is the subsequent geocoding, that is, assigning it the appropriate geographical coordinates. Despite the fact that there is still no (and probably this is not achievable) method that would flawlessly enable the conversion of, for example, addresses to appropriate points in space, specialised servers using the API are able to automate this process and achieve quite good results now. This chapter will indicate the possibility of performing this activity automatically directly in the R program. The growing importance of using data accessed by the API is also associated with the fact that more and more data is placed on the network on the basis of open access, which greatly facilitates the process of obtaining data for analysis.

3.1 What is an application programming interface (API)?

The basic method of creating data that is used for further analysis is to create personal databases. This applies to both traditional data and the spatial analysis data resources (including, for example, vector maps). Of course, it is not possible to create data by yourself from scratch, so it is natural to use existing resources. The difficulty in the traditional approach is the fact that data from external sources is downloaded once and reflects the state of the used set (or in the case of cartographic data, maps) for a given day. Due to the constantly growing Internet resources and their constant update (including in particular the publication of data referred to as public information), a useful idea was to create a mechanism that would allow downloading data from external sources, updating it (e.g. on demand by the user or automatically when the program code is executed) and using such collections many times. In this case, each use provides a different result based on the currently downloaded data. The mechanism that meets the idea described here is the so-called API service used in IT applications. In general, the API defines the way programs can communicate with each other, ensuring interaction. In other words, it is a way to build new applications using the already-created software and activities that these programs generate.

For the purposes of data analysis (including spatial data analysis), a special type of API called WebAPI is important. Each WebAPI – specific to a given service located on a server in the network – defines the form of communication between such server and computers/programs connecting to it and also indicates a specific structure of generated information. To put it differently, each server on the Internet with a WebAPI has a specific form and principles according to which another program can "ask" for information and a specific format of data/information provided for this request.

R has packages that allow one to connect to servers/network resources in popular protocols, that is, methods of communication. Selected packages are also created for specific services – WebAPI servers for the given service. The functions in these packages enable data acquisition without the need to know the technical aspects of constructing queries. Also, the results returned by the remote server are converted to one of the standard data classes (e.g. *data.frame*). Users obtaining data from such a source do not need to have knowledge about the structure of standard data exchange formats in the network (e.g. XML or JSON). Manual "translation" of the server's response to a data class ready for further processing is also avoided. Of course, R also has a set of functions that allow one to create direct queries in the form of, for example, "http" addresses with the appropriate parameters. There are also packages that help to independently analyse and transform a portion of data obtained from the server. The latter method gives the user more flexibility and greater possibilities to select the desired information but also requires a little more knowledge and creating more code.

The possibilities of the R program in the context of using WebAPI services can be used on all three levels of using the data indicated in the introduction to this chapter. First, it will be obtaining geolocated maps in the form of raster images to supplement the graphics with such a background. Second, the collection of spatial data in vector form whilst keeping it in the appropriate classes of spatial objects, which will allow it to be analysed and used in modelling phenomena taking into account the spatial factor. Finally, WebAPI services can be used to download non-spatial data – without location – which can complement the existing dataset with new variables. It is also worth mentioning that the latter type of data can be supplemented with location using API services that will provide the so-called geolocation function, that is, indication of the geographical location on the basis of the textual description of the given record (most often, it will be the address of the occurrence of a given).

3.2 Creating background maps with use of an application programming interface

The first and probably simplest option of using WebAPI services in spatial analysis is the visualisation of the data in such a way that background maps are taken from the Internet. Visualisation is, of course, possible without the use of API servers. In this situation, the cartographic elements are usually limited to vector data on administrative boundaries, areas with a given characteristic – for example, land use and sometimes also objects such as roads or points marking major cities. The use of current map data downloaded from the Internet makes the visualisation possibilities more attractive, but most importantly, it shows the analysed phenomenon in the context of full cartographic information available on the used background. Furthermore, there are different types (physical, topographic, tourist, road and other), and it is possible to use satellite images as a background. In this way, the so-called contextual analysis of the studied phenomena occurs – each time when dynamically downloading a map resource from a given server, access to the latest data on a given server is also provided. Of course, instead of the background maps downloaded by the API services, it is possible to use raster images (e.g. scanned maps), to which georeferences can be added in external GIS programs (e.g. QGIS); that is, the correct coordinates are assigned (in the selected cartographic reference system). Such an image can then also be loaded as a background. However, the advantage of dynamically downloading current data (already with georeferences) from selected API servers seems to be obvious.

Google Maps is undoubtedly the most popular map service on the web. This site offers both a map (which could be called a road map) and access to satellite images. With the release of this service (and Web API servers), additional packages have appeared in R that allow the use of these primers in contextual maps. One of the most suggested packages ensuring R cooperation with Google Maps is the RgoogleMaps:: package. This package is being developed on a regular basis. As stated in the description of this package, it contains both the function of downloading static map from Google servers, as well as using it as a background and allowing layering on it with other data. The basic command is **GetMap.bbox**(), which allows one to load (and save to a local disk) maps from Google by pointing to the centre of the downloaded area and using zoom or giving the coordinates (rectangle) of the area to be downloaded. Another command often used for similar purposes is **MapBackground**(). The presented example will show a fairly simple way to create a contextual map using the first command. A layer of administrative boundaries from the vector file will be added to the graphics. Both the background and the vector borders will be drawn using the **PlotPolysOnStaticMap**() function.

The first step will be to load the necessary additional packages and load vector data from shapefile containing administrative boundaries. The Mazowieckie voivodeship will be selected from the voivodeship boundary layer. Because objects loaded with the **readOGR**() function or created by selecting only a given province are of the *SpatialPolygonsDataFrame* class from the sp:: package and thus are objects containing spatial information, they have the attribute of the coordinate system used (together with parameters, including cartographic mapping). This is information on how the earth's surface was moved to the plane and in what units one measures the location. One can check this with the function **proj4string**(). The last part of the code defines the variables denoting the codes of the three systems used. The first one is global for the whole world and uses geographical coordinates (WGS84), the second is Google maps (as well as OpenstreetMap) for rendering raster maps and the third one is used in Poland. The following vector files describe administrative boundaries. The **CRS**() function gives the full parameters of these systems.

```
library(RgoogleMaps)
library(PBSmapping)
library(maptools)
library(rgdal)

pl<-readOGR("./Data", "Panstwo", use_iconv=TRUE, encoding='Windows-1250')
voi<-readOGR("./Data", "wojewodztwa", use_iconv=TRUE,
encoding='Windows-1250')
pov<-readOGR("./Data", "powiaty", use_iconv=TRUE, encoding='Windows-1250')

mazowieckie<-voi[voi@data$jpt_nazwa_=='mazowieckie',]

class(mazowieckie)
#[1] "SpatialPolygonsDataFrame"
attr(,"package")
#[1] "sp"

proj4string(mazowieckie)
```

```
[1] "+proj=tmerc +lat_0=0 +lon_0=18.99999999999998 +k=0.9993 +x_0=500000
    +y_0=-5300000 +ellps=GRS80 +towgs84=0,0,0 +units=m +no_defs"
```

```
latlong<-"+init=epsg:4326"
google<-"+init=epsg:3857"
polish_borders<-"+init=epsg:2180"

CRS(latlong)
```

```
CRS arguments:
+init=epsg:4326 +proj=longlat +datum=WGS84 +no_defs +ellps=WGS84
+towgs84=0,0,0
```

```
CRS(google)
```

```
CRS arguments:
+init=epsg:3857 +proj=merc +a=6378137 +b=6378137 +lat_ts=0.0
+lon_0=0.0 +x_0=0.0 +y_0=0 +k=1.0 +units=m +nadgrids=@null
+no_defs
```

```
CRS(polish_borders)
```

```
CRS arguments:
+init=epsg:2180 +proj=tmerc +lat_0=0 +lon_0=19 +k=0.9993
+x_0=500000 +y_0=-5300000 +ellps=GRS80 +towgs84=0,0,0,0,0,0,0
+units=m +no_defs
```

Currently, using the GoogleMap API server requires the creation of an access key. To create it one should use the following website: https://console.cloud.google.com/getting-started. There is a platform that allows one to create such a key and manage other Google services available to users. There are many instructions on the web on how to create an API key for GoogleMaps services. Although registration and key creation require providing data regarding the form of payment (e.g. credit card), the created key allows one to obtain a specific number of free queries to the GoogleMaps service in a month. This is sufficient even for the needs of basic analysis. The created key has been assigned to the APIKEY variable subsequently for easy use. Of course, a fake code is shown subsequently (for presentation purposes).

```
# assigning an individual Google API access key to the variable
APIKEY<-"AbCdEfGhIjKlMnOpRsTuWxYz" # false key
```

In order to download the background map from GoogleMaps, one must specify the geographical coordinates of corners of the rectangle that marks the area acquired. One can enter them manually, but one can set them on the basis of another vector (spatial) object – for example, voivodeship borders. To do this, one can use two functions: **qbbox**() from the Rgooglemaps:: package or **bbox**() from the sp:: package. Both functions provide the extreme geographical coordinates of a given object, except that the first requires a separate specification of the latitude vector and longitude; hence, the maz.polyset object is created first, giving easier access to the geographical coordinate vectors used in **qbbox**(). For the **bbox**() function, the argument can be any *Spatial* object. In the example, the problem is that the Masovian object is in the EPSG coordinate system: 2180 (Cartesian coordinates), while for later download of maps from Google, geographical coordinates based on EPSG: 4326 are required. The Masovian object will first be converted to the desired coordinate system using the **spTransform**() function from the rgdal:: package using pre-defined codes specifying different coordinate systems.

```
# boundaries of Mazovian region - method I
mazowieckie.lonlat<-spTransform(mazowieckie, CRS(latlong))
maz.polyset<-SpatialPolygons2PolySet(mazowieckie.lonlat) #using maptools::
maz_box<-qbbox(lat=maz.polyset[,"Y"], lon=maz.polyset[,"X"]) #RgoogleMaps::

# boundaries of the download area - method II
maz_box2<-bbox(mazowieckie.lonlat) #using sp::

maz_box
$latR
#[1] 51.00077 53.49415
$lonR
#[1] 19.23987 23.14775
maz_box2
```

```
        min      max
x 19.25921 23.12841
y 51.01311 53.48181
```

The advantage of the **qbbox**() function is that it generates area boundaries with a margin. The standard margin value can be changed in options. This downloads maps which are slightly larger than the expected area but improves their readability. It is also worth paying attention to a different order of results of both obtained boundaries of the area that will be downloaded.

GetMap.box() is the basic function for downloading maps in raster form via the Google WebAPI service. The result is an image (in PNG, JPG or GIF format) in the *StaticMap* class. The feature requires specifying the latitude and longitude range of the area or centre and zoom. Currently, one must also provide the Google API key, and one can specify the name of the file to which the downloaded map will be saved to disk. After downloading the map to the Maz.map object, it was drawn along with the outline of the Masovian Voivodeship. The **PlotPolysOnStaticMap**() function was used, which imposes a vector layer on the raster image in polygonal format. The easiest way is to use an object with the *PolySet* class obtained as a result of the **SpatialPolygons2PolySet**() function, although it is also possible to use objects from the *SpatialPolygons* class from the sp:: package.

```
maz.map<-GetMap.bbox(maz_box$lonR, maz_box$latR, destfile="MAZ_Google.png",
API_console_key=APIKEY, SCALE=2)
class(maz.map)
#[1] "staticMap"

# drawing a background and adding a contour map, Figure 3.1a
par(mar=c(0,0,0,0)) # erasing the margins
PlotPolysOnStaticMap(Maz.map, maz.polyset, lwd=2, col=rgb(0.25, 0.25, 0.25,
0.025), border="red", add=FALSE)
```

Figure 3.1 Contours with a raster background downloaded from GoogleMaps: a) Mazowieckie voivodeship, b) regions in Poland

Source: Own study using R and the RgoogleMaps:: package

Figure 3.1 (Continued)

The saved map in the form of a raster file can be reloaded (as an object in the *StaticMap* class), without sending the query to the API server again, using the **GetMap.box()** function. However, then it is necessary to specify the file name and use the *NEWMAP=FALSE* parameter (set to *TRUE* by default, which automatically sends a query to the server). The same effect can be achieved by using the **ReadMapTile()** function by providing only the name of the previously downloaded file.

Obtainable map backgrounds are not limited to Google roadmaps. In the maptype option for the **Getmap. bbox()** function, one can indicate the following types of sleepers: roadmap, mobile, satellite, terrain, hybrid, mapmaker-roadmap and mapmaker-hybrid. It is worth trying what they look like and choosing the best one for given applications. Slightly more options for choosing the type of background map will be available in the next method of preparing graphics containing maps described subsequently. The last point to keep in mind when using the GoogleMaps WebAPI service is that the downloaded maps have quite limited resolution. The maximum size of a standard downloaded image is 640 × 640 pixels. It can be increased by using the *SCALE=2* parameter. The result is an image with 1280 × 1280 pixels. For presentation purposes, this is sufficient, while on the printouts, it can be seen that this is quite limited quality.

In the example subsequently, a similar operation was performed – the provincial map of Poland was superimposed on the base map of Europe. The **GetMap.bbox()** command from the RgoogleMaps:: package retrieves a background map. The coordinates defining the geographical coverage of the base were obtained from the shapefile contour map for provinces with the command **SpatialPolygons2PolySet**(). They have been transformed into a bounding box by the **qbbox**() command. The **GetMap.bbox**() command uses the bounding box coordinates and saves the acquired image as *.jpg in the Working Directory, and the map itself is in the *staticMap* class. The zoom option is not available because it is in fact defined by the bounding box.

```
library(RgoogleMaps)
library(PBSmapping)
library(maptools)
library(rgdal)


# contour and bounding box
woj.lonlat<-spTransform(woj, CRS(latlong))
woj.polyset<-SpatialPolygons2PolySet(woj.lonlat) # using maptools::
bb<-qbbox(lat=woj.polyset[,"Y"], lon=woj.polyset[,"X"]) # RgoogleMaps::


# RgoogleMaps - generates a map in the "staticMap" class
MyMap<-GetMap.bbox(bb$lonR, bb$latR, destfile="DC.jpg", API_console_
key=APIKEY)


# drawing a background and adding a contour map, Figure 3.1b
par(mar=c(0,0,0,0))
PlotPolysOnStaticMap(MyMap, shp, lwd=1, col=rgb(0.25, 0.25, 0.25, 0.025),
add=FALSE)
```

An alternative to the Google service may be to use servers with open data. The most popular and fastest-growing free and open resource of cartographic data is the OpenStreetMap project. In fact, this project is not a map but a collection (database) of spatial data in vector format. Each of the spatial objects is described by attributes clearly indicating its type and possessed features. Data for this database is collected by the user community. At the same time, the quality of this data is mutually controlled and improved by them. The OSM Association and its local sections also ensure an unequivocal description of the data and also its quality. Due to the efforts of the Association, in many countries, official map data has been released and transferred to the OSM website (it is also updated frequently). OSM currently binds data from official resources as well as local knowledge of its users. Therefore, data from areas that do not have official maps is often correct. Based on this database, a map is built in the form of raster images. The basic map is available at: https://www.openstreetmap.org/. Due to the constantly increasing data resource, other map services are available that provide thematic maps, for example, tourist maps. Additionally, services and applications enabling route finding and navigation are created on the OSM database. The OSM database has official API services for both downloading raster images and vector data and an address geocoding API server. These servers are available on a free access basis, sometimes with restrictions on the number of queries so that they are not overloaded and blocked. Other API servers have also been created on the OSM database, which sometimes share data even without the limits mentioned previously.

For example, a similar operation scheme as for Google servers will be presented, that is, downloading a raster map and supplementing it with vector layers of administrative boundaries. The operation of downloading a map from the OpenStreetMap WebAPI service can be performed manually, although it is easier to use a dedicated package in R (OpenStreetMap::), which will provide the required data and download the appropriate raster images from the OSM server. The OSM server does not require obtaining or providing any user key. The function for downloading maps is **openmap**(). It requires first of all geographical coordinates (latitude, longitude) of the upper-left and lower-right corner of download areas. By default, the OSM-style raster map is downloaded (see Figure 3.2).

```
library(OpenStreetMap)
maz_box # reminder of the border values of the Mazowieckie area
```

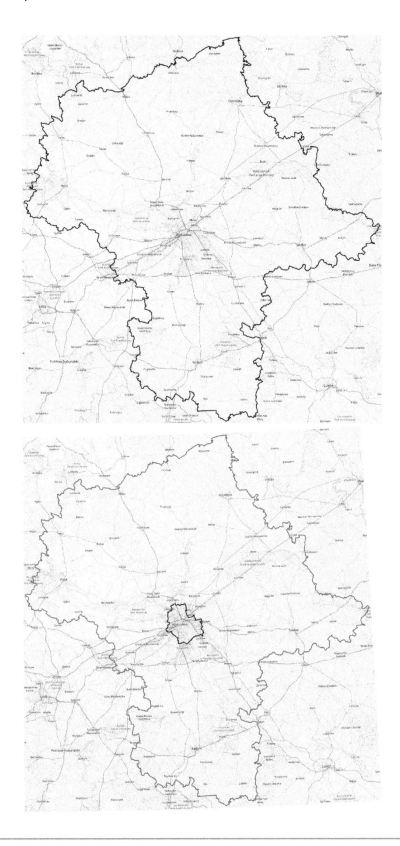

Figure 3.2 Contour of the Mazowieckie voivodeship with the default OpenStreetMap raster background: a) method I – original system, b) method II – EPSG coordinate system: 2180 according to the vector layer

Source: Own study using R and the OpenStreetMaps:: package

```
$latR
#[1] 51.00077 53.49415
$lonR
#[1] 19.23987 23.14775
```

```
# limits entered manually
maz_osm<-openmap(c(53.49415,19.23987),c(51.00077,23.14775))

# I method - in the original layout of the OSM underlay - Figure 3.2a
# change in the reference system of the Masovian district to native OSM
plot(maz_osm)
plot(spTransform(mazowieckie, osm()), add=TRUE)

# II method - in a reference system consistent with administrative bounda-
ries
maz_osm.2180<-openproj(maz_osm, CRS(polish_borders)) # Figure 3.2b

plot(maz_osm.2180)
plot(mazowieckie, lwd=2, border="red", add=TRUE)

waw<-pow[pow@data$jpt_name_=="powiat Warszawa",]
polygon(waw@polygons[[1]]@Polygons[[1]]@coords) # adding Warsaw contour
```

The resulting file is an OpenStreetMap image. An important advantage of this result is the fact that it has a coordinate system stored in the object (EPSG code with parameters) – the native OSM system is identical to that used by Google (EPSG: 3857). The OpenStreetMap:: package has a function that allows for changing the coordinate system of the downloaded map and adapting it to the user's own data. The advantage, unlike the example of Google map use shown earlier, is the fact that on the obtained OSM background, one can apply subsequent layers using the standard functions **plot**(), **lines**(), **points**() and so on, without changing OSM map to the simple general raster object. The distortion of the background image in Figure 3.2b is due to the change in its coordinate system.

Files downloaded from the OpenStreetMap service can also be easily printed using the ggplot2:: package and its capabilities. The **autoplot**() function automatically calls the appropriate commands from the ggplot2:: package; in addition, according to the syntax based on the functions present in this package, a layer with the border of the Mazowieckie voivodeship has been added. The syntax used by the functions of the ggplot2:: package will be discussed in more detail in other parts of the book. It should be noted that the coordinates in the figure are in the Cartesian system and the map background is slightly rotated. This is due to the fact that the map downloaded from the OSM is adapted to the EPSG: 2180 system, in which there is a vector file. Of course, one can draw a similar graph in the geographical coordinate system, which, however, will cause slightly different transformations of the background, which were already observed in earlier examples.

```
library(ggplot2)
maz.f<-fortify(mazowieckie)
# figure similar to previous ones
autoplot(Maz_osm.2180) + geom_polygon(aes(x=long, y=lat, group=group),
data=maz.f, colour='red', alpha=0, size=0.3)
```

The **openmap**() function can be used not only to download OSM maps in the default style. By using the type option additionally, the user can choose the desired map style from several different servers. The currently available style list is as follows: "*osm*", "*osm-bw*", "*maptoolkit-topo*", "*waze*", "*bing*", "*stamen-toner*", "*stamen-terrain*", "*stamen-watercolor*", "*osm-german*", "*osm-wanderreitkarte*", "*mapbox*", *"esri"*, *"esri-topo"*, *"nps"*, *"apple-iphoto"*, *"skobbler"*, *"hillshade"*, *"opencyclemap"*, *"osm-transport"*, *"osm-public-transport"*, *"osm-bbike"* and *"osm-bbike-german"*, although not all servers are always available. The following example will be based on the code contained in the help for the **openmap()** command. For comparison, the same map fragment will be displayed in nine selected styles; the map fragments contain the names of the styles used.

```
library(OpenStreetMap) # Figure 3.3
nm< - c("osm", "bing", "stamen-toner", "stamen-terrain", "stamen-watercolor",
"esri", "esri-topo", "nps", "apple-iphoto")
par(mfrow=c(3,3))
for(i in 1:length(nm)){
 par(mar=c(1,0,1,0))
   map<-openmap(c(53.49415,19.23987),c(51.00077,23.14775),type=nm[i])
plot(map)
title(main = nm[i])}
```

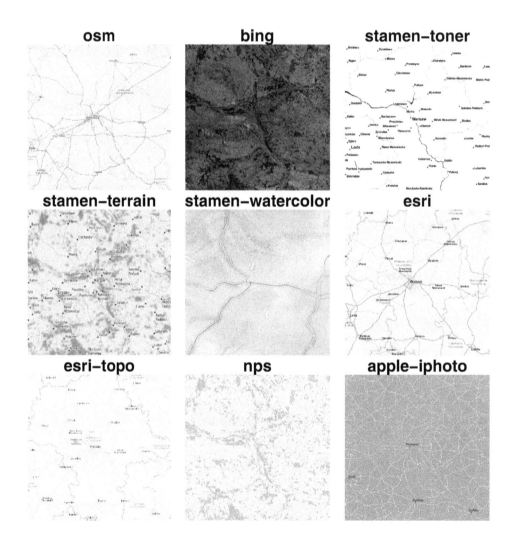

Figure 3.3 Selected map styles available for download using the **openmap()** function from the OpenStreetMap:: package

Source: Own study using R and the OpenStreetMap:: package

There is another map downloading scheme available in R. Using the ggmap:: package and the **get_map()** command, one can download attractive background maps. They are the so-called Static Map in *ggmap* class and *raster* class. When using this package, a Google API key must be first registered with the command **register_google()**. A detailed discussion of this command is provided in Section 3.6. To draw a map, one needs three elements: defining the location, choosing the map type and code for drawing the map itself. They are discussed subsequently.

1   *Defining the location* – this is possible in several ways. First, the location command, for example, the geocode ("Olsztyn, Poland") is defined descriptively in the **geocode()** localisation command. The result of the **geocode()** command is in the *data.frame* and *tibble* classes.

```
library(ggmap)
register_google(key=APIKEY) # previously defined key
ggmap_hide_api_key()
ggmap(get_map(geocode("Olsztyn, Poland"), maptype="roadmap",
source="google"))


gc<-geocode("Olsztyn, Poland")
gc
```

```
# A tibble: 1 x 2
lon lat
<dbl> <dbl>
1 20.5 53.8
```

Second, in the **get_map()** map download command, the location option specifies the country, for example, *location = 'Poland'*.

```
ggmap(get_map(location="Poland", maptype="roadmap", source="google",
zoom=6))
```

Third, in the map download command **get_map()**, the location option specifies geographical coordinates, for example, location = c (22.57, 51.24) – these are for Lublin – they are then the centre of the map.

```
ggmap(get_map(location=c(22.57, 51.24), source="stamen",
maptype="watercolor", crop=FALSE, zoom=10))
```

2   *Map type selection* – the following map type can be used: *maptype = c ("terrain", "terrain-background", "satellite", "roadmap", "hybrid", "toner", "watercolor", "terrain-labels", "terrain-lines", "toner-2010", "toner-2011", "toner-background", "toner-hybrid", "toner-labels", "toner-lines", "toner-lite")*. Maps can come from three sources: *source = c ("google", "osm", "stamen")*, and source selection is always automatic. Stamen is the name of the company providing the maps (http://maps.stamen.com). There can be three types of maps: watercolour, toner and terrain. The Google API key is required for each option to specify the map source. It is also possible to control zoom – the zoom option requires entering consecutive natural numbers. Usually, *zoom=3* is continent, *zoom=10* is city, *zoom=21* is building. In OpenStreetMap maximal value for *zoom* is *18*. The available *crop=FALSE* option prevents light scaling of the map from Stamen. Figure 3.4 shows the example of available map types.

3   *Drawing a map* – the **ggmap()** command draws aesthetic maps with a margin and coordinates. The **plot()** command also performs such graphics, but maps typically have no margin and may be slightly blurred. The following examples show the contour map of Poland in various approximations (see

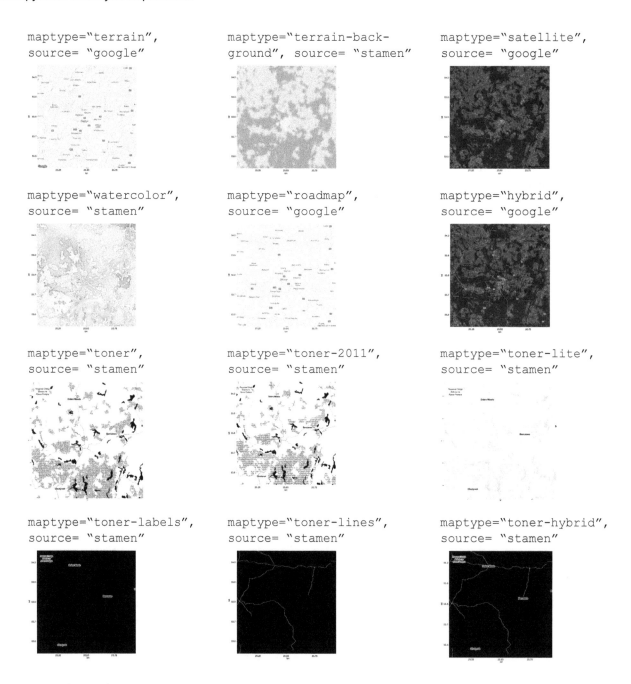

Figure 3.4 Available map types in Google and Stamen

Source: Own study using R and the ggmap:: package

Figure 3.5) and the road system in the watercolour style for the Lublin area in different approximations (see Figure 3.6).

```
# drawing with the plot() command
MyMap2<-get_map(location='Poland', zoom=4) # from the ggmap package
plot(MyMap2) # Figure 3.5a
MyMap2<-get_map(location='Poland', zoom=6) # from the ggmap package
plot(MyMap2) # similar to Figure 3.5a, but higher zoom
```

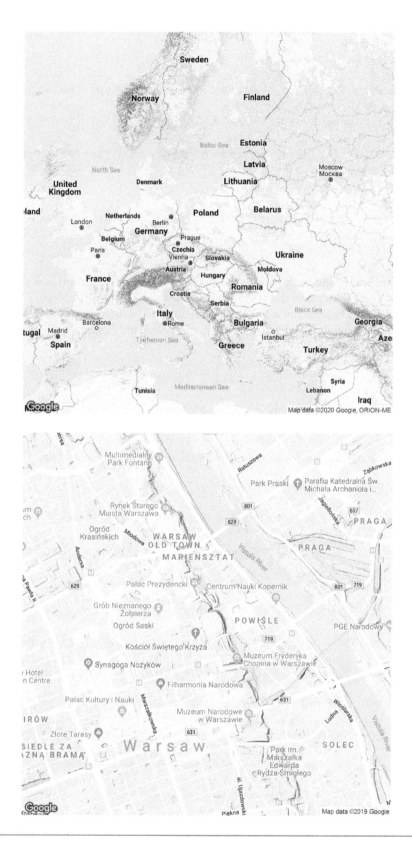

Figure 3.5 Selected maps with different zoom download from the ggmap:: package: a) map of Europe downloaded with **get_map()** function, b) map of Warsaw downloaded using the **qmap()** function

Source: Own study using R and the ggmap:: package

```
# Road system around Lublin - using different zoom
# drawing with the ggmap() command
loc<-c(22.57, 51.24)

Lublin<-get_map(location=loc, source="stamen", maptype="watercolor",
crop=FALSE, zoom=10, scale=2)
ggmap(Lublin) # Figure 3.6a

Lublin<-get_map(location=loc, source="stamen", maptype="watercolor",
crop=FALSE, zoom=12, scale=2)
ggmap(Lublin) # Figure 3.6b

Lublin<-get_map(location=loc, source="stamen", maptype="watercolor",
crop=FALSE, zoom=13, scale=2)
ggmap(Lublin) # Figure 3.6c
```

Note that the **qmap()** command from the ggmap:: package replaces the **ggmap()** and **get_map()** commands together. It allows for quick drawing of the desired location without intermediate steps. This command allows both localisation specified in English and a localisation language. The location of the University of Warsaw is shown in Figure 3.5 using the following code, indicating it in Polish and English.

```
qmap(location='University of Warsaw', zoom=14) # from the ggmap package
qmap(location='Uniwersytet Warszawski', zoom=13) # Fig.3.5b
```

The examples shown previously are only an introduction to the possibilities offered by downloading raster tiles from map services available on the web. Emphasis was placed on downloading raster images from publicly available servers and printing them. The supplementation of the obtained charts with the presentation of the examined variables depends on the needs of the specific analysis and the available data.

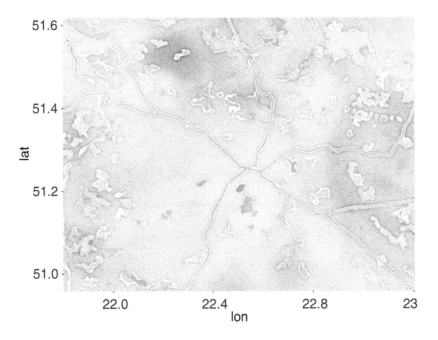

Figure 3.6 **Selected maps of the road system around Lublin with different zoom down-loaded using the get_map() function from the ggmap:: package**

Source: Own study using R and the ggmap:: package

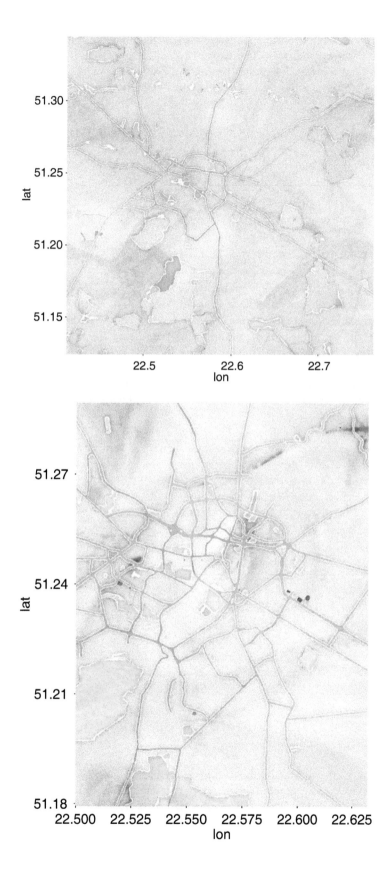

Figure 3.6 (Continued)

In addition to background maps, outline maps are also available online. The raster:: package includes the **getData**() function, which downloads shapefile maps and data at various levels of aggregation for the whole world, drawing from several sources. The **ccodes**() command displays the names of all available countries along with their codes and geographical assignment. For example, Poland is in row 181.

```
library(raster)
ccodes()   # full list
ccodes()[181,]   # the list is limited to one country
```

```
        NAME  ISO3  ISO2  NAME_ISO  NAME_FAO  NAME_LOCAL  SOVEREIGN
181  Poland  POL   PL    POLAND    Poland      Polska      Poland
          UNREGION1    UNREGION2   continent
181  Eastern Europe      Europe      Europe
```

To download the selected contour, the source (one of: 'GADM', 'countries', 'SRTM', 'alt', 'worldclim'), country code and aggregation level (most often from 0 to 3, where 0 is the country and 1, 2, 3 successive lower levels of territorial aggregation). It should be noted that the vector objects downloaded from GADM are assigned the EPSG:4326 coordinate system. On the project website (https://gadm.org/index. html), GADM has ready-to-download objects in the *sp* and *sf* classes. The **getData**() command automatically gets data into the *sp* class; later conversion to the *sf* class is possible. Attributes of the downloaded object can be examined as always with the **attributes**() command, including displaying the names of regions from the regional map.

```
voi<-getData('GADM', country='PL', level=1) # from the raster package::
attributes voi) # checking attributes and slot names
voi$NAME_1 # names of subregions
dim(voi) # length of set - number of regions
pov<-getData('GADM', country='PL', level=2) # other level of aggregation
gm<-getData('GADM', country='PL', level=3) # different level of aggregation
pol<-getData('GADM', country='PL', level=0) # other level of aggregation

plot(pol, lwd=3) # multi-layer figure for two top aggregate levels, Fig-
ure 3.7a
plot(voi, add=TRUE) # country NTS0 & regions NTS2

plot(gm, border='grey70') # multi-layer figure for two other aggregate lev-
els
plot(pov, add=TRUE) # municipalities NTS5 & poviats NTS4
```

3.3 Ways to visualise spatial data – maps for point and regional data

One can apply point charts (the so-called bubble map) on the background maps. An interesting gallery of such charts is available at: https://www.r-graph-gallery.com/bubble-map/. There are several schemes of such charts, because they use different commands. They will be presented subsequently.

Scheme 1 – with bubbleMap() from the RgoogleMaps:: package

From the RgoogleMaps:: package, one can use the **bubbleMap**() command, which draws points in selected locations. The command gives references to two objects: maps in the *map* option and dataset. The object with data is given as the first argument; in the *coords* option, the names of columns in which the geographical coordinates are given are given and in the *zcol* option, the name of the variable that will vary the size of the circles is given. The background map was taken with the **GetMap.bbox**() command from the package from RgoogleMaps::.

Figure 3.7 Multilayer maps: a) contour maps from GADM – levels 0 and 1, raster::, b) map of Poland with a population layer with **GetMap.bbox()** and **bubbleMap()**, RgoogleMaps::

Source: Own study using R and raster:: and RgoogleMaps:: packages

One can apply a legend to the figure. Unfortunately, one usually has to work on its aesthetics, and one can't always adapt it to expectations.

In the example, the *populationxy.csv* data was loaded, which contains data on the population for the largest 30 cities in Poland, as well as the number of women and men, city coordinates and belonging to the voivodeship.

```
# loading prepared data
populationxy<-read.table("populationxy.csv", sep=";", dec=".",header=TRUE)
populationxy
```

LP	City	Voi.	total	MEN	WOMEN	xx	yy
1 1	M.st.Warszawa	MAZOWIECKIE	1744351	800800	943551	52.25900	21.02000
2 2	Kraków	MAŁOPOLSKIE	761069	354954	406115	50.10220	19.96223
3 3	Łódź	ŁÓDZKIE	700982	318978	382004	51.77636	19.48418
4 4	Wrocław	DOLNOŚLĄSKIE	635759	296654	339105	51.11240	17.07775
5 5	Poznań	WIELKOPOLSKIE	542348	252870	289478	52.40637	16.92517
6 6	Gdańsk	POMORSKIE	462249	218901	243348	54.25925	18.66179

```
# preparation of map components
library(sp)
library(rgdal)
library(RgoogleMaps)
voi<-readOGR(".", "wojewodztwa") # 16 units
voi<-spTransform(voi, CRS(latlong))
shp<-SpatialPolygons2PolySet(voi) # z pakietu maptools
APIKEY<-"AbCdEfGhIjKlMnOpRsTuWxYz" # false Google API key
bb<-qbbox(lat=shp[,"Y"], lon=shp[,"X"]) # z pakietu RgoogleMaps
MyMap<-GetMap.bboxMK(bb$lonR, bb$latR, destfile="DC.jpg", API_console_
key=APIKEY, SCALE = 2)

# basic figure based on RgoogleMaps::, Figure 3.7b
bubbleMap(populationxy, coords=c("yy","xx"), map=MyMap, zcol='total', key.
entries = 10000000) # from the RgoogleMaps

# figure with a legend, similar to Figure 3.7b
bubbleMap(populationxy, coords=c("yy","xx"), map=MyMap, zcol='total', key.
entries=populationxy$Women, LEGEND=T, do.sqrt=TRUE, max.radius=100000)
```

Scheme 2 – with ggmap() from the ggmap:: package

Using the ggmap:: package, one can put points on the background map with the **get_map**() command from the above-mentioned package. The ggplot syntax is used – the background map is drawn with the **ggmap**() command, and the points are applied with the **geom_point**() command, where their location (*aes*), variable for drawing (*date*), transparency (*alpha*), colour (*color*) and size are defined (*scale_size*). The legend appears automatically next to it. As one can see from the two examples subsequently, there may be several thousand drawn points. The drawing scheme remains the same.

```
# map of Poland with the population in large cities - Figure 3.8a
library(ggmap)
register_google(key="AbCdEfGhIjKlMnOpRsTuWxYz") # False Google API key
MyMap2<-get_map(location='Poland', zoom=6)

# based on ggmap:: package
ggmap(MyMap2) + geom_point(aes(x=yy, y=xx, size=sqrt(total)),
data=populationxy, alpha=.5, color="darkred")+ scale_size(range=c(3,12))
```

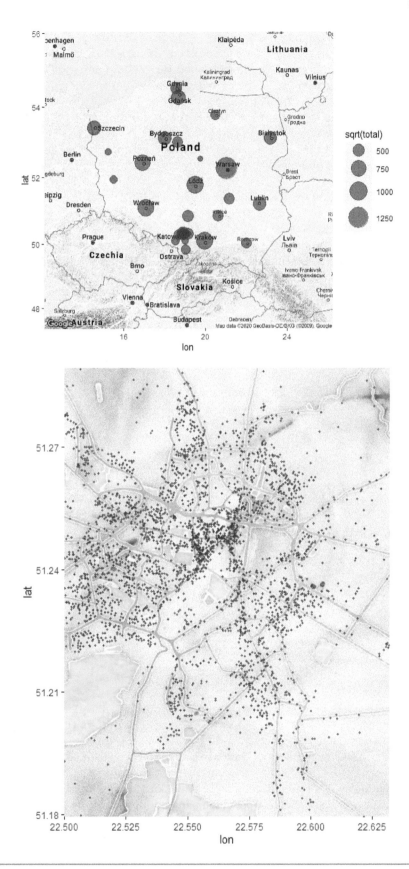

Figure 3.8 Background maps with a point layer made using get_map() and ggmap() from the ggmap:: package

Source: Own study using R and the ggmap:: package

```
# map of Lublin with company locations - Figure 3.8b
# loading point data regarding the location of firms in Lubelskie
firms<-read.csv("geoloc_data_firms.csv", header=TRUE, dec=",", sep=";")
firms$ones<-as.data.frame(rep(1, times=dim(firms)[1]))

# Lublin city map with location of firms (geom_point)
loc<-c(22.57, 51.24)
Lublin<-get_map(location=loc, source="stamen", maptype="watercolor",
crop=FALSE, zoom=13) # based on ggmap::

ggmap(Lublin) + geom_point(aes(x=coords.x1, y=coords.x2, size=ones/5),
data=firms, alpha=0.5, color="darkred", size=0.75) # based on ggmap::
```

In the ggmap:: package, it is also possible to make a layered drawing on the map with location of objects, labels, administrative contours, phenomenon density and isolines. The syntax for the **ggmap**() command is similar to the **ggplot**() command. Subsequently, these opportunities for the Lublin voivodeship will be presented.

```
# background map
p<-ggmap(get_googlemap(center=c(lon=22.85, lat=51.30), zoom=8, scale=2,
maptype ='terrain', color = 'color'))
p # background map with an administrative outline (geom_polygon)

voi<-readOGR(".", "wojewodztwa") # 16 units
voi<-spTransform(voi, CRS(latlong))
lub.voi<-voi[voi$jpt_nazwa_=="lubelskie",]
lub.f<-fortify(lub.voi) # converted for ggmap::

p # background map with transparent contour
p + geom_polygon(aes(x=long, y=lat, group=group), data=lub.f, size=0.3,
fill='grey80', colour='white', alpha=0.3) #background + filled contour
```

Having point data, as in this example on the location of firms, one can easily visualize them, as well as illustrate their spatial density.

```
# background map with contour (geom_polygon) and points (geom_point)
p + geom_polygon(aes(x=long, y=lat, group=group), data=lub.f,
colour='white', fill='grey80', alpha=0.3, size=0.3) + geom_
point(aes(x=coords.x1, y=coords.x2, size=zatr/5), data=firms, alpha=0.5,
color="darkred", size=0.75)

# background map with contour (geom_polygon) and point density (stat_densi-
ty2d)
p + geom_polygon(aes(x=long, y=lat, group=group), data=lub.f,
colour='white', fill='grey80', alpha=0.3, size=0.3) + stat_
density2d(aes(x=coords.x1, y=coords.x2, fill =.. level..,
alpha=0.25),size=0.01, bins=30, data=firms, geom="polygon")
```

Subsequently, an additional set of data on cities in the Lubelskie voivodeship has been loaded. It includes the 10 largest cities and their coordinates, names, population and position in the ranking of cities in the province. The data will be used to apply labels on the location of cities on the map. To create an aesthetic label, city position and name were glued together in one object with the **paste**() command.

```
# loading city data
cities.lub<-read.table("cities of lubelskie.csv", sep=";", dec=",",
header=TRUE)
cities.lub
```

	xx	yy	city	population	position
1	51.24806	22.57028	Lublin	339682	1
2	50.72056	23.25861	Zamość	64354	2
3	51.13222	23.47778	Chełm	63734	3
4	52.03333	23.11667	Biała Podlaska	57303	4
5	51.41639	21.96917	Puławy	48114	5
6	51.21972	22.70000	Świdnik	39913	6
7	50.92139	22.22083	Kraśnik	34821	7
8	51.92722	22.38333	Łuków	30310	8
9	50.54083	22.72194	Biłgoraj	26724	9
10	51.46222	22.60861	Lubartów	22138	10

```
cities.lub$label<-paste(cities.lub$position, cities.lub$city, sep=".")
# background map with contour (geom_polygon),
# point density (stat_density2d),
# and labels (geom_point, geom_label_repel)
library(ggrepel)

# Figure 3.9a
p + geom_polygon(aes(x=long, y=lat, group=group), data=lub.f,
colour='white', fill='grey80', alpha=0.3, size=0.3) + stat_
```

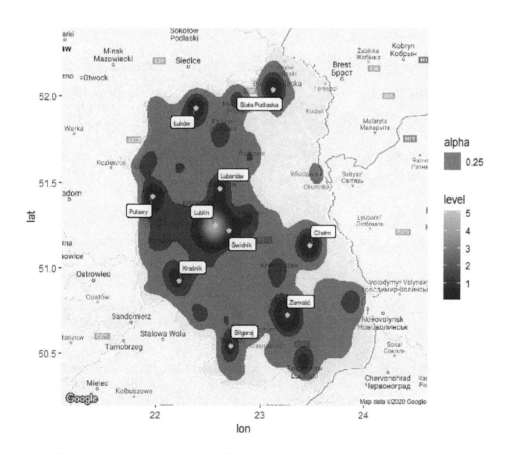

Figure 3.9 Background maps a) with city labels, b) with isolines and point density

Source: Own study using R and the ggmap:: package

107

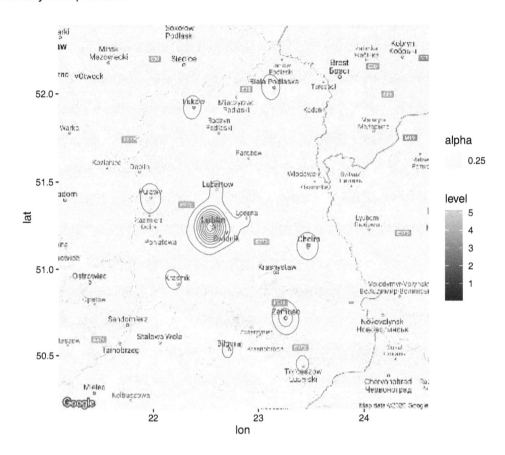

Figure 3.9 (Continued)

```
density2d(aes(x=coords.x1, y=coords.x2, fill =.. level..,
alpha=0.25),size=0.01, bins=30, data=firms, geom = "polygon") + geom_
point(aes(x=yy, y=xx, stroke=2), colour='grey80', data=cities.lub,
size=0.5) + geom_label_repel(aes(x=yy, y=xx, label=city), data=cities.
lub, family='Times', size=2, box.padding=0.2, point.padding=0.3, segment.
color='grey50')

# Figure 3.9b
# background map with contour (geom_polygon),
# point density (stat_density2d) and isolines (geom_density2d)
p + geom_polygon(aes(x=long, y=lat, group=group), data=lub.f,
colour='white', fill='grey80', alpha=0.3, size=0.3) + stat_
density2d(aes(x=coords.x1, y=coords.x2, fill =.. level.., alpha= 0.25),
size=0.01, bins=30, data=firms, geom="polygon") + geom_density2d(data=firms,
aes(x=coords.x1, y=coords.x2), size=0.3)
```

When using the ggmap:: package, it's worth mentioning the tmap:: package, which is an overlay on ggmap:: and is designed to help in drawing maps and contains several datasets for the whole world (maps and data – world, rivers, metro). The qtm() command enables fast graphics. The package easily defines map styles and options (style_catalogue(), theme_ps(), tmap-element(), tmap_icons(), tmap_options(), tmap_style(), tm_add_legend(), tm_compass(), tm_credits(), tm_fill(), tm_grid(), tm_iso(), tm_layout(), tm_lines(), tm_logo(), tm_raster()) and can generate interactive maps (tmap_leaflet(), tmap_mode()) and so on.

Scheme 3 – with PlotOnStaticMap() from the RgoogleMaps:: package

The RgoogleMaps:: package allows for applying point layers to the ground map in another way. By downloading the background map with the **MapBackground**() command, one can apply the layer with the **PlotOnStaticMap**() command. The background map download is based on the coordinates of the dataset. The downloaded map can be terrain or satellite but also roadmap or hybrid (as *maptype* options), as already described in the previous examples.

```
# based on RgoogleMap
library(RgoogleMaps)
Lat<-as.vector(populationxy$xx)
Lon<-as.vector(populationxy$yy)

# default map – roadmap, map from Google with cities
MyMap1<-MapBackground(lat=Lat, lon=Lon, zoom=10, API_console_key=APIKEY,
SCALE=2)
PlotOnStaticMap(MyMap1, Lat, Lon, cex=4, pch=21, bg="lightcoral")

# satellite map, map from Google with cities
MyMap2<-MapBackground(lat=Lat, lon=Lon, zoom=9, maptype="satellite",
API_console_key=APIKEY)
PlotOnStaticMap(MyMap2, Lat, Lon, cex=4, pch=21, bg="lightcoral") #Bubbles

# terrain map, map from Google with cities
MyMap3<-MapBackground(lat=Lat, lon=Lon, zoom=10, maptype="terrain",
API_console_key=APIKEY, SCALE=2)
PlotOnStaticMap(MyMap3, Lat, Lon)

# hybrid map, map from Google with land coverage and cities
MyMap4<-MapBackground(lat=Lat, lon=Lon, zoom=10, maptype="hybrid",
API_console_key=APIKEY, SCALE=2)
PlotOnStaticMap(MyMap4, Lat, Lon)
```

Scheme 4 – with RGoogleMaps:: GetMap() and conversion of staticMap into a raster

As can be seen from previous diagrams, drawing maps in different classes and layering on them is not always compatible between packages. The following code retrieves the map in the *staticMap* class using the **GetMap**() command from the RgoogleMaps:: package and converts it to a raster using its own function. One can draw raster maps using the **rasterImage**() command from the graphics:: package. The code was taken from the book by Brunsdon and Comber (2015, p. 316). Such a transformation of the map class allows applying subsequent layers with typical commands such as **points**() or **polygon**(). It should be noted that after conversion, the map is centred at (0,0), and its range in each direction (down/up/right/left) is 320 points. This necessitates the scaling of geographical coordinate data to the map scale. This is enabled by the **LatLon2XY.centered**() command from the RgoogleMaps:: package, which specifies the map to which it is scaled and the coordinates to be scaled.

```
# draws staticMap scaled in (0,0) as a raster
# The own backdrop() function works here, which transforms staticMap
backdrop<-function(gmt){
limx<-c(-320,320)
limy<-c(-320,320)
par(mar=c(0,0,0,0))
plot(limx, limy, type='n', asp=1, xlab='', ylab='', xaxt='n', yaxt='n',
bty='n')
box()
rasterImage(gmt$myTile, -320,-320,320,320)
} # end of function
```

```
# downloads and draws a map, using RgoogleMaps::
LivMap<-GetMap(center=c(51.25,22.55), zoom=11, API_console_key=APIKEY,
SCALE=2)
backdrop(LivMap) # own function calling

# scaling geographical coordinates to the map scale with RgoogleMaps::
firms.XY<-LatLon2XY.centered(LivMap, firms[,24], firms[,23])

# drawing maps and layers
backdrop(LivMap) #background map of Lublin with white points
points(firms.XY$newX, firms.XY$newY, pch=16, col='darkred', cex=0.7)
points(firms.XY$newX, firms.XY$newY, pch=16, col='white', cex=0.5)

backdrop(LivMap) #background map of Lublin with red points
points(firms.XY$newX, firms.XY$newY, pch=16, col=rgb(0.7, 0,0,0.15), cex=0.7)
```

3.4 Spatial data in vector format – example of the OSM database

Using the services offered by API servers may have even greater value if one can access dynamically generated data containing objects in vector form. Such data, whether containing information only about geometry and location or often containing attributes of downloaded objects, in the case of the possibility of saving as an object in one of the basic classes used in spatial analyses in R, can become direct source data for this type of research. Of course, not based on API services on the network, there is quite a large resource of open spatial data (including those from public registers) made available on the basis of ready-to-download files. For Poland, an example can be data provided by the Central Office of Geodesy and Cartography, available at http://www.gugik.gov.pl/pzgik/Data-udostepniane-bez-oplat. There exists, among others, data from the state register of borders and territorial division, register of geographical names, data on general geographical objects and selected data from the digital terrain model. An example on a European Union scale is the Geographic Information System of the Commission (GISCO) data, an office operating under EUROSTAT and providing access to selected geographical data. On the GISCO website, it is possible to access data such as territorial division according to various criteria, population, information on land height, land use and transport networks.

As mentioned previously, these examples were for downloading previously placed files. However, API services allow for choosing the range of data to download. Such a response is created as if "on the fly" at the request of the user. In order to present the possibility of downloading vector data, the OpenStreetMap project database will be used. As indicated, it combines data created by users (authors) based on official websites and using the individual work of these authors and their knowledge of the local area. Often, data appears in this database faster (or is corrected faster) than on official or commercial websites.

The Overpass API service is used to download data from the OSM database. Basic information about this API is available at http://overpass-api.de/. This service is installed on several publicly available instances. This service is optimised for downloading data from the OSM database (in contrast to the main API service, which is used to build and edit the OSM database by the community). Overpass has its own language in which queries sent to the server are created. When creating queries directly (on the website), it is recommended to use a dedicated user interface called Overpass Turbo, available at http://overpass-turbo.eu/. The website also provides help for constructing queries; through the menu, one can also choose the format in which the results will be returned. Constructing queries according to Overpass API syntax (or also in XML according to the schema accepted by this server) requires some skill and some knowledge of the keys and attribute values in the OSM database (as objects are called in the OSM database). It is possible to directly send queries using R and process the results, but the more sensible solution seems to be using the dedicated osmdata:: package.

This package is actively developed; the current version is dated February 2020. The results obtained using this package are converted depending on the selected function to the Simple Features *sf* class or to the appropriate spatial object classes defined by the sp:: package. The package can be installed from the default R package repository, that is, using the **install.packages("osmdata")** command. Of course, one should load it with the **library()** command before using it.

```
library(osmdata)
library(sp)
library(sf)
library(rgdal)
```

The presented instructions will show only part of the possibilities of the osmdata:: package and obtaining OSM vector data from Overpass API. In the subsequent example, information about gas stations will be taken from the rectangle covering Warsaw. This data will then be drawn onto the city's outline, and several corrections related to the previously obtained set will be presented, in particular cutting the set only to stations within the city limits.

When obtaining data from the OpenStreetMap database, it is necessary to build the appropriate query. It is usually based on the area in which one searches for data marked with geographical coordinates and an indication of the objects (keys and attribute values) one is looking for. Initially, the problem may be a good definition of objects that need to be acquired. The keys and values used in OSM can be found at https://taginfo.open-streetmap.org/projects/nominatim#tags or an even more extensive set at https://wiki.openstreetmap.org/wiki/Map_Features. It can also be helpful to use the **available_features()** function from the osmdata:: package, which provides the most important keys – object types – used in the database.

```
head(available_features(),20)
```

```
 [1]  "4wd only"                "abandoned"
 [3]  "abutters"                "access"
 [5]  "addr"                    "addr:city"
 [7]  "addr:conscriptionnumber" "addr:country"
 [9]  "addr:district"           "addr:flats"
[11]  "addr:full"               "addr:hamlet"
[13]  "addr:housename"          "addr:housenumber"
[15]  "addr:inclusion"          "addr:interpolation"
[17]  "addr:place"              "addr:postcode"
[19]  "addr:province"           "addr:state"
```

Each type can have a specific value. Another function is to check the most commonly used values for a given key. In the example, the values for the key "amenity" for public utilities and the like and the general key "building" for building will be shown.

```
head(available_tags ("building"),20)
```

```
 [1]  "apartments"    "bakehouse"
 [3]  "barn"          "bridge"
 [5]  "bungalow"      "bunker"
 [7]  "cabin"         "carport"
 [9]  "cathedral"     "chapel"
[11]  "church"        "civic"
[13]  "commercial"    "conservatory"
[15]  "construction"  "cowshed"
[17]  "detached"      "digester"
[19]  "dormitory"     "entrance"
```

```
head(available_tags ("amenity"),20)
```

```
 [1]  "animal boarding"           "animal shelter"
 [3]  "arts centre"               "atm"
 [5]  "baby hatch"                "baking oven"
 [7]  "bank"                      "bar"
 [9]  "bbq"                       "bench"
[11]  "bicycle parking"           "bicycle rental"
[13]  "bicycle repair station"    "biergarten"
[15]  "boat rental"               "boat sharing"
[17]  "brothel"                   "bureau de change"
[19]  "bus station"               "cafe"
```

The next code will show how to build a query. The first step is to identify the area from which the objects will be taken. The following are two ways to identify such a "rectangle". The first of them is based on already-available vector data on administrative division. Please note that a file with all poviats for Poland has been loaded, and then a new **WAW** object has been created that contains only Warsaw (code 1465). The **bbox**() function from the sp:: package gives the outermost points of the area as a matrix. In the second case, the **getbb**() function from the osmdata:: package was used. It finds the area based on location data from the OpenStreetMap database. By default, areas smaller than the country and larger than the street are considered. This can be changed in the corresponding function options. In the case of Warsaw, both methods gave the same result.

```
latlong <- "+init=epsg:4326"
pov<-readOGR(".", "powiaty") # 380 units
WAW<-pov[pov$jpt_kod_je==1465,]
WAW<-spTransform (WAW, CRS (latlong)) # transformation coordinate to WGS84
WAW_box<-bbox(WAW) # reading the extreme points of the Warsaw area
WAW_box
```

```
      min       max
x   20.85169   21.27115
y   52.09785   52.36815
```

```
WAW_box2<-getbb("Warszawa, Polska") # osmdata::
WAW_box2
```

```
      min       max
x   20.85169   21.27115
y   52.09785   52.36815
```

The basic function that prepares data to be sent to the API server is **opq**(). It creates an object of the *overpass_query* class containing the boundaries of the area from which data will be retrieved. The previously identified area will be used for this purpose. In addition, the **add_osm_feature**() function adds filters based on the selected keys and values to the question created by **opq**(). In the example, as indicated, the "amenity" key with the value "fuel" for gas stations will be used. It is worth noting that the so-called pipe operator%>% allows for passing the result of the function just executed to the next without assigning it to an auxiliary object. Using the%>% operator gives the same results as executing the command subsequently, but the latter requires the creation of an additional object.

```
query<-opq(WAW_box)
query_final<-add_osm_feature(query, key='amenity', value='fuel')
```

```
# using pipe operator
query<-opq(WAW_box)%>% add_osm_feature(key='amenity', value='fuel')
```

After building the question, it can be sent to the Overpass API service, and the results will be generated. It is important that the result can be converted to an object in one of the classes that is used to store spatial data, that is, primarily to an object of the class *sf* or *sp* and also to an object in xml format. The osmdata_... family of functions is used for this purpose, where *sf*, *sp* or *xml* appear as the suffix depending on the expected format. The characteristics of the object obtained can be viewed by entering its name.

```
petrol_waw.sf<-osmdata_sf(query) # output in sf class
petrol_waw.sp<-osmdata_sp(query) # output in sp class

petrol_waw.sf
```

```
Object of class 'osmdata' with:
            $bbox : 52.0978496125492,20.8516883368428,52.368153944595,21.2711512942955
    $overpass_call : The call submitted to the overpass API
            $meta : metadata including timestamp and version numbers
      $osm_points : 'sf' Simple Features Collection with 1200 points
       $osm_lines : 'sf' Simple Features Collection with 2 linestrings
    $osm_polygons : 'sf' Simple Features Collection with 142 polygons
   $osm_multilines : NULL
$osm_multipolygons : NULL
```

```
petrol_waw.sp
```

```
Object of class 'osmdata' with:
            $bbox : 52.0978496125492,20.8516883368428,52.368153944595,21.2711512942955
    $overpass_call : The call submitted to the overpass API
            $meta : metadata including timestamp and version numbers
      $osm_points : 'sp' SpatialPointsDataFrame with 1200 points
       $osm_lines : 'sp' SpatialLinesDataFrame with 2 lines
    $osm_polygons : 'sp' SpatialPolygonsDataFrame with 142 polygons
   $osm_multilines : 'sp' SpatialNADataFrame with 0 multilines
$osm_multipolygons : 'sp' SpatialPolygonsDataFrame with 0 multipolygons
```

The next step may be to draw the object with the borders of the city of Warsaw as a background. Of course, only one of the downloaded objects will be drawn.

```
par(mar=c(0,0,0,0))
plot(WAW)
plot(petrol_waw.sp$osm_points, add=TRUE) # plot of raw data
```

Sometimes it can happen that one object is represented by several points. These points overlap, which can be seen as bold markers. This is due to the fact that some objects can also be represented not only as points but also as areas – drawn buildings. In the osmdata package, the **unique_osmdata()** function can be useful, which reduces such repeated objects. It works better for objects of the *osmdata_sf* class.

```
petrol_waw_clear.sf<-unique_osmdata(petrol_waw.sf)
par(mar=c(0,0,0,0))
plot(WAW)
```

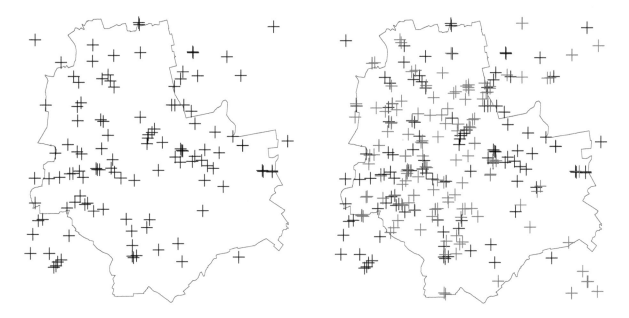

Figure 3.10 Location of gas stations in Warsaw – data from the OSM database downloaded using the API service: a) cleaned data, b) buildings transformed into point data

Source: Own study using R and the OpenStreetMap:: package

```
plot(petrol_waw_clear.sf$osm_points$geometry, pch=3, add=TRUE) # Fig-
ure 3.10a
```

It is also worth noting that the downloaded data from the OSM server did not contain only point objects. This is due to the map creation characteristics of the users. Sometimes a given object is marked in points, while sometimes it can be entered into the database as a building and given attributes assigned to that building. When displaying the characteristics of the data (after cleaning any duplicate values), it can be seen that on the day of creating this example, there were 121 point objects and 141 objects in the form of polygons (two objects in the form of lines can be omitted, which in the case of gas stations data seem to be incorrectly marked). In this case, it is worth supplementing the previously created drawings with polygon objects. It should also be added that both types of points and polygons are spatial (vector) objects and can be used for further analysis.

```
par(mar=c(0,0,0,0))
plot(WAW)
plot(petrol_waw_clear.sf$osm_points$geometry, pch=3,add=TRUE)
plot(petrol_waw_clear.sf$osm_polygons$geometry, add=TRUE)
```

If it is not desirable to use polygonal objects for further analysis, it is possible to convert such objects into point data. The easiest way is to find their geometric centre. In the sf package, there is a function, **st_centroid**(), that allows finding the centres of such polygons. In the next example, both the points and the centres of the polygons (instead of these objects) will be drawn on the Warsaw contour. The latter will be in the picture in red.

```
# Figure 3.10b
petrol_waw_clear_center<-st_centroid(petrol_waw_clear.sf$osm_polygons)
par(mar=c(0,0,0,0))
```

```
plot(WAW)
plot(petrol_waw_clear.sf$osm_points$geometry, pch=3, add=TRUE)
plot(petrol_waw_clear_center, pch=3, col="red", add=TRUE) # red centers
```

In previous examples, it was not pointed out that in fact some of the objects go beyond the area of the city being analysed. Of course, this is due to the fact that the area for receiving data from the Overpass server is in the form of a rectangle. For further analysis, it is worth removing from the object those observations that are outside the area delineated by the borders of Warsaw. The osmdata package offers the function **trim_osmdata()** for this purpose. It requires that the area indicating the data be in the format of a polygon obtained using the **getbb()** function with the option *format_out = 'polygon'*.

```
WAW_box2.poly<-getbb("Warszawa, Polska", format_out = 'polygon')
head(WAW_box2.poly[[1]])
```

```
          [,1]       [,2]
[1,]  20.85169   52.20098
[2,]  20.85170   52.20084
[3,]  20.85171   52.20072
[4,]  20.85172   52.20068
[5,]  20.85177   52.20028
[6,]  20.85188   52.19983
```

```
# object limited to the borders of Warsaw
petrol_waw_clear2.sf<-trim_osmdata(petrol_waw_clear.sf,WAW_box2.poly)
par(mar=c(0,0,0,0))
plot(WAW)
plot(petrol_waw_clear2.sf$osm_points$geometry, pch=3,add=TRUE)
```

The data obtained from the Overpass API was filtered only due to one attribute. There are no obstacles to getting data due to two attributes. Another example will show how to choose data for gas stations in Warsaw but only those with the name Shell. In this case, the conjunction (combination) of two attributes will be used. The *value_exact=FALSE* option ensures that all records containing the searched value are selected, not just exactly the one. Using this option, one can specify the value in the form of so-called regular expressions (regexp). A similar option (*key_exact=FALSE*) is for key definition, but it is not recommended to use it because of the possibility of quite wide results that also contain unexpected data.

```
query1<-opq(WAW_box)%>%
add_osm_feature(key='amenity', value='fuel')%>%
add_osm_feature(key='name', value='Shell', value_exact=FALSE)
petrol_waw_Shell.sf<-osmdata_sf(query1)
petrol_waw_Shell.sf<-trim_osmdata(petrol_waw_Shell.sf, WAW_box2.poly)
par(mar=c(0,0,0,0))
plot(WAW)
plot(petrol_waw_Shell.sf$osm_points$geometry, pch=3, add=TRUE)
plot(petrol_waw_Shell.sf$osm_polygons$geometry, add=TRUE)
```

Objects of the *sf* or *sp* class have quite a complex structure. However, it is not difficult to transform data about downloaded objects, e.g. to the *data.frame* class. Of course, in this case, the possibility of direct further analysis is lost, taking into account the spatial specificity (e.g. information about the data/ projection reference system is lost) and relevant functions based on objects in specific classes containing spatial data.

```
petrol_waw_Shell.sf # first three observations
```

```
Object of class 'osmdata' with:
              $bbox : 52.0978496125492,20.8516883368428,52.368153944595,21.2711512942955
     $overpass_call : The call submitted to the overpass API
              $meta : metadata including timestamp and version numbers
        $osm_points : 'sf' Simple Features Collection with 132 points
         $osm_lines : NULL
      $osm_polygons : 'sf' Simple Features Collection with 15 polygons
    $osm_multilines : NULL
$osm_multipolygons : NULL
petrol_waw_Shell.sf$osm_point
Simple feature collection with 132 features and 21 fields
geometry type:  POINT
dimension:      XY
bbox:           xmin: 20.89199 ymin: 52.14051 xmax: 21.15521 ymax: 52.36473
epsg (SRID):    4326
proj4string:    +proj=longlat +datum=WGS84 +no_defs
First 5 features:
                osm_id    name addr.city  addr.housenumber    addr.postcode
261666238    261666238   Shell      <NA>              <NA>             <NA>
334630863    334630863    <NA>      <NA>              <NA>             <NA>
334631719    334631719   Shell      <NA>              <NA>             <NA>
            addr.street  amenity     brand   brand.wikidata brand.wikipedia
261666238          <NA>     fuel      <NA>             <NA>            <NA>
334630863          <NA>     <NA>      <NA>             <NA>            <NA>
334631719          <NA>     fuel      <NA>             <NA>            <NA>
            fuel.diesel  fuel.electricity  fuel.lpg   fuel.octane_100
261666238          <NA>              <NA>      <NA>              <NA>
334630863          <NA>              <NA>      <NA>              <NA>
334631719          <NA>              <NA>      <NA>              <NA>
            fuel.octane_95  fuel.octane_98 opening_hours      operator  shop
261666238             <NA>            <NA>          <NA>         Shell  <NA>
334630863             <NA>            <NA>          <NA>          <NA>  <NA>
334631719             <NA>            <NA>          <NA> Shell Polska  <NA>
               source   source.addr                      geometry
261666238        <NA>          <NA>   POINT (21.01877 52.14051)
334630863        <NA>          <NA>   POINT (20.91907 52.19834)
334631719        <NA>          <NA>   POINT (20.94446 52.19259)
```

```
# information about point objects converted to data.frame class

petrol_waw_Shell.df <-as.data.frame (petrol_waw_Shell.sf$osm_points)
# similarly can be done with polygons
```

This section only presents an example of obtaining spatial data from a selected database available through the API. Attention was paid to OpenStreetMap resources due to the fact that they are published on an open and free license, have a fairly extensive API service and are available through the actively created package in R. They also contain an amount and scope of data that seem to be absent in any other single database. Of course, due to the fact that this database is created by the user community, there is a possibility that in the selected range, the data may not be completely accurate. However, it is usually quickly repaired by other authors of this resource. Observed inaccuracies can also be corrected in the OSM project (contributing to its development) or reported as incorrect data. It must be admitted that the resources of public institutions are also not error-free.

3.5 Access to non-spatial internet databases and resources via application programming interface – examples

Two trends contribute to the fact that the amount of data and access to it via the Internet seem to grow very quickly – apart from the previously discussed options for downloading maps and spatial (vector) data. Somewhat perversely, one could say that the creation of WebAPI services has simply become fashionable and more and more institutions are trying to stay current. On the other hand, there is more and more data available also under open and free licenses (without restrictions) – which also results from legal regulations, and access to them should be ensured. The R environment offers a lot in terms of access to such data. The general functions and tools from the basic and additional packages used to "communicate" with any servers on the network are complemented by packages with functions that provide access (and facilitate it) to specific WebAPI service providers that offer data. It is impossible to point out even the most important sources that should be noted because of their multiplicity and diversity. For this reason, this subsection will focus on showing access to sample data using specialised packages and through queries created "manually" to selected WebAPIs. Knowing the way presented in the examples, R user will be able to easily transfer them to acquiring data from desired sources.

The first to be presented is access to the DB NOMICS website, which is considered one of the largest economic databases on the internet. This website has access to its resources using an interface in a web browser, which is available at https://db.nomics.world/ with the option of downloading the indicated range of data, for example, to a csv or xlsx file. At the same time, it provides the WebAPI service through which one can download the requested data after formulating the appropriate query. It is important to note that choosing data through the browser interface, DB NOMICS also provides the already-built query to the API server, which can be used to download the indicated data. An R user who does not undertake the construction of queries by himself can therefore access ready queries. An additional advantage of using this site in R is the fact that a dedicated additional rdbnomics:: package has been created, available in official repositories to facilitate the use of this API. The example of this package will show the use of DB NOMICS in R. The scope of data offered by this website is best described by the information on the main page of the website. As of July 26, 2019, it offers data collected from 62 providers that contain almost 20,000 datasets and over 600 million data series. In this package, there are two functions that can be used to inform about the content of the DB NOMICS website. The first one is **rdb_providers**(), which gives a list of data sources used by the service, and the second one (**rdb_last_updates**()) allows one to get an idea of what data has recently been updated. The latter gives the last 100 updates by default.

```
library (rdbnomics)
#Visit <https://db.nomics.world>.
library (ggplot2) # package used for creating charts

rdb_prov<-rdb_providers() # data providers for the DB NOMICS website
head(rdb_prov) # first four only
```

```
      code         converted_at             created_at              indexed_at
1:    AFDB   2019-02-18  14:28:44   2018-12-21  17:04:45   2019-02-18   14:28:48
2:   AMECO   2019-05-15  08:02:53   2017-03-03  16:31:12   2019-05-15   08:02:59
3:     BCB   2019-07-27  05:58:33   2018-06-06  09:52:03   2019-07-27   06:02:59
4:   BCEAO   2019-07-25  05:47:47   2017-12-12  14:55:17   2019-07-25   07:32:17

json_data_commit_ref
1: c97a17d4e72f7171282f09d76c181ef58500d8af
2: f08b34fd8221e14849f9ea0afed42d979258fe8b
3: af28e5c88a5f1d87870c95d20f540e53d3040109
4: 82eaa482574d2d77268644c2add1807d4dae4a10
name
1: African Development Group
2: Annual macro-economic database of the European Commission's Directorate
General for Economic and Financial Affairs
3: Banco Central do Brasil
4: Banque Centrale des Etats de l'Afrique de l'Ouest
```

```
         region     slug
1:       Africa     afdb
2:           EU     ameco
3:           BR     bcb
4: West Africa      bceao


                                             terms_of_use
1:                   https://www.afdb.org/en/terms-and-conditions/
2: https://ec.europa.eu/info/legal-notice_en#copyright-notice
3:                                                          <NA>
4:                   http://www.bceao.int/Mentions-legales.html
website
1: https://www.afdb.org/en/
2: https://ec.europa.eu/info/business-economy-euro/indicators-statistics/eco-
   nomic-databases/macro-economic-database-ameco_en
3: http://www.bcb.gov.br
4: http://www.bceao.int
```

```
rdb_upd <-rdb_last_updates() # list of the last 100 updates
head (rdb_upd, 4) # for example, a list of the last 4 updates is shown
```

```
      code           converted_at            created_at            indexed_at
1: CISP  2019-07-27  09:50:34  2019-04-03 12:10:05  2019-07-27  09:55:48
2:   CL  2019-07-27  09:50:34  2019-04-03 12:10:05  2019-07-27  09:55:48
3:   CP  2019-07-27  09:50:34                 <NA>  2019-07-27  09:55:48
4:   CS  2019-07-27  09:50:34  2019-04-03 12:10:05  2019-07-27  09:55:48

                    json_data_commit_ref
1: 4fe75d9e643e6b8fa36097961f2d7e7cf5113b90
2: 4fe75d9e643e6b8fa36097961f2d7e7cf5113b90
3: 4fe75d9e643e6b8fa36097961f2d7e7cf5113b90
4: 4fe75d9e643e6b8fa36097961f2d7e7cf5113b90
                                                                  name
1:                 Investment: Country Investment Statistics Profile
2: Food Balance: Food Supply - Livestock and Fish Primary Equivalent
3:                                   Prices: Consumer Price Indices
4:                                   Macro-Statistics: Capital Stock
   nb_series  provider_code
1:      4790           FAO
2:     42269           FAO
3:       358           FAO
4:      3335           FAO
                                            provider_name updated_at
1: Food and Agriculture Organization of the United Nations 2018-12-19
2: Food and Agriculture Organization of the United Nations 2018-01-17
3: Food and Agriculture Organization of the United Nations 2019-06-18
4: Food and Agriculture Organization of the United Nations 2018-09-11
```

Looking at the list of data sources, it is worth pointing out that much of this data comes from official, reliable sources. It is worth paying attention in the first received set to the data source code and in the second to the code of the updated dataset itself, because it can help indicate the code that is necessary to download the required data using WebAPI.

The basic function that retrieves the selected data packet is **rdb**() and requires a minimum identifier (code) of the desired data. Built-in help for this function explicitly states that the best way to identify this code is to find it on the DB NOMICS website (https://db.nomics.world/), because the construction of such an identifier is quite complex. Generally, it has the scheme: "source code/dataset code/series code". When specifying the entire identifier as an argument to the **rdb()** function, no other parameters are required. Alternatively, use the source code, dataset code and the so-called "dimension". In this case, the dimension should be given as a list and explicitly indicate the name of the parameter: *dimensions* = . . . Examples will be shown subsequently.

In downloaded data (on July 26, 2019) on updates, the following record was found on position no 99:

```
# execution in the following days may give a different result each time
rdb_upd[99]
```

```
              code                    converted_at  created_at description
indexed_at                 json_data_commit_ref
1: apro_mk_pobta  2019-07-26   02:20:54     <NA>       <NA>   2019-07-26
02:46:10    f9f407d9db16e214fccb94c2c6b039443df038ed
    name  nb_series   provider_code    provider_name    updated_at
1: Milk  collection  (all  milks) and  dairy products   obtained      -
annual    data   7438 Eurostat    Eurostat     2019-07-25  13:19:08
```

The record indicates EUROSTAT data on annual milk production. Searching the received code on the DB NOMICS website and providing further parameters, one can find the series identifier. Figure 3.11 shows the result of such a search.

After finding the code, one can proceed to download the data. The series will be saved to the object of *data. frame* and *data.table*.

```
# download data series from DB NOMICS
milk_PL_dbn<-rdb('Eurostat/apro_mk_pobta/A.D2100.PRO.PL')
class(milk_PL_dbn)
#[1] "data.table" "data.frame"
tail(milk_PL_dbn) # Last 6 records retrieved, 4 in output
```

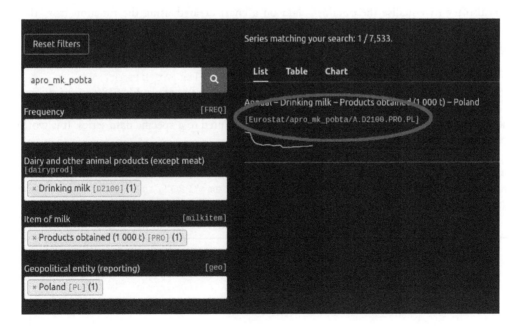

Figure 3.11 **Results of the data series code search on the DB NOMICS page**

Source: Own study

```
      @frequency dataset_code
1:       annual apro_mk_pobta
2:       annual apro_mk_pobta
3:       annual apro_mk_pobta
4:       annual apro_mk_pobta
                                                          dataset_name
1: Milk collection (all milks) and dairy products obtained - annual data
2: Milk collection (all milks) and dairy products obtained - annual data
3: Milk collection (all milks) and dairy products obtained - annual data
4: Milk collection (all milks) and dairy products obtained - annual data
            indexed_at    observations_attributes original_period      period
1: 2019-07-26 02:46:10              OBS_STATUS,             2013  2013-01-01
2: 2019-07-26 02:46:10              OBS_STATUS,             2014  2014-01-01
3: 2019-07-26 02:46:10              OBS_STATUS,             2015  2015-01-01
4: 2019-07-26 02:46:10              OBS_STATUS,             2016  2016-01-01
      provider_code       series_code
1:        Eurostat   A.D2100.PRO.PL
2:        Eurostat   A.D2100.PRO.PL
3:        Eurostat   A.D2100.PRO.PL
4:        Eurostat   A.D2100.PRO.PL
                                                    series_name    value
1: Annual - Drinking milk - Products obtained (1 000 t) - Poland 1616.05
2: Annual - Drinking milk - Products obtained (1 000 t) - Poland 1596.98
3: Annual - Drinking milk - Products obtained (1 000 t) - Poland 1638.65
4: Annual - Drinking milk - Products obtained (1 000 t) - Poland 1655.08
      FREQ dairyprod  geo milkitem
1:    A      D2100   PL      PRO
2:    A      D2100   PL      PRO
3:    A      D2100   PL      PRO
4:    A      D2100   PL      PRO
```

One can also try to visualise the obtained data on a chart created using the function from the ggplot2:: package.

```
ggplot(milk_PL_dbn, aes(x=period, y=value, color=series_code)) +
geom_line(size=1) +
theme(legend.position="bottom", legend.title=element_blank())
```

Another example also downloads one data series, this time for the entire EU (28 countries); however, the choice of data is made slightly differently. Instead of the batch identifier, the data provider parameters, file code and "dimensions" given as the list object were used, which pointed to a specific data series. It is worth comparing the codes used subsequently and previously.

```
milk_EU28_dbn<-rdb(provider_code = 'Eurostat',
dataset_code = 'apro_mk_pobta',
dimensions = list(dairyprod="D2100", milkitem="PRO", geo="EU28"))
tail(milk_EU28_dbn) # below only 3 lines
```

```
      @frequency    dataset_code
1:       annual     apro_mk_pobta
2:       annual     apro_mk_pobta
3:       annual     apro_mk_pobta
                                                          dataset_name
1: Milk collection (all milks) and dairy products obtained - annual data
2: Milk collection (all milks) and dairy products obtained - annual data
3: Milk collection (all milks) and dairy products obtained - annual data
```

```
            indexed_at observations_attributes  original_period         period
1:2019-07-26  02:46:10             OBS_STATUS,c            2012    2012-01-01
2:2019-07-26  02:46:10             OBS_STATUS,c            2013    2013-01-01
3:2019-07-26  02:46:10             OBS_STATUS,c            2014    2014-01-01
    provider_code          series_code
1:      Eurostat     A.D2100.PRO.EU28
2:      Eurostat     A.D2100.PRO.EU28
3:      Eurostat     A.D2100.PRO.EU28
                                                                  series_name
1: Annual-Drinking milk-Products obtained (1 000 t)- European Union - 28 coun-
tries
2: Annual-Drinking milk-Products obtained (1 000 t)- European Union - 28 coun-
tries
3: Annual-Drinking milk-Products obtained (1 000 t)- European Union - 28 coun-
tries
    value  FREQ   dairyprod   geo   milkitem
1:     NA     A       D2100  EU28        PRO
2:     NA     A       D2100  EU28        PRO
3:     NA     A       D2100  EU28        PRO
```

When retrieving data from DB NOMICS, there is no need to be limited to one data series. An example will show sending a request for milk production for Poland and Romania (see Figure 3.12).

```
milk_PLRO_dbn<-rdb(provider_code='Eurostat', dataset_code='apro_mk_pobta',
dimensions=list(dairyprod="D2100", milkitem="PRO", geo=c("PL","RO")))
```

A similar effect can be obtained by downloading two data series directly by indicating the batch codes. A graph will be produced comparing both sets of data.

```
milk_PLRO_dbn<-rdb(c('Eurostat/apro_mk_pobta/A.D2100.PRO.PL', 'Eurostat/
apro_mk_pobta/A.D2100.PRO.RO'))
tail(milk_PL_dbn) # below 3 lines only
```

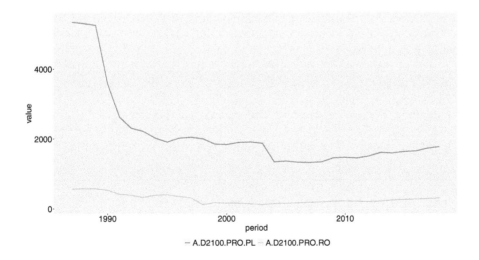

Figure 3.12 Milk production in Poland and Romania – data taken from the WEBAPI DB
         NOMICS service

Source: Own study using R and rdbnomics:: and ggplot2:: packages

```
    @frequency    dataset_code
1:      annual    apro_mk_pobta
2:      annual    apro_mk_pobta
3:      annual    apro_mk_pobta
                                                      dataset_name
1: Milk collection (all milks) and dairy products obtained - annual data
2: Milk collection (all milks) and dairy products obtained - annual data
3: Milk collection (all milks) and dairy products obtained - annual data
            indexed_at  observations_attributes original_period      period
1: 2019-07-26  02:46:10             OBS_STATUS,            2013  2013-01-01
2: 2019-07-26  02:46:10             OBS_STATUS,            2014  2014-01-01
3: 2019-07-26  02:46:10             OBS_STATUS,            2015  2015-01-01
    provider_code         series_code
1:       Eurostat        A.D2100.PRO.PL
2:       Eurostat        A.D2100.PRO.PL
3:       Eurostat        A.D2100.PRO.PL
                                                 series_name    value
1: Annual - Drinking milk - Products obtained (1 000 t) - Poland 1616.05
2: Annual - Drinking milk - Products obtained (1 000 t) - Poland 1596.98
3: Annual - Drinking milk - Products obtained (1 000 t) - Poland 1638.65
    FREQ  dairyprod   geo  milkitem
1:    A      D2100     PL       PRO
2:    A      D2100     PL       PRO
3:    A      D2100     PL       PRO
```

```
ggplot(milk_PLRO_dbn, aes(x=period, y=value, color=series_code)) +
geom_line(size=1) +
theme(legend.position="bottom", legend.title=element_blank())
```

When using the previously mentioned method, one can simultaneously download completely different data (more than two series) – each series from a different provider and a different type of data. This is achieved by indicating the specific series codes found in the search engine on the DB NOMICS website.

Downloading a larger number of data series with a similar code (but from one supplier and one type) can be implemented using the *mask* = . . . option. The example will show how to download the previously analysed data type for milk production for all countries in the database. Looking at the codes found in the DB NOMICS search engine, it can be seen that, depending on the country/region, they differ only in the last term, for example, for Albania and Austria, they are as follows: A.D2100.PRO.AL, A.D2100.PRO.AT. The code that will retrieve data for all countries is as follows:

```
milk_ALL_dbn<-rdb(provider_code='Eurostat', dataset_code='apro_mk_pobta',
mask='A.D2100.PRO.')
dim(milk_ALL_dbn) # dimensions of the obtained data set
#[1] 1173 15
table(milk_ALL_dbn$geo) # number of observations from given country
```

```
AL   AT   BE   BG   CH   CY   CZ   DE   DK   EE   EL   ES  EU28   FI   FR
 5   36   59   32   10   20   32   59   59   32   39   32    14   34   59
HR   HU   IE   IS   IT   LT   LU   LV   ME   MT   NL   NO    PL   PT   RO
19   27   59    8   59   32   58   29    7   18   59    7    32   33   31
RS   SE   SI   SK   TR   UK
 5   46   24   32    7   59
```

It is clear that the acquired dataset has 1173 rows. The number of observations from each country/region was identified using the **table**() function.

The *mask* = . . . parameter has many more possibilities. A more comprehensive description can be found in the help for this function in the R environment.

It is worth mentioning one more function from the rdbnomics:: package, i.e. **rdb_by_api_link**(). It allows for sending a query from the R environment to the WebAPI DB NOMIC service directly using the "http" address that points to the searched resource. As mentioned, such an address can be found directly in the data search engine on the website (see Figure 3.13)

After copying the link from the search engine page, the sample download code would look like this:

```
milk_AL_dbn<-rdb_by_api_link('https://api.db.nomics.world/v22/series/Euro-
stat/apro_mk_pobta/A.D2100.PRO.AL?observations=1')
```

Although access to the DB NOMICS website presented previously can be a good scheme of using additional specialised packages for downloading data from a specific WebAPI, one more example of access to a data service using a dedicated solution in R will be presented. In this part, attention will be paid to access to the recently launched Local Data Bank (*pl* Bank Danych Lokalnych – BDL) server of the Central Statistical Office (*pl* Główny Urząd Statystyczny – GUS) in Poland. The reason for describing this service and downloading data in R is that the additional package that allows for submitting queries to this API is not yet available in the official R repository and it is available to download from the creator's repository, which changes the initial steps of working with this package. Undoubtedly, BDL is also an important source of data in Poland (with great potential for spatial research); hence, it is worth pointing out the possible ways of downloading data from it. It is also, as indicated, a service recently released.

It should also be mentioned that the GUS API portal provides access to three official data registers: REGON, TERYT and BDL. The first is a database on entities of the national economy; the second contains constantly updated (and historical) official data on the territorial division of Poland and the third is a collection of data from the area of economic, social and environmental phenomena. As for access to the REGON file, it is limited and requires filling out an application and obtaining an API key. Access also varies depending on the entity, as this register also contains personal and other sensitive data. The TERYT database is entirely public. At the moment, there does not appear to be a specialised additional package providing access to these resources. However, the construction of API queries for the TERYT service is quite simple; hence it is possible to construct queries and communicate with the server directly from the R without using dedicated packages. In the next part of the chapter, the "manual" method of communication with WebAPI servers will be described, which can also be a good example for using the TERYT API. The WebAPI Local Data Bank service has access open to all users.

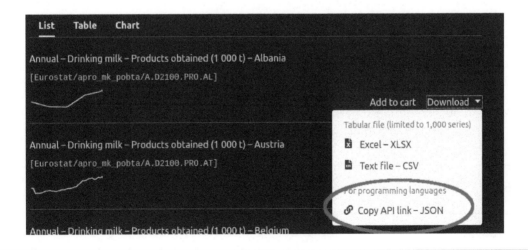

Figure 3.13 Direct http address to the API resource

Source: Own study

It generally does not require a key, although it is possible to use dedicated access with personal key. The registered user has less restrictive limits for sending queries to the server. The subsequent examples will be presented without using an access key.

This part presents an example of access to BDL data using the bdl:: package by Krzysztof Kania. The installation of this package is different in R because it is available in the author's repository on the well-known GitHub programming platform. Installation of the bdl:: package requires, as suggested by the author, prior installation of the remotes:: package. The full code to install the bdl:: package looks like this:

```
library(remotes)
remotes::install_github("statisticspoland/R_Package_to_API_BDL", upgrade=
"always")
```

During installation, several or even a dozen other packages will be downloaded that are required for the bdl:: package to function.

Indication of data to be downloaded from the Local Data Bank requires, as in the case of DB NOMICS services, operating on identifiers of both administrative and variable units. Unfortunately, the interface on the BDL website does not have such a convenient identifier finder. However, the bdl package contains functions that allow for searching for the appropriate identifiers. The **search_subjects**() and **search_variables**() functions are used to search for an issue or specific variable using a text query.

It is also worth pointing out that most functions from the previously mentioned package bdl:: also allow for getting results in English when using the *lang=en* option, which means that the package can be used globally. The standard answer language (without parameter specification) is Polish.

```
library (bdl)
search_subjects("unempl", lang="en")
```

```
# A tibble: 43 x 6
   id    parentId name                   hasVariables children     levels
   <chr> <chr>    <chr>                  <lgl>        <list>       <list>
 1 G12   K4       REGISTERED UNEMPL . . .FALSE        <chr [2. . .<int [. . .
 2 P1364 G12      Registered unempl . . .FALSE        <chr [0. . .<int [. . .
 3 P1944 G12      Registered unempl . . .FALSE        <chr [0. . .<int [. . .
 4 P1946 G12      Registered unempl . . .FALSE        <chr [0. . .<int [. . .
 5 P1947 G12      Registered unempl . . .FALSE        <chr [0. . .<int [. . .
 6 P1948 G12      Registered unempl . . .FALSE        <chr [0. . .<int [. . .
 7 P1949 G12      Registered unempl . . .FALSE        <chr [0. . .<int [. . .
 8 P2061 G290     Registered in          FALSE        <chr [0. . .<int [. . .
                  the District . . .
 9 P2241 G301     Unempl by the period   FALSE        <chr [0. . .<int [. . .
                  of the search . . .
10 P2358 G380     Unemployed for a       FALSE        <chr [0. . .<int [. . .
                  long time (13. . .
# . . . with 33 more rows
```

```
search_variables("unempl", lang="en")
```

```
# A tibble: 707 x 10
     id subjectId n1         n2         level measureUnitId measureUnitName  n3
  <int>     <chr> <chr>      <chr>      <int>         <int> <chr>           <chr>
 1 1812     P2058 with . . . using . . .    6            26 person           <NA>
 2 1813     P2058 with . . . using . . .    6            26 person           <NA>
```

```
 3 1870      P2068 adde . . . <NA>           6         26 person        <NA>
 4 4244      P1385 acti . . . all . . .       3         23 thous. people <NA>
 5 4245      P1385 acti . . . male . . .      3         23 thous. people <NA>
 6 4246      P1385 acti . . . fema . . .      3         23 thous. people <NA>
 7 4252      P1386 acti . . . 15-29           3         23 thous. people <NA>
 8 4256      P1386 acti . . . 30-39           3         23 thous. people <NA>
 9 4260      P1386 acti . . . 40-49           3         23 thous. people <NA>
10 4264      P1386 acti . . . 50 and . . .    3         23 thous. people <NA>
# . . . with 697 more rows, and 2 more variables: n4 <chr>, n5 <chr>
```

The first search shows that unemployment data is marked with a collective code (higher level G12 – parentID). Knowing this code, one can search for specific issues/topics covering this group using **get_subjects()**. Similarly, one can search for variables in a given issue using the **get_variables()** function. Examples subsequently.

```
get_subjects(parentId = "G12", lang="en")
```

```
# A tibble: 27 x 6
   id     parentId name          hasVariables children      levels
   <chr>  <chr>    <chr>         <lgl>        <list>        <list>
 1 P1364  G12      Registered    TRUE         <list [. . .  <int [. . .
                   unempl . . .
 2 P1365  G12      Job offers by TRUE         <list [. . .  <int [. . .
                   groups of people . . .
 3 P1944  G12      Registered    TRUE         <list [. . .  <int [. . .
                   unempl . . .
 4 P1946  G12      Registered    TRUE         <list [. . .  <int [. . .
                   unempl . . .
 5 P1947  G12      Registered    TRUE         <list [. . .  <int [. . .
                   unempl . . .
 6 P1948  G12      Registered    TRUE         <list [. . .  <int [. . .
                   unempl . . .
 7 P1949  G12      Registered    TRUE         <list [. . .  <int [. . .
                   unempl . . .
 8 P2392  G12      Registered    TRUE         <list [. . .  <int [. . .
                   unemployment rate . . .
 9 P2511  G12      Registered    TRUE         <list [. . .  <int [. . .
                   unempl . . .
10 P2512  G12      Registered    TRUE         <list [. . .  <int [. . .
                   unempl . . .
# . . . with 17 more rows
```

```
# subjectID based on previous search
get_variables(subjectId = "P1385", lang="en")
```

```
# A tibble: 12 x 7
  id     subjectId n1           n2        level measureUnitId measureUnitName
  <int> <chr>     <chr>        <chr>     <int> <int>         <chr>
1 4238  P1385     population   ogół. . . 3     26            person
                  fakty . . .
```

```
 2 4239   P1385    population męż c . . . 3      26          person
                   fakty . . .
 3 4240   P1385    population kobi . . . 3       26          person
                   fakty . . .
 4 4241   P1385    aktywni    ogół. . . 3        23          tys. people
                   zawod . . .
 5 4242   P1385    aktywni    męż c . . . 3      23          tys. people
                   zawod . . .
 6 4243   P1385    aktywni    kobi . . . 3       23          tys. people
                   zawod . . .
 7 4244   P1385    aktywni    ogół. . . 3        23          tys. people
                   zawod . . .
 8 4245   P1385    aktywni    męż c . . . 3      23          tys. people
                   zawod . . .
 9 4246   P1385    aktywni    kobi . . . 3       23          tys. people
                   zawod . . .
10 4247   P1385    bierni     ogół. . . 3        23          tys. people
                   zawodo . . .
11 4248   P1385    bierni     męż c . . . 3      23          tys. people
                   zawodo . . .
12 4249   P1385    bierni     kobi . . . 3       23          tys. people
                   zawodo . . .
```

It is worth noting the differences between these two types of BDL search. The first gives the name of the issue, and the second gives a specific variable. When searching for an issue, it is important to pay attention to the parentID and children columns, which allow for finding out where in the data hierarchy the user is. Similarly, the hasVariables column informs the user if he or she has already reached the search level that contains numeric data.

From previous searches, it is known that the topic "P1944" is defined as "Registered unemployed persons by sex in communes". Variables from this topic can be displayed with the following code:

```
get_variables(subjectId = "P1944", lang="en")
```

```
# A tibble: 3 x 6
   id      subjectId n1     level measureUnitId measureUnitName
   <int>   <chr>     <chr>  <int> <int>         <chr>
1  10514   P1944     total  6     26            person
2  10515   P1944     women  6     26            person
3  33484   P1944     men    6     26            person
```

The variable denoting the total data at the level of communes (level 6) is denoted by id = 10514. This code will allow for getting data for this variable using the **get_data_by_variable()** function.

```
unempl_gm<-get_data_by_variable(varId="10514", lang="en")
tail(unempl_gm, 5) # last 5 records
```

```
# A tibble: 10 x 5
   id           name        year      val    attrId
   <chr>        <chr>       <chr>    <dbl>    <int>
```

```
1  071427338052  Wiskitki  2009  455  1
2  071427338052  Wiskitki  2010  497  1
3  071427338052  Wiskitki  2011  499  1
4  071427338052  Wiskitki  2012  604  1
5  071427338052  Wiskitki  2013  634  1
```

```
unempl_gm[unempl_gm$name=="POLSKA",] # Aggregate data at the country level
```

```
# A tibble: 16 x 5
    id            name    year   val      attrId
    <chr>         <chr>   <chr>  <dbl>    <int>
 1  000000000000  POLSKA  2003   3175674  1
 2  000000000000  POLSKA  2004   2999601  1
 3  000000000000  POLSKA  2005   2773000  1
 4  000000000000  POLSKA  2006   2309410  1
 5  000000000000  POLSKA  2007   1746573  1
 6  000000000000  POLSKA  2008   1473752  1
 7  000000000000  POLSKA  2009   1892680  1
 8  000000000000  POLSKA  2010   1954706  1
 9  000000000000  POLSKA  2011   1982676  1
10  000000000000  POLSKA  2012   2136815  1
11  000000000000  POLSKA  2013   2157883  1
12  000000000000  POLSKA  2014   1825180  1
13  000000000000  POLSKA  2015   1563339  1
14  000000000000  POLSKA  2016   1335155  1
15  000000000000  POLSKA  2017   1081746  1
16  000000000000  POLSKA  2018   968888   1
```

Note the download of this collection takes a long time because data is obtained for each administrative level up to the level of municipalities for each year. The result has over 47,000 records.

The package includes functions for obtaining useful charts. For example, a line chart will be shown for the previously searched variables meaning "Registered unemployment" – here broken down into men and women. The data will be limited to the country level by the unitID option.

```
line_plot(data_type="unit", unitId="000000000000", varId=
c("10515","33484"), lang="en") # Figure 3.14a
```

The chart looks promising; the only drawback is that with long descriptions of variables, the legend is not quite printed correctly. One can construct a similar graph, for example, using the tools from the ggplot2:: package as shown for other issues in this chapter.

A very interesting option available in the package is the ability to automatically generate a map based on data downloaded from BDL. An example of such a map is shown in Figure 3.14b. The map is made for the already-analysed total unemployment variable (*varID*="10514"). Source data is at the level of municipalities, while on the map, it has been aggregated to the level of provinces (option *unitlevel=2*). The map reveals all its usefulness after performing it in the R environment, because it contains interactive elements. It is possible to zoom and change the background layer. For this function to work properly, however, a map file should be downloaded separately. It is available at: https://github.com/statisticspoland/R_Package_to_API_BDL/releases/download/1.0.0/bdl.maps.RData. This file automatically (after execution by doubleclick) adds the required maps to the environment in the current session. If one executes the **generate_map()** function without loading the maps first, the appropriate instruction will appear.

```
generate_map(varId="10514", year="2017", unitLevel=2, lang="en")
#Figure 3.14b
```

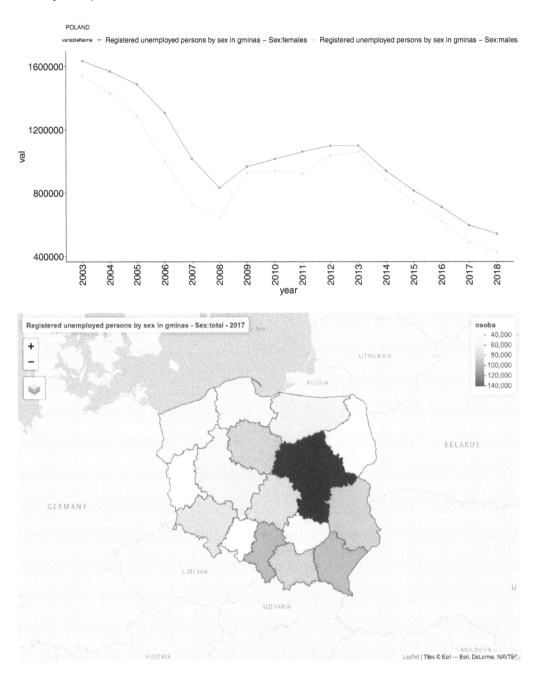

Figure 3.14 Visualisation of data from BDL: a) time series generated with **line_plot()**, b) map of regions generated by **generate_map()**

Source: Own study using R and bdl:: packages

In the examples shown previously, there was sometimes an option limiting downloading data to the selected unit or restricting the download area (unitID). Territorial division unit codes are also available using the functions in the bdl package. Similar to searching for subjects and variables for units, the **search_units**() and **get_units**() functions are available. Their use is similar to previously used functions. For example, the search for the Barcin commune (in the Kuyavian-Pomeranian voivodeship) will be presented. Provided below code and results should be fully understandable.

```
search_units("barcin", lang="en")
```

```
# A tibble: 3 x 6
  id           name                  parentId          level kind hasDescription
  <chr>        <chr>                 <chr>             <int><chr> <lgl>
1 040416719013 Barcin                0404167190. . .  6     3    FALSE
2 040416719014 Barcin - city         0404167190. . .  6     4    FALSE
3 040416719015 Barcin - rural area . . . 0404167190. . . 6  5    FALSE
```

```
# subunit search for the parent unit 040416719000
get_units(parentId = "040416719000", lang="en")
```

```
# A tibble: 14 x 6
   id           name               parentId        level kind hasDescription
   <chr>        <chr>              <chr>           <int> <chr> <lgl>
1  04041671. . . Barcin            040416719. . .  6     3    FALSE
2  04041671. . . Barcin - city     040416719. . .  6     4    FALSE
3  04041671. . . Barcin - rural area 040416719. . . 6   5    FALSE
4  04041671. . . Gąsawa            040416719. . .  6     2    FALSE
5  04041671. . . Janowiec          040416719. . .  6     3    FALSE
                 Wielkopolski
6  04041671. . . Janowiec          040416719. . .  6     4    FALSE
                 Wielkopolski -. . .
7  04041671. . . Janowiec          040416719. . .  6     5    FALSE
                 Wielkopolski -. . .
8  04041671. . . Łabiszyn          040416719. . .  6     3    FALSE
9  04041671. . . Łabiszyn - city   040416719. . .  6     4    FALSE
10 04041671. . . Łabiszyn - rural  040416719. . .  6     5    FALSE
                 area . . .
11 04041671. . . Rogowo            040416719. . .  6     2    FALSE
12 04041671. . . Żnin             040416719. . .  6     3    FALSE
13 04041671. . . Żnin - city      040416719. . .  6     4    FALSE
14 04041671. . . Żnin - rural area 040416719. . . 6     5    FALSE
```

After installing the bdl:: package in the R environment, it is worth turning to the built-in help for this package, where more detailed data acquisition and graphic options are described.

An example of direct access to the WebAPI service – without intermediation of additional packages dedicated to the given service – will be shown in the example of the POL-on service. This is the official website of the Ministry of Science and Higher Education in Poland, providing access to an extensive resource of data on higher education in Poland. Full access requires having an account on the website and appropriate permissions; however, there is quite a large amount of data on Polish science available to the public. In addition to the records available on the site, this site has a WebAPI service that allows one to interact with the site. The website supports basic access methods (GET, POST, DELETE, PUT) that allow downloading, creating, updating and deleting data – of course, an unlogged user (without the appropriate permissions) can only perform GET queries on publicly available resources. A website description and specification of queries and their parameters can be found at: https://polon.nauka.gov.pl/opi-ws/api/api-docs. To send a properly prepared query, one can use the **GET**() function from the httr:: package. The result will be received in JSON – it is currently one of the most popular file-sharing formats on the web. The resulting query result should be converted to the desired format (e.g. *list* or *data.frame* object) using the appropriate parser. This last step will be provided by the functions available in the jsonlite:: add-on package. Of course, this is not the only package that allows analysis of this type of format.

In the example, a list of universities operating in Poland (resources available without restrictions) along with their identifiers and status will be obtained. To obtain such a list – in accordance with the server specification – the WebAPI POL-on address should be supplemented with the ending "/academicInstitutions". One can specify additional parameters that will not be used in the example. As a result, an object in the *data.frame* class containing the expected information was obtained.

```
library("httr")
library("jsonlite")
U1_result<-GET("https://polon.nauka.gov.pl/opi-ws/api/
academicInstitutions") # query result assigned to variable
U1_parsed <- fromJSON(content(U1_result,"text"), simplifyVector = TRUE,
flatten = TRUE) # transformed object to readable form in the list class
U1.df<-as.data.frame(U1_parsed$institutions) # result content in the data.
frame class object
head(U1.df) # first four only
```

```
                              uid
1 bEc0DbHHbpYEHJgPTQBH5yA
2 b2DEgwUdIFsd73ADJul4MIA
3 bUs0WvJiBCm7L_89_bwdf2g
4 bEArsU8PLvAEqrb5LhSYBPg

                                                               name
1 College of Management in Legnica in liquidation with headquarters in Leg-
nica
2 University of Entrepreneurship and Regional Rozvoiu in liquidation based
in Falenty
3 College of International Studies in Łódź
4 State Higher Vocational School in Chełm
           status
1   IN_LIQUIDATION
2       LIQUIDATED
3        OPERATING
4        OPERATING
```

```
nrow(U1.df) # number of rows in the result
#[1] 532
```

The used **fromJSON()** function transformed the object from the JSON format (in the *response* class) to a readable format in the *list* class based on the proper JSON format identifiers. Before using this function, the received response had to be converted to a text format – the default POL-on server sends data in the "raw" format, which is a string of numbers. This was done using the **content()** function from the same package. Finally, from the transformed result the variables were extracted and saved in the *data.frame* class object. By the way, it was checked that there are 532 universities in Poland in the POL-on database (as on July, 2019).

It is also worth paying attention to the information Status: 200 present in the response from the server. This status means that the query has been successfully completed and the results returned; any other status means that there were errors in the query or response. This value can be used for further analysis and automatic data processing (e.g. error handling).

```
U1_result
```

```
Response [https://polon.nauka.gov.pl/opi-ws/api/academicInstitutions]
  Date: 2019-07-27 14:05
  Status: 200
  Content-Type: application/json;charset=UTF-8
  Size: 65.8 kB
```

```
U1_result$status_code # direct reference to the status of the response
#[1] 200
```

The object identifier (*uid* variable) obtained in the first example can be used to further query the server for detailed information about the object. To do this, add the additional path "institutions/{uid}" to the main server address, where {uid} is the parameter received in the previous query. In the example, the data on the first 10 universities from the previously obtained list will be downloaded. The following code may seem a bit complex. It is worth noting that the query for the first object is made separately so as to create a result object to which data about subsequent universities will be added. Further queries are already called in the *for*() loop. The two lines that use the **setdiff**() function from the base package (automatically loaded at R startup) check that the existing object and the currently downloaded (temporary) object have the same columns (for some universities, not all variables exist) and supplement the columns which are missing by assigning them NA (not available). This then allows for combining the query result with the previously obtained with **rbind**() function. The final result is saved to a csv file.

```
# clearing variables from any previous code executions
rm(output,output_temp,parsed,resp,i,q_url)

q_url<-paste("https://polon.nauka.gov.pl/opi-ws/api/institutions/",U1.
df$uid[1], sep="")

resp<-GET(q_url)
parsed<-jsonlite::fromJSON(content(resp,"text"), simplifyVector=TRUE,
flatten=TRUE)
output<-as.data.frame(parsed)

for(i in 2:10) # replacing 10 with nrow (U1_df) would retrieve information
about all schools.
{
 q_url<-paste("https://polon.nauka.gov.pl/opi-ws/api/institutions/",U1.
 df$uid[i],sep="")
 resp<-GET(q_url)
 parsed<-fromJSON(content(resp,"text"), simplifyVector=TRUE, flatten=TRUE)
 output_temp<-as.data.frame(parsed) # temporary object for the current
 query
 output[setdiff(names(output_temp), names(output))]<-NA
 output_temp[setdiff(names(output), names(output_temp))]<-NA
 output<-rbind(output,output_temp) # adding a temporary object to
 previously downloaded data
}
 output # received data, 4 lines only
```

```
   errorCode   version                       uid              mainUid
1          0       1.0   bEcODbHHbpYEHJgPTQBH5yA   bEcODbHHbpYEHJgPTQBH5yA
2          0       1.0   b2DEgwUdIFsd73ADJul4MIA   b2DEgwUdIFsd73ADJul4MIA
3          0       1.0   bUsOWvJiBCm7L_89_bwdf2g   bUsOWvJiBCm7L_89_bwdf2g
4          0       1.0   bEArsU8PLvAEqrb5LhSYBPg   bEArsU8PLvAEqrb5LhSYBPg
                                                                      name
1 College of Management in Legnica in liquidation with headquarters in Leg-
nica
2 University of Entrepreneurship and Regional Rozvoiu in liquidation based
in Falenty
3 College of International Studies in Łódź
4 State Higher Vocational School in Chełm
                                                                  fullName
1 College of Management in Legnica in liquidation with headquarters in Legnica
```

```
3 College of International Studies in Łódź
4 State Higher Vocational School in Chełm

        code                supervisingInstitutionName
1    WSML   Ministerstwo Nauki i Szkolnictwa Wyższego
2  WSPRRF   Ministerstwo Nauki i Szkolnictwa Wyższego
3   WSSML   Ministerstwo Nauki i Szkolnictwa Wyższego
4   PWSZC   Ministerstwo Nauki i Szkolnictwa Wyższego

                      kind    dateFrom  liquidationStart             status
1 NONPUBLIC_UNIVERSITY  1997-07-31        2017-02-24   IN_LIQUIDATION
2 NONPUBLIC_UNIVERSITY  2002-07-05        2012-12-11        LIQUIDATED
3 NONPUBLIC_UNIVERSITY  1997-07-21              <NA>         OPERATING
4    PUBLIC_UNIVERSITY  2001-09-01              <NA>         OPERATING

      regon                            www                       email
1 390571111        http://www.wsm.edu.pl     sekretariat@wsm.edu.pl
2 015194107      http://www.wspirr.edu.pl       wspirr@wspirr.edu.pl
3 471582365       http://www.wssm.edu.pl     sekretariat@wssm.edu.pl
4 110607010        http://pwsz.chelm.pl       rektorat@pwsz.chelm.pl

    address.country        address.city          address.street
1           Polska            Legnica      Voiciecha Korfantego
2           Polska            Falenty            Aleja Hrabska
3           Polska               Łódź                 Brzozowa
4           Polska              Chełm                 Pocztowa

    address.buildingNumber  address.postalCode  address.postalCity
1                        4             59-220             Legnica
2                        3             05-090    Raszyn, Falenty
3                      5/7             93-101                Łódź
4                       54             22-100               Chełm

    address.voivodeship isLaboratory  isLibrary isScientific     dateTo
1         dolnośląskie        FALSE      FALSE        FALSE       <NA>
2          mazowieckie        FALSE      FALSE        FALSE 2012-12-15
3             łódzkie        FALSE      FALSE         TRUE       <NA>
4           lubelskie        FALSE      FALSE         TRUE       <NA>

    phoneNumber        faxNumber          universityType        nip
1         <NA>             <NA>                    <NA>       <NA>
2  22 720 05 35      226283763                    <NA>       <NA>
3     426897210      426897213                    <NA>       <NA>
4  82 565 88 95   82 565 88 94   VOCATIONAL_UNIVERSITY       <NA>
```

```
write.csv2(output,file="./Data2/highschools_PL.csv",quote=TRUE, row.
names = FALSE)
```

It is worth noting that the result contains address information that can be used for so-called geocoding, that is, obtaining geographical coordinates based on address. Discussion of this issue will be the subject of the next part of the chapter.

3.6 Geocoding of data

When downloading data from online resources or collecting data in a different way, very often the individual observations lack the most important information, i.e. unambiguous location – geographical coordinates. Very often shared databases already have fields ready for such information, but practice shows that these fields are rarely filled. Carrying out spatial analyses, it is necessary to assign the collected observations (data) to the correct geographical location. It is also worth adding that this problem is most important in the case of point phenomena, where precise location is important – in the case of observations for areas (e.g. administrative), one does not need to know the specific location in order to assign appropriate observations to a given region and present them on maps or to use them further for quantitative spatial techniques.

The process described previously is called geocoding. In a broad sense, it is not only about the indication of geographical coordinates for a given phenomenon (observation), the term means the assignment of unambiguous codes, identifiers and so on that would precisely indicate the place. Any marking that describes a location at a specific level of accuracy can be used as a location identifier – for example, zip code, telephone area code, car license plate, NUTS codes, TERYT, country codes and so on. Of course, the use of a given identifier is justified depending on the desired degree of precision. In spatial analyses, the most precise and unambiguous indication is usually needed (in particular in relation to point phenomena); hence, geocoding is usually the process of exchanging the description of a place (usually a postal address) for specific geographical coordinates, most often in the latitude/longitude (or other dependent from the reference system used – determined by the coordinate system of the vector maps owned). Sometimes it is necessary to perform a reverse operation, that is, the geographical coordinate should be assigned a location name, for example, address or the name of a company, building and so on. This process is called reverse geocoding or re(geo)coding. Reverse geocoding can be used to check if the results of geocoding are correct, but it should be noted that reverse geocoding, especially in highly urbanised areas, usually gives different results than the primary data (for example, it may be a different name of a company, institution or address nearby). Getting a result similar to the original one also depends on the precision of previous geocoding.

Geocoding in R can be carried out in two basic ways: using previously created packages for geocoding (and sometimes for regeocoding) or using directly selected servers through their API services by manually creating queries. The first way is easier, but of course it limits the user's flexibility in some ways; the second is a bit more complicated and requires knowledge about creating queries for WebAPI services. It is worth noting that until recently, geocoding (using R packages) was essentially limited to querying Google Maps; now packages are created using other – including free and open – datasets and servers. In addition to the Google Maps website mentioned previously (which is of course a commercial website), Nominatim and Photon are also noteworthy open services enabling geocoding. Both of these services are based on data from the OpenStreetMap database. Example queries for all three services will be presented subsequently.

The most common way of geocoding is to use the **geocode**() command from the ggmap:: package. Due to the fact that this command uses the Google service, it also requires entering the API key, which was already used in the previous examples. The following code will load the ggmap:: package into the session. At the same time, the next two commands (already available in the loaded package) will register private key to Google API services and set the key to be not visible when executing subsequent commands that need it – it will not be explicitly shown in the console (for security). It is worth pointing out that key registration is active in a specific R session, that is, until it is closed. The API key must be reactivated at the next session. It is similar with the **ggmap_hide_api_key**() function – it works only during the current session. By using the *write = TRUE* option in **register_google(key = ". . .", write = TRUE)**, one can save the used key permanently in one's own R environment. It will be loaded each time when running this package. It is a solution that facilitates the use of the key, although it should be noted that the API key, after its registration in R, is potentially available for all functions – functions from packages from an unknown source can use this key and transfer it elsewhere.

```
library(ggmap)
register_google(key="AbCdEfGhIjKlMnOpRsTuWxYz" # false key
ggmap_hide_api_key()
```

The first of the examples presented is based on the list of cities in Poland with the largest number of inhabitants taken from the Central Statistical Office. This and other tables can be downloaded in a convenient csv format. It is worth noting that geocoding will be carried out for very generally described locations, because the file only contains the name of the city. The object loaded from the csv file will be appended with two variables generated during geocoding denoting the latitude and longitude – the centre of each city. The loop will be used in such a way as to obtain a location for all observations in dataset. Two columns have been added to the population object with the latitude and longitude provided by the server.

```
population<-read.csv("population.csv", sep=",", stringsAsFactors=FALSE)
population$lon<-NA
population$lat<-NA
for(i in 1:nrow(population)){
xy<-geocode(paste(population[i,2], "Polska"))
population$lon[i]<-xy$lon
population$lat[i]<-xy$lat}
head(population)
```

lp	name	name_voi	total	men	women	lon	lat
1 1	Warszawa	MAZOWIECKIE	1777972	817660	960312	21.01223	52.22968
2 2	Kraków	MAŁOPOLSKIE	771069	359865	411204	19.94498	50.06465
3 3	Łódź	ŁÓDZKIE	685285	312328	372957	19.45598	51.75925
4 4	Wrocław	DOLNOŚLĄSKIE	640648	299190	341458	17.03854	51.10789
5 5	Poznań	WIELKOPOLSKIE	536438	250244	286194	16.92517	52.40637
6 6	Gdańsk	POMORSKIE	466631	221187	245444	18.64664	54.35203

```
# writing to a new result object file after geocoding
write.csv(population, file="ludnoscxy.csv", row.names=FALSE)
```

It's worth trying a simple code to geocode a single location, for example, Warsaw: geocode ("Warsaw, Poland"). The displayed result seems to be accurate to one decimal place, but in fact the number returned by the **geocode**() function has an accuracy of five decimal places, as seen in the previous example looking at the first few rows of the table.

The **geocode**() function can also be used to find the coordinates of several addresses at once. In this case, enter the searched location as a vector of text data. One can then avoid using loops. In the following example, the search for geographical coordinates for cities from the set used previously will be presented again – the searched data will be limited to the first five observations.

```
geocode(population$name[1:5])
```

```
# A tibble: 5 x 2
     lon    lat
   <dbl>  <dbl>
1   21.0   52.2
2   19.9   50.1
3   19.5   51.8
4   17.0   51.1
5   16.9   52.4
```

The additional parameter *output* = *c* (*"latlon"*, *"latlona"*, *"more"*, *"all"*) in **geocode**() determines the content of the results returned by the Google server – one of the following can be selected. The default value – the first in the list – returns only the coordinates. Another setting is interesting, that is, *latlona*. It returns – in addition to the coordinates – the address found by the server being queried. It can be used to compare the address searched for with the one found by Google. This will allow for checking the results for any misidentified queries. The *more* option provides somewhat broader information, including the boundaries of the location found. In the case of the whole city, these will be the coordinates of the border point in the northeast and southwest of the area for the given address which Google assigns to it. Option *all* returns all results obtained from the API server in the form of a fairly complex object in the *list* class. A comparison of the results of the various options is given subsequently.

```
geocode("Pacanów, Polska")
```

```
Source : https://maps.googleapis.com/maps/api/geocode/json?address=Pacan%C3%B
3w,+Polska&key=xxx
# A tibble: 1 x 2
      lon     lat
    <dbl>   <dbl>
1    21.0    50.4
```

```
geocode("Pacanów, Polska", output="latlona")
```

```
Source : https://maps.googleapis.com/maps/api/geocode/json?address=Pacan%C3%B
3w,+Polska&key=xxx
# A tibble: 1 x 3
    lon    lat    address
  <dbl>  <dbl>    <chr>
1  21.0   50.4    28-133 pacanów, poland
```

```
geocode("Pacanów, Polska", output="more")
```

```
Source : https://maps.googleapis.com/maps/api/geocode/json?address=Pacan%C3%B
3w,+Polska&key=xxx
# A tibble: 1 x 9
  lon   lat   type     loctype      address         north south  east  west
 <dbl> <dbl> <chr>    <chr>         <chr>           <dbl> <dbl> <dbl> <dbl>
1 21.0  50.4  locality approxima . . . 28-133 pacanów,. . . 50.4  50.4  21.1  21.0
```

```
# the result assigned to the temp object due to the length of the result
temp<-geocode("Pacanów, Polska", output="all")
temp
```

```
Source : https://maps.googleapis.com/maps/api/geocode/json?address=Pacan%C3%B
3w,+Polska&key=xxx
```

```
temp$results[[1]]$address_components[[1]] # only a fragment of the result-
ing object is shown
```

```
$long_name
[1] "Pacanów"

$short_name
[1] "Pacanów"

$types
$types[[1]]
[1] "locality"

$types[[2]]
[1] "political"
```

Of course, geocoding can refer to more specific addresses, or one can try to identify the location only by the name of the institution (object). The following is an example of obtaining geographical coordinates by address and an attempt when asking for the name of the institution only.

```
addresses.df<- data.frame(Lp=integer(), adres=character(), lon=numeric(),
lat=numeric(), stringsAsFactors=FALSE) # empty dataframe
addresses.df[1,]<-c(1,"Długa 44/50, Warszawa, Polska",NA, NA)
addresses.df[2,]<-c(2,"Wydział Nauk Ekonomicznych, Uniwersytet Warszawski,
Warszawa, Polska",NA,NA)
result<-geocode(addresses.df[,2]) # geocoding of two addresses
```

```
Source : https://maps.googleapis.com/maps/api/geocode/json?address=D%C5%82uga
+44/50,+Warszawa,+Polska&key=xxx
Source : https://maps.googleapis.com/maps/api/geocode/json?address=Wydzia%C5%
82+Nauk+Ekonomicznych,+Uniwersytet+Warszawski,+Warszawa,+Polska&key=xxx
```

```
# adding geocoding results to the source object
addresses.df[,3:4]<-result
addresses.df
```

```
   Lp                                                              address
1  1                                       Długa 44/50, Warszawa, Polska
2  2  Wydział Nauk Ekonomicznych, Uniwersytet Warszawski, Warszawa, Polska
        lon       lat
1  21.00394  52.24649
2  21.00381  52.24652
```

The so-called reverse geocoding is the reverse action for already described geocoding. The result is the name or address of the object identified from given geographical coordinates. The following example appends the result of reverse geocoding to the source table as the *regeo* variable. In both cases, the result is an address. It is also worth seeing the result of regeocoding in the console, because it contains information about finding many objects under given coordinates. As a result of the **revgeocode**() function, the first of them is selected.

```
addresses.df$regeo<-NA
addresses.df$regeo[1]<-revgeocode(c(addresses.df$lon[1],addresses.
df$lat[1]))
```

```
Source : https://maps.googleapis.com/maps/api/geocode/json?latlng=52.2464855,
21.0039365&key=xxx
Multiple addresses found, the first will be returned:
  Długa 44/50, 00-241 Warszawa, Poland
  OSK YETI. Szkoła nauki jazdy, Długa 44/50, 00-241 Warszawa, Poland
  00-241 Warsaw, Poland
  Muranów, 00-001 Warsaw, Poland
  Śródmieście, 00-001 Warsaw, Poland
  Warsaw, Poland
  Warszawa, Poland
  Masovian Voivodeship, Poland
  Poland
```

```
addresses.df$regeo[2]<-revgeocode(c(addresses.df$lon[2],addresses.
df$lat[2]))
```

```
Source : https://maps.googleapis.com/maps/api/geocode/json?latlng=52.2465181,
21.0038102&key=xxx
Multiple addresses found, the first will be returned:
  Długa 44, 00-241 Warszawa, Poland
  OSK YETI. Szkoła nauki jazdy, Długa 44/50, 00-241 Warszawa, Poland
  00-241 Warsaw, Poland
  Muranów, 00-001 Warsaw, Poland
  Śródmieście, 00-001 Warsaw, Poland
  Warsaw, Poland
  Warszawa, Poland
  Masovian Voivodeship, Poland
  Poland
```

addresses.df

```
   Lp                                                                      address
1  1                                                    Długa 44/50, Warszawa, Polska
2  2      Wydział Nauk Ekonomicznych, Uniwersytet Warszawski, Warszawa, Polska
       lon       lat                                         regeo
1  21.00394   52.24649    Długa 44/50, 00-241 Warszawa, Poland
2  21.00381   52.24652       Długa 44, 00-241 Warszawa, Poland
```

The **geocode**() function was also originally used to query the Data Science Toolkit: http://www.datascience-toolkit.org/ server, which provided geocoding functionality on an open access basis. Unfortunately, this service is no longer developed and is even marked as "dead" in the help for the **geocode**() function.

When using Google data, it's a good idea to use the ggmap:: package. The **mapdist**() command from the ggmap:: package allows for calculating the distance between addresses using the Google Distance Matrix API. The daily limit of free calculations is 2500 queries. The command allows for taking into account various means of communication: by car (*mode="driving"*), on foot (*mode="walking"*), by bicycle (*mode="bicycling"*) and transit (*mode="transit"*). When entering data as geographical coordinates, the command regeocodes in the background, replacing the coordinates with an approximate address. This operation is possible using the **revgeocode**() command from the ggmap:: package.

In the following example, the distances between the main campus of the University of Warsaw and the Faculty of Economic Sciences of the University of Warsaw were examined. The address was given as *from* and *to*. The **mapdist**() command has determined distances in various ways.

```
library(ggmap)
register_google(key="AbCdEfGhIjKlMnOpRsTuWxYz" # false key
ggmap_hide_api_key()
from<-"Krakowskie Przedmieście 26/28, Warszawa, Polska"
to<-"Długa 44/50, Warszawa, Polska"
distAB<-mapdist(from, to, mode="bicycling", output="all")
distAB
```

```
$'Krakowskie Przedmieście 26/28, Warszawa, Polska'
$'Krakowskie Przedmieście 26/28, Warszawa, Polska'$'Długa 44/50, Warszawa,
Polska'
$'Krakowskie Przedmieście 26/28, Warszawa, Polska'$'Długa 44/50, Warszawa,
Polska'$distance
$'Krakowskie Przedmieście 26/28, Warszawa, Polska'$'Długa 44/50, Warszawa,
Polska'$distance$text
[1] "1.6 km"
```

```
$'Krakowskie Przedmieście 26/28, Warszawa, Polska'$'Długa 44/50, Warszawa,
Polska'$distance$value
[1] 1632

$'Krakowskie Przedmieście 26/28, Warszawa, Polska'$'Długa 44/50, Warszawa,
Polska'$duration
$'Krakowskie Przedmieście 26/28, Warszawa, Polska'$'Długa 44/50, Warszawa,
Polska'$duration$text
[1] "5 mins"

$'Krakowskie Przedmieście 26/28, Warszawa, Polska'$'Długa 44/50, Warszawa,
Polska'$duration$value
[1] 319

$'Krakowskie Przedmieście 26/28, Warszawa, Polska'$'Długa 44/50, Warszawa,
Polska'$status
[1] "OK"
```

```
from<-"Krakowskie Przedmieście 26/28, Warszawa, Polska"
to<-"Długa 44/50, Warszawa, Polska"
distAB<-mapdist(from, to, mode="bicycling", output="simple")
distAB
```

```
# A tibble: 1 x 9
        from        to       m      km miles seconds minutes hours      mode
       <chr>     <chr>   <int>   <dbl> <dbl>   <int>   <dbl> <dbl>     <chr>
1 Krakowskie Przedmi~ Długa 44/50~  1632   1.63   1.01     319  5.320.0886 bicyc~
```

It is also possible to map a route between two locations. The **route**() and **trek**() commands allow this, both from the ggmap:: package. The starting and destination addresses are given as arguments, and the output can be in various formats. The structure defined as "route" allows easy visualisation of the path by generating a base map with the **qmap**() command. In the **route**() and **trek**() commands, as in **mapdist**(), car, bicycle, pedestrian and transit modes are available.

```
from<-"Krakowskie Przedmieście 26/28, Warszawa, Polska"
to<-"Długa 44/50, Warszawa, Polska"
wayAB1<-route(from, to, structure="legs")
wayAB2<-route(from, to, structure="route")
wayAB3<-route(from, to, alternatives=TRUE)
wayAB4<-trek(from, to, structure="route")
wayAB1
```

```
# A tibble: 9 x 11
    m      km  miles seconds minutes    hours start_lon start_lat end_lon end_lat
<int><dbl>  <dbl>  <int>   <dbl>   <dbl>    <dbl>     <dbl>   <dbl>   <dbl>
1 288 0.288  0.179    124    2.07  0.0344     21.0      52.2    21.0    52.2
2 139 0.139 0.0864     28   0.467 0.00778     21.0      52.2    21.0    52.2
3 430  0.43  0.267     94    1.57  0.0261     21.0      52.2    21.0    52.2
4 519 0.519  0.323     56   0.933  0.0156     21.0      52.2    21.0    52.2
5 235 0.235  0.146     47   0.783  0.0131     21.0      52.2    21.0    52.2
6 301 0.301  0.187     48     0.8  0.0133     21.0      52.2    21.0    52.2
7 168 0.168  0.104     48     0.8  0.0133     21.0      52.2    21.0    52.2
```

```
8  68  0.068 0.0423      18      0.3   0.005       21.0      52.2   21.0      52.2
9 119  0.119 0.0739      29    0.483 0.00806       21.0      52.2   21.0      52.2
# . . . with 1 more variable: route <chr>
```

wayAB2

```
# A tibble: 10 x 9
     route       m      km     miles  seconds minutes     hours     lon      lat
     <chr>   <int>   <dbl>     <dbl>    <int>   <dbl>     <dbl>   <dbl>    <dbl>
 1 A           288   0.288     0.179      124    2.07    0.0344    21.0     52.2
 2 A           139   0.139    0.0864       28   0.467   0.00778    21.0     52.2
 3 A           430    0.43     0.267       94    1.57    0.0261    21.0     52.2
 4 A           519   0.519     0.323       56   0.933    0.0156    21.0     52.2
 5 A           235   0.235     0.146       47   0.783    0.0131    21.0     52.2
 6 A           301   0.301     0.187       48     0.8    0.0133    21.0     52.2
 7 A           168   0.168     0.104       48     0.8    0.0133    21.0     52.2
 8 A            68   0.068    0.0423       18     0.3     0.005    21.0     52.2
 9 A           119   0.119    0.0739       29   0.483   0.00806    21.0     52.2
10 A            NA      NA        NA       NA      NA        NA    21.0     52.2
```

wayAB3

```
# A tibble: 18 x 11
      m     km  miles seconds minutes    hours start_lon start_lat end_lon end_lat
  <int>  <dbl>  <dbl>   <int>   <dbl>    <dbl>     <dbl>     <dbl>   <dbl>   <dbl>
 1 288  0.288  0.179      124    2.07   0.0344      21.0      52.2    21.0    52.2
 2 139  0.139 0.0864       28   0.467  0.00778      21.0      52.2    21.0    52.2
 3 430   0.43  0.267       94    1.57   0.0261      21.0      52.2    21.0    52.2
 4 519  0.519  0.323       56   0.933   0.0156      21.0      52.2    21.0    52.2
 5 235  0.235  0.146       47   0.783   0.0131      21.0      52.2    21.0    52.2
 6 301  0.301  0.187       48     0.8   0.0133      21.0      52.2    21.0    52.2
 7 168  0.168  0.104       48     0.8   0.0133      21.0      52.2    21.0    52.2
 8  68  0.068 0.0423       18     0.3    0.005      21.0      52.2    21.0    52.2
 9 119  0.119 0.0739       29   0.483  0.00806      21.0      52.2    21.0    52.2
10 288  0.288  0.179      124    2.07   0.0344      21.0      52.2    21.0    52.2
11 139  0.139 0.0864       28   0.467  0.00778      21.0      52.2    21.0    52.2
12 128  0.128 0.0795       30     0.5  0.00833      21.0      52.2    21.0    52.2
13 199  0.199  0.124       43   0.717   0.0119      21.0      52.2    21.0    52.2
14 149  0.149 0.0926       22   0.367  0.00611      21.0      52.2    21.0    52.2
15 220   0.22  0.137       42     0.7   0.0117      21.0      52.2    21.0    52.2
16 182  0.182  0.113       52   0.867   0.0144      21.0      52.2    21.0    52.2
17 498  0.498  0.309       92    1.53   0.0256      21.0      52.2    21.0    52.2
18 292  0.292  0.181       67    1.12   0.0186      21.0      52.2    21.0    52.2
# . . . with 1 more variable: route <chr>
```

wayAB4

```
# A tibble: 64 x 3
     lat    lon  route
   <dbl>  <dbl>  <chr>
 1  52.2   21.0      A
```

```
 2   52.2   21.0       A
 3   52.2   21.0       A
 4   52.2   21.0       A
 5   52.2   21.0       A
 6   52.2   21.0       A
 7   52.2   21.0       A
 8   52.2   21.0       A
 9   52.2   21.0       A
10   52.2   21.0       A
# . . . with 54 more rows
```

```
qmap("Uniwersytet Warszawski, Polska", zoom=13, SCALE=2) +
geom_path(aes(x=lon, y=lat), colour="red", size=1.5, data=wayAB2,
lineend="round") # Figure 3.15a
```

Figure 3.15 Visualisation of the route between the main campus of the University of Warsaw and the Faculty of Economic Sciences of the University of Warsaw using the **route()** and **track()** commands

Source: Own study using R and the ggmap:: package

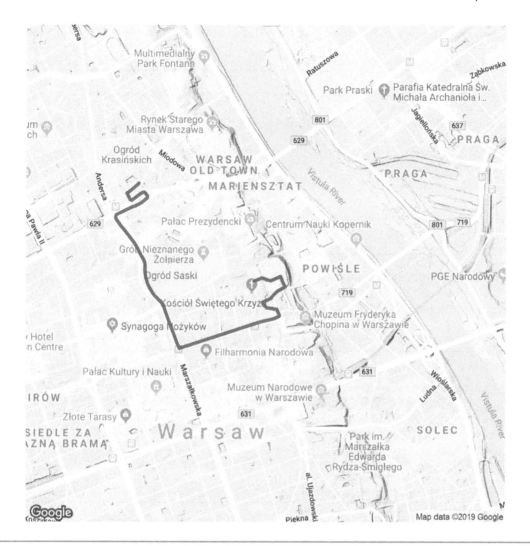

Figure 3.15 (Continued)

```
qmap("Uniwersytet Warszawski, Polska", zoom=14, SCALE=2) +
geom_path(aes(x=lon, y=lat), colour="red", size=1.5, data=wayAB4,
lineend="round") # Figure 3.15b
```

It is also worth mentioning that in the gmapsdistance:: package, one can determine the distance and travel time between two points based on a Google map. The **get.api.key()** and **set.api.key()** commands allow for managing the API. The **gmapsdistance()** command finds the distance or travel time.

The aforementioned alternative to Google may be the use of geocoding API services based on OSM data. The basic server offering the geocoding (and reverse geocoding) service is Nominatim, available in a web version at: https://nominatim.openstreetmap.org/. This website also offers an API service. The most important condition for using the services is to limit the frequency of queries to one per second and limit the number of queries to a reasonable number. It is also recommended to save the results to avoid repeated requests for the same object. Queries to the Nominatim server can be built directly in R, but one can also use packages already created for this service. The following uses the **geocode_OSM()** function from the tmaptools:: package. There is one more package osmar:: in the official R resources that performs similar functions and names it as OSM. However, this package has not been developed for several years.

```
library(tmaptools)
library(rgdal)
library(OpenStreetMap)
library(sf)
```

```
# the simplest form of a query
geocode_OSM("Warszawa, Polska")
```

```
$query
[1] "Warszawa, Polska"

$coords
        x               y
21.00673       52.23192

$bbox
    xmin      ymin      xmax      ymax
20.85169   52.09785   21.27115   52.36815
```

```
# address request, detailed result (additional attributes),
# result as an object of the data.frame class
geocode_OSM("Koński Jar 2, Warszawa, Polska", details=TRUE, as.data.frame=
TRUE)
```

```
                          query      lat       lon   lat_min   lat_max
1 Koński Jar 2, Warszawa, Polska 52.16252 21.02625 52.16229 52.16274
    lon_min    lon_max   place_id osm_type     osm_id  place_rank
1 21.02616   21.02654   97221023      way   99896914          30
                                              display_name    class type
1 2, Koński Jar, Ursynów, Warszawa, mazowieckie, 02-785, RP building yes
importance icon
1 0.421 <NA>
```

```
# an example of geocoding and saving several firms from Warsaw
firms<-read.csv2("firms_warszawa.csv", stringsAsFactors=FALSE)
# creating a variable with full address
firms$Address_full<-paste(firms$Address, firms$Postcode, firms$City,
firms$Country, sep=",")

firms_geo<-geocode_OSM(firms$Address_full)
firms<-cbind(firms, firms_geo[,2:3]) # adding geocode to firms object
# writing to a new result object file after geocoding
write.csv(firms, file="./Data2/firms_warszawaxy.csv", row.names=FALSE)

#defining the codes for used georeference systems
latlong<-"+init=epsg:4326"
google<-"+init=epsg:3857"
polish_borders<-"+init=epsg:2180"

pov<-readOGR(".", "powiaty") # 380 units

# selection of the area of Warsaw (city with poviat status)
WAW<-pov[pov$jpt_kod_je==1465,]
WAW_latlong<-spTransform(WAW, CRS(latlong))
WAW_box<-bbox(WAW_latlong)
WAW_osm.map<-openmap(c(WAW_box[2,2], WAW_box[1,1]), c(WAW_box[2,1],
WAW_box[1,2]))
par(mfrow=c(1,2)) # split of the graphic window into two active areas
```

```
# 1st figure
# OSM map layout changed to geographical coordinates
WAW_osm2.map<-openproj(WAW_osm.map, CRS(latlong))
plot(WAW_osm2.map)
plot(WAW_latlong, add=TRUE) # borders of Warsaw area
# geographical coordinates of firms
crds<-as.data.frame(cbind(firms$lon, firms$lat))
points(crds, pch=16, col="red")

# 2nd figure
par(mar=c(0,0,0,0))
plot(WAW_osm.map) # drawing a map in native OSM system
WAW_osm<-spTransform(WAW, osm()) # transformation of city boundaries into
OSM
plot(WAW_osm, add=T)

# transformation of company coordinate matrix into OSM
# use of the sp_project() function from the sf package
crds_osm<-sf_project(as.character(CRS(latlong)), as.character(osm()),crds)
points(crds_osm, pch=16, col="red")
```

As a result of executing this code (except for two sample queries at the beginning), the geographical coordinates of 11 selected firms from Warsaw were obtained. The data was added to the loaded company database and then saved in a new file. To illustrate the results, the location of these firms – on the contour of Warsaw together with a base map from the OpenStreetMap website – was marked. Basically, each of the elements applied was originally in a different georeference system – this required the conversion of objects to one selected system. A comparison of figures in two coordinate systems has been provided. The first one is in the geographical coordinate system (EPSG: 4326), and the second one is in the native OSM (and Google) map system. Instead of the **CRS**() function, **osm**() was used from the OpenStreetMap:: package, which directly provides the appropriate parameters for the OSM system. It is worth noting that the first drawing required conversion of the raster image, which caused its distortion. The second kept the originally downloaded map and required coordinate conversion of vector objects. This did not distort the final results. When constructing the second figure, the **sf_project**() function was used, which enables the conversion of an ordinary matrix containing coordinates in a given system to the selected target system. The only disadvantage of using this function is the need to change the coordinate system parameters received using the previously used **CRS**() function into simple text.

For the sake of order, it should also be pointed out that the **geocode_OSM**() function may have additional options that affect the resulting object. It is worth paying attention to the projection option, which will determine in which coordinate system geocoding results should be returned. The following example will show the geocoding of the first two firms from the previous set straight to the original system in which OSM raster underlays are downloaded. In this way, one can also avoid transforming the raster backgrounds into another coordinate system, which can cause distortion as shown. In the projection option, instead of entering projection parameters, the result of **osm**() function was used again.

```
firms<-read.csv2("./Data2/firms_warszawa.csv", stringsAsFactors=FALSE,
encoding = "UTF-8")
firms$Address_full<-paste(firms$Address,firms$Postcode, firms$City,
firms$Country, sep=", ")
osm() # parameters of the original reference system used by OSM maps

CRS arguments:
+proj=merc +a=6378137 +b=6378137 +lat_ts=0.0 +lon_0=0.0 +x_0=0.0
+y_0=0 +k=1.0 +units=m +nadgrids=@null +no_defs

firms_geo<-geocode_OSM(firms$Address_full[1:2], projection =osm())
firms_geo # the result in x and y is from the Cartesian coordinate system
```

```
                                        query       x          y     y_min     y_max
1 BALETOWA  29,   02-867, WARSZAWA,  Poland 2338513  6822503  6822490  6822515
2 CELULOZY  17B,  04-986, WARSZAWA,  Poland 2357938  6829324  6829315  6829333
        x_min        x_max
1     2338496      2338530
2     2357933      2357944
```

The geocoding result from the variables *x* and *y* corresponds to the same points obtained in the previous example, in which the results were given as latitude and longitude (in degrees).

Another parameter worth attention is *as.sf*. Setting this parameter to *TRUE* causes the resulting object to be in the *sf* class defined in the additional package with the same name. This class (and package) allows simple and standardised storage of spatial vector objects. This package refers to some of the spatial packages mentioned earlier and allows for manipulating and drawing localised vector data as well. The following example will use the option mentioned previously and mark the two selected locations from the company database on the contour of Warsaw borders. It is worth noting that the resulting object has an *sf* class, while the contour is in the class *SpatialPolygonsDataFrame* from the package sp::. Both objects are spatial objects; hence they could be placed without problems in one figure.

```
library(sf)
# geocoding result directly as an sf object
firms_geo.sf<-geocode_OSM(firms$Address_full[2:3], as.sf=TRUE)
firms_geo.sf
```

```
Simple feature collection with 2 features and 7 fields
geometry type:  POINT
dimension:      XY
bbox: xmin:     21.14353 ymin: 52.1612 xmax: 21.18172 ymax: 52.22836
epsg (SRID):    4326
proj4string:    +proj=longlat +datum=WGS84 +no_defs


                                    query      lat       lon    lat_min
1 CELULOZY 17B, 04-986, WARSZAWA,  Poland 52.16120  21.18172  52.16115
2 MINERSKA 3,   04-506, WARSZAWA,  Poland 52.22836  21.14353  52.22831
  lat_max    lon_min   lon_max                         geometry
1 52.16125  21.18167  21.18177  POINT  (21.18172   52.1612)
2 52.22841  21.14348  21.14358  POINT  (21.14353  52.22836)
```

```
class(firms_geo.sf)
#[1] "sf" "data.frame"
plot(WAW_latlong) # contour map with two points in sf class
plot(firms_geo.sf, add=TRUE)
```

As in the case of Google services, Nominatim based on OSM also offers a reverse geocoding service, that is, address search based on coordinates. This task is performed by the function **rev_geocode_OSM()** from the already-used OpenStreetMap:: package. In general, the options for this function are very similar to the **geocode_OSM()** shown previously. Of course, the argument in this case will be the geographical coordinates of the point or points to be re-coded – by default, as latitude and longitude. However, after setting the appropriate coordinate system in options, one can also enter coordinates in other systems. Input data can give latitude and longitude vectors separately or as one argument of the *SpatialPoints* class. In the example, the previously saved file with geocoded data about firms from Warsaw will be used. The result of reverse geocoding should be the original addresses.

```
library(OpenStreetMap)
firms<-read.csv("./Data2/firms_warszawaxy.csv", stringsAsFactors=FALSE,
encoding = "UTF-8")
```

```
firms_rev<-rev_geocode_OSM(firms$lon,firms$lat)

# combining the result of regeocoding with the previous data
firms<-cbind(firms,firms_rev[,3:25])

# shown address originally forwarded for geocoding
# and the address received after reverse geocoding

firms[,c(8,11)]
```

```
                                                          Address_full
 1                            BALETOWA 29, 02-867, WARSZAWA, Poland
 2                            CELULOZY 17B, 04-986, WARSZAWA, Poland
 3                             MINERSKA 3, 04-506, WARSZAWA, Poland
 4                    ABRAHAMA ROMANA 18, 03-982, WARSZAWA, Poland
 5   KONSTANTEGO ILDEFONSA GAŁCZYNSKIEGO 4, 00-362, WARSZAWA, Poland
 6                       NOWOGRODZKA 47 A, 00-695, WARSZAWA, Poland
 7                         RENESANSOWA 7B, 01-905, WARSZAWA, Poland
 8                                SOLEC 85, 00-382, WARSZAWA, Poland
 9                           SPEDYCYJNA 24, 03-191, WARSZAWA, Poland
 10                             STALOWA 52, 03-429, WARSZAWA, Poland
 11            STANISŁAWA MONIUSZKI 1A, 00-014, WARSZAWA, Poland
name
 1 29, Baletowa, Nowe Jeziorki, Ursynów, Warszawa, mazowieckie, 02-867, RP
 2 17B, Celulozy, Nadwiśle, Wawer, Warszawa, mazowieckie, 04-986, RP
 3 3, Minerska, Marysin Wawerski Południowy, Wawer, Warszawa, mazowieckie,
04-506, RP
 4 18, Generała Romana Abrahama, Gocław, Praga-Południe, Warszawa, mazowieck-
ie, 03-982, RP
 5 Ministerstwo Rodziny, Pracy i Polityki Społecznej, 4, Konstantego Ildefon-
sa Gałczyńskiego, VII, Śródmieście, Warszawa, mazowieckie, 00-362, RP
 6 1, Jana Pankiewicza, VIII, Śródmieście, Warszawa, mazowieckie, 00-696, RP
 7 7B, Renesansowa, Wawrzyszew, Bielany, Warszawa, mazowieckie, 01-905, RP
 8 85, Solec, X, Śródmieście, Warszawa, mazowieckie, 00-382, RP
 9 24, Spedycyjna, Żerań, Białołęka, Warszawa, mazowieckie, 03-191, RP
 10 52, Stalowa, Nowa Praga, Praga-Północ, Warszawa, mazowieckie, 03-429, RP
 11 Moniuszki Tower, 1A, Stanisława Moniuszki, VI, Śródmieście, Warszawa, ma-
zowieckie, 00-014, RP
```

The last line of the previous code allows for comparing the original addresses with those obtained after regeocoding. It is worth noting that only in one example was the address changed (Nowogrodzka 47A was changed to Pankiewicza 1). It should be pointed out that these addresses are at one intersection in Warsaw, and their location is very near; hence, a result seems acceptable assuming a reasonable margin of error. It is worth adding that the result of regeocoding from the OSM database contains much more information, including some attributes of the objects from OSM base. This enables their subsequent analysis and filtering.

The last presented geocoding option will be based on an alternative WebAPI service based on OpenStreet-Map data. The Photon project is considered faster than the Nominatim website described previously. The Photon service also seems slightly more resistant to typos in the names of geocoded places, supports names in many languages (if they appear in the OSM database) and searches the current version of the OSM database (any corrections in OSM data are included in searches with only a short delay). It is worth noting that Photon also supports reverse geocoding. Using Photon also allows one to filter results using OSM attributes. This gives the ability to search for objects of a specific type only. Photon installation is available at http://photon.komoot. de/; it also has a WebAPI, which allows sending queries from external applications. There is no formal limit to queries; it is only general so that it is not overly intense.

There is currently no package available in the official R package repositories that would use the Photon service in geocoding. In 2017, a package was created that is available at: https://github.com/rCarto/photon.

It should be noted once again that this is not an official R package. Interested users can install this package (to their R package library) according to the instructions on the indicated page. To do this, follow these code:

```
require(devtools)
devtools::install_github(repo = 'rCarto/photon')
```

After installation, two functions are available in the package: **geocode**() and **reverse**(). Because this is not an official package, direct querying of the Photon server using http address and manual process of the results from the server will be presented. It is worth mentioning that the result returned by the Photon service is in the geoJSON format, which, according to the home page of this standard, is used to describe (exchange of information) geographical (spatial) data. Together with the presentation of the structure of the object, it is possible to include non-spatial features (attributes). To send a properly prepared request for geocoding, for example, a given address, the previously used **GET**() function from the httr:: package will be used. The obtained query result in the mentioned geoJSON format will be converted to the desired format (e.g. *list* or *data.frame* class object) in the same way as in the previously mentioned example using functions from the jsonlite:: package.

```
library(httr)
library(jsonlite)

# simplest location request (address)
resp<-GET("https://photon.komoot.de/api/?q='Długa+44/50+Warszawa'")
parsed<-fromJSON(content(resp,"text"), flatten = TRUE)
parsed$features # result converted to data.frame class
```

```
    type    geometry.coordinates   geometry.type    properties.osm_id
1 Feature     21.00290, 52.24607           Point         5677460723
2 Feature     21.00367, 52.24659           Point           34497643
  properties.osm_type      properties.country    properties.osm_key
1                   N                  Poland               leisure
2                   W                  Poland               amenity
    properties.housenumber     properties.city    properties.street
1                     44/50             Warsaw                Długa
2                     44/50             Warsaw                Długa
    properties.osm_value     properties.postcode
1                  dance               00-241
2             university               00-241
                                  properties.name      properties.state
1                    Szkoła Tańca ABRA STUDIO    Masovian Voivodeship
2 Faculty of Economic Sciences, University of Warsaw    Masovian Voivodeship
                      properties.extent
1                               NULL
2 21.00319, 52.24687, 21.00403, 52.24636
```

```
parsed$features$geometry.coordinates
```

```
[[1]]
[1] 21.00290 52.24607
[[2]]
[1] 21.00367 52.24659
```

```
parsed$features$geometry.coordinates[[1]][[1]] # longitude
#[1] 21.0029
parsed$features$geometry.coordinates[[1]][[2]] # latitude
```

```
#[1] 52.24607
# limiting the results to the first
resp<-GET("https://photon.komoot.de/api/?q='Długa+44/50+Warszawa'&limit=1")
parsed<-fromJSON(content(resp,"text"), flatten = TRUE)
parsed$features
```

```
        type   geometry.coordinates    geometry.type       properties.osm_id
1   Feature        21.00290, 52.24607           Point             5677460723
    properties.osm_type      properties.country      properties.osm_key
1                     N                  Poland                 leisure
    properties.housenumber      properties.city      properties.street
1                   44/50               Warsaw                 Długa
    properties.osm_value  properties.postcode         properties.name
1          dance 00-241        Szkoła Tańca          ABRA STUDIO
        properties.state
1   Masovian Voivodeship
```

```
# limiting the results to a specific value of the OSM attribute (here: uni-
versity)
resp<-GET("https://photon.komoot.de/api/?q='Długa+44/50+Warszawa'&osm_
tag=:university")
parsed<-fromJSON(content(resp,"text"), flatten = TRUE)
parsed$features
```

```
        type   geometry.coordinates   geometry.type    properties.osm_id
1 Feature              21.00367,          52.24659        Point    34497643
      properties.osm_type         properties.extent
1                       W       21.00319, 52.24687, 21.00403 52.24636
    properties.country   properties.osm_key   properties.housenumber
1           Poland               amenity                    44/50
    properties.city        properties.street    properties.osm_value
1        Warsaw                  Długa                university
  properties.postcode                       properties.name
1     00-241 Faculty of Economic Sciences, University of Warsaw
        properties.state
1   Masovian Voivodeship
```

One of the drawbacks of returned results is the fact that geographical coordinates are returned in one column. Therefore, the first example shows a way to obtain latitude and longitude values. Please also note that the address request (parameter q =) must not contain any spaces. They must be encoded with either the + sign or the %20 code. The next two examples show a way to limit results. The first automatically limits the returned locations to the first, while the next allows one to choose (if they are in the results) locations with a specific value of the data attribute in OpenStreetMap. A list of available attributes and their values can be found on the previously provided page: https://taginfo.openstreetmap.org/projects/nominatim#tags. These are also the values used by the Nominatim website. The API Photon API service also has other parameters to limit the search to a given area (rectangle) or near a given location. More on the project website indicated previously.

The following example shows a slightly more advanced implementation of the Photon API service. Part of the company database will be used again as the source data. It will be supplemented with a variable that contains the full address for searching and three variables for latitude, longitude and address returned by the Photon Service. The result object is used to check the results and contains the next check number, number of results received, status (OK if the address was identified, BAD if the service did not return the address) and which of the results was used (saved). This last information matters, because the type of the found object is checked in the code. Due to the fact that building addresses are most expected, if a building type object appears in the results obtained from the server (this is one of the attributes of the objects from the OSM database), this location will be used to indicate the coordinates; otherwise, the first result is taken. The proposed scheme results in better quality of collected data. Two more functions need clarification. The first is **URLencode**() from the default utils:: package. This command converts the web address to a format that uses the %20 character described previously to replace the space. In addition, it converts some other national and special characters (to hexadecimal codes) in such a way that network service servers do not have a problem reading the query. The second function is **Sys.sleep**(). This function suspends code execution for a specified number of seconds. In the example shown, it provides a minimum interval of 1 second between queries. This is due to the request of the server owners so as not to overload the service. Of course, for 11 queries, it doesn't matter, but with more geocoding addresses, this ensures that the Photon service is used fairly.

```
firms<-read.csv2("./Data2/firms_warszawa.csv", stringsAsFactors=FALSE,
encoding = "UTF-8")
firms$Address_full<-paste(firms$Address, firms$Postcode, firms$City,
firms$Country, sep=" ")
firms$lat<-NA
firms$lon<-NA
firms$geoadres<-NA
Result<-data.frame(id=numeric(length=nrow(firms)), result=numeric(length=
nrow(firms)), status=character(length=nrow(firms)), whichone=numeric(length=
nrow(firms)), stringsAsFactors=FALSE)

for(i in 1:nrow(firms))
{
 q_url<-paste("https://photon.komoot.de/api/?q='",firms$Address_
full[i],"'",sep="")
 q_url<-URLencode(q_url)
 resp<-GET(q_url)
 parsed<-jsonlite::fromJSON(content(resp,"text"), flatten =TRUE)
 parsed<-parsed$features
 print(paste(i, nrow(parsed)))
 Result$id[i]<-i
 if (!is.null(nrow(parsed))) {
   Result$result[i]<-nrow(parsed)
   if (length(which(parsed$properties.osm_key=="building"))!=0) {
     j<-min(which(parsed$properties.osm_key=="building"))
   } else {
     j<-1
   }
   firms$lat[i]<-as.numeric(parsed$geometry.coordinates[[j]][2])
   firms$lon[i]<-as.numeric(parsed$geometry.coordinates[[j]][1])
   firms$geoadres[i]<-as.character(paste(parsed$properties.street[[j]],
parsed$properties.housenumber[[j]], parsed$properties.postcode[[j]],
parsed$properties.city[[j]],sep=" "))
   Result$status[i]<-"OK"
   Result$whichone[i]<-j
   } else {
   Result$result[i]<-NA
   Result$status[i]<-"BAD"
   }
   Sys.sleep(1)
}
```

```
[1] "1 2"
[1] "2 1"
[1] "3 2"
[1] "4 11"
[1] "5 1"
[1] "6 4"
[1] "7 1"
[1] "8 1"
[1] "9 1"
[1] "10 3"
[1] "11 1"
```

Result

	id	result	status	whichone
1	1	2	OK	1
2	2	1	OK	1
3	3	2	OK	1
4	4	11	OK	2
5	5	1	OK	1
6	6	4	OK	4
7	7	1	OK	1
8	8	1	OK	1
9	9	1	OK	1
10	10	3	OK	3
11	11	1	OK	1

```
rm(resp,parsed,q_url,i,j)
```

Above, only the basic possibilities of geocoding using the Photon website and basic processing of results are presented. One can be encouraged to build a slightly more complex code that will transform the data obtained from the server into an even more convenient form or build a more specific query.

## Chapter 4

Spatial weights matrix, distance measurement, tessellation, spatial statistics

Katarzyna Kopczewska and
Maria Kubara

This chapter will present methods of spatial statistics. The most important element of spatial analysis is the spatial weights matrix presented in 4.2. In detail, for the general framework for creating a spatial weights matrix, 4.2.2 gives rules for the selection of the neighbourhood matrix, 4.2.3 describes the contiguity (common boundary) matrix, 4.2.4 discusses the matrix of *k*-nearest neighbours, 4.2.5 describes the matrix based on the distance criterion (neighbours within *d* km), 4.2.6 describes the inverse distance matrix, 4.2.7 summarises the weights matrix and its editions, 4.2.8 describes spatial lags and higher-order neighbourhoods and 4.2.9 discusses creating the matrix of weights based on the group membership.

The following sections present other important elements of statistical analyses: distance measurement and spatial data aggregation (4.3) and tessellation (4.4). Finally in 4.5, spatial, global (4.5.1) and local (4.5.2) statistics are discussed. The chapter ends with correlation analyses: spatial cross correlations (4.6) and correlogram (4.7).

The vast majority of this chapter is based on the spdep:: package, which is dedicated to the analysis of spatial dependencies. The spatstat:: and spatialEco:: packages were also used. Due to the limited space, the lctools:: package, which contains a rich library of spatial statistics measures for point data, is not discussed.

4.1 Introduction to spatial data analysis

Spatial analysis allows for a better understanding of the real characteristics of phenomena. Adding a geographical aspect to the study allows one to see previously undiscovered patterns. This enables the extraction of information on the spatial dependence of regions and interactions between the values of the variables studied in different locations.

The aim of spatial analysis is to discover information about the spatial dependence of regions and interactions between the values of the studied variables in different locations. Spatial analysis of data makes it possible to determine the similarity and differences between regions in general and in individual terms. Thanks to these methods, groups (clusters) of regions similar to each other can be distinguished, as well as finding regions significantly different from the neighbouring ones. Thanks to the estimation of models that take into account the spatial factor, it is possible to determine the spatial relationship between observations in different locations, as well as to prove that there is a non-measurable spatial factor differentiating the studied phenomenon between locations. Knowledge and understanding of space allow one to anticipate changes and better plan, for example, development policy.

Spatial methods are increasingly used in economic sciences, including in the analysis of spatial variability of economic processes, spatial patterns of trend and location, economic concentration, trade, international processes, valuation of real estate and land, location values, local public finance, the city economy (urban economics), environment, resources or market organisation (industrial organisation) and so on. Spatial methods are also useful in many other areas of science, including sociology (behaviour in space, ethnic patterns, crime, demographic trends), political science (voting patterns), urban planning and geography (patterns of interpersonal interactions, reduction of distance, land use), transport (transport systems, analysis of accidents and traffic), history (socioeconomic changes in time and space), health care and epidemiology (spreading and castration of diseases) and others (Anselin, 2001).

Space analysis can take place on many planes. These may include:

1 Basic location analysis – situation in individual locations
2 Spatial interactions like communication, information, migration, role of distance
3 Spatial point patterns (clustering of point observations, relationships between them, the importance of the direction)
4 Scale effects
5 Diffusion of economic processes, innovation and culture
6 Spatial autocorrelation – the quantitative relationship between locations

Spatial effects have different origins. They can be divided into two groups: spatial heterogeneity and spatial dependence. Spatial heterogeneity refers to structural relationships that change with the location of the object. These changes can be abrupt (e.g. countryside–town) or continuous. Spatial dependence, or spatial autocorrelation, refers to systematic spatial changes that are observed as clusters of similar values or a systematic spatial pattern. The term is borrowed from the analysis of time series and autocorrelation as relationships in time. Positive spatial autocorrelation means that the objects observed in the region and are more similar to objects in neighbouring regions than would result from the random distribution of these values. In the case of negative autocorrelation, the observed objects are more diverse.

Practical evidence shows that spatial heterogeneity and dependence are not fully distinguishable. As an example, Florax and Nijkamp (2003) cite high value clusters in cities and low values outside the city, which can be interpreted as both spatial dependence (clusters) but also as spatial heterogeneity (abrupt changes). The appearance of spatial heterogeneity does not mean that more information is available. In the case of spatial autocorrelation, there is an improvement in prediction, because the values of variables partly depend on the values of these variables in neighbouring areas. The purpose of using spatial quantitative methods is to detect and measure spatial autocorrelation.

Spatial statistics include a spatial observation component in the measurement. A lack of spatial autocorrelation means spatial randomness. The values observed in a given area do not depend on the values observed in neighbouring areas, and the observed spatial pattern is as likely as any other spatial pattern. Moreover, the location of variable values can be changed without disturbing the content of the set information. In the situation of spatial autocorrelation, one can talk about positive and negative autocorrelation (see Figure 4.1). Positive autocorrelation means that the values are focused in space and the neighbouring areas are similar. It is a process comparable to diffusion. However, it should be remembered that positive spatial autocorrelation does not directly mean diffusion, although diffusion usually gives a positive autocorrelation. In the case of negative spatial autocorrelation, the neighbouring areas are different, more than it would appear from the random distribution. This is called a "checkerboard pattern".

Several groups of tools are used in spatial analysis. These include: map visualisation, spatial statistics, neighbourhood modelling, spatial modelling, variogram, correlogram and so on. All these tools will be discussed in the following sections. Testing spatial dependencies with the help of these tools takes place on several planes and aims to answer the question: Does spatial dependence exist at all, and if so, what is its scope and structure? Testing the direct existence of spatial effects means studying the existence of spatial dependence, spatial heterogeneity and the existence of spatial regimes. Global and local Moran and Geary statistics for a given variable and the Moran test for the remainders of the ordinary least squares linear model are used for this type of test. The range of spatial effects, also understood as process diffusion, can be examined by analysing the number of lags in the spatial process, that is, variogram and correlogram analysis. Finally, the spatial relationship structure is examined by testing and selecting a spatial weights matrix defined according to various criteria, as well as choosing a spatial model, error or lag based on Lagrange multiplier tests and information criteria.

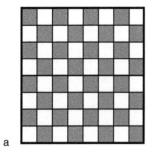 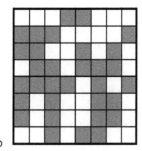 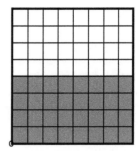

a   b   c

Figure 4.1 **Autocorrelation spatial pattern: a) negative autocorrelation, b) random distribution, c) positive autocorrelation**

Source: Own study based on Anselin, 1988

The spatial statistics tools described in this book are generally static. Methods that work directly on variables over different periods of time are developing. A simple and effective method is to set measures separately for individual time periods and compare them with each other. The current trend in such data indicates changes over time. This type of analysis method is possible in both statistics and spatial econometrics.

4.2 Spatial weights matrix

4.2.1 General framework for creating spatial weights matrices

Understanding neighbourhood structure is a key element of spatial modelling. It is the basis for later analysis of regional dependencies and modelling of neighbourly interactions. It is customary to present it in the form of a spatial weights matrix. This is a table in which information about the occurrence and strength of dependencies between objects i and j is stored.

The starting point for creating such an object is to build a neighbourhood matrix. It is defined as a matrix W with the dimensions $n \times n$, where n is the number of objects observed. Non-zero elements in a given row inform about the relationship of the object and region j. Diagonal usually remains empty – according to the principle that the object is not a neighbour to itself.

There are many criteria for defining a neighbourhood relationship. However, the most common way of modelling it is the approach recognising a common border of regions as a neighbourhood criterion (contiguity or adjacency criterion). The starting point is a binary matrix of zero–one elements, where the value is 1 if the regions have a common boundary and 0 if they do not. This matrix is symmetrical and square, which simplifies many calculation procedures. Elements of the matrix W are in forms of:

$$\begin{cases} w_{ij} = 1 \text{ when object } i \text{ is a neighbour of object } j, \text{ i.e. they have a common border} \\ w_{ij} = 0 \text{ when object } i \text{ is not a neighbour of object } j, \text{ i.e. they have no common border} \\ w_{ij} = 0 \text{ diagonal elements of the matrix} \end{cases} \quad (4.1)$$

Neighbourhood can be first-order, second-order or higher. A first-order neighbourhood means that only immediate neighbours of the examined object are considered (according to the contiguity criterion), while in the case of the second-order neighbourhood matrix, neighbours' neighbours are also included (also according to the contiguity criterion). This type of matrix is suitable when the data sample is homogeneous over time (pooled data) and when spatial diffusion over time is assumed.

Similarly, one can create a matrix W based on the distance criterion – a matrix of neighbours in a radius of d km. Then the elements w_{ij} of the matrix W are in the form of:

$$\begin{cases} w_{ij} = 1 \text{ when object } j \text{ is away from object } i \text{ by } d \text{ or less} \\ w_{ij} = 0 \text{ when object j is away from object } i \text{ more than } d \\ w_{ij} = 0 \text{ diagonal elements of the matrix} \end{cases} \quad (4.2)$$

The distance between regions is measured based on Euclidean measures applied to the coordinates of the centres of the areas. This matrix can also be modified by using different approaches in defining distance, for example, by taking squared distance, pover distance, exponential distance or inverse distance.

It is noteworthy that neighbourhood matrices only provide information about the existence of a neighbourhood relationship. However, they do not take into account the weights – the strength of spatial interaction. It is easy to imagine that the selected object A is affected not only by the fact of having neighbour B but also by its distance, size or surroundings by other neighbouring regions. Therefore, it is necessary to standardise the relationship between objects. Weighing the neighbourly relationship is most often done in two ways:

- weighing with distance;
- standardising with lines to one, so that for each line i $\sum_j w_{ij} = 1$ (after this transformation, each w_{ij} element is $1/n$ when the region has n neighbours).

Distance weighing is characteristic of matrices based on distance between regions. The most common is the inverse matrix with the following weight form:

$$w_{ij} = \left(1 / d_{ij}\right)^k \tag{4.3}$$

where d_{ij} is the Euclidean distance between region centroids (Duncan, White, & Mengersen, 2017). The addition of an exponent k different from one allows for non-linear weighting of the distance – the higher its value, the greater the relative influence of closer objects (Earnest et al., 2007; Getis i Altstadt, 2008).

Other examples of weight-creating functions based on distance d are, for example, the exponential lag function:

$$w_{ij} = \exp\left(-\lambda d_{ij}\right) \tag{4.4}$$

where $\lambda > 0$ is the decay parameter and the Gaussian distribution function:

$$w_{ij} = \exp\left(-\frac{d_{ij}^2}{2b^2}\right) \tag{4.5}$$

where $b > 0$ is the bandwidth.

The conversion of binary elements into scales facilitates subsequent calculations and the modelling process. This procedure also leads to the spatial lag operator. For example, the spatially lagged value of the variable y for location 1 (y_1) represented as Wy_1 (for a matrix standardised by weights) is a weighted average of adjacent observations. More information about lags can be found in Chapter 5.

There are also other criteria than distance or neighbourhood in creating spatial weights matrices. Going beyond geographical definitions, one can define environmental distance matrices (Dormann et al., 2007), social distance (Doreian, 1980) or economic distance (Case, Rosen, & Hines, 1993; Conley & Tsiang, 1994; Conley, 1999). Such matrices are based on climate characteristics, trade relations, capital flows and migrations between areas. The approach behind them indicates that the geographic distance between regions alone is less important than their actual characteristics. This is due to the observation that regions with similar properties face similar risks (Duncan et al., 2017). Such matrices are based on the observation of the value of interdependent variables (e.g. consideration of the climatic properties of a given area typical for life sciences for the study of habitat behaviour of a rare species or taking into account the strength of the impact of trade flows on economic connections between neighbours). For example, for the environmental distance, neighbours are defined as regions with similar properties. The relationship is expressed using the parameter δ, calculated as the absolute difference of the dependent variable in the regions studied ($\delta_{ij} = |x_i - x_j|$). However, non-geographical matrices are not popular in the literature – they are difficult to implement because the weights must be exogenous to the model.

Although the approach based strictly on non-geographical distances is not often used, it can be combined with the traditional distance. This creates hybrid spatial weights matrices (Duncan et al., 2017). Kuhnert (2003) and Earnest et al. (2007) proposed the following form of hybrid scales $w_{ij} = \dfrac{1}{d_{ij}\delta_{ij}}$, where d_{ij} and δ_{ij} represent geographical and environmental distances, respectively. This approach allows one to set aside the assumption of similarity arising only from distance and better reflects the actual connections between remote but related or similar regions (Duncan et al., 2017).

In the literature, one can also meet the matrices proposed by Cliff and Ord (1981) and Dacey (1968). The former is constructed by using a combination of distance measure and the relative length of the border between objects. This matrix is asymmetrical. W_{ij} elements are the form $w_{ij} = (d_{ij})^{-a}(\beta_{ij})^{b}$, where $d_{ij} x$ is the distance between the centres i and j, β_{ij} is the ratio of the border from j to the entire border, while a and b are parameters. Dacey (1968) takes into account the relative surface of objects. The w_{ij} matrix has the form $w_{ij} = d_{ij} \alpha_i \beta_{ij}$, where d_{ij} is the binary neighbourhood factor, α_i the share of the area of the object and in the total examined area and β_{ij} is a measure of the boundary (as shown previously). In a situation where there are islands in the analysed sample, binary matrices do not fulfil their role, as individual rows can only consist of zero elements. The geographical distance criterion (the so-called great circle between the centres of the regions) is then used. A broad discussion on the spatial weights matrix can be found in Cliff and Ord (1981), Upton and Fingleton (1985), Anselin (1988) and Anselin and Bera (1998). Economic scales were introduced by Case et al. (1993).

Depending on the methods of obtaining them, the weighting matrices may be symmetrical or asymmetrical. The asymmetric group includes the k-nearest neighbours matrix. The implications of asymmetry are disturbances in the results of the Moran's I and Geary's C statistics. However, symmetrical matrices, although convenient in calculations, may be inappropriate for large samples, in particular for hierarchical diffusion or core–periphery relations).

Technically, many types of matrices can be defined in R. All statistical and econometric procedures require two classes of objects for calculations: *nb* or *listw*. In addition to the basic neighbourhood matrix according to the contiguity criterion, one can create spatial weights matrices based on the distance between areas (in particular the inverse of distance), k nearest neighbours and neighbours within a given radius of d km. For these three types of matrices, computational operations are performed on the coordinates of the centres of the areas (so-called centroids). In addition to calculating the matrix, they can also be analysed graphically. For this purpose, an empty contour map is drawn first, and then the spatial connections are applied as an object of the *nb* class. The general rule is as follows: *nb* class objects are used for drawing, while *listw* class objects are used for calculations. Subsequently one will find more detailed information about matrices with examples.

4.2.2 Selection of a neighbourhood matrix

The multiplicity of spatial weights matrices allows flexible matching of an appropriate matrix to data, but can also be a source of problems. The choice of the spatial weights matrix determines the results of the analysis. Usually, the W matrix is exogenously determined, which can cause specification problems (Florax & Rey, 1995). Most commonly used are neighbourhood matrices defined according to the cotiguity criterion. Equally popular are matrices using the nearest neighbours (5 or 10), neighbours in a given radius and inverse distance. Each time, however, the researcher must ensure that the selected matrix is appropriate for the problem being analysed.

There is still discussion in the literature about whether the matrix of spatial weights W should be assumed a priori or estimated on the basis of data. Proponents of the W-based matrix explain their approach as testing hypotheses about the extent of spatial interactions between individuals. This also justifies model theoretical interactions – for example, only local. Supporters of the calculation-based W matrix show a bias of spatial coefficients in the situation of a wrongly specified spatial weights matrix W (e.g. Corrado & Fingleton, 2012). The approach of the estimated spatial weights matrix W is closer to filtering the spatial relation. The approach of using the a priori matrix W is closer to testing hypotheses about spatial relationships.

Having prior knowledge about the form of weights is a good basis for choosing the optimal matrix (Benjanuvatra & Burridge, 2015). Other researchers suggest observing panel data and tracking their relationships to find better weight matching (Bhattacharjee & Jensen-Butler, 2013; Beenstock & Felsenstein, 2012). The appropriate matrix can also be selected using optimisation – by selecting the best matrix from a predetermined set of potential candidates. This is done by maximising the Moran's I statistic (Kooijman, 1976), comparing spatial dependency statistics or information criteria, that is, Akaike information criterion (AIC), BIC (Bayesian), as in Duncan et al. (2017). However, these approaches are strongly arbitrary – even the optimisation method requires the initial narrowing of the group of evaluated matrices.

Examples of matrix selection using analytical tools are also known in the literature. The W matrix can be estimated for panel data with a wide time range and a small number of dependent variables (lower T and increased n cause calculation problems and reduce estimation efficiency). The system of equations with matrix weights is determined assuming matrix symmetry and homoscedasticity. Estimation is done using the generalised method of moments (GMM).

One can also use bootstrap in the matrix estimation process. The procedure is then two stage: first, the best type of matrix is found, and then the best bootstrapped matrix is selected (for many types of matrices, the bootstrapping procedure is performed and several best matrices are obtained according to various criteria) (Herrera, Mur, & Ruiz, 2019). More about the bootstrap method in Chapter 9.

The problem of choosing the best possible spatial weights matrix often appears in the literature. However, it is difficult to find simple tips that could clearly indicate the optimal course of action in a choice situation. The majority of authors are still convinced that the selection of the matrix is crucial in the estimation process – incorrect specification of the W matrix may lead to overestimation of spatial dependence parameters or even indicate the opposite direction of interactions than that occurring in reality; Importantly, a particularly high bias arises when studying extreme direct and indirect effects (Florax & Rey, 1995; Franzese Jr & Hays, 2007; Lee & Yu, 2012; Debarsy & Ertur, 2016). However, there are more and more doubts in this regard. LeSage and Pace (2014) indicate that only drastic changes in the spatial relationship matrix can lead to significant differences in the estimated values or significantly disrupt the model (the so-called biggest myth).

From a practical point of view, it is worth specifying the model using several selected weight matrices, for example, the most popular and most commonly used in the literature in similar studies. For the final model, it is best to use the matrix W, which shows the best values of the information criterion. In this situation, it can be assumed that the given matrix most accurately maps the real neighbour relations. In case of specification problems, it is also worth remembering that the spatial weights matrices differ in subtleties, and the greatest value is already introduced by taking into account the spatial relationship in the model. To quote Herrera, Riuz, and Mur (2013): "*Any useful result is welcomed in this field but, overall, we need practical, clear guides to solve the problem*".

4.2.3 Neighbourhood matrices according to the contiguity criterion

Neighbourhood is most often modelled on the basis of a common border – contiguity. In this case, the spatial weights matrix is created starting with the binary matrix representing the neighbourhood. Neighbourhood can be first order, second order or higher. A first-order neighbourhood means that only neighbours of the examined object are considered (according to the contiguity criterion), while in the case of the second-order neighbourhood matrix, neighbours' neighbours are also included (also according to the contiguity criterion). Then standardisation and/or weighting of values is carried out. The general form of such a matrix of weights, for neighbours of the nth order, is as follows:

$$w_{ij} = \begin{cases} \omega_k \text{ if } i \text{ and } j \text{ are } k\text{'s neighbours}, k = 1, \ldots, n \\ 0 \text{ otherwise} \end{cases} \tag{4.6}$$

In this approach, $\omega = (\omega_1, \ldots, \omega_n)$ is a weight vector. Higher weights are usually assigned to closer geographical regions. The weights can also be defined by means of functions, for example, $\omega_k = \exp\left(\dfrac{k-1}{n-1}\right)$ (Duncan et al., 2017). Unfortunately, the border approach does not take into account the variation in the size of neighbouring regions. Fahrmeir and Kneib (2011) indicate that the Euclidean distance indicates isotropy – that is, the same pover of influence on objects in different directions (Duncan et al., 2017).

This matrix is symmetrical, which facilitates testing procedures. Furthermore, this type of matrix is suitable when the data sample is homogeneous over time (pooled data) and when the presence of spatial diffusion over time is assumed.

Technically, the matrix of spatial weights according to the contiguity criterion is determined by two commands: first, with the command **poly2nb()** from the spdep:: package, a neighbourhood structure is created in the *nb* class on the basis of available polygons (polygons) specified by a map in the *SpatialPolygons* class, and then the **nb2listw()** command converts the neighbourhood matrix to a spatial weights matrix, the result being in the *listw* class. In the options of this command, one can declare the type of matrix in the *style* option (e.g. *style*="W"). There are several types of matrices to choose from:

- B – basic binary coding (zero-one)
- W – first-order matrix, row standardised
- C – globally standardised matrix
- U – matrix C divided by the number of neighbours (sums everything up to 1)
- S – variance-stabilising matrix
- minmax – matrix standardised by the smallest of the maximum sums of weights in the rows and columns of the original matrix (similar to the C and U matrix) (after: Kelejian and Prucha, 2010)

Most often, analysis uses the W matrix. This is a first-order matrix, which means that the neighbours are only the areas bordering the studied region. One can also create a higher-order matrix. Then the neighbourhood is determined by successive layers of neighbours. The S matrix is used in the percentage analysis when the populations in the regions studied are different.

Subsequently, a spatial weights matrix will be created based on the contiguity criterion for the poviat map (NTS4) containing 380 territorial units. This most popular variant of the W matrix in the *listw* class, suitable for further calculations, and its equivalent suitable for visualisation will be prepared.

```
# loading packages
library(rgdal)
library(spdep)
# poviat map
pov<-readOGR(".", "powiaty") # 380 units
pov<-spTransform(pov, CRS("+proj=longlat +datum=NAD83"))

# matrix of weights according to the contiguity criterion
cont.sp<-as(pov, "SpatialPolygons") # conversion of polygons to class sp
cont.nb<-poly2nb(cont.sp, queen=T) # conversion of sp to class nb
```

When creating an *nb* class object, note the *queen=T* option. This refers to the method of defining neighbourly connections, constructed in analogy to the movements of chess figures. For the default option *queen=T* objects will be considered neighbours if they have at least one border point (just like the figure of the queen can move on oblique fields connected at only one point). When changing the option to *queen=F*, one can create a neighbourly relationship between areas where at least two border points will be required (in analogy to the movements of the tower <<rook>>, having access to the fields vertically and horizontally). This option is particularly important when there are four regions touching at one point. With *queen=T*, all will be considered adjacent; otherwise, oblique objects will not be considered immediate neighbours.

```
# creation of spatial weights matrix row-standardised to 1 (option W)
cont.listw<-nb2listw(cont.nb, style="W")
cont.listw # displays a summary of the weights matrix
```

```
Characteristics of weights list object:
Neighbour list object:
Number of regions: 380
Number of nonzero links: 2006
Percentage nonzero weights: 1.389197
Average number of links: 5.278947

Weights style: W
Weights constants summary:
    n    nn  S0      S1       S2
W 380 144400 380 169.7589 1632.39
```

```
# poviats' coordinates of measures
crds.pov<-coordinates(pov)
colnames(crds.pov)<-c("cx", "cy")
plot(pov) # contour map Figure 4.2a
plot(cont.nb, crds.pov, add=TRUE) # neighbourly layer
```

In the spatial weights matrix according to the contiguity criterion, the percentage of non-zero relationships is not high – it is 2006 cells from a total of $380^2 = 144,400$ fields, or 1.39%. That is why it is a rare matrix. One poviat has on average about five neighbours. The neighbourhood or spatial weights matrix can be converted to the *matrix* class with the command **nb2mat()**, which allows displaying the W matrix in typical matrix notation.

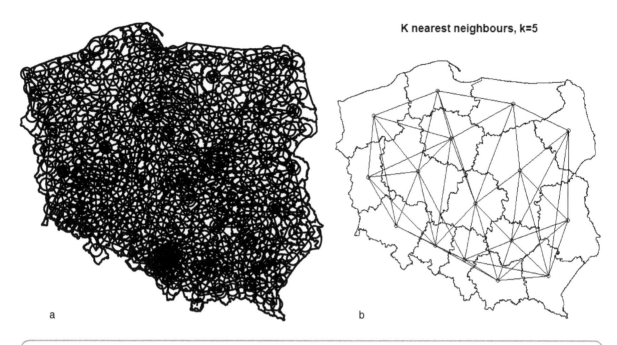

K nearest neighbours, k=5

Figure 4.2 Illustration of the neighbourhood matrix: a) according to the contiguity criterion, b) five nearest neighbours criterion

Source: Own study using R and the spdep:: package

```
# conversion to matrix class
cont.mat<-nb2mat(cont.nb)
cont.mat[1:5, 1:5]
```

```
    [,1]       [,2]       [,3] [,4]       [,5]
0  0.0 0.0000000 0.0000000  0.0 0.1666667
1  0.0 0.0000000 0.1428571  0.0 0.1428571
2  0.0 0.1428571 0.0000000  0.0 0.1428571
3  0.0 0.0000000 0.0000000  0.0 0.2000000
4  0.1 0.1000000 0.1000000  0.1 0.0000000
```

The map operations for voivodeships – NTS2 – will be carried out subsequently. The voivodeship shapefile contains 16 territorial units whose names have been saved as *names_voi*. Conversion to the *nb* class can be done as before using the **poly2nb**() command, with row-region names given in the *row.names* option. In summaries, for example, by displaying an matrix in the *matrix* class, one can get the names of regions.

```
voi<-readOGR(".", "wojewodztwa") # 16 units
voi<-spTransform(voi, CRS("+proj=longlat +datum=NAD83"))
names_voi<-as.character(voi$jpt_nazwa_) # region names vector
names_voi
```

```
 [1] "opolskie"            "świętokrzyskie"     "kujawsko-pomorskie"
 [4] "mazowieckie"         "pomorskie"          "śląskie"
 [7] "warmińsko-mazurskie" "zachodniopomorskie" "dolnośląskie"
[10] "wielkopolskie"       "łódzkie"            "podlaskie"
[13] "małopolskie"         "lubuskie"           "podkarpackie"
[16] "lubelskie"
```

```
cont.voi.nb<-poly2nb(as(voi, "SpatialPolygons"), row.names=names_voi)
cont.voi.mat<-nb2mat(cont.voi.nb) # conversion to matrix class
cont.voi.mat[1:5, 1:5]
```

```
                     [,1]       [,2]      [,3]       [,4] [,5]
opolskie                0 0.0000000 0.0000000 0.0000000  0.0
świętokrzyskie          0 0.0000000 0.0000000 0.1666667  0.0
kujawsko-pomorskie      0 0.0000000 0.0000000 0.2000000  0.2
mazowieckie             0 0.1666667 0.1666667 0.0000000  0.0
pomorskie               0 0.0000000 0.2500000 0.0000000  0.0
```

This arrangement of the object enables the displaying of the names of regions neighbouring the examined area. To do this, use the **which**() command to specify which element of the name vector is the region being investigated. This value will become an argument in conditioning the weights list in the spatial weights matrix in the *nb* class. This in turn will return the numbers of the regions that are neighbours. These numbers have been substituted into the region names vector to display the names of neighbours.

```
names_voi[cont.voi.nb[[which(names_voi=="lubuskie")]]]
#[1] "zachodniopomorskie" "dolnośląskie"       "wielkopolskie"
```

The previous print shows that the neighbours of the Lubuskie voivodeship are the Zachodniopomorskie, Dolnośląskie and Wielkopolskie voivodeships, which is true.

4.2.4 Matrix of *k* nearest neighbours (knn)

The matrix *k* nearest neighbours (knn neighbours, knn) is usually constructed for point data, because unlike the contiguity matrix, it only examines point data (without referring to areas). One can also create a knn matrix for area data by first determining the area centroids (centres of gravity of regions – spatial geometries) and operating on these points. In the case of point data, the knn matrix is a natural analytical solution, although determining the number of neighbours is most often based on modelling or random experience. In the case of area data, the structure of the neighbourhood (for example, indication *k* = 5 nearest neighbours) strongly depends on the shape and surface of the regions for which centroids have been designated. A narrow and long region may have a centroid significantly distant from another centroid and therefore may not be one of the closest neighbours, although it will be a neighbour in light of the contiguity criterion.

Subsequently, the matrix *k* of nearest neighbours has been determined for regional area data (voivodeships, NTS2, 16 territorial units). Centroids were calculated using the **coordinates**() command from the sp:: package.

```
crds.voi<-coordinates(voi) # centroids of NTS 2 voivodeships
head(crds.voi)
```

```
       [,1]     [,2]
0 17.89988 50.64711
1 20.76909 50.76339
2 18.48822 53.07270
3 21.09645 52.34576
4 17.98619 54.15424
5 18.99410 50.33108
```

It should be remembered that the nearest neighbour is the area whose centre lies closest to the centre of the given area according to Euclidean measures, which depends on the shape of the examined and neighbouring regions. The knn matrix usually uses 5 to 10 nearest neighbours. Calculations are performed using the **knearneigh**() command from the spdep:: package, which creates a *knn* object.

Spatial weights matrices obtained from *knn* class of objects are asymmetrical, which may disturb the results of spatial tests (e.g. Moran or Geary) or even make calculations impossible. The symmetry of the matrix is checked with the command **is.symetric.nb()** from the spdep:: package. As the name of the command indicates, it can be applied to objects of the *nb* class, which requires conversion from the *knn* class to this format, using the **knn2nb()** command from the spdep:: package. The asymmetric matrix can be "symmetrical" by using the **make.sym.nb()** command from the spdep:: package.

```
# matrix k=5 nearest neighbours for area data
voi.knn<-knearneigh(crds.voi, k=5) # a knn object is created
voi.knn.nb<- knn2nb(voi.knn)

#Figure 4.2b - connections of k nearest neighbours
plot(voi, main="K nearest neighbours, k=5")      # regional outline
plot(voi.knn.nb, crds.voi, add=TRUE) # relationship layer

# checking the symmetry of the matrix
print(is.symmetric.nb(voi.knn.nb))
#[1] FALSE
voi.knn.sym.nb<-make.sym.nb(voi.knn.nb)
print(is.symmetric.nb(voi.knn.sym.nb))
#[1] TRUE

# creating a listw class object
voi.knn.sym.listw<-nb2listw(voi.knn.sym.nb)
```

In Figure 4.2b, each region is connected to the five closest regions. When neighbouring regions are the same distance away, the region is connected to more regions. The visualisation of the matrix after the symmetry is the same; only the matrix W changes, in which the relations are automatically agreed and when A is the neighbour B, then B becomes the neighbour A.

In the following code, the knn matrix for point data has also been determined. A business dataset that contains geolocations of observations was loaded and *n* = 100 rows drawn as a subset using the **sample_n()** command from the dplyr:: package. For this, it was necessary to convert data from the *data.frame* class to the *tibble* (tbl) class with the **as_tibble()** command from the dplyr:: package – the target subset of the *company.sub* is in the *tbl* and *data.frame* classes. To determine the W matrix, no shapefile is necessary – it can be used as a bounding box (bbox) for points for better visualisation. The summary of the spatial weights matrix in the *listw* class with the **summary()** command shows that the number of neighbours after adjustments is higher than 5 (it is 6.16), which shows the mechanism of correction of connections up.

```
library(dplyr)
firms<-read.csv("geoloc_data_firms.csv", header=TRUE, dec=",", sep=";")
firms.sub<-sample_n(as_tibble(firms), size=100, replace=FALSE)
names(firms.sub)
```

```
 [1] "ID"          "ADDRESS"   "STREET"     "STREET.NO"   "ZIP"
 [6] "CITY_POST" "CITY"      "region2"    "poviat"      "region3"
[11] "subreg"      "coords.x1" "coords.x2" "LEGAL_FORM1" "LEGAL_FORM2"
[16] "OWNERSHIP" "PKD7"      "SEC_PKD7" "GR_EMPL"     "empl"
```

```
class(firms.sub)
#[1] "tbl_df"     "tbl"        "data.frame"

# matrix k = 5 nearest neighbours for point data
firms.knn<-knearneigh(as.matrix(firms.sub[,12:13]), k=5) # knn class object
firms.knn.nb<- knn2nb(firms.knn)
```

```
# figure of k nearest neighbours
plot(voi[voi$jpt_nazwa_=="lubelskie",]) # fragment of a map
plot(firms.knn.nb, as.matrix(firms.sub[,12:13]), add=TRUE) # relationship layer

print(is.symmetric.nb(firms.knn.nb))
#[1] FALSE
firms.knn.sym.nb<-make.sym.nb(firms.knn.nb)
print(is.symmetric.nb(firms.knn.sym.nb))
#[1] TRUE

# creating a listw class object
firms.knn.sym.listw<-nb2listw(firms.knn.sym.nb)
summary(firms.knn.sym.listw)
```

```
Characteristics of weights list object:
Neighbour list object:
Number of regions: 100
Number of nonzero links: 616
Percentage nonzero weights: 6.16
Average number of links: 6.16
Link number distribution:

 5   6   7   8  9 11
39 24 23 12  1  1
39 least connected regions:
2 7 8 11 12 17 20 22 23 24 26 28 32 33 35 41 44 45 49 52 54 60 61 63 64 68 71
72 75 76 77 80 81 84 89 90 97 98 99 with 5 links
1 most connected region:
15 with 11 links

Weights style: W
Weights constants summary:
    n    nn   S0        S1       S2
W 100 10000 100 33.10909 403.301
```

4.2.5 Matrix based on distance criterion (neighbours in a radius of d km)

Using the coordinates of the centres of the areas, one can also create a matrix of spatial weights according to the criterion of neighbourhood in a radius of d km. This means that the neighbour will be an object whose centre is not more than d km away in a straight line. A special case of such a matrix is the inclusion of all areas as neighbours.

To construct a matrix of neighbours within a radius of d km, use the **dnearneigh**() command from the spdep:: package, which returns an object of class *nb*. The arguments for the command are: matrix of coordinates of areas centres, minimum distance, maximum distance, row names, type of coordinates in the *longlat* option. Setting the *longlat=TRUE* option determines the distance measurement in kilometres (according to the Great Circle method) and *longlat = FALSE* or omitting results in expressing the distance in degrees.

In the following example, poviat neighbours (NTS4) are sought within a 0–30 km radius. Previously created centroids (*crds.pov*) were used for the map of the area As can be seen from Figure 4.3, not all poviats have a neighbour – this applies mainly to large territorial units for which the nearest neighbouring centroid is more than 30 km away.

```
# Figure 4.3 - a matrix of neighbours within a radius of d km
conti.d.30<-dnearneigh(crds.pov, 0, 30, longlat=TRUE)
plot(pov) # background map
plot(conti.d.30, crds.pov, add=TRUE) # neighbourly relations
```

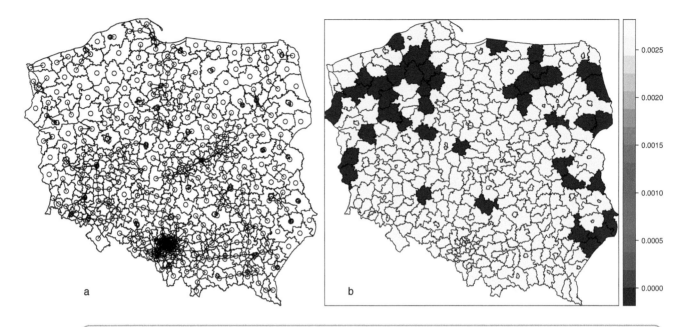

Figure 4.3 Spatial relationships in a spatial weights matrix W according to the distance criterion (in a radius *d* = 30 km): a) links between spatial units, b) in dark – regions without a link

Source: Own study using the R program and spdep:: and sp:: packages

In matrices with no connections (too small a distance for neighbours within a radius of *d* km or an island in the contiguity matrix), it is worth setting the option *zero.policy=TRUE*. This allows the creation of zero-length vectors. This is especially important when converting a neighbourhood matrix in which there are elements not connected to others. If *zero.policy=FALSE* is used, the program will stop and display an error when there is no connection between the area and others.

Conversion to the *matrix* class is useful when the goal is colour visualisation on the regional map of the number of connections – in particular the lack of connections. In the following example, the W matrix was converted to the *matrix* class with the **nb2mat()** command and the average weights for a given region were calculated in rows. The **colMeans()** command calculates the means in columns, while the t() command transposes the matrix so that the means in the rows are effectively calculated. A vector of these values has been added to the shapefile database, so one can use the **spplot()** command to visualise this mean vector. Areas marked in black in Figure 4.3b have no neighbours.

```
# convert nb to matrix
conti.d.30.m<-nb2mat(conti.d.30, zero.policy=TRUE)
a<-colMeans(t(conti.d.30.m)) # averages in rows
pov$a<-a # appending a value vector to shapefile
spplot(pov, "a") # Figure 4.3b
```

Arrays of this type are particularly vulnerable to the lack of connection problem. If the selected distance *d* km is too small and the regions have no neighbour, then such a neighbourhood matrix cannot be converted to a *listw* class object and thus used in further calculations.

```
conti.d.30.listw<-nb2listw(conti.d.30)
#Error in command 'nb2listw(conti.d.30) #Empty neighbour sets found
```

The following is the code to create a spatial weights matrix in which each region has at least one neighbour. The logic of the code is as follows. The *kkk* object (after conversion is in the *nb* class) contains connections to *k=1* (omitted in the code) of neighbours (i.e. to one nearest neighbour). The **nbdists()** command allows one to determine the distance between neighbours in an *nb* class object – the result has been converted from a list to a vector by the **unlist()** command, so it can be an argument of the **max()** command. The object *all* contains the maximum distance between one-neighbour relations, that is, it is the de facto minimum distance for the **dnearneigh()** command so that all regions have at least one neighbour. Next, a relationship was created in the *everyone.nb* object so that each region has a neighbour. Summary by the **summary()** command confirms that the least connected regions have one neighbour (*2 least connected regions: 53 & 216 with 1 link*), while the most strongly connected region has as many as 28 connections (*1 most connected region: 322 with 28 links*). The average number of links is 7.3, and the matrix is rare because it contains only 1.92% of non-empty links. The summary also gives the distribution of the number of neighbours (*Link number distribution*). These connections can be visualised on the map.

```
# matrix of neighbours within a radius of d km, so that each region has a neighbour
kkk<-knn2nb(knearneigh(crds.pov))
all<-max(unlist(nbdists(kkk, crds.pov)))
all.nb<-dnearneigh(crds.pov, 0, all)
summary(all.nb, crds.pov)
```

```
Neighbour list object:
Number of regions: 380
Number of nonzero links: 2772
Percentage nonzero weights: 1.919668
Average number of links: 7.294737
Link number distribution:

 1  2  3  4  5  6  7  8  9 10 11 12 13 15 16 18 19 20 21 22 23 24 25 26 27 28
 2 27 22 53 51 48 48 49 26 13  7  3  2  1  1  1  2  2  2  2  3  5  3  3  3  1
2 least connected regions:
53 216 with 1 link
1 most connected region:
322 with 28 links
Summary of link distances:
   Min. 1st Qu.  Median    Mean 3rd Qu.    Max.
0.01929 0.28109 0.36955 0.35014 0.43826 0.49580
```

```
plot(pov, border="grey") # neighbours by distance
plot(all.nb, crds.pov, add=TRUE)
```

4.2.6 Inverse distance matrix

Another type of spatial weights matrix very often used in practice is the matrix based on the distance criterion, and in particular the inverse of the distance. The weight is defined as $w_{ij} = 1/d_{ij}$ for all pairs of neighbouring regions. The so-called square weights are $w_{ij} = 1/d^2_{ij}()$. Technically, in R, first one needs to create a **knnneigh()** *knn* class object containing all other regions as neighbours with the **knearneigh()** command and convert it to the *nb* class using the **knn2nb()** command. The next step is to create a distance matrix between all neighbours with the **nbdists()** command. The **lapply()** command converts it to a distance inverse matrix. In the final step, the **nb2listw()** command converts the *nb* class object to the *listw* class with the *dist* option, which specifies the inverse of the distance from the **lapply()** command.

In the example, a matrix of inverse distances between poviats was determined, assuming that all regions are neighbours (hence $k = 379$ for 380 regions). Summary by the **summary()** command confirms that the regions have an average of 379 neighbours, and the matrix is significantly filled with connections (99.7%).

```
# inverse matrix
poviats.knn<-knearneigh(crds.pov, k=379)
poviats.nb<-knn2nb(poviats.knn)
dist<-nbdists(poviats.nb, crds.pov)
dist1<-lapply(dist, function(x) 1/x) # list class object
poviats.dist.listw<-nb2listw(poviats.nb, glist=dist1)
summary(poviats.dist.listw)
```

```
Characteristics of weights list object:
Neighbour list object:
Number of regions: 380
Number of nonzero links: 144020
Percentage nonzero weights: 99.73684
Average number of links: 379
Non-symmetric neighbours list
```

This method of distance weighing allows the mapping of the decreasing influence of further neighbours. Other than simple inverse, distance-dependent functions are also possible. Assuming that the influence decreases non-linearly, it is easy to transform the distance matrix according to a given mathematical function using the **lapply()** command. In this way, power matrices can arise ($w_{ij} = d_{ij}^{-\alpha}$, for α usually equal to 1 or 2) or exponential distance ($w_{ij} = \exp(-\alpha d_{ij})$, where α is any positive exponent). Examples of these transformations are presented subsequently.

```
# power matrix
dist2<-lapply(dist, function(x) x**(-2))
poviats.dist.listw2<-nb2listw(poviats.nb, glist=dist2)

# exponential matrix
dist3<-lapply(dist, function(x) exp(-1.5*x))
poviats.dist.listw3<-nb2listw(poviats.nb, glist=dist3)
```

4.2.7 Summarising and editing spatial weights matrix

There are several commands that can summarise the calculated spatial weights matrices. The **summary()** command applied to *listw* and *nb* class objects produces similar results. Similar, though poorer, results can be obtained using the **print()** command. The number of regions, the number and percentage of non-zero associations, the average number of associations, the distribution of the number of associations and the most and least associated region are displayed.

```
summary(conti.d.30) # summary of the nb class neighbourhood matrix
```

```
Neighbour list object:
Number of regions: 380
Number of nonzero links: 1168
Percentage nonzero weights: 0.8088643
Average number of links: 3.073684
41 regions with no links:
18 24 51 53 57 60 67 69 109 120 125 126 141 153 165 187 188 194 201 213 216 230
231 242 243 247 255 257 258 262 264 266 267 273 289 294 297 338 364 375 377
Link number distribution:

 0  1  2  3  4  5 6 7 9 10 11 13 14 15 16 17 18 19
41 80 82 78 47 25 1 3 1  2  3  3  1  3  3  4  2
80 least connected regions:
17 22 26 27 30 31 32 33 36 38 43 50 52 54 59 62 63 64 72 90 95 107 111 128 129
130 131 132 134 144 150 154 163 172 173 175 189 190 198 199 200 204 205 210 212
215 217 220 221 222 224 227 228 234 235 237 244 253 256 261 263 268 271 274 284
285 287 288 290 292 293 295 296 298 334 339 340 352 368 369 with 1 link
2 most connected regions:
276 283 with 19 links
```

To find out the weight statistics of a given matrix, use the **summary()** command on the *listw* class object, additionally extracting list elements to the vector with the **unlist()** command. It is important that the transformed matrix not contain empty elements (such as those in the *conti.d.30* matrix).

```
all.listw<-nb2listw(all.nb)
summary(unlist(all.listw$weights))
```

```
   Min. 1st Qu.  Median    Mean 3rd Qu.    Max.
0.03571 0.08333 0.12500 0.13709 0.16667 1.00000
```

There is a possibility of statistical analysis of connections for each region. Simply apply the **summary()** command to a "classic" matrix, that is, to an object of the *matrix* class. Both *nb* and *listw* class objects can be converted to the *matrix* class using **nb2mat()** and **listw2mat()**, respectively. The result is statistics for each row (i.e. object) – the example displays the first four of 380. The weights displayed (in max statistics) reveal how many neighbours a given region has – in the case of V3 max (w_{ij}) = 0.25 = 1/4, which means that region 3 (V3) has four neighbours. Similarly, region 4 (V4) has six neighbours (1/6 = 0.166), region V2 has three neighbours (1/3 = 0.333) and region V1 has five neighbours (1/5 = 0.2).

```
all.mat<-nb2mat(all.nb)
summary(all.mat)
```

```
      V1                 V2                 V3                 V4
Min.    :0.000000   Min.    :0.00000   Min.    :0.000000   Min.    :0.000000
1st Qu.:0.000000    1st Qu.:0.00000    1st Qu.:0.000000    1st Qu.:0.000000
Median :0.000000    Median :0.00000    Median :0.000000    Median :0.000000
Mean    :0.002701   Mean    :0.00239   Mean    :0.003775   Mean    :0.003281
3rd Qu.:0.000000    3rd Qu.:0.00000    3rd Qu.:0.000000    3rd Qu.:0.000000
Max.    :0.200000   Max.    :0.33333   Max.    :0.250000   Max.    :0.166667
```

To display only the distribution of the number of neighbours of all areas, use the **card()** command in conjunction with the **table()** command. The **card()** command requires an *nb* class object as input. As one can see in the following example, most regions have between 9 and 15 neighbours. One of the regions has as many as 20 neighbours, but there are also regions with two or three neighbours.

```
table(card(all.nb))
```

```
 1  2  3  4  5  6  7  8  9 10 11 12 13 15 16 18 19 20 21 22 23 24 25 26 27 28
 2 27 22 53 51 48 48 49 26 13  7  3  2  1  1  1  2  2  2  2  3  5  3  3  3  1
```

To view scales in the form of a matrix, convert *nb* or *listw* class objects to *matrix* class objects, using the **nb2mat()** and **listw2mat()** commands, respectively. In such a matrix, area connections and weights assigned to them can be examined. The full spatial weights matrix *knn = 5* for 16 voivodeships is shown subsequently. One can check that, for example, area 4 connects to areas 2, 7, 11, 12 and 16.

```
voi.knn.mat<-nb2mat(voi.knn.nb)
print(voi.knn.mat)
```

```
      [,1]  [,2]  [,3]  [,4]  [,5]  [,6]  [,7]  [,8]  [,9]  [,10] [,11] [,12] [,13] [,14] [,15] [,16]
1    0.0   0.0   0.2   0.0   0.0   0.2   0.0   0.0   0.2   0.2   0.2   0.0   0.0   0.0   0.0   0.0
2    0.0   0.0   0.0   0.2   0.0   0.2   0.0   0.0   0.0   0.0   0.2   0.0   0.2   0.0   0.2   0.0
3    0.2   0.0   0.0   0.0   0.2   0.0   0.2   0.0   0.0   0.2   0.2   0.0   0.0   0.0   0.0   0.0
4    0.0   0.2   0.0   0.0   0.0   0.0   0.2   0.0   0.0   0.0   0.2   0.2   0.0   0.0   0.0   0.2
5    0.0   0.0   0.2   0.0   0.0   0.0   0.2   0.2   0.0   0.2   0.2   0.0   0.0   0.0   0.0   0.0
6    0.2   0.2   0.0   0.0   0.0   0.0   0.0   0.0   0.0   0.2   0.2   0.0   0.2   0.0   0.0   0.0
7    0.0   0.0   0.2   0.2   0.2   0.0   0.0   0.0   0.0   0.0   0.2   0.2   0.0   0.0   0.0   0.0
8    0.0   0.0   0.2   0.0   0.2   0.0   0.0   0.0   0.2   0.2   0.0   0.0   0.0   0.2   0.0   0.0
9    0.2   0.0   0.0   0.0   0.0   0.2   0.0   0.2   0.0   0.2   0.0   0.0   0.0   0.2   0.0   0.0
10   0.2   0.0   0.2   0.0   0.2   0.0   0.0   0.0   0.2   0.0   0.0   0.0   0.0   0.2   0.0   0.0
11   0.2   0.2   0.2   0.2   0.0   0.2   0.0   0.0   0.0   0.0   0.0   0.0   0.0   0.0   0.0   0.0
12   0.0   0.2   0.0   0.2   0.0   0.0   0.2   0.0   0.0   0.0   0.0   0.0   0.0   0.0   0.2   0.2
13   0.2   0.2   0.0   0.0   0.0   0.2   0.0   0.0   0.0   0.0   0.2   0.0   0.0   0.0   0.2   0.0
14   0.2   0.0   0.2   0.0   0.0   0.0   0.0   0.2   0.2   0.2   0.0   0.0   0.0   0.0   0.0   0.0
15   0.0   0.2   0.0   0.2   0.0   0.2   0.0   0.0   0.0   0.0   0.0   0.0   0.2   0.0   0.0   0.2
16   0.0   0.2   0.0   0.2   0.0   0.0   0.0   0.0   0.0   0.0   0.0   0.2   0.2   0.0   0.2   0.0
attr(,"call")
nb2mat(neighbours = voi.knn.nb)
```

The weights assigned to each observation can also be displayed by calling slot weights from the *listw* class object. This way one can also view a summary of how many neighbours each region has.

```
cont.listw$weights
```

```
[[1]]
[1] 0.1666667 0.1666667 0.1666667 0.1666667 0.1666667 0.1666667

[[2]]
[1] 0.1428571 0.1428571 0.1428571 0.1428571 0.1428571 0.1428571 0.1428571

[[3]]
[1] 0.1428571 0.1428571 0.1428571 0.1428571 0.1428571 0.1428571 0.1428571

[[4]]
[1] 0.2 0.2 0.2 0.2 0.2

[[5]]
[1] 0.1 0.1 0.1 0.1 0.1 0.1 0.1 0.1 0.1 0.1
```

When working outside RStudio (in the console), one can interactively edit the neighbour matrix. A map appears on the screen with a connection diagram saved in the *nb* class file. Using the **edit.nb()** command, one can use the mouse to add or remove links between objects. In addition, by applying these connections to the contour map, one can easily manage the relationships between the studied objects.

After running the **edit.nb()** command, which defines the W spatial weights matrix (here *voi.knn.nb*), the coordinates of the centroids (here *crds.voi*) and the contour map (here *voi*), select the mouse by clicking the two selected locations – when there is no connection between them, the program will inform one about it and ask if one wants to create a connection ("add contiguity?"), to which one can answer YES (*y*) or NO (*n*). After performing the operation, one can leave the interface (*q*), refresh the drawing (*r*) or continue editing (*c*). When there is a connection between the indicated regions, R will inform the user about it and ask whether to delete the connection ("Delete this line?"), to which one can answer YES (*y*) or NO (*n*).

```
# launching the editing interface
edit.nb(voi.knn.nb, crds.voi, polys=voi)
```

The spdep:: package contains several other useful commands that are tools for managing matrix matrices. The **droplinks**() function gives interesting options. It is, in a way, an automated version of the interactive editing of neighbours mentioned previously. This command allows the deleting of connections in the *nb* object according to a given rule. As arguments, it accepts the *nb* object being edited and the *drop* element, informing about connections to be removed. The *drop* can be in the form of a vector with logical values, names or IDs of regions that are to be erased from connections. The logical argument *sym* specifies whether symmetric connections will be removed (from and to the region, *TRUE* by default) or only outgoing connections (*FALSE* option).

In the following example, operations are carried out on a matrix for voivodeships (NTS2), in which there are 16 territorial units. The matrix *cont.voi.nb* is a matrix according to the contiguity criterion. By calling its name, the statistics of this matrix are displayed. Data were prepared for removing associations in three ways: 1) *logic* vector, which defines what to do with each subsequent region (leave *F* = *FALSE*, delete *T* = *TRUE*); 2) vector of region names to be deleted and 3) region number to be deleted. The effect of all three formulas is the same. The effect of the **droplinks**() command is an object of the *nb* class.

```
cont.voi.nb
```

```
Neighbour list object:
Number of regions: 16
Number of nonzero links: 68
Percentage nonzero weights: 26.5625
Average number of links: 4.25
```

```
logic<-c(F,F,F,T,F,F,F,F,F,F,F,F,F,F,F,F)
name<-c("mazowieckie")
id<-c(4)
change1<-droplinks(cont.voi.nb, logic)
change2<-droplinks(cont.voi.nb, name)
change3<-droplinks(cont.voi.nb, id)
change1          # others give identical results
```

```
Neighbour list object:
Number of regions: 16
Number of nonzero links: 56
Percentage nonzero weights: 21.875
Average number of links: 3.5
1 region with no links:
Mazowieckie
```

Another useful function is the **subset.nb**() command. It allows one to get a new neighbourhood matrix that is a subset of the previous one. As arguments, it takes the original object of the *nb* class and a logical vector that designates regions to a subset (those for which the vector assumes T values will be selected).

```
subset<-c(F,T,F,T,F,F,F,F,F,F,T,F,F,F,F,F)
subset.voi<-subset.nb(cont.voi.nb, subset)
subset.voi
```

```
Neighbour list object:
Number of regions: 3
Number of nonzero links: 6
Percentage nonzero weights: 66.66667
Average number of links: 2
```

The **include.self()** command allows one to enter into the spatial weights matrix connections of a given object with itself (the matrix diagonal will then be filled with ones). It is very simple to use – as the only argument it accepts is an object of type *nb* and transforms it. The effect of the action can be seen in the number of non-zero links (*number of nonzero links*) – from 68 increased to 84, exactly 16 (as many as regions), and the average number of links increased by 1, from 4.25 to 5.25.

```
cont.voi.nb.self<-include.self(cont.voi.nb)
cont.voi.nb.self
```

```
Neighbour list object:
Number of regions: 16
Number of nonzero links: 84
Percentage nonzero weights: 32.8125
Average number of links: 5.25
```

There are also functions that treat neighbourhood matrices from the point of view of operations on sets. The **setdiff.nb()** command shows the difference of two sets, **insersect.nb()** sets the common part of two matrices, **union.nb()** gives the sum of connections and **complement.nb()** completes the single neighbourhood matrix with all possible missing links (see Figure 4.4).

```
par(mfrow=c(1,3)) # 1 x 3 split window
par(mar=c(5.1, 2, 4.1, 1))

plot(voi, main="complement.nb()")
plot(complement.nb(change1), crds.voi, add=T)

plot(voi, main="intersect.nb()")
plot(intersect.nb(cont.voi.nb, change1), crds.voi, add=T)

plot(voi, main="setdiff.nb()")
plot(setdiff.nb(cont.voi.nb, change1), crds.voi, add=T)

par(mfrow=c(1,1)) # 1 x 1 split window
```

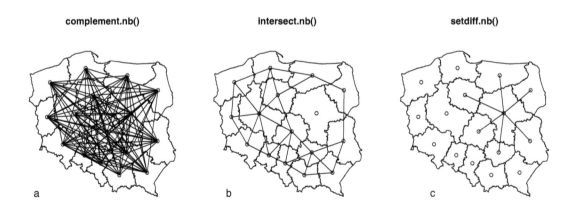

complement.nb()   **intersect.nb()**   **setdiff.nb()**

a   b   c

Figure 4.4 Neighbour connections resulting from subsequent operations: a) completed links, b) common part of two matrices, c) difference of two sets

Source: Own study using the R program and the spdep:: package

4.2.8 Spatial lags and higher-order neighbourhoods

A higher-order neighbourhood (i.e. neighbour neighbours) can be used to model diffusion, that is, spread shock. A very useful command is **lag.listw**(), which creates spatial lags of the variable, and the **nblag**() command that works in a pair, which creates lags in the spatial weights matrix, and **nblag_cumul**() that accumulates lags and saves them in the *nb* class.

The following example shows the use of spatial weights matrices for modelling lags and shock diffusion. The impact of selected voivodeship cities was adopted as a shock, and subsequent lags distributed by lower-order neighbours were presented on the maps.

Technically, the spelling of poviat names was checked with the **levels**() command to select key cities in Poland. The *SHOCK* variable was added to the database in shapefile, which initially contains only values of 0 and further conditionally assigned values of 1 for selected locations. Figure 4.5a presents these shock centres.

```
head(levels(pov$jpt_nazwa_))
```

```
[1] "powiat aleksandrowski" "powiat augustowski" "powiat bartoszycki"
[4] "powiat bełchatowski"   "powiat będziński"   "powiat bialski"
```

```
pov$SHOCK<-rep(0,380)
pov$SHOCK[pov$jpt_nazwa_=="powiat Łódź"]<-1      #Łódź
pov$SHOCK[pov$jpt_nazwa_=="powiat Warszawa"]<-1  #Warszawa
pov$SHOCK[pov$jpt_nazwa_=="powiat Gdańsk"]<-1    #Gdańsk
pov$SHOCK[pov$jpt_nazwa_=="powiat Poznań"]<-1    #Poznań
spplot(pov, "SHOCK")                             #shock centres Figure 4.5a
```

The **lag.listw**() command allows one to create a spatial lags vector. As arguments, it takes a *listw* class matrix and a numerical vector (lagged variable) of the same length as the number of neighbours. As can be seen from Figure 4.5b, the original shock centres were not included, because in principle the studied region is not its own neighbour.

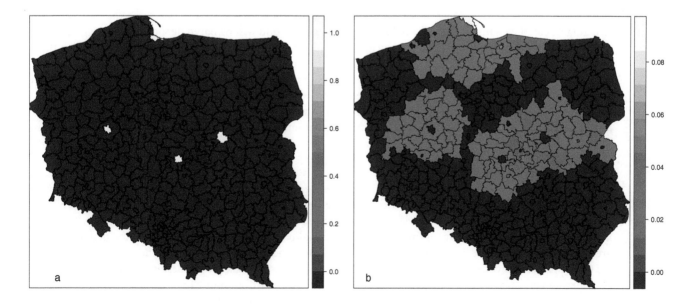

Figure 4.5 Diffusion of shocks: a) centres, b) spatial lag of third order

Source: Own study using the R program and spdep:: and sp:: packages

```
# spatial lag of the SHOCK variable - first row
pov$lagg<-lag.listw(cont.listw, pov$SHOCK)
summary(pov$lagg)
```

```
    Min.  1st Qu.   Median     Mean 3rd Qu.     Max.
0.000000 0.000000 0.000000 0.008103 0.000000 0.500000
```

```
spplot(pov, "lagg")                    # first order lag
```

Use the **nblag**() and **nblag_cumul**() commands to obtain higher order lags. The first of these creates a list of higher-order neighbours for lagged objects. Its arguments are the *nb* class neighbourhood matrix and the *maxlag* numeric variable representing the number of lags to be constructed. The **nblag_cumul**() command creates one consistent list of associations from the original table and newly created lags. The results obtained in this way should be merged with the dataset with the already known command **lag.listw**() in order to be able to visualise them on a **spplot**() chart.

```
# spatial lag of the SHOCK variable - second row
poviats.2.list<-nblag(cont.nb, 2)
poviats.2.nb<-nblag_cumul(poviats.2.list)
pov$lagg<-lag.listw(nb2listw(poviats.2.nb), pov$SHOCK)
spplot(pov, "lagg") # second order lag

# spatial lag of the SHOCK variable - third row
poviats.3.list<-nblag(cont.nb, 3)
poviats.3.nb<-nblag_cumul(poviats.3.list)
pov$lagg<- lag.listw(nb2listw(poviats.3.nb), pov$SHOCK)
spplot(pov, "lagg") # Figure 4.5b - third order lag
```

4.2.9 Creating weights matrix based on group membership

One can think of local relationships not only in the context of location in space but also on the basis of belonging to the same groups or similarity of a set of features. One can imagine that two regions with similar economic indicators will respond in a similar way to the shocks affecting them, even if they are in a different spatial environment. One can think of categorical variables in a similar way – by assigning them to one group, two entities, even if separated from each other, will have common features. Such a similarity can also be the basis for constructing a spatial weights matrix. The definition of neighbourhood will be based here on categorical variables, not distances in space. Importantly, such a neighbourhood matrix can be built for a-spatial data in which there are group assignments or clear clusters of a given feature that could not be taken into account otherwise. The use of this type of object in the modelling process (instead of the standard matrix) can improve the quality of analyses by filtering non-random similarity of elements.

The **mat2listw**() function from the spdep:: package allows one to convert a user-created weights matrix to the *listw* class. The transformed object can be used for calculations analogous to standard spatial weights matrices.

Example

To construct a matrix of weights based on group membership, the poviat dataset (NTS4) for $n = 380$ units will be used. The set contains the categorical variable *region_name* – belonging to a voivodeship (NTS2). In the matrix, the neighbours will be poviats that are located in one province. In each row of the newly created neighbourhood matrix with dimensions $n \times n$, the diagonal has zero elements, and outside the diagonal, there are values equal to 1 at the intersection of poviats from the same voivodeship and 0 at the intersection of poviats from different voivodeships. If there are 42 poviats in the Mazowieckie voivodeship, each poviat from this voivodeship will have 41 neighbours, and the weights, after standardisation by rows to 1, will be $1/41 = 0.024$. The number of poviats in a given voivodeship was summarised by the **table**() command.

```
data<-read.csv("data_nts4_2019.csv", header=TRUE, dec=",", sep=";")
sub<-data[data$year==2017,]
names(sub)
```

```
 [1]  "ID_MAP"          "code_GUS"        "poviat_name1"  "poviat_name2"
 [5]  "subreg72_name"   "subreg72_nr"     "region_name"   "region_nr"
 [9]  "year"            "core_city"       "dist"          "XA01"
[13]  "XA02"            "XA03"            "XA04"          "XA05"
[17]  "XA06"            "XA07"            "XA08"          "XA09"
[21]  "XA10"            "XA11"            "XA12"          "XA13"
[25]  "XA14"            "XA15"            "XA16"          "XA17"
[29]  "XA18"            "XA19"            "XA20"          "XA21"
[33]  "XA22"            "XA23"
```

```
table(sub$region_name)
```

Dolnośląskie	Kujawsko-Pomorskie	Lubelskie	Lubuskie
30	23	24	14
Łódzkie	Małopolskie	Mazowieckie	Opolskie
24	22	42	12
Podkarpackie	Podlaskie	Pomorskie	Śląskie
25	17	20	36
Świętokrzyskie	Warmińsko-Mazurskie	Wielkopolskie	Zachodniopomorskie
14	21	35	21

To build a weights matrix, follow these steps. First, prepare an empty matrix with dimensions of $n \times n$. Then assign the ones at the intersection of the observation numbers belonging to the same province. This is equivalent to specifying in which rows there are poviats from a given voivodeship, and then in the weights matrix in those rows and appropriate columns (the same as the numbers of the rows), specify the neighbourhood marker. Finally, the diagonal elements must be zero, which can be done with the **diag**() command.

```
matt<-matrix(0, nrow=380, ncol=380) # creating an empty matrix

names<-levels(sub$region_name)
names
```

```
 [1]  "Dolnośląskie"    "Kujawsko-Pomorskie" "Lubelskie"     "Lubuskie"
 [5]  "Łódzkie"         "Małopolskie"        "Mazowieckie"   "Opolskie"
 [9]  "Podkarpackie"    "Podlaskie"          "Pomorskie"     "Śląskie"
[13]  "Świętokrzyskie"  "Warmińsko-Mazurskie" "Wielkopolskie" "Zachodniopomorskie"
```

```
for(i in 1:16){
c1<-which(sub$region_name==names[i])
matt[c1, c1]<-1}
diag(matt)<-0
```

The matrix prepared in this way should be standardised with rows to one, divided by sums in rows, and then transformed into a *listw* class. Totals in rows were obtained by the **rowSums**() command. The operation of dividing the matrix by a vector means that each element of the line is divided by the corresponding element of the vector (i.e. all elements of the first line by the first element of the vector). Visualisation of correctness of operation can be achieved by drawing maximum values by rows, that is, de facto weights (because weights are equal) for a given poviat. As one can see in Figure 4.6a, the colours are in line with the voivodeship's outline. Assigning the same colour to neighbouring voivodeships results from a similar number of poviats in the voivodeship and consequently similar weights.

```
vec<-rowSums(matt) # sums in rows - de facto number of neighbours
matt.W<-matt/vec # matrix standardisation
matt.listw<-mat2listw(matt.W) # conversion to listw
matt.listw
```

```
Characteristics of weights list object:
Neighbour list object:
Number of regions: 380
Number of nonzero links: 9702
Percentage nonzero weights: 6.718837
Average number of links: 25.53158

Weights style: M
Weights constants summary:
    n     nn  S0      S1    S2
M 380 144400 380 33.59688 1520
```

```
# visualisation of values in the weight matrix
a<-apply(matt.W, 1, max)     # max in rows
library(GISTools)            # Figure 4.6a
choropleth(pov, a)           # value of weights in the poviat
plot(voi, add=TRUE, lwd=2)   # voivodship outline
```

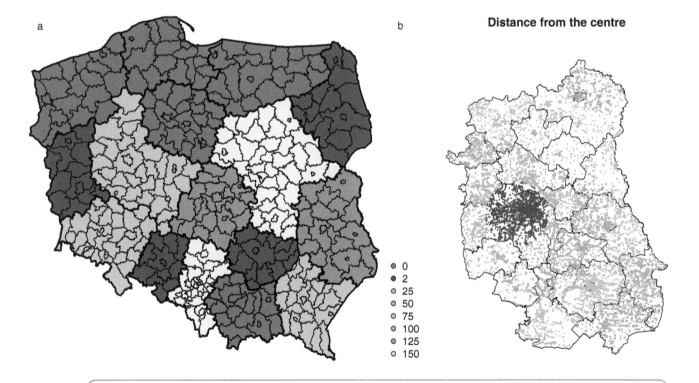

Figure 4.6 Visualisation of W matrix: a) category-based matrix, b) distances for inverse distance matrix

Source: Own study using the R software and spdep:: and GISTools:: packages

Example

The second weights matrix based on group membership will be built on the results of clustering. The same data as in the previous example will be used. Clustering was carried out on several selected economic and social variables using the *k*-means algorithm (more in Chapter 7). Assuming the *silhouette* statistic values as a criterion for a good division, the cluster membership (cf. Figure 4.7a) was determined. A spatial weights matrix was created in the same way as in the previous example.

The set of clustered *sub.k* variables includes three variables: *XA10* – population per 1/km² (column 26), *XA21* – unemployment rate (column 37), *dist* – distance from the voivodeship city (column 16). Clustering was performed with the **eclust()** command from the factoextra:: package, specifying the *kmeans* option and Euclidean distances. The clustering result, the cluster vector, is in the *cluster* slot. As can be seen from Figure 4.7b, the joint cluster includes the smallest poviats territorially – these are generally regional cities, while the remaining poviats are in the second cluster.

```
sub<-data[data$year==2017,]
names(sub)
```

```
 [1] "ID_MAP"         "code_GUS"      "poviat_name1"   "poviat_name2"
 [5] "subreg72_name"  "subreg72_nr"   "region_name"    "region_nr"
 [9] "year"           "core_city"     "dist"           "XA01"
[13] "XA02"           "XA03"          "XA04"           "XA05"
[17] "XA06"           "XA07"          "XA08"           "XA09"
[21] "XA10"           "XA11"          "XA12"           "XA13"
[25] "XA14"           "XA15"          "XA16"           "XA17"
[29] "XA18"           "XA19"          "XA20"           "XA21"
[33] "XA22"           "XA23"
```

```
sub.k<-sub[,c(21,32,11)]

library(factoextra)
library(RColorBrewer)
fviz_nbclust(sub.k, kmeans, method="silhouette") # Figure 4.7a
clus<-eclust(sub.k, "kmeans", hc_metric="euclidean",k=2, graph=FALSE)

brks<-c(0,1,2,3)
cols<-brewer.pal(3, "Purples")
plot(pov, col=cols[findInterval(clus$cluster, brks)]) # Figure 4.7b
```

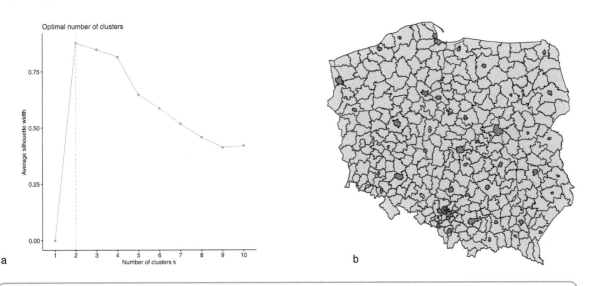

a   b

Figure 4.7 Analysis of *k*-means: a) silhouette statistics, b) map of cluster membership

Source: Own study using the factoextra:: package

The next steps are similar to the previous case. First, an empty matrix with dimensions $n \times n$ and a vector with observations assigned to clusters are created. Then, two vectors are created in the loop (each for a given cluster) containing the observation numbers from that cluster, after which the loops will be iterated. The resulting matrix should be standardised and transformed into a *listw* class. Finally, draw the maximum weights (as in the previous example) to check whether it was possible to recover the spatial structure as in the clustering itself.

```
matt<-matrix(0, nrow=380, ncol=380) # creating an empty matrix
for(i in 1:2){
c1<-which(clus$cluster==i)
matt[c1, c1]<-1}
diag(matt)<-0

vec<-rowSums(matt) # sums in rows - de facto number of neighbours
matt.W<-matt/vec # matrix standardisation
matt.listw<-mat2listw(matt.W) # conversion to listw
matt.listw
```

```
Characteristics of weights list object:
Neighbour list object:
Number of regions: 380
Number of nonzero links: 107732
Percentage nonzero weights: 74.60665
Average number of links: 283.5053

Weights style: M
Weights constants summary:
    n     nn    S0       S1   S2
M 380 144400 380 4.042556 1520
```

```
# visualisation of values in the weight matrix
a<-apply(matt.W, 1, max)      # max in rows
library(GISTools)
choropleth(pov, a)            # value of weights in the poviat
plot(voi, add=TRUE, lwd=2)    # voivodeship outline
```

4.3 Distance measurement and spatial aggregation

Calculations can be made on spatial objects. One of the typical operations is determining the distance between points. This action can be performed in several ways; the commands differ in the way the result is presented, the class of the result object and so on.

The basic distance command is **spDistN1()** from the sp:: package. It determines the distance between the distance vector (first argument) and a single point (second argument). As arguments, it accepts objects in the *matrix* class, as well as objects of the *SpatialPoints* class.

The fossil:: package contains three more functions that perform a similar task. The **deg.dist()** function calculates haversine distances (according to the formula: $\frac{1}{2} * \left(1 - \cos(x)\right)$: haversine distance formula), the **earth. dist()** command calculates geographical distance and the **ecol.dist()** command creates a matrix of distances between any number of locations. In the gmapsdistance:: package, one can determine the distance and travel time between two points based on a Google map. The **get.api.key()** and **set.api.key()** commands allow one to manage the API (more in Chapter 3). The **gmapsdistance()** command sets the distance or travel time.

The following example specifies the coordinates of Lublin as the centre. Based on point data for firms (*firms* object) from spatial coordinates, the **SpatialPoints()** command was created with the *SpatialPoints* class object, which will be the input for the **spDistsN1()** command. The **spDistsN1()** command determined the distance vector between each observation and the indicated centre – Lublin. The distances assigned to the points were plotted on the map in a colour point layer. The palette from the wesanderson:: package was used.

```
firms<-read.csv("geoloc_data_firms.csv", header=TRUE, dec=",", sep=";")
crds.Lublin<-c(22.4236877, 51.2180254) # coordinates of Lublin

# xy coordinates of firms - SpatialPoints class object
loc<-SpatialPoints(firms[,12:13],          proj4string=CRS("+proj=longlat
+datum=NAD83"))
head(loc)
```

```
SpatialPoints:
     coords.x1 coords.x2
[1,] 22.00263  51.40935
[2,] 22.34528  51.35417
[3,] 22.96182  50.31523
[4,] 21.83905  50.95910
[5,] 23.37778  51.96064
[6,] 22.25986  51.74347
Coordinate Reference System (CRS) arguments: +proj=longlat +datum=NAD83
+ellps=GRS80 +towgs84=0,0,0
```

```
# vector distance of all points from the centre - Lublin
dist<-spDistsN1(loc, crds.Lublin, longlat=TRUE)

# colour palette
library(wesanderson)
cols<-wes_palette(n=8, name="Darjeeling1", type="continuous")
cols    # Figure of palette colours

# a point map of the distance between the point and the centre
summary(dist)
```

```
   Min.   1st Qu.   Median     Mean   3rd Qu.      Max.
 0.4886   27.5838  56.6484  56.4965   81.9026  141.1772
```

```
brks<-c(0, 2, 25, 50, 75, 100, 125, 150)
lubelskie.voi<-voi[voi$jpt_nazwa_=="lubelskie",] # creating a map subset
lubelskie.pov<-pov[50:73,] # creating a map subset
plot(lubelskie.voi) # Figure 4.6b
plot(lubelskie.pov, add=TRUE)

points(firms[,12:13], col=cols[findInterval(odle, brks)],
pch=21, bg=cols[findInterval(dist, brks)], cex=0.2)

legend("bottomleft", legend=brks, pt.bg=cols, bty="n", pch=21)
title(main="Distance from the centre")
savePlot(filename="dists", type="jpg")
```

Aside from counting distances, another important activity is assigning points to areas. For a set of geolocated points (as in analysed firms previously), one can check their belonging to geographical areas – for example, poviats. The **over()** command from the sp:: package shows in which region (poviat) the point is drawn. As input, it requires objects in the *sp* class – for example, *SpatialPolygons* and *SpatialPoints*. Initially, a length dataset is created, like a point set, and its content is like an area set. There are poviat names in the shapefile file, and they will be assigned to the points. The result set from the **over** () command does not contain information from the point set – trust that it is in the same order as the original point set. Both sets – point and result – can be glued with the **cbind**() command.

```
# checking the affiliation of points to areas
lubelskie.pov<-pov[50:73,] # creating a map subset
loc<-SpatialPoints(firms[,12:13],           proj4string=CRS("+proj=longlat
+datum=NAD83"))

# rewriting points to areas (poviats)
firms.over<-over(loc, lubelskie.pov) # from sp:: package
dim(firms.over)
#[1] 37378    29

names(firms.over) # the same result will be obtained from names (pov)
```

```
 [1] "iip_przest" "iip_identy" "iip_wersja" "jpt_sjr_ko" "jpt_kod_je"
 [6] "jpt_nazwa_" "jpt_nazw01" "jpt_organ_" "jpt_orga01" "jpt_jor_id"
[11] "wazny_od"   "wazny_do"   "jpt_wazna_" "wersja_od"  "wersja_do"
[16] "jpt_powier" "jpt_kj_iip" "jpt_kj_i01" "jpt_kj_i02" "jpt_kod_01"
[21] "id_bufora_" "id_bufor01" "id_technic" "jpt_opis"   "jpt_sps_ko"
[26] "gra_ids"    "status_obi" "opis_bledu" "typ_bledu"  "a"
```

```
head(firms.over$jpt_nazwa_)
```

```
[1] powiat     puławski powiat lubelski     powiat biłgorajski
[4] powiat kraśnicki  powiat bialski     powiat łukowski
370 Levels: powiat aleksandrowski . . .  powiat żywiecki
```

An interesting command, useful in point analyses, is the **zerodist**() from the sp:: package, which finds pairs of points with the same coordinates in the same set, and the **zerodist2**() command, which finds duplicates in two sets. Duplicate locations may result from replication of observations or be the same address of a building but of different premises. If there are such duplicates, there are two strategies. One of them may assume the removal of duplicate coordinates – one can then use the **remove.duplicates**() command from the sp:: package. The second may, on the contrary, assume the distribution of these coordinates by epsilon – one can use the **jitter**() command or add a random value from the distribution, for example, normal with *mean = 0* and a small standard deviation.

The **zerodist**() command displays a matrix of point pairs as a result. It is also possible to enter an error margin in which the objects will be considered the same. A properly selected parameter can allow one to find elements in the closest neighbourhood.

```
the.same<-zerodist(loc)
head(the.same)
```

```
      [,1] [,2]
[1,]   48   53
[2,]   73   83
[3,]    3   84
[4,]   73   95
[5,]   83   95
[6,]   73  209
```

Example

The following problem is presented subsequently: points were drawn on the country's surface (*pl* object). The goal is to calculate the distance between these points and the central regional cities by region; that is, if the point is in region X, the distance between this point and the main city in region X is calculated. The following task requires several steps:

- Points randomisation – the **spsample**() command from the sp:: package is used, which draws points inside a given contour map from various spatial distributions (including random, regular, stratified, unbalanced, hexagonal, clustered draw and Fibonacci distribution) (more in Chapter 9).
- Assigning points to regions – the **over**() command from the sp:: package is used, which checks the affiliation of a point to data regions with a regional map (shapefile) and returns all variables from the database in shapefile for a given point.
- Selecting central city coordinates – centrodium coordinates for all areas are determined by the **coordinates**() command from the sp:: package. The *dataset3* poviat dataset contains a dummy (zero-one) variable for provincial cities. To create a subset of data containing only the name of a poviat that is a voivodeship city and the coordinates of this location, one must first add the coordinate matrix to the poviat set as columns, and then conditionally select interesting rows.
- Correction of spelling of regions' names – in the examined datasets, voivodeship names were spelled differently: with uppercase and lowercase letters. One needs to agree on the spelling to combine the files according to this identifier. All names were converted to lowercase letters using the **tolower**() command from the R database.
- Connection of point data (together with the identifier of the voivodeship) with the coordinates of central cities in the given voivodeship. The **merge**() command was used, in which sorting was deactivated (*sort=FALSE*) to keep the structure of the file as in the original. Thanks to this procedure, a dataset was obtained containing the coordinates of a given point and coordinates of the central city, as well as the name of the voivodeship in which the given point lies.
- Determining distance for coordinate pairs – the **pointDistance**() command from the raster:: package allows one to calculate the distance between two points. It was implemented in a loop for each subsequent row. The result of the calculations was saved as the next column.
- Drawing point values – a simple command that draws the value of a variable at a given spatial point in colour is **quilt.plot**() from the fields:: package (see Figure 4.8). The coordinates of points and the value in points are given as arguments.

```
# draw points on the map, options: random | regular | stratified
# | nonaligned | hexagonal | clustered | fibonacci

pl<-readOGR(".", "Panstwo") #outline map of Poland without internal divisions
pl<-spTransform(pl, CRS("+proj=longlat +datum=NAD83"))

newpoints<-spsample(pl, 20000, type="stratified")
newpoints.df<-as.data.frame(newpoints)      # class conversion
newpoints.m<-as.matrix(newpoints.df)         # class conversion

# checking the affiliation of a point to a region and choosing a key variable
where<-over(newpoints, voi)
newpoints.df$voi<-where$jpt_nazwa_  # name of the region
head(newpoints.df)
```

```
        x1       x2          voi
1 22.64712 49.06419 podkarpackie
2 22.69542 49.06649 podkarpackie
3 22.75105 49.06346 podkarpackie
4 22.79052 49.08243 podkarpackie
5 22.81472 49.05017 podkarpackie
6 22.85597 49.07044 podkarpackie
```

```
dim(newpoints.df)
#[1] 19977 3

regions<-levels(newpoints.df$voi)    # voivodeship names
regions
```

```
 [1] "dolnośląskie"          "kujawsko-pomorskie"  "lubelskie"
 [4] "lubuskie"              "łódzkie"             "małopolskie"
 [7] "mazowieckie"           "opolskie"            "podkarpackie"
[10] "podlaskie"             "pomorskie"           "śląskie"
[13] "świętokrzyskie"        "warmińsko-mazurskie" "wielkopolskie"
[16] "zachodniopomorskie"
```

```
# loading poviat data that has the identifier of region centres
unempl<-read.csv("unemp2018.csv", header=TRUE, dec=",", sep=";")

# adding centroid coordinates to the data set
unempl$crds<-coordinates(pov) # added as a matrix in the data frame

# selection of relevant column - name voivodeship and geographical coordinates
capitals.xy<-unempl[unempl$core_city==1, c(6,102)]

# case correction
capitals.xy$voivodeship<-tolower(capitals.xy$voivodeship)
capitals.xy<-capitals.xy[1:16,]
head(capitals.xy)
```

```
      voivodeship   crds.1   crds.2
 13  podkarpackie  22.77714  49.78058
 71     lubelskie  22.55662  51.22575
 94   małopolskie  19.98507  50.05316
119       łódzkie  19.48026  51.76649
122      lubuskie  15.53183  51.92588
151   mazowieckie  21.04774  52.22933
```

```
# joining data sets by key-variable
dist<-merge(newpoints.df, capitals.xy, by.x="voi", by.y="voivodeship", all.
x=TRUE, sort=FALSE)
names(dist)
#[1] "voi" "x1" "x2" "crds"

head(dist)
```

```
           voi       x1       x2   crds.1   crds.2
1 podkarpackie  22.64712  49.06419  22.77714  49.78058
2 podkarpackie  22.69542  49.06649  22.77714  49.78058
3 podkarpackie  22.75105  49.06346  22.77714  49.78058
4 podkarpackie  22.79052  49.08243  22.77714  49.78058
5 podkarpackie  22.81472  49.05017  22.77714  49.78058
6 podkarpackie  22.85597  49.07044  22.77714  49.78058
```

```
# counting the distance between coordinate pairs
library(raster)
for(i in 1:dim(dist)[1]){
dist$dist[i]<-pointDistance(dist[i,2:3],dist$crds[i,], lonlat=FALSE)}
head(odle)
```

```
           voi       x1       x2   crds.1   crds.2       dist
1 podkarpackie 22.68015 49.04960 22.77714 49.78058 0.7373881
2 podkarpackie 22.71904 49.06112 22.77714 49.78058 0.7217979
3 podkarpackie 22.76122 49.05907 22.77714 49.78058 0.7216895
4 podkarpackie 22.81219 49.04100 22.77714 49.78058 0.7404076
5 podkarpackie 22.85409 49.04517 22.77714 49.78058 0.7394244
6 podkarpackie 22.50673 49.09419 22.77714 49.78058 0.7377317
```

```
# value graph in point
library(fields)
quilt.plot(dist[,2:3], dist$dist) # Figure 4.8a
```

Another option for operations on distances and point affiliation is to group points based on clustering results. This allows one to determine, for example, in which poviats there are points, for example, from the first cluster (grouping, e.g. according to CLARA) or how many points there are in a given poviat. One can also specify the density of points in regions (poviats) per area (poviat) by colour. For more information on spatial clustering and spatial cluster analysis, see Chapter 7.

In the following example, spatial coordinates were clustered from point data for firms, using the **clara**() command from the cluster:: package. Based on this grouping, statistics were made concerning in which poviats the points from the selected cluster lie, using the **unique**() command. It was also checked with the **poly.counts**() command from the GISTools:: package how many points are within counties. The counties area was counted with the **poly.areas**() command from the GISTools:: package, which enabled determination and visualisation of point density per unit of area (see Figure 4.8b).

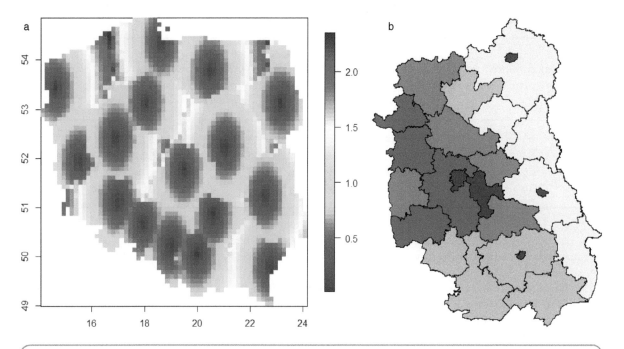

Figure 4.8 Data aggregation visualisation: a) graph of the value of the distance function for randomly drawn points, b) point density in the cluster in relation to the poviat surface

Source: Own study using R and the spdep:: package

```
library(cluster)
library(factoextra)
library(wesanderson)
library(GISTools)

# coordinate clustering
c2<-clara(firms[,12:13], 5, metric="euclidean", sampsize=1000)

# a combination of clustering and membership verification results
points<-cbind(firms[,12:13], firms.over$jpt_nazwa_, c2$clustering)
head(points)
names(points)<-c("xx", "yy", "poviat", "cluster")
head(points)
```

```
        xx       yy              poviat cluster
1 22.00263 51.40935     poviat puławski       1
2 22.34528 51.35417     poviat lubelski       2
3 22.96182 50.31523 poviat biłgorajski       3
4 21.83905 50.95910    poviat kraśnicki       1
5 23.37778 51.96064     poviat bialski       4
6 22.25986 51.74347     poviat łukowski       1
```

```
# summary of points belonging to poviats according to clusters from CLARA
# i.e. what poviats are points from the first CLARA cluster
unique(points$poviat[points$cluster==1])
```

```
[1] puławski poviat kraśnicki   poviat łukowski   poviat rycki    poviat
[5] opolski  poviat lubartowski poviat radzyński poviat lubelski poviat
```

```
# checking how many points there are within polygons (poviats)
how.many.pts.in.pov<-poly.counts(loc, lubelskie.pov)# from GISTools::
```

```
  49   50   51   52  53   54   55   56   57  58   59   60  61   62   63  64
2022 1969 1446 1361 975 1303 1646 1683 2755 928 2180 1179 658 1857 1195 966
  65   66  67  68  69   70  71   72
1060 1756 754 702 2132 5035 810 1006
```

```
# point density in relation to the surface - Figure 4.8b
choropleth(lubelskie.pov, how.many.pts.in.pov/poly.areas(lubelskie.pov))
```

Similar aggregation and window assignments can be performed on data that has a numerical value in addition to location. In the following example, regional totals from the drawn points were counted. This time, points (xy locations) were drawn using the **spsample**() command from the sp:: package for the internally shared area – the poviat map. Additionally, numerical values (z dimension) assigned to the location (xy dimension) were drawn. The assignment of points to poviats and voivodeships was checked in turn, and all information was combined into one object. Aggregation with the **aggregate**() command was performed, and the aggregated data was combined by common identifiers with the **merge**() command – it resulted from different orders of data in the map and in the set after aggregation. The results were drawn with the **plot**()

command using division into intervals with the **classIntervals**() command – the drawing mechanism is discussed in detail in Chapter 2.

```
NN<-1500 # number of observations
samm<-spsample(pov, n=NN, type="random") # spatial sample
val<-rnorm(NN, mean=100, sd=20) # drawn values
samm.df<-as.data.frame(samm)
samm.df$val<-val

head(samm.df)
```

```
          x        y       val
1 22.45018 52.93122 125.08417
2 20.56552 51.83059 128.39009
3 19.78832 51.82504 112.56964
4 21.74917 50.10530 113.07227
5 19.79134 53.01229  79.64336
6 23.48443 52.01306  75.69600
```

```
brks<-c(40,60,80,100,120,140,160)
cols<-wes_palette(n=8, name="BottleRocket2", type="continuous")
plot(pov, border="grey80")
points(samm.df[,1:2], pch=".", cex=5, col=cols[findInterval(samm.df[,3], brks)])

# checking the point's affiliation with the area
samm.over.pov<-over(samm, pov) # do poviats
samm.over.voi<-over(samm, voi) # do voivodeship
samm.df$pov<-samm.over.pov$jpt_nazwa_
samm.df$voi<-samm.over.voi$jpt_nazwa_

head(samm.df)
```

```
          x        y       val                            pov          voi
1 22.45018 52.93122 125.08417       poviat wysokomazowiecki     podlaskie
2 20.56552 51.83059 128.39009                  poviat rawski       łódzkie
3 19.78832 51.82504 112.56964               poviat brzeziński     łódzkie
4 21.74917 50.10530 113.07227 poviat ropczycko-sędziszowski podkarpackie
5 19.79134 53.01229  79.64336               poviat żuromiński mazowieckie
6 23.48443 52.01306  75.69600                 poviat bialski     lubelskie
```

```
# aggregation of point data
result<-aggregate(samm.df$val, by=list(samm.df$voi), sum)
voi.df<-as.data.frame(voi)
voi.df$id<-1:16
voi.df<-voi.df[,c(6,30)]
voi.df.m<-merge(voi.df, result, by.x="jpt_nazwa_", by.y="Group.1")

# aggregate data sorting
library(doBy)
result.sort<-orderBy(~id, data=voi.df.m)

# colourful voivodeship map
library(RColorBrewer)
```

```
library(classInt)
variable<-result.sort$x
summary(variable)
intervals<-8
colors<-brewer.pal(intervals, "BuPu") # colour selection
class<-classIntervals(variable, intervals, style="pretty")
color.table<-findColours(class, colors)
plot(voi, col=color.table)
legend("bottomleft", legend=names(attr(color.table,
"table")), fill=attr(color.table, "palette"), cex=0.75, bty="n")
title(main="Sums according to voivodeship")
```

In the current dataset of point data and around distances and locations, it is quite easy to identify several nearest neighbours of each point (*k* nearest neighbours) and specify values in these locations. The **knearneigh**() command from the spdep:: package builds neighbour relations, indicating which is whose neighbour by this criterion. Averaged values in neighbourly locations can be determined by the already known command **lag.listw**() from the sp:: package, which calculates the spatial lag. For randomly drawn points, this happens in the same way as for real data. The scatterplot of observed values and values in the neighbourhood (spatial lag) can be generated as in the following.

```
# closest neighbours of generated points
samm.knn<-knearneigh(samm, k=2) # class knn object k=2
samm.nb<-knn2nb(samm.knn)
samm.listw<-nb2listw(samm.nb) # conversion class nb to listw
samm.lag<-lag.listw(samm.listw, samm.df$val) # spatial lag
dev.off()
plot(samm.df$val, samm.lag)
```

4.4 Tessellation

Dirichlet tessellation is the process of artificially dividing space into smaller areas. The algorithm works according to the following rule: the distance from the centre of the object to its border is exactly half the distance to the next point or centre in a given direction (Sibson, 1980; Halls, Bulling, White, Garland, & Harris, 2001). In practice, by creating a division between two points, the border is carried out exactly halfway between them. This approach allows the space to be divided into completely filling tiles/tiles that do not overlap (Bowyer, 1981; Halls et al., 2001). It is often used for the allocation of areas, the allocation of public buildings to communities or the construction of new units (Halls et al., 2001). Applications in the field of epidemiology and crystallography are also known (Bowyer, 1981). The extension of this approach to the *n*-dimensional case (Vornoi's tessellation) and further von Theissen studies allowed the method to be adapted to meteorological purposes (Halls et al., 2001).

The implementation of the Dirichlet tessellation algorithm in R is available in the spatstat:: package, which works fully after loading the maptools:: package. To divide the space, one needs to prepare a contour object of the *owin* class, defining the borders of the entire territory (created by the command **as.owin**() from the rgdal:: package) and unique coordinates of points serving as the centre of the tile. Then, using the **ppp**() command from the spatstat:: package, one creates a spatial object of the *ppp* class, which is the basis for building target areas using the **dirichlet**() function. The resulting object can be drawn using the basic **plot**() command. It is also possible to create a *SpatialPolygons* object from the obtained tessellation, using the **as**() command with the *SpatialPolygons* option. An important element is map projection. The **as.owin**() command works only for shapes in planar projection – that is, unlike commands in the *sp* class, which require spherical projection objects.

The following code runs the Dirichlet's tessellation for the Lubelskie voivodeship. A sample of *n* = 800 firms in this area was used for this purpose. They served as reference points for creating a division of space. Planar projection for the analysed data can be obtained by loading the shapefile map without changing the projection, as the file itself contains a planar projection. With a loaded and converted object, one can change the projection type to planar by using *merc* instead of *longlat* in the projection syntax. Planar coordinates of the map require reconciliation to the same format of point data coordinates. The easiest way is to convert the coordinates of points to the *SpatialPoints* object and change the projection with the **spTransform**() command. Referencing the coordinates in *sp* class object requires using "@" to get to the *coords* socket. The tessellation result object, saved as *SpatialPolygons*, can also be transformed into a spherical projection. When adding the projection with the **proj4string**() command, one should use the planar (*merc*) projection, and then change with **spTransform**() to the spherical (*longlat*) projection.

```
library(spatstat)
library(rgdal)
library(maptools)


# option A - reading a map with planar projection
voi<-readOGR(".", "wojewodztwa") # 16 units
lub.voi<-voi[voi$jpt_nazwa_=="lubelskie",] # creating a fragment of the map
lub.owin<-as(lub.voi, "owin") # from rgdal::


# option B - conversion of projection from spherical to planar
#voi<-readOGR(".", "wojewodztwa") # 16 units
#voi<-spTransform(voi, CRS("+proj=longlat +datum=NAD83")) #spherical
voi<-spTransform(voi, CRS("+proj=merc +datum=NAD83")) # planar
lub.voi<- voi[voi$jpt_nazwa_=="lubelskie",] # creating a fragment of the map
lub.owin<-as(lub.voi, "owin") # from rgdal::


# extraction of unique points and random sampling
cord<-as.matrix(cbind(firms$coords.x1, firms$coords.x2))
cord1<-unique(cord)
x<-sample(1:length(cord1), 100)
x<-order(x)
cord2<-cord1[x,]


cord2.sp<-SpatialPoints(cord2) # points in sp class spherical
proj4string(cord2.sp)<-CRS("+proj=longlat +datum=NAD83") # spherical
cord2.sp<-spTransform(cord2.sp, CRS("+proj=merc +datum=NAD83")) #planar


# construction of the spatial object and tessellation
lub.ppp<-ppp(x=cord2.sp@coords[,1], y=cord2.sp@coords[,2], window=lub.owin)
lub.tes<-dirichlet(lub.ppp) # Dirichlet tessellation
plot(lub.tes, main="Tessellation for n = 100 points") # Figure 4.9a
plot(lub.ppp, add=TRUE, pch=".", col="darkblue", cex=2)
degAxis(1)
degAxis(2)


tes.poly<-as(lub.tes, "SpatialPolygons")
proj4string(tes.poly)<-CRS("+proj=merc +datum=NAD83") # spherical
tes.poly<-spTransform(tes.poly, CRS("+proj=longlat +datum=NAD83"))
```

As a result of the algorithm, the space of the Lubelskie Voivodeship was divided into tiles, the centre of which are drawn objects. The system of regions is irregular and is only the result of mathematical calculations. A particularly high density of tiles is found in the neighbourhood of Lublin (more points, smaller result areas). Tiles in peripheral areas occupy a larger area. It is worth noting that even in more dense places, the tiles do not overlap, and their arrangement tightly fills the entire territory.

Tessellation is also possible in the tripack:: package. The basic difference is that the **voronoic.mosaic**() command works only on coordinates, without reference to the bounding area; hence the tessellation division is on an unlimited area (see Figure 4.9b). The resulting object is in the *voronoi.mosaic* class, which specifies the split

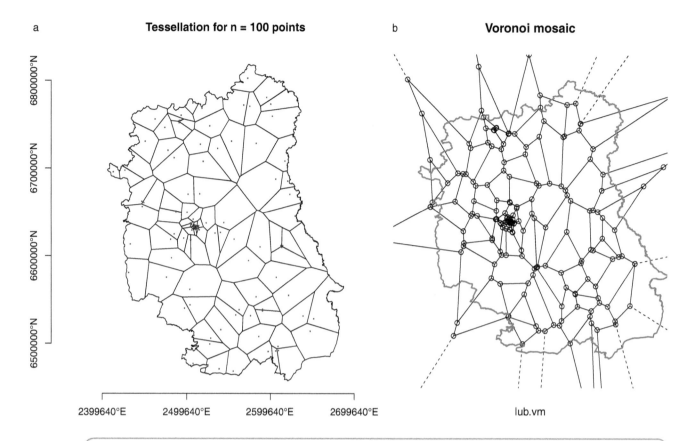

a **Tessellation for n = 100 points**

b **Voronoi mosaic**

2399640°E   2499640°E   2599640°E   2699640°E

lub.vm

> **Figure 4.9** Tessellation for randomly selected points from the Lubelskie voivodeship: a) using the **dirichlet()** command from the spatstat:: package, b) using the **voronoi. mosaic()** command from the tripack:: package
>
> *Source*: Own study using R and spatstat:: and tripack:: packages

vertices. It is also possible to convert to the *voronoi.polygons* class with the **voronoi.polygons**() command, thanks to which the coordinates of all vertices of all tiles are specified within the list. Commands from the tripack:: package work on objects in spherical projection.

```
library(tripack)
lub.vm <- voronoi.mosaic(cord2[,1], cord2[,2])
plot(lub.vm) # Figure 4.9b
voi<-readOGR(".", "wojewodztwa") # 16 units
voi<-spTransform(voi, CRS("+proj=longlat +datum=NAD83")) #spherical
library(tripack)
lub.vm<-voronoi.mosaic(cord2[,1], cord2[,2])
plot(lub.vm)
voi<-readOGR(".", "wojewodztwa") # 16 units
voi<-spTransform(voi, CRS("+proj=longlat +datum=NAD83")) #spherical
lub.voi<-voi[voi$jpt_nazwa_=="lubelskie",] # fragment of a map
plot(lub.voi, add=TRUE, lwd=2, border="red")

lub.vp<-voronoi.polygons(lub.vm)
head(lub.vp)
```

```
[[1]]
            x        y
[1,] 22.80549 50.49007
[2,] 22.68587 50.72253
[3,] 22.52926 50.56334
[4,] 22.52546 50.55790

[[2]]
            x        y
[1,] 23.47407 51.48259
[2,] 23.48531 51.60096
[3,] 23.35592 51.71397
[4,] 23.10375 51.48598
[5,] 23.12194 51.44968
[6,] 23.15742 51.40901
[7,] 23.35110 51.33569
[8,] 23.38466 51.34000
```

4.5 Spatial statistics

Spatial statistics methods are also called explorative spatial data analysis (ESDA). Spatial statistics are a convenient and effective method of testing the existence of spatial autocorrelation processes. Spatial autocorrelation measures are used to assess the correlation of variables with respect to spatial location. Spatial autocorrelation means that geographically close observations are more similar to each other than distant ones. This makes it possible to predict values at a given point, knowing the values in other locations. The spatial autocorrelation measure can be used in several ways: to test the existence of spatial autocorrelation and population characteristics; to determine the degree of autocorrelation, for example, the distance above which observations are independent; to determine the theoretical model appropriate for the observed spatial structure or to conclude on the spatial process. Based on the conclusions from ESDA, the hypothesis of spatial randomness can be rejected, which opens the way to the search for spatial regimes (Anselin, 1999). The conclusions from ESDA are also the basis for the spatial specification of models.

Two types of measures are used in spatial statistics: global and local measures. Global measures are a single-digit indicator of spatial autocorrelation or general similarity of regions. Their advantage is synthetic, and their disadvantage is averaging. In turn, thanks to local statistics, which are determined for each region, one can answer the question of whether the region is surrounded by regions with high or low values or whether it is similar/different to neighbouring regions. The values of local statistics are not generalised for the entire surveyed area, and information about the position of each region relative to neighbours is obtained.

The use of ESDA techniques is the first step in the analysis of spatial data and allows the detection of global and local spatial autocorrelation patterns in data. Global autocorrelation can be measured by global Moran's I (Moran, 1950; Cliff & Ord, 1981) and Geary's C statistics. The most commonly used local measures include local statistics Moran's II and local statistics Geary's C_i, which are part of the so-called local indicator of spatial association (LISA) (Anselin, 1995) and local G Getis-Ord statistics (Ord & Getis, 1995). Global statistics can identify clusters and spatial relationships only for the entire system. However, these statistics can be disaggregated to local statistics, thanks to which patterns of local spatial relationships between regions and their neighbours can be detected. LISA calculation allows detection of which of the regions strongly influence the formation of local clusters and examination of non-random local clusters, so-called hot spots, local instability, significant atypical observations and spatial regimes (Ministeri, 2003). The listed statistics are summarised in Table 4.1. They are discussed in detail later in this chapter.

Spatial statistics are based on several assumptions. The most important of them is stationarity, which means that the data analysed should have a normal distribution with a constant average and variance. However, spatial autocorrelation structures are not always constant throughout the entire study area. However, satisfying poor stationarity is sufficient, where the mean and variance are constant, while autocorrelation depends only on the distance between the tested objects. Spatial patterns should be isotropic; that is, they should not depend on the direction of the study. In addition, the shape of the studied area affects the intensity, scope and type of estimated

Table 4.1 Characteristics of global and local spatial statistics

Spatial Statistics	Statistics Type	Type of measure	Interpretation	functions in R
Moran's I	Global	A measure of spatial autocorrelation	$I > 0$: positive spatial autocorrelation, observation values at distance d are similar $I < 0$: negative spatial autocorrelation, observation values at distance d are different $I = 0$: observation values at distance d are randomly distributed	**moran ()** **moran.test ()** **moran.mc ()** **moran.plot ()**
Geary's C	Global	A measure of spatial autocorrelation	$0 < C < 1$: positive spatial autocorrelation, observation values at distance d are similar $1 < C < 2$: negative spatial autocorrelation, observation values at distance d are different	**geary ()** **geary.test ()** **geary.mc ()**
Join-count	Global in groups	A measure of spatial autocorrelation	H_0 assumes no spatial autocorrelation, p-value > 0.05 confirms this hypothesis H_1 assumes the existence of spatial autocorrelation, p-value < 0.05 confirms this hypothesis	**joincount.test ()** **joincount.mc ()** **joincount.multi ()**
Local Moran's I_i	Local	Together with Local Geary, it creates LISA, a measure of spatial relationships	p-value < 0.05: surrounded by relatively high values, a high values cluster p-value > 0.95: an area surrounded by relatively low values, a cluster of low values	**local.moran ()** **localmoran.sad ()**
Local Geary's C_i	Local	Together with Local Moran, it creates LISA, a measure of spatial diversity	p-value > 0.95: positive, region surrounded by similar observations p-value < 0.05: negative, region surrounded by different observations	—
Local G_i, G^*	Local	Measure of spatial relationships	$G_i > 0$: area surrounded by relatively high values, high-value cluster $G_i < 0$: area surrounded by relatively low values, low-value cluster	**localG ()**
Local H_i, LOSH	Local	Measure of spatial relationships	Interpreted in conjunction with G^* statistics. High values indicate a heterogeneous environment; low values indicate local homogeneity.	**LOSH()**
Empirical Bayes index	Local	Measure of spatial relationships	Interpreted as Moran's I statistics, used for interest (e.g. number of cases/population)	**EBImoran.mc()**

Source: Own study

spatial process, as well as its significance. In addition, there are so-called edge effects (Fortin, Dale, & Ver Hoef, 2002). This problem is revealed during observations on the border of the study area. These regions have fewer neighbours than the objects located in the middle. As a result of this effect, differences in estimates may appear depending on the adopted neighbourhood matrix (e.g. first or second order) (Cressie, 1993b).

A common element of spatial statistics analysis is the so-called spatial lag, which is determined for the examined variable. The spatial lag operator is a weighted average of the variable value in neighbouring regions, in accordance with the declared spatial weights matrix W. When using the first-order neighbourhood matrix for the calculation according to the contiguity criterion, the spatial lag will be the average of the values in the regions bordering the examined area. If the second-order matrix were adopted, the spatial lag would be determined from the values of the neighbours of these neighbours. An analogous interpretation occurs when using the distance matrix or matrix k of the nearest neighbours.

The following are examples of spatial, global and local statistics. The illustration will use a set of data on monthly unemployment rates registered in 380 poviats in 2011–2018. The order of collection is in accordance with the *pov* map of the poviats. For most statistics, a spatial weights matrix is needed – the matrix according to the contiguity criterion, the object *cont.listw* will be used.

```
# reading poviat data
unempl<-read.csv("unemp2018.csv", header=TRUE, dec=",", sep=";")
names(unempl)
```

```
  [1]        "code1"        "code2"          "poviat"
  [4]        "ID_MAP"       "subregion"      "region"
  [7] "subregion_nr"    "region_nr"         "core_city"
 [10]         "dist"    "population_2006" "population_2013"
 [13]    "X2011.01"     "X2011.02"        "X2011.03"
 [16]    "X2011.04"     "X2011.05"        "X2011.06"
 [19]    "X2011.07"     "X2011.08"        "X2011.09"
 [22]    "X2011.10"     "X2011.11"        "X2011.12"
 [25]    "X2012.01"     "X2012.02"        "X2012.03"
 [28]    "X2012.04"     "X2012.05"        "X2012.06"
 [31]    "X2012.07"     "X2012.08"        "X2012.09"
 [34]    "X2012.10"     "X2012.11"        "X2012.12"
 [37]    "X2013.01"     "X2013.02"        "X2013.03"
 [40]    "X2013.04"     "X2013.05"        "X2013.06"
 [43]    "X2013.07"     "X2013.08"        "X2013.09"
 [46]    "X2013.10"     "X2013.11"        "X2013.12"
 [49]    "X2014.01"     "X2014.02"        "X2014.03"
 [52]    "X2014.04"     "X2014.05"        "X2014.06"
 [55]    "X2014.07"     "X2014.08"        "X2014.09"
 [58]    "X2014.10"     "X2014.11"        "X2014.12"
 [61]    "X2015.01"     "X2015.02"        "X2015.03"
 [64]    "X2015.04"     "X2015.05"        "X2015.06"
 [67]    "X2015.07"     "X2015.08"        "X2015.09"
 [70]    "X2015.10"     "X2015.11"        "X2015.12"
 [73]    "X2016.01"     "X2016.02"        "X2016.03"
 [76]    "X2016.04"     "X2016.05"        "X2016.06"
 [79]    "X2016.07"     "X2016.08"        "X2016.09"
 [82]    "X2016.10"     "X2016.11"        "X2016.12"
 [85]    "X2017.01"     "X2017.02"        "X2017.03"
 [88]    "X2017.04"     "X2017.05"        "X2017.06"
 [91]    "X2017.07"     "X2017.08"        "X2017.09"
 [94]    "X2017.10"     "X2017.11"        "X2017.12"
 [97]    "X2018.01"     "X2018.02"        "X2018.03"
[100]    "X2018.04"     "X2018.05"
```

```
# map of poviats and spatial weights matrix
pov<-readOGR(".", "powiaty") # 380 units
pov<-spTransform(pov, CRS("+proj=longlat +datum=NAD83"))

cont.nb<-poly2nb(as(pov, "SpatialPolygons"))      # from spdep::
cont.listw<-nb2listw(cont.nb, style="W")   # from spdep::
```

4.5.1 Global statistics

Global measures in spatial statistics allow the determination of the general spatial trend in the region. Their analysis allows a synthetic view of the data. They also provide a good basis for subsequent more detailed evaluation of the phenomenon.

4.5.1.1 Global Moran's I statistics

Global Moran's I statistics are probably the oldest spatial measure, introduced by Moran in 1950 (Moran, 1950). It is used to test global spatial autocorrelation (Cliff & Ord, 1975a, 1981; Upton & Fingleton, 1985; Haining, 1991). It depends on the difference between the test value and the average – it is a covariance-based test, similar to the Pearson test. Moran's I statistic is expressed by the formula:

$$I = \frac{\sum_i \sum_j w_{ij}(x_i - \bar{x})(x_j - \bar{x})}{S^2 \sum_i \sum_j w_{ij}} \tag{4.7}$$

and

$$S^2 = \frac{1}{n}\sum_i (x_i - \bar{x})^2 \tag{4.8}$$

where x_i is the observation in the region i, $\bar{x}$ is the average of all regions studied, n is the number of regions and w_{ij} is part of the spatial matrix W. The spatial weights matrix should be standardised by rows up to 1 (row-standardised spatial weights matrix).

Tests of significance of Moran's I statistic can be based on theoretical moments of statistics or a permutation approach. The theoretical average (expected value) of Moran's statistic is $\bar{I} = \frac{-1}{(n-1)}$ and is close to 0, which can be interpreted as randomness. To determine the theoretical variance, two spatial data distribution patterns are assumed: normal and randomised (Cliff & Ord, 1981; Goodchild, 1986). If the normal distribution is assumed, the expected value and variance depend only on the spatial weights. With the randomisation approach, these moments also depend on the value of the examined variable. Empirical variance is also determined using the permutation distribution based on a Monte Carlo simulation with approximately 10,000 permutations. The null hypothesis in this approach assumes that any observed value can appear in any location with equal probability (Anselin, 1995). The significance of standardised Moran's I statistics is tested with the U test, based on the distribution function of the normal distribution, in which the null hypothesis assumes the lack of spatial autocorrelation, that is, random distribution of values.

Moran's I statistics indicate whether there is a spatial effect of agglomeration. Positive and significant values of the I statistic (precisely, greater than) mean the existence of positive spatial autocorrelation, that is, similarity of the examined objects at a certain distance d. In this situation, similar values come into contact more often than randomly. Negative values of the I statistic (less than) mean negative autocorrelation, that is, differentiation of the examined objects (similar values touch less often than randomly). Positive autocorrelation means the existence of clusters of similar values (high or low), while negative values of the I statistic are interpreted as so-called hot spots or islands of definitely different values (Goodchild, 1986).

Usually, Moran's I statistics are interpreted as correlation coefficients, although its value is not limited in the range [–1, 1]. This correlation occurs between the value of the variable x in the location and the values of the variable x in the neighbouring locations, analysed as the spatial lag of the tested variable W_x. Moran's statistic informs about spatial autocorrelation regardless of the distance between objects. Determination of spatial correlation taking into account distances is possible using a correlogram.

The spatial autocorrelation coefficient can be interpreted in a similar way as the linear correlation coefficient. In linear correlation, the square of the correlation coefficient is an approximation of the model's determination coefficient (R^2). By examining the interdependence of the X and Y variables, one can check to what extent the information contained in both variables is common. For example, when the linear correlation coefficient is 0.75, then the variation of X in approximately 56% explains the variability of Y. In the case when the correlation coefficient is 1 (or -1), information about X is enough to fully determine Y, and no additional information is necessary. The same applies to spatial autocorrelation, with the difference that the relationship between X and space is examined. When the spatial correlation coefficient is, for example, about 0.8, the location explains the variation of X in about 65%. In addition, obtaining data from neighbourly locations only partially provides new information. For new observations to give 100% new information, they must come from remote locations where the correlation is 0.

For correct Moran statistics estimates, the analysed variable must have a constant variance. This problem occurs when estimating the percentage, especially when the populations are of different sizes. When this condition is not met, the result of spatial autocorrelation can be misleading (Cressie and Chan, 1989; Cressie, 1993b). In a situation of changing (non-stationary) variance, a variance stabilising approach should be used - a conditional randomisation approach is used (Anselin, 1995) or Bayesian statistics (EBI – empirical Bayes index) (Assuncao & Reis, 1999).

The graphic presentation of the global Moran's I statistic is the scatterplot of Moran's I statistics (Moran scatterplot). It is used to visualise local spatial relationships (clusters) and atypical observations (outliers), as well as spatial instability (Anselin, 1995, 1996). This graph on the x-axis set aside has the analysed standardised variable (expressed in the number of deviations from the sample mean and divided by the standard deviation) and on the y-axis the tested standardised variable with spatial lag.

The chart is divided into four quadrants, relative to point (0,0). Points in the lower-left quadrant and upper-right quadrant indicate positive (positive) spatial autocorrelation, while points in the upper-left and lower-right quadrants indicate negative spatial autocorrelation. HH and LL squares mean clustering of regions with similar low or high values. The slope of the regression line at these points is the same as the global Moran's I statistic for the row-standardised matrix. If the observations are quite evenly distributed between the four quadrants of the graph, then it can be concluded that there is no spatial autocorrelation. The distribution of squares on the graph and their interpretation is presented in Table 4.2.

The Moran scatterplot can be a tool for diagnosing unusual regions in relation to the global trend. These are the observations in the HL and LH squares. It may also indicate polarisation tendencies of the studied regions and spatial heterogeneity. The Moran scatterplot is derived for given spatial weights matrix W. Depending on its form, the results may vary. When the results coincide, regardless of W, it indicates the resistance of the results to the choice of weights matrix.

The directional coefficient in the regression line (slope) on the Moran scatterplot is the global Moran I statistic. For example, a value of Moran statistic $I = 0.15$ means that the correlation is quite weak (15%) and only about 2.25% of the phenomenon in the region i results from the value of the phenomenon in neighbouring regions. Because the variable x is standardised, statistical interpretation (e.g. the 2σ rule) can be used to evaluate atypical values. The Moran scatterplot can be used for any variable, including the rest of the regression model (Florax & Nijkamp, 2003).

To calculate the global Moran statistics in R, four commands are used: **moran()**, **moran.test()**, **moran.mc()** and **moran.plot()**. They will be discussed briefly subsequently.

moran() – the poorest of commands; only calculates the global Moran's I statistic without testing its significance. The first argument is the variable for which Moran statistics are to be calculated; the second argument is the spatial weights matrix, the *listw* class object and the third argument is the number of observations – it can

---

Table 4.2 Relationship between the region and neighbours for Moran scatterplot

	Low values in the region i (L)	High values in the region i (H)
High values in neighbouring regions (H)	LH square Negative spatial autocorrelation	HH square Positive spatial autocorrelation
Low values in neighbouring regions (L)	LL square Positive spatial autocorrelation	HL square Negative spatial autocorrelation

Source: Own study

be expressed as the length of the data vector in the *nb* class object. The fourth element is the sum of the weights and is an object of the *listw* class.

```
moran(unempl$X2011.06, cont.listw, length(cont.nb), Szero(cont.listw))
```

```
$I
[1] 0.4713219
$K
[1] 2.755576
```

moran.test() – calculates the global Moran's *I* statistic and also tests its significance using a traditional approach. In the options, one can specify the calculation method for variance. Specifying *randomisation=TRUE* (default option), the randomisation approach is selected, while the *FALSE* option uses the approximation with a normal distribution. By default, the greater (*"greater"*) alternative hypothesis assumes positive spatial autocorrelation. This can be changed by writing *alternative = "less"* for the "smaller" or "two-sided test" (*two.sided*). Syntax is like in the **moran**() command.

In the following examples, H_0 with spatial randomness of residuals is rejected in favour of H_1 with positive autocorrelation due to the very low *p*-value. A Moran's *I* statistic with a value of 0.47 means a moderately strong spatial autocorrelation; that is, unemployment rates are similar in neighbouring regions. The test result is the same regardless of how the variance is determined.

```
moran.test(unempl$X2011.06, cont.listw)
```

```
        Moran I test under randomisation
data:     unempl$X2011.06
weights: cont.listw
Moran I statistic standard deviate = 13.914, p-value < 2.2e-16
alternative hypothesis: greater
sample estimates:
Moran I statistic      Expectation        Variance
    0.471321919      -0.002638522      0.001160394
```

```
m3<-moran.test(unempl$X2011.06, cont.listw, randomisation=FALSE)
m3
```

```
        Moran I test under normality
data:     unempl$X2011.06
weights: cont.listw
Moran I statistic standard deviate = 13.918, p-value < 2.2e-16
alternative hypothesis: greater
sample estimates:
Moran I statistic      Expectation        Variance
    0.471321919      -0.002638522      0.001159688
```

Approach of randomisation or approximation by the normal distribution is revealed in the value of variance. Statistics and Moran, as well as its expectation, are the same in both cases. The result of different variances is the different value of test statistics (Moran's *I* statistic standard deviate), which is Moran's *I* statistic standardised with the formula (X-EX)/SX. The *p*-value is the probability of a distribution function of normal distribution. One can get them "manually" by entering:

```
#pvalue for Moran's I (normal distribution)
pval.norm<-1-pnorm(m3$statistic, mean=0, sd=1)
pval.norm
```

```
Moran I statistic standard deviate
                                   0
```

Moran's *I* statistics are significant and indicate moderate positive spatial autocorrelation. It can be assumed that the observations in the neighbouring regions are similar to each other.

moran.mc() – calculates the Moran's *I* statistic and tests its significance with a permutation test based on Monte Carlo simulation, usually with 99 iterations. Placing an iteration number of 99 (or 999) gives a "nice" *p*-value of 0.01 (0.001). Increasing the number of iterations by an order decreases the *p*-value by an order. Syntax is like in the **moran()** command. The expected value and variance of statistics are not determined based on theoretical moments, as in the command **moran.test()**, but based on numerical simulation.

```
moran.mc(unempl$X2011.06, cont.listw, 99)
```

```
          Monte-Carlo simulation of Moran I

date:      unempl$X2011.06
weights:   cont.listw
number of simulations + 1: 100

statistic = 0.47132, observed rank = 100, p-value = 0.01
alternative hypothesis: greater
```

moran.plot() – draws all individual observations (including outliers) and compares them with global Moran statistics, expressed as the angle of the regression line. The first argument is the variable being tested. It can be extracted from the dataset with the **spNamedVec()** command so that the variable values as well as the corresponding region names are imported. The second argument is the spatial weights matrix, the *listw* class object; the third argument defines the variable from which the observation names are to be taken and the last argument defines the type of individual observation tag. It is preferable to carry out operations on a standardised variable, because then it is easy to evaluate atypical observations, for example, using the three-sigma principle (see Figure 4.10a).

```
variable<-unempl$X2011.06
variable.std<-((variable-mean(variable))/sd(variable))
moran.plot(variable.std, cont.listw, labels=as.character(unempl$poviat),
pch=19, quiet=F) # Figure 10a
```

An additional result of this command is the list of the most outliers.[1]

Potentially influential observations of			lm(formula = wx ~ x) :			
	dfb.1_	dfb.x	dffit	cov.r	cook.d	hat
Poviat m.Krosno	0.18	-0.25	0.31_*	0.95_*	0.05	0.01
Poviat przysuski	-0.01	-0.03	-0.03	1.02_*	0.00	0.01
Poviat radomski	-0.02	-0.06	-0.07	1.02_*	0.00	0.02_*
Poviat szydłowiecki	-0.01	-0.05	-0.05	1.04_*	0.00	0.03_*
Poviat m.Kraków	0.01	-0.02	0.02	1.02_*	0.00	0.01
Poviat m.Radom	0.18	0.20	0.27_*	0.95_*	0.04	0.01
Poviat m. st. Warszawa	-0.01	0.02	-0.03	1.02_*	0.00	0.01

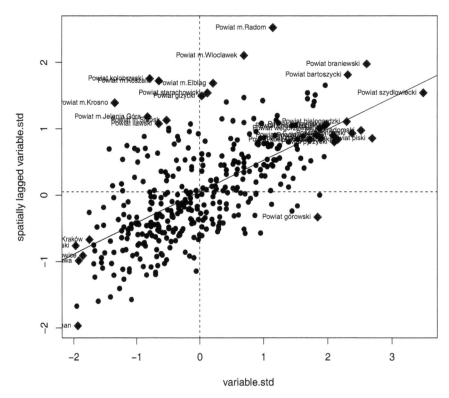

a

Regions belonging to quarters from the Moran scatter plot

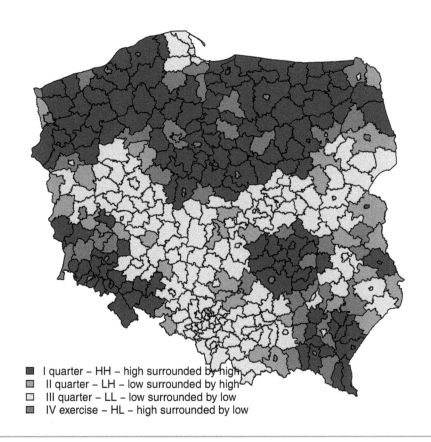

- ■ I quarter – HH – high surrounded by high
- ■ II quarter – LH – low surrounded by high
- □ III quarter – LL – low surrounded by low
- ■ IV exercise – HL – high surrounded by low

b

Figure 4.10 Moran scatterplot: a) using **moran.plot()**, b) visualised on the map
Source: Own study using R and the spdep:: package

The graph shows that most of the outliers highlighted are more than two standard deviations from the mean. Their position relative to the regression line is indicated by the values from the first column of results, dfb.1. Observations with negative dfb.1 values (Przysuski, Radom and Szydłowiecki poviats) are located below the regression line. The values of the *X2011.06* variable in these locations outweigh the values in the neighbouring regions much more than it would appear from the general spatial pattern, that is, hot spots. These are regions with a much more difficult labour market (high unemployment) than their neighbours. Observations above the regression line are regions in a relatively good situation, whose neighbours have higher-than-average unemployment rates. Regions whose values of the standardised variable are below zero are the regions with the lowest unemployment rates relative to the entire population, not just their neighbours.

All statistics have the *zero.policy* option. The default is always *zero.policy=FALSE*, which means that all calculations are made assuming that each region has a neighbour. This means that islands are not allowed to exist. If the analysed set contains islands, enter *zero.policy=TRUE*.

The parameters of the regression line, free expression and directional coefficient from the Moran chart can be calculated manually. To this end, it is necessary to extract the analysed variable, standardise it and determine the spatial lag. Sequences of the spatially lagged variable on the examined variable should be carried out in turn. In the last step, it is enough to draw a regression line and, using the properties of the *htest* class object, use the regression results. The graph is the same as that obtained using **moran.plot**().

```
x<-unempl$X2011.06                # extracting the variable
zx<-scale(x)                      # variable standardisation
mean(zx)                          # average control
sd(zx)                            # standard deviation control
wzx<-lag.listw(cont.listw, zx)    # spatial lag of x
morlm<-lm(wzx~zx)                 # regression
slope<-morlm$coefficients[2]      # directional coefficient
intercept<-morlm$coefficients[1]  # constant term
par(pty="s")                      # square chart window
plot(zx, wzx, xlab="zx",ylab="spatial lag zx", pch="*")
abline(intercept, slope)          # regression line
abline(h=0, lty=2)                # horizontal line at y = 0
abline(v=0, lty=2)                # vertical line at x = 0
```

The Moran scatterplot chart only shows the relationship between the region and its neighbours, not a wider location context. This chart does not show where the points on the map are located, which lie in the next quarters of the chart. Hence, an alternative possibility to visualising the obtained data is to transfer the values from the next quarters of the Moran scatterplot to the colour map.

In the following example, **ifelse**() commands created a variable with the values 1, 2, 3, 4, depending on the location in quadrants – that is, depending on the value of the variable *x* and the value of its spatial lag *wzx*. The graph (see Figure 4.10b) was prepared in four shades of gray.

```
# map of belonging to the quarters of the Moran scatterplot
# creating a variable for analysis
x<-unempl$X2011.06 # creating a variable for analysis
zx<-scale(x) # variable standardisation
wzx<-lag.listw(cont.listw, zx) # spatial lag of x
cond1<-ifelse(zx>=0 & wzx>=0, 1,0) # I quarter
cond2<-ifelse(zx>=0 & wzx<0, 2,0) # II quarter
cond3<-ifelse(zx<0 & wzx<0, 3,0) # III quarter
cond4<-ifelse(zx<0 & wzx>=0, 4,0) # IV quarter
cond.all<-as.data.frame(cond1+cond2+cond3+cond4)

# chart - colour map, Figure 10b
brks<-c(1,2,3,4)
cols<-c("grey25", "grey60", "grey85", "grey45")
```

```
par(mar=c(5.1,1,4.1,1))
plot(pov, col=cols[findInterval(cond.all$V1, brks)])

legend("bottomleft", legend=c("I quarter - HH - high surrounded by high",
"II quarter - LH - low surrounded by high", "III quarter - LL - low
surrounded by low", "IV exercise - HL - high surrounded by low"),
fill=cols, bty="n", cex=0.8)
title(main="Regions belonging to quarters from the Moran scatterplot")
```

4.5.1.2 Global Geary's C statistics

Geary's statistic (Geary, 1954), like Moran's statistic, is an indicator of spatial autocorrelation and is a development of join-count statistics. It is expressed by the formula:

$$C(d) = \frac{(n-1)}{(2\sum_i^n \sum_j^n w_{ij})} \cdot \frac{\sum_i^n \sum_j^n w_{ij}(x_i - x_j)^2}{\sum_i^n (x_i - \bar{x})^2} \qquad (4.9)$$

where all formula components are defined as in Moran's I statistic.

The value of Geary's statistic is always positive and asymptotically normal and fluctuates in the range [0, 2]. The expected value of Geary's statistic is 1, assuming no spatial autocorrelation. In the calculations, as in Moran's I, a standardised first-order matrix is used. C values above 1 indicate differentiation of the observed objects (negative autocorrelation), while C values less than 1 (and greater than 0) indicate similarity of regions (positive autocorrelation). The moments of Geary's statistics were derived by Cliff and Ord (1981). However, as Cliff and Ord (1981) indicate, more effective than Geary's C statistic is Moran's I statistic, which is partly due to the high sensitivity of C statistic variance to sample distribution. Geary statistic significance tests are based on the assumption of symmetry of the spatial weights matrix. For asymmetrical matrices (such as k nearest neighbours), Geary's C statistic may be disturbed.

Spatial autocorrelation statistics (Moran's I and Geary's C) have, in addition to their advantages, a lot of disadvantages (Cliff & Ord, 1981; Anselin, 1988). First, the spatial weights matrix is invariant when the size, shape and relative strength of links between areas change. Second, in the weights matrix, all neighbourhood connections have the same weight, which does not always reflect reality. Third, the level of aggregation of the studied variables is usually arbitrary, but it affects the size of statistics.

To calculate Geary's global statistic in R, three commands are used: **geary()**, **geary.test()** and **geary.mc()**. They will be discussed briefly subsequently.

geary() – only calculates Geary's statistic; this is the poorest and least-used command. The first argument is the examined variable, the second the spatial weights matrix W expressed by the *listw* class object, the third and fourth determine the number of observations n and the last is the sum of the weights.

```
geary(unempl$X2011.06, cont.listw, length(cont.nb), length(cont.nb)-1,
Szero(cont.listw))
```

```
$C
[1] 0.5331133
$K
[1] 2.755576
```

geary.test() – calculates Geary's statistic and also tests it assuming randomisation or normal distribution, similar to **moran.test()**. The command arguments are the same as the first two arguments of the **geary()** command. The form of displaying the result is the same as in the case of Moran's statistic.

```
geary.test(spNamedVec("X2011.06", unempl), cont.listw)
```

```
        Geary C test under randomisation

data:  spNamedVec("X2011.06", unempl)
weights: cont.listw

Geary C statistic standard deviate = 12.061, p-value < 2.2e-16
alternative hypothesis: Expectation greater than statistic
sample estimates:
Geary C statistic      Expectation          Variance
      0.533113331      1.000000000       0.001498602
```

Geary's C statistic is significantly less than 1, which means positive autocorrelation. The result is consistent with Moran's I statistic.

geary.mc() – calculates Geary's statistic and tests it with a permutation test based on Monte Carlo simulation. The first argument is the variable studied; the second is the spatial weights matrix and the third is the number of interactions, usually taken as 99 (as well as 999, 9999 or 99,999).

```
geary.mc(unempl$X2011.06, cont.listw, 99)
```

```
        Monte-Carlo simulation of Geary C

data: unempl$X2011.06
weights: cont.listw
number of simulations + 1: 100

statistic = 0.53311, observed rank = 1, p-value = 0.01
alternative hypothesis: greater
```

Geary's C statistic for unemployment in June 2011 (*X2011.06*) is 0.53 and is significant. By examining changes in statistics between periods, an increase in Geary's C means a decrease in spatial autocorrelation, which can be interpreted as increasing variation in the value of a variable between regions.

4.5.1.3 Join-count statistics

One of the types of statistics testing spatial relationships are join-count statistics. The idea of the test is to divide the variable value into groups, assign a colour to them and apply colours to the map. The number of contacts (connections) of the same colours relative to the total number of contacts is tested. Comparison of the observed frequency with the expected values is the essence of the spatial autocorrelation test. As spatial autocorrelation is considered more frequent than expected in probability contact of objects of the same colour. Tests are carried out for each colour group. The basis is a two-colour black and white system, which is used in the test terminology (BW, black-white).

The highlighted colour (e.g. black) is the examined feature, for example, a high variable level or the presence of a phenomenon. Joint-count statistics determine the contact frequencies of black-black (BB), white-black (BW) and white-white (WW) areas. Hence, J_{BB} and J_{WW} statistics examine positive spatial autocorrelation, while J_{BW} statistics examine negative spatial autocorrelation.

In testing the significance of BW statistics, the observed BW value is compared with its distribution assuming randomness. In the analysis of spatial autocorrelation, they use two definitions of randomness: free sampling and non-free sampling (randomisation). In the case of free sampling, it is assumed that each region may be white or black (or a different colour in multidimensional cases). In the randomisation approach, the number of regions of a given colour is fixed; only the colour assignment to the region is random. In R, tests use the non-free sampling approach on the **joincount.test()** command and free-sampling as Monte Carlo on the **joincount. mc()** command.

Joint-count statistics have been extended to multidimensional data (i.e. taking more than two values). Join count tests examine the existence of positive spatial autocorrelation in groups of "colour" objects. In other words, by dividing the values of objects, for example, into four groups, and assuming the minimum, standard deviation, average and max as the criterion for division, the join count tests will examine whether there is positive spatial autocorrelation in each of the four groups. Expected values are calculated based on the hypergeometric distribution.

There are several uses for join count tests. First, it can be used as an alternative to "non-colour" tests for spatial autocorrelation, that is, the Moran and Geary tests. Second, the test works for qualitative data, because it abstracts from values and only examines the connections of regions. Third, the test is used to analyse the randomness of residuals from the regression model. Model errors are divided into two groups: positive values resulting from forecast underestimation and negative values resulting from forecast revaluation. The randomness of their occurrence is examined. In a situation where spatial autocorrelation occurs, positive and negative errors will be spatially grouped. The non-free sampling approach is used to study regression residuals.

The distribution of test statistics has an asymptotic normal distribution. The moments were derived by Cliff and Ord (1981). The null hypothesis assumes that the distribution of events in space is random and the correlation coefficient equal 0. As an alternative hypothesis, it is assumed that the correlation coefficient is greater than 0.

In R software, there are three commands for join-count tests. They all work for multidimensional divisions (i.e. more than two colour groups). All three tests have similar syntax. To perform the test, one needs to prepare the data in the appropriate format, that is, specify the variable, specify grouping intervals and specify group labels using the **cut**() command, as well as reformatting the data from vector to factor using the **factor**() command. The division into groups of values can be visualised for point and mapped data (see Figure 4.11).

```
summary(unempl$X2011.06)
```

```
 Min.   1st Qu.   Median   Mean   3rd Qu.    Max.
 3.40     10.55    13.90   14.72    18.70    34.80
```

```
head(unempl$X2011.06)
# [1] 23.0 18.7 18.0 18.8 12.6 11.9

variable.f<-factor(cut(unempl$X2011.06,    breaks=c(0,10,    20,    40),
labels=c("low", "medium", "high")))

head(variable.f)
#[1] high average medium average medium average
#Levels: low medium high

# graphic parameters
brks1<-c(0, 10, 20, 40)
cols<-c("green", "blue", "red")

# scatterplot of subsequent values of the tested variable - Figure 4.11a
plot(unempl$X2011.06, bg=cols[findInterval(unempl$X2011.06, brks1)], pch=21)
abline(h=c(10,20,40), lty=3)

# spatial distribution of values into three distinguished groups - Figure 4.11b
plot(pov, col=cols[findInterval(unempl$X2011.06, brks1)])
plot(voi, add=TRUE, lwd=2)
title(main="Unemployment in June 2011")
legend("bottomleft", legend=c("low", "medium", "high"), leglabs(brks1),
fill=cols, bty="n")
```

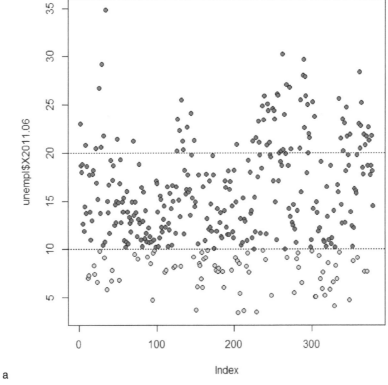

a

Unemployment in June 2011

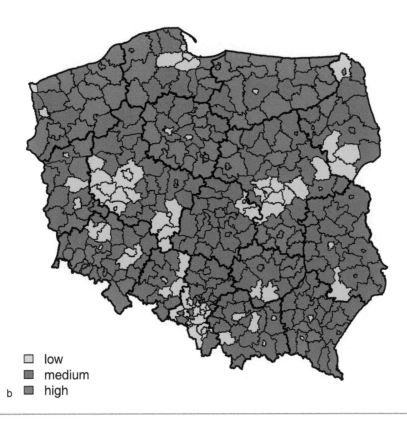

b

Figure 4.11 Graph of variable *X2011.06*: a) scatterplot (*x* = ordinal number, *y* = variable value), b) spatial distribution of values according to value groups

Source: Own study using R and the spdep:: package

The tests described subsequently are carried out on such prepared data.

joincount.test() – tests hypotheses about spatial autocorrelation based on the assumption of randomisation (non-free sampling). It provides the statistic value (standard deviation), its mean, variance, expected value and *p*-value of the test. One test displays *k* tests (for *k* colour groups). The first argument is the variable studied, and the second argument is the spatial weights matrix. In the following printouts, information about which level it is a test for is presented in the phrase *Std. deviate for* (underlined).

```
joincount.test(variable.f, cont.listw)
```

```
        Join count test under nonfree sampling
data: variable.f
weights: cont.listw
Std. deviate for low = 6.137, p-value = 4.204e-10
alternative hypothesis: greater
sample estimates:
Same colour statistic          Expectation          Variance
         16.070527                8.762533          1.418012

        Join count test under nonfree sampling
data: variable.f
weights: cont.listw
Std. deviate for medium = 4.2663, p-value = 9.935e-06
alternative hypothesis: greater
sample estimates:
Same colour statistic          Expectation          Variance
         75.282233               65.899736          4.836438

Join count test under nonfree sampling
data: variable.f
weights: cont.listw
Std. deviate for high = 8.5773, p-value < 2.2e-16
alternative hypothesis: greater
sample estimates:
Same colour statistic          Expectation          Variance
         16.495833                7.126649          1.193161
```

The printout shows that the H_0 hypotheses about randomness in each colour group should be rejected (due to low *p*-value). This means that the observations in all three groups of the values of the examined variable are focused more often than it would appear from the random distribution.

joincount.mc() – tests the existence of spatial autocorrelation based on the assumption of free sampling using Monte Carlo simulation. It provides the join-count statistic value, its mean and variance and the *p*-value of the test. Similar to the **joincount.test()** command, all *k* tests are displayed for individual colour groups. Syntax as previously; only the number of iterations of the simulation is added. The result is very similar to the one from the **joincount.test()** command.

```
joincount.mc(variable.f, cont.listw, 99)
```

joincount.multi() – tests spatial autocorrelation in and between colour groups. Combinations of individual colours (e.g. red-white, white-white, black-black), as well as all combinations (e.g. red-white, red-black, white-black), are tested. The printout includes the statistic's value, its variance, its expected value and *z*-value. The *z*-value should be compared with the distribution tables of the normal distribution.[2] The syntax is the same as in the **joincount.test()** command.

```
joincount.multi(variable.f, cont.listw)
```

```
              Joincount  Expected  Variance  z-value
low:low         16.0705    8.7625    1.4180   6.1370
medium:medium   75.2822   65.8997    4.8364   4.2663
high:high       16.4958    7.1266    1.1932   8.5773
medium:low      39.4716   48.4644    5.8712  -3.7114
high:low         4.9365   16.0106    2.6733  -6.7730
high:medium     37.7433   43.7361    5.3545  -2.5898
Jtot            82.1514  108.2111    8.3075  -9.0414
```

The previous result means that the statistics of all colour combinations are significant. It should be interpreted that the connections of different colour groups are more frequent or rarer than the results from a random distribution. This is due to the clustering of the values of highlighted colour groups and more frequent than random connections between them and, as a consequence, less frequent than random connections between observations from different colour groups.

4.5.2. Local spatial autocorrelation statistics

Local indicators of spatial relationships (LISA) were proposed by Anselin (1995). LISA includes local Moran's I_i and local Geary's C_i statistics. Local Moran allows the identification of spatial agglomeration effects (similar to G Getis-Ord statistics), while local Geary shows spatial similarities and differences. Local Moran statistics give information on low- or high-value clusters, while Geary's local statistics show average differences between the object and neighbours, which helps to find outliers and similarity/difference patterns. These statistics are related to their global counterparts and can be used to estimate the impact of individual statistics on their global counterparts.

Thanks to LISA, it is possible to identify so-called hot spots, that is, centres with high values surrounded by low values, as well as local clusters in the absence of global autocorrelation. High-value islands can be interpreted not only as hot spots but also as outliers. Then local statistics are an indicator of local instability and local deviations from the global autocorrelation pattern.

4.5.2.2 Local Moran's I statistics (local indicator of spatial association)

Local Moran's I statistics measure whether a region is surrounded by neighbouring regions with similar or different values of the examined variable in relation to the random distribution of these values in space. I_i is a smoothed index for individual observations, so it can be used to find the so-called hot spots and local clusters. Local Moran's I statistics are proportional to Moran's global statistics. It is expressed by the formula:

$$I_i = \frac{(x_i - \bar{x})\sum_{i=1}^{n} w_{ij}(x_j - \bar{x})}{\sum_{i=1}^{n}(x_i - \bar{x})^2 / n}$$

(4.10)

where the w_{ij} elements come from a spatial weights matrix row standardised to one. Statistical significance tests are based on the distribution resulting from conditional randomisation or permutation (Anselin, 1995), and hypotheses are verified on the basis of a pseudo level of significance. Standardised local Moran statistics assume significantly negative values when the object i is surrounded by relatively low values in neighbouring objects and significantly positive values when the object i is surrounded by relatively high values. A low p-value ($p < 0.05$) means that the region i is surrounded by relatively high values, while a high p-value ($p > 0.95$) means that the region i is surrounded by relatively low values (Bao, 2000).

The combination of information from the Moran scatterplot and the significance of LISA statistics can be represented on the map. The four colours then indicate the belonging of a given object to a given square of the Moran scatterplot (Anselin and Bao, 1997). Thanks to such a map, it is quite easy to visually identify spatial regimes.

In R, local Moran statistics are calculated using the **local.moran**() command. Its arguments are: the examined variable, *listw* class spatial weights matrix and options. As a result, local statistics are obtained for each region (I_i), expected value for the entire area ($E.I_i$), variance for the given area ($Var.I_i$), test statistics ($Z.I_i$) and p-value

($Pr(z > 0)$). It is also possible to determine the approximation of Moran's I statistic with a saddlepoint by using the **localmoran.sad**() command. This modification of Moran's statistic has a different variance because it is based on the assumption of randomisation.

In the example subsequently, local Moran statistics for the unemployment rate were determined in June 2011. The code generating the list of local statistics with the name of the region was used for this. The **localmoran**() command from the spdep:: package was used. Unfortunately, in this case, the command does not work – the names of regions are duplicated for 10 poviats, which causes an error.

```
locM<-localmoran(spNamedVec("X2011.06", unempl), cont.listw)
oid1<-order(unempl$ID_MAP)
printCoefmat(data.frame(locM[oid1,],        row.names=unempl$poviat[oid1]),
check.names=FALSE)
```

```
Error in command 'data.frame(locM[oid1,], row.names = unempl$poviat[oid1])':
duplicate row.names: Poviat tomaszowski, Poviat krośnieński, Poviat brzeski,
Poviat ostrowski, Poviat opolski, Poviat grodziski, Poviat nowodworski,
Poviat bielski, Poviat średzki, Poviat świdnicki
```

One can work around this situation as follows. One must create a vector for duplicate poviat names and slightly modify the second instance of the name by adding an apostrophe. After replacing the vector names with new ones (with an apostrophe), the **localmoran**() command works fine.

```
# duplicate names vector
dupli<-c("Powiat tomaszowski", "Powiat krośnieński", "Powiat brzeski",
"Powiat ostrowski", "Powiat opolski", "Powiat grodziski", "Powiat
nowodworski", "Powiat bielski", "Powiat średzki", "Powiat świdnicki")

# change data type from factor to character
unempl$poviat<-as.character(unempl$poviat)

# loop that adds an apostrophe to the second instance of names
for(i in 1: length(dupli)){
a<-which(unempl$poviat==dupli[i])[2]
unempl$poviat[a]
unempl$poviat[a]<-paste0(as.character(unempl$poviat[a]), "'")}

# restoring the original class variable
unempl$poviat<-as.factor(unempl$poviat)

# code generating a list of local Moran statistics with region names assigned
locM<-localmoran(spNamedVec("X2011.06", unempl), cont.listw)
oid1<-order(unempl$ID_MAP)
locMorMat<-printCoefmat(data.frame(locM[oid1,],
row.names=unempl$poviat[oid1]), check.names=FALSE)
```

It is important to remember that p-value < 0.05 and p-value > 0.95 are assessed for significance. As many as 99 regions have p-value < 0.05 and local Moran statistics significantly positive, which means that these areas are surrounded by relatively high values of the unemployment rate. These areas are therefore local "employment centres". On the contrary, only four regions have p-values > 0.95 and show values of local Moran statistics significantly less than 0, which means that they are surrounded by relatively low values and are local "unemployment centres".

```
# printing of the first six significant values <0.05
head(locMorMat[locMorMat$Pr.z . . . 0.<0.05,])
```

```
                          Ii         E.Ii      Var.Ii     Z.Ii Pr.z . . . 0.
Powiat otwocki        1.2937341 -0.002638522 0.1969754 2.920948  1.744840e-03
Powiat piaseczyński   1.7993769 -0.002638522 0.1969754 4.060247  2.451037e-05
Powiat płocki         0.7804189 -0.002638522 0.1399479 2.093199  1.816571e-02
Powiat pruszkowski    2.1734146 -0.002638522 0.2468745 4.379569  5.945712e-06
Powiat przysuski      1.8772801 -0.002638522 0.1399479 5.025230  2.514154e-07
Powiat radomski       2.4565547 -0.002638522 0.1221268 7.036995  9.821510e-13
```

```
# printing significant values> 0.95
locMorMat[locMorMat$Pr.z . . . 0.>0.95,]
```

```
                          Ii         E.Ii      Var.Ii      Z.Ii Pr.z . . . 0.
Powiat m.Krosno      -1.8692118 -0.002638522 0.9953609 -1.870918    0.9693218
Powiat suwalski      -0.7238298 -0.002638522 0.1637093 -1.782435    0.9626608
Powiat iławski       -0.6975976 -0.002638522 0.1637093 -1.717602    0.9570654
Powiat kołobrzeski   -1.3825399 -0.002638522 0.1969754 -3.109153    0.9990619
```

Significant local Moran statistics are plotted on the map, that is, for which the *p*-value is below 0.05 or above 0.95. To facilitate data operations, the variable header has been changed from *Pr.z. . . > 0)* to *Prob* with the **names()** command.

```
# map of the significance of Moran's local statistics
names(locMorMat)[5]<-"Prob"
brks<-c(min(locMorMat[,5]), 0.05000, 0.95000, max(locMorMat[,5]))
cols<-c("grey30", "grey90", "grey60")

plot(pov, col=cols[findInterval(locMorMat[,5], brks)])
legend("bottomleft", legend=c("surrounded by relatively high values,
locM>0", "insignificant", "surrounded by relatively low values, locM<0"),
fill=cols, bty="n", cex=0.75)
title(main=" Local Moran statistics ", cex=0.7)
plot(voi, add=TRUE, lwd=2)
```

Regions marked in dark gray are relative employment centres, because they are surrounded by regions with significantly higher values of the examined variable (see Figure 4.12a). There are also four areas on the map that can be described as statistically significant unemployment centres, that is, regions surrounded by regions with significantly lower values of the studied variable.

4.5.2.3 Local Geary's C statistics

The local version of Geary's statistic is expressed by the formula:

$$C_i(d) = \sum_{j \neq i}^{n} w_{ij}(Z_i - Z_j)^2 \tag{4.11}$$

where Z_i and Z_j are standardised values, and w_{ij} are from the matrix W, row-standardised. The student's *t* test assumes the null hypothesis $C_i = 0$. A value of $C_i < 0$ means positive spatial relationships and similarities of the region and with neighbours, while $C_i > 0$ means neighbour diversity and negative spatial autocorrelation. Anselin (2019) proposes to expand this statistic. Unfortunately, the statistics for one are not programmed in R taking into account the spatial weights matrix. There is the usdm:: **lisa()** command that allows the user to calculate this statistic, but for rasters with a roaming neighbourhood.

4.5.2.4 Local Getis-Ord G_i statistics

In detecting the existence of spatial relationships in the studied variable, the local G statistic is used, derived by Ord and Getis (1995). It is expressed by the formula:

$$G_i(d) = \left(\sum_{j,j\neq i}^{n} w_{ij}x_j\right) / \left(\sum_{j,j\neq i}^{n} x_j\right) \tag{4.12}$$

where n is the number of observations, x_i is the value observed in the region i and w_{ij} are elements of a symmetrical binary spatial weights matrix. The G_i statistic compares the sum of values in neighbouring areas at a distance with the sum of values in all areas.

The student's t test assumes the null hypothesis that $G_i = 0$. Positive and significant statistical values in the region mean a grouping of high-value regions (region and neighbouring regions), the so-called high-value cluster. Negative values of the G statistic indicate the existence of a cluster of regions with low values of the examined variable, that is, the region i is surrounded by similar low-value regions. As in the case of local Moran statistics, the p-value is used. Positive and significant Getis-Ord statistics have a p-value below 0.05, while negative and significant statistics have a p-value above 0.95. G can be used in the analysis of agglomeration effects and high- or low-value clusters.

G_i^* (Gstar) statistics are also used. It differs from the G statistics by the construction of the spatial weights matrix. In G_i^*, the diagonal of the matrix W contains non-zero elements, which are interpreted as a region that borders on itself; therefore, the measure G^* can be used to detect clusters – high $Gi\,s$ results occur for clusters of elevated values of a variable. Statistics then compare the sum of the values in the neighbourhood to the sum for all observations.

In R, the **localG()** command is used to count these statistics. As a result, the standardised value of the G (or G^*) statistic is displayed. The critical values of the test at the 5% significance level depend on the sample size and are: for $n = 1$ 1.645, for $n = 50$ 3.083, for $n = 100$ 3.289 and for $n = 1000$ 3.886 (according to the description of the **localG()** command of the spdep:: package). The only difference between G and G^* is the *listw* object specification. To determine G^* (*locGstar*), one must modify the spatial weights matrix so that the studied region is one's own neighbour – just write *nb2listw (include.self (cont.nb))*.

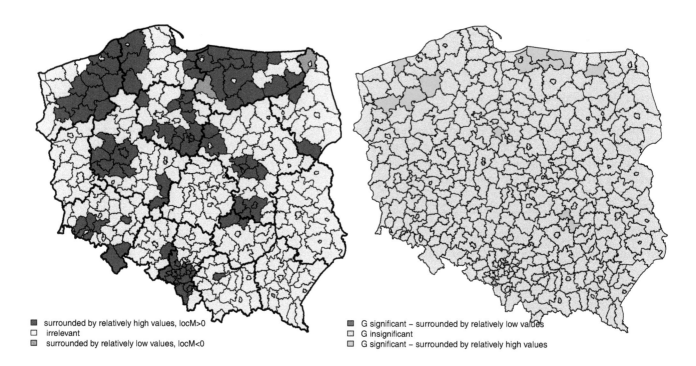

a **Local Moran statistics**　　　　　　b **relevant statistics G**

- ■ surrounded by relatively high values, locM>0
- □ irrelevant
- ▨ surrounded by relatively low values, locM<0

- ▨ G significant – surrounded by relatively low values
- □ G insignificant
- □ G significant – surrounded by relatively high values

Figure 4.12 Significant local statistics: a) local Moran, b) local Getis-Ord
Source: Own study

In the example subsequently, localG and localGstar and their significance were determined, and information on significant values was displayed and visualised on the map (see Figure 4.12b).

```
# local Gi statistics
locG<-localG(unempl$X2011.06, cont.listw)
# locGstar<-localG(unempl$X2011.06, nb2listw(include.self(cont.nb)))
summary(locG)
```

```
 Min.    1st Qu.   Median   Mean 3rd Qu.    Max.
-3.23502 -1.09334 -0.05330 0.09059 1.29399 4.13362
```

```
# significance t-student test for n=100
sig<-ifelse(locG<=-3.289 | locG>=3.289, "*", " ")
which(sig=="*")
#[1] 34 226 243 247 251 253 256 288 289 294 298 348

unempl[which(sig=="*"), c(3,6,10,18)]
```

```
                 poviat        voivodeshipo       odle X2011.06
 34     Poviat szydłowiecki      Mazowieckie   46.26772    34.8
226   Poviat starachowicki    Świętokrzyskie   40.36918    15.4
243   Poviat szczecinecki  Zachodniopomorskie 134.55000    25.1
247    Poviat stargardzki  Zachodniopomorskie  40.94362    19.6
251     Poviat elbląski  Warmińsko-Mazurskie   67.19430    24.4
253     Poviat giżycki   Warmińsko-Mazurskie   93.28757    14.9
256   Poviat lidzbarski  Warmińsko-Mazurskie   37.02356    26.0
288  Poviat bartoszycki  Warmińsko-Mazurskie   55.09402    28.0
289   Poviat braniewski  Warmińsko-Mazurskie   65.06196    29.7
294      Poviat drawski  Zachodniopomorskie   91.59537    25.0
298   Poviat kołobrzeski  Zachodniopomorskie   96.11545    10.2
348 Poviat aleksandrowski  Kujawsko-Pomorskie  26.00010    18.7
```

```
# graph of significant locG statistics
brks<-c(-10,-3.289, 3.289, 10)
cols<-c("grey40", "grey85", "grey55")
cols<-c("red", "grey85", "green")
plot(pov, col=cols[findInterval(locG, brks)])
legend("bottomleft", legend=c("G significant - surrounded by relatively low
values", "G insignificant", "G significant - surrounded by relatively high
values"), fill=cols, bty="n", cex=0.75)
title(main="Significant G statistics")
```

As can be seen from Figure 4.12b, in most regions, the local Getis-Ord statistics are insignificant. Green areas are local employment centres, as they are regions surrounded by neighbours with relatively high values of the examined variable. G_i^* statistics have more variance than G_i statistics. In addition, more local statistics have proved to be significant. These statistics show more local employment centres than the basic Getis-Ord statistics. There are also areas surrounded by relatively low values – unemployment centres, marked in red.

4.5.2.5. Local spatial heteroscedasticity

Extending the analysis of neighbourly similarity by the stability of a variable in space is possible using the local spatial heteroscedasticity (LOSH) (H_i) measure (local spatial heteroscedasticity), complementary to the localG

(G_i) statistics described earlier. The combined analysis of both indicators allows the measuring of the similarity and instability of a variable in space. H_i statistics measure the variance of a variable in the neighbourhood, indicating areas with uniform and varied variability, in an analogous way to the measure of autocorrelation in the G_i indicator (Ord & Getis, 2012). The local measure of spatial heteroscedasticity LOSH (H_i) is given by the formula:

$$H_i(d) = \frac{\sum_j w_{ij}(d) \mid e_j(d) \mid^a}{\sum_j w_{ij}(d)} \qquad (4.13)$$

where $e_j(d) = x_j - \bar{x}_j(d)$ for $j \epsilon N(i,d)$ plays the role of the rest, and $\bar{x}_i(d) = \left[\sum_j w_{ij}(d) x_j \right] \Big/ \left[\sum_j w_{ij}(d) \right]$ is

the determinant of the local average. Therefore, $H_i(d)$ is a measure of the absolute deviation for $a = 1$ and the variance for $a = 2$. High H_i values indicate heterogeneous surroundings (differentiation), while low values indicate local homogeneity (similarity). The spatial weights matrix can be defined as the matrix of neighbours in a radius of d km (for point data) but also as a matrix based on the contiguity criterion (for area data and rasters). H_i and G_i statistics can be calculated for individual and aggregated data, which changes the level of analysis.

The combined analysis of both statistics allows the determination of the relative level of the phenomenon (G_i) and its local diversity (H_i) (Ord & Getis, 2012). Both indicators, as local measures, are helpful in determining the stability and similarity of spatial patterns. G_i statistics allow the detection of high-value clusters (by tracing spatial patterns and distribution of values on the map), and the H_i measure allows one to determine to what extent the pattern is constant (or varied) in space.

The H_i measure can be calculated using the spdep:: package and the **LOSH()** command from the spdep:: package. The command gives two basic arguments – a vector of the variable describing the phenomenon and a matrix of weights determining the spatial relationships. Then the options for the specification of the indicator are determined. The default values are $a = 2$ (determining the form of the function for residuals) and *var_hi=T* (moments and test values will be calculated for each location separately). The next two arguments shape behaviour towards missing values. It is convenient to use the options *zero.policy=T* and *na.action=na.exclude*, which allow one to skip missing items.

In the example subsequently, a joint analysis of G_i and H_i statistics for unemployment data was carried out in June 2011. For this purpose, two colour maps were made, indicating areas with high and low values of both measures. A comparative analysis of both drawings will allow the determination of spatial patterns and their stability.

The G_i statistics were calculated analogously to the previous example. The changes were made to the distribution of values on the chart. This time, a different colour palette and division based on quartiles was used.

```
# local Gi statistics
locG<-localG(unempl$X2011.06, cont.listw)
a<-summary(locG)
```

```
     Min.  1st Qu.   Median    Mean 3rd Qu.     Max.
  -3.23502 -1.09334 -0.05330 0.09059 1.29399 4.13362
```

```
# graph of Gi statistics, Figure 13a
brks<-c(a[1], a[2], a[3], a[5], a[6])
colfunc<-colorRampPalette(c("royalblue", "springgreen", "yellow", "red"))
coli<-colfunc(5)
plot(pov, col=coli[findInterval(locG, brks)], main=" Gi statistics")
legend("bottomleft", legend=c("Very low", "Low", "Medium", "High", "Very
high"), fill=coli, bty="n")
```

LOSH statistics were calculated for the same arguments. The division into value ranges was also based on positional statistics of the calculated measure.

```
#local Hi LOSH
locH<-LOSH(unempl$X2011.06, cont.listw, a=2, var_hi=TRUE, zero.policy=TRUE,
na.action=na.exclude)
summary(locH)
```

```
       Hi                  E.Hi           Var.Hi              Z.Hi
Min.   :0.000753   Min.    :1    Min.    :0.2289    Min.    : 0.000582
1st Qu.:0.331463   1st Qu.:1    1st Qu.:0.3635    1st Qu.: 1.343089
Median :0.695751   Median :1    Median :0.5117    Median : 2.738944
Mean   :0.909153   Mean    :1    Mean    :0.6784    Mean    : 3.714242
3rd Qu.:1.245673   3rd Qu.:1    3rd Qu.:0.6413    3rd Qu.: 5.207994
Max.   :4.987379   Max.    :1    Max.    :2.5857    Max.    :19.520694

    x_bar_i              ei
Min.   : 3.40    Min.    :   0.00062
1st Qu.:11.95    1st Qu.:   1.53880
Median :14.59    Median :   7.86378
Mean   :15.02    Mean    :  19.85061
3rd Qu.:18.04    3rd Qu.:  23.32132
Max.   :29.20    Max.    : 246.49000
```

```
b<-summary(locH[,"Hi"])
b
```

```
    Min.   1st Qu.   Median     Mean  3rd Qu.      Max.
0.000753 0.331463 0.695751 0.909153 1.245673 4.987379
```

```
# LOSH statistics mapped, Figure 4.13b
brks<-c(b[1], b[2], b[3], b[5], b[6])
plot(pov, col=coli[findInterval(locH[,"Hi"], brks)])
legend("bottomleft", legend=c("Very low", "Low", "Medium", "High", "Very
high"), fill=coli, bty="n")
title(main="Hi statistics")
```

The first graph (Figure 4.13a) illustrates spatial autocorrelation. One can see several areas with very high values (marked in red) and wide areas of high values. Comparing them with the LOSH statistics map (Figure 4.13b) allows the determination of the stability of the pattern, based on variance in space. A grouping of high G_i values is visible in the northern part of the map (Warmian-Masurian, Kuyavian-Pomeranian, Pomeranian and West Pomeranian voivodeships). In the analogous area, different levels of local and H_i diversity are visible: low in the middle of the band and high in the extreme eastern and western parts. This means that the central-northern part is a relatively stable cluster of high unemployment rates, while the northeast and northwest areas have a high level of locally differentiated unemployment rate. Other patterns are observed in the central-southern part of Poland. Low levels of the phenomenon (navy blue) have a low level of diversity in Silesia (central southern part) and in the neighbourhood of Warsaw. The Świętokrzyskie voivodeship is quite heterogeneous – it has a high level of unemployment, and at the same time, it's high diversity. Increased LOSH suggests spatial discontinuity. It is a cluster of distinct values with a heterogeneous environment – neighbouring areas show different values of a characteristic.

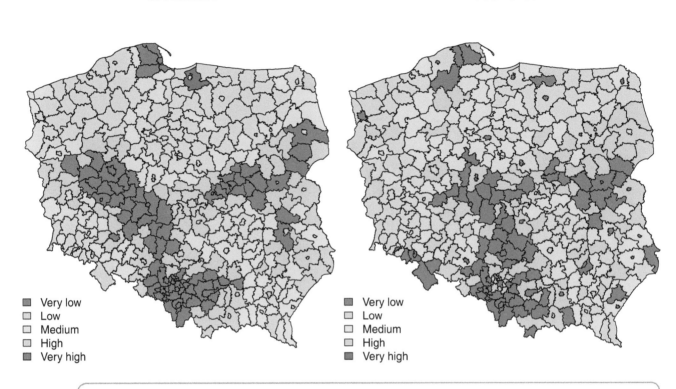

| a | **Gi statistics** |
| b | **Hi statistics** |

■ Very low
□ Low
□ Medium
□ High
■ Very high

■ Very low
□ Low
□ Medium
□ High
■ Very high

Figure 4.13 Common analysis of local statistics: a) measure of spatial correlation G_i, b) measure of spatial heteroscedasticity

Source: Own study

4.6 Spatial cross-correlations for two variables

The spatialEco:: package has a command available to determine local cross-correlations (for two variables) using local Moran statistics (LISA). This is the implementation of the Chen method (2015). He introduced the global spatial cross-correlation index (GSCI), which is expressed by the formula:

$$R_c = x^T W y \tag{4.14}$$

and is a statistic scaled between [–1, 1]; x and y are the variables studied, and W is the spatial weights matrix. This is an indirect measure that tests correlation due to distance.

Chen (2015) shows a relationship between the traditional correlation of two variables and its GSCI as:

$$R_0 = R_c + R_p \tag{4.15}$$

where R_0 is a typical Pearson correlation coefficient, where dependence consists of variable relations and relations in space; R_c is the GSCI coefficient – an indirect measure based on effects in space and R_p is a partial spatial cross-correlation, which expresses the direct influence of variables without space effects.

The global measure is supplemented by local measures, the so-called local spatial cross-correlation index (LSCI), based on the concept of local Moran statistics – LISA – expressed as:

$$R_i^{(xy)} = x_i \sum_{j=1}^{n} w_{ij} y_j \tag{4.16}$$

$$R_j^{(yx)} = y_i \sum_{j=1}^{n} w_{ij} y_j \tag{4.17}$$

where *i* and *j* are different locations studied in pairs, and w_{ij} are the spatial weights assigned to these locations.

As Chen (2015) writes, global SCI reflects the sum of cross-correlations between any two (all) elements, while local SCI measures the cross-correlation between a given element and all others in the system.[3]

The **crossCorrelation**() command from the spatialEco:: package requires two variables, as well as an argument for the spatial structure: spatial weights matrix in the *matrix* class – a typical neighbourhood matrix in the *nb* class converted to the *matrix* class with the **nb2mat**() command or coordinates from which there are distances later to be calculated – for example, determined as poviat centroids with the **coordinates**() command. The options should specify the calculation details, including the type of measure in the *type* option (local LSCI for each region/point or global GSCI as a single value for all observations), the number of simulations to perform permutation tests in the *k* option, standardisation of variables in the option *scale.xy*, standardisation of statistics for the range [–1, 1] in the *scale.partial* option, standardisation of spatial weights matrices by lines up to 1 in the *scale.matrix* option and level of significance in the *alpha* option.

```
library(spatialEco) # calculations, crossCorrelation()
library(GISTools) # graphics, choropleth()

cont.mat<-nb2mat(cont.nb) # W matrix converted to class matrix

x1<-unempl[,101]
x2<-unempl[,100] # date month to month

# version with matrix W
ii<-crossCorrelation(x1, x2, w=cont.mat, type=c("LSCI", "GSCI"), k=99,
scale.xy=FALSE, scale.partial=TRUE, scale.matrix=TRUE, alpha=0.05)
ii
```

```
Moran's-I under randomisation assumptions. . .
   First-order Moran's-I: 0.1971509
   First-order p-value: 0
Chen's SCI under randomisation assumptions. . .

Summary statistics of local partial cross-correlation [xy]
   Min. 1st Qu. Median Mean 3rd Qu. Max.
-1.0000-0.9111-0.8102-0.7204-0.6451 1.0000

   p-value based on 2-tailed t-test: 0.2727273
   p-value based on 2-tailed t-test observations above/below CI: 0.5454545
```

```
# version with coordinates and distance
iii<-crossCorrelation(x1, x2, coords=crds.pov, type = c("LSCI", "GSCI"),
k=99, dist.function="inv.pover", scale.xy=FALSE, scale.partial=TRUE, scale.
matrix=FALSE, alpha=0.05)
iii
```

```
Moran's-I under randomisation assumptions. . .
   First-order Moran's-I: 0.167682
   First-order p-value: 0.9292929
Chen's SCI under randomisation assumptions. . .

Summary statistics of local partial cross-correlation [xy]
   Min. 1st Qu. Median Mean 3rd Qu. Max.
-1.0000-0.7392-0.5738-0.5226-0.3542 1.0000

   p-value based on 2-tailed t-test: 0.3030303
   p-value based on 2-tailed t-test observations above/below CI: 0.6060606
```

```
cor(x1,x2) # Pearson correlation
#[1] 0.9984963

head(iii$SCI[,"lsci.xy"]) # accessing the result
#[1] 0.1171843-0.2141666-0.2460691-0.1858888-0.3914828-0.6620357

# mapping of result ii
choropleth(pov, ii$SCI[,"lsci.xy"])
shades<-auto.shading(ii$SCI[,"lsci.xy"])
choro.legend(14, 50.25, shades, cex=0.65, bty="n")
title(main="Contiguity matrix, date month to month")

# mapping of result iii
choropleth(pov, iii$SCI[,"lsci.xy"])
shades<-auto.shading(iii$SCI[,"lsci.xy"])
choro.legend(14, 50.25, shades, cex=0.65, bty="n")
title(main="Inverse square distance matrix, date month to month")
```

The result of the spatial cross-correlation shows that the global cross-correlation is GSCI = 0.197 and is significant (p-value= 0.00); that is, both variables are correlated with each other and in space. This is the correct result – the unemployment rate shows strong dependence in time (path dependence) and strong dependence in space. With Pearson's correlation coefficient $R_0 = 0.99$, it means that the space effect is 20% (0.197/0.99), while the effect of variable association is (0.99–0.20)/0.99 = 80%. The effect of the phenomenon trajectory is about four times stronger than the effect of interdependence (similarity) in space.

Local statistics for *lsci.xy* should be interpreted as the influence of x on y (i.e. variable $x1$ on variable $x2$). In the example, they are mainly negative, which results from the change of all interest rates in one direction. When plotted, clusters of high and low values are visible. Light areas (strong negative spatial cross-correlation) mean much stronger spatial differentiation (process relation and spatial effects) than darker fields (mean negative spatial cross-correlation). Using a square inverse distance to model relationships increases diversity (more light fields).

4.7 Correlogram

The correlogram is a graph showing the changes in spatial autocorrelation that occur with the change in distance between neighbours (Oden, 1984). In its basic form, it can be a table whose elements are subsequent statistics' values along with their significance test. Autocorrelation is used in a wide spectrum of fields – from time series analysis (to determine the randomness of a variable and carry out the Box-Jenkins procedure) (Harvey & Todd, 1983), through epidemiology (Cliff & Ord, 1981) and to study natural resource allocation or diffusion estimation innovation (Oden, 1984).

The spatial correlogram is used to analyse the degree of spatial correlation depending on the distance between objects or the degree of spatial lag. The correlogram is conceptually similar to the variogram, except that it measures spatial correlation, not variance (or semi-variance). There are two methods for estimating the correlogram: the correlation method, where the classic correlation coefficient for each group is calculated, and the Moran method, where Moran's *I* statistics are calculated. Autocorrelation, positive or negative, exists when the value in the area under study can be determined based on the value of observations in neighbouring areas. Lack of autocorrelation (i.e. the correlation coefficient or Moran's *I* do not differ significantly from 0) means randomness. The entire correlation is significant when at least one correlation coefficient or Moran's *I* value is significant (in one of the spatial lags).

In the correlogram interpretation, several spatial lags (usually about five or six) are assumed in the initial model, and their significance is checked. The continuity of subsequent lags is important from the first lag. If there is a "continuity gap"; that is, the first and fourth lags are significant, then this may indicate data instability. In this situation, only the first spatial lag is considered. The direction of changes in the spatial autocorrelation coefficient is also important. In the event of an autocorrelation coefficient increasing to the second or third lag, this may mean the existence of macro-regions whose operation is based on economies of scale or on a similar climate or similar natural conditions in, for example, agriculture. Otherwise, if the autocorrelation drops since the first lag, such a concentric pattern can mean the existence of several leading regions whose relationship with neighbours decreases with distance. The reason may be high transport costs or negative range benefits. This spatial pattern is confirmed by the core-peripheral model. Completely different patterns may also appear, for example, negative correlation between close neighbours and so on.

To estimate the correlation in R, use the **sp.correlogram**() command from the spdep:: package. To estimate the correlation method using the Moran method, write *method="I"* in the options and *method="corr"* for the correlation method. The result of the operation is an object of the *spcor* class. The disadvantage of this command is displaying only the spatial correlation coefficient, without its significance. The significance of the classic correlation coefficient *r* can be calculated using the basic formula:

$$\frac{r}{\sqrt{1-r^2}}\sqrt{n-2} \qquad (4.18)$$

whereas in the case of the Moran's *I* autocorrelation coefficient, standardisation according to:

$$\frac{I - E(I)}{S(I)} \qquad (4.19)$$

The following example estimates the correlation for the poviat unemployment rate recorded in May 2018. The classic correlation coefficient (for which the significance test was calculated) and the Moran spatial autocorrelation coefficient were used.

```
# correlogram based on the classic correlation coefficient
corr.classic<-sp.correlogram(cont.nb,      unempl$X2018.05,    order=6,
method="corr")
print(corr.classic)
```

```
Spatial correlogram for unempl$X2018.05
method: Spatial autocorrelation
             1            2            3            4            5            6
0.62644416   0.43323902   0.23791504   0.07287527   0.01866754  -0.03291839
```

```
stat<-(dim(unempl)[1]-2)^0.5*corr.classic$res/(1-corr.classic$res^2)^0.5
pvalue<-1-pnorm(abs(stat), mean=0, sd=1)
corr.classic.sig<-cbind(corr.classic$res, stat, pvalue)
options(scipen=999, digits=2)
corr.classic.sig
```

```
        stat      pvalue
1 0.626  15.63  0.00000000
2 0.433   9.35  0.00000000
3 0.238   4.76  0.00000096
4 0.073   1.42  0.07771150
5 0.019   0.36  0.35830178
6-0.033  -0.64  0.26097131
```

```
corr.moran<-sp.correlogram(cont.nb, unempl$X2018.05, order=6, method="I")
print(corr.moran)
```

```
Spatial correlogram for unempl$X2018.05
method: Moran's I
            estimate  expectation  variance  standard deviate  Pr(I) two sided
1 (380)   0.45134152  -0.00263852  0.00115823         13.3395       < 2.2e-16 ***
2 (380)   0.23214989  -0.00263852  0.00045363         11.0236       < 2.2e-16 ***
3 (380)   0.10150598  -0.00263852  0.00029137          6.1012       1.053e-09 ***
4 (380)   0.02667282  -0.00263852  0.00022042          1.9743         0.04835 *
5 (380)   0.00608868  -0.00263852  0.00017886          0.6526         0.51404
6 (380)  -0.00961604  -0.00263852  0.00015663         -0.5575         0.57717
---
Signif. codes: 0 '***' 0.001 '**' 0.01 '*' 0.05 '.' 0.1 ' ' 1
```

The correlogram can be drawn using the basic **plot**() command, declaring the result of the **sp.correlogram**() command as an argument.

```
plot(corr.classic)
plot(corr.moran) # Figure 4.14a
```

Only six spatial lags could be created for the studied variable. It should be remembered that the contiguity spatial weights matrix was used for the calculations. It can be seen that the first four spatial lags are sequentially significant (see Figure 4.14a). This means that the spatial relationship occurs not only between neighbours sharing a common border but also further neighbours. From the fifth lag, the variable distribution is random. The significance of the sixth lag may indicate that data is not stationary. Spatial autocorrelation between nearest neighbours is moderate (0.45). For unemployment variability in May 2018, as much as 20.25% (0.45^2) can be explained by location. Further neighbourhood also has an impact on the labour market, albeit to a lesser extent. The second row of neighbourhoods translates only in 5.3% volatility.

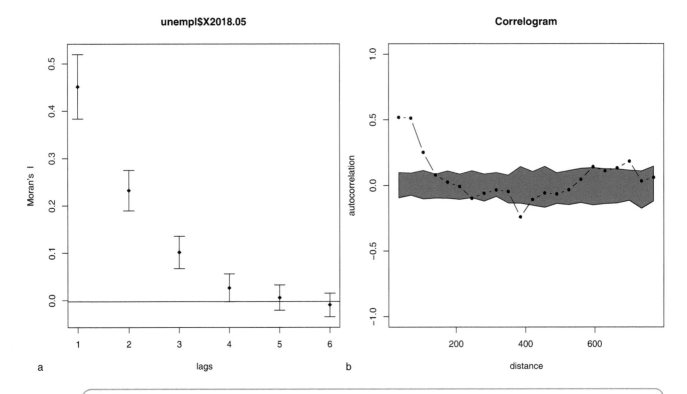

Figure 4.14 Correlogram for the variable unemployment rate in May 2018: a) sixth-order correlogram, b) correlation graph

Source: Own study using the spatialEco:: package

The correlogram can also be determined by the **correlogram()** command from the spatialEco:: package. The command works on objects in the *SpatialPointsDataFrame* class. As a result, variable autocorrelation is obtained in given ranges, for example, every 35 km. Distances for subsequent lags are calculated in km when the option *latlong=TRUE*. The command immediately displays the confidence interval for correlation = 0 (coloured corridor). The line with points shows the correlation – when it goes beyond the designated corridor, the correlation is significantly different from 0.

```
library(spatialEco)
unempl$crds<-crds.pov # adding xy coordinates to the data set
unempl.lim<-unempl[,c(102,101)] # data.frame coordinates + 1 one variable
class(unempl.lim)       # data.frame class
coordinates(unempl.lim)<-unempl.lim[,1] # definition of coordinates
class(unempl.lim)       # already the SpatialPointsDataFrame class

a<-correlogram(x=unempl.lim,  v=unempl.lim@data[,'X2018.05'],  dist=35,
ns=99, latlong=TRUE, dmatrix=TRUE)
plot(a)
attributes(a)
```

```
$names
[1] "autocorrelation" "CorrPlot" "dmatrix"


a$autocorrelation


     autocorrelation    dist    lci   uci
1             0.5186     35   -0.101 0.143
2             0.5123     70   -0.137 0.101
3             0.2525    105   -0.150 0.109
4             0.0796    140   -0.183 0.128
5             0.0239    175   -0.160 0.125
6            -0.0099    210   -0.143 0.102
7            -0.0987    245   -0.153 0.090
8            -0.0604    280   -0.170 0.097
9            -0.0359    315   -0.134 0.134
10           -0.0475    350   -0.132 0.150
11           -0.2402    385   -0.163 0.132
12           -0.1082    420   -0.136 0.121
13           -0.0589    455   -0.107 0.117


a$CorrPlot
```

The previous chart shows that the first three values are significantly different from 0 – that is, for the first three distances, here: 35, 70 and 105 km. The fourth value (for a distance of 140 km) is on the border of significance (or simply insignificant). So it can be said that at a distance of up to about 120 km, there are clear similarities in the value of the variable, but they disappear later. In the autocorrelation print, the columns *lci* and *uci* are the limits of the confidence interval: *lci* – lower confidence interval and *uci* – upper confidence interval.

Notes

1 One can read more about outliers in the help files under the heading *?influence.measures*
2 In a situation where the absolute value of z exceeds 1.64, the hypothesis of the existence of spatial autocorrelation may be adopted at the significance level of 0.05 (with z-value = 2.33 at the significance level of 0.01, respectively).
3 "The GSCI is used to reflect the summation of cross-correlation between any two elements, while the LSCI is utilised to measure the cross-correlation between a given element and all other elements in a geographical system."

Applied spatial econometrics

Katarzyna Kopczewska

This chapter presents methods of estimating popular spatial models – the cross-sectional (5.2) and panel (5.4) models. For cross-sectional models, in addition to typical models, unidirectional spatial interaction models (5.3.1), cumulative models (5.3.2), bootstrapped models (5.3.3) and grid data models (5.3.4) will be shown. The principles of estimation (5.2.1), ways of assessing the quality of models (5.2.2), aspects of spatial weights matrices (5.2.3), forecasting issues (5.2.4) and causality problems (5.2.5) are presented. In order to comply with the concept of the whole book, the theory is limited to the minimum necessary, while the emphasis is on the applied approach and elements that the researcher must pay attention to when estimating models (5.1).

In R, there is a rich collection of packages and commands for estimating spatial models. The most important of these is the spatialreg:: package, on which this chapter is based. It contains basic commands for estimation and diagnostics of these models. It should be noted that this is a relatively fresh package and is based on spdep::. The authors of spdep:: (Roger Bivand and others) decided to leave the spdep:: commands related to the study of spatial dependencies (mainly in relation to area data; the name of the spdep package is from spatial dependence) and decided to create a new spatialreg:: package for transparency to which econometric commands were transferred. Spatial panels are estimated in the splm:: package.

Due to space limitations, specific estimation methods and less common packages are not discussed. These include the packages spsur::, which estimates seemingly independent regressions (SUR) for spatial specifications (including SLX (spatially lagged X), SLM, SEM, SDEM) according to the methodology of Mur, López, and Herrera (2010) and López, Mur, and Angulo (2014); sphet::, which estimates Cliff-Ord models in heteroscedasticity (Piras, 2010); spGARCH::, which allows for taking into account temporal and spatio-temporal conditional heteroscedasticity according to the methodology of Otto, Schmid, and Garthoff (2018); ProbitSpatial:: and spatialprobit::, which estimate binary selection models (probit, logit) taking into account spatial dependencies; HSAR::, which estimates hierarchical autoregressive spatial models and CARbayes::, which estimates Bayesian generalised mixed models.

5.1 Added value from spatial modelling and classes of models

Econometric spatial modelling is a process that should be considered on many levels. First, it is an estimation theory in which estimator classes and their properties are considered. Second, it is the construction of theoretical models based on the theory of economics, most often consisting of the selection of variables but also determining the form of the model. Third, it is a technical way of carrying out the estimation together with the assessment of the fit and quality of the estimated models, the selection of the best model and the implementation of the forecast on new data. Fourth, it is the interpretation of models, on the one hand, consisting of examining the size and significance of the obtained econometric model coefficients but, on the other hand, translating quantitative results into phenomena and mechanisms discussed in theory. This is a very wide area, significantly exceeding the possibility of one chapter. For this reason, the topics selected in this chapter are arbitrary and will be addressed. The chapter, as well as the entire book, are strictly applied. For this reason, the theory of estimation was omitted and the construction of theoretical models was substantially limited, and the focus was on presenting the method of estimation and interpretation. Books are a good source of estimation theory: Arbia, G. (2014). *A Primer for Spatial Econometrics: With Applications in R. Springer*; Kelejian, H., & Piras, G. (2017). *Spatial Econometrics. Academic Press* or LeSage, J., & Pace, R. K. (2009). *Introduction to Spatial Econometrics. Chapman and Hall/CRC*. In turn, the alternative publication in terms of use is an open-access book entitled *Handbook of Spatial Analysis. Theory and Application with R* (INSEE/EUROSTAT, 2018).

Starting econometric spatial modelling, in which dependent and exogenous variables necessarily occur, one should consider sources of spatial dependence and spatial heterogeneity. Their first source is the existence of spatial spillovers, which appear as interregional flows of knowledge, trade or production factors. Second, they may result from omitted variables, which are most often unobserved location factors, such as various location amenities or clusters, affecting a dependent and spatially correlated variable. Their third source are measurement errors or unobserved heterogeneity, resulting from administrative boundaries that are not consistent with the nature of the spatial processes studied.

When thinking about spatial models, one can ask the question: What are the consequences of omitting the spatial dimension when the data is spatial? It can be assumed that, in principle, all data that can be geolocated shows a spatial nature; that is, there are spatial autocorrelation, heterogeneity and heteroscedasticity. In economics, therefore, it is basically all data collected by statistical offices for areas separated by administrative boundaries and point data to which locations can be assigned – relating to the life of the population, activities of firms and traces in virtual reality (logging of mobile phones, cars, IP computers). The lack of consideration of the spatial dimension in modelling data of this nature (i.e. when observations in location i are related to observations in location j) is most important for the dependent variable, because it breaks the assumptions of the classical linear regression model on independence, and thus errors in the model do not have zero covariance, because they most often show spatial error. The omission of the significant spatial variable, for example, the spatial lag of the dependent variable ρWy or of the explanatory variable θWy translates as a consequence into a model specification error (misspecification). In the situation where the econometric model requires the addition of the spatial lag of the dependent variable (y) and is estimated using a typical a-spatial least squares method (OLS), the estimators are biased and overestimated (overbiased) and inconsistent. When there is spatial autocorrelation of residuals in the model that is not addressed, OLS estimates are inefficient, though asymptotically consistent.

In the presence of dependence or spatial heterogeneity, there are several ways to deal with this problem. First of all, it is possible to use data on a different spatial scale, that is, data at a lower or higher level of aggregation, point or aggregated data or data for administrative division or in a grid or raster. Second, it is possible to include variables that approximate unobserved spatial phenomena – they can be dummy (zero-one) variables for specific characteristics or locations (e.g. regional capitals, cluster membership) or variables with a very similar spatial distribution to the error distribution. Third, it is possible to include geographical coordinates as explanatory variables – these are the expansion models. Fourth, it is possible to estimate econometric models in groups (clubs) that show internal homogeneity and interclub diversity. Fifth, the right spatial estimation method should be chosen – by selecting the best possible spatial weights matrix, appropriate model type and class and good model specification.

In recent years, spatial econometrics has changed the paradigm of perception and modelling of spatial interactions (see Figure 5.1). Until 2010, it was assumed that in spatial models, only one spatial interaction was

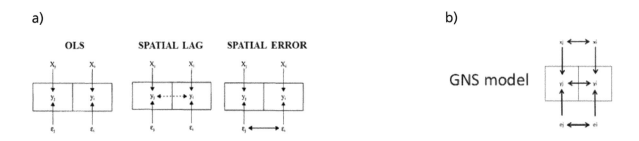

Figure 5.1 Paradigm of estimating spatial models: a) until 2010, b) from 2010 to present

Source: Own figure on the basis of Baller, Anselin, Messner, Deane, and Hawkins (2001)

taken into account, either at the level of the dependent variable (y) under spatial lag models or at the level of model error under spatial error models, estimated by the maximum likelihood method (MLE).

After 2010, with the appearance of the book *Introduction to Spatial Econometrics* (LeSage & Pace, 2009) and its review – the article *Applied Spatial Econometrics: Raising the Bar* (Elhorst, 2010) in spatial econometrics, bold estimations of models appeared, containing all three possible ways of spatial interaction: in dependent variables, in error and in explanatory variables (so-called Durbin component). A model with three components, the so-called GNS, is rarely estimated as a target because of overspecification, and one of the models with two spatial factors is selected. What is important is the fact that most often, one matrix of spatial weights is still used for all three spatial processes. This solution is recommended by LeSage and Pace in the article *The Biggest Myth in Spatial Econometrics* (2014).

The construction of a good econometric model requires consideration of several aspects, which are discussed in detail subsequently. These are: 1) model class, resulting from the purpose of the study and data availability; 2) estimation method, taking into account spatial variables and 3) model specification, including identification of variables tested in the hypothesis and control. In the iterative search for the best model, in the initial phase, it is necessary to take into account assumptions that elude from theories and expectations. Further, in the matching phase, changes are possible in both the model class, estimation method and model specification.

In economic applications, in particular in regional problems, several important classes of econometric models can be distinguished.

First of all, the basic spatial dependence model is based on cross-sectional data, most often for one period or taking into account time or space-time lags. A continuous dependent variable can be expressed as:

$$y_i = f\left(X_i\right) \tag{5.1}$$

where y_i is the support of the dependent variable in locations i, and X_i is a set of explanatory variables x_i, also in locations i. These types of models are used as a spatial alternative to typical linear regression models in which the level of the dependent phenomenon is explained by the level of the phenomenon from the exogenous assumption. Details of these models are provided in Section 5.2. In a special case, the dependent variable y_i can be binary (a so-called dummy; takes the values 0 and 1); then probit or logit models are estimated.

Second, one can estimate a spatial interactions model that is unidirectional by design. Such a model assumes the flow of the phenomenon from the core to the periphery in a way that extends with distance (decay). The model can be unconditional when the level of the phenomenon of y depends only on the distance (*dist*) and conditional when it also depends on other phenomena. It can be estimated in the static (unconditional model) or dynamic form when, in the conditional model, one of the conditioning variables is the level of a dependent variable from an earlier period. Such a model can be expressed as:

$$\frac{y_{i,t=1}}{\overline{y}_{t=1}} = f\left(\frac{y_{i,t=0}}{\overline{y}_{t=0}}, dist_{ij}, X_i\right) \tag{5.2}$$

where $y_{i,t=1}$ and $y_{i,t=0}$ are the values of the dependent variable at the location i in later periods ($t = 1$) and earlier ($t = 0$); $\overline{y}_{t=1}$ and $\overline{y}_{t=0}$ are the mean of the dependent variable in later periods ($t = 1$) and earlier; *dist* is the distance between locations i and j ($i \neq j$); one of the locations is the core and the other is the periphery and X_i is a set of explanatory variables x_i, also in the locations i. These types of models are used in modelling centrifugal diffusion resulting from natural processes or cohesive policy. By expressing the dependent variable as a relativised average, its values oscillate around 1 (>1 for values above the average and <1 for values below the average). More about these models is discussed in Section 5.3.1.

Third, one can estimate a convergence model in which the change (in time) of the level of the dependent variable Δy_i is translated by the initial state of the dependent variable $y_{i,t=0}$ in the unconditional model and additionally other exogenous explanatory variables X_i in the conditional model. The model is expressed as:

$$\Delta y_i = f\left(y_{i,t=0}, X_i\right) \tag{5.3}$$

Convergence models were very popular in the 1990s and in the first decade of the 21st century. There is a justification for combining the convergence model and the spatial interaction presented previously. Convergence models are also estimated in groups, studying the so-called club convergence. The dependent variable in convergence models is usually income or another monetary value of cumulative nature. This is due to the fact that the model is assumed to examine the growth rate and implicitly assumes that the variable increases over time.

Fourth, cross-sectional time-panel models can be estimated in which the studied units are geolocated and there is a spatial relationship between them. The model is expressed as:

$$y_{it} = f\left(X_{it}\right)$$ (5.4)

where y_{it} is a dependent variable at the location i in time t, and X_{it} is a set of dependent variables at the location i in time t. While y_{it} should be a continuous variable, varied in time and space, the explanatory variables can meet this assumption, as well as constant variables in time or constant variables in separated areas (hierarchical system). More about these models is discussed in Section 5.4.

Fifth, one can estimate spatial cumulative models, estimated in an expanding window. These types of models are based on the financial concept of net present value, the discounted sum of the payment stream. The model assumes that the base year is chosen arbitrarily ($t = 0$), and the variables are divided into three types: 1) monetary variables that can be accumulated (e.g. inflows, expenses, income, investments), 2) state variables that change every year and are non-cumulative (e.g. unemployment rate, number of children in nurseries) and 3) characteristics of the quasi-permanent location over time (e.g. forest area, commune type, seat of the authorities). For each year, it sets a set of variables to be used in the equation for a given year: for monetary variables, it is the discounted sum of values from the base year to the audited year; for state variables, it is the value from a given year and for quasi-constant variables, it is simply the value of the phenomenon. The model can be represented as:

$$\sum_{t=0}^{T} y_{it} = f\left(\sum_{t=0}^{T} x_{it}, x_{i,t=T}\right)$$ (5.5)

where $\sum_{t=0}^{T} y_{it}$ and $\sum_{t=0}^{T} x_{it}$ are the discounted sum of streams between the year $t = 0$ and T for dependent and explanatory variables, while $x_{i,t=T}$ is the state variable or the characteristics of the place in period T. The model is estimated for each year. The interpretation of coefficients from such models is extremely interesting. The value of the coefficient with the cumulative dependent variable is interpreted as a multiplier (e.g. investment multiplier, income multiplier), while the change (trend) of these coefficients determines the degree of saturation (e.g. investment, income). More about this class of models can be found in Section 5.3.2 and in Kopczewska (2016).

The effect of the estimation of most spatial models is the quantification of diffusion (spillover) between regions or geolocated units. Why control diffusion in econometric modelling? First, when diffusion and spatial interactions are ignored, individuals are treated autarctically, which limits information and changes the understanding of processes. This is the problem of perceiving data as independent, despite the existence of spatial autocorrelation, as well as uncontrolled spatial variance (spatial heterogeneity). Second, controlling diffusion makes it possible to assess to what extent the positive and negative effects of endogenous and esgogenic processes or the implications of policies with territorial impact are transmitted to other spatial units. This in turn is important in assessing the absorption capacity of units. Third, simply controlling and testing diffusion does not mean that this diffusion always exists. It happens that there is a very strong internalisation of effects. Fourth, despite the fact that positive autocorrelation and similarity of neighbouring units are most often observed, the so-called backwash effect, understood as negative diffusion, results in increasing differentiation.

5.2 Basic cross-sectional models

5.2.1 Estimation

The current spatial econometrics paradigm assumes that in the model there may be an element of spatial autocorrelation at three different levels – the lag of the dependent variable ρWy (spatial lag), the spatial lag of the dependent variables θWx (Durbin component) and the spatial error λWu (spatial error). In each of these lags, the key role is played by W – a matrix of spatial weights, constructed according to a given criterion. In current practice of spatial econometrics, it is recommended to use the same spatial weights matrix in all lags (LeSage, 2014).

Elhorst (2010), in the classic article entitled "Applied Spatial Econometrics: Raising the Bar" (*Spatial Economic Analysis*, 5: 1, 9–28), presented the classification of spatial models, which is currently the estimation standard. The same chart, supplemented with the names of commands and options from the spatialreg:: package used to estimate these models, is shown in Figure 5.2. Starting from the left, the most complete model, GNS, assumes the existence of all three spatial components, though in practice, it is extremely rarely estimated due to overspecification. Imposing the successive zero restriction on the spatial components of the model leads to

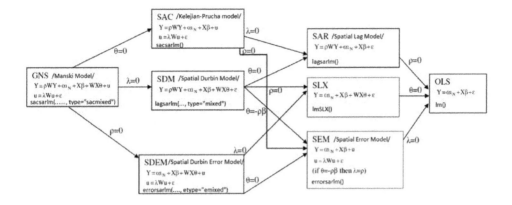

Figure 5.2 Structure of spatial models – classification since 2010

Source: Classification adopted from Elhorst (2010), supplemented with R codes from the spatialreg:: package

its reduced forms. Two-component models, SAC (so-called Keleijan-Prucha), spatial Durbin model (SDM) and spatial Durbin error model (SDEM) are among the most commonly chosen for estimation. The most popular of them include both Durbin models, although SAC sometimes gives better matches. Among the models with one spatial component, two models were intensively used until 2010 – spatial error model (SEM) and spatial autoregressive model (SAR; spatial lag model) and the relatively poorly used SLX model, which contains only explanatory variables spatially lagged. The specification of these models can be represented as follows:

Model with three spatial components:

$$\text{GNS}: Y = \beta_0 + \rho WY + X\beta + WX\theta + u \; oraz \; u = \lambda Wu + e \tag{5.6}$$

Models with two spatial components:

$$\text{SAC}: Y = \beta_0 + \rho WY + X\beta + u \; and \; u = \lambda Wu + e \tag{5.7}$$
$$\text{SDM}: Y = \beta_0 + \rho WY + X\beta + WX\theta + e \tag{5.8}$$
$$\text{SDEM}: Y = \beta_0 + X\beta + WX\theta + u \; and \; u = \lambda Wu + e \tag{5.9}$$

Models with one spatial component:

$$\text{SAR}: \quad Y = \beta_0 + \rho WY + X\beta + e \tag{5.10}$$
$$\text{SEM}: Y = \beta_0 + X\beta + u \; and \; u = \lambda Wu + e \tag{5.11}$$
$$\text{SLX}: \; Y = \beta_0 + X\beta + WX\theta + e \tag{5.12}$$

In spatial econometrics, general to specific modelling is practiced, which means that the three-component GNS model or Manski model is a good starting point for many estimations.

In models where there is a spatial lag of the explained variable (W_y), there is a problem of simultaneity that limits the possibility of forecasting, as well as the interpretation of coefficients in the model. In such models, y at location j influences y at location i (for $i \neq j$), with the inverse relationship being satisfied simultaneously. For this reason, in the final model, instead of β coefficients, effects (impacts) both indirect and direct are interpreted. The direct effect is understood as the influence of the explanatory variable x in the location i on the variable y in the location i. An indirect effect occurs when the explanatory variable x in the location affects the explained variable in the location j (where $i \neq j$) – this is interpreted as diffusion (spillover). The total effect is the sum of direct and indirect effects and means the total impact of the explanatory variable x in the location i and j on the explained variable y in the location i and j. Table 5.1 presents the method of determining the effects. As can be seen, in relation to models without the spatial lag y, the beta and theta coefficients are interpreted directly as impacts.

Table 5.1 Direct and indirect effects in spatial models

	Direct effect	Indirect – spillover effect
OLS/SEM	β_k	0
SAR/SAC	Diagonal elements from the matrix $(I-\rho W)^{-1}\beta_k$	Non-diagonal elements from the matrix $(I-\rho W)^{-1}\beta_k$
SXL/SDEM	β_k	θ_k
SDM/GNS	Diagonal elements from the matrix $(I-\rho W)^{-1}[\beta_k +W\theta_k]$	Non-diagonal elements from the matrix $(I-\rho W)^{-1}[\beta_k +W\theta_k]$

Source: Own table based on LeSage and Pace (2009)

Spatial components can be characterised and interpreted as below.

Spatial lag of the dependent variable (spatial lag of y) – usually referred to as ρWy; is interpreted as the average level of the dependent variable in neighbouring regions, where the weighing is carried out with weights from the spatial weights matrix W, and the neighbourhood criterion is defined by the structure of the matrix W itself. ρWy is the autoregressive factor y, most often used in a situation of significant autocorrelation of the studied variable, that is, when the values of y are systematically linked to the values of y in neighbouring areas and it expresses the similarity (difference) of regions perceived as neighbours. The model containing ρWy allows for the assessment of the degree of spatial dependence of the examined variable, controlling other explanatory variables, and on the contrary, for assessing the significance of other non-spatial variables, controlling spatial processes. Its inclusion allows to some extent filter spatial autocorrelation and requires estimation of effects (impacts) due to the simultaneity of the variable y on both sides of the equation. Omitting the lag of the dependent variable generates a bias of coefficients. This occurs in the GNS, SAC, SDM and SAR models.

Spatial lag of the explanatory variable (spatial lag of x) – usually referred to as θWx; is the so-called Durbin factor. Its inclusion in the model results from the assumption that the explanatory variables x are spatially correlated, and as a consequence, a change in the explanatory variable x in a neighbour affects the change of the explained variable y in the examined location. Most often, a set of all spatially lagged explanatory variables is added to the model, although only a few of them can be included. In models without ρWy, that is, SDEM and SLX (but taking into account θWx), it plays the role of indirect effect, without the need for further transformations. Interestingly, in models with the Durbin component, when block spatial lags of x variables are included (i.e. including all X), the structure of direct/indirect effects is constant, and when some variables are enabled, the structure changes. The advantage of models with the Durbin component is that the estimates from such models are unbiased, regardless of whether the correct model should additionally include the spatial lag of the error or the explained variable. This occurs in the GNS, SDM, SDEM and SLX models. It is assumed that θWx are local externalities; however, since X variables are exogenous, θWx are also exogenous, and models can be estimated without obstacles.

Spatial lag of the error u (spatial lag of u) – usually referred to as λWu; is the autoregressive factor of error u, treated as a residual disorder. It is interpreted as an average error from neighbouring locations. Its inclusion allows modelling of invisible or hardly measurable supra-regional characteristics, occurring more broadly than regional boundaries (e.g. cultural variables, soil fertility, weather, pollution, etc.). λ is used when region features are correlated with unobserved characteristics of neighbours but also for modelling unobserved shocks. It controls the spatial dependence of errors, that is, the unmodified effects of the explained and explanatory variables, making it more resistant. According to Rey and Montouri (1999), in the situation of modelling spatial error, an exogenous shock in a given region will affect not only the situation in this region (e.g. economic growth) but also the situation in neighbouring regions, precisely due to the presence of the spatial dependence of the error. This occurs in the GNS, SAC, SDEM and SEM models.

Diffusion and external effects (spillover) – are most often derived from direct and indirect effects, measuring the impact of exogenous variables on a given location and neighbourhood. They show the interdependence of spatial units – when it is positive, it is most often interpreted as cooperation and when negative, as leaching.

The impact of absolute and relative location – while the mechanisms of spatial lags discussed previously de facto examine the effect of relative location (location near someone), it is also possible to study the absolute location (location in a specific place), the implementation of which involves taking into account dummy (zero-one) variables for specific locations (e.g. location in the business district, location in the main regional city, etc.).

Spatio-temporal lag – in a situation where variables are correlated in time and space, one can include time-space lag in the model. This is the spatial lag W_y for the variable in the earlier period. The interpretation is based on path dependence related to process persistence. A very common example of the significance of such a variable is the unemployment rate in autoregressive models due to the relative durability of the spatial and temporal patterns of the phenomenon.

In R, spatial dependency models can be estimated using the spatialreg:: package. The following commands are included:

For the estimation of basic spatial models (they return an object in the *sarlm* class):
- **sacsarlm()** with option *type="sacmixed"* under ML_models – GNS model
- **sacsarlm()** under ML_models – SAC model
- **errorsarlm()** under ML_models – SEM model
- **errorsarlm()** with option *etype="emixed"* under ML_models – SDEM model
- **lagsarlm()** under ML_models – SAR model
- **lagsarlm()** with option *type="mixed"* under ML_models – SDM model
- **lmSLX()** – SLX model

Used in estimating impacts:
- **impacts()** and **intImpacts()** – estimation of direct and indirect effects in spatial lag and SLX models

Used in testing (they return objects in the *htest* class):
- **LR.sarlm()** – likelihood ratio test
- **logLik()**
- **LR1.sarlm()** and **Wald1.sarlm()** – tests for the existence of spatial dependence in spatial models that are part of **summary.sarlm()**
- **Hausman.test()** – Hausman test for spatial error models (**errorsarlm()**)
- **anova()** (also **anova.sarlm()**)
- **bptest.sarlm()**

Used in post-estimation:
- **summary()** under ML_models – summary of the estimation result
- **print()** under ML_models – displaying the estimation result
- **residuals()** as part of ML_models – the remainder of the econometric model
- **deviance()** under ML_models – identical to sum of squared errors (SSE)
- **coef()** under ML_models – model coefficients
- **vcov()** under ML_models – a matrix of variance of main parameters from the fitted model
- **fitted()** within ML_models – fitted values from the model
- **predict.sarlm()** – for prediction in spatial simultaneous autoregressive linear models

There is also a large collection of other commands not discussed in this book.

The following are commands for estimating other spatial models: **lagmess()** – matrix exponential spatial lag model, an alternative to SAR; **gstsls()** – SAC model estimated using generalised moments (spatial simultaneous autoregressive SAC model estimation by GMM); **spautolm()** – spatial conditional and simultaneous autoregression model estimation; **stsls()** – generalised two-stage spatial lag model with possible correction of heteroscedasticity (generalised spatial two stage least squares); **spBreg_lag()** – Bayesian MCMC model of spatial simultaneous autoregressive model (Bayesian MCMC spatial simultaneous autoregressive model estimation) and **GMerrorsar()** – spatial simultaneous autoregressive error model estimated GMM.

There are also commands used in spatial filtering: **griffith_sone()** – eigenvalues of the spatial weights matrix, **lextrB()** – extreme eigenvalues of the binary symmetrical spatial weights matrix, **ME()** – GLM filtering with Moran's own vectors and **SpatialFiltering()** – semi-parametric spatial filtering.

There are also commands used to manage the spatial weights matrix: **as.spam.listw()** – to convert the spatial weights matrix from the *listw* class to the *spam* class, **set.ZeroPolicyOption()** – to check the spatial object ID, **similar.listw()** – to create a symmetrical similar list of weights and **trW()** – for determining traces of spatial weights matrices, as well as commands supporting parallel calculations – in the **set.mcOption()** group, MCMC sampling from matched values – **MCMCsamp()** or highest posterior density (HPD) intervals for parameters in the MCMC sample – **HPDinterval()**.

Example

Cross-sectional models will be presented on data for poviats. In the set data_pow_2019.csv, there are panel data for 2006–2017 for 380 NTS4 territorial units for 23 variables and 15 identifiers. An attempt will be made to determine the determinants of average wage in the poviat (in Poland = 100%) with various socioeconomic and

---

Table 5.2 **Specification of cross-sectional model**

Variable symbol	Variable code	Description of the variable
Y	XA14	Salaries Poland =100%
X1	XA08/XA09	Percentage of working-age population in relation to post-working age population
X2	XA13	Number of firms per 10,000 working age residents
X3	(XA05/XA06) /mean(XA05/XA06)	Public investments per capita (Poland = 100)
X4	(XA18+XA19+XA20)/XA15	Share of employment in services
X5	XA16/XA15	Share of employment in agriculture
X6	XA15/XA06	Share of employees relative to population
X7	dist	Distance of the poviat from the voivodeship city
X8	XA10/mean(XA10)	Population density per 1 km^2 (Poland = 100)
X9	XA21	Unemployment rate

Source: Own work

---

population variables. The location of 380 territorial units, poviats, is defined by shapefile for poviats, defined as the object of the poviat. The order of the dataset according to the order of units in shapefile is specified by the *ID_MAP* variable. The model specification is in Table 5.2.

```
#eq1<-average salary (Poland=100) ~
    + population_in_production_age_to_post_production_age
    + number_of_firms_to_population_in_production_age
    + per_capita_public_investment(Poland=100)
    + share_of_employment_in_services
    + share_of_employment_in_agriculture
    + share_of_employed_to_population
    + distance
    + population_density(Poland=100)
    + unemployment rate

#eq1<-XA14 ~ XA08/XA09 + XA13 + (XA05/XA06)/mean(XA05/XA06) + (XA18+XA19+XA20)/
XA15 + XA16/XA15 + XA15/XA06 + dist + XA10/mean(XA10) + XA21
```

The previous code shows which variables will be taken into account in the econometric model. To simplify the process, the variables used received new codes – these variables were created in the *sub* subset for the selected year.

```
data<-read.csv("data_nts4_2019.csv", header=TRUE, dec=",", sep=";")
sub<-data[data$year==2017,]

sub$y<-sub$XA14
sub$x1<-sub$XA08/sub$XA09
sub$x2<-sub$XA13
sub$x3<-(sub$XA05/sub$XA06)/mean(sub$XA05/sub$XA06)
sub$x4<-(sub$XA18+sub$XA19+sub$XA20)/sub$XA15
sub$x5<-sub$XA16/sub$XA15
sub$x6<-sub$XA15/sub$XA06
sub$x7<-sub$dist
sub$x8<-sub$XA10/mean(sub$XA10)
sub$x9<-sub$XA21
eq1<-y~x1+x2+x3+x4+x5+x6+x7+x8+x9 # model equation
```

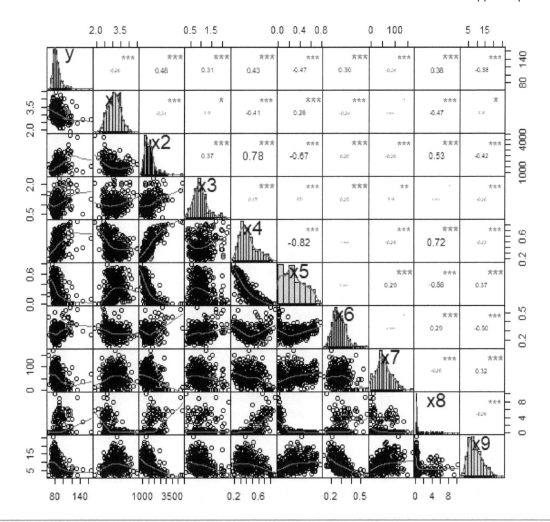

Figure 5.3 Examination of correlations between variables selected for the model

Source: Own study using R and the PerformanceAnalytics:: package.

For the purposes of correlation analysis, a *regdata* subset was created that contains only variables selected for analysis. This type of file is required for matrix correlation analysis – it was performed subsequently with the command **chart.Correlation()** from the PerformanceAnalytics:: package, but there are many other commands, such as **cor()** from the stats:: package or **corrgram()** from the corrgram:: package. The correlation graph shows that the distributions of variables are generally right skewed, the variables are usually weakly or moderately correlated and there are several pairs of variables with clear correlations:

```
cor(x2,x4)=0.78, cor(x4,x5)=-0.82, cor(x4,x8)=0.72.

regdata<-sub[,c("y", "x1", "x2", "x3", "x4", "x5","x6", "x7", "x8", "x9")]

library(PerformanceAnalytics)
chart.Correlation(regdata, histogram=TRUE, pch=19) # Figure 5.3
```

The following are the basic results of linear regression (OLS model, estimated using the least squares method), estimated with the **lm()** command (short for linear model).

```
model.lm<-lm(eq1, data=sub)
summary(model.lm)
```

```
Call:
lm(formula = eq1, data = sub)

Residuals:
 Min 1Q Median 3Q Max
-27.040 -4.659 -1.156 3.342 75.219

Coefficients:
   Estimate Std. Error t value Pr(>|t|)
(Intercept) 89.453667 7.225117 12.381 < 2e-16 ***
x1  -3.868098 1.377277 -2.809 0.00524 **
x2  0.002061 0.001811 1.138 0.25581
x3  4.547280 1.540568 2.952 0.00336 **
x4  -1.479978 8.569714 -0.173 0.86298
x5  -20.336263 4.559384 -4.460 1.09e-05 ***
x6  38.753277 9.610871 4.032 6.71e-05 ***
x7  -0.062031 0.015654 -3.963 8.90e-05 ***
x8  -0.346144 0.440771 -0.785 0.43277
x9  -0.015593 0.150482 -0.104 0.91753
---
Signif. codes: 0 '***' 0.001 '**' 0.01 '*' 0.05 '.' 0.1 ' ' 1

Residual standard error: 9.011 on 370 degrees of freedom
Multiple R-squared: 0.3949, Adjusted R-squared: 0.3802
F-statistic: 26.83 on 9 and 370 DF, p-value: < 2.2e-16
```

The previous model was estimated on the nominal values of created variables. Regression coefficients (from the *Estimate* column) indicate how much the level of the dependent variable y will change when the specified explanatory variable x changes by 1. Standard errors (*Std.Error* column) specify in plus and in minus the corridor of regression coefficient fluctuation – when it is relatively narrow, the model is considered well fitted and the variable x a significant determinant of the variable y. A wide corridor, that is, when the standard error is greater than the estimate, indicates the uncertainty of the variable x in explaining y. When the error is greater than the estimated regression coefficient (β), the corridor includes the possible positive, negative and equal to 0 values of the regression coefficient – this is not an attractive situation, because there is no certainty whether the relationship exists (when $\beta = 0$, there is no relationship) or, if it exists, what its direction is (positive when $\beta > 0$ or negative when $\beta < 0$). The *t value* column is the value of the $t = estimate/std.error$ test statistic in the test for equality of the coefficient (H_0: $\beta = 0$) and de facto means the estimation of β is greater than its error. In the sense of standard deviations, this multiplicity is confronted with the properties of statistical distributions – it is assumed that the chance of an event smaller or larger by three standard deviations in the Student's t distribution is small (the limits of probability distribution are reached). A value of 4 means that the chance that $\beta = 0$ is very small, because it would be necessary to have an event at the extreme end of the distribution, four standard deviations away from the mean – and this confirms the high chance that the real parameter is close to the estimated regression coefficient. This opportunity is confirmed by the value from the *Pr (> | t |)* column, the so-called *p*-value. Residual standard error (RSE) determines the forecast fluctuation. In this case, the y values range between 65 and 166; that is, the error at the RSE = 9 level is from 13.8% to 5.4% of the y value. The goodness-of-fit measured with R2 and multiple R2 is moderate (approximately 0.39), although in cross-sectional models rather typical. The F test for the combined significance of all coefficients of the H_0 model: $\beta1 = . . . = \beta n$ confirms that they are all significant together.

The basic spatial model requires the preparation of a spatial weights matrix. As presented in Chapter 4, the most basic, and also the most popular, spatial weights matrix according to the common border criterion (contiguity matrix) is created as:

```
library(spdep)
library(spatialreg)
library(rgdal)
pov<-readOGR(".", "powiaty") # 380 units
```

```
pov<- spTransform(pov, CRS("+proj=longlat +datum=NAD83"))
cont.nb<-poly2nb(as(pov, "SpatialPolygons"))       # from spdep::
cont.listw<-nb2listw(cont.nb, style="W")           # from spdep::
```

The most general spatial model with three spatial coefficients, GNS or the so-called *Manski* model, which usually has overspecification, is estimated using the **sacsarlm()** command with the option *type = "sacmixed"*. By adding the *Nagelkerke = TRUE* option in the **summary()** command, one can display the pseudo-R^2.

```
# Manski model (all three spatial coefficients)
# spatial lag Y (rho)
# spatial lags X (theta)
# spatial autocorrelation of error (lambda)

model.GNS<-sacsarlm(eq1, data=sub, listw=cont.listw, type="sacmixed")
summary(model.GNS, Nagelkerke=TRUE)
```

```
Call:sacsarlm(formula=eq1, data=sub, listw=cont.listw, type="sacmixed")

Residuals:
      Min      1Q  Median      3Q      Max
 -23.3217 -4.6444 -1.1510  3.4150  72.2148

Type: sacmixed
Coefficients: (asymptotic standard errors)
              Estimate   Std. Error   z value    Pr(>|z|)
(Intercept)  34.7861563  17.2804009    2.0130    0.04411
x1           -3.2421965   1.6222431   -1.9986    0.04565
x2           -0.0029298   0.0023321   -1.2563    0.20902
x3            2.9950610   1.5363298    1.9495    0.05124
x4            4.0067192   8.9755505    0.4464    0.65531
x5          -20.2754767   5.0660168   -4.0023  6.274e-05
x6           59.8812715  10.5973033    5.6506  1.599e-08
x7           -0.0719491   0.0296636   -2.4255    0.01529
x8           -0.2500320   0.5307138   -0.4711    0.63755
x9           -0.0484336   0.1717766   -0.2820    0.77798
lag.x1       -0.8009749   2.3748686   -0.3373    0.73591
lag.x2        0.0038153   0.0035660    1.0699    0.28467
lag.x3        1.8476196   3.1501470    0.5865    0.55753
lag.x4       22.3269181  16.5363273    1.3502    0.17696
lag.x5       12.6373117   8.0485410    1.5701    0.11638
lag.x6      -10.0757565  21.6853828   -0.4646    0.64219
lag.x7        0.0593318   0.0422366    1.4047    0.16010
lag.x8       -0.1929782   0.9774510   -0.1974    0.84349
lag.x9        0.2389813   0.2461626    0.9708    0.33163

Rho:          0.40246
Asymptotic standard error: 0.19302
    z-value: 2.085, p-value: 0.037069
Lambda: -0.24525
Asymptotic standard error: 0.24441
    z-value: -1.0034, p-value: 0.31566

LR test value: 32.79, p-value: 0.00056905
Log likelihood: -1353.154 for sacmixed model
ML residual variance (sigma squared): 69.431, (sigma: 8.3325)
Nagelkerke pseudo-R-squared: 0.44494
Number of observations: 380
Number of parameters estimated: 22
AIC: 2750.3, (AIC for lm: 2761.1)
```

```
# logLik of the OLS model and degrees of freedom of the LR test
model.GNS$logLik_lm.model
#[1] 'log Lik.' -1369.549 (df=11)
model.GNS$LL                        # logLik of the GNS model
#[1] -1353.154
```

The previous example shows that in the GNS model, the significance is similar to in the OLS model, although the *p*-value has changed. The estimates of the coefficients are evolutionarily different (they are not revolutionary changes), which results from the introduction of new parameters for the OLS model: the spatial lags of the explanatory variables X_1: X_9, designated as *lag.x₁*, *lag.x₂*, . . . *lag.x₉*, as well as the spatial lag of the explained variable defined as Rho and the spatial error lag specified as Lambda. The values of *p*-value *lag.x* and *Lambda* error lags are all high, which may be due to excessive specification or large variance in spatial distributions (then the spatial lags have a high variance and, consequently, a high standard error). The lowest *p*-value is at the spatial lag *y* (*rho*). The likelihood ratio (LR) test, comparing models with factor constraints (here a-spatial) and without constraints (GNS), indicates the choice of H_1 and the adoption of an unrestricted (full) model. LR statistics were designated as:

$$LR = -2\left(ln\frac{Lik_{limited}}{Lik_{unlimited}} \right) = -2\left[\ln\left(Lik_{limited}\right) - \ln\left(Lik_{unlimited}\right) \right] == -2\left[logLik_{MNK} - logLik_{GNS}\right] \quad (5.13)$$

which in practice can be determined using operations on the attributes of the regression result:

$$-2 * model.GNS\$logLik\_lm.model - model.GNS\$LL \quad (5.14)$$

The statistics have an χ^2 distribution, which has as many degrees of freedom as the parameters are limited – in practice, it is the number of spatial coefficients θ, ρ and λ (determined in the attribute *model.GNS\$logLik_lm.model*). This does not mean that GNS is the best model but only better than the a-spatial model. Similar conclusions are prompted by the assessment of the AIC, where the AIC of the spatial model is better (because lower) than the AIC of the a-spatial model (OLS). Nagelkerke's pseudo-R^2 shows that the fit of the model is moderately good.

In the estimation process, the best-fitted model is sought, in accordance with the principles described in Sections 5.2.2 and 5.1. Starting from the most complete model (as previously), one of the elements of searching for the best specification (regardless of other activities) is the reduction of spatial coefficients. Subsequently are the estimates of the popular SDM model.

```
# SDM model (two spatial coefficients)
# spatial lag Y (rho ρ)
# spatial lags X (theta θ)

model.SDM<-lagsarlm(eq1, data=sub, listw=cont.listw, type="mixed", tol.
solve=1.0e-20, method="LU")
summary(model.SDM, Nagelkerke=TRUE)
```

```
Call:lagsarlm(formula=eq1, data=sub, listw=cont.listw, type="mixed",
 tol.solve=1e-20)


Residuals:
 Min 1Q Median 3Q Max
-23.7766 -4.6834 -1.2621 3.5488 72.8687


Type: mixed
Coefficients: (asymptotic standard errors)
            Estimate   Std. Error  z value    Pr(>|z|)
(Intercept) 50.2649819 12.6889515   3.9613    7.454e-05
x1          -3.2411465  1.5815537  -2.0493    0.040429
x2          -0.0029229  0.0022533  -1.2972    0.194570
```

```
x3              3.1224077    1.5189078    2.0557   0.039812
x4              5.9059077    8.7910066    0.6718   0.501703
x5            -19.5892389    4.9620588   -3.9478   7.887e-05
x6             59.6131558   10.5585061    5.6460   1.642e-08
x7             -0.0765630    0.0296326   -2.5837   0.009774
x8             -0.2270201    0.5173852   -0.4388   0.660818
x9             -0.0650779    0.1683102   -0.3867   0.699012
lag.x1         -2.0205824    2.2413951   -0.9015   0.367331
lag.x2          0.0038880    0.0036319    1.0705   0.284387
lag.x3          2.7588659    3.0739733    0.8975   0.369457
lag.x4         24.5049579   16.9616167    1.4447   0.148534
lag.x5          7.7160839    8.0366415    0.9601   0.336998
lag.x6          3.6485921   20.2354682    0.1803   0.856912
lag.x7          0.0590442    0.0425570    1.3874   0.165315
lag.x8         -0.3765914    1.0244115   -0.3676   0.713159
lag.x9          0.3015723    0.2555540    1.1801   0.237971


Rho: 0.2044, LR test value: 6.72, p-value: 0.0095338
Asymptotic standard error: 0.070603
 z-value: 2.8951, p-value: 0.0037906
Wald statistic: 8.3815, p-value: 0.0037906


Log likelihood: -1353.37 for mixed model
ML residual variance (sigma squared): 72.034, (sigma: 8.4873)
Nagelkerke pseudo-R-squared: 0.4443
Number of observations: 380
Number of parameters estimated: 21
AIC: 2748.7, (AIC for lm: 2753.5)
LM test for residual autocorrelation
test value: 0.24981, p-value: 0.61721
```

Due to computational accuracy when reversing the asymptotic covariance matrix, it is necessary to increase the computational tolerance, which was declared with the *tol.solve = 1.0e-20* option. In the previous model, evolutionary changes in *p*-value and regression coefficient values can be seen again. The coefficients θ of spatial lags *x* have maintained high standard errors, but ρ is significant. The LR test also indicates that the spatial model is more favourable than the a-spatial one. The conclusions of the Wald test are similar, because the distance between the parameters of the restricted and unlimited models is large, which suggests adopting H_1. The alignment measured by the pseudo-R^2 is similar to the previous GNS model, and the information criteria also indicate that the spatial model is better. The LM test result shows that the rest of the SDM model does not have spatial autocorrelation and the entire spatial relationship has been filtered out.

Regardless of the model type chosen, it is important to note that various Jacobian counting methods have been implemented in the spatial model commands (*method* option), the default *method = "eigen"* is much slower than the alternative method specified as *method = "LU"*. Estimations of other types of models are presented in the examples later in the book. Subsequently are implementations of all models specified in Figure 5.3. The commands contain, in turn, a regression formula, a data object, a spatial weights matrix of *listw* class, and options: a method for estimating a weights matrix, calculation accuracy and more. A summary of estimation results is obtained by the **summary**() command – its options can be specified, among others *Nagelkerke = TRUE*, which displays the pseudo-R^2.

```
# Manski model (all three spatial coefficients)
# spatial lag Y (rho)
# spatial lags X (theta)
# spatial autocorrelation of error (lambda)
model.GNS<-sacsarlm(eq1, data=sub, listw=cont.listw, type="sacmixed")
summary(model.GNS, correlation=TRUE)
```

```
# SAC model (two spatial coefficients)
# spatial lag Y (rho)
# spatial autocorrelation of error (lambda)
model.SAC<-sacsarlm(eq1, data=sub, listw=cont.listw, method="LU", tol.
solve=1.0e-20)

# SDM model (two spatial coefficients)
# spatial lag Y (rho ρ)
# spatial lags X (theta θ)
model.SDM<-lagsarlm(eq1, data=sub, listw=cont.listw, type="mixed", tol.
solve=1.0e-20, method="LU")

# SDEM model (two spatial coefficients)
# spatial lags X (theta)
# spatial autocorrelation of error (lambda)
model.SDEM<-errorsarlm(eq1, data=sub, listw=cont.listw, etype="emixed",
method="LU", tol.solve=1.0e-20)

# SAR model (one spatial factor)
# spatial lag Y (rho)
model.SAR<-lagsarlm(eq1, data=sub, listw=cont.listw)

# SLX model (one spatial factor)
# spatial lags X (theta)
model.SLX<-lmSLX(eq1, data=sub, listw=cont.listw)

# SEM model (one spatial factor)
# spatial autocorrelation of error (lambda)
model.SEM<-errorsarlm(eq1, data=sub, listw=cont.listw)
```

To the previous basic models, one can attach the *space-time lag*, which should be understood as the spatial lag determined for the variable in another period (e.g. a year earlier). The role of such variables is to absorb dependencies that penetrate slowly, with periodic lag. One should be aware that the spatial interactions modelled previously, especially in Durbin models, were immediate. Spatial lags are created by the **lag.listw**() command from the spdep:: package. Subsequently, for the model estimated on data from 2017, spatial lags were created from the variable in 2016.

```
# creating time-space lag
sub0<-data[data$year==2016,]
sub$y.stlag<-lag.listw(cont.listw, sub0$XA14)

eq1<-y~x1+x2+x3+x4+x5+x6+x7+x8+x9 # form of the regression equation
eq2<-y~x1+x2+x3+x4+x5+x6+x7+x8+x9+y.stlag # form of the regression equation

# SDM model (two spatial factors rho and theta)
# additionally time-space lag y (y.stlag)
model.SDMst<-lagsarlm(eq2, data=sub, listw=cont.listw, type="mixed", tol.
solve=1.0e-20, method="LU")
summary(model.SDMst)
```

```
Call:lagsarlm(formula = eq2, data = sub, listw = cont.listw, type = "mixed",
    method = "LU", tol.solve = 1e-20)

Residuals:
      Min        1Q     Median        3Q       Max
-34.882274  -3.740133  -0.051281   3.336532  49.633717
```

```
Type: mixed
Coefficients: (asymptotic standard errors)
                Estimate    Std. Error  z value    Pr(>|z|)
(Intercept)  -6.7543e+01  1.3379e+01   -5.0484   4.455e-07
x1           -2.2941e+00  1.2825e+00   -1.7888    0.073647
x2            3.6910e-04  1.8322e-03    0.2015    0.840342
x3            2.9828e+00  1.2290e+00    2.4269    0.015229
x4           -3.6600e+00  7.1211e+00   -0.5140    0.607274
x5           -1.3000e+01  4.0304e+00   -3.2255    0.001258
x6            4.3705e+01  8.6038e+00    5.0797   3.780e-07
x7           -5.6209e-02  2.4051e-02   -2.3370    0.019437
x8           -2.7610e-02  4.1811e-01   -0.0660    0.947351
x9            3.1213e-02  1.3605e-01    0.2294    0.818543
y.stlag      -4.7036e-01  1.1042e-01   -4.2596   2.048e-05
lag.x1        3.2276e+00  1.8469e+00    1.7476    0.080536
lag.x2       -6.5557e-04  2.9474e-03   -0.2224    0.823987
lag.x3        5.4103e-01  2.5183e+00    0.2148    0.829889
lag.x4        1.0580e+01  1.3808e+01    0.7662    0.443559
lag.x5       -3.0899e+00  6.5710e+00   -0.4702    0.638192
lag.x6        4.3179e+01  1.7255e+01    2.5025    0.012332
lag.x7        6.3403e-02  3.4589e-02    1.8331    0.066794
lag.x8       -1.2621e+00  8.3097e-01   -1.5188    0.128812
lag.x9        5.9513e-01  2.0743e-01    2.8690    0.004117
lag.y.stlag   1.8550e+00  1.3403e-01   13.8401   < 2.2e-16

Rho: 0.027698, LR test value: 0.057589, p-value: 0.81035
Asymptotic standard error: 0.07208
 z-value: 0.38427, p-value: 0.70078
Wald statistic: 0.14766, p-value: 0.70078

Log likelihood: -1270.304 for mixed model
ML residual variance (sigma squared): 46.89, (sigma: 6.8476)
Number of observations: 380
Number of parameters estimated: 23
AIC: 2586.6, (AIC for lm: 2584.7)
LM test for residual autocorrelation
test value: 57.95, p-value: 2.6867e-14
```

In the previous model, the time-space lag of the dependent variable (*y.stlag*) took over the significance of the *rho* time lag. The parameter at *lag.y.stlag* is also significant, which should be interpreted as a second order spatial lag (neighbours of the neighbours). Relative to the non-dynamic model, the significance of the spatial lags of the explanatory variables has improved. Another option in the estimation would be to check the SDEM (which does not contain *rho*) or SLX.

As mentioned earlier, in models where there is a spatial lag of the dependent variable, due to simultaneity (variable *y* on both sides of the equation), effects should be estimated. The direct effect assumes the influence *x* in the location *i* and on *y* in the location *i*. The indirect effect assumes the influence *x* in the location *i* and on *y* in location *j*. The determination of impacts is possible with the command **impacts()** from the spatialreg:: package. It is worth noting that in the case of models without the spatial lag *y*, the **impacts()** command actually rewrites the coefficients β and θ.

```
# impacts dla SDM
model.SDM<-lagsarlm(eq1, data=sub, listw=cont.listw, type="mixed",
tol.solve=1.0e-20, method="LU")
W.c<-as(as_dgRMatrix_listw(cont.listw), "CsparseMatrix")
trMat<-trW(W.c, type="mult")
model.SDM.imp<-impacts(model.SDM, tr=trMat, R=2000)
summary(model.SDM.imp, zstats=TRUE, short=TRUE)
```

```
Impact measures (mixed, trace):
         Direct       Indirect        Total
x1  -3.34889332  -3.264645517   -6.613538841
x2  -0.00279077   0.004003842    0.001213072
x3   3.25882788   4.133425022    7.392252906
x4   6.93845512  31.285376603   38.223831724
x5 -19.44017091   4.516640532  -14.923530376
x6  60.24899940  19.265554860   79.514554256
x7  -0.07482039   0.052800715   -0.022019674
x8  -0.24400666  -0.514680857   -0.758687514
x9  -0.05350155   0.350754647    0.297253095
============================================================
Simulation results (asymptotic variance matrix):
============================================================
Simulated standard errors
          Direct       Indirect        Total
x1   1.556928222   2.506392650   2.322959638
x2   0.002223777   0.004145669   0.004080375
x3   1.505590315   3.632002975   3.854545228
x4   8.912368913  20.786038133  23.010583857
x5   4.925475256   9.483979764  10.136426423
x6  10.661009587  24.454880814  28.469951308
x7   0.028391785   0.046122663   0.030719628
x8   0.504153619   1.190088793   1.088112717
x9   0.166924893   0.298026064   0.314198447

Simulated z-values:
         Direct     Indirect       Total
x1  -2.1662828   -1.2918135   -2.8457398
x2  -1.2334592    0.9581134    0.3012182
x3   2.1570606    1.1386793    1.9154882
x4   0.7890414    1.5074446    1.6673210
x5  -3.9545693    0.4671418   -1.4845242
x6   5.6394549    0.8365522    2.8303550
x7  -2.6453467    1.1656210   -0.6948185
x8  -0.5234587   -0.4146976   -0.6960957
x9  -0.3390107    1.2143165    0.9717064

Simulated p-values:
       Direct   Indirect      Total
x1  0.0302896    0.19642  0.0044308
x2  0.2174045    0.33801  0.7632481
x3  0.0310009    0.25484  0.0554303
x4  0.4300878    0.13170  0.0954506
x5  7.6673e-05   0.64040  0.1376699
x6  1.7059e-08   0.40284  0.0046496
x7  0.0081607    0.24377  0.4871691
x8  0.6006551    0.67836  0.4863689
x9  0.7346017   0.22463  0.3311966
```

The use of the previous result allows for interpreting the modelling results using the SDM specification. Due to their simultaneity, coefficients from the pure SDM model cannot be used to determine the impact of x changes on y changes. For this purpose, indirect, direct and total effects are examined. To some extent, the effects are similar to the β and θ coefficients and usually keep the same direction; however, the differences between β and *impacts* can reach tens of percent.

There are three types of relationships in the previous result:

- when both effects are positive ($x3$, $x4$ and $x6$), that is, an increase of x_i in region i and an increase of x_j in the neighbourhood of j translate into an increase of y_i in region i – this means there exists an effect of mutual full support of processes, regardless of location;
- when both effects are negative ($x1$ and $x8$), that is, an increase in x_i in region i and an increase in x_j in the neighbourhood of j translate into a decrease in y_i in region i – this means there exists an effect of mutual weakening of processes, regardless of location; and
- when the direct effects are negative and the indirect effects are positive ($x2$, $x5$, $x7$ and $x9$), that is, an increase of x_i in region i translates into a decrease in y_i in region i, while an increase in x_j in the neighbourhood of j translates into an increase in y_i in region i, this means that factors interact differently, often due to the saturation effect and the appearance of scale disadvantages or to the apparent heterogeneity of the study and neighbouring areas.

There was no last relationship possible:

- when direct effects are positive and indirect effects are negative, that is, an increase in x_i in region i translates into an increase of y_i in region i, while an increase of x_j in neighbourhood j translates into a decrease of y_i in region i – this is the leaching relation, when resources from one location work in favour of another location, weakening their own location.

The proportion of direct and indirect effects is another issue. The sum of both effects gives the total effect. One can easily test the effect ratio. All effects are separated subsequently, and the share of the direct effect in the total (a/c) and the relation of the strength of the direct and indirect effects (abs (a)/abs (b)) are calculated in absolute terms using the **abs()** command.

```
a<-model.SDM.imp$res$direct   # direct effects only
b<-model.SDM.imp$res$indirect # only indirect effects
c<-model.SDM.imp$res$total    # only total effects
a/c                           # share of direct effect in the total
```

```
        x1          x2         x3          x4          x5          x6          x7
0.5063693  -2.3005799  0.4408437   0.1815217   1.3026523   0.7577103   3.3978882
        x8          x9
0.3216168  -0.1799865
```

```
abs(a)/abs(b)            # relation of direct to indirect effects
```

```
        x1          x2         x3          x4          x5          x6          x7
1.0258061   0.6970229  0.7884086   0.2217795   4.3041218   3.1272912   1.4170336
        x8          x9
0.4740931   0.1525327
```

The result, in the case of the same direction of effects, is easy to interpret. When both are positive or negative, it is easy to determine to what extent the effects are internalised (direct effect, impact stays in place), and to what extent diffusion occurs (spillover, external effect for the location, when the action in the studied place translates into the effect elsewhere). In the studied case, the compliance of the directions of influence was shown by the variables $x1$, $x3$, $x4$, $x6$ and $x8$. The share of the direct effect is from 18% ($x4$) to 75% ($x6$), that is, from rather marginal (approximately 1/6) to dominant (approximately 3/4).

Effects with different directions of impact are in fact cancelled out; hence the relation to the total effect is less informative, and the relative relationship of both effects should be studied. In the examined case, the compliance of the directions of influence was shown by the variables $x2$, $x5$, $x7$ and $x9$. By examining the ratio of direct and indirect effects, it can be determined that the direct effect prevails when the result > 1. The ratio of 4.3 (for $x5$) means that the effects are roughly 4:1 (or 9:2), and this means strong internalisation (four times stronger than spillover). On the contrary, a score of <1, for example, a quotient of 0.15 (for $x9$), means that the effects are like 1:6, and internalisation is repressed in favour of a very strong spillover.

5.2.2 Quality assessment of spatial models

The assessment of the quality of spatial models has slightly different rules than the assessment of other types of econometric models. First of all, *the normality* or typical heteroscedasticity of residuals and variables *is not generally studied*, since spatial models do not have such assumptions.[1] Second, *when outliers are detected, they are not removed* in models on area data, but rather are modelled, for example, with dummy (zero-one) variables. Their removal could cause spatial discontinuity and data deficiencies in the areas on the related map and as a consequence an incompatible spatial weights matrix W. Third, an important element of diagnostics is the *assessment of spatial autocorrelation* of residuals using Moran's I statistics, which allows determining whether a spatial model is necessary when residuals originating from OLS are tested or whether the spatial model has successfully filtered the spatial relationship when residuals from the spatial model are examined. Fourth, in assessing the fit of spatial models, due to the MLE estimation, the Akaike information criterion, Bayesian (BIC) or LogLik criterion is most often used, while pseudo-R^2 is used to a much lesser extent. Fifth, the search for the best model consists of finding such specification and taking into account the model class, estimation method and selection of variables so that the *significance of the coefficients is the most advantageous*. The American Statistical Association recommendations of 2016 regarding understanding *p*-value and its use in assessing the quality of regression are applicable (Wasserstein & Lazar, 2016). Sixth, *spatial heteroscedasticity tests* based on scanning statistics (scan test) for point data (Le Gallo, López, & Chasco, 2019) or so-called LOSH are an extension of local G spatial autocorrelation measures (Ord & Getis, 1995, 2012). They allow tracking the formation of spatial clusters of residuals from models and thus infer local unobserved characteristics.

Details of these issues will be discussed subsequently.

5.2.2.1 Information criteria and pseudo-R^2 in assessing model fit

The following measures are available for spatial models: Akaike, Bayesian, logLik and pseudo-R^2. In regression models, AIC is expressed by the formula:

$$AIC = -2logLik + 2k = A + N \log\left(\frac{RSS}{N}\right) + 2k \tag{5.15}$$

where N is the number of observations, residual sum of squares (RSS) is expressed as $RSS = \sum_{i=1}^{n} \varepsilon_i^2 = \sum_{i=1}^{n} \left(y_i - f(x_i)\right)^2$,

k is the number of estimated parameters of the free model parameters and A is a constant. Similarly, BIC is expressed by the formula:

$$BIC = -2logLik + k\log(N) = A + N \log\left(\frac{RSS}{N}\right) + k\log(N) \tag{5.16}$$

Both AIC and BIC depend on the size of sample B; hence they are incomparable for the same specification estimated on different length of datasets.

Among the pseudo-R^2, econometrics has a choice of several measures that work differently in selected model classes and with selected estimation methods. Generally, among pseudo-R^2 measures are: Efron, McFadden, McFadden, Cox-Snell, Nagelkerke, McKelvey and Zavoin, count and corrected count. In the case of spatial models, the pseudo-R^2 Nagelkerke is most often used, given by the formula:

$$R^2 = \frac{1 - \left\{ L\left(M_{intercept}\right) \middle/ L\left(M_{Full}\right) \right\}^{2/N}}{1 - L\left(M_{intercept}\right)^{2/N}} \tag{5.17}$$

where N is the number of observations and $L(M)$ is the conditional probability of the dependent variable assuming independent variable data, understood as the sum of error squares (SSE), where it is estimated for the model with the constant L only ($M_{intercept}$) or for the full L model (M_{Full}).

Using the information criteria, a model is chosen for which the AIC and/or BIC are the lowest (also for negative values). On the contrary, when assessing logLik itself, the highest value is sought. As a general rule, the model with the highest LogLik (log likelihood, log probability) or the lowest information criterion (AIC) should be preferred. When examining pseudo-R^2, as in the case of traditional R^2, a model with the highest value of this measure is sought. In all cases, this is a way to compare models with the same dependent variable estimated on a sample of the same size.

Technically, information criteria for estimated models can be obtained in several ways. One can use the **AIC()** and **BIC()** commands. The AIC for the appropriate linear model can be viewed by referring to the *AIC_lm.model* of the regression object – attributes can be displayed by the **attributes()** command. LogLik can be obtained with the **logLik()** command or as an LL attribute. Pseudo-R^2 is displayed in the regression summary after entering *Nagelkerke = TRUE* as the option.

```
# SAR model (one spatial factor rho)
model.SAR<-lagsarlm(eq1, data=sub, listw=cont.listw)

AIC(model.SAR)
#[1] 2748.24

BIC(model.SAR)
#[1] 2795.522

attributes(model.SAR)
```

```
$names
 [1] "type"          "dvars"           "rho"            "coefficients"
 [5] "rest.se"       "LL"              "s2"             "SSE"
 [9] "parameters"    "logLik_lm.model" "AIC_lm.model"   "method"
[13] "call"          "residuals"       "opt"            "tarX"
[17] "tary"          "y"               "X"              "fitted.values"
[21] "se.fit"        "similar"         "ase"            "rho.se"
[25] "LMtest"        "resvar"          "zero.policy"    "aliased"
[29] "listw_style"   "interval"        "fdHess"         "optimHess"
[33] "insert"        "trs"             "LLNullLlm"      "timings"
[37] "f_calls"       "hf_calls"        "intern_classic"

$class
[1] "sarlm"
```

```
model.SAR$AIC_lm.model
#[1] 2761.098

logLik(model.SAR)
#'log Lik.' -1362.12 (df=12)

model.SAR$LL
```

```
   [,1]
[1,] -1362.12
```

Information criteria for a single model interpreted in isolation from other values do not make sense, as they are a comparative tool. Models can be compared using only one criterion or all three at a time. The AIC and BIC criteria can be displayed for many models together using the **AIC()** or **BIC()** command and specifying further model objects as arguments.

```
AIC(model.lm, model.SAR, model.SDM) # AIC information criteria
```

```
          df       AIC
model.lm  11 2761.098
model.SAR 12 2748.240
model.SDM 21 2748.741
```

```
BIC(model.lm, model.SAR, model.SDM) # BIC information criteria
```

```
          df       BIC
model.lm  11 2804.440
model.SAR 12 2795.522
model.SDM 21 2831.484
```

LogLik and AIC values can be viewed in bulk, along with LR tests, using the **anova.sarlm()** command. Unfortunately, the **anova.sarlm()** command does not display the Bayesian BIC criterion. However, they can be easily added to the printout.

```
anova.sarlm(model.lm, model.SAR, model.SDM) # comparison of AIC and logLik
```

```
          Model df   AIC  logLik Test L.Ratio  p-value
model.lm      1 11 2761.1 -1369.5    1
model.SAR     2 12 2748.2 -1362.1    2  14.857 0.000116
model.SDM     3 21 2748.7 -1353.4    3  17.500 0.041443
```

```
out1<-anova.sarlm (model.lm, model.SAR, model.SDM )
out2<-BIC(model.lm, model.SAR, model.SDM)
out3<-cbind(out1, out2) # combination of objects
out3 # displaying the result
```

```
          Model df   AIC     logLik   Test L.Ratio   p-value      df BIC
model.lm      1 11 2761.098 -1369.549   NA      NA        11  2804.440
model.SAR     2 12 2748.240 -1362.120 1 vs 2 14.85734 0.0001159559 12 2795.522
model.SDM     3 21 2748.741 -1353.370 2 vs 3 17.49965 0.0414427685 21 2831.484
```

5.2.2.2 Test for heteroscedasticity of model residuals

In the diagnostics of linear and spatial models, heteroscedasticity of residuals can be tested. The well-known Breusch-Pagan test (BP test) or the localG-related LOSH test can be used. The LOSH test, described in Chapter 4, measures the variance of a neighbourhood variable, indicating areas with uniform and varied variability. High H_i values indicate a heterogeneous environment, while low values indicate local homogeneity.

The Breusch-Pagan test assumes in H_0 that the variance of residuals is constant (homoscedastisticity, $Var(\varepsilon_i) = \sigma^2$) and in H_1 that the variance of residuals is unstable (heteroscedasticity) and in particular depends on other factors that can be described by the function ($Var(\varepsilon_i) = \sigma_i^2 = \sigma^2 f(z_i)$). In the BP test, the variance of the residuals from the estimated model is determined, and a new variable is constructed – for each observation of the residual, the square of the residual value divided by the variance of these residuals is counted. Then the regression is performed as in the base model, with the difference that the explained variable is replaced by a newly constructed variable (based on residuals). The BP test statistic is 1/2 explained sum of squares (ESS; calculated as the sum of squares of differences between the fitted values from the model and the mean of the explained variable) and has a distribution χ^2 with $m - 1$ degrees of freedom, where m is the number of variables in the second regression.

In the case of statistically significant heteroscedasticity, models for spatial heterogeneity should be used. Spatial heterogeneity means that processes are unstable in space. Heterogeneity may result from poor functional specification of the model or from incorrect specification of the model, for example, by omitting a variable. Then the spatial distribution of model errors is like the distribution of the omitted variable. Another source of spatial heterogeneity is parameter instability, for example, structural instability or group heteroscedastic patterns (e.g. southern/non-southern regions, village/city). Such patterns are called **spatial regimes** (Anselin, 1988). If the residuals are heteroscedastic in a linear model, the output is usually zero-one modelling of the spatial regime or the use of spatial models. Unfortunately, in practice, based on the rest of the linear regression model, it is often difficult to distinguish spatial dependence on spatial heterogeneity.

In spatial models, heteroscedasticity tests also respond to autocorrelation, so a test result rejecting homoscedasticity can simply mean the existence of spatial autocorrelation. One way to distinguish these spatial effects is to control one of them when testing the other. This is possible, for example, by introducing a dummy (zero-one variable, e.g. south/non-south) that would suggest a horizontal shift. Another solution is modelling by *switching spatial regimes*, *geographically weighted regression* and so on.

The Breusch-Pagan test for heteroscedasticity of residuals can be used for both linear and spatial models. In both cases, however, other commands are used. For linear models, use the **bptest()** command from the lmtest:: package, and for spatial models, use the **bptest.sarlm()** command from the spatialreg:: package (as part of the **LR.sarlm()** command). The null hypothesis in both cases assumes homoscedasticity of residuals.

```
library(lmtest)
model.lm<-lm(eq1, data=sub)
bptest(model.lm)                # BP test for residuals from the OLS model
```

```
        studentized Breusch-Pagan test

data: model.lm
BP = 21.902, df = 9, p-value = 0.009194
```

```
# SAC model (two spatial coefficients rho and lambda)
model.SAC<-sacsarlm(eq1, data=sub, listw=cont.listw, method="LU")
bptest.sarlm(model.SAC)         # BP test for residuals from the SAC model
```

```
        studentized Breusch-Pagan test

data:
BP = 22.754, df = 9, p-value = 0.006774
```

```
# LOSH statistics for the rest of the model
losh.stat.SAC<-LOSH(model.SAC$residuals, cont.listw, a=2, var_hi=TRUE,
zero.policy=TRUE, na.action=na.exclude)
```

```
          Hi                  E.Hi      Var.Hi            Z.Hi
Min.   : 0.000182 Min.   :1 Min.   : 1.565 Min.   :0.000021
1st Qu.: 0.220844 1st Qu.:1 1st Qu.: 2.485 1st Qu.:0.111049
Median : 0.368830 Median :1 Median : 3.498 Median :0.230046
Mean   : 0.915260 Mean   :1 Mean   : 4.638 Mean   :0.588099
3rd Qu.: 0.813641 3rd Qu.:1 3rd Qu.: 4.385 3rd Qu.:0.512103
Max.   :13.244467 Max.   :1 Max.   :17.678 Max.   :7.648379
```

```
     x_bar_i                 ei
Min.   :-7.4247 Min.   :    0.001
1st Qu.:-2.1943 1st Qu.:    4.207
Median :-0.1456 Median :   18.850
Mean   : 0.3261 Mean   :   85.484
3rd Qu.: 2.4172 3rd Qu.:   54.844
Max.   :15.3843 Max.   :5195.863
```

```
losh.stat.LM<-LOSH(model.lm$residuals, cont.listw, a=2, var_hi=TRUE, zero.
policy=TRUE, na.action=na.exclude)

library(GISTools)
choropleth(pov, losh.stat.SAC[,1], main="Spatial distribution of LOSH
statistics for residuals from the SAC model")
shades<-auto.shading(losh.stat.SAC[,1])
choro.legend(14, 50.25, shades, cex=0.65, bty="n")

choropleth(pov, losh.stat.LM[,1], main="Spatial distribution of LOSH sta-
tistics for residuals from the LM model ")
shades<-auto.shading(losh.stat.LM[,1])
choro.legend(14, 50.25, shades, cex=0.65, bty="n")
```

The previous result shows that both residuals from the OLS and SAC model show spatial heterogeneity. This is confirmed by the BP test (H$_1$ accepted), as well as the very different results of the LOSH statistics. With a figure (not presented, code as previously) one can confirm that the highest diversity of the rest of the model is around significant urban centres: Warsaw, Łódź, the Silesian metropolis and Wrocław.

5.2.2.3 Residual autocorrelation tests

Spatial modelling aims to improve the specification of the econometric model. The spatial factor may be relevant in explaining variability. When modelling time series, autocorrelation of errors over time is examined. In spatial modelling, the first step in model diagnosis is to study the residuals from a linear model for the occurrence of spatial autocorrelation. For this purpose, Moran's I for residuals is used, as well as join-count statistics.

The existence of spatial autocorrelation in residuals does not necessarily mean the use of spatial models. It may result from the effect of estimating a non-linear relationship with a linear model. In this situation, it is worth checking the remainders from the OLS estimated model on variable logarithms (Cliff & Ord, 1981). If, despite this, spatial autocorrelation persists, the reason for this state may be the variables omitted in the model. It is worth testing the rest in terms of their spatial distribution. For this purpose, the Moran test, graphical analysis and join-count test are carried out. The visualisation of variables and colour test in groups can help to find a variable that may affect the studied phenomenon and has not been included in the model. Omitting the variable in the model may result in spatial heterogeneity. In a situation where spatial autocorrelation occurs despite regression on logarithms and the addition of a variable explaining spatial variability, it is necessary to estimate spatial models.

In a situation where the null hypothesis about the randomness of residuals of the OLS model is rejected, a spatial approach should be used in the estimation, because the OLS estimators are biased and inconsistent, as the data is non-random and there is an incorrect specification of the model. In a situation of spatial correlation, R^2 measures and the significance tests of parameters in OLS models are unreliable.

The Moran's I test is the best-known and oldest test used for regression residuals to investigate the existence of spatial autocorrelation. Unfortunately, based on this test, it is not possible to determine which form of spatial autocorrelation, error, explanatory variables or lag of the dependent variable is the reason for the incorrect specification of the OLS model. Moments of statistics distribution and details of estimation can be found in Cliff and Ord (1981) and Anselin (1988). The matrix test has the form:

$$I = \frac{N}{W} \frac{\sum_i \sum_j w_{ij} (u_i - \bar{u})(u_j - \bar{u})}{\sum_i (u_i - \bar{u})^2} \qquad (5.18)$$

where u_i is a residual in the studied region, and u_j is a residual in another region (in particular in the neighbouring region). The interpretation of Moran's statistic is similar to the interpretation of the Pearson's linear correlation coefficient. Values of $0 < I < 1$ mean positive spatial autocorrelation of residuals, and values of $-1 < I < 0$ negative. $I = 0$ is interpreted as the lack of spatial dependence of the residuals.

Moran's I statistic tests the spatial autocorrelation of model residuals. The test is run by the **lm.morantest()** command from the spdep:: package. This requires as an argument the object of class *lm* (regression result) and the spatial weights matrix – the object of the *listw* class. The H_1 alternative hypothesis assumes the existence of spatial autocorrelation. The test can be two sided or one sided and assumes an alternative hypothesis with positive or negative autocorrelation. Usually a two-sided test is performed, verifying only the existence of autocorrelation. To determine the Moran's I statistic, it is necessary to estimate the linear model with the result assigned to the object and successively calculate the Moran's I statistic using objects of the *lm* and *listw* class, resulting in the object of the *htest* class. By default, the one-sided hypothesis is tested. To test two-sided hypotheses, it is necessary to multiply by two significance (*p*-value) or change the settings by defining options. The same test can be performed manually, by creating a model residual object and testing it with a regular Moran test, run by the **moran.test()** command from the spdep:: package.

```
model.lm<-lm(eq1, data=sub)
lm.morantest(model.lm, cont.listw)        # Moran test for residuals from the OLS model
```

```
        Global Moran I for regression residuals
data:
model: lm(formula = eq1, date = sub)
weights: cont.listw
Moran I statistic standard deviate = 3.2048, p-value = 0.0006758
alternative hypothesis: greater
sample estimates:
Observed Moran I      Expectation        Variance
      0.09723969        -0.01034565      0.00112697
```

```
# SAC model (two spatial coefficients rho and lambda)
model.SAC<-sacsarlm(eq1, data=sub, listw=cont.listw, method="LU")
moran.test(model.SAC$residuals, cont.listw)# Moran test for residuals from SAC
```

```
        Moran I test under randomisation

data: model.SAC$residuals
weights: cont.listw

Moran I statistic standard deviate = 0.11066, p-value = 0.4559
alternative hypothesis: greater
sample estimates:
Moran I statistic      Expectation        Variance
      0.001031651      -0.002638522      0.001100089
```

The previous result quite clearly shows that in the a-spatial linear regression model, there is a fairly pronounced spatial autocorrelation of residuals ($I_{Moran} = 0.097$, *p*-value = 0.00067), and in the spatial SAC model, it was filtered out ($I_{Moran} = 0.001$, *p*-value = 0.456).

The autocorrelation of residuals is also worth visualising. The following are residuals broken down by mean and standard deviation, plus by positive and negative values. Positive residual mapping shows clear clustering of these values.

```
# spatial distribution of residuals from the OLS linear model
res<-model.lm$residuals
brks<-c(min(res), mean(res)-sd(res), mean(res), mean(res)+sd(res),
max(res))
cols<-c("steelblue4","lightskyblue","thistle1","plum3")
plot(pov, col=cols[findInterval(res,brks)])
title(main=" Rest in the OLS model ")
legend("bottomleft", legend=c("<mean-sd", "(mean-sd, mean)", "(mean,
mean+sd)", ">mean+sd"), leglabs(brks1), fill=cols, bty="n", cex=0.8)

# quick map of residuals divided into positive and negative
pov$res<-res
rng<-c(-100,0,100)
cls<-brewer.pal(3, "PuBuGn")
spplot(pov, "res", col.regions=cls, at=rng)
title(main=" Positive and negative residuals in the OLS model ")
```

The spatial relationship of the positive and negative residuals can be confirmed by the join-count test. The test requires variable factorisation with the **factor**() command. The result is the following: negative residuals are rather randomly distributed in space (although local clusters are optically visible; there are also lonely areas), and positive residuals show positive spatial autocorrelation, that is, they are located next to each other more often than would result from random distribution.

```
# join.count test for residuals (positive vs. negative)
resid<-factor(cut(res, breaks=c(-100, 0, 100),
      labels=c("negative", "positive")))
joincount.test(resid, cont.listw)
```

```
        Join count test under nonfree sampling
data: resid
weights: cont.listw
Std. deviate for negative = 1.2466, p-value = 0.1063
alternative hypothesis: greater
sample estimates:
Same colour statistic   Expectation   Variance
          66.87745         64.14248    4.81364

        Join count test under nonfree sampling
data: resid
weights: cont.listw
Std. deviate for positive = 2.9562, p-value = 0.001557
alternative hypothesis: greater
sample estimates:
Same colour statistic  Expectation  Variance
          38.820635    33.142480    3.689208
```

5.2.2.4 Lagrange multiplier tests for model type selection

In a situation where the test for spatial autocorrelation of residuals from the OLS model shows the existence of a spatial relationship, one can use the Lagrange multiplier (LM) test to find a priori better specification of spatial models, lag and/or spatial error. LM tests examine the existence of spatial relationships in linear models. Five LM tests can be performed: two normal for lag and error models (LMerr and LMlag); two robust, also for lag and error models (RLMlag, RLMerr) and the combined SARMA test for the SAC model. The LMerr test tests the spatial dependence of an error, while the LMlag test tests the significance of a spatially lagged dependent variable. For both ordinary LM tests, there are versions that are robust (LM robust) of local poor model specification (existence of spatial error when the presence of spatial lag is tested and vice versa) (Anselin, 1998; Anselin, Bera, Florax, & Yoon, 1996). All LM tests are used for the a-spatial linear regression model estimated by OLS. The results of these tests are the basis for decisions as to the choice of the form of spatial dependence (Anselin & Bera, 1998) – based on the smallest p-value, it is decided which specification is the most appropriate (Anselin and Florax, 1995).[2] Spatial autocorrelation in residuals may occur in the situation of incorrect specification of the model or when the errors are from a non-normal distribution, when there is a non-linear relationship, when a weak set of explanatory variables was selected or when the spatial weights matrix was chosen incorrectly.

The LM test statistics for the error model (LMerror) have the form:

$$LM_{ERROR} = \frac{(e'We / S^2)^2}{T} \tag{5.19}$$

where e is a vector of residuals from the OLS, S^2 is the estimated standard error, $T = [tr(W + W')W]$ and tr is the matrix trace operator. Statistics have an asymptotic distribution $\chi^2(1)$. LM statistics are based on the assumption of normality; however, as demonstrated by Anselin and Kelejian (1997) this is not necessary, and asymptotically the test is equivalent to the Moran test used for OLS residuals. The LM test statistics for the lag model are expressed by the formula:

$$LM_{LAG} = \frac{(e'Wy / S^2)^2}{[(WXb)'M(WXb) / S^2 + T]} \tag{5.20}$$

where $M = I - X(X'X)^{-1}X'$, b is an estimate of the OLS and $T = [tr(W + W')W]$, where tr is the matrix trace operator. Statistics also have an asymptotic distribution $\chi^2(1)$.

Technically, tests are run with the **lm.LMtests()** command from the spdep:: package. Two objects are used as arguments: the linear regression result and spatial weights matrix – *listw* class object. The command's options specify which of the tests is being run. One can use the option to display all tests (*test = "all"*). The level of significance is decisive in the tests. The best model is indicated by tests with the lowest p-value. Robust tests are conclusive when ordinary tests fail. Summary statistics can be viewed with the **summary()** command, whose argument is the result of the **lm.LMtests()** command.

```
lm.LMtests(model.lm, cont.listw, test="all")
```

```
        Lagrange multiplier diagnostics for spatial dependence
data:
model: lm(formula = eq1, data = sub)
weights: cont.listw
LMerr = 8.0431, df = 1, p-value = 0.004568

        Lagrange multiplier diagnostics for spatial dependence
data:
model: lm(formula = eq1, data = sub)
weights: cont.listw
LMlag = 14.342, df = 1, p-value = 0.0001524

        Lagrange multiplier diagnostics for spatial dependence
data:
model: lm(formula = eq1, data = sub)
weights: cont.listw
RLMerr = 1.6084, df = 1, p-value = 0.2047

        Lagrange multiplier diagnostics for spatial dependence
data:
model: lm(formula = eq1, data = sub)
weights: cont.listw
RLMlag = 7.9074, df = 1, p-value = 0.004923
```

```
        Lagrange multiplier diagnostics for spatial dependence
data:
model: lm(formula = eq1, data = sub)
weights: cont.listw
SARMA = 15.95, df = 2, p-value = 0.0003439
```

```
summary(lm.LMtests(model.lm, cont.listw, test="all"))
```

```
        Lagrange multiplier diagnostics for spatial dependence
data:
model: lm(formula = eq1, data = sub)
weights: cont.listw

        statistic parameter   p.value
LMerr      8.0431          1 0.0045678 **
LMlag     14.3421          1 0.0001524 ***
RLMerr     1.6084          1 0.2047233
RLMlag     7.9074          1 0.0049233 **
SARMA     15.9505          2 0.0003439 ***
---
Signif. codes:  0 '***' 0.001 '**' 0.01 '*' 0.05 '.' 0.1 ' ' 1
```

It follows from the previous that the *LMlag* specification has the lowest *p*-value, but the robust *RLMlag* specification is no longer as good. The second best is SARMA, which combines lag and error specifications and is probably the most appropriate.

5.2.2.5 Likelihood ratio and Wald tests for model restrictions

LR, Wald and LM tests (also called score test) can help in choosing the best model. Each of them is based on a comparison of the likelihood function;[3] that is, it works for estimation results for which logLik is determined.

The *LR test* is used for testing the nested models. Assuming that the unrestricted logLik model is higher than the limited model, the LR test tests whether the logLik difference is significant. If so, then the restricted model should be discarded. If the difference is insignificant, choose a simpler or limited model, because the unlimited model does not bring added value. A limited model is one in which one of the parameters is zeroed, for example, spatial coefficient or variable. H_0 assumes that $logLik_{limited} = logLik_{unlimited}$, while H_1 *implicitly* assumes that $logLik_{limited} < logLik_{unlimited}$. Adopting H_0 means choosing the limited (simpler) model, and adopting H_1 the unlimited (fuller) model. Character test statistics $LR = -2\ln\left(Lik_{limited} / Lik_{unlimited}\right)$ have a distribution χ^2 with as many degrees of freedom as there are limits.

Equivalent models are compared in an interesting way, for example, SEM and SAR (i.e. when none is a limited version of the other). In such a system, both models are compared to OLS (because OLS is a limited version of SEM and SAR by imposing restrictions, respectively, $\lambda = 0$ and $\rho = 0$). The test examines how much better SEM is over OLS and SAR over OLS, and comparing their dominance, one is searching for greater increase in the likelihood function. The more dominant model over OLS automatically becomes a better model in a pair of equivalent models.

The *Wald test*, like the LR test, compares the distances between the parameters of both models, although only one, limited model is used in the test itself. H_0 assumes that the distances between the parameters of the models are small; hence the limited model is better. In H_1, it is assumed that the distance between the parameters is large, and the greater the distance between the parameters, the less likely that the restrictions are appropriate; that is, the unlimited model is better. The restricted model is the OLS model, which assumes that the spatial coefficients are equal to 0.

The *LM test*, evaluates the slope of the tangent to the logLik function for the limited model. The logLik function is the inverse of u-shaped ("hill-shaped"); hence the tangent slope at maximum is 0, and it increases as one moves away from the max. point. When the bounded and unlimited models (max. LogLik locates this model at the apex of the distribution) are similar, the tangent slope will be close to 0.

The LR, LM and Wald diagnostic tests are displayed along with the model estimation. However, one can call them using separate commands for single models or for several at once. The **LR.sarlm()** command from the spatialreg:: package can be used to compare two spatial models. As a result, of course, LR and logLik statistics appear for both models tested. Both command arguments are objects of the *sarlm* class, recorded with the results of estimation of spatial models. In this family of commands, **LR1.sarlm()** and **Wald1.sarlm()** are still available, testing the existence of spatial dependence. The command results are objects of the *htest* class.

```
# unlimited model
# Manski model (three spatial coefficients - lambda, rho, theta)
model.GNS<-sacsarlm(eq1, data=sub, listw=cont.listw, type="sacmixed")

# restricted model
# SDM model (two spatial coefficients rho and theta)
model.SDM<-lagsarlm(eq1, data=sub, listw=cont.listw, type="mixed", tol.
solve=1.0e-20, method="LU")

LR.sarlm(model.GNS, model.SDM) # comparison of the two indicated models
```

```
        Likelihood ratio for spatial linear models
data:
Likelihood ratio = 0.43276, df = 1, p-value = 0.5106
sample estimates:
Log likelihood of model.GNS Log likelihood of model.SDM
                -1353.154                    -1353.370
```

```
Wald1.sarlm(model.SDM) # comparison of the indicated model with OLS
```

```
        Wald diagnostics for spatial dependence
data:
Wald statistic = 8.3815, df = 1, p-value = 0.003791
sample estimates:
        rho
0.2044004
```

```
LR1.sarlm(model.SDM) # comparison of the indicated model with OLS
```

```
        Likelihood Ratio diagnostics for spatial dependence
data:
Likelihood ratio = 6.72, df = 1, p-value = 0.009534
sample estimates:
Log likelihood of spatial lag model  Log likelihood of OLS fit y
                -1353.37                              -1356.73
```

The previous results should be interpreted as follows: the **LR.sarlm()** test compared the limited model (SDM) with the unlimited model (GNS), and the difference in logLik turned out to be insignificant, so the simpler, unlimited model, SDM, should be chosen. The **Wald1.sarlm()** test compared the distances between the SDM and OLS models. Distances are significantly large; hence the unlimited model (i.e. SDM) is better than the limited model (OLS). The **LR1.sarlm()** test compares the slopes of the tangents to the logLik curve, and the difference in these slopes is significant, so the unrestricted model (SDM) is better than the OLS base model.

5.2.3 Selection of spatial weights matrix and modelling of diffusion strength

Chapter 4 discussed the selection of a spatial weights matrix, which is current not only in the context of spatial statistics but also in econometric modelling.

The choice of the spatial weights matrix W is important for the result of spatial modelling. This is because the matrix W determines which regions and to what extent the neighbours of the regions are studied. Based on this information, spatial lags are determined, that is, the average of values in neighbouring regions, weighted with spatial weights. For this reason, along with another matrix of spatial weights, spatial lags with different values go differently into the model, which translates into different parameter estimates, with explanatory variables and spatial components.

The analysis of the sensitivity of the model to the selection of the W matrix is particularly important in the case of the matrix based on the criterion *k* nearest neighbours, because the selection of the number of neighbours is usually arbitrary. This allows for assessing the strength and direction of spatial interactions depending on the adopted weights matrix W. It also allows for detecting the diffusion strength depending on the neighbour distance. This is an important element of the research in the process of determining the internalisation of processes and externalities.

The following example examines diffusion in a selected SAC econometric model. The variability of spatial coefficients *rho* and *lambda* resulting from the change of the spatial weights matrix W was analysed. Ten models with W matrices with an increasing number of neighbours were estimated, from *knn = 10* to *knn = 100* neighbours. In addition to the spatial coefficients, the AIC was used for the comparison.

```
crds<-coordinates(pov)
pov.k.sym.listw<-nb2listw(make.sym.nb(knn2nb(knearneigh(crds, k=10))))
model.SAC.10<-sacsarlm(eq1, data=sub, listw=pov.k.sym.listw, method="LU")

pov.k.sym.listw<-nb2listw(make.sym.nb(knn2nb(knearneigh(crds, k=20))))
model.SAC.20<-sacsarlm(eq1, data=sub, listw=pov.k.sym.listw, method="LU")

pov.k.sym.listw<-nb2listw(make.sym.nb(knn2nb(knearneigh(crds, k=30))))
model.SAC.30<-sacsarlm(eq1, data=sub, listw=pov.k.sym.listw, method="LU")

pov.k.sym.listw<-nb2listw(make.sym.nb(knn2nb(knearneigh(crds, k=40))))
model.SAC.40<-sacsarlm(eq1, data=sub, listw=pov.k.sym.listw, method="LU")

pov.k.sym.listw<-nb2listw(make.sym.nb(knn2nb(knearneigh(crds, k=50))))
model.SAC.50<-sacsarlm(eq1, data=sub, listw=pov.k.sym.listw, method="LU")

pov.k.sym.listw<-nb2listw(make.sym.nb(knn2nb(knearneigh(crds, k=60))))
model.SAC.60<-sacsarlm(eq1, data=sub, listw=pov.k.sym.listw, method="LU")

pov.k.sym.listw<-nb2listw(make.sym.nb(knn2nb(knearneigh(crds, k=70))))
model.SAC.70<-sacsarlm(eq1, data=sub, listw=pov.k.sym.listw, method="LU")

pov.k.sym.listw<-nb2listw(make.sym.nb(knn2nb(knearneigh(crds, k=80))))
model.SAC.80<-sacsarlm(eq1, data=sub, listw=pov.k.sym.listw, method="LU")

pov.k.sym.listw<-nb2listw(make.sym.nb(knn2nb(knearneigh(crds, k=90))))
model.SAC.90<-sacsarlm(eq1, data=sub, listw=pov.k.sym.listw, method="LU")
```

```
pov.k.sym.listw<-nb2listw(make.sym.nb(knn2nb(knearneigh(crds, k=100))))
model.SAC.100<-sacsarlm(eq1, data=sub, listw=pov.k.sym.listw, method="LU")
out<-anova.sarlm(model.SAC.10, model.SAC.20, model.SAC.30, model.SAC.40,
model.SAC.50, model.SAC.60, model.SAC.70, model.SAC.80, model.SAC.90,
model.SAC.100)

out<-cbind(out, lambda=c(model.SAC.10$lambda, model.SAC.20$lambda,
model.SAC.30$lambda, model.SAC.40$lambda, model.SAC.50$lambda, mod-
el.SAC.60$lambda, model.SAC.70$lambda, model.SAC.80$lambda, model.
SAC.90$lambda, model.SAC.100$lambda), rho=c(model.SAC.10$rho, model.
SAC.20$rho, model.SAC.30$rho, model.SAC.40$rho, model.SAC.50$rho, model.
SAC.60$rho, model.SAC.70$rho, model.SAC.80$rho, model.SAC.90$rho, model.
SAC.100$rho))

out
```

```
               Model df      AIC     logLik     lambda          rho
model.SAC.10       1 13 2756.633 -1365.317 0.1819736   0.12783495
model.SAC.20       2 13 2755.119 -1364.559 0.3091111   0.15602136
model.SAC.30       3 13 2759.045 -1366.523 0.2834714   0.15805861
model.SAC.40       4 13 2760.381 -1367.190 0.1924195   0.24104748
model.SAC.50       5 13 2760.919 -1367.460 0.3169217   0.17301562
model.SAC.60       6 13 2762.646 -1368.323 0.3005130   0.11784694
model.SAC.70       7 13 2763.342 -1368.671 0.3339067   0.04653375
model.SAC.80       8 13 2764.143 -1369.072 0.3272610  -0.10777881
model.SAC.90       9 13 2764.070 -1369.035 0.3437959  -0.21799251
model.SAC.100     10 13 2762.173 -1368.087 0.3595670  -0.70966129
```

```
# graph of spatial parameters in subsequent models - Figure 5.4a
plot((1:10)*10, out[,5], type="l", ylim=c(-0.8,0.4), xlab="knn number of
neighbours", ylab="spatial parameters")
lines((1:10)*10, out[,6], lwd=2)
legend("bottomleft", legend=c("lambda", "rho"), lty=c(1,1), lwd=c(1,2), bty="n")
abline(h=(-8:4)/10, lty=3, col="grey80")

# AIC information criteria chart in subsequent models - Figure 5.4b
plot((1:10)*10, out[,3], type="l", ylim=c(2750,2770), xlab = "number of
neighbours knn", ylab="AIC", lwd=2)
legend("bottomleft", legend=c("AIC"), lty=1, lwd=2, bty="n")
abline(h=c(2750,2755,2760, 2765, 2770), lty=3, col="grey80")
```

The printout of the *out* object shows the changes in the resulting estimation parameters in models with an increasingly wide neighbourhood. Figure 5.4 illustrates these changes. In relation to the AIC, the minimum achieved in the model with *knn=20* neighbours and the increase in AIC with neighbourhood expansion (except the last model) is visible. This would suggest choosing W for *knn=20* for the final model. The spatial parameters *rho* and *lambda* are negatively correlated, and usually an increase in one causes a decrease in the other. This property suggests a mechanism for balancing spatial effects. The *rho* parameter increases to approximately 40 neighbours and then decreases, while negative *rho* values appear for W, approximately 75 neighbours. This pattern shows expiring neighbourhood effects over 40 neighbours, suggesting that this is the end of the poviat interaction range. Further neighbourhoods than *knn=40* are subject to other central cities and build spatial interactions around other centres. This suggests that the model that is to cover all functional relationships should have a matrix W not wider than 40 neighbours, and to maximise fit, select *knn=20*.

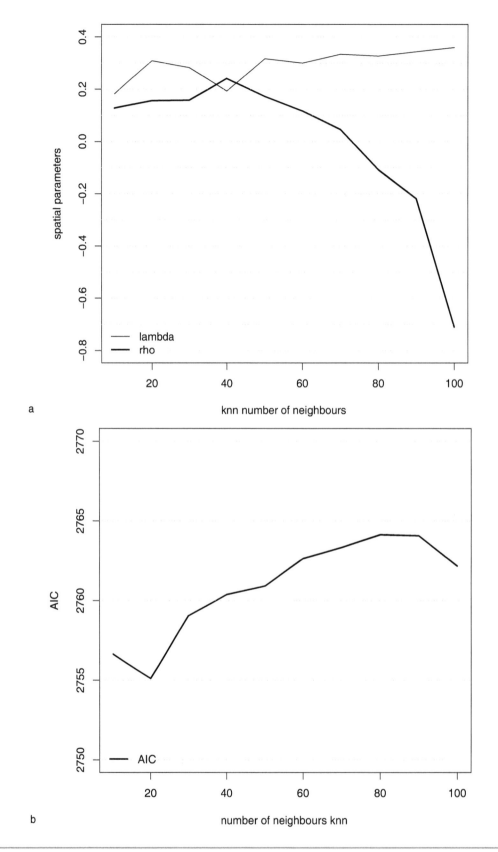

a

b

Figure 5.4 Parameters in the SAC model depending on the number of neighbours in the spatial weights matrix: a) rho and lambda, b) AIC

Source: Own study using R and the spatialreg:: package

5.2.4 Forecasts in spatial models

There are several degrees of difficulty in spatial forecasting, depending on the data used in modelling. The biggest challenge lies in the spatial weights matrix W for which the model is calibrated and which contains only specific spatial units. In the simplest forecasting scenario, the values of the explained variable y are determined for the same spatial units specified in W and the same input data (x). In a slightly more advanced scenario, for the same spatial units specified in W, new values of the explanatory data are determined for new values of the explained variable. In the most advanced scenario, the dependent variable values are forecast for new spatial units and new observations of explanatory variables.

The basis for forecasting is the Haining methodology (1990). Developments for more advanced cases can be found in Cressie (1993b), Bivand (2002), Kelejian and Prucha (2010) and Goulard, Thibault, and Thomas-Agnan (2017). In part of the strategy, spatial forecasting is an attempt to determine the value of a variable in neighbouring locations, knowing the process value in a given region or estimating the value of a dependent variable based on known values of explanatory variables based on the estimated model and the adopted structure of spatial dependencies. There are three components of the forecast: trend, signal and noise. The trend is an a-spatial smoothing factor and is expressed by a pattern $X \cdot \beta$, while the signal is a spatial factor and is expressed by a formula $\lambda Wy - \lambda WX\beta$ for error models and ρWy for lag models. The sum of the trend and signal factors is called fit. The forecast error is noise. For an explanation of how to create forecasts, see the help for forecasting functions, the command **predict.sarlm()** from the spatialreg:: package. While the trend can be estimated in any model, signal estimation, in particular for autoregressive models, is indirectly possible, including as a difference between prediction and a trend, or its value is assumed to be 0. The problem of forecasting is also still spatial homogeneity, stationarity, edge effect and so on.

The quality of the forecast can be checked in many ways. Popular measures include forecast bias (FB) (5.21 in nominal terms and 5.22 in percentage terms), mean deviation (MAD) and mean absolute percentage error (MAPE), as well as mean absolute error (MAE), mean square error (RMSE, root mean square error) and the correlation coefficient between the forecast and the actual data. There is one rule: the smaller the error, the better the model. Forecast quality measures can be defined as assuming that n is the number of observations, y_i is the observed value, $\overline{y_i}$ is the predicted value and $\overline{y_i}$ is the average of the observed values:

$$FB = \Sigma\left(\overline{y_i} - y_i\right) \tag{5.21}$$

$$FB = \Sigma \overline{y_i} / \Sigma y_i \tag{5.22}$$

$$MAD = \frac{1}{n}\Sigma\left|\overline{y_i} - y_i\right| \tag{5.23}$$

$$MAPE = \frac{1}{n}\Sigma\frac{\left|\overline{y_i} - y_i\right|}{y_i} \tag{5.24}$$

$$MAE = \Sigma\frac{\left|\overline{y_i} - y_i\right|}{n} \tag{5.25}$$

$$RMSE = \sqrt{\frac{\Sigma(\overline{y_i} - y_i)^2}{n}} \tag{5.26}$$

From the previous measures, MAPE is independent of the scale, while the other measures are not relativised. MAD, MAPE and MAE are based on the absolute difference between forecast and observation, while RMSE is based on the squares of these differences. MAD and FB (according to 5.21) are not averaged (by reference to the number of observations), and FB (according to 5.22) gives a percentage measure.

The following is the quality of matching to empirical data in several spatial models – the simplest scenario was used. To assess the quality of the forecasts, the Metrics:: package was used, which contains effective functions determining the measures discussed previously. These are **bias()**, **percent_bias()**, **mae()**, **mape()** and **rmse()**. Three forecast components were also shown – the fit component is the sum of the *trend* and *signal* components. By analogy with models of seasonal fluctuations, the trend shows the average level of the phenomenon, while the signal corrects this value by the influence of spatial factors.

```
# SDM model (two spatial coefficients, rho and theta)
model.SDM<-lagsarlm(eq1, data=sub, listw=cont.listw, type="mixed", tol.
solve=1.0e-20, method="LU")

# SDEM model (two spatial coefficients, theta and lambda)
model.SDEM<-errorsarlm(eq1, data=sub, listw=cont.listw, etype="emixed",
method="LU")

# SAR model (one spatial factor rho)
model.SAR<-lagsarlm(eq1, data=sub, listw=cont.listw)

# SEM model (one lambda spatial coefficient)
model.SEM<-errorsarlm(eq1, data=sub, listw=cont.listw)

model.SDM.p<-predict.sarlm(model.SDM)
model.SDEM.p<-predict.sarlm(model.SDEM)
model.SAR.p<-predict.sarlm(model.SAR)
model.SEM.p<-predict.sarlm(model.SEM)

model.SEM.p
```

```
            fit      trend         signal
4181 77.04326 77.62578 -0.5825122016
4182 72.96587 73.90672 -0.9408548636
4183 81.61913 82.24747 -0.6283384426
4184 82.94573 84.14776 -1.2020344168
4185 84.91456 86.13334 -1.2187844362
4186 87.48227 87.34561  0.1366678687
```

```
library(Metrics)
vec<-c("model.SDM.p", "model.SDEM.p", "model.SAR.p", "model.SEM.p")
metrics<-matrix(0, nrow=4, ncol=5)
rownames(metrics)<-vec
colnames(metrics)<-c("bias", "bias%", "MAE","MAPE", "RMSE")
for(i in 1:4){
     metrics[i,1]<-bias(sub$y, get(vec[i]))
     metrics[i,2]<-percent_bias(sub$y, get(vec[i]))
     metrics[i,3]<-mae(sub$y, get(vec[i]))
     metrics[i,4]<-mape(sub$y, get(vec[i]))
     metrics[i,5]<-rmse(sub$y, get(vec[i]))}
metrics
```

```
                     bias        bias%       MAE      MAPE      RMSE
model.SDM.p    1.866789e-16 -0.007670291 5.513246 0.06198692 8.487296
model.SDEM.p  -1.127488e-16 -0.007690391 5.537118 0.06226533 8.491335
model.SAR.p   -2.612328e-16 -0.008023855 5.547210 0.06208778 8.661793
model.SEM.p    6.348677e-16 -0.008221258 5.633744 0.06303802 8.729438
```

The previous result shows the quality of matching spatial models in the simplest scenario. The bias is close to 0, with two being positive and two negative. The percentage bias (%) is negative everywhere and amounts to approximately 0.7%–0.8%. The mean absolute error is 5, which should be related to the scale of the forecasted phenomenon (for y: average = 85.56, standard deviation = 11.44) – that is, it is about 5% of the average and

0.4 standard deviation. The mean absolute percentage error is around 0.06 or 6%, while the root mean square error is around 8.5, which also needs to be related to the level of the phenomenon (around 10%). All measures show that the models are fairly well fitted to the variables, and there are no significant differences in the quality of the forecast, regardless of the model used.

5.2.5 Causality

The literature distinguishes between two basic goals of econometric modelling: explanation and prediction (Shmueli, 2010). The development of machine learning methods, dedicated to predictions in models, deepens the differences with regard to the econometric methods, treated as a tool for explaining these relationships. The problem of causality is deeper than the problem of dependence – when there is a relationship between rainfall and crop yields, the causality is only one-way: rainfall affects yields, and it is difficult to expect that the amount of yield affects rainfall. While in predictive models, more and more often defined in supervised learning as a black box, causality is not a problem as long as the forecast is good; in econometric models focused on explaining dependencies, causality is a key issue. In spatial and a-spatial models, the approach to the problem of causality is similar. Kolak and Anselin (2019) present a good overview of problem in the paper "*A Spatial Perspective on the Econometrics of Program Evaluation*".

As tools for studying the causality in econometric models, one typically indicates instrumental variables, regression discontinuity, the difference-in-difference method (also *diff-in-diff*), the propensity score matching method and various natural and designed experiments (e.g. randomized experiment). These approaches are operationalised with respect to both traditional and spatial methods. Kolak and Anselin (2019) impose on this the problem of spatial dependence (SD) and spatial heterogeneity (SH). They start with the stable unit treatment value assumption (SUTVA) problem about the constant unit impact of intervention, introduced by Rubin (1974). SUTVA assumes that individuals under the influence of the intervention do not affect those non-affected by the intervention and that the impact results are similar at the individual and population levels, independent of the intervention towards other entities and identical in form. In spatial terms, SUTVA assumptions are not met when diffusion and spillover appear. Kolak and Anselin (2019) point out the problems associated with the spatial effects in counterfactual methods. First, they consider how the intervention is attributed – is it according to the spatial regime (SD problem)? The effectiveness of the impact depends on the distance from the intervention (SD problem); in some locations (with specific features), intervention is more likely (SH problem). Second, it is necessary to examine what the potential sources of variability are in variables under the influence of intervention – whether there are interactions within or between individuals as a spatial lag or spillover (SD problem); there is a measurement/aggregation scale error (SD problem); spatial patterns result from exogenous factors, which generates spatial heterogeneity (SH problem). Third, one is to assess what effects are estimated – spatial effects (neighbourhoods, diffusion), spatial dependence (SD problem), spatial heterogeneity (SH problem). Recognition of the previous problems translates into the choice of method, which implementations and assumptions Kolak and Anselin (2019) describe.

There are empirical studies in the literature that include causality research in various ways. With reference to instrumental variables, Chong, Qin, and Chen (2019) try to address the problem of endogenicity in the model explaining the relationship between economic growth and transport infrastructure. They use as instrumental variables the spatial lag of the explained (dependent) variable and the historical transport network. The use of the spatial lag of a dependent variable also appeared in Bottasso, Conti, Ferrari, and Tei (2014) and Rodríguez-Pose and Peralta (2015), who believe that these lags are not affected by dependent variables, so that there is no correlation between lags and error term. In turn, Le Gallo and Páez (2013) use as instrumental variables eigenvectors obtained from the spatial weights matrix. It is also possible to use two-stage estimation and a 2SLS estimator. Betz, Cook, and Hollenbach (2019), in the first stage, perform regression of the endogenous predictor x on the z instrument, which gives fitted x *hat* values, while in the second stage, they perform regression of the dependent variable y on fitted x *hat* values. As instruments, one typically uses geographical, meteorological or economic variables, such as natural disasters, rain and shocks in commodity prices – their common feature is that there is a spatial relationship over borders, since these processes do not stop at the territorial boundaries. With respect to *diff in-diff* estimation, Dubé et al. (2014) derive spatial estimators and implement real-estate modelling in relation to public transport, while Delgado and Florax (2015) show how to model *diff-in-diff*, taking into account local spatial interactions. The spatial approach to *propensity score matching* can be found in Chagas, Toneto, and Azzoni (2012), who model sugarcane crops; in Gonzales, Aranda, and Mendizabal (2017), who use a Bayesian approach to estimate the regional effects of microfinance or in De Castris and Pellegrini (2015), who model the neighborhood effects in the context of SUTVA.

In R, there is a choice of implementations addressing the problem of causality. The spatialreg:: package includes the **stsls()** function, which performs a generalized two-step OLS estimation. This function assumes spatial lag model $y = \rho W y + X\beta + \varepsilon$, which uses the Durbin component (spatially lagged X variables) as instruments for spatially lagged dependent variable y. The package sphet:: includes function **stslshac()**, also based on the S2SLS approach, whereby an instrument matrix H = (X, WX, $W2X$) is available. There is also the S2sls:: package dedicated to two-stage estimation with instrumental variables. The McSpatial:: package includes the **matchprop()** function, which allows for connecting the data using the *propensity score matching* method.

5.3 Selected specifications of cross-sectional spatial models

5.3.1 Unidirectional spatial interaction models

Spatial interaction models have been present in the literature for many years, although their first formalisation and theoretical foundations did not appear until Wilson (1971). As indicated by Fotheringham (2001), four main currents in the history of the development of spatial interaction models can be distinguished, associated with the so-called social physics (on a strong foundation of gravity models, mainly for modelling bidirectional flows such as trade or migration), statistical mechanics, a-spatial information processing and spatial information processing. While the first models based on gravitational models took into account the importance of mass and distance in attraction, in the next phases, the main factor explaining flows was distance and neighbouring relations.

The models as described subsequently were introduced by Fotheringham and O'Kelly (1989). These models assume that there are flows between the indicated locations. These flows depend mainly on the distance – the greater the distance, the weaker the flow. Two-way flows (from A to B and from B to A) are assumed in the widest form – this applies to commercial models. It is then assumed that there exists a matrix T of flows T_{ij} between locations M and N (5.27) and a matrix of distance d between these units (5.28). Then the matrix of flows is as shown in (5.29):

$$T = \{T_{ij}\}_{i,j=1}^{M,N} \tag{5.27}$$

$$d = \{d_{ij}\}_{i,j=1}^{M,N} \tag{5.28}$$

$$\begin{bmatrix} A & B & C & D \\ E & F & G & H \\ I & J & K & L \end{bmatrix} \tag{5.29}$$

However, in modelling socioeconomic processes, in fact, flows are often one way, for example, migrations from weak to strong regions, diffusion of development incentives from centres to the periphery. Flows are then given as a vector (5.30).

$$\begin{bmatrix} M \\ N \\ O \\ P \end{bmatrix} \tag{5.30}$$

This is especially useful in the implementation for the theory of central places, which assumes unidirectional impact to map the hierarchical (spatially discontinuous) process. A broader description of the model and applications in the analysis of the spatial extent of local governments in the delivery of public goods can be found in the article: Kopczewska, K. (2013). "The Spatial Range of Local Governments: Does Geographical Distance Affect Governance and Public Service?", *The Annals of Regional Science*, 51 (3), 793–810. This arrangement allows for studying the processes of expansion and relocation.

In the literature, several functional forms of spatial interaction models are most often used – polynomial, exponential or power, which allows for non-linearity of flows depending on spatial separation. In all these models, the main explanatory variable is the distance between the units tested, measured as Euclidean distance (simplest), road distance, travel time and travel cost (the most advanced measure). These models can be saved as:

$$\text{Distance polynomial model}: x_i = \beta_0 + \varphi_1 D_i^1 + \varphi_2 D_i^2 + \varphi_3 D_i^3 + \ldots + e \tag{5.31}$$

$$\text{Exponential model}: lnx_1 = \beta_0 + \beta_1 D + e \quad \text{because} \quad T_{ij} = \gamma_0 e^{\gamma_1 d_{ij}} \tag{5.32}$$

$$\text{Power model}: lnx_1 = \beta_0 + \beta_1 lnD + e \quad \text{because} \quad T_{ij} = \gamma_0 d_{ij}^{\gamma_1} \tag{5.33}$$

The interpretation of these models is always the same – they determine how the distance between the core and the periphery affects the level of the phenomenon.

There are several issues to consider in the final selection of the model. These are:

- *Choosing the level of data aggregation* – having available data at different levels of aggregation, one should consider whether the model is to be estimated on more detailed or more generalised data. A higher level of aggregation usually blurs local variations and averages the results. There is no clear translation that any model at a lower level of aggregation will definitely be better.
- *Choice of functional form of the model* – the three functional forms presented previously perform estimation in different ways. In research, it is worth considering each of the functional forms and choosing the best suited.
- *Choice of estimation method* – models of spatial interaction were typically estimated in an a-spatial way. However, there are no obstacles to include in the estimation the spatial weights matrix W, which naturally complements the location-related data. It should be treated, then, as if the explanatory variables related to distance explain the absolute location (relative to the central centre), and the spatial weights matrix explains the relative location (relative to the characteristics of the neighbours).

Evaluating models estimated by both OLS and MLE is possible using *standardised root mean square error* (SRMSE), where SRMSE is given by formula (5.34) as:

$$SRMSE = \frac{\sqrt{\dfrac{\sum_{i,j=1}^{M,N}(T_{ij} - T_{ij}')^2}{M \cdot N}}}{\dfrac{\sum_{i,j=1}^{M,N}T_{ij}}{M \cdot N}}$$

where the interpretation of *SRMSE* is as follows: *SRMSE* <0.5 means a very good fit, *SRMSE* ~ 0.75 means a decent fit and captured the main trends, while *SRMSE* > 1 means a weak fit and the existence of off-scale observation.

Modelling of unidirectional spatial interactions, where one indicates the core regions being the source of flowing stimuli, and the peripheral regions absorbing these stimuli, needs determining two issues: 1) selection of core regions and 2) examining whether the core-peripheral relation along with distance is present. For the purposes of regional policy, stimulus flows resulting from regional policy or natural diffusion processes related to, for example, agglomeration or urbanisation, are most often modelled in this way. As these are unobserved processes, their effects are examined, for example, the number of firms per capita or the unemployment rate in poviats or communes. On this basis, it is possible to draw conclusions on the spatial impact of the policy pursued.

Determining central locations is usually arbitrary – that is, based on administrative functions (e.g. headquarters of regional authorities) or according to population potential. In the example, it was assumed that the core centres are voivodeship cities – the main cities of the NUTS2 regions (see Figure 5.5) and development stimuli flow from them towards the borders of regions. It is assumed, then, that the processes flow from the centre to the periphery, which means that within the region (NUTS2), there is one "transmitting" region and several regions "receiving" these stimuli.

The following code generates a graph (see Figure 5.5a) of possible directions of regional policy impact. The **arrows()** command from the basic graphics:: package was used to draw the arrows. The first two arguments are the locations (*x, y*) of the core cities that are the starting point, and the next two arguments are the target coordinates (*x, y*), randomly generated based on the normal distribution.

```
# drawing of core cities and diffusion directions Figure 5.9a
voi<-readOGR(".", "wojewodztwa") # 16 units
voi<-spTransform(voi, CRS("+proj=longlat +datum=NAD83"))
par(mar=c(2,1,1,1)) # setting narrow margins
bins<-c(0,1)
variable<-data$core_city[data$year==2017]
cols<-c("white", "red")
plot(pov, col=cols[findInterval(variable, bins)])
plot(voi, add=TRUE, lwd=2)

# directional arrows
crds<-coordinates(pov)
```

```
city.id<-which(data$core_city==1 & data$year==2006)
city.crds<-crds[city.id,]

for(i in 1:4){
arrows(city.crds[,1], city.crds[,2], city.crds[,1]+rnorm(16,0,0.35), city.
crds[,2]+rnorm(16,0,0.35), angle=15, length=0.10, lwd=1, col="blue")}
```

In conjunction, it is worth assessing the spatial distribution of the core-peripheral distance for a defined relationship and location hierarchy. The following example (see Figure 5.5b) presents Euclidean distances between centroid regions, counted from the central hub to others within each region – voivodeship (NUTS2).

```
library(GISTools)
library(RColorBrewer)

# map of distances between peripheral and central locations Figure 5.5a
variable<-data$dist[data$year==2017]
shading<-auto.shading(variable, n=6, cols=rev(brewer.pal(6, "Spectral")))
choropleth(pov, variable, shading=shading)
choro.legend(15, 50, shading, cex=0.65, bty="n")
plot(voi, add=TRUE, lwd=2)
par(mar=c(5,4,4,2))     # return to typical margin settings
```

Identifying whether there is a core–periphery relationship and whether it decays with distance is an important element in assessing indications for modelling spatial interactions. Studying these distribution properties is possible in several ways, including panel charts, phenomenon over time in groups by distance, or a point chart of the phenomenon depending on distance.

The panel chart (see Figure 5.6a) allows illustration of the changes in the phenomenon over time in selected groups, in particular by location relative to major regional cities. In the following code, long panel data (380 units for 12 years – in 2006–2017, data for subsequent years are in the following lines) are divided into subsets by relative location – the distance of the poviat centroid from the centroid of the regional centre. Distance groups were separated every 25 or 50 km. For these groups, averages were determined in subsequent years and

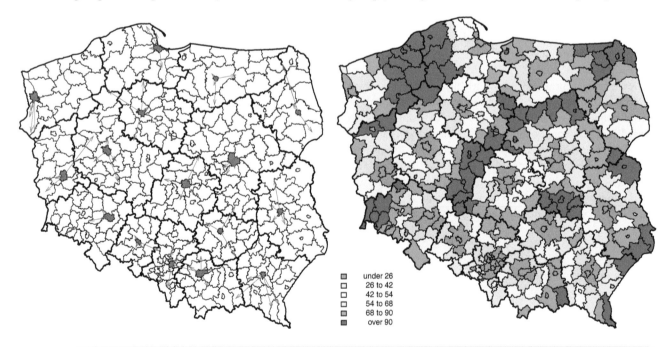

Figure 5.5 Directions of the influence of cores and distances between core and peripheral locations

Source: Own development using R and GISTools:: and spdep:: packages

combined into one object. The means by group in years were drawn as line charts. The panel chart allows for tracking the dynamics of various groups of units and assessing convergence/divergence processes.

The point plot of the phenomenon depending on the distance (see Figure 5.6b) allows for making a static approximation of relations in space. For the sake of clarity, the values of the unemployment rate were presented for one voivodeship (Mazowieckie). The size of points depends on the number of inhabitants of a given poviat.

```
# panel chart Figure 5.6a
data$variable<-data$XA21 # unemployment rate

# subsets by distance
sub1<-data[data$core_city==1,]              # capital of the region
sub2<-data[data$dist>=2 & data$dist<25,]    # distance up to 25 km
sub3<-data[data$dist>=25 & data$dist<50,]   # dist. between 25 and 50 km
sub4<-data[data$dist>=50 & data$dist<100,]  # dist. between 50 and 100 km
sub5<-data[data$dist>=100,]                 # dist. over 100 km

# average values by years in subsets by distance
msub1<-aggregate(sub1$variable, by=list(sub1$year), mean)
msub2<-aggregate(sub2$variable, by=list(sub2$year), mean)
msub3<-aggregate(sub3$variable, by=list(sub3$year), mean)
msub4<-aggregate(sub4$variable, by=list(sub4$year), mean)
msub5<-aggregate(sub5$variable, by=list(sub5$year), mean)

# combination of average values into one object
sub<-cbind(msub1, msub2$x, msub3$x, msub4$x, msub5$x)
minsub<-min(sub[,2:5], na.rm=TRUE)
maxsub<-max(sub[,2:5], na.rm=TRUE)

# panel chart
plot(msub1, type="n", ylim=c(0,21), xlab=" ", ylab=" ")
lines(msub1[1:12,], lwd=2)
lines(msub2[1:12,], lwd=2, lty=2)
lines(msub3[1:12,], lty=1)
lines(msub4[1:12,], lty=2)
lines(msub5[1:12,], lty=3)

title(main="Unemployment rate in poviats")
legend("bottom", legend=c("core - main regional cities","poviats located up to
25 km from the core","poviats located between 25 and 50 km from the core","poviats
located between 50 and 100 km from the core","poviats located more than 100 km
from the core"), lty=c(1,2,1,2,3), lwd=c(2,2,1,1,1), bty="n", cex=0.8)

# diagram of the phenomenon depending on the distance Figure 5.6b
variable<-data$XA21[data$region_name=="Mazowieckie" & data$year==2017]
dist<-data$dist[data$region_name=="Mazowieckie" & data$year==2017]
population<-data$XA06[data$region_name=="Mazowieckie" & data$year==2017] /
mean(data$XA06[data$region_name=="Mazowieckie" & data$year==2017])

brks.pop<-c(0, 0.5,0.75, 1.00, 1.25, 2, 5, 20) #intervals for the population
size<-brks.pop*1.6      # dot size scaling
cols<-"chartreuse3"     # dot color
plot(dist, variable, xlim=c(0,120),ylim=c(0,21), ylab="unemployment rate",
xlab="distance of the poviat from the voivodeship city", col=cols, bg=cols,
cex=size[findInterval(population, brks.pop)], pch=21)

title(main="Unemployment rate in poviats")

abline(h=(0:8)*5, lty=3, col="grey80")
abline(v=(0:10)*20, lty=3, col="grey80")
```

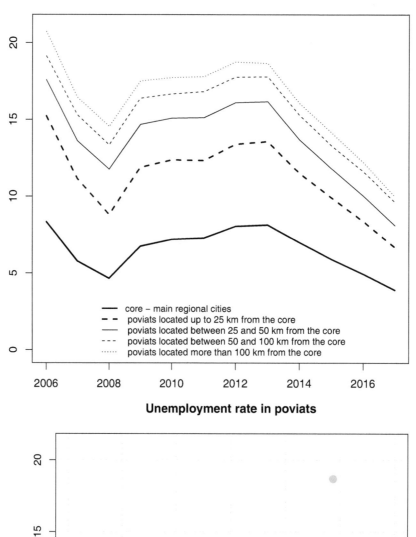

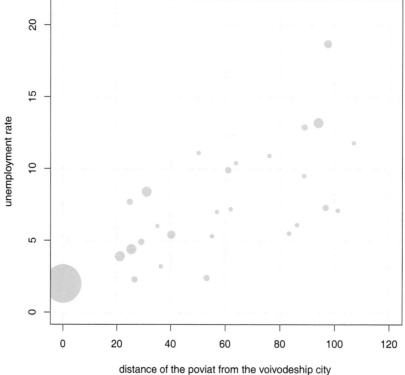

Figure 5.6 Dependence of the unemployment rate on the distance from the core city

Source: Own study using R

The previous charts show quite precisely that the value of the phenomenon – the unemployment rate – is strongly and permanently dependent on the absolute location, that is, the distance of the poviat from the regional city. The further the district (poviat) is located, the higher the unemployment rate. The relationship is stable over time, although its spatial diversity decreases as the phenomenon level decreases. The unemployment rate also does not depend on the number of inhabitants.

When starting to estimate models, one should consider all three specifications estimated in a-spatial (*asp, a-spatial*) and spatial (*sp, spatial*) ways, which gives a total of six models to evaluate. Subsequently are all these models. The unemployment rate is modelled in relative terms Poland = 100% – that is, the index value equal to 1 means the average level of the phenomenon. Creating the power of a dependent variable uses the operator **I**(). One can also use the **poly**() command in the equation specification, specifying the variable and the highest degree of polynomial as its arguments. When there is a problem in determining the natural logarithm of the value <1, it can be solved by the command **log1p**(), which instead of log (x) estimates the value of log ($x + 1$).

```
sub<-data[data$year==2017,]
sub$variable<-sub$XA21/mean(sub$XA21, na.rm=TRUE)


# visualization of the distance - phenomenon relation
plot(log(sub$dist),log(sub$variable), main="log x, log y")
plot(log(sub$dist), sub$variable, main="x, log y")
plot(sub$dist, sub$variable, main="x, y")


# matrix of spatial weights according to the contiguity criterion
cont.nb<-poly2nb(as(pov, "SpatialPolygons")) # from spdep::
cont.listw<-nb2listw(cont.nb, style="W")# from spdep::


# model of polynomial distance (multinominal distance)
mod.multi.asp<-glm(variable~dist+I(dist^2)+I(dist^3)+ I(dist^4), data=sub)
mod.multi.sp<-errorsarlm(variable~dist+I(dist^2)+ I(dist^3)+ I(dist^4),
data=sub, cont.listw, tol.solve=2e-40)


# power model
mod.power.asp<-glm(log1p(variable)~log1p(dist), data=sub)
mod.power.sp<-errorsarlm(log1p(variable)~log1p(dist), data=sub, cont.listw)


# exponential model
mod.exp.asp<-glm(log1p(variable)~dist, data=sub)
mod.exp.sp<-errorsarlm(log1p(variable)~dist, data=sub, cont.listw)


# goodness-of-fit measures
out<-matrix(0, nrow=2, ncol=6)
colnames(out)<-c("multi.asp", "multi.sp", "power.asp", "power.sp", "exp.
asp", "exp.sp")
rownames(out)<-c("SRMSE", "lambda")


a<-mean(sub$variable)
b<-dim(sub)[1]
c<-sub$variable


out[1,1]<-sqrt(sum((mod.multi.asp$fitted.values-c)^2)/b)/a
out[1,2]<-sqrt(sum((mod.multi.sp$fitted.values-c)^2)/b)/a
out[1,3]<-sqrt(sum((mod.power.asp$fitted.values-c)^2)/b)/a
out[1,4]<-sqrt(sum((mod.power.sp$fitted.values-c)^2)/b)/a
out[1,5]<-sqrt(sum((mod.exp.asp$fitted.values-c)^2)/b)/a
out[1,6]<-sqrt(sum((mod.exp.sp$fitted.values-c)^2)/b)/a
```

```
out[2,2]<- mod.multi.sp$lambda
out[2,4]<- mod.power.sp$lambda
out[2,6]<- mod.exp.sp$lambda
out
```

```
        multi.asp  multi.sp  power.asp  power.sp    exp.asp     exp.sp
SRMSE   0.4708004  0.3842633  0.5860224  0.5437467  0.5889698  0.5501229
lambda  0.0000000  0.6310112  0.0000000  0.6426204  0.0000000  0.6299987
```

```
# visualization of model fit - Figure 5.7
plot(sub$dist, sub$variable, main="OLS, multinominal, fitted values")
points(sub$dist, mod.multi.asp$fitted.values, col="red")
abline(h=1, lty=3)

plot(sub$dist, sub$variable, main="SEM, multinominal, fitted values")
points(sub$dist, mod.multi.sp$fitted.values, col="red")
abline(h=1, lty=3)

plot(log1p(sub$dist), log1p(sub$variable), main="OLS, power, fitted values")
points(log1p(sub$dist), mod.power.asp$fitted.values, col="red")
abline(h=1, lty=3)

plot(log1p(sub$dist), log1p(sub$variable), main="SEM, power, fitted values")
points(log1p(sub$dist), mod.power.sp$fitted.values, col="red")
abline(h=1, lty=3)

plot(sub$dist, log1p(sub$variable), main="OLS, exponential, fitted values")
points(sub$dist, mod.exp.asp$fitted.values, col="red")
abline(h=1, lty=3)

plot(sub$dist, log1p(sub$variable), main="SEM, exponential, fitted values")
points(sub$dist, mod.exp.sp$fitted.values, col="red")
abline(h=1, lty=3)
```

The previous estimation shows that from a technical point of view, the spatial polynomial model is best suited – its SRMSE is definitely the lowest. This result is often obtained for the data type presented. Other models fare worse, and their fit is often weaker.

Interpretation of unidirectional spatial interaction models is mainly based on the assessment of the intersection of the fit and the extension line $y = 1$. The fit is a distance-dependent function. Point $x = 0$ represents core centres, while observations x km away are peripheral centres. The values on the y-axis are an index of the phenomenon, and the values on the x-axis are a distance from the core.

In the case of the analysed unemployment rate, the level of the phenomenon (and index) is significantly lower in the core measures and higher in the peripheral centres. The intersection of the fit curve with $y = 1$ at x^* means that the average level is reached, and consequently, observations farther than x^* km have a higher than average phenomenon level. The assessment of the cut-off point x^* allows for checking how far the influence of the core city goes and where it disappears.

Subsequently, for a full overview of the capabilities of these models, two spatial polynomial models were estimated for two different periods, 2006 and 2017, and the intersection point was examined (see Figure 5.8). The estimation with **errorsarlm()** uses the **poly()** function to efficiently create a polynomial. The results show that the polynomial is well chosen, because all the subsequent powers of the explanatory variable are significant. Based on the AIC, it can be concluded that spatial estimation (AIC = 130.81) is significantly better than a-spatial (AIC = 303.15). The intersection of the fit curve with the $y = 1$ line

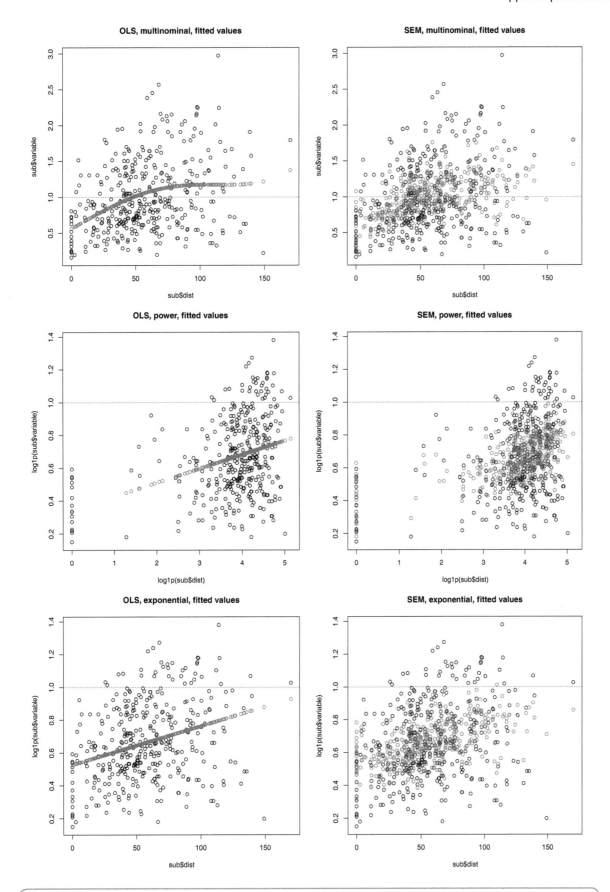

Figure 5.7 Visualisation of the quality of matching spatial interaction models to empirical data

Source: Own study using R and the spatialreg:: package

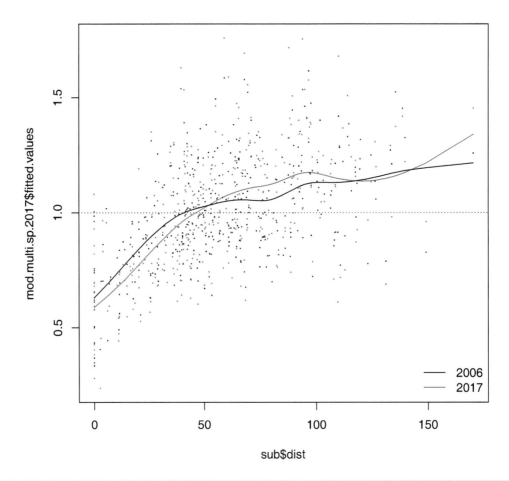

Figure 5.8 Smoothed fitting of polynomial models in 2006 and 2017
Source: Own study using R and the spatialreg:: package

moved away about 5 km, which means an increase in the range of impact of core cities. The visualisation uses the **smooth.spline**() smoothing command from the stats:: package with a very high smoothing factor (*spar=0.99*).

```
# spatial polynomial models
sub<-data[data$year==2017,]
sub$variable<-sub$XA21/mean(sub$XA21, na.rm=TRUE)
mod.multi.sp.2017<-errorsarlm(variable~poly(dist,4), data=sub, cont.listw,
tol.solve=2e-40)
sqrt(sum((mod.multi.sp.2017$fitted.values-sub$variable)^2)/dim(sub)[1])/
mean(sub$variable)
#[1] 0.3842633

sub<-data[data$year==2006,]
sub$variable<-sub$XA21/mean(sub$XA21, na.rm=TRUE)
mod.multi.sp.2006<-errorsarlm(variable~poly(dist,4), data=sub, cont.listw,
tol.solve=2e-40)
sqrt(sum((mod.multi.sp.2006$fitted.values-sub$variable)^2)/dim(sub)[1])/
mean(sub$variable)
#[1] 0.2653964
```

```
summary(mod.multi.sp.2006)
```

```
Call:
errorsarlm(formula = variable ~ poly(dist, 4), data = sub, listw = cont.listw,
    tol.solve = 2e-40)

Residuals:
      Min         1Q     Median         3Q        Max
-0.7816167 -0.1565798 -0.0087304  0.1689882  0.7595056

Type: error
Coefficients: (asymptotic standard errors)
                Estimate Std. Error z value  Pr(>|z|)
(Intercept)     0.943175   0.047283 19.9473 < 2.2e-16
poly(dist, 4)1  1.240573   0.459602  2.6992  0.006950
poly(dist, 4)2 -1.647014   0.335285 -4.9123 9.002e-07
poly(dist, 4)3  0.853938   0.287480  2.9704  0.002974
poly(dist, 4)4 -0.502295   0.273232 -1.8383  0.066012

Lambda: 0.71185, LR test value: 174.34, p-value: < 2.22e-16
Asymptotic standard error: 0.043105
    z-value: 16.514, p-value: < 2.22e-16
Wald statistic: 272.72, p-value: < 2.22e-16

Log likelihood: -58.40521 for error model
ML residual variance (sigma squared): 0.070435, (sigma: 0.2654)
Number of observations: 380
Number of parameters estimated: 7
AIC: 130.81, (AIC for lm: 303.15)
```

```
# matching visualization
plot(sub$dist, mod.multi.sp.2017$fitted.values, col="red", pch=".", cex=1.5)
lines(smooth.spline(sub$dist, mod.multi.sp.2017$fitted.values, spar=0.99),
col="red")
points(sub$dist, mod.multi.sp.2006$fitted.values, col="black", pch=".", cex=1.3)
lines(smooth.spline(sub$dist, mod.multi.sp.2006$fitted.values, spar=0.99),
col="black")
abline(h=1, lty=3)
legend("bottomright", legend=c("2006", "2017"), col=c("black", "red"),
lty=c(1,1), bty="n")
```

5.3.2 Cumulative models

Cumulative models are mainly used in modelling cash flows in the economy, for example, income, investments, costs and so on and are based on the net present value approach (NPV). A broader description of the model and applications in the analysis of multiplier effects in local government investments can be found in the article: Kopczewska, K. (2016). "Efficiency of Regional Public Investment: An NPV-Based Spatial Econometric Approach". *Spatial Economic Analysis*, 11(4), 413–431.

As a starting point, it is assumed that the distribution over the cash flow does not matter, and above all, the total amount of funds obtained or spent in a given period counts. For example, if a local government builds a viaduct and a road that improves the conditions for entrepreneurs, then such an action should be analysed in the context of all investment expenditure. A secondary factor is whether the expenditure was made in December and is reported in year t or after a month in January, in year $t + 1$. A several-year investment is perceived in such models as the sum of funds spent, including discounting. On the other hand, if the local government assumes that the construction of the viaduct and road will translate into increased economic activity in a few years, it is again of secondary importance whether firms will start paying higher taxes (on higher profits) in December t or January $t + 1$ – the sum of taxes paid in the examined period is significant.

In cumulative modelling, the following steps are performed:

1   The base period is determined – the starting point from which all cash flows accumulate
2   Variables are prepared for estimation:

● *flow-type variables* – monetary, which are accumulated. The sum of discounted financial flows starting from the base year is counted for each period. Financial variables in the area of the public sector that are cumulated are own income, investment expenditure, income from PIT, income from CIT and so on. From an accounting perspective, they are variable as in the profit and loss account.

● *stock-type variables* – defining the state (resource) for a given day. Their accumulation does not make sense. In the economic area, these are the number of employed, business entities, unemployment rate, property value, number of patents and so on. From an accounting perspective, they are variable as from the balance sheet.

● *control variables* – specifying characteristics that are constant throughout the test period. These are, for example, the type of commune, distance from the core city and tourist attractions but often also the percentage of forested area, the number of large enterprises and so on.

3   A separate equation is estimated for each year in which these three types of variables occur. It has the form for cumulative variables:

$$\sum_{t=1}^{t=k} y_t = \beta_0 + \sum_{t=1}^{t=k} x_t \beta + \ldots + u_n \tag{5.35}$$

For example,
In the model for year 1: y_1 is explained by x_1
In the model for year 2: $y_1 + y_2$ is explained by $x_1 + x_2$
In the model for year 3: $y_1 + y_2 + y_3$ is explained by $x_1 + x_2 + x_3$
In the model for the year k: $y_1 + y_2 + y_3 + \ldots + y_k$ is explained by $x_1 + x_2 + x_3 + \ldots + x_k$
where subscripts $1, 2, \ldots k$ represent years. The full model, for example, assuming the SDEM specification, is as follows:

$$y = \alpha + X\beta + WX\theta + u \text{ and } u = \lambda Wu + e \tag{5.36}$$

where (when i is the territorial unit and k is the time unit for cumulative variables): $x = x_{k,i} = \sum_{t=1}^{t=k} x_{t,i} d_t$, $y = y_{k,i} = \sum_{t=1}^{t=k} y_{t,i} d_t$, where dt is the deflator for year t; for state variables: $x = x_{k,i}$, $y = y_{k,i}$ and for constant characteristics over time, $x = x_i$.

4   For each period, the coefficients are interpreted – with cumulative variables as marginal multipliers (e.g. investment or income), with the remaining variables – in the traditional way, changes in marginal multipliers – as the saturation effect. In the model in which there is a flow type variable on both sides of the equation, the multiplier (coefficient for the cumulative variable) determines how much the total result, for example, of income, for the next investment per unit will change. If the multipliers increase from year to year, it means that subsequent expenditure increases the multiplier effect (state of unsaturation); when the multipliers fall from period to period, it means that the effect increases are weaker, which indicates saturation.

This approach has several important properties. First, it smoothes cash flows that are reported at different times, often random. Second, the problem of arbitrary selection of time lags disappears, which often appears in typical models. Third, the results of regression – coefficients with monetary variables have an interpretation of multipliers, for example, investment or income ones. Fourth, the change from period to period of factors interpreted as multipliers allows assessment of saturation of, for example, the local investment economy, as it shows the marginal response of the dependent variable. Typical time-lagged models do not have these

> Table 5.3 **Specification of cumulative model**

Variable symbol	Variable code	The nature of the variable	Description of the variable
Y	XA01/XA06	flow	Own income per capita
X1	XA05/XA06	flow	Investment expenditure per capita
X2	XA14	stock	Salaries Poland = 100%
X3	XA08/XA09	stock	Percentage of working-age population in relation to post-working-age population
X4	XA13	stock	Number of firms per 10,000 working-age residents
X5	(XA18+XA19+XA20)/XA15	stock	Share of employment in services
X6	dist	Control (fixed value)	Distance of the poviat from the voivodeship city
X7	XA10/mean(XA10)	Control (fixed value)	Population density per 1 km^2 (Poland = 100)
X8	XA21	stock	Unemployment rate

Source: Own work

properties. More detailed analytical details and construction of economic models can be found in Kopczewska (2016).

The details of creating cumulative variables follow. There are a few things to keep in mind: first, panel data is in long form; that is, each observation block for the following year is in the following lines; second, spatial model commands require that all variables used in the model come from one data object specified in the data option. In this solution, 2010 was chosen as the base year (arbitrarily). A separate subset was created for each year from 2010–2017. Cumulative variables were created from the final years of accumulation; that is, the cumulative variable in 2010–2015 was saved in the subset of 2015. The model specification is presented in Table 5.3.

```
# creating simple quotients of two variables (or without changes)
data$y<-data$XA01/data$XA06
data$x1<-data$XA05/data$XA06
data$x2<-data$XA14
data$x3<-data$XA08/data$XA09
data$x4<-data$XA13
data$x5<-(data$XA18+data$XA19+data$XA20)/data$XA15
data$x6<-data$dist
data$x8<-data$XA21

# subsets for each year
sub10<-data[data$year==2010,]
sub11<-data[data$year==2011,]
sub12<-data[data$year==2012,]
sub13<-data[data$year==2013,]
sub14<-data[data$year==2014,]
sub15<-data[data$year==2015,]
sub16<-data[data$year==2016,]
sub17<-data[data$year==2017,]

# variables Poland=100%, referring to the average from a given year
sub10$x7<-sub10$XA10/mean(sub10$XA10, na.rm=TRUE)
sub11$x7<-sub11$XA10/mean(sub11$XA10, na.rm=TRUE)
sub12$x7<-sub12$XA10/mean(sub12$XA10, na.rm=TRUE)
sub13$x7<-sub13$XA10/mean(sub13$XA10, na.rm=TRUE)
sub14$x7<-sub14$XA10/mean(sub14$XA10, na.rm=TRUE)
```

```
sub15$x7<-sub15$XA10/mean(sub15$XA10, na.rm=TRUE)
sub16$x7<-sub16$XA10/mean(sub16$XA10, na.rm=TRUE)
sub17$x7<-sub17$XA10/mean(sub17$XA10, na.rm=TRUE)

# cumulative dependent variable
sub10$y.cum<-sub10$y
sub11$y.cum<-sub10$y+sub11$y
sub12$y.cum<-sub10$y+sub11$y+sub12$y
sub13$y.cum<-sub10$y+sub11$y+sub12$y+sub13$y
sub14$y.cum<-sub10$y+sub11$y+sub12$y+sub13$y+sub14$y
sub15$y.cum<-sub10$y+sub11$y+sub12$y+sub13$y+sub14$y+sub15$y
sub16$y.cum<-sub10$y+sub11$y+sub12$y+sub13$y+sub14$y+sub15$y+sub16$y
sub17$y.cum<-sub10$y+sub11$y+sub12$y+sub13$y+sub14$y+sub15$y+sub16$y+
sub17$y

# cumulative explanatory variable
sub10$x1.cum<-sub10$x1
sub11$x1.cum<-sub10$x1+sub11$x1
sub12$x1.cum<-sub10$x1+sub11$x1+sub12$x1
sub13$x1.cum<-sub10$x1+sub11$x1+sub12$x1+sub13$x1
sub14$x1.cum<-sub10$x1+sub11$x1+sub12$x1+sub13$x1+sub14$x1
sub15$x1.cum<-sub10$x1+sub11$x1+sub12$x1+sub13$x1+sub14$x1+sub15$x1
sub16$x1.cum<-sub10$x1+sub11$x1+sub12$x1+sub13$x1+sub14$x1+sub15$x1 +sub16$x1
sub17$x1.cum<-sub10$x1+sub11$x1+sub12$x1+sub13$x1+sub14$x1+sub15$x1
+sub16$x1+ sub17$x1

# model equation
eq<-y.cum~x1.cum+x2+x3+x4+x5+x6+x7+x8 # form of a regression equation

# contiguity spatial weights matrix
cont.nb<-poly2nb(as(pov, "SpatialPolygons"))
cont.listw<-nb2listw(cont.nb, style="W")

# model estimation for subsequent years
m10<-errorsarlm(eq, data=sub10, cont.listw, etype="emixed", tol.solve=1e-20)
m11<-errorsarlm(eq, data=sub11, cont.listw, etype="emixed", tol.solve=1e-20)
m12<-errorsarlm(eq, data=sub12, cont.listw, etype="emixed", tol.solve=1e-20)
m13<-errorsarlm(eq, data=sub13, cont.listw, etype="emixed", tol.solve=1e-20)
m14<-errorsarlm(eq, data=sub14, cont.listw, etype="emixed", tol.solve=1e-20)
m15<-errorsarlm(eq, data=sub15, cont.listw, etype="emixed", tol.solve=1e-20)
m16<-errorsarlm(eq, data=sub16, cont.listw, etype="emixed", tol.solve=1e-20)
m17<-errorsarlm(eq, data=sub17, cont.listw, etype="emixed", tol.solve=1e-20)

# combining model results into one printout
options(scipen=999, digits=2)
result<-cbind(m10$coefficients, m11$coefficients, m12$coefficients,
m13$coefficients, m14$coefficients, m15$coefficients, m16$coefficients,
m17$coefficients)
colnames(result)<-paste(rep("mod",times=8),2010:2017)
lambda<-cbind(m10$lambda, m11$lambda, m12$lambda, m13$lambda, m14$lambda,
m15$lambda, m16$lambda, m17$lambda)
AIC<-cbind(AIC(m10), AIC(m11), AIC(m12), AIC(m13), AIC(m14), AIC(m15),
AIC(m16), AIC(m17))
result<-rbind(result,lambda, AIC)
rownames(result)[19]<-"AIC"
result
```

	mod 2010	mod 2011	mod 2012	mod 2013	mod 2014	mod 2015	mod 2016	mod 2017
(Intercept)	-2194.100	-4058.154	-6151.450	-7408.9011	-8925.061	-11549.683	-13676.84	-15494.58
x1.kum	0.482	0.468	0.520	0.6066	0.625	0.602	0.68	0.76
x2	15.039	27.231	42.419	52.0092	64.047	86.867	105.82	122.86
x3	76.110	162.769	269.351	263.8917	286.254	335.079	219.94	6.22
x4	0.395	1.012	1.349	1.7142	2.123	2.603	2.84	3.34
x5	530.276	658.899	1198.526	2224.5390	3149.084	3477.850	4900.45	6009.79
x6	1.019	1.175	2.190	1.8414	1.846	3.194	5.11	7.49
x7	66.757	155.381	240.926	299.9315	381.852	473.157	505.47	545.72
x8	-12.677	-21.297	-29.730	-38.1687	-47.485	-52.631	-64.81	-67.64
lag.x1.kum	-0.026	0.011	0.016	-0.0061	0.015	0.059	0.10	0.11
lag.x2	8.532	18.428	30.870	41.7750	51.480	68.705	81.41	89.10
lag.x3	3.685	-9.211	-71.265	-211.4286	-321.146	-503.737	-610.56	-639.14
lag.x4	-0.126	-0.676	-1.018	-1.5882	-2.052	-2.518	-2.80	-3.42
lag.x5	1100.336	3429.408	5207.142	8454.0844	10861.525	12672.098	13665.05	15923.38
lag.x6	-0.413	0.494	0.911	5.4253	8.472	11.065	12.33	12.69
lag.x7	-13.232	-107.943	-207.638	-356.0689	-474.167	-641.011	-680.72	-708.00
lag.x8	0.232	-11.166	-19.100	-34.5966	-57.298	-65.118	-68.85	-89.60
lambda	0.277	0.258	0.281	0.2448	0.251	0.252	0.28	0.31
AIC	5372.642	5886.789	6197.956	6395.4097	6570.994	6700.178	6815.28	6919.73

summary(m17)

```
Call:errorsarlm(formula = eq1, data = sub17, listw = cont.listw,
etype = "emixed",
    tol.solve = 0.00000000000000000001)


Residuals:
     Min       1Q  Median       3Q      Max
-5472.22 -1153.88 -145.48  888.94 14287.89


Type: error
Coefficients: (asymptotic standard errors)
              Estimate  Std. Error z value              Pr(>|z|)
(Intercept) -15494.577211 3363.550829 -4.6066       0.00000409280555
x1.cum          0.759572    0.076229  9.9644 < 0.00000000000000022
x2            122.856806   11.985850 10.2502 < 0.00000000000000022
x3              6.215773  355.742140  0.0175             0.9860595
x4              3.340953    0.501809  6.6578    0.00000000002779
x5           6009.788684 1561.538870  3.8486             0.0001188
x6              7.492074    6.985071  1.0726             0.2834579
x7            545.722166  112.806688  4.8377       0.00000131366287
x8            -67.644780   34.624655 -1.9537             0.0507415
lag.x1.cum      0.106004    0.187616  0.5650             0.5720705
lag.x2         89.095604   28.243930  3.1545             0.0016077
lag.x3       -639.135049  567.708369 -1.1258             0.2602435
lag.x4         -3.416341    0.862165 -3.9625       0.00007416503575
lag.x5      15923.376945 3185.467652  4.9988       0.00000057701231
lag.x6         12.693322   10.362924  1.2249             0.2206210
lag.x7       -707.995440  249.822091 -2.8340             0.0045970
lag.x8        -89.602032   59.256997 -1.5121             0.1305105
```

```
Lambda: 0.31, LR test value: 15, p-value: 0.000121
Asymptotic standard error: 0.068
    z-value: 4.5, p-value: 0.0000055744


Wald statistic: 21, p-value: 0.0000055744
Log likelihood: -3441 for error model
ML residual variance (sigma squared): 4211300, (sigma: 2052)
Number of observations: 380
Number of parameters estimated: 19
AIC: 6920, (AIC for lm: 6932)
```

```
# ratio of direct and indirect effects
abs(result[2:9,])/abs(result[10:17,])
```

	mod 2010	mod 2011	mod 2012	mod 2013	mod 2014	mod 2015	mod 2016	mod 2017
x1.cum	18.71	41.58	32.87	99.39	41.55	10.27	6.82	7.1655
x2	1.76	1.48	1.37	1.24	1.24	1.26	1.30	1.3789
x3	20.66	17.67	3.78	1.25	0.89	0.67	0.36	0.0097
x4	3.13	1.50	1.32	1.08	1.03	1.03	1.01	0.9779
x5	0.48	0.19	0.23	0.26	0.29	0.27	0.36	0.3774
x6	2.47	2.38	2.40	0.34	0.22	0.29	0.41	0.5902
x7	5.05	1.44	1.16	0.84	0.81	0.74	0.74	0.7708
x8	54.75	1.91	1.56	1.10	0.83	0.81	0.94	0.7549

The previous results of the cumulative model should be interpreted as follows:

● Models are quite well fitted to the data – the vast majority of variables in the models are significant, AIC in the spatial model is lower than in the OLS model and there are significant spatial parameters
● Coefficient with the cumulative variable $x1.cum$ is below 1 and is increasing. This means that in this period, there is a clear positive connection between investment expenditure of municipalities (aggregated by poviats) and their own income, although investment expenditure is not fully refunded – that is, for each zloty spent on investments, own income increases by about 0.5–0.75 PLN. The multiplier is strongly growing (from 0.48 in 2010 to 0.76 in 2017), which means that local economies are not saturated with investments and benefit from economies of scale – the larger the investment, the higher the proportionate returns.
● The lambda spatial coefficient is significant and increasing, which means increasing spatial interactions between territorial units. Theta coefficients are significant. In the case of the SDEM model, in which there is no spatial lag of the explained variable, the Durbin component is interpreted directly as a spillover. In this model, the financial multiplier effects are very internalised, as the effect ratio varies between 6.8 and 99.39 (i.e. the direct effect is stronger than the indirect effect from 6 times to 100 times). In recent years, the direct effect has weakened, which indicates growing neighbouring interactions and externalisation of the investment impact (cross-border effects occur). Variables $x2$, $x4$ and $x5$ have very stable parity of direct and indirect effects, which means that they are anchors of local development. The quotients below the value of 1 mean that the indirect (external) effect is stronger than the internal one, which for example, occurs constantly with the variable $x5$, that is, the share of employees in services.

Technically, in part of the code to combine model results into one printout, the **options**() command was used with the option *scipen = 999* and *digits = 2*, which limits the display of results in 1e-02 notation for decimal value benefits as 0.01 and limits the number of significant decimal places to 2. Column headers were also created automatically by combining the **paste**() constant element with the year number from 2010 to 2017. The **colnames**() command requires a vector name; hence it was necessary to replicate the constant part *"mod"* with the **rep()** command. The results of several estimations were combined with columns (**cbind()**) and rows (**rbind()**), which facilitates cross-sectional evaluation of results.

5.3.3 Bootstrapped models for big data

In a situation where the dataset is larger than the computational capabilities, it is necessary to resort to statistical methods that allow for achieving high-quality calculations at a relatively low cost of these calculations. One of the old, well-known methods for improving the quality of calculations is the bootstrap method (more in Chapter 8). It is possible to use the bootstrap method to reduce the dataset, as opposed to typical applications of dataset reproduction.

The algorithm presented subsequently allows bootstrapping the econometric model – that is, multiple model estimation on a randomly selected subsample, followed by the selection of the best model based on a combined analysis of factor estimates and an indication of the most typical (middle) combination of factors using partitioning around medoid (PAM). The model thus estimated (i.e. known observations from the best model and model coefficients) can be used in prediction for out-of-sample data. To this end, the space is tessellated based on the points on which the best model was estimated. For out-of-sample data, a new point is assigned to the tessellation tile, so that the new point is associated with the old spatial arrangement. This enables forecasting for a new point using neighbourhood information from the data used to calibrate the model. The out-of-sample forecast uses $n - 1$ old points and 1 new point – spatial lags are determined based on the old points. This simple procedure allows one to quickly estimate a spatial model on large datasets. The article entitled "Spatial Bootstrapped Microeconometrics: Forecasting for Out-of-Sample Geo-Locations in Big Data" (Kopczewska, 2020) presents the details of modelling and examining the properties of models and forecasts estimated using this algorithm. This article clearly demonstrates that bootstrapped models are no worse than models on a full sample of data, and their computational efficiency is much higher.

Example

The following example uses point data for business locations. The data is supplemented with information about the distance from the company's location to the regional city – Lublin. The sector of activity and the employment size class are also known. For the data loaded, the information in the dataset of the sector of activity based on the section has been supplemented (i.e. section A is the agriculture sector; sections B, C, D and E are the production sector; section F is the construction sector and sections G to S are the sector services). The average expected profitability by sectors and sections (hypothetical values) and the premium for the central location were determined. The created tag set was connected to the base set by the **merge()** command. The merger was carried out twice, assigning sectors to the section for the first time and section profitability, assigning profitability to the sector the second time. Based on the average yields thus generated, μROA was drawn from the normal distribution $N(\mu_{ROA}, \sigma^2_{ROA} = 0.045)$ with the **rnorm()** command, which gives individual yields for each observation. A dummy (zero-one) variable was created for each sector with the **ifelse()** command. Two actions were performed in relation to the location: the **spDistsN1()** command was calculated from the spdep:: Euclidean distance between each point and the city centre of Lublin (main regional city) location, and locations were scattered by random epsilon to avoid overlapping points, adding small geographic coordinates to the value drawn from the normal distribution $N (0, 0.0005)$. Finally, the data was sorted randomly by adding as a value the ID from the uniform distribution (obtained from the **runif()** command) and sorting by this value in ascending order using the **orderBy()** command from the doBy:: package and divided into two subsamples: training data (*in*) and test data (*out*).

```
# loading data on firms
firms<-read.csv("geoloc_data_firms.csv", header=TRUE, dec=",", sep=";")
voi<-readOGR(".", "wojewodztwa") # 16 units
voi<- spTransform(voi, CRS("+proj=longlat +datum=NAD83"))

# set of additional parameters for non-aggregated sectors
param<-data.frame(SEC_PKD7=c("A", "B", "C", "D", "E", "F", "G",
"H", "I", "J", "K", "L", "M", "N", "O", "P", "Q", "R", "S"),
SEK_agg=c("agri", "prod", "prod", "prod", "prod", "constr",
"serv", "serv", "serv", "serv", "serv", "serv", "serv",
```

```
"serv", "serv", "serv", "serv", "serv", "serv"), roa_ind
=c(2,2.5,3,3.5,4,4.5,5,5.5,6,6.5,7,7.5,8,8.5,9,9.5,10,10.5,11))
firms1<-merge(firms, param, by="SEC_PKD7") # merging parameters and data

# set of additional parameters for aggregated sectors
param2<-data.frame(SEK_agg=c("agri", "prod", "constr", "serv"), roa_
sec=c(2,3.5,5,8))
firms1<-merge(firms1, param2, by="SEK_agg") # merging parameters and data
# premium for central location
firms1$roa_geo<-ifelse(firms1$poviat=="powiat Lublin", 1.5,0)
firms1$roa_param<-firms1$roa_sec+firms1$roa_geo # final ROA with premium

# drawing ROA profitability based on the assumed parameter
for(i in 1:dim(firms1)){
firms1$roa[i]<-rnorm(1, firms1$roa_param[i], 0.045)}
# dummy variables for sectors
firms1$agri<-ifelse(firms1$SEK_agg=="agri",1,0)
firms1$prod<-ifelse(firms1$SEK_agg=="prod",1,0)
firms1$constr<-ifelse(firms1$SEK_agg=="constr",1,0)
firms1$serv<-ifelse(firms1$SEK_agg=="serv",1,0)

library(spdep)
# Euclidean distance between a point and the centre of Lublin
coords<-as.matrix(data.frame(x=firms1$coords.x1, y=firms1$coords.x2))
core<-c(22.5666700, 51.2500000) # coordinates of central Lublin
firms1$dist<-spDistsN1(coords, core, longlat=TRUE)

# randomly scattered locations (by epsilon) of observations
epsilon.x<-rnorm(dim(firms1)[1], mean=0, sd=0.015)
epsilon.y<-rnorm(dim(firms1)[1], mean=0, sd=0.015)
firms1$xxe<-firms1[,24]+epsilon.x
firms1$yye<-firms1[,25]+epsilon.y

# random ordering of the firms set
library(doBy)
firms1$los<-runif(dim(firms1)[1], 0,1)
firms1<-orderBy(~los, data=firms1)

# division of firms into training (in) and test (out)
firms1.in<-firms1[1:30000,]
firms1.out<-firms1[30001:37374,]
```

Bootstrapped models can be estimated on such prepared data. In the bootstrap preparation phase, specify the number of bootstrap iterations and the size of the random sample – the greater the number of iterations (model estimation) and the larger the sample, the greater the chance of a better fit of the model to the data. Increasing both parameters slows the estimation process due to the increasing time consumption. Above 1000 iterations and 10,000 observations, increasing these parameters usually does not significantly increase the model quality.

```
# simulation parameters (to be changed by the researcher)
# parameters of simulation - change here the parameters
n.col<-50 # number of iterations
n.row<-800 # number of obs in a sample
```

Another element of preparation is the drawing of points for each iteration. Due to the fact that the observation IDs are saved, it is possible to reconstruct the observations from which the model was estimated, which allows subsequent operations, tessellation and forecasting. Observation IDs are stored in the selector matrix. According to the literature, spatial sampling requires the use of an appropriate sampling model, and one of the most effective is the sampling of irregular shapes obtained from the division into groups using the *k*-means algorithm. For this reason, the sample in (based solely on geographical coordinates) was divided into clusters around *k*-means with the **kmeans**() command from the underlying stats:: package, and randomised stratification from these clusters with the command **strata**() from the sampling:: package was used, allowing the selection of *n* observations from each group. The clustering vector, that is, containing information about the assignment of observations to the cluster, can be recovered as a cluster slot from the resulting object of the *k*-means algorithm. The *method* = *"srswor"* option from the **strata**() command means simple random sampling without replacement.

```
# division of observations into k groups by k-means method
firms1.in.crds<-firms1.in[,14:15] # geographical coordinates
groups<-kmeans(firms1.in.crds, n.row/100) # based on geographic coordinates
firms1.in$kmean<-groups$cluster # clustering vector

library(sampling)
# ID matrix of observations drawn for each iteration
# randomly selected with the strata() command from the sampling:: package
# randomized observations from groups determined by k-means
selector<-matrix(0, nrow=n.row, ncol=n.col)
for(i in 1:n.col){
vec<-sample(1:dim(firms1.in)[1], n.row, replace=FALSE)
x<-strata(firms1.in, "kmean", size=rep(100, times=n.row/100),
method="srswor") # from sampling::
selector[,i]<-x$ID_unit}
```

Turning to the correct estimation, specify the model specification that will be estimated. The following example adopts SDM. For bootstrap estimation in the **for**() loop, one needs to prepare objects for saving key results. A new different spatial weights matrix is created for each iteration due to a different set of points in the sub-sample – the criterion *k* = *5* nearest neighbours was adopted, and the commands **knearneigh**(), **knn2nb**(), **make.sym.nb**() and **nb2listw**() were used.

```
# objects for saving estimation results
coef.sdm<-matrix(0, nrow=n.col, ncol=11) # matrix of model coefficients
error.sdm<-matrix(0, nrow=n.col, ncol=11) # matrix of standard errors
fitted.sdm<-matrix(0, nrow=n.row, ncol=n.col) # fitted values
roa<-matrix(0, nrow=n.row, ncol=n.col) # matrix of y values
other.sdm<-matrix(0, nrow=n.row, ncol=4) # AIC.sdm, BIC.sdm, time.sdm, rho.sdm

eq<-roa~empl+prod+constr+serv+dist # model structure

library(spdep) # necessary for spatial regression
library(spatialreg) # necessary for spatial regression

# estimation in the loop of models with saving the results to objects
for(i in 1:n.col) { # n.col defines the number of iterations
datax<-firms1.in[selector[,i],] # selection of observation id for given iteration
roa[,i]<-datax$roa # write y for each iteration

# creating a matrix of spatial weights W for a subset of points
crds<-as.matrix(firms1.in[selector[,i],14:15])
pkt.knn<-knearneigh(crds, k=5, longlat = NULL) # planar system, knn=5
pkt.k.nb<-knn2nb(pkt.knn)
pkt.k.sym.nb<-make.sym.nb(pkt.k.nb) # matrix symmetry
pkt.k.sym.listw<-nb2listw(pkt.k.sym.nb)

# SDM model
start.time <- Sys.time() # system time measurement
model.sdm<-lagsarlm(eq, data=datax, pkt.k.sym.listw, method="LU",
type="mixed")
```

```
end.time <- Sys.time()
time.sdm<- difftime(end.time, start.time, units="secs")

# saving results
coef.sdm[i,]<-model.sdm$coefficients
error.sdm[i,]<-model.sdm$rest.se
fitted.sdm[,i]<-model.sdm$fitted.values
other.sdm[i,1]<-AIC(model.sdm) # AIC.sdm
other.sdm[i,2]<-BIC(model.sdm) # BIC.sdm
other.sdm[i,3]<-time.sdm
other.sdm[i,4]<-model.sdm$rho
}

print(head(coef.sdm), digits=3)
```

```
        [,1]       [,2] [,3] [,4] [,5]      [,6]       [,7]  [,8]  [,9] [,10]   [,11]
[1,] 0.284  0.001468 1.50 3.02 6.02 -0.00771 -0.003967 -1.17 -2.54 -5.16 0.00717
[2,] 0.216 -0.000126 1.53 2.99 6.01 -0.00433 -0.001538 -1.34 -2.74 -5.35 0.00398
[3,] 0.263  0.000178 1.50 3.00 6.04 -0.00387 -0.000435 -1.34 -2.69 -5.25 0.00333
[4,] 0.328 -0.000414 1.54 2.98 6.02 -0.00526 -0.000387 -1.35 -2.40 -5.05 0.00465
[5,] 0.230 -0.000062 1.49 3.04 6.00 -0.00521  0.001090 -1.21 -2.75 -5.26 0.00472
[6,] 0.188 -0.000239 1.55 3.02 6.02 -0.00290 -0.001127 -1.26 -2.80 -5.41 0.00253
```

```
print(head(other.sdm), digits=3)
```

```
       [,1] [,2] [,3]  [,4]
[1,]  -696 -636 2.83 0.871
[2,]  -891 -830 2.93 0.902
[3,]  -889 -828 2.98 0.881
[4,]  -652 -591 3.19 0.850
[5,]  -829 -769 3.12 0.890
[6,] -1020 -959 2.88 0.911
```

With these calculations, the *coef.sdm* matrix is obtained, in which the number is equal to the number of iterations and the number of columns is equal to the number of estimated coefficients (there are 11 – beta and theta coefficients for five variables + constant). In the remaining objects, there is stored information about: standard error (*error.sdm*), fitted values (*fitted.sdm*), the values of the dependent variable y (*roa*) and information criteria, the time of the estimation and spatial coefficient *rho* (*other.sdm*).

Based on the previous data, choose the best model. In this approach, it is assumed that the total distribution of all model coefficients is examined and a model whose coefficients are most central is selected, that is, they are a medoid in the grouping process using the partitioning around medoid algorithm. One cluster is assumed in the PAM algorithm because the coefficients do not show heterogeneity. The justification of extracting only one cluster can be tested with the Hopkins statistics (**hopkins**() command from the clustertend:: package). PAM grouping was performed using the **pam**() command from the cluster:: package. *Coef* coefficients combined with the *rho* coefficient from the *other.sdm* object served as input. The result of grouping can be examined by entering the slots: *clustering* (shows the clustering vector – in this case only the values 1 as the assignment to cluster No. 1 are the result), *medoids* (shows the coefficients from the medoid model) and *id.med* (shows the model number which is medoid – the most central set of coefficients).

```
library(cluster)
library(clustertend)
c1.sdm<-pam(cbind(coef.sdm, other.sdm[1:50,4]),1) # z pakietu cluster::
summary(c1.sdm)
```

```
c1.sdm$clustering # clustering vector
c1.sdm$medoids # medoid model coefficients
c1.sdm$id.med # medoid model number
hopkins(cbind(coef.sdm, other.sdm[1:50,4]), n=nrow(coef.sdm)-1) #
```

Knowing the medoid number, one can view the results of the best SDM model (as previously), but the resulting regression object was overwritten in the loop, hence the need to reassess. The RAMSE for the medoid model is determined subsequently (fitted and observed values are compared there – they are all stored in objects from bootstrap). As can be seen from the printout, all regression coefficients are significant, in addition to the spatial component *rho*. The Akaike information criterion for the spatial model is significantly better than for the linear model. The Moran test for residuals from the SDM model clearly shows that there is no spatial autocorrelation in the residuals (for *knn = 5*).

```
# re-estimation of the best model
# selection of data on which the medoid model was scored
data.x<-firms1.in[selector[,c1.sdm$id.med],]
# comparison of empirical and fitted y values
RAMSE.med.sdm<-(sum((firms1.in[selector[,c1.sdm$id.med],33]-fitted.sdm[,c1.
sdm$id.med])^2)/n.row)^(0.5)
RAMSE.med.sdm
#[1] 0.1217029


crds<-as.matrix(data.x[,14:15]) # xy coordinates to the W matrix
pkt.knn<-knearneigh(crds, k=5) # knn object
pkt.k.nb<-knn2nb(pkt.knn)
pkt.k.sym.nb<-make.sym.nb(pkt.k.nb)
pkt.k.sym.listw<-nb2listw(pkt.k.sym.nb)


eq<-roa~unempl+prod+constr+serv+dist
model.sdm<-lagsarlm(eq, data=data.x, pkt.k.sym.listw, method="LU",
type="mixed")
summary(model.sdm)
```

```
Call:lagsarlm(formula = eq, data = data.x, listw = pkt.k.sym.listw,
type = "mixed", method = "LU")

Residuals:
        Min          1Q      Median          3Q         Max
-1.1570981  -0.0423921  -0.0020206   0.0421809   1.0246296

Type: mixed
Coefficients: (asymptotic standard errors)
              Estimate Std. Error   z value   Pr(>|z|)
(Intercept)  2.6256e-01 3.6501e-02    7.1933 6.324e-13
empl         1.7845e-04 7.9227e-05    2.2524   0.02430
prod         1.4987e+00 2.2415e-02   66.8621 < 2.2e-16
constr       3.0040e+00 1.9047e-02  157.7181 < 2.2e-16
serv         6.0396e+00 1.0468e-02  576.9740 < 2.2e-16
dist        -3.8691e-03 1.8089e-03   -2.1389   0.03244
lag.zatr    -4.3490e-04 2.0294e-04   -2.1430   0.03212
lag.prod    -1.3361e+00 5.8699e-02  -22.7624 < 2.2e-16
lag.constr  -2.6857e+00 6.3593e-02  -42.2322 < 2.2e-16
lag.serv    -5.2469e+00 9.6399e-02  -54.4293 < 2.2e-16
lag.dist     3.3266e-03 1.8297e-03    1.8182   0.06904
```

```
Rho: 0.88134, LR test value: 1108.4, p-value: < 2.22e-16
Asymptotic standard error: 0.014872
 z-value: 59.261, p-value: < 2.22e-16
Wald statistic: 3511.9, p-value: < 2.22e-16

Log likelihood: 457.5348 for mixed model
ML residual variance (sigma squared): 0.014812, (sigma: 0.1217)
Number of observations: 800
Number of parameters estimated: 13
AIC: -889.07, (AIC for lm: 217.32)
LM test for residual autocorrelation
test value: 29.989, p-value: 4.3443e-08
```

```
moran.test(model.sdm$residuals, pkt.k.sym.listw)
```

```
        Moran I test under randomisation

data: model.sdm$residuals
weights: pkt.k.sym.listw

Moran I statistic standard deviate = -3.0882, p-value = 0.999
alternative hypothesis: greater
sample estimates:
Moran I statistic    Expectation       Variance
    -0.0611745090 -0.0012515645 0.0003764973
```

Choosing the best set of regression coefficients allowed model calibration. Because the observations that were the basis for estimating this model are known, surface tessellation can be performed. Thanks to tessellation, the point distribution will be approximated by the distribution of polygons, and the area of analysis will be continuously divided into tiles. For tessellation, it is necessary to load a contour map for the studied area – in the example subsequently one province, Lubleskie, has been selected as the study area. Tessellation is possible in the spatstat:: package, which works fully after loading the maptools:: package. Tessellation is performed by the **dirichlet**() command on the prominent *ppp* classes, which is formed from a combination of points (coordinates) and a contour in *owin* class. The *ppp* class object is created by the **ppp**() command, and the *owin* class object is created by the **as.owin**() command. It is also possible to create a *SpatialPolygons* object from the obtained tessellation, using the **as**() command with the "*SpatialPolygons*" option.

```
library(rgdal)
library(spatstat)
library(maptools)

voi<-readOGR(".", "wojewodztwa") # 16 units
voi<-spTransform(voi, CRS("+proj=longlat +datum=NAD83")) # spherical
voi<-spTransform(voi, CRS("+proj=merc +datum=NAD83")) # planar
region<-voi[voi@data$jpt_nazwa_=="lubelskie",] # one region only
region.owin<-as.owin(region) # rgdal:: requires planar coordinates

points<-data.frame(x=firms1.in[selector[,c1.sdm$id.med],14],
y=firms1.in[selector[,c1.sdm$id.med],15])
points.sp<-SpatialPoints(points) # new points in sp class - spherical
proj4string(points.sp)<-CRS("+proj=longlat +datum=NAD83") # spherical
points.sp<-spTransform(points.sp, CRS("+proj=merc +datum=NAD83")) # planar
```

```
region.ppp<-ppp(x=points.sp@coords[,1], y=points.sp@coords[,2],
window=region.owin) # points of ppp class

region.tes<-dirichlet(region.ppp) # Dirichlet tessellation
tes.poly<-as(region.tes, "SpatialPolygons")
proj4string(tes.poly)<-CRS("+proj=merc +datum=NAD83")
tes.poly<-spTransform(tes.poly, CRS("+proj=merc +datum=NAD83")) #planar

plot(region) # Figure 5.9a
points(points.sp, pch=".")

plot(region.tes, main=" ") # tessellation plot, Figure 5.9b
plot(region.ppp, add=TRUE, pch=".", col="darkblue", cex=2)
```

In the last step, one can make a forecast for out-of-sample data. The *nnew* object subsequently defines how many new observations the forecast will be made for. The target object for forecasts was also defined, in which forecasted, actual values (available from the out object) and coordinates of new locations are saved. The new points have been converted to the *SpatialPoints* class by the **SpatialPoints()** command so that they can be input to the **over()** command. In the **over()** command from the sp:: package, the spatial superimposition of the points specified in the first argument on the areas (tiles from the tessellation in the *SpatialPolygons* class) specified in the second argument is checked. As a result, points are assigned to tile numbers. It was also necessary to complete the dataset by adding points for observations for which the assignment to tiles failed (*NA*). The forecast for 100 new points was made in a loop. In each iteration, only one observation was listed from the base model, and the remaining $n - 1$ were model calibration data, so that the model

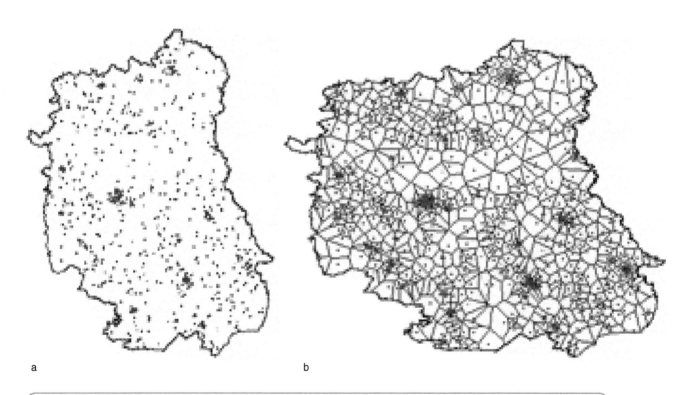

a   b

Figure 5.9 Observations underlying the best model: a) in a spherical projection as points, b) in a planar projection after tessellation

Source: Own study using R and spatialreg:: and sp:: packages

is not weakened. The forecast uses the command **predict()** (in fact **predict.sarlm()**), whose arguments are the model object, new data and the spatial weights matrix (as in estimation). Prediction results have been saved to *forecast1* objects. RAMSE was determined for all new points (without taking into account data from the modelling stage).

```
nnew<-100 # number of new points in the forecast

forecasts1<-matrix(0, nrow=nnew, ncol=5)
colnames(forecasts1)<-c("predicted y","real y","crds x","crds y", "diff")

points.pred<-SpatialPoints(firms1.out[1:nnew, 14:15]) # new sp points
proj4string(points.pred)<-CRS("+proj=longlat +datum=NAD83") # spherical
points.pred<-spTransform(points.pred, CRS("+proj=merc +datum=NAD83"))
a1<-over(points.pred, tes.poly) # assigning points to tessellation tiles
head(a1)
```

```
27268 18360 11155 37205 22892 23198
  773   352   322   436   384   624
```

```
# completing the draw when NA occurs
# determining the number of new points to be drawn (as from-to)
a2<-nnew+1 # from . . .
a3<-which(is.na(a1))
a4<-a2+length(a3)-1 # to . . .
points.pred2<-SpatialPoints(firms1.out[a2:a4, 14:15]) # new points
proj4string(points.pred2)<-CRS("+proj=longlat +datum=NAD83") # spherical
points.pred2<-spTransform(points.pred2, CRS("+proj=merc +datum=NAD83"))
points.pred2
a5<-over(points.pred2, tes.poly) # putting new points on the tile
a5
a1[which(is.na(a1))]<-a5 # overwriting with new points

# loop for forecasts for new points
# there is a separate match for each point
for(i in 1:nnew){
    # point by point - assigning new data to the old data set
    data.x.new<-data.x
    xxx<-firms1.out[i,]
    data.x.new[a1[i],]<-xxx
    rownames(data.x.new)<-1:dim(data.x.new)[1]

    # prediction for out-of-sample calibrated SDM model
    pred<-predict(model.sdm, newdata=data.x.new, listw=pkt.k.sym.listw,
    legacy.mixed=TRUE)
    pred[a1[i]] # prediction for a new point
    xxx[,33] # empirical y value of the new point

forecasts1[i,1]<- pred[a1[i]] # predicted value of y
forecasts1[i,2]<- xxx[,33] # empirical value of y
forecasts1[i,3]<-xxx[,39] # x coordinates
forecasts1[i,4]<-xxx[,40] # y coordinates
}
forecasts1[,5]<-(forecasts1[,1]-forecasts1[,2])^2
RAMSE.sdm<-(mean(forecasts1[,5]))^0.5
```

```
head(forecasts1)
```

```
           [,1]     [,2]      [,3]      [,4]          [,5]
 [1,]  7.945700  8.037431  23.11310  52.03856  8.414524e-03
 [2,]  2.107821  1.988089  22.09031  51.74640  1.433584e-02
 [3,]  2.120041  1.955649  22.42842  51.72729  2.702459e-02
 [4,]  7.932718  7.982926  23.68576  51.02439  2.520775e-03
 [5,]  4.906223  4.977703  21.94728  51.96171  5.109477e-03
 [6,]  3.771481  3.506796  22.19967  51.26922  7.005784e-02
```

```
RAMSE.sdm
#[1] 0.387511
```

The previous printout of the forecast object presents matches and empirical values of the variable *y* and locations *xy* of the examined points. The quality of the fit should be assessed highly. The article entitled "Spatial Bootstrapped Microeconometrics: Forecasting for Out-of-Sample Geo-Locations in Big Data" (Kopczewska, 2020) presents all methodological details of modelling and forecasts as well as comparisons with estimation on the full dataset or other spatial models. The quality of estimation with the previous method is comparable to the estimation on a full sample, while the properties of bootstrapped model estimators are in line with expectations (effective, unbiased, consistent).

5.3.4 Models for grid data

Spatial data can be aggregated in different ways, which increases the possibilities of using different data sources. In a situation where point and grid data are available, it is possible to aggregate point data according to the existing grid and estimate the econometric model on the grid. The following example shows such an operation.

Example

The following example uses point data for business locations prepared in the same way as for the bootstrap model (see 5.3.3). A population grid was also used (see 2.7), originally containing 315,857 cells. Point data are only for the Lubelskie voivodeship (NTS2), while the grid with the population is for all of Poland. The example presents the method of aggregating point data according to the existing grid to maintain the compatibility of the spatial system and estimation of the spatial model on such integrated data.

First, it was recalled from Section 2.7 how to load the grid with the **readOGR()** command and decompose the object into a dataset in the *data.frame* class (*pop.df*) (with the **as.data.frame()** command) and the grid (*pop. grid*), which is the equivalent of a contour map (with the **as()** command and "*SpatialPolygons*" argument). The data has been converted into numbers in the **for()** loop.

```
#a reminder of the loaded grid data for the population
#loading grid for population and converting projections
#pop<-readOGR(".", "PD_STAT_GRID_CELL_2011")
#pop<-spTransform(pop, CRS("+proj=longlat +datum=NAD83"))
#pop.df<-as.data.frame(pop) # extracting data to data.frame
#pop.grid<-as(pop, "SpatialPolygons") # extracting grid

#conversion to numerical data of subsequent columns of the data set
#for(i in 1:12){
#pop.df[,i]<-as.numeric(as.character(pop.df[,i]))}
```

Second, the loaded grid for the whole of Poland was cut off to the Lubelskie voivodeship, according to the location of point data. The voivodeship contour was extracted from the map of Poland (object *woj.lub*) using conditional operation on the contour map. Next, using the **over()** command, the regional contour and grid for Poland were overlaid – as a result, a *data.frame* class object was obtained, with the structure of variables as in *woj.lub*. Grids belonging to the contour of the province were assigned with variable values as in the shapefile file province, while the rest obtained NA values. The summary shows that 25,753 grid cells were classified within the voivodeship's contour, while the remaining 290,104 cells remained outside it. The **which**() command specified which of the rows in the grid were qualified for the contour. Subsequently, the conditional operation cut the grid to the contour of the voivodeship, drew the result and examined its properties. Using the vector result obtained from **which**(), the dataset assigned to the grid was truncated.

```
##cutting the grid according to the contour of the lubelskie region
#cutting the contour map
voi.lub<-voi[voi@data$jpt_nazwa_=="lubelskie",]
plot(voi.lub)

lim<-over(pop.grid, voi.lub) # overlay of grid and contour
summary(lim)
```

```
         iip_przest                                iip_identy
PL.PZGIK.200: 25753  a1e8cdc7-d26d-4982-946f-ff3e5082eecd: 25753
NA's        :290104  42d2335f-bd81-491e-b93c-effe1bcab872:     0
                     45e7cfc6-6d8b-42ff-acbd-ef1d0a53dacb:     0
                     4b6c492a-eb04-441d-a92a-f44359c06de7:     0
                     53ad7aea-d9d3-40c9-9a5c-ff737d5b076e:     0
                     (Other)                             :     0
                     NA's                                :290104
              iip_wersja       jpt_sjr_ko      jpt_kod_je
2012-09-27T13:45:12+02:00: 25753  WOJ : 25753   06      : 25753
2012-09-27T13:45:13+02:00:     0  NA's:290104   02      :     0
2013-07-09T11:04:17+02:00:     0                04      :     0
NA's                     :290104                08      :     0
                                                10      :     0
                                                (Other):     0
                                                NA's    :290104
               jpt_nazwa_       jpt_nazw01     jpt_organ_      jpt_orga01
lubelskie            : 25753  NA's:315857   NA's:315857   NZN : 25753
dolnośląskie         :     0                              NA's:290104
kujawsko-pomorskie   :     0
lubuskie             :     0
łódzkie              :     0
(Other)              :     0
NA's                 :290104
```

```
a<-which(lim$jpt_nazwa_=="lubelskie") # rows which fulfill the criteria
head(lim[a,])

# cutting grid to regional contour
pop.grid.lub<-pop.grid[lim$jpt_nazwa_=="lubelskie",]
class(pop.grid.lub) # sp class
length(pop.grid.lub)
```

```
[1] 25753
```

```
## grid data limited as grid shapefile
pop.df.lub<-pop.df[a,]
```

Third, the effects of data preparation should be visualised. One can draw the obtained grid with the **plot**() command and check whether the cut in accordance with the province's contour has been made correctly (see Figure 5.10). One can also draw the variable values on the limited grid. Unfortunately, due to the large number

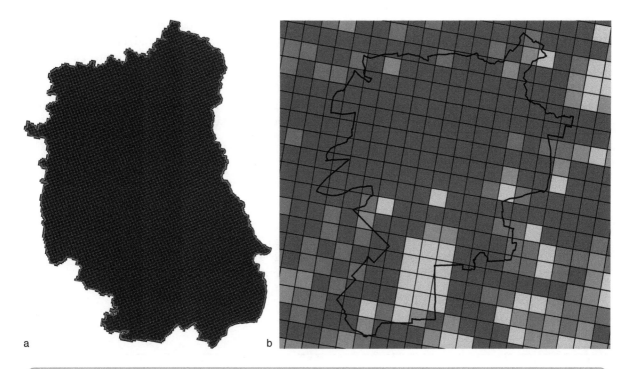

Figure 5.10 Grid: a) grid limited to the contour of the province, b) zoom of the grid for a limited area

Source: Own study using R and the GISTools:: package

of cells, the variable values are illegible. One can use a smart trick to zoom in the figure. By drawing the selected single poviat (NTS4) as the first layer (here the poviat of Lublin), limited coordinates of the figure are determined. As the next layer, one can apply a grid, which will be drawn only within the limited coordinates. Again, one can apply the poviat's contour to the grid (the first and third layers are the same).

```
# Figure 5.10a - administrative contour and grid
plot(pop.grid.lub)
plot(voi.lub, add=TRUE, border="red")

# Figure - values of the examined variable - the whole province
library(GISTools)
choropleth(pop.grid.lub, pop.df.lub$TOT)
plot(voi.lub, add=TRUE)

# Figure 5.10b - values of the examined variable - zoomed poviat
library(GISTools)
plot(pow[pow@data$jpt_nazwa_=="powiat Lublin",])
choropleth(pop.grid.lub, pop.df.lub$TOT, add=TRUE)
plot(pow[pow@data$jpt_nazwa_=="powiat Lublin",], add=TRUE, lwd=2)
```

Fourth, the key element of the process is aggregation of point data according to the existing separated grid. The point data (37,374 obs) was converted to the *SpatialPoints* class by the **coordinates()** command, and the projection of these data and the grid were agreed with the **proj4string()** and **spTransform()** commands. The **over()** command was used to assign points to grid cells. The *locs.lim* object is of *integer* class, its length is as in the point dataset (37,374 obs.), and it contains the row IDs of the grid object. Its headers are the numbers of the original grid set, while the values are the ID of the limited grid set. This vector has been added to the original point dataset (*grid* variable in *data1* object). With the **lapply()** and **unlist()** commands, one can check the ID structure in sets – it is crucial in further ordering the data. This study confirms that the variable ID from with values from 1 to *n* (here *n* = 25,753) should be added to the restricted grid dataset. It will be used to merge datasets – aggregated points and truncated grid data. Statistics in the **summary()** command output indicate that some of the grid cells are empty – there are no points from the point dataset (in particular the first and last 10 cells). This forces **merge()** to link the data – aggregated with the use of **aggregate()** command according to the grid ID and the limited grid data. NA has been replaced with 0 with the **is.na()** command. In the limited grid dataset, it was necessary to change the name of the variable appended with the **merge()** command. The code shows how to change only the last variable name, keeping all others unchanged.

```
## assignment and aggregation of point data according by grid cells
#data reminder - points from REGON
#dane<-read.csv("geoloc data.csv", header=TRUE, dec=",", sep=";")

dane.sp<-dane1
coordinates(dane.sp)<-c("coords.x1","coords.x2") # change of object class
proj4string(dane.sp)<-CRS("+proj=longlat +datum=NAD83")
pop.grid.lub<-spTransform(pop.grid.lub, CRS("+proj=longlat +datum=NAD83"))
dane.sp<-spTransform(dane.sp, CRS("+proj=longlat +datum=NAD83"))

# assigning points to grid
locs.lim<-over(dane.sp, pop.grid.lub)
head(locs.lim)
```

```
30140 27559  2219 12104 22254  7971
21201  7061 17503  3835  7699 19705
```

```
summary(locs.lim)
```

```
 Min. 1st Qu. Median   Mean 3rd Qu.  Max.
   10    6201   8979  11277   17513 25743
```

```
dane1$grid<-locs.lim # adds a new (truncated) grid ID
head(dane1)

# preparation of objects with ordered data
# summary of slot & ID number assignment
aaa1<-lapply(pop.grid.lub@polygons, slot, "ID")
head(aaa1)
```

```
[[1]]
[1] "241239"

[[2]]
[1] "241321"

[[3]]
[1] "241858"

[[4]]
[1] "241964"

[[5]]
[1] "241984"

[[6]]
[1] "241990"
```

```
# list of slot numbers
aaa2<-unlist(lapply(pop.grid.lub@polygons, slot, "ID"))
```

```
[1] "241239" "241321" "241858" "241964" "241984" "241990"
```

```
pop.df.lub$ID<-1:25753

# data aggregation by grid
roa.ag<-aggregate(dane1$roa, by=list(dane1$grid), mean, na.rm=TRUE)
pop.df.lub<-merge(pop.df.lub, roa.ag, by.x="ID", by.y="Group.1", all.x=TRUE)
pop.df.lub$x[is.na(pop.df.lub$x)]<-0
choropleth(pop.grid.lub, pop.df.lub$x)

# change of the name of the added variable (by merge())
ccc<-colnames(pop.df.lub)
n<-length(ccc)
colnames(pop.df.lub)<-c(ccc[1:n-1], "roa")
colnames(pop.df.lub)
```

```
 [1]  "ID"          "TOT"         "TOT_0_14"   "TOT_15_64"  "TOT_65__"
 [6]  "TOT_MALE"    "TOT_FEM"     "MALE_0_14"  "MALE_15_64" "MALE_65__"
[11]  "FEM_0_14"    "FEM_15_64"   "FEM_65__"   "FEM_RATIO"  "SHAPE_Leng"
[16]  "SHAPE_Area"  "CODE"        "roa"
```

In the last, fifth step, one can proceed to the estimation of the spatial model. The key element is to create a spatial weights matrix – based on rules, as for area data. In the example, a spatial weights matrix was created on the basis of contiguity rule, with the use of the typical commands **poly2nb()** and **nb2listw()**. In order to estimate the model, an additional variable was created based on the grid data: percentage of the working age population, from which NA values (with the **is.na()** command) and infinity (with the **is.infinite()** command) were eliminated, replacing them with 0. An example presents the estimated model in which the average profitability of companies (*roa*) is explained by the percentage of women in the population (*FEM_RATIO*) and the percentage of the population of working age (*pop.prod*). The presented model is only a technical illustration of the possibilities.

```
# spatial weights matrix as contiguity matrix for grid
cont.nb<-poly2nb(pop.grid.lub, queen=T) #conversion from sp to nb class
cont.listw<-nb2listw(cont.nb, style="W")
cont.listw #displays summary of matrix

# variables included in the model
pop.df.lub$pop.prod<-pop.df.lub$TOT_15_64/pop.df.lub$TOT
pop.df.lub[is.na(pop.df.lub)]<-0
pop.df.lub[is.infinite(pop.df.lub)]<-0
pop.df.lub$pop.prod[which(pop.df.lub$pop.prod ==Inf)] <- 0

# estimation of model on grid
model<-errorsarlm(roa~FEM_RATIO+pop.prod, data=pop.df.lub, nb2listw(cont.
nb), zero.policy=TRUE, method="LU")
summary(model, Nagelkerke=TRUE)
```

```
Call:errorsarlm(formula = roa ~ FEM_RATIO + pop.prod, data = pop.df.lub,
    listw = nb2listw(cont.nb), method = "LU", zero.policy = TRUE)

Residuals:
     Min      1Q    Median      3Q      Max
-4.82677 -0.99575 -0.10497 0.35233 8.23679

Type: error
Coefficients: (asymptotic standard errors)
             Estimate Std. Error z value  Pr(>|z|)
(Intercept) 0.20672453 0.02373418  8.7100 < 2.2e-16
FEM_RATIO   0.00137579 0.00023533  5.8463 5.026e-09
pop.prod    1.44667644 0.04315606 33.5220 < 2.2e-16

Lambda: 0.51947, LR test value: 3342.8, p-value: < 2.22e-16
Approximate (numerical Hessian) standard error: 0.0080423
    z-value: 64.593, p-value: < 2.22e-16
Wald statistic: 4172.3, p-value: < 2.22e-16

Log likelihood: -47514.61 for error model
ML residual variance (sigma squared): 2.2489, (sigma: 1.4996)
Nagelkerke pseudo-R-squared: 0.25066
Number of observations: 25753
Number of parameters estimated: 5
AIC: 95039, (AIC for lm: 98380)
```

5.4 Spatial panel models

Spatial panel models are quite an advanced estimation tool due to the relatively large number of parameters to be determined within the model. The selection of appropriate parameters is an art and requires determination of the panel model estimation strategy.

The literature is typically divided into theory-based studies, related to the properties of estimators, and practice-based studies, related to software and estimation strategies. With regard to the former category, that is, theory-based literature, it is worth recommending the texts: Elhorst, J. P. (2014b). Spatial Panel Data Models. *In Spatial Econometrics (pp. 37–93). Springer, Berlin*; Heidelberg and Elhorst, J. P. (2014a). Dynamic Spatial Panels: Models, Methods and Inferences. *In Spatial Econometrics (pp. 95–119). Springer, Berlin, Heidelberg.* With regard to the second category, that is, practice-based studies, the authors of the splm:: package estimating spatial panels deserve special attention (Millo, G., & Piras, G. (2012). splm: Spatial Panel Data Models. *In R. Journal of Statistical Software, 47(1), 1–38*). It is also worth referring to the description of estimation strategies in the article Kopczewska, K., Kudła, J., & Walczyk, K. (2017). Strategy of Spatial Panel Estimation: Spatial Spillovers Between Taxation and Economic Growth. *Applied Spatial Analysis and Policy, 10(1), 77–102*, on which the following chapter is based.

In panel modelling, decisions should be made about several model elements. They are described subsequently.

First, decide what types of specific effects will be included in the model. As in traditional models, they can be fixed effects (FE), random effects (RE) and models without specific effects (pooled). In the group of permanent effects, one can estimate individual effects one way (for units), time (for periods) or both (two way). The choice between FE and RE models in the Cliff-Ord class uses the Hausman test (Mutl & Pfaffermayr, 2008). RE and FE may result from theory: FE when the population is tested and RE when the sample is tested.

Specific effects – time and individual – introduced to the model allow for controlling the characteristic features of the cross-sections studied. Individual effects for each of the observed regions express spatial characteristics that are unchanged over time and affect the dependent variable. Failure to do so may result in biased estimators. Similarly, time effects are introduced into the model for each period of time, invariable in space that may represent, for example, business cycles or institutional and legal changes. Permanent effects should be understood as dummy variables (with 0–1 values) for each unit of the extracted dimension. Random effects are understood as IID random variables.

On RE models, the *phi* parameter and its standard error are estimated. In FE models, the unit coefficients can be extracted – beta. When FE and RE are insignificant, so-called pooled model is estimated without isolating the effects.

The choice between the fixed effects model and random effects further determines how to treat variables. In FE models, it is necessary to transform demeaning – subtracting the mean of observations for each cross-section in time, which eliminates constants, constant effects and variables with time-constant values. Thanks to this, it is possible to estimate individual effects for units, time or two-way. The RE models assume that the effects come from a distribution with an average of 0 and variance σ^2.

Second, the model type is selected between the Cliff-Ord model (containing spatial lags y and/or ε) and the Durbin model (containing spatial lags x and, if necessary, spatial lags y and/or ε). The starting point is the full Manski model (GNS). The spatial lag of the explained variable y is treated as a global spillover or long-term relation. The Durbin component (spatial lag of the explanatory variables x) absorbs the endogenous nature of regressors and is treated as a local spillover. The spatial lag of error ε is treated as local characteristics, exogenous shocks not being modelled by variables in the model.

There are a number of pros and cons to estimating with the Durbin and Cliff-Ord models. The estimation of Cliff-Ord models may lead to uncertainty of parameters or their incomplete interpretation; however, it allows correction of heteroscedasticity, which causes the bias of standard errors (Romero & Burkey, 2011). Choosing a Durbin model leads to lower estimation efficiency. Durbin models cause a loss of degrees of freedom when adding spatially lagged explanatory variables, so they can be effectively used in models with a limited set of explanatory variables. In the Durbin model, it is not possible to model the spatial autocorrelation of an error. Ignoring this component leads to ineffective estimates and erroneous inferences.

The path for turning variables on and off can be illustrated as follows.

All approaches come from the classic linear regression model – the additive linear model, where the explained variable is translated by a set of explanatory variables and the error is to be independent and IID.

$$Y = \alpha\iota + X\beta + \varepsilon \tag{5.37}$$

where Y is the explained variable, and X is the set of explanatory variables.

In the non-dynamic spatial approach according to Millo and Piras (2012), in line with the Cliff-Ord approach, the full model takes into account the spatial lag of the explained variable, the spatial autocorrelation of the residual component and specific effects (permanent or random) are also spatially correlated. It does not contain spatially lagged explanatory variables. Disturbances are spatially modelled precisely, most often according to the approach of Baltagi, Song, and Koh (2003) and Kapoor, Kelejian, and Prucha (2007).

$$Y = \lambda WY + X\beta + u, \; where \; u = f\left(\mu, \iota, W\right) + \varepsilon \tag{5.38}$$

where W is a spatial weights matrix according to the selected neighbourhood criterion.

In the non-dynamic approach with the Durbin component (Elhorst, 2010), the full model assumes taking into account the spatial lag of the explanatory variable, the spatial lag of the explanatory variables and the spatial autocorrelation of the error.[4]

$$Y = \rho WY + \alpha\iota + X\beta + WX\theta + u, \; where \; u = \lambda Wu + \varepsilon \tag{5.39}$$

where ι is a vector of specific effects.

Due to the potential inability to fully interpret all three interaction effects, Elhorst (2010) proposes disabling spatial autocorrelation of error, which leads to the so-called Durbin model, that is, a model only with endogenous (WY) and exogenous (WX) interactions:

$$Y = \rho WY + \alpha\iota + X\beta + WX\theta + \varepsilon \tag{5.40}$$

Third, it is necessary to determine what lag variables will be included in the model. It is possible to enable:

1   a time-lagged explained variable
2   time-space-lagged explained variable
3   time-lagged explanatory variables
4   time-space-lagged explanatory variables

In Cliff-Ord models (i.e. those modelling the error disturbance and omitting the spatial lag of independent variables), an important element of the estimation strategy is to include (or not) the spatial lag of the dependent variable as an explanatory variable. This introduces a kind of endogenicity to the model. In practice, this allows for absorbing the variability arising from clusters. The model with spatial lag has a generalised form:

$$y = \lambda \left(I_T \otimes W_N \right) y + X\beta + u \tag{5.41}$$

where y is a vector of the dependent variable with dimensions $NT \times 1$, W_N is the $n \times n$ matrix of spatial weights, X is the matrix $NT \times k$ of explanatory variables and u is the residual component.

In the dynamic approach according to Elhorst (2010), the Durbin standard non-dynamic model is supplemented with the time lag of the explained variable and the time-spatial lag of the explained variable, as well as specific cross-sectional effects $\alpha \iota\_N$ and time $\xi \iota_T$.

$$Y_t = \tau Y_{t-1} + \rho W Y_t + \eta W Y_{t-1} + \alpha \iota_N + \xi \iota_T + X_t \beta + W X_t \theta + \varepsilon \tag{5.42}$$

It is possible to determine the direct and indirect long- and short-term effects on the basis of the variance-covariance matrix (LeSage & Pace, 2009).

Fourth, the error structure should be defined. In the literature on models that take into account the spatial error found in Clif-Ord models, two approaches are proposed: error modelling according to the approach of Kapoor et al. (2007) or according to the approach of Baltagi et al. (2003). The Durbin model does not include spatial error. The first approach, according to Baltagi et al. (2003) assumes that error can be decomposed into individual effects, group effects independent of time and space and idiosyncratic error, with spatially correlated shocks (innovations) (Millo & Piras, 2012). Then the residual component takes the form:

$$u = \left(\iota_T \otimes I_N \right) \mu + \varepsilon \tag{5.43}$$
$$\varepsilon = \rho \left(I_T \otimes W_N \right) \varepsilon + \upsilon$$
$$\upsilon_{it} \sim IID\left(0, \sigma_\upsilon^2 \right)$$

where ι_T is a unit vector, I_N a unit matrix $N \times N$, μ a vector of specific effects (time-constant and spatially correlated), and ε an autoregressive vector of shocks (spatially correlated innovations). In the second approach, proposed by Kapoor et al. (2007), in the residual component, both individual effects and the rest of the error are spatially correlated. Then the residual component takes the form:

$$u = \rho \left(I_T \otimes W_N \right) u + \varepsilon \tag{5.44}$$
$$\varepsilon = \left(\iota_T \otimes I_N \right) \mu + \upsilon$$
$$\upsilon_{it} \sim IID\left(0, \sigma_\upsilon^2 \right)$$

where μ is a vector of specific effects[5], and υ is a vector of shocks that change over time and across (Millo & Piras, 2012).

Fifth, decide on the spatial weights matrix W. In modelling, several types of spatial weights matrix can be used, according to different neighbourhood criteria: common neighbourhood matrix, inverse distance and inverse square (also polynomial) distance. Arrays of k nearest neighbours or neighbours within a radius of d km are also available. Most studies use a weights matrix based on the *contiguity criterion* standardised by lines to unity. In some studies, an inverse distance or inverse square distance matrix appears as an alternative (e.g. Piras and Arbia, 2007). Each of the spatial weights matrices W expresses different spatial relations.

The contiguity matrix allows the assessment of local effects without taking into account global contexts. The neighbourhood is expressed as zero-one and takes the value = 1 for a pair of regions that have a common border, while in other cases it is 0. The weights result from standardisation by rows to 1 and are equal for all neighbours, regardless of the length of the border. This means that in a region surrounded by n neighbours, each weight will be $1/n$. The advantage of using this matrix is the ease of construction and interpretation of results. However, the matrix does not take into account spatial relations further than the first-order neighbours, that is, with neighbours' neighbours.

The reverse distance matrix takes into account neighbouring relations with all territorial units. They are linear, which means that the strength of the relationship changes proportionally to the distance. The matrix allows testing the effects of global spatial interactions, without taking into account local clusters.

In turn, *the inverse square distance* matrix assumes that neighbourly relations are non-linear and weaken more than proportionally while increasing the distance. This allows modelling local clusters, in which neighbouring countries are more closely connected for cultural, historical or economic reasons than results from the distance alone. The matrix takes into account global effects due to non-zero weights for all territorial units.

The matrix *k of nearest neighbours* very often turns out to be asymmetrical, which requires symmetry, usually equalising the number of neighbours up, which disturbs the nature of the matrix.

The matrix of *neighbours within a radius of d km* requires precise selection of radius *d* so that each region has at least one neighbour. Lack of neighbours means that the object is in the situation of an island, and it causes computational complications. The matrix is strongly dependent on the shape of territorial units. Searching for regions within a reasonable radius means searching for centres of gravity of regions not more than a kilometre away from the Euclidean distance, where the centres of gravity of geometric figures representing the given areas do not have to coincide with the actual centre, for example, main city, region.

Geographical distance-based matrices are exogenous in the model, thanks to which they are identified (Manski, 1993). In a generalised form, the spatial weights matrix W based on distance takes the form (Anselin, 2002):

$$W(i,j) = 1 / d(i,j)_m^\gamma \tag{5.45}$$

where $d(i,j)_m$ is the distance between the *m*th nearest region *j* and *i*, and γ is the decay parameter. Most often, it is assumed that in the sample *N* regions, region *i* has $m = N - 1$ of the nearest neighbouring regions. Then the inverse distance matrix with the parameter $\gamma = 1$ takes into account neighbourly relations with all territorial units. They are linear, which means that the strength of the relationship changes inversely in proportion to the distance. The matrix allows testing the effects of global spatial interactions without taking into account local clusters. The matrix of inverse square distance with the parameter $\gamma = 2$, also in relation to all units, assumes that neighbourly relations are non-linear and weaken more than proportionally to the distance. It allows modelling of local clusters, in which neighbouring countries are more closely connected for cultural, historical or economic reasons than results from the distance alone. The matrix takes into account global effects due to non-zero weights for all territorial units.

Sixth, choose the method of estimation and the form of the model. There are several schools of conduct. First, it is possible to choose based on theory, both in relation to the matrix of weights and the variables contained in the model. Second, it is possible to follow the data to get the highest estimation efficiency or the best fit. The criterion can be a comparison of spatial advanced models with a simple OLS model on pooled data. The comparative measure may be LogLik. Pace and LeSage (2008) developed a test based on the Hausman test used to assess the significance of differences in β estimates from OLS and SEM models. It is known that OLS estimates in the presence of a spatial lag spatial process are biased and inconsistent, and OLS estimates in the case of spatial error are inefficient (but inefficient and asymptotically consistent). Pace and LeSage (2009) indicate that parameter estimates in OLS and SEM models should be similar if the model specification is correct, while spatial estimation allows for robust estimates. In the case of significant discrepancies between OLS and SEM, there is a problem of erroneous specification: omitted variables and omitted spatial lag of the explained variable. Testing the correctness of the SEM model specification consists of assessing the differentiation of OLS and SEM parameters for each parameter separately (Pace, LeSage, & Zhu, 2012). Test statistics *t* take the form:

$$t = \frac{(\beta_{SEM} - \beta_{OLS})}{\sigma_{SEM}} \tag{5.46}$$

The spatial panel model of error should control spatial correlation and heterogeneity. The Breusch-Pagan LM tests developed by Baltagi et al. (2003) allow testing the existence of spatial autocorrelation and random individual effects (LMH test), as well as marginal effects: the existence of spatial autocorrelation assuming random individual effects (LM2) and the existence of individual random effects assuming spatial autocorrelation (LM1).

Seventh, remember the restrictions. Panel estimation allows the estimation of compatible model coefficients for the entire period studied. This advantage is also a disadvantage, because it does not allow for assessing changes over time, including under the influence of cyclical changes such as the financial crisis. While it is possible to extract individual effects in FE (time and group) models, it is impossible to assess the spatial correlation changes for the spatially lagged dependent variable and the residual component. The assessment of changes in the strength of spatial correlations can allow conclusions about the strength of neighbourly interactions, the strength of economic phenomena transfer between neighbours and the interdependence of phenomena (cluster effects) from period to period. In a simplified approach, the extraction of these neighbourly effects is possible due to the estimation for each year of a static simple spatial lag model taking into account the spatially lagged dependent variable. The estimation may also include an approach using heteroscedastic innovations (Arraiz, Drukker, Kelejian, & Prucha, 2010; Kelejian & Prucha, 2007b, 2010; Piras, 2010).

It should also be remembered that the panel model in the result gives for a given variable a one-directional coefficient for the entire period for all observations. This requires the assumption that the relationship between x and y be constant, that is, that an increase in x is accompanied by an increase in y in at a constant proportion. On the other hand, if this relation is variable in time (e.g. there is a constant increase in x and cyclicity of y), then the panel model will give an average directional coefficient, which will very poorly reflect real relations. In the event of such relationship changes, a cumulative model (for monetary variables) or a one-period model with time lags is recommended, estimated separately for each year.

Consideration of all the previous modelling elements is necessary to obtain a good model. However, it should be emphasised that the order in which these criteria are resolved is important. According to the recommendations contained in Kopczewska et al. (2017), the order given subsequently for deciding on spatial components should be used. It should be emphasised that the following procedure is for a selected matrix of spatial weights and a selected form of the model (interaction model, convergence model etc.). When changing the matrix of spatial weights W or model form, the following procedure should be fully performed again.

Phase I: Starting from the full model with the initial two-way specific effects and all spatial components

Round 1 from the full model – reduction of specific effects (FE or RE), maintaining all spatial components
Round 2 from a reduced model – reduction of spatial components for specific effects.

Control of Phase I: Starting from a full model with initial two-way specific effects and all spatial components

Round 1 from the full model – reduction of spatial components (θ, λ, ρ) while maintaining two-way specific effects
Round 2 from a reduced model – reduction of specific effects (FE or RE) for specific spatial components.

Regardless of the order of Rounds 1 and 2, the best model estimated (including only these components) should be similar. The order of these two phases should not matter for the result.

Phase 2: Starting from a reduced model with specific target spatial components and specific effects:
Round 3: Reduction/consideration of time and space-time lags while maintaining other variables
Round 4: Final adjustments of variables included in the model

Round 5: Adjustments in the error structure.

In R, the estimation of spatial panel models is possible in the splm:: package (Millo & Piras, 2012). The package assumes Cliff-Ord models, that is, without the Durbin component turned on automatically but taking into account error distortions according to the Kapoor or Baltagi approach. Other spatial operations, construction of the spatial weights matrix and spatial lags use the packages: sp:: (Pebesma & Bivand, 2005), spdep:: (Bivand, 2013) and maptools:: (Bivand and Lewin-Koh, 2017).

In the commands for estimating panel models with splm:: like **spml()** or **spgm()**, specify the elements mentioned previously as part of the option. Table 5.4 shows the available diagrams – combinations of panel model options.

Example###

The example uses panel data for 380 poviats (NTS4) from 2006–2017. Panel *databases_pow_2019.csv* contains panel data for 23 variables and 15 identifiers. An attempt will be made to estimate a panel model with variable specifications as in the example for cross-sectional models – that is, an attempt to determine the determinants of average wage in the poviat (in Poland = 100%) by various socio-economic-population variables. The order of the dataset according to the order of units in the shapefile is specified by the *ID_MAP* variable. The location of 380 territorial units – poviats – is defined by the shapefile for poviats, defined as the object of

Table 5.4 Available combinations of panel model options

Option model=c("within", "random", "pooling"),	effect=c("time", "individual", "twoways"),	lag=FALSE	spatial.error=c("b", "kkp", "none")
FE: model="within"	Dummies for units: effect="individual"	spatial lag of Y – YES lag=TRUE	Spatial lag of e – YES (choose) spatial.error="b" spatial.error=kkp"
FE: model="within"	Dummies for units: effect="individual"	spatial lag of Y – YES lag=TRUE	Spatial lag of e – NO spatial.error="none"
FE: model="within"	Dummies for units: effect="individual"	spatial lag of Y – NO lag=FALSE	Spatial lag of e – YES (choose) spatial.error="b" spatial.error=kkp"
FE: model="within"	Dummies for units: effect="individual"	spatial lag of Y – NO lag=FALSE	Spatial lag of e – NO spatial.error="none"
FE: model="within"	Dummies for time: effect="time"	spatial lag of Y – YES lag=TRUE	Spatial lag of e – YES (choose) spatial.error="b" spatial.error=kkp"
FE: model="within"	Dummies for time: effect="time"	spatial lag of Y – YES lag=TRUE	Spatial lag of e – NO spatial.error="none"
FE: model="within"	Dummies for time: effect="time"	spatial lag of Y – NO lag=FALSE	Spatial lag of e – YES (choose) spatial.error="b" spatial.error=kkp"
FE: model="within"	Dummies for time: effect="time"	spatial lag of Y – NO lag=FALSE	Spatial lag of e – NO spatial.error="none"
FE: model="within"	Dummies for both: effect="twoways"	spatial lag of Y – YES lag=TRUE	Spatial lag of e – YES (choose) spatial.error="b" spatial.error=kkp"
FE: model="within"	Dummies for both: effect="twoways"	spatial lag of Y – YES lag=TRUE	Spatial lag of e – NO spatial.error="none"
FE: model="within"	Dummies for both: effect="twoways"	spatial lag of Y – NO lag=FALSE	Spatial lag of e – YES (choose) spatial.error="b" spatial.error=kkp"
FE: model="within"	Dummies for both: effect="twoways"	spatial lag of Y – NO lag=FALSE	Spatial lag of e – NO spatial.error="none"
RE: model="random"	–	spatial lag of Y – YES lag=TRUE	Spatial lag of e – YES (choose) spatial.error="b" spatial.error=kkp"
RE: model="random"	–	spatial lag of Y – YES lag=TRUE	Spatial lag of e – NO spatial.error="none"
RE: model="random"	–	spatial lag of Y – NO lag=FALSE	Spatial lag of e – YES (choose) spatial.error="b" spatial.error=kkp"
RE: model="random"	–	spatial lag of Y – NO lag=FALSE	Spatial lag of e – NO spatial.error="none"
No effects model="pooling"	–	spatial lag of Y – YES lag=TRUE	Spatial lag of e – YES (choose) spatial.error="b" spatial.error=kkp"
No effects model="pooling"	–	spatial lag of Y – YES lag=TRUE	Spatial lag of e – NO spatial.error="none"
No effects model="pooling"	–	spatial lag of Y – NO lag=FALSE	Spatial lag of e – YES (choose) spatial.error="b" spatial.error=kkp"
No effects model="pooling"	–	spatial lag of Y – NO lag=FALSE	Spatial lag of e – NO spatial.error="none"

Source: Own work

Table 5.5	**Specification of panel model**

Variable symbol	Variable code	Description of the variable
Y	XA14	Poland salaries = 100%
X1	XA08/XA09	Percentage of working-age population in relation to post-working-age population
X2	XA13	Number of firms per 10,000 working age residents
X3	(XA05/XA06)/mean(XA05/XA06)	Public investments per capita (Poland = 100)
X4	(XA18+XA19+XA20)/XA15	Share of employment in services
X5	XA16/XA15	Share of employment in agriculture
X6	XA15/XA06	Share of employees relative to population
X7	XA10/mean(XA10)	Population density per 1 km^2 (Poland = 100)
X8	XA21	Unemployment rate

Source: Own study

the poviat time is specified by the variable *year*. In the example, the relationship between variables and panel modelling will be determined. The specification of the model is presented in Table 5.5.

Table 5.5 shows which variables will be taken into account in the econometric model. To simplify the process, the variables used received new codes – these variables were created in the dataset. A little more transformation was required for variables based on the mean, since it is necessary to divide the value by the mean of the observation year. There are many solutions to this issue – here the average annual numbers were calculated using the **aggregate**() command, the vector of average annual lengths of the dataset was created using the **rep**() command and the examined observations were divided by the appropriate average values. The variables were also standardised using the **scale**() command. Panel data requires group operation (by *year*), which is why the **transformBy**() command from the doBy:: package was used. Finally, the equations for estimation were prepared – on the basis of nominal variables and on the basis of scaled variables. The dataset was also ordered to ensure the order of variables required by the package (variable #1 – region ID, variable #2 – time id). Two spatial weights matrices were prepared – according to the contiguity criterion (*cont.listw*) and an inverse distance matrix (*dist.listw*).

```
library(spdep)
library(rgdal)
# loading data
data<-read.csv("data_nts4_2019.csv", header=TRUE, dec=",", sep=";")
pov<-readOGR(".", "powiaty") # 380 units
pov<- spTransform(pov, CRS("+proj=longlat +datum=NAD83"))

# creating variables as described
data$y<-data$XA14
data$x1<-data$XA08/data$XA09
data$x2<-data$XA13
data$x4<-(data$XA18+data$XA19+data$XA20)/data$XA15
data$x5<-data$XA16/data$XA15
data$x6<-data$XA15/data$XA06
data$x8<-data$XA21

# variables based on periodic average
a1<-data$XA05/data$XA06 # variable for analysis
a2<-aggregate(a1, by=list(data$year), mean, na.rm=TRUE)
a3<-rep(a2$x, each=380) # periodic average assigned to observations
data$x3<-a1/a3 # variable index (Poland = 100%)

b2<-aggregate(data$XA10, by=list(data$year), mean, na.rm=TRUE)
b3<-rep(b2$x, each=380)
data$x7<-data$XA10/b3
```

```
# standardization of variables according to temporal parameters (μ, σ)
library(doBy)
data$y.sc<-transformBy(~year, data=data, y=scale(y))$y
data$x1.sc<-transformBy(~year, data=data, x1=scale(x1))$x1
data$x2.sc<-transformBy(~year, data=data, x2=scale(x2))$x2
data$x3.sc<-transformBy(~year, data=data, x3=scale(x3))$x3
data$x4.sc<-transformBy(~year, data=data, x4=scale(x4))$x4
data$x5.sc<-transformBy(~year, data=data, x5=scale(x5))$x5
data$x6.sc<-transformBy(~year, data=data, x6=scale(x6))$x6
data$x7.sc<-transformBy(~year, data=data, x7=scale(x7))$x7
data$x8.sc<-transformBy(~year, data=data, x8=scale(x8))$x8

# changing the order of variables (No. 1 region, No. 2 year)
data<-data[,c(1,9,2:52)]

# contiguity spatial weights matrix
cont.nb<-poly2nb(as(pov, "SpatialPolygons"))
cont.listw<-nb2listw(cont.nb, style="W")

# matrix W according to the inverse criterion
crds<-coordinates(pov)
pov.knn<-knearneigh(crds, k=379) # knn=380-1, there are 380 counties
pov.nb<-knn2nb(pov.knn)
dist<-nbdists(pov.nb, crds)
dist1<-lapply(dist, function(x) 1/x) # listw class object
dist.listw<-nb2listw(pov.nb, glist=dist1) # listw class object
```

The following example shows an estimation of the panel model based on the **spml**() command from the splm:: package. The SAC model with specific unidirectional – individual effects was estimated. The **options**() command with the *scipen* option allows the result to be displayed without scientific notation (e.g. 1e-15). In the model, both spatial coefficients, *rho* and *lambda*, are significant, as well as most variables. Fixed effects obtained with the **effects**() command are in most units significant. They can be visualized as a density distribution (using **plot**() and **density**()) or a spatial distribution (using e.g. **choropleth**()).

```
library(splm)
# model with permanent effects (FE), this is the SAC model
# there are lambda coefficients (spatial lag y)
# rho coefficients (spatial lag of error) occur
# there is a Baltagi error

# equations for the model
# equation on nominal variables and the Poland=100 index variables
eq1<-y~x1+x2+x3+x4+x5+x6+x7+x8

model.spml<-spml(eq1, data=data, listw=cont.listw, model="within", spatial.
error="b", lag=TRUE, effect="individual", rel.tol=2e-40)
options(scipen=999, digits=2)
summary(model.spml)
```

```
Spatial panel fixed effects sarar model

Call:
spml(formula = eq1, data = data, listw = cont.listw, model = "within",
     effect = "individual", lag = TRUE, spatial.error = "b",
     rel.tol = 0.00000000000000000000000000000000000000002)
```

```
Residuals:
      Min.  1st Qu.   Median 3rd Qu.     Max.
-13.87206 -1.22670 -0.00539 1.18006 19.75253

Spatial error parameter:
 Estimate Std. Error t-value  Pr(>|t|)
rho -0.4774 0.0501 -9.53 <0.0000000000000002 ***

Spatial autoregressive coefficient:
 Estimate Std. Error t-value  Pr(>|t|)
lambda 0.5185 0.0361 14.4 <0.0000000000000002 ***

Coefficients:
     Estimate Std. Error t-value          Pr(>|t|)
x1 -1.184415   0.113792  -10.41 < 0.0000000000000002 ***
x2 -0.001627   0.000402   -4.05          0.000050934 ***
x3  0.359111   0.088818    4.04          0.000052725 ***
x4 -5.693622   1.637570   -3.48              0.00051 ***
x5 -7.134217   1.190050   -5.99          0.000000002 ***
x6  9.890663   2.330517    4.24          0.000021959 ***
x7  0.346749   0.274777    1.26              0.20697
x8  0.045371   0.010792    4.20          0.000026188 ***
---
Signif. codes: 0 '***' 0.001 '**' 0.01 '*' 0.05 '.' 0.1 ' ' 1
```

```
eff<-effects(model.spml)  # fixed effects
eff
```

```
Intercept:
            Estimate Std. Error t-value          Pr(>|t|)
(Intercept)     48.5        1.5    32.4 <0.0000000000000002 ***

Spatial fixed effects:
  Estimate Std. Error t-value          Pr(>|t|)
1  -8.4189     1.5323   -5.49 0.00000003926353661 ***
2  -3.7549     1.6617   -2.26             0.02384 *
3  -6.1596     1.5826   -3.89 0.00009944453409089 ***
4  -2.0936     1.5446   -1.36             0.17528
5  -3.4361     1.5653   -2.20             0.02815 *
6  -4.3420     1.5718   -2.76             0.00574 **
```

```
attributes(eff)  # attributes of specific effects object
```

```
$'names'
[1] "INTTable" "SETable" "effects"

$class
[1] "effects.splm"
```

```
plot(density(eff$SETable[,1]))  # specific effect density distribution

library(GISTools)
choropleth(pov, eff$SETable[,1], main="Fixed effects")  # fixed effects mapped
```

```
shades<-auto.shading(eff$SETable[,1])
choro.legend(14, 50.25, shades, cex=0.65, bty="n")
```

The same model can be estimated on standardised data obtained as part of the **transformBy()** command. The significance and properties of the model remain unchanged, though the results differ in interpretation. In a typical "change of x by 1" model, necessary for interpreting the beta coefficient, it is on a variable scale and requires knowledge of whether this type of change is possible. The coefficients obtained in the model are not directly comparable (larger/smaller impact), as their volume depends on the scale of the phenomenon studied. In the case of scaled variables, "change of x by 1" is on the scale of standard deviations of the variable, so it is much easier to interpret such a change. Beta coefficients are directly comparable (larger/smaller) because they refer to changes scaled by standard deviation. The following code shows that the variables $x7$ and $x8$ have the greatest impact on the studied phenomenon, while the variable $x2$ would have the weakest if it were significant.

```
# equation for scaled variables
eq1.sc<-y.sc~ x1.sc+ x2.sc+ x3.sc+ x4.sc+ x5.sc+ x6.sc+ x7.sc+ x8.sc

model.spml<-spml(eq1.sc, data=data, listw=cont.listw, model="within", spa-
tial.error="b", lag=TRUE, effect="individual", rel.tol=2e-40)
options(scipen=999, digits=2)
summary(model.spml)
```

```
Spatial panel fixed effects sarar model

Call:
spml(formula = eq1.sc, data = data, listw = cont.listw, model = "within",
    effect = "individual", lag = TRUE, spatial.error = "b",
    rel.tol = 0.0000000000000000000000000000000000000002)

Residuals:
    Min.  1st Qu.   Median 3rd Qu.    Max.
-1.11463 -0.10091 -0.00162 0.09100 1.48698

Spatial error parameter:
  Estimate Std. Error t-value Pr(>|t|)
rho 0.2647     0.0868    3.05   0.0023 **

Spatial autoregressive coefficient:
     Estimate Std. Error t-value Pr(>|t|)
lambda -0.190     0.098   -1.94   0.052.

Coefficients:
        Estimate Std. Error t-value          Pr(>|t|)
x1.sc -0.07113    0.01056   -6.74 0.00000000001627 ***
x2.sc -0.00601    0.02137   -0.28          0.77835
x3.sc  0.01547    0.00402    3.85          0.00012 ***
x4.sc -0.06647    0.02294   -2.90          0.00376 **
x5.sc -0.14260    0.02378   -6.00 0.00000000201694 ***
x6.sc  0.04005    0.01492    2.68          0.00729 **
x7.sc -3.90307    0.55373   -7.05 0.00000000000181 ***
x8.sc  2.22776    0.30884    7.21 0.00000000000055 ***
---
Signif. codes: 0 '***' 0.001 '**' 0.01 '*' 0.05 '.' 0.1 ' ' 1
```

In testing existing relationships, to choose the best model, it is worth using the available tests. The BSJK test, run by the **bsjktest()** command, has two versions: version C1, which conditionally assumes that there is a spatial relationship in error provided that random effects and serial autocorrelation (in groups by units) are assumed, and version J, assuming that there is one of the phenomena: spatial dependence in error, random effects or serial autocorrelation. The printouts show that in both tests it is necessary to assume H_1 about the existence of the tested phenomena. This means that the model requires diagnostics of residuals and finding a way to filter the spatial relationship.

```
# BSJK test - version C1 (conditional)
bsjktest(eq1, data=data, listw=cont.listw, test="C.1")
```

```
        Baltagi, Song, Jung and Koh C.1 conditional test
data: y ~ x1 + x2 + x3 + x4 + x5 + x6 + x7 + x8
LM = 18, df = 1, p-value = 0.00002
alternative hypothesis: spatial dependence in error terms, sub RE and serial
corr.
```

```
# test BSJK - wersja J (join, łączna)
bsjktest(eq1, data=data, listw=cont.listw, test="J")
```

```
        Baltagi, Song, Jung and Koh joint test (J)
data: y ~ x1 + x2 + x3 + x4 + x5 + x6 + x7 + x8
LM = 21063, df = 3, p-value <0.0000000000000002
alternative hypothesis: random effects or serial corr. or spatial dependence
in error terms
```

Other tests within the **bsktest()** command are random effects test (LM1), test for spatial autocorrelation (LM2) and the test for both effect (LMH). As the results show, in each case, there are no grounds for rejection of H_1, which confirms that the phenomena studied occur.

```
bsktest(eq1, data=data, listw=cont.listw, test="LMH", standardize=TRUE)
```

```
        Baltagi, Song and Koh LM-H one-sided joint test
data: y ~ x1 + x2 + x3 + x4 + x5 + x6 + x7 + x8
LM-H = 21048, p-value <0.0000000000000002
alternative hypothesis: Random Regional Effects and Spatial autocorrelation
```

```
bsktest(eq1, data=data, listw=cont.listw, test="LM1", standardize=TRUE)
```

```
        Baltagi, Song and Koh SLM1 marginal test
data: y ~ x1 + x2 + x3 + x4 + x5 + x6 + x7 + x8
SLM1 = 146, p-value <0.0000000000000002
alternative hypothesis: Random effects
```

```
bsktest(eq1, data=data, listw=cont.listw, test="LM2", standardize=TRUE)
```

```
        Baltagi, Song and Koh LM2 marginal test
data: y ~ x1 + x2 + x3 + x4 + x5 + x6 + x7 + x8
SLM2 = 14, p-value <0.0000000000000002
alternative hypothesis: Spatial autocorrelation
```

The best-known panel econometric test, the Hausman test, is also available for spatial models. The interpretation is according to typical principles, where H_1 suggests FE and H_0 suggests RE. The test is performed with the **sphtest()** command. The result subsequently suggests that the fixed effects model will be better.

```
sphtest(eq1, data=data, listw=cont.listw, spatial.model="error",
method="GM")
```

```
        Hausman test for spatial models
data: x
chisq = 30, df = 8, p-value = 0.0001
alternative hypothesis: one model is inconsistent
```

Similar to cross-sectional models, it is possible (and necessary) to determine impacts when there is a spatial lag of the dependent variable y in the model. Their interpretation is the same as in cross-sectional models. In the result subsequently, for most impacts, the value and sign of the direct effect is the same as the indirect effect, suggesting that 50% of the effects are internalised.

```
# impacts with simulation
W.c<-as(as_dgRMatrix_listw(cont.listw), "CsparseMatrix")
trMat<-trW(W.c, type="mult")
imp<-impacts(model.spml, tr=trMat, R=20000)
summary(imp)
```

```
Impact measures (lag, trace):
     Direct Indirect    Total
x1 -1.1908  -1.2692  -2.4601
x2 -0.0016  -0.0017  -0.0034
x3  0.3611   0.3848   0.7459
x4 -5.7245  -6.1013 -11.8259
x5 -7.1730  -7.6451 -14.8180
x6  9.9444  10.5989  20.5433
x7  0.3486   0.3716   0.7202
x8  0.0456   0.0486   0.0942
========================================================
Simulation results (variance matrix):
Direct:

Iterations = 1:20000
Thinning interval = 1
Number of chains = 1
Sample size per chain = 20000

1. Empirical mean and standard deviation for each variable,
   plus standard error of the mean:

        Mean       SD   Naive SE Time-series SE
x1 -1.19063 0.114632 0.00081057     0.00081057
x2 -0.00164 0.000406 0.00000287     0.00000289
x3  0.36014 0.089370 0.00063194     0.00064769
x4 -5.70541 1.661001 0.01174505     0.01174505
x5 -7.15893 1.194376 0.00844552     0.00844552
x6  9.94662 2.369132 0.01675229     0.01662090
x7  0.34693 0.276717 0.00195668     0.00195668
x8  0.04558 0.010796 0.00007634     0.00007592
```

```
2. Quantiles for each variable:

        2.5%       25%       50%       75%       98%
x1 -1.41573 -1.26795 -1.19059 -1.11390 -0.965293
x2 -0.00246 -0.00191 -0.00164 -0.00137 -0.000841
x3  0.18631  0.29984  0.36012  0.42069  0.533767
x4 -8.96468 -6.83042 -5.70495 -4.59289 -2.433215
x5 -9.48380 -7.96645 -7.16313 -6.35236 -4.816475
x6  5.35647  8.34310  9.93655 11.54115 14.665007
x7 -0.18912  0.16080  0.34220  0.53213  0.896992
x8  0.02441  0.03833  0.04552  0.05273  0.067026
==============================================================
Indirect:

Iterations = 1:20000
Thinning interval = 1
Number of chains = 1
Sample size per chain = 20000

1. Empirical mean and standard deviation for each variable,
   plus standard error of the mean:

       Mean       SD      Naive SE Time-series SE
x1 -1.28246 0.22345 0.00158001      0.00158001
x2 -0.00177 0.00051 0.00000361      0.00000361
x3  0.38806 0.11285 0.00079799      0.00079799
x4 -6.14571 2.01579 0.01425376      0.01425376
x5 -7.71253 1.71890 0.01215447      0.01177918
x6 10.71377 3.00735 0.02126519      0.02137231
x7  0.37429 0.30570 0.00216166      0.00216166
x8  0.04907 0.01361 0.00009626      0.00009570

2. Quantiles for each variable:

        2.5%       25%  50%  75%  98%
x1  -1.76885 -1.42080 -1.26578 -1.12529 -0.893652
x2  -0.00287 -0.00209 -0.00173 -0.00141 -0.000852
x3   0.18723  0.30954  0.38136  0.45958  0.626348
x4 -10.49423 -7.39840 -6.02802 -4.75088 -2.520203
x5 -11.44931 -8.77173 -7.59060 -6.50912 -4.700629
x6   5.33736  8.60314 10.50479 12.61525 17.196390
x7  -0.20558  0.17015  0.36210  0.56931  1.007415
x8   0.02468  0.03959  0.04815  0.05770  0.077916
==============================================================
Total:

Iterations = 1:20000
Thinning interval = 1
Number of chains = 1
Sample size per chain = 20000
```

```
1. Empirical mean and standard deviation for each variable,
   plus standard error of the mean:

        Mean        SD      Naive SE Time-series SE
x1  -2.47309 0.301925 0.00213493       0.00213493
x2  -0.00341 0.000883 0.00000624       0.00000624
x3   0.74820 0.194945 0.00137847       0.00140289
x4 -11.85112 3.572242 0.02525957       0.02525957
x5 -14.87146 2.732814 0.01932391       0.01932391
x6  20.66039 5.172472 0.03657490       0.03678780
x7   0.72122 0.578614 0.00409142       0.00409142
x8   0.09465 0.023453 0.00016584       0.00016511


2. Quantiles for each variable:

         2.5%       25%        50%        75%        98%
x1  -3.10946  -2.66535  -2.45834  -2.26409 -1.92337
x2  -0.00523  -0.00398  -0.00338  -0.00281 -0.00172
x3   0.37691   0.61523   0.74435   0.87939  1.14149
x4 -19.08509 -14.21592 -11.76516  -9.42112 -5.02028
x5 -20.50477 -16.65450 -14.79930 -12.99992 -9.74196
x6  10.86676  17.09803  20.52456  24.07600 31.12299
x7  -0.39520   0.33267   0.70971   1.10560  1.88048
x8   0.04967   0.07882   0.09409   0.11009  0.14172
```

Notes

1 Baltagi and Yang (2013) developed a version of the LM test resistant to lack of normality and heteroscedasticity of residuals.

2 Until 2010, when it was decided to choose the error or lag specification, there was a procedure for selecting the appropriate form of spatial relationship when all LM tests are significant. There were several approaches called specification strategy. The first approach (according to Haining, Getis, Griffith and Tiefelsdorf) recommended filtering the variables to be free from spatial dependencies and estimation of such a model by the OLS. The so-called approach classic recommended starting the OLS estimation, performing LM tests and choosing the model with the most relevant test. Hedry's approach recommended the first estimation with a very general model containing spatial effects and gradual reduction of the model based on significance. Per Florax, Raymond, and Nijkamp (2003), it can be proved by the Monte Carlo method that the classic approach is better than the Hendry approach. However, there is no comparison with the filtering method (first approach) (Florax & Nijkamp, 2003). Currently, both specifications can be included in the model; hence these procedures have lost their relevance.

3 In the maximum likelihood method, likelihood is the probability of obtaining specific data at the adopted parameters (coefficients) of the model (reliability depends on the parameters and data). The goal of the method is to find a set of coefficients that maximize the reliability function, that is, make the data most likely. For practical reasons, log likelihood (logLik) is used; however, it causes logLik to be negative, and the closer to zero, the better the model.

4 It is worth noting that in spatial models in the literature, there is no uniform notation. In Millo and Piras's notation, the coefficient at spatial lag is λ, and the error autocorrelation coefficient is ρ. In Elhorst's notation and spatial static models, most often these symbols are defined inversely.

5 In the applied approach, the parameter for specific effects is most commonly φ.

Geographically weighted regression – modelling spatial heterogeneity

Piotr Ćwiakowski

The models discussed in previous chapters assume the stability of parameters throughout the sample. When considering spatial phenomena, this means that over the entire area from which the sample comes, the impact of a given independent variable on the explained variable is the same. Thus, the average effect of the variable in the whole sample is estimated. This assumption simplifies the calculations and the complexity of the model, but it is often an unrealistic assumption, easy to subvert in the context of spatial phenomena. When modelling spatial point pattern data it can be suspected that parameter values of regression equation will depend on the observation location – points/areas lying close together will have similar marginal effects of observation characteristics. It can be demonstrated on the case of the housing market. Depending on whether the property (apartment) is located in the centre or in the suburbs, a specific attribute can be a huge advantage or, on the contrary, be a price-reducing factor. One can imagine that the higher the apartment is located in the centre, the better (better view from the window, skyscrapers are more prestigious than ordinary blocks of flats) while in the suburbs, it will rather disturb potential buyers. Similarly, underground parking – a huge advantage in dense buildings – and much less necessary in suburban districts.

6.1 Geographically weighted regression

The assumption of spatially varying coefficients is associated with the belief that there are homogeneous clusters in space in which model parameter values are constant. The heterogeneity of the spatial sample can be modelled using several different approaches, which have been summarised by, among others, Gargallo, Miguel, and Salvador (2018). All of them assume the existence of segments (clusters) in the sample, within which the variance is smaller and the parameters of the model are stable (the same effects for each observation). One of the most popular models in this approach is the geographically weighted regression (GWR). Its most important advantages are: flexibility (can adjust the individual model even to small clusters of points, although this has some negative consequences, about which later), lack of need to define segment affiliation in advance (the model itself chooses the optimal number of nearest neighbours) and the simplicity of the functional form – all equations are linear; that is, interpretation of each of the effects is possible and straightforward.

GWR is an adaptation of the local regression model in spatial econometrics, first formulated in the famous work of Brunsdon, Fotheringham, and Charlton (1996). The idea of local regression is to estimate value at a given point based on its immediate surroundings (nearest neighbours). The closest neighbours are determined based on the distance from the point at which the estimation is carried out. While in the classical approach, the distance between points is determined on the basis of the values of explanatory variables (X matrix), in geographical weighted regression, a more intuitive approach can be applied and the distance between the location of two points in geographical space is calculated. According to the notation adopted in the literature (Fotheringham, Brunsdon, & Charlton, 2003), the GWR model can be written as:

$$y_i = \beta_0(u_i, v_i) + \sum_{k=1}^{p} x_{i,k} \beta_k(u_i, v_i) \qquad (6.1)$$

where i is the observation number, $i = \{1, \ldots, N\}$, k is the feature number in the set, $k = \{1, \ldots, P\}$, y_i is the value of the explained variable of the ith observation, $x_{i,k}$ is the value of the kth feature of the ith observation, u_i, v_i are the geographical coordinates of the observation, and $\beta_k(u_i, v_i)$ is the value of the effect of the kth feature for given graphic coordinates.

In this model, a separate equation and distinct coefficients are estimated for each observation. In addition, each observation enters the equation with the weight, calculated on the basis of a certain function assuming the inversely proportional relationship of the weights and distance from the point. In this simple way, the a-spatial model can be used to model spatial heterogeneity, that is, spatial non-stationarity of parameters. Geographically weighted regression, originally designed for property valuation, is a very popular analysis tool in other areas, such as crime (Wheeler, 2008) and epidemiology (Wheeler & Tiefelsdorf, 2005). It has also been used to study the demographic effects of famine in Ireland (Fotheringham, Kelly, & Charlton, 2013).

The introduction of the GWR model into spatial econometrics has allowed the improvement of existing techniques of modelling spatial heterogeneity (e.g. moving window regression) with little complexity of the functional form, though correct estimation of the GWR model required that several problems be solved, among others:

- selecting the functional form to weigh the observation,
- determining by cross-validation or information criteria the optimal range (bandwidth) for the weighing function,
- solving the problem of collinearity of variables in local equations; many spatial filtering techniques are presented in the literature (e.g. Griffith, 2008b or Gargallo et al., 2018), such as L1/L2 regularisation and least-angle regression (see Wheeler, 2007, 2009; Vidaurre, Bielza, & Larrañaga, 2012) and *Bayesian spatial varying coefficient* (see Bárcena, Ménendez, Palacios, & Tusell, 2014; Wheeler, Páez, Spinney, & Waller, 2014) – in this chapter, the L1/L2 regularisation approach is presented,
- determining methods for calculating the distance between observations (Euclidean distance is the default, but not the only solution; see Lu et al., 2014, 2016).

Among the critical issues in the correct estimation of GWR, Páez, Farber, and Wheeler (2011) list collinearity, range selection for local equations (related to the previous one) and sample size (according to the simulations carried out, it should not be less than 160 observations).

In R software, GWR regression can be estimated in four packages: spgwr::, gwrr::, GWmodel:: and McSpatial::. In the first one, basic procedures were programmed to estimate the geographically weighted regression equation. Available functions allow estimating the linear equation and the generalised linear equation (available compound functions and residual distributions analogous to the **glm()** function), selecting the optimal bandwidth for local regression using cross-validation, estimating weights for observations using various kernel functions and estimating weighted descriptive statistics, including spatial statistics (local Moran's I for *gwr* class objects). However, despite the fairly recent update (the moment the book was written on CRAN, package version 29.10.2019 is available), the package does not contain many GWR extensions, described in this chapter. The gwrr:: package contains the procedures needed to diagnose and tackle collinearity in GWR models using ridge regression (also known as, L2-type regularisation) and least absolute shrinkage and selection operator regression (LASSO, also known as L1-type regularisation) – however, it has not been updated since 2013[23] and does not contain most current GWR regularisation methods. The most-developed, systematically updated and most current package is GWmodel::.[24] The most important functionalities not available in the previous two packages are mainly the ability to estimate mixed GWR, robust GWR model which takes into account the spatial as well as temporal distribution of the observations (geographically and temporally weighted regression [GWTR]), nonparametric tests of significance parameters and assumptions of GWR using Monte Carlo and bootstrap (see Gollini, Lu, Charlton, Brunsdon, & Harris, 2015). An alternative to the mentioned above approach is the McSpatial:: package, which contains tools for non-parametric spatial data analysis. Because it is a separate family of methods, with a completely different bibliography, and is less often used in the mainstream of spatial econometrics, for the sake of clarity and brevity the focus is put on packages using the more prevalent, parametric approach. This chapter presents three packages for estimation of parametric GWR – at the beginning, for simple examples, the short spgwr:: package was used; then, for presenting advanced and more recent models, the gwrr:: and GWmodel:: packages were employed.

6.2 Basic estimation of geographically weighted regression model

In the application example, poviat data for Poland (NTS4 level) from 2012 will be used. The dependent variable will be average variable salary in the poviat normalised by the average salary in the country (y). Explanatory variables are: the share of workers employed in services (x_1), percentage of unemployment (x_2), count of the companies in the poviat normalised by the average number of companies in each poviat in Poland (x_3) and the indexed population density in the poviat (x_4). The model has no theoretical basis and will serve only as an application example in which geographically weighted regression will be presented in practice.

6.2.1 Estimation of the reference ordinary least squares model

First, the reference linear regression model (OLS) of the following form was estimated:

$$y_i = \beta_0 + \beta_1 x_1 + \beta_2 x_2 + \beta_3 x_3 + \beta_4 x_4 \qquad (6.2)$$

This model should be treated only as a preliminary exploratory data analysis and the preliminary model that will be referred to later – it will be used as a benchmark for various variants of the geographically weighted regression model. The first steps are: loading data, preparing the data set for analysis and loading into memory the packages needed at this stage:

```
# loading packages and data
library(car) # car:: includes VIF statistics
mydata<-read.csv2('data_nts4_2019.csv') # loading data
mydata12<-mydata[mydata$year==2012,] # selecting data for year 2012

# Y - average salary Polish = 100% (XA14)
mydata12$y<-mydata12$XA14
# X1 share of people employed in services (XA19/XA15)
mydata12$x1<-mydata12$XA19/mydata12$XA15
# X2 unemployment rate (XA21)
mydata12$x2<-mydata12$XA21
# X3 number of companies in Poland = 100% (XA13/mean(XA13))
mydata12$x3<-mydata12$XA13/mean(mydata12$XA13, na.rm=TRUE)
# X4 indexed population density by location      (XA10/mean(XA10))
mydata12$x4<-mydata12$XA10/mean(mydata12$XA10, na.rm = TRUE)

# Linear regression model
lm.model<-lm(y ~ x1 + x2 + x3 + x4, data=mydata12)
summary(lm.model)
```

```
Call:
lm(formula = y ~ x1 + x2 + x3 + x4, data = mydata12)

Residuals:
Min 1Q Median    3Q    Max
-79.234-5.309-1.229    3.144 83.677

Coefficients:
Estimate Std. Error t value Pr(>|t|)
```

```
(Intercept) 79.2339    2.6816 29.548 < 2e-16 ***
x1   47.2289   9.6889     4.875     1.61e-06 ***
x2   -0.1977   0.1023    -1.933     0.0540.
x3   -0.5083   0.6842    -0.743     0.4580
x4    1.4555   0.7021     2.073     0.0389 *
---
Signif.    codes: 0      '***' 0.001 '**' 0.01 '*' 0.05 '.' 0.1 ' ' 1

Residual standard error: 10.99 on 375 degrees of freedom
Multiple R-squared: 0.2643,   Adjusted R-squared: 0.2564
F-statistic: 33.67 on 4 and 375 DF, p-value: < 2.2e-16
```

The estimated model has an *R*-square at a relatively low level – about 26% – two variables are significant at 5%, one at 10% and one statistically insignificant (the population density turns out to be significant). As already mentioned, the results will not be interpreted on the basis of economic theory – they are only used as an example to present the interface of R packages for modelling spatial heterogeneity and model diagnostics. In the context of the latter, subsequently it was checked whether the linear model had a problem of collinearity (variance inflation factor (VIF) were used). According to the literature, collinearity testing is a crucial diagnostic tool for GWR models.

```
# Computing VIF statistics vif(lm.model)
```

```
      x1        x2        x3        x4
2.182999 1.274839 6.362061 7.740968
```

None of the VIF statistics exceeds 10, thus collinearity is not a significant problem in this OLS model.

6.2.2 Choosing the optimal bandwidth for a dataset

Before GWR model can be estimated observations in the dataset needs to be geolocated. A non-controversial convention adopted in the literature devoted to regional analysis is the use of geographical coordinates of poviat centroids as representatives of the geographical location of the poviat. The relevant code is given subsequently:

```
# libraries
library(rgdal)
library(maptools)
library(spgwr)

# loading the map
nts4<-readOGR("#R8_0 Data", "poviats") # 380 units
nts4<-spTransform(nts4, CRS("+proj=longlat +datum=NAD83"))
# poviat centroids coordinates
crds<-coordinates(nts4)
colnames(crds)<-c("cx", "cy")
```

The most important step in estimating the GWR model is to choose the optimal range within which the nearest neighbours are defined for each location. In GWR, this parameter is called *bandwidth* and is a parameter of kernel weighing functions for determining the weight decreasing rate with increasing distance. There are two ways to define bandwidth. Either it can be fixed for each location (*fixed spatial kernel*) or be fitted in a way that for each local regression, the same percentage of observations from the sample is used, regardless of the absolute distance *(adaptive spatial kernel)*. If a fixed kernel is used, it may happen that some equations for locations in dense clusters will be calculated for hundreds or thousands of observations and others located on the periphery for very few only. On the other hand, the adaptive kernel is especially useful when the density of observations is different in different areas of space – then each model will be calculated on the same sample (uneven distribution in space is typical for geolocated points).

Since there is no theory saying what bandwidth value to choose in a particular set, it is treated as the so-called hyperparameter, that is, a parameter of the model that must be set manually based on the statistics of the model's fit to the data. In the GWR model, the optimal bandwidth is calibrated using well-known error metrics. The best bandwidth parameter value can be selected based on *mean square prediction error* (MSPE) or *root mean square prediction error* (RMSPE) calculated using cross-validation (in the *leave-one-out cross validation*, LOOCV)[1] or optimising the Akaike information criterion. LOOCV is the default, recommended option to search for optimal bandwidth. In the case of the AIC, information criterion only includes a penalty for taking into account insignificant parameters, not overfitting of the model on training data. Therefore, although theoretically the AIC allows for correctly comparing the fit of models with different numbers of parameters, the LOOCV criterion is more useful tool for optimising the value of the bandwidth parameter.

The spgwr:: package uses the **gwr.sel()** function to determine its optimal bandwidth. The regression equation should be specified in the *formula* argument and the dataset in the *data* argument (as *data.frame* or *SpatialPolygonsDataFrame*). If the *data* argument specifies an object of the *data.frame* class instead of *SpatialPolygonsDataFrame*, one must additionally supplement the *coord* argument with a coordinate matrix for each point. When estimating bandwidth, one needs to opt for specific kernel function. There are several kernel functions available in the spgwr:: package: **gwr.gauss()** (Gaussian curve, default), **gwr.bisquare()** and **gwr.tricube()**. Formulas for specific kernel functions can be found in the package documentation or in Fotheringham et al. (2003). The choice of a particular form of function has been well studied theoretically in, among other works, Abramson (1982) and Wand and Jones (1995). However, if there are no prerequisites to use a specific *kernel*, it is worth staying with the default values, which are selected by the package authors in accordance with the best practices in the field. The other way out is, of course, test everything and choose the one that best fits the data (based on the information criterion or cross-validation). The weighting function in latter approach is treated as a hyperparameter (like bandwidth). All possibilities should be taken under consideration, and the best one should be chosen in cross-validation.

The value of bandwidth also gives preliminary information about the spatial heterogeneity of the parameters in the sample. If the model does not detect significant spatial effects, the adaptive kernel will return a value close to 1 and the kernel with a constant radius a very large value (which means all or almost all points will be included in each regression with large weights). The results and meaning of the model will then be close to linear regression (but still different because of the weighting observations). The selection of the optimal radius can be compared to the *bias variance trade-off* problem known from machine learning. Deciding on a very small radius, only the closest observations will take part in the estimation of the equations, each of which will have a large impact on the estimated parameters. This can cause overfitting of the model and a large forecast error on out-of-sample observations. On the opposite, the very large radius of local regression will result in obtaining similar parameters as in linear regression; that is, it will simplify/generalise the model, reducing the risk of overfitting the model and a large variation of forecasts. Somewhere between these extremes is the optimal bandwidth value.

Because it was decided to cross-validate, one can additionally choose in the **gwr.sel()** procedure which metric will be reported in the results. If *TRUE* is selected in the *RMSE* argument, the average RSME error in the cross-validation (CV) will be printed. Otherwise, the sum of residual squares (RSS) will be presented. In general, RMSE is a better choice, since it is a more intuitive and very popular metric. In the function interface, one must also decide whether the width is to be defined in the fixed spatial kernel approach (*adapt = FALSE*) or in the adaptive spatial kernel approach (parameter *adapt = TRUE*). Below the code and results of the procedure on the data from the application example is presented.

```
# replace data.frame with SpatialPointsDataFrame
map<-SpatialPointsDataFrame(data=mydata12, coords=crds)

# Choosing the optimal number of nearest neighbors
bw<-gwr.sel(y~x1+x2+x3+x4, data=map, adapt=T, RMSE=T)
```

Adaptive	q:	0.381966	CV	score:	11.17969
Adaptive	q:	0.618034	CV	score:	11.16121
Adaptive	q:	0.763932	CV	score:	11.16143
Adaptive	q:	0.6874321	CV	score:	11.15952
Adaptive	q:	0.6887939	CV	score:	11.15966
Adaptive	q:	0.6593341	CV	score:	11.16028
Adaptive	q:	0.6766996	CV	score:	11.15962
Adaptive	q:	0.6825634	CV	score:	11.1594
Adaptive	q:	0.6828774	CV	score:	11.1594
Adaptive	q:	0.680851	CV	score:	11.15946
Adaptive	q:	0.6824914	CV	score:	11.1594
Adaptive	q:	0.6818648	CV	score:	11.15939
Adaptive	q:	0.6814776	CV	score:	11.1594
Adaptive	q:	0.6819892	CV	score:	11.15939
Adaptive	q:	0.6817169	CV	score:	11.15939
Adaptive	q:	0.6816255	CV	score:	11.15939
Adaptive	q:	0.681569	CV	score:	11.15939
Adaptive	q:	0.6816662	CV	score:	11.15939
Adaptive	q:	0.6816255	CV	score:	11.15939

The function automatically proposes subsequent values to be checked in cross-validation (proceeds in accordance with the *Golden Section Search* algorithm; cf. Greig, 1980). The algorithm at the beginning "jumps" between large and small values (for example, the fit for 38.1966% of the nearest observations is checked first, then 76.3932%) finally ends searching more thoroughly the area around 68%. The algorithm stops working if, in several subsequent iterations, the value of the error metric (RMSE) ceases to improve. The procedure returns the value of the optimal bandwidth, which for convenience should be saved to a variable (in the example, the variable is called *bw*). Next, let's check what the results will be for the fixed bandwidth (if the calculation result should be given in kilometres instead of longitude and latitude degrees, set the option *longlat = TRUE*):

```
# Choosing the optimal bandwidth
bw<-gwr.sel(y~x1+x2+x3+x4, data=map, adapt=F, RMSE=T)
```

Bandwidth:	4.203304	CV	score:	11.1723
Bandwidth:	6.794299	CV	score:	11.16863
Bandwidth:	8.395621	CV	score:	11.1688
Bandwidth:	7.449065	CV	score:	11.16866
Bandwidth:	6.654513	CV	score:	11.16864
Bandwidth:	6.893061	CV	score:	11.16863
Bandwidth:	6.860616	CV	score:	11.16863
Bandwidth:	6.862158	CV	score:	11.16863
Bandwidth:	6.861731	CV	score:	11.16863
Bandwidth:	6.861691	CV	score:	11.16863
Bandwidth:	6.861772	CV	score:	11.16863
Bandwidth:	6.861731	CV	score:	11.16863

Estimation is similar to adaptive kernel. The bandwidth values are given in the same units as the coordinates – in this case, in degrees. In the north-south dimension, Poland stretches over 649 km, that is, about 5 degrees and 50 minutes. On the east-west line, it is 689 km or 10 degrees and 2 minutes. That means, in the case of a fixed kernel only for border counties lying on the edge of the space, there is a chance to estimate local regressions on a limited dataset. All in all, more localised bandwidth was obtained using a adaptive kernel (68% sample). This will allow to fit a more spatially differentiated model. Therefore, in the example, this approach was decided for estimating the GWR model.

After presenting how the **gwr.sel()** function works, the goodness-to-fit of different kernels was checked to select the optimal kernel weighting function. Code presented below generates results for each kernel in the *fixed* and *adaptive* approach. Notice that the **bw.gwr()** function from the GWmodel:: package was used because it contains more kernels (six instead of three).

```
# Auxiliary variables
bw_adapt<-numeric(5)
bw_fixed<-numeric(5)
i=1


# A loop to find an optimal bandwidth for different kernels
for(kernel in c('gaussian', 'exponential', 'bisquare', 'tricube', 'boxcar')){
bw_adapt[i]<-bw.gwr(y~x1+x2+x3+x4, data=map, kernel=kernel, adaptive=T)
bw_fixed[i]<-bw.gwr(y~x1+x2+x3+x4, data=map, kernel=kernel, adaptive=F)
i = i + 1}


results<-data.frame (Kernel=rep(c('gaussian','exponential','bisquare','tric
ube','boxcar'),2),
    Type=c(rep('adaptive', 5), rep('constant', 5)),
    RSS=c(47322.13, 47400.55, 47334.44,47383.18,47542.08,47458.59,
        47542.08,47458.59, 47334.89, 47405.07))
```

Table 6.1 shows results based on the previous calculations. Unfortunately, the procedures only return bandwidth values, so errors must be rewritten from printouts.

The kernel with the lowest error value is the Gaussian kernel in the adaptive approach – this approach will be used in all GWR models estimated on the exemplary dataset.

Table 6.1 **Summary of LOOCV errors for different kernels**

Kernel	Type	RSS
Gaussian	adaptive	47322.13
exponential	adaptive	47400.55
bisquare	adaptive	47334.44
tricube	adaptive	47383.18
boxcar	adaptive	47542.08
Gaussian	constant	47458.59
exponential	constant	47542.08
bisquare	constant	47458.59
tricube	constant	47334.89
boxcar	constant	47405.07

Source: Own study using the GWmodel:: package.

6.2.3 Local geographically weighted statistics

When the optimal bandwidth for certain dataset is selected, local statistics and regression equations can be finally estimated. Through the **gw.cor()** procedure, local statistics can be computed, such as the mean, standard error of the mean and standard deviation in the sample using the weighting function with the estimated bandwidth. The spatial distribution of selected results is presented in Figure 6.1, and the data-generating code is shown subsequently.

```
# computing spatial statistics
stats<-spgwr::gw.cov(map, 'y', adapt=T, bw=0.6816255, gweight=gwr.gaussian,
longlat=FALSE)


# list boxes with results (e.g. average, standard deviation, standard error)
names(stats$SDF)
```

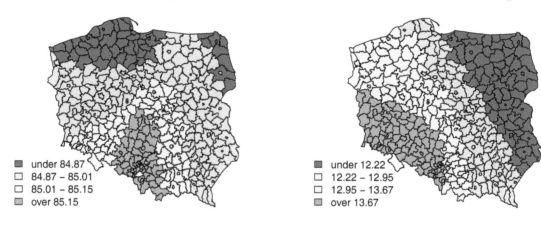

Spatial distribution of the geographically weighted mean of the average salary

Spatial distribution of the geographically weighted standard deviation of the average salary

- under 84.87
- 84.87 – 85.01
- 85.01 – 85.15
- over 85.15

- under 12.22
- 12.22 – 12.95
- 12.95 – 13.67
- over 13.67

Figure 6.1 Example of local statistics calculated in the spgwr:: package

Source: Own study using the graphics:: package

```
[1] "mean.V1" "sd.V1""sem.V1" "diff.V1" "cx"        "cy"
```

```
# Saving vectors with parameters to variables
b1<-stats$SDF$mean.V1

# Setting colors for different values in the distribution of parameters x1
brks<-c(min(b1), mean(b1) - sd(b1), mean(b1), mean(b1) + sd(b1), max(b1))

# color palette
cols<-c("steelblue4", "lightskyblue", "thistle1", "plum3")

# Graph (map, legend title)
plot(nts4, col=cols[findInterval(b1, brks)])
title(main="Spatial distribution of the geographically weighted\n mean of
the average salary")
legend("bottomleft", legend=leglabs(round(brks, 2)), fill=cols, bty="n")

# Saving vectors with parameters to variables
b1<-stats$SDF$sd.V1

# Graph (map, legend title)
plot(nts4, col=cols[findInterval(b1, brks)])
title(main="Spatial distribution of the geographically weighted\nstandard
deviation of the average salary")
legend("bottomleft", legend=leglabs(round(brks, 2)), fill=cols, bty="n")
```

The figure clearly shows the spatial correlation of the levels of the explained variable for both the weighted average and the weighted standard deviation.[2] On the other hand, the variance of the explained variable relatively small. Probably this is the reason, why the statistical model does not have a fit with the data well – small differences between poviats are difficult to explain by the deterministic features of poviats and are mostly random. Geographically weighted regression should improve the results, but one should not observe big differences in goodness to fit statistics.

6.2.4 Geographically weighted regression estimation

Model estimation (see formula 6.2) in the spgwr:: package is available through the **gwr()** procedure. Below, the call of the function for exemplary dataset and its result are presented:

```
# Estimation of the GWR model
gwr.model<-gwr(y~x1+x2+x3+x4, data=map, adapt=bw)

# Summary print
gwr.model
```

```
Call:
gwr(formula = y ~ x1 + x2 + x3 + x4, data = map, adapt = bw)

Kernel function: gwr.Gauss

Adaptive quantile: 0.6816255 (about 259 of 380 data points)

Summary of GWR coefficient estimates at data points:
                 Min.     1st Qu.    Median    3rd Qu.       Max.    Global
X.Intercept. 76.03646    77.02730  78.72201   81.11485   81.82793   79.2339
x1           41.47916    44.45846  51.42708   55.05053   58.18604   47.2289
x2           -0.30463    -0.27620  -0.20480   -0.13769   -0.10383    0.1977
x3           -0.70034    -0.60362  -0.50744   -0.41489   -0.30896   -0.5083
x4            1.11830     1.25738   1.35965    1.49669    1.60388    1.4555
```

At the beginning of the printout, one can read information about the functional form of the model, the type of kernel function weighting the observations and the percentage of nearest neighbours used to estimate local regression for each point (about 68% of observations). Next, information about the distribution of individual parameter values in the sample is given. The distribution characteristics: minimum value, first quartile, median, third quartile and maximum are reported for each parameter. The last column is the OLS estimates. Based on the results, it can be concluded that the parameters differ in the sample between different locations. For example, according to the OLS an additional 1 pp. of people employed in services cause an average increase in average remuneration (normalised) of 47.22 units *ceteris paribus*. On the other hand, GWR model indicates that in some poviats, the average increase in the share of the services sector in the employment structure by 1 pp. gives an average increase in the average salary (normalised) only by 41–42 units. However, there are also local government units where the effect is much stronger then OLS estimate and equals 57–58 units. The distribution of parameters should also be analysed spatially. Below one may find the code, which generates the maps of parameter values. The results are visualised in Figure 6.2.

```
par(mfrow=c(2, 2)) # 2x2 graphic window
cols<-c("steelblue4", "lightskyblue", "thistle1", "plum3") # color palette

# Saving vectors with parameters to variables
b1<-gwr.model$SDF$x1
b2<-gwr.model$SDF$x2
b3<-gwr.model$SDF$x3
b4<-gwr.model$SDF$x4

# Setting color ranges for different x1 parameter values
brks<-c(min(b1), mean(b1)-sd(b1), mean(b1), mean(b1)+sd(b1), max(b1))

# Graph (map, legend title)
plot(nts4, col=cols[findInterval(b1, brks)])
```

```
title(main="The average impact of the percentage of employees in the
services sector on the average wage in the poviat")
legend("bottomleft", legend=leglabs(round(brks, 2)), fill=cols, bty="n")

# Setting color ranges for different x2 parameter values
brks<-c(min(b2), mean(b2)-sd(b2), mean(b2), mean(b2)+sd(b2), max(b2))

plot(nts4, col=cols[findInterval(b2, brks)]) # Graph (map, legend title)

title(main="Impact of the unemployment rate on average wages in poviat")
legend("bottomleft", legend=leglabs(round(brks, 2)), fill=cols, bty="n")

# Setting color ranges for different x3 parameter values
brks<-c(min(b3), mean(b3)-sd(b3), mean(b3), mean(b3)+sd(b3), max(b3))

plot(nts4, col=cols[findInterval(b3, brks)]) # Graph (map, legend title)

title(main="Impact of the percentage of companies registered in the poviat
on the average remuneration in the poviat")
legend("bottomleft", legend=leglabs(round(brks, 2)), fill=cols, bty="n")

# Setting color ranges for different x4 parameter values
brks<-c(min(b4), mean(b4)-sd(b4), mean(b4), mean(b4)+sd(b4), max(b4))

plot(nts4, col=cols[findInterval(b4, brks)]) # Graph (map, legend title)
title(main="Impact of population density in the poviat on average salary in poviat")
legend("bottomleft", legend=leglabs(round(brks, 2)), fill=cols, bty="n")
par(mfrow=c(1, 1)) # Return to default settings (1 chart in 1 window)
```

By analysing each of the charts, one can observe spatial patterns in the distribution of parameters. For example, the unemployment rate has the strongest impact in the eastern part of the country and relatively the lowest in the west. On the other hand, the percentage of companies registered in the poviat has the strongest impact on the average salary in the northern voivodeships and relatively the lowest in the southern voivodeships.

6.2.5 Basic diagnostic tests of the geographically weighted regression model

In the context of the analysis of parameters variability over space, two questions appear to be relevant: 1) if the set of spatially weighted equations is really significantly better than the OLS? 2) Whether the variance of equation parameters between points in geographical space is statistically significant? Statistical tests, designed to answer both questions, and additional diagnostic tools were proposed in the article by Leung, Mei, and Zhang (2000a). The answer to the first question is given by the $F(1)$ test of goodness-of-fit. Test statistics have a Fisher-Snedecor distribution and the following formula (Leung et al., 2000a):

$$F_1 = \frac{\dfrac{RSS_0 - RSS_g}{v_1}}{\dfrac{RSS_0}{n - p - 1}}$$

(6.3)

where RSS_g is the unexplained part of the phenomenon in the GWR model, and RSS_0 is an analogous value for linear regression. Both measures are divided by the number of degrees of freedom for corresponding models (in linear regression, this is the number of observations n minus the number of parameters, and the GWR parameter is calculated in a slightly more complex way; see Leung et al., 2000a). The null hypothesis states that the models are not significantly different, with the alternative that the GWR model is better (F statistics must be significantly less than 1). Printouts from the procedure are presented subsequently – according to them, the GWR model is not significantly better than OLS.

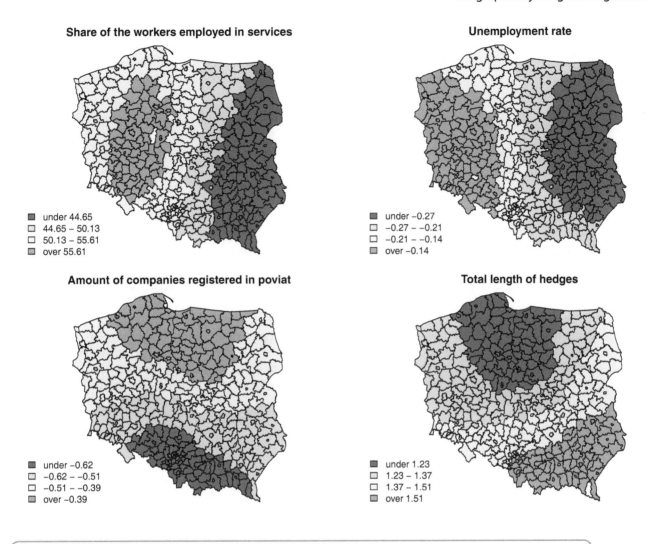

Figure 6.2 Spatial distribution of parameter values in geographically weighted regression
Source: Own study using the graphics:: package

```
LMZ.F1GWR.test(gwr.model)
```

```
        Leung et al. (2000) F(1) test

data: gwr.model
F = 0.98996, df1 = 374.26, df2 = 375.00, p-value = 0.4612
alternative hypothesis: less sample estimates:
SS OLS residuals SS GWR residuals
    45252.96      44433.58
```

Another way to test the significance of the GWR model is the $F(2)$ test:

$$F_2 = \frac{\frac{RSS_0 - RSS_g}{v_1}}{\frac{RSS_0}{n - p - 1}}$$

(6.4)

where the designations are the same as in formula 6.3, and v_1 is the number of degrees of freedom for the difference between RSS_0 and RSS_g. In other words, this time, it is tested whether the increase in explained variability is significantly better than OLS. Statistical analysis in this case clearly indicates that yes – the p-value of the test allows rejection of the null hypothesis about the statistical insignificance of spatial heterogeneity estimated by the GWR method at a significance level of 5%.

```
LMZ.F2GWR.test(gwr.model)
```

```
     Leung et al. (2000) F(2) test

data: gwr.model
F = 2.2214, df1 = 12.347, df2 = 375.000, p-value = 0.009671
alternative hypothesis: greater sample estimates:
SS OLS residuals SS GWR improvement
    45252.9613   819.3773
```

The ambiguous conclusions of both tests may be considered puzzling. The ambiguity of the results may result from, amongst other things, relatively low spatial heterogeneity of the equation parameters and therefore from a little improvement in fit using the GWR model. One of the tests ($F2$) is more sensitive, hence it reports the significance of even small improvements. The earlier publication of Brunsdon, Fotheringham, and Charlton (1999) also includes another version of the F test, namely the ANOVA test, which also examines the significance of the GWR model relative to OLS:

```
BFC99.gwr.test(gwr.model)
```

```
     Brunsdon et al. (1999) ANOVA

data: gwr.model
F = 2.2439, df1 = 142.05, df2 = 374.26, p-value = 4.975e-10
alternative hypothesis: greater sample estimates:
SS GWR improvement       SS GWR residuals
          819.3773       44433.5841
```

Another available diagnostic test for the GWR model is the test of significance of variance of estimated local parameters.[3] The null hypothesis states that the variance is statistically insignificant, and therefore parameters should be constant. Large values of the $F(3)$ statistic provide evidence for rejection of the null hypothesis in favour of the alternative, assuming the significance of parameters. The results of the statistical procedure are presented below:

```
LMZ.F3GWR.test(gwr.model)

Leung et al. (2000) F(3) test
```

	F statistic	Numerator d.f.	Denominator d.f.	Pr(>)	
(Intercept)	9.00180	67.32271	374.26	<2e-16	***
x1	5.55581	38.48513	374.26	<2e-16	***
x2	6.77117	89.64911	374.26	<2e-16	***
x3	0.69455	23.91588	374.26	0.8573	
x4	0.91317	28.84494	374.26	0.5983	

Signif.	codes: 0	'***' 0.001 '**'	0.01 '*' 0.05	'.' 0.1 ' ' 1	

Based on the results of the tests, it can be stated that the constant and f parameters are significantly non-stationary in space, whereas $x3$ and $x4$ should remain constant (in this case, mixed GWR can be used). Perhaps this is the reason (stationary of some parameters), why tests $F(1)$ and $F(2)$ give ambiguous results.

An important element of assessing the results of the model is the issue of spatial dependency between individual observations. In the article by Leung, Mei, and Zhang (2000b), Moran's I statistics for GWR were proposed, which allows checking the spatial autocorrelation of residuals in the GWR model. Similar to traditional Moran statistics, the null hypothesis assumes a lack of spatial autocorrelation. The test for the fitted model is carried out below (for this, one needs to load the spdep:: library needed to prepare the *listw* class objects). The 50 nearest neighbours were selected as the representative surroundings of the testing point:

```
library(spdep)
neib<-spdep::knearneigh(crds, k=50) # Computing the closest neighbors
neib.nb<-spdep::knn2nb(neib)
spgwr::gwr.morantest(gwr.model, spdep::nb2listw(neib.nb))
```

```
    Leung et al., 2000 three moment approximation for Moran's I

data: GWR residuals
statistic = 49.298, df = 29.404, p-value = 0.0122
sample estimates:

I 0.0160325
```

Unfortunately, the GWR model has not eliminated spatial autocorrelation (Moran statistic is significant). However, it can be seen that Moran's statistics are lower than in the case of linear regression – by modelling spatial heterogeneity, some spatial autocorrelation is also reduced (see Fotheringham et al., 2003, p. 115). Following is the Moran's I analysis for the linear regression model – therefore we can compare results between OLS and GWR:

```
spdep::lm.morantest(lm.model, spdep::nb2listw(neib.nb))
```

```
    Global Moran I for regression residuals

data:
model: lm(formula = y ~ x1 + x2 + x3 + x4, data = mydata12) weights:
spdep::nb2listw(neib.nb)

Moran I statistic standard deviate = 2.3933, p-value = 0.008349 alterna-
tive hypothesis: greater
sample estimates:
Observed Moran I    Expectation        Variance
   1.800861e-02    -3.231065e-03    7.876079e-05
```

In addition to traditional parametric tests, the GWmodel:: package includes nonparametric tests based on bootstrap and Monte Carlo simulations. In order to discuss them, the identical GWR model in the GWmodel:: package will be estimated first. Then, optimal bandwidth has to be found (almost identical to previous analysis – 259 observations, about 2/3 of the set) – and finally, the model can be estimated. The procedure has slightly more options (more weighting functions are available, and a generalised Minkowski[4] distance metric can be used instead of the Euclidean distance, which is a special case of Minkowski distance for $p = 2$).

```
# Model estimation
bw1<-bw.gwr(y~x1+x2+x3+x4, data=map, kernel="gaussian", adaptive=T)
```

```
# Model estimation with Euclidean distance
gwr.res<-gwr.basic(y~x1+x2+x3+x4,    data=map,    bw=bw1,      p=2,
adaptive=T, kernel='gaussian')

# Results printout

gwr.res
```

```
******************************************************************
*    Results of Geographically Weighted Regression     *
******************************************************************


*******************Model calibration information*******************
Kernel function: gaussian
Adaptive bandwidth: 259 (number of nearest neighbours) Regression points:
the same locations as observations are used. Distance metric: Euclidean
distance metric is used.
*****************Summary of GWR coefficient estimates:*****************
        Min. 1st Qu. Median 3rd Qu. Max.
Intercept 76.03611 77.02718 78.72194 81.11523 81.8291

  x1    41.47600    44.45828    51.42739    55.05110    58.1891
  x2    -0.30465    -0.27620    -0.20480    -0.13768    -0.1038
  x3    -0.70036    -0.60363    -0.50745    -0.41489    -0.3089

  x4    1.11827    1.25736    1.35964    1.49669    1.6039

************************Diagnostic information************************
Number of data points: 380
Effective number of parameters (2trace(S) - trace(S'S)): 8.056998 Effec-
tive degrees of freedom (n-2trace(S) + trace(S'S)): 371.943 AICc (GWR book,
Fotheringham et al., 2002, p. 61, eq 2.33): 2903.562
AIC (GWR book, Fotheringham et al., 2002,GWR p. 96, eq. 4.22): 2894.497 Re-
sidual sum of squares: 44433.5
R-square value: 0.2775847
Adjusted R-square value: 0.2618936


******************************************************************
Program stops at: 2019-06-27 00:27:36
```

The estimated model is almost identical to one computed with the spgwr:: package, while the printout of the GWmodel:: object contains much more information – amongst other things, one can read the AIC and AICc statistics for GWR. If the value of the argument *F123.test = TRUE* is also set in the **gwr.basic()** procedure, parametric diagnostic tests for *F*(1), *F*(2) and *F*(3) are obtained. The test results are identical to those obtained in the spgwr:: package.

Nonparametric tests are available through the **gwr.bootstrap()** function (test for stationarity of the parameters under four null hypotheses: OLS [multiple regression analysis, MRA], ERR [simultaneous autoregressive error model], SMA [moving average error model] and SLA [simultaneous autoregressive lag model]). The test was proposed in the article by Harris, Brunsdon, Lu, Nakaya, and Charlton (2017). In the following example, a slightly modified function is used (a minor technical modifications were introduced to adjust code to the dataset[5]). Below, the code and results with statistical interpretation are presented. Please notice that procedure lasts long even if the calculations are limited to 99 iterations.

```
source('#R8 - functions.R')
gwr.bootstrap2(y ~ x1 + x2 + x3 + x4, data=map, p=2, R=99, approach='CV',
k.nearneigh=5, adaptive=T, kernel='gaussian', longlat=F, bw=bw)
```

```
*********************************************************************
***          Modified test statistic          ***
*Comparison with a multiple linear regression model (MLR):

Modified statistic for MLR at 95% level:
Intercept  x1  x2  x3  x4
 1.28925 0.79070 0.81892 0.55273 0.6016

p value to accept null hypothese(MLR):
Intercept  x1  x2  x3  x4
   0.14 0.09 0.06 0.31 0.2

*Comparison with a simultaneous autoregressive error model (ERR):

Modified statistic for ERR at 95%:
Intercept  x1  x2  x3  x4
1.79742 1.01182 0.94518 1.00788 1.1042

p value to accept null hypothese(ERR):
Intercept  x1  x2  x3  x4
0.32 0.21 0.16 0.44 0.4

*Comparison with a moving average error model (SMA):

Modified statistic for SMA at 95%:
Intercept  x1  x2  x3  x4
1.65664 0.89852 1.08106 0.85842 0.7272

p value to accept null hypothese(SMA):
Intercept  x1  x2  x3  x4
0.36 0.22 0.14 0.47 0.42

*Comparison with a simultaneous autoregressive lag model (LAG):

Modified statistic for LAG at 95%:
Intercept  x1  x2  x3  x4
2.64993 0.94971 1.36872 1.01744 1.0063

p value to accept null hypothese(LAG):
Intercept  x1  x2  x3  x4
0.35 0.22 0.25 0.46 0.41
*********************************************************************
*** Localized test statistic ***
            Min. 1st Qu. Median 3rd Qu. Max.
Intercept_MLR_p 0.0100 0.0375 0.8300 0.9600 0.99
x1_MLR_p   0.0100 0.0400 0.0700 0.9100 0.94
x2_MLR_p   0.0000 0.0400 0.6300 0.9700 0.99
x3_MLR_p   0.0600 0.2100 0.5350 0.8300 0.94
x4_MLR_p   0.0600 0.3000 0.7900 0.8800 0.95
x1_ERR_p   0.0200 0.0900 0.1300 0.8025 0.89
x2_ERR_p   0.1300 0.2600 0.8400 0.9600 0.98
```

```
x3_ERR_p   0.4400 0.6400 0.7200 0.7600 0.84
x4_ERR_p   0.1100 0.2000 0.2500 0.3600 0.58
Intercept_SMA_p 0.0100 0.0300 0.5450 0.9500 0.97
x1_SMA_p   0.0100 0.0200 0.0800 0.8725 0.97
x2_SMA_p   0.0600 0.1500 0.7900 0.9500 0.97
x3_SMA_p   0.3500 0.6200 0.7750 0.8700 0.91
x4_SMA_p   0.0500 0.1100 0.2300 0.4500 0.74
Intercept_LAG_p 0.6600 0.8100 0.8600 0.8900 0.92
x1_LAG_p   0.0000 0.0700 0.1300 0.8800 0.97
x2_LAG_p   0.0800 0.1500 0.4100 0.8400 0.91
x3_LAG_p   0.3400 0.4700 0.6200 0.7125 0.82
x4_LAG_p   0.1400 0.3100 0.5150 0.6400 0.77
*** Note the '_p' means the p value from the localised pseudo-t statistic.
*******************************************************************
```

The results returned by the procedure are quite extensive and contain a lot of information. First, the following printouts presents results of the test for the spatial stability of each parameter, under different null hipothesis. It can be seen that each time there are no grounds for rejecting the null hypothesis, thus parameters are homogeneous. This means that the parameters are indeed geographically stable and no GWR model is needed. The second part of the printout gives pseudo-test statistics for reference models – which do not have a strict statistical interpretation.

The second non-parametric test available in the GWmodel:: package, the parameter stability test, is available in the **gwr.montecarlo()** procedure, as suggested by Fotheringham et al. (2003, p. 93). The Monte Carlo test is based on repeated permutations of geographical coordinates in the dataset and re-estimation of the model – if the parameter variance obtained in the original set is in the critical areas of the distribution of statistics obtained as a result of the set of experiments, then the heterogeneity of the variable is considered statistically significant. The model proposed in the application example, the Monte Carlo simulation, was repeated 1000 times and did not provide evidence of rejection of the null hypothesis in any of the four parameters:

```
gwr.montecarlo(y~x1+x2+x3+x4, data=map, nsims=999, bw=bw1, adaptive=T)
```

```
Tests based on the Monte Carlo significance test
            p-value
(Intercept)    0.270
x1          0.390
x2          0.124
x3          0.741
x4          0.819
```

6.2.6 Testing the significance of parameters in geographically weighted regression

Until now, only the significance of the variability of GWR coefficients has been tested, which is a justification of estimating a GWR model instead of model with fixed parameters. However, nothing is still known about the statistical significance of the coefficients in the GWR model, that is, whether the parameter value is significantly different from zero. Only the pseudo-statistic t for local regressions is obtained as a result of the algorithm – no statistical test has been created so far to reliably test the hypothesis on the significance of variables in the GWR model globally. The only statistical procedure is the Fotheringham and Byrne procedure (Byrne et al., 2009) developed in 2009, which addresses only Lovell's bias in repeated testing, not the statistical properties and reliability of these procedures. That is why in the article about the GWmodel:: package (Gollini et al., 2015) one can read that *t* statistics in these regressions can be treated only as an element of exploratory data analysis. Nevertheless, the Fotheringham and Byrne procedure is available in the GWmodel:: package in the **gwr.t.adjust()** function. The following is an example of this function and local test statistics with a *p*-value adjustment. The only argument of the function is the object with the estimated GWR model.

```
# Significance tests of local regression parameters with p-value correction
# by Fotheringham - Byrne
gwr.t.adjust(gwr.res)$results$fb
```

```
Intercept_p_fb x1_p_fb x2_p_fb x3_p_fb x4_p_fb
2281      0         0.027      0.263       1          0.561
2282      0         0.027      0.308       1          0.525
2283      0         0.027      0.290       1          0.543
2284      0         0.027      0.308       1          0.543
2285      0         0.027      0.299       1          0.534
2286      0         0.027      0.335       1          0.507
2287      0         0.027      0.253       1          0.579
2288      0         0.027      0.317       1          0.525
2289      0         0.027      0.272       1          0.579
```

6.2.7 Selection of the optimal functional form of the model

The issue of significance of variables is related to the selection of the appropriate functional form of the model – that is, the selection of variables and their transformations that significantly improve the fit of the model to the data. From this perspective, it can be stated that a significant variable in GWR is one whose inclusion in the model improves the information criterion.[6] The GWR models use the AIC_c. Procedure can be summarized in three steps::

$$AIC_c(b) = 2nln(\sigma)+nln(2\pi) + n\left\{\frac{n + tr(s)}{n-2-tr(s)}\right\} \tag{6.5}$$

where *b* is the bandwidth, *n* is the local sample size for the assumed bandwidth, $\hat{\sigma}$ (hat matrix) is an estimator of the standard deviation of the random error, and $tr(S)$ is the trace of the S matrix, with ($\hat{y} = Sy$). In GWR, the most common procedure for selecting variables to the model is "forward stepwise regression", which is as follows:

1   Estimate the GWR model with each explanatory variable separately.
2   Find the variable for which GWR has the best AIC_c and include this variable permanently in the model.
3   Repeat steps 1 and 2 until all variables are included in the model.

The GWmodel:: package has a procedure that performs forward selection automatically – it is called **model. selection.gwr()**. In addition, **model.sort.gwr** allows to sort generated models according to the AIC (within subsets of models with the same number of variables) and visualise the entire process on the chart (see Figure 6.3). Below is the code for generating exemplary charts.

```
# Forward regression fitting
model.sel<-model.selection.gwr(DeVar='y', InDeVars=c('x1', 'x2', 'x3',
'x4'), data=map, kernel="gaussian", adaptive=TRUE, bw=259)

# Sorting models by AICc
sorted.models<-model.sort.gwr(model.sel, numVars=4, ruler.vector= model.sel[[2]][,3])

# Save list of sorted models
model.list<-sorted.models[[1]]

# Illustration of the steps of step regression
model.view.gwr('y', c('x1', 'x2', 'x3', 'x4'), model.list=model.list)
```

View of GWR model selection with different variables

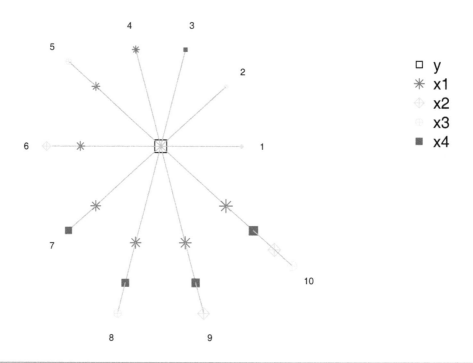

Figure 6.3 Illustration of the GWR step regression process

Source: Own study using the GWmodel:: package

The chart works well for a small number of variables. The disadvantage of visualisation from Figure 6.3 is the inability to read the AIC_c for individual models, in particular checking whether the full model is the best suggestion or it is profitable to reduce equation. Therefore, the previous interesting visualisation should be complemented by printing actual statistics. The following is a printout of sorted model statistics (according to AIC_c):

```
# Statistics print of sorted models
sorted.models[[2]]
```

	[,1]	[,2]	[,3]	[,4]
[1,]	259	2974.686	2979.612	55435.41
[2,]	259	2956.357	2961.101	52849.16
[3,]	259	2935.070	2939.903	49958.90
[4,]	259	2907.618	2912.549	46465.62
[5,]	259	2903.738	2909.934	45848.38
[6,]	259	2902.026	2908.421	45619.71
[7,]	259	2900.181	2906.462	45411.79
[8,]	259	2900.315	2907.882	45283.90
[9,]	259	2894.423	2902.185	44565.88
[10,]	259	2894.497	2903.562	44433.50

Unfortunately, the columns are not described, but according to the documentation, they are, starting from the left: bandwidth, AIC, AIC_c and RSS (sum of squared residuals). Analysing the information criteria, the conclusion is that the full model (10th row) is worse than model 9, which is the best of all estimated equations.

6.2.8 Geographically weighted regression with heteroscedastic random error

In packages devoted to geographically weighted regression, it is not possible to formally test heteroscedasticity. Estimating standard GWR model, we implicitly assume that the variance of residuals is constant throughout the sample. This assumption does not directly affect the statistical inference regarding model parameters because traditional significance tests and confidence intervals are not used (any local pseudo-statistics t are also descriptive and informative only). However, violation of this assumption may have a negative impact on the estimation of prediction error – observations in two poviats may have the same expected value but different error variance. If there is a strong suspicion regarding the instability of residuals, heteroscedastic GWR can be used, which additionally treats observations with large residuals by assigning smaller weights. The modification of the heteroscedastic GWR changes the assumption about the random component – instead of $\epsilon_i \sim N(0, \sigma^2)$, $\epsilon_i \sim N(0, \sigma^2(u_i, v_i))$ is accepted, where u_i and v_i are the geographical coordinates of the ith point (see Fotheringham et al., 2003). The model is iteratively fitted (in subsequent rounds, the weights are modified until the difference between the weight values before and after the modification is not less than the accepted tolerance threshold) and, according to the literature, converts quickly and without problems (Fotheringham et al., 2003). The heteroskedastic GWR is available in the **gwr.hetero**() procedure, which has syntax analogous to the **gwr.basic**() procedure. In this case, it can be programmed as follows:

```
gwr_het<-gwr.hetero(y~x1+x2+x3+x4, data=map, regression.points=map,
longlat=T, bw=bw, adaptive=TRUE, kernel='gaussian')
```

```
Iteration Delta
================
1    0.1307
2    0.0013
```

```
summary(gwr_het)
```

```
Object of class SpatialPointsDataFrame Coordinates:
     min     max
cx 14.32621 23.87668
cy 49.29806 54.72892
Is projected: NA proj4string : [NA] Number of points: 380 Data
attributes:
  Intercept       x1        x2        x3
 Min. :77.08   Min.  :41.26 Min. :-0.3007   Min. :-0.6697  Min. :0.9733
 1st Qu.:78.14  Qu.:1.11591st Qu.:45.10 1st Qu.:-0.2667 1st Qu.:-0.5650
 Median :79.31  Median :49.27  Median :-0.2227  Median :-0.4019 Median
 :1.2540
 Mean :79.43   Mean :48.96 Mean  :-0.2204 Mean  :-0.4081  Mean :1.2866
 3rd Qu.:80.86   3rd Qu.:52.87   3rd Qu.:-0.1791   3rd Qu.:-0.2579  3rd
 Qu.:1.4911
Max. :81.88  Max. :55.81  Max. :-0.1367  Max. :-0.1530  Max. :1.6103
```

The model converged in two iterations. Model estimates have changed only a little – this can be recognised by slightly different parameter distributions in the results.

To sum up, in the series of procedures, estimation and basic diagnostics of the GWR model were presented. Based on parametric and non-parametric tests as well as goodness-to-fit statistics (R-squared and AIC), it can be stated that the heterogeneity of parameters in the application example is weak, and the use of models with constant parameters does not involve a significant loss of fit. However, due to the complex structure of the model itself (380 linear regression equations), it is necessary to diagnose the results thoroughly. The next section presents one of the most important problems in the estimation of GWR: collinearity of covariates in local regressions of GWR and methods of tackling this issue.

6.3 The problem of collinearity in geographically weighted regression models

One of the most important problems that prevents proper interpretation of the results is collinearity (correlation) of explanatories variables. In the case of a large correlation between variables, parameter estimates are unstable. The issue of diagnostics and elimination of collinearity with GWR has been thoroughly studied in a series of publications (see Wheeler & Tiefelsdorf, 2005; Wheeler, 2007, 2009).

6.3.1 Diagnosing collinearity in geographically weighted regression

At the beginning, it is worth looking at the correlations between the parameters of individual variables (see Figure 6.4) – for this purpose, a correlogram was generated (using the GGally:: package) for local regression parameters. This graph is the basic preliminary instrument for diagnosing collinearity. Following is the code for creating the parameters and the chart itself.

```
library(GGally)
# Saving the revision coefficients as data.frame columns
wsp<-data.frame(x1=gwr.model$SDF$x1, x2=gwr.model$SDF$x2,
        x3=gwr.model$SDF$x3, x4=gwr.model$SDF$x4)
ggpairs(wsp) # Generating a correlogram
```

In Figure 6.4, one can notice a correlation between local regression parameters – some of them are linear (which is confirmed by Pearson's correlation) and non-linear (in the graph, the points are arranged in a circle/ellipse with few observations in the centre).

The correlogram obtained in the example is a typical result in GWR models (see Wheeler & Tiefelsdorf, 2005). The correlation between coefficients suggests collinearity between variables, which manifests itself in local regressions even if it is not present in the full dataset. In the immediate vicinity of the given point, there

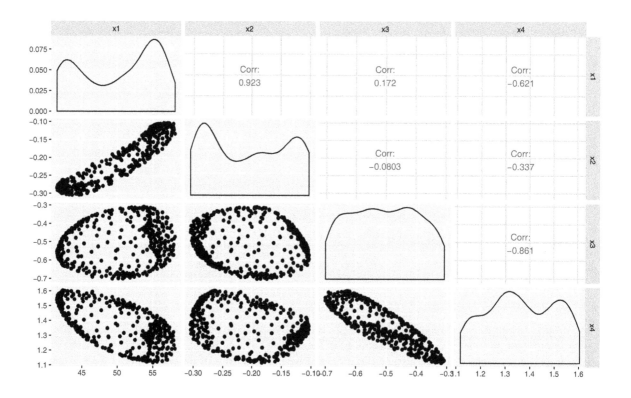

Figure 6.4 Correlation between parameters in geographically weighted regression

Source: Own study using **ggpairs()** command from GGally::

are usually points with similar characteristics, which causes the collinearity of explanatories variables of local regression. This can be detected by looking for high values of collinearity metrics calculated on local X matrices or globally, looking for high correlation between GWR regression parameter vectors. Interpretation of GWR results and drawing conclusions as to the spatial heterogeneity of the phenomenon are incorrect in these cases – the explanatory variables cease to be independent of each other.

Statistics needed to diagnose the problem are available through the **gwr.collin.diagno**() procedure, in which one should define the form of the equation and the database similar to other functions in the *formula* and *data* arguments. The local regression radius should be defined in the argument *bw* (the number of nearest neighbours, as can be seen in the code below); kernel type should be set to adaptive (*adaptive = TRUE*) and preferred kernel is gaussian. An example procedure call looks as follows:

```
library(GWmodel)

# Computing diagnostic statistics in GWR
diag_gwr<-gwr.collin.diagno(y~x1+x2+x3+x4, map, bw=floor(bw*380),
kernel="gaussian", adaptive=T)
```

The object returned by the function is a *list* object. For each of the 380 regression equations, statistics were obtained informing about the level of collinearity between variables in the GWR equations: spatial correlation matrix between variables (*corr.mat* attribute), VIF statistics (*VIF* attribute), conditional number from each regression (*local_CN* attribute) and the variance decomposition proportion (*VDP* attribute). The object also includes a *SpatialPointsDataFrame* the dataset itself.

```
summary(diag_gwr)
```

	Length	Class	Mode
corr.mat	3800	-none-	numeric
VIF	1520	-none-	numeric
local_CN	380	-none-	numeric
VDP	1900	-none-	numeric
SDF	380	-none-	numeric
			SpatialPointsDataFrame S4

The correlation matrix in each row contains a matrix of correlation between variables, calculated only for the observations used to estimate the given equation. There are no column names in the matrix, but the convention is to write the coefficients in the order in which the parameters appear in the regression equation. First, a constant with all variables $x1$, $x2$, $x3$, $x4$ (naturally, the correlation cannot be calculated because the variance of the constant is 0), then $x1$ with $x2$, $x3$, $x4$ and so on. At first glance, the correlation is very large, especially for variables $x3$ and $x4$. This explains the low significance of both variables in the model:

```
diag_gwr$corr.mat
```

	[,1]	[,2]	[,3]	[,4]	[,5]	[,6]	[,7]	[,8]	[,9]	[,10]
[1,]	NaN	NaN	NaN	NaN	-0.4502027	0.5853955	0.6689448	-0.2741994	-0.3059002	0.9249623
[2,]	NaN	NaN	NaN	NaN	-0.4505014	0.5842491	0.6684790	-0.2749180	-0.3071836	0.9229327
[3,]	NaN	NaN	NaN	NaN	-0.4503250	0.5846144	0.6686608	-0.2746240	-0.3066660	0.9237852
[4,]	NaN	NaN	NaN	NaN	-0.4506427	0.5851141	0.6674735	-0.2744164	0.9230547	-0.3064479

The content of the matrix is illustrated in Figure 6.5; each non-zero column is presented on a separate map. Local correlation of the coefficients is observed for each pair of variables. Following is the code that generates Figure 6.5.

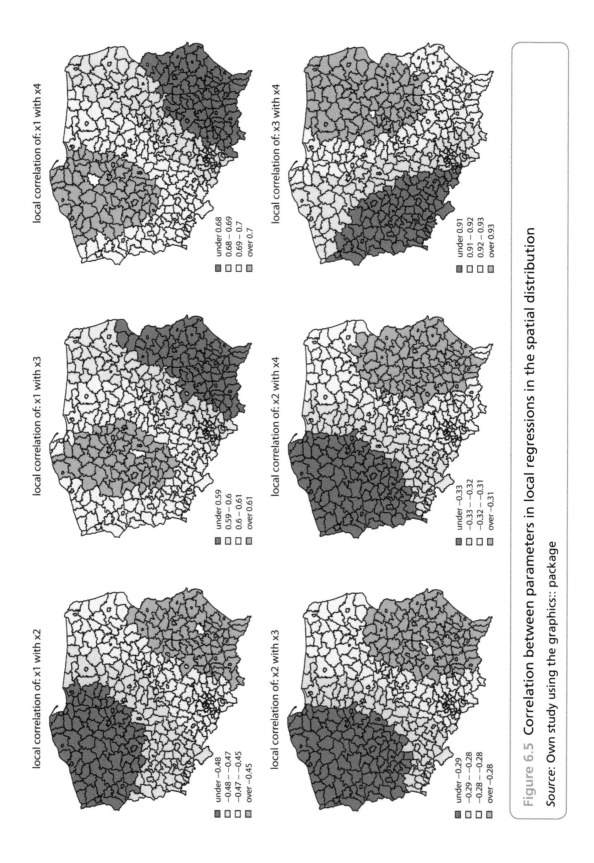

local correlation of: x1 with x2

under −0.48
−0.48 − −0.47
−0.47 − −0.45
over −0.45

local correlation of: x1 with x3

under 0.59
0.59 − 0.6
0.6 − 0.61
over 0.61

local correlation of: x1 with x4

under 0.68
0.68 − 0.69
0.69 − 0.7
over 0.7

local correlation of: x2 with x3

under −0.29
−0.29 − −0.28
−0.28 − −0.28
over −0.28

local correlation of: x2 with x4

under −0.33
−0.33 − −0.32
−0.32 − −0.31
over −0.31

local correlation of: x3 with x4

under 0.91
0.91 − 0.92
0.92 − 0.93
over 0.93

Figure 6.5 Correlation between parameters in local regressions in the spatial distribution

Source: Own study using the graphics:: package

```
par(mfrow=c(2, 3)) # 2x2 graphic window
# vector with headers for individual maps
variables=c('x1 & x2','x1 & x3','x1 & x4','x2 & x3','x2 & x4','x3 & x4')

# generating graphs in a loop
for (i in 5:10){
  print(i)
  b1<-diag_gwr$corr.mat[, i] # Saving vector with parameters
# Setting colors for different values in the distribution of parameters x1
  brks<-c(min(b1), mean(b1)-sd(b1), mean(b1), mean(b1)+sd(b1), max(b1))
  cols<-c("steelblue4", "lightskyblue", "thistle1", "plum3")# color palette
plot(nts4, col=cols[findInterval(b1, brks)]) # Graph (map, legend title)
  title(main=paste0("local correlation of:", variables[i - 4]))
  legend("bottomleft", legend=leglabs(round(brks, 2)), fill=cols, bty="n")}
par(mfrow=c(1, 1))
```

It is worth comparing the correlation distribution of variables in local regressions with the correlation across the entire set. The results of the calculations are presented subsequently – it is clear that the coefficients are stable between the models, and the problem does not increase in the GWR model – in particular, there are no "islands" with much higher collinearity, which Wheeler (2005) pointed out. Below, the correlation matrix for the whole set is computed and presented.

```
# Correlation matrix for the whole set
cor(mydata12[, c('x1', 'x2', 'x3', 'x4')])
```

	x1	x2	x3	x4
x1	1.0000000	-0.4640218	0.5942380	0.6843961
x2	-0.4640218	1.0000000	-0.2826421	-0.3179679
x3	0.5942380	-0.2826421	1.0000000	0.9168866
x4	0.6843961	-0.3179679	0.9168866	1.0000000

Another measure of collinearity is VIF statistics – the results for the first few equations are presented subsequently. Similar to the correlation coefficients, the statistics do not differ much from the results for ordinary linear regression throughout the set.

```
diag_gwr$VIF
```

	[,1]	[,2]	[,3]	[,4]
[1,]	2.086239	1.254677	7.023213	8.353976
[2,]	2.080999	1.255108	6.839153	8.142345
[3,]	2.083144	1.254859	6.915825	8.231932
[4,]	2.074401	1.255223	6.839878	8.111258
[5,]	2.079055	1.255117	6.870415	8.164133

A histogram or box graph can be used to detect alarmingly high VIF values. Alternatively, one can use simple statistical functions such as **summary**(). The following results show that despite a large correlation between variables, this measure never exceeds the critical value of 10 for any variable or equation. Only VIF for variables V3 and V4 is moderately high – same as in the correlation analysis.

```
summary(diag_gwr$VIF)
```

```
       V1               V2              V3              V4
 Min.   :2.071   Min.    :1.255  Min.    :5.713  Min.    :6.941
 1st Qu.:2.129   1st Qu.:1.262   1st Qu.:6.014   1st Qu.:7.309
 Median :2.219   Median :1.274   Median :6.571   Median :7.957
 Mean   :2.214   Mean    :1.277  Mean    :6.597  Mean    :7.998
 3rd Qu.:2.291   3rd Qu.:1.292   3rd Qu.:7.150   3rd Qu.:8.611
 Max.   :2.377   Max.    :1.310  Max.    :7.791  Max.    :9.390
```

One disadvantage of VIF is that it does not take into account the constant in the model and does not return information about the details of the correlation, for example, with which variables the given column is correlated (Bárcena et al., 2014). This information can be extracted thanks to lesser known measures of collinearity available in results of the procedure **gwr.collin.diagno**(): VDP and *conditional number* (CN; also called the *conditional index*) are obtained on the basis of the decomposition of the X matrix (matrix with explanatory variables) using the *singular value composition* (SVD) model in accordance with the Belsley proposal (1991). VDP is the share of variance of a given regression coefficient in a given component obtained from SVD decomposition. In other words, CN is the proportion of the largest and smallest singular values in a diagonal matrix in SVD decomposition. The higher the conditional index value, the greater the collinearity of variables (it is easier to aggregate them into one, the largest component). In the literature cited previously, it is assumed that if the CN value is greater than 10 (according to conservative criteria above 30), the VDP should be checked. If VDP is greater than 0.5, given observation should be treated more carefully. The tables obtained thanks to the **gwr.collin.diagno**() function provide statistics for the largest observation. Each equation exceeds the threshold 10:

```
head(diag_gwr$local_CN, 30)
```

```
          [,1]      [,2]      [,3]      [,4]      [,5]      [,6]      [,7]
 [1]   11.15615  11.12789  11.14030  11.10566  11.12144  11.09619  11.16123
 [8]   11.10051  11.14481  11.10074  11.09179  11.11788  11.13320  11.14871
[15]   11.36723  11.67495  11.70696  11.53659  11.44083  11.47025  11.77074
[22]   11.76221  11.53335  11.81808  11.27455  11.44616  11.87662  11.68834
```

VDP values for these observations are also high:

```
head(diag_gwr$VDP)
```

```
         [,1]        [,2]        [,3]        [,4]        [,5]
[1,]  0.9707929   0.5851334   0.6967117   0.05591056  0.11819541
[2,]  0.9744490   0.5810699   0.7031131   0.04823426  0.10689003
[3,]  0.9729299   0.5825705   0.7003589   0.05165994  0.11182572
[4,]  0.9767093   0.5793764   0.7107943   0.04091082  0.09685160
[5,]  0.9749275   0.5810626   0.7055839   0.04603180  0.10409105
[6,]  0.9775288   0.5764431   0.7116271   0.04019817  0.09487157
```

Given the restrictive thresholds of the VDP and CN values, there is a need to regularise the GWR model. Regularisation is a technique used in machine learning to reduce the risk of overfitting a model. This is achieved by reduction of values of certain model parameters (shrinkage). This usually results in a deterioration in the fit of the model on the training set but reduces the forecast variance, which in the case of collinear parameters can be very large – the linear regression algorithm based on the training sample cannot be "sure" how distribute the effect between two – highly correlated variables. The regularisation algorithm imposes a limit on the total size of coefficients, reducing their value. Smaller parameter values (in extreme cases equal to 0) reduce the collinearity between variables. There are a lot of regularisation algorithms that achieve this goal in different ways. The most important of them is LASSO (also called L1) and ridge regression (also called L2). Both techniques modify the linear regression loss function, only in slightly different ways. The optimisation task in LASSO (see formula 6.6) and ridge regression (see formula 6.7), in which the optimal vectors of parameters β^{L1} and β^{L2} are selected, looks as follows:

$$\beta^{L1} = \arg\min_{\beta}\left\{\frac{1}{N}\sum_{i=1}^{N}\left(y_i - \beta_0 - \sum_{k=1}^{K}x_{i,k}\beta_k\right)^2 + \lambda\sum_{k=1}^{K}|\beta_k|\right\} \tag{6.6}$$

$$\beta^{L2} = \arg\min_{\beta}\left\{\frac{1}{N}\sum_{i=1}^{N}\left(y_i - \beta_0 - \sum_{k=1}^{K}x_{i,k}\beta_k\right)^2 + \lambda\sum_{k=1}^{K}\beta_k^2\right\} \tag{6.7}$$

where *lambda* is a some positive value that controls the size of the parameters in the regression. It should be noted that for a sufficiently large lambda, the vector β will consist of only zeros. Conversely, when lambda is zero, the linear regression equation will be estimated. If the model is overfitted and the collinearity of the X matrix exists, the optimal choice will be some $\lambda > 0$. The higher the lambda, the lower the values of the model parameters – the model will be worse suited to the training data but will have less variance and a smaller error on the test set (simulating new data) – this gives hope that the equation found corresponds more closely to reality and is less overfitted (and thus better for predictions).

Both techniques reduce parameter values, starting from those that least affect the value of the loss function, but differ slightly in the manner of reducing those parameters. It is due to the design of the penalty term, and therefore the L1 and L2 algorithms are used for different purposes. Ridge regression imposes a non-linear, quadratic penalty, while LASSO has a linear cost of increasing parameter values (see formulas 6.6 and 6.7). Hence, in ridge regression, for significant variables, simultaneous gradual contraction of all parameters is usually observed. In LASSO, on the other hand, the parameters with the least significant variables are reduced first. Thus, LASSO not only reduces the risk of overfitting by reducing coefficients but is also an automatic tool for selecting variables in local regression (coefficient shrinkage and local model selection). However, in contrast to step regression (described in Section 6.1), the selection of variables is not discrete (including or excluding the variable from the model) but continuous. In the GWR approach, the reduction of collinearity should help better identify the phenomenon of spatial heterogeneity (Gollini et al., 2015). Both GWR regularisation techniques were implemented first by Wheeler (2007) introducing them into the GWR model as geographically weighted ridge regression (GWRR) and geographically weighted LASSO (GWL). In R, these models are implemented in two packages: gwrr:: and GWmodel::. GWL is only available in the gwrr:: package. However, GWRR has at least two implementations. The first version of the model has been implemented in the gwrr:: package, which allows setting only a global constant value of λ for all local regressions. Improvement of this approach was proposed by the authors of the GWmodel:: package (see Gollini et al., 2015; Brunsdon, Charlton, & Harris, 2012), where they implemented a model called geographically weighted ridge regression with locally compensated ridge term (GWR-LCR). The model according to their proposal estimates the equation adjusted with a different parameter λ for each point and only for points with a CN greater than the given level (e.g. 10 or 30). Let's look more closely at the implementation of the previous models in both packages.

GWL is available in the gwrr:: package under the **gwl.est**() procedure. In this function, the hyperparameter λ, like bandwidth is determined in the process of cross-validation.

According to the simulations carried out by Wheeler (2007), despite the interaction between hyperparameters, the procedure which determines bandwidth first and then λ gives only slightly worse results than checking different combinations of both hyperparameters simultaneously, being at the same time, a much more time-efficient procedure. Thus, the sequential algorithm was chosen in the estimation procedure of the GWL.

```
library(gwrr)  # Loading packages

gwr_lasso<-gwl.est(y~x1+x2+x3+x4, locs=crds, kernel="gauss", data=
as.data.frame(map))  # Model estimation
```

The object returned by the **gwl.est**() function has a selected bandwidth (*phi*), goodness-to-fit statistics (RMSE, RMSPE, *R*-square) and a matrix of beta coefficients in individual models. Unfortunately, the λ parameter value is not returned in the procedure. The disadvantage is also that bandwidth is estimated only in the *fixed kernel* approach.

```
summary(gwr_lasso) # Results
```

```
        Length  Class   Mode
phi        1    -none-  numeric
RMSPE      1    -none-  numeric
beta    1900    -none-  numeric
yhat     380    -none-  numeric
RMSE       1    -none-  numeric
```

The results of matching statistics and optimal *bandwidth* looks encouraging. First, the range of local regression (although constant) is smaller than in the ordinary GWR – it is 5.30 instead of 6.86, which was obtained by estimating the bandwidth for the basic GWR model.

```
gwr_lasso$phi           # Optimal bandwidth for GWL
#[1] 5.302233
gwr_lasso$rsquare       # GWL R-square
#[1] 0.4890978
gwr_lasso$RMSE          # RMSE (in-sample) GWL
#[1] 9.093664
gwr_lasso$RMSPE         # RMSE (out-of-sample) GWL
#[1] 9.51557
```

This means that the reduction of collinearity actually allowed capturing spatial heterogeneity to a greater extent – the model is also better matched to the data than a regular GWR – the *R*-squared is higher, and the RMSPE and RMSE are lower than the corresponding measures for GWR (the first one can be read with bandwidth estimation – it is over 11). The in-sample error for OLS, GWR and GWL is compared below.

```
# Model estimation with in-sample prediction option
gwr.model<-gwr(y~x1+x2+x3+x4, data=map, adapt=.68,
        hatmatrix=TRUE, predictions=TRUE)
# Comparison of in-sample prediction errors
Metrics::rmse(mydata12$y, gwr.model$SDF$pred)
#[1] 10.81322
# Linear Regression
Metrics::rmse(mydata12$y, predict(lm.model, mydata12))
#[1] 10.91269
# GWR (in-sample)
Metrics::rmse(mydata12$y, gwr.model$SDF$pred)
#[1] 10.81322
# GWL (in-sample)
Metrics::rmse(mydata12$y, gwr_lasso$yhat)
#[1] 9.093664
```

In the next step, the ridge regression model was estimated, using the GWR-LCR variant, which is more flexible and is included in the newer GWmodel:: package. Model estimation was carried out using the **gwr.lcr()** function. First of all, the bandwidth in the adaptive kernel version was once again used using the Gaussian function to weight the observation by distance (in order to compare results with GWR estimated in spgwr:: package).

```
# Bandwidth estimation
bw<-bw.gwr.lcr(y~x1+x2+x3+x4, data=map, kernel="gaussian", adaptive=TRUE)
```

```
Adaptive bandwidth(number of nearest neighbours): 242 CV score: 47328.38
Adaptive bandwidth(number of nearest neighbours): 158 CV score: 47474.96
Adaptive bandwidth(number of nearest neighbours): 296 CV score: 47343.37
```

```
Adaptive bandwidth(number of nearest neighbours): 211 CV score: 47362.27
Adaptive bandwidth(number of nearest neighbours): 264 CV score: 47323.63
Adaptive bandwidth(number of nearest neighbours): 274 CV score: 47328.81
Adaptive bandwidth(number of nearest neighbours): 253 CV score: 47329.89
Adaptive bandwidth(number of nearest neighbours): 265 CV score: 47325.19
Adaptive bandwidth(number of nearest neighbours): 257 CV score: 47324.06
Adaptive bandwidth(number of nearest neighbours): 262 CV score: 47325.09
Adaptive bandwidth(number of nearest neighbours): 259 CV score: 47322.13
Adaptive bandwidth(number of nearest neighbours): 262 CV score: 47325.09
Adaptive bandwidth(number of nearest neighbours): 263 CV score: 47325.92
Adaptive bandwidth(number of nearest neighbours): 262 CV score: 47325.09
Adaptive bandwidth(number of nearest neighbours): 262 CV score: 47325.09
Adaptive bandwidth(number of nearest neighbours): 261 CV score: 47322.71
Adaptive bandwidth(number of nearest neighbours): 261 CV score: 47322.71
Adaptive bandwidth(number of nearest neighbours): 260 CV score: 47322.38
Adaptive bandwidth(number of nearest neighbours): 260 CV score: 47322.38
Adaptive bandwidth(number of nearest neighbours): 259 CV score: 47322.13
```

After selecting the bandwidth, one can proceed to estimating the GWR-LCR model. The novel arguments, crucial for the correct estimation of GWR-LCR, are: *lambda* (possibility of setting the global lambda parameter, 0 by default), *lambda.adjust* (logical parameter: if *FALSE* – GWRR with global λ is estimated, if *TRUE* – the local model is estimated) and *cn.thresh* (mandatory parameter, if *lambda.adjust = TRUE*, threshold of the minimum conditional number for which the local parameter λ is fitted in local regression). At the beginnig, the reference GWR model without regularisation was estimated:

```
gwrr_model<-gwr.lcr(y~x1+x2+x3+x4, data=map, bw=bw, kernel="gaussian",
adaptive=TRUE, lambda.adjust=F, lambda=0)
gwrr_model
```

```
*********************************************************************
*           Package GWmodel           *
*********************************************************************
Program starts at: 2019-06-27 04:08:31
Call:
gwr.lcr(formula = y ~ x1 + x2 + x3 + x4, data = map, bw = bw, ker-
nel = "gaussian", lambda = 0, lambda.adjust = F,
adaptive = TRUE)

Dependent (y) variable: y Independent variables: x1 x2 x3 x4
*********************************************************************
*     Results of Ridge Geographically Weighted Regression     *
*********************************************************************


*******************Model calibration information*******************
Kernel function: gaussian
Adaptive bandwidth: 259 (number of nearest neighbours) Regression points:
the same locations as observations are used. Distance metric: Euclidean
distance metric is used.
Lambda(ridge parameter for gwr ridge model): 0
*************Summary of Ridge GWR coefficient estimates:*************
                Min. 1st Qu. Median 3rd Qu. Max.
Intercept 76.03611 77.02718 78.72194 81.11523 81.8291
x1  41.47600    44.45828    51.42739    55.05110    58.1891
x2  -0.30465    -0.27620    -0.20480    -0.13768    -0.1038
```

```
x3   -0.70036    -0.60363    -0.50745    -0.41489    -0.3089
x4    1.11827     1.25736     1.35964     1.49669     1.6039
***********************Diagnostic information***********************
Number of data points: 380
Effective number of parameters (2trace(S) - trace(S'S)): 8.056998 Effec-
tive degrees of freedom (n-2trace(S) + trace(S'S)): 371.943 AICc (GWR book,
Fotheringham et al., 2002, p. 61, eq 2.33): 2914.543
 AIC (GWR book, Fotheringham et al., 2002,GWR p. 96, eq. 4.22): 2914.05 Re-
sidual sum of squares: 44433.5
********************************************************************
Program stops at: 2019-06-27 04:08:32
```

Goodness-to-fit statistics and distribution of the parameters are the same as in regular GWR. The reader can now analyse the difference between a regularised and non-regularised model for his or her own. Because the procedure **gwr.lcr()** does not have a built-in cross-validation mechanism, the parameter must be set manually. In this situation, one should try different lambda parameter values and check how the AIC behaves. With higher values of lambda, parameters in the equations should shrink and the AIC should rise. The evolution of the model parameters looks slightly different if local regularisation is allowed. In this approach, only local regressions for which the conditional number statistic is greater than the defined threshold (which in the **gwr.lcr()** function is set in the argument *cn.thresh* – the threshold should be between 20 and 30) have regularised parameters in the equation. Thus lambda is set locally and varies in space between locations. As long as GWR model does not exceed this threshold in any equation, no regularisation is needed in this case.

The subject of regularisation and removal of collinearity in GWR is still being developed and discussed. Fotheringham and Oshan (2016) conducted simulations which showed that only in the case of small samples and very high correlation between variables (at the level of approximately 99%) do the results of the GWR model give inconsistent and non-interpretable results – therefore, the problem of collinearity is not in any way more severe for GWR than for other algorithms. At the same time, work is being carried out on implementation in R Elastic.Net regularisation algorithm for GWR, which would combine the advantages of LASSO and ridge regression. Elastic.Net is the generalised case of L1 and L2. The linear regression loss function is modified by the weighted average of both algorithms according to the formula:

$$y_i = \beta_0\left(u_i, v_i\right) + \sum_{k=1}^{P} x_{i,k}\beta_k\left(u_i, v_i\right) + \sum_{l=1}^{M} x_i\beta_l \qquad (6.8)$$

where the model's hyperparameters are λ (penalty term) and α (mixing parameter). It should be noted that in the case of α = 1, the equation is simplified to LASSO, and when α = 0, ridge regression is estimated. Determining $\alpha \in (0,1)$ allows for taking advantage of the properties of both forms of regularisation. Theoretical work has already been carried out. The analytical solution and empirical analysis of the model were published by Li and Lam (2018) – they named the model GWEN (geographically weighted elastic net) – unfortunately, this algorithm has not yet been introduced to R. In addition, there is relatively new work of Yoneoka et al. (2016) and Comber and Harris (2018) regarding the adaptation of Elastic.Net to geographically weighted logistic regression. The Yoneoki model has a package in R GWLelast:: available on the CRAN repository.

6.4 Mixed geographically weighted regression

Since the beginning of the research on the GWR model, a mixed generalised model was considered in which some parameters would be fixed and others determined locally (cf. Brunsdon et al., 1996, 1999). Mixed

GWR is a semi-parametric model (global and local parameters are set separately) and can be written as follows:

$$y_i = \beta_0(u_i, v_i) + \sum_{k=1}^{p} x_{i,k}\, \beta_k(u_i, v_i) + \sum_{l=1}^{M} x_i \beta_l \tag{6.9}$$

where M is the number of fixed coefficients and rest of the notation is the same as in previous formulas. Let's apply this model to exemplary dataset. Computed previously parametric tests indicated $x3$ and $x4$ stationary variables, while nonparametric tests considered all stationary. Since F tests showed a slight improvement over OLS, it is worth trying on this basis to estimate a mixed GWR model with fixed parameters $x3$ and $x4$.

The procedure needed to estimate the mixed model is called **gwr.mixed()**, and its syntax is similar to other discussed functions from the GWmodel:: package. Additionally, the *fixed.vars* argument is available, in which stationary variable names should be defined as character vector and the logical parameter *intercept.fixed*, through which it is separately determined whether the constant is to be estimated globally or locally (in the example, the application will be estimated locally, so the value of this parameter is set to *FALSE*). If the user wants to additionally calculate the fitting statistics and the complexity of the model (AIC$_c$, *Df* and residual squares), the value *TRUE* should be set in the *diagnostic* argument. Relevant codes and modelling results are presented below.

```
model.mixed<-gwr.mixed(y~x1+x2+x3+x4, data=map, fixed.vars=c('x3',
'x4'), bw=bw, diagnostic=TRUE, intercept.fixed=FALSE, adaptive=TRUE,
kernel='gaussian')
```

```
***********************************************************************
*           Package GWmodel          *
***********************************************************************
Program starts at: 2019-07-11 22:49:35
Call:
gwr.mixed(formula = y ~ x1 + x2 + x3 + x4, data = map, fixed.vars = c("x3",
"x4"), intercept.fixed = FALSE, bw = bw, diagnostic = TRUE,
kernel = "gaussian", adaptive = TRUE)

********************Model calibration information********************
Mixed GWR model with local variables : Intercept x1 x2
Global variables : x3 x4 Kernel function: gaussian
Adaptive bandwidth: 259 (number of nearest neighbours) Regression points:
the same locations as observations are used. Distance metric: Euclidean
distance metric is used.

***************Summary of mixed GWR coefficient estimates:***************
Estimated global variables :
              x3      x4
Estimated global coefficients: -0.50484 1.4486 Estimated GWR variables :
    Min. 1st Qu. Median 3rd Qu. Max.
Intercept 76.38679 77.32653 78.97028 81.13948 81.8153
x1   42.29535 44.05636 49.38435 52.66055 54.7082
x2   -0.30491-0.27618-0.20538-0.13822-0.1039
*********************Diagnostic information*********************
Effective D.F.:   6.197
Corrected AIC:   2904
Residual sum of squares:       44569
***********************************************************************
```

The AIC, without rounding, is on the results list.

```
model.mixed$aic
#[1] 2903.662
```

The first visible issue is the fact that the model almost did not deteriorate its fit – the AIC_c in the model with constants $x1$ and $x2$ is only worse by about 0.1 AIC_c. Unfortunately, the difference cannot be tested with the F test (as in Brunsdon et al., 1999), because so far, an adequate procedure has not been developed in this package.[7] Based only on the information criteria of the model stability tests (see Section 6.1), it can be stated that the simplification of the model is justified (the simpler model is only slightly worse). In addition, the distribution of local parameter $x1$ slightly changed in relation to the basic GWR (previously it had values from roughly 40 to 58, now from about 42 to 54). On the other hand, global parameters are close to linear regression, but not identical; let's recall the linear equation for comparison:

```
lm(y~x1+x2+x3+x4, data=map)
```

```
Call:
lm(formula = y ~ x1 + x2 + x3 + x4, data = map)
Coefficients:
(Intercept)         x2      x3       x4
79.2339     47.2289 -0.1977    -0.5083        1.4555
```

The mixed model is a useful tool if the theory and statistical tests suggest that some parameters are spatially stable.

6.5 Robust regression in the geographically weighted regression model

Linear regression, in particular local estimation with a small number of observations, may have biased parameter values due to outliers in the sample – the closest neighbours are selected only on the basis of geographical distance. In small local sub samples, the impact of outliers increases, so there is a risk of obtaining estimators for a specific value deviating from true parameters, although consistent and unbiased. To counteract this, local regression equations that are robust to outliers are estimated. In the case of weighted regression, it is called robust GWR. In robust GWR the removal of influential observations can occur in one of two ways: either by removing observations based on their studentised residuals[8] (in absolute terms greater than 3) or weight reduction for the largest (non-studentised) residuals in the iterative local regression fitting procedure (the weights for the largest residuals are reduced and the model is re-estimated). Robust regression has been proposed in the GWR fundamental handbook (Fotheringham et al., 2003), and later literature has attempted to use the model in practice, although there were relatively few attempts (cf. Harris et al., 2010).

In the GWmodel:: package, it is possible to estimate robust regression with the **gwr.robust()** function. With the *filtered = TRUE* parameter, the cases are filtered based on the absolute values of *studentised* residuals, and the cut-off threshold is set in the *cut.filter* argument (the default value is 3). However, if the user wants to use the iterative approach (the value of parameter *filtered = FALSE*), he must specify additional arguments: *maxiter* (maximum number of iterations, default 20), *cut1* and *cut2* (cutoff thresholds above which the observation weights are reduced[9] – defaults 2 and 3 are used for this in estimation). If user does not have expert knowledge, he/she should leave the default values of these parameters. An exemplary function call in the filtering approach for large studentised residuals is presented below – in the application example, there are no large outliers, so the results differ only slightly but nevertheless should be more stable.

```
gwr.robust(y~x1+x2+x3+x4, data=map, bw=bw, adaptive=TRUE,
    kernel='gaussian', F123.test=FALSE, filtered=TRUE, cut.filter=3)
```

```
*****************************************************************
*    Results of Geographically Weighted Regression      *
*****************************************************************

********************Model calibration information********************
Kernel function: gaussian
Adaptive bandwidth: 259 (number of nearest neighbours) Regression points:
the same locations as observations are used. Distance metric: Euclidean
distance metric is used.

***************Summary of GWR coefficient estimates:**************** Min.
1st Qu.    Median 3rd Qu.      Max.
Intercept 78.60795 79.27565 79.57950 80.40868 80.7363
x1        40.21512    42.60578 43.71079 45.88571 48.6081
x2        -0.26702    -0.25596 -0.22719 -0.20696 -0.1921
x3         0.29273     0.34116  0.41124  0.54277  0.6129
x4         0.37659     0.45231  0.53393  0.64129  0.7093
```

```
***********************Diagnostic information************************
Number of data points: 380
Effective number of parameters (2trace(S) - trace(S'S)): 7.901105 Effective
degrees of freedom (n-2trace(S) + trace(S'S)): 372.0989 AICc (GWR book,
Fotheringham et al., 2002, p. 61, eq 2.33): 2910.64
AIC (GWR book, Fotheringham et al., 2002,GWR p. 96, eq. 4.22): 2901.589
Residual sum of squares: 45272.23
R-square value: 0.2639483
Adjusted R-square value: 0.2482769
```

The model is worse fitted, than GWR – the AIC$_c$ is larger (it is 2910.64) than in the basic GWR (where it was 2603.52).

6.6 Geographically and temporally weighted regression

The concept of parameters heterogeneity in GWR modelling over space and time was initially proposed in the work of Huang, Wu, and Barry (2010). The equation for the spatially and temporally weighted regression model (*geographically and temporally weighted regression*) can be denoted as follows:

$$y_i = \beta_0(u_i, v_i, t_i) + \sum_{k=1}^{p} x_{i,k}\, \beta_k(u_i, v_i, t_i) \tag{6.10}$$

where all designations are analogous to those in the other formulas, and the additional parameter t_i relates to the moment of occurrence of the ith observation in time. Thus, the difference between GWR and GTWR is that the distance matrix is calculated after three dimensions instead of two. Of course, in order to be able to calculate the distance in time, one must define time units and calculate the numerical distances. Here, an additional problem arises – in the space-time coordinate system, two axes are measured in degrees/distances, the third, depending on the data, can be measured in years, months, days and hours. The variance of this third dimension may therefore be accidentally wrongly selected and intepreted; thus, the Euclidean distance in the distance matrix must be weighted. In the model proposed by Huang et al. (2010), the distance between points i and j is defined as follows:

$$\left(d_{ij}^{ST}\right)^2 = \lambda\left[\left(u_i - u_j\right)^2 + \left(v_i - v_j\right)^2\right] + \mu\left(t_i - t_j\right)^2 \tag{6.11}$$

In practice, to reduce the number of hyper-parameters in the cross-validation process, the μ value is set, and the λ value is permanently equal to 1 (for more detailed explanations, see Huang et al., 2010). This approach was then developed in the study by Wu et al. (2014), which used GTWR to model real estate transactions. The structure proposed in these two articles has been implemented in the GWmodel:: package. An alternative approach to the estimation of the GTWR Fotheringham model (Fotheringham, Crespo, & Yao, 2015) has also appeared in the literature. They proposed a new approach consisting of a separate estimation of spatial and temporal bandwidth; however, it has not yet been implemented in R.

In the GWmodel:: package, the GTWR model is returned by the **gtwr()** function. However, to estimate a model, one first needs to prepare panel data. Following is the code that prepares the data for analysis:

```
# Computing the time variable
mydata$time<-as.Date(as.character(mydata$year), format='%Y')

# Separating columns from the set
mydata_t<-mydata[, c('y', 'x1', 'x2', 'x3', 'x4', 'time',)]

# Coordinates vector for each observation in the panel
crds2<-do.call("rbind", replicate(10, crds, simplify=FALSE))
map<-SpatialPointsDataFrame(data=mydata, coords=crds2)
```

The next step is to find the optimal bandwidth for spatial-temporal space, using another function from the GWmodel:: package – **bw.gtwr()**. Because the procedure returned errors for *Date* and *POSIXlt type of* data it was decided to use the modified function **st.dist()** in estimation,[10] which is used in **bw.gtwr()** procedure to calculate the spatial temporal distance matrix separately and subsequently define this matrix in argument *st. dMat* while calling procedure **bw.gtwr**. The **st.dist()** function contains two additional parameters – a *lambda* (for weighing the spatial and temporal distance according to Huang's proposal (2010) and *ksi*, which was introduced to the GTWR model in Wu (2014). The example uses the default values of both parameters, which is set optimally in the cross-validation process.

```
x<-st.dist(coordinates(map), obs.tv=mydata$time, p=2, theta=0,
    longlat=F, lamda=0.05, t.units = "auto", ksi=0)

bw_st<-bw.gtwr(y~x1+x2+x3+x4, map, mydata$time, approach="CV",
    kernel="gaussian", adaptive=T, verbose=T, st.dMat=x)

Take a cup of tea and have a break, it will take a few minutes.
-----A kind suggestion from GWmodel development group
Adaptive bandwidth: 2356 CV score: 438714.5
Adaptive bandwidth: 1464 CV score: 438133.9
Adaptive bandwidth: 912 CV score: 440052.1
Adaptive bandwidth: 1804 CV score: 438014.5
Adaptive bandwidth: 2015 CV score: 438306.1
Adaptive bandwidth: 1674 CV score: 438014.5
Adaptive bandwidth: 1593 CV score: 438014
Adaptive bandwidth: 1543 CV score: 438014.2
Adaptive bandwidth: 1623 CV score: 438014.1
Adaptive bandwidth: 1573 CV score: 438014.1
Adaptive bandwidth: 1604 CV score: 438014.3
Adaptive bandwidth: 1585 CV score: 438014
Adaptive bandwidth: 1581 CV score: 438014.1
Adaptive bandwidth: 1588 CV score: 438013.9
Adaptive bandwidth: 1589 CV score: 438013.9
Adaptive bandwidth: 1586 CV score: 438013.9
Adaptive bandwidth: 1588 CV score: 438013.9
```

After finding the optimal *bw* value, one can proceed to estimate the GTWR equations and use the **gtwr()** procedure. The procedure parameters are similar to the **bw.gtwr()** function, except that in addition is necessary to provide a unique list of measurement moments for each observation in the panel (*reg.tv* argument).If the user wants to obtain information diagnostic, the *regression.points* argument must be set to *NULL* as instructed in the documentation. Below the code for estimating the model and printouts of the results are presented:

```
model<-gtwr(y~x1+x2+x3+x4, map, regression. points=NULL, mydata$time,
unique(mydata$time), bw_st, kernel="bisquare", adaptive=FALSE, p=2, theta=0,
longlat=F, lamda=0.05, t.units="auto", ksi=0, st.dMat = x)
```

```
***********************************************************************
*           Package GWmodel           *
***********************************************************************
Program starts at: 2019-07-25 23:36:04
Call:
gtwr(formula = y ~ x1 + x2 + x3 + x4, data = map, regression.points = NULL,
obs.tv = mydata$time, reg.tv = unique(mydata$time), st.bw = bw_st, ker-
nel = "bisquare", adaptive = FALSE, p = 2, theta = 0,
longlat = F, lamda = 0.05, t.units = "auto", ksi = 0, st.dMat = x)

Dependent (y) variable: y
Independent variables: x1 x2 x3 x4
Number of data points: 3800
***********************************************************************
*           Results of Global Regression           *
***********************************************************************

Call:
lm(formula = formula, data = data)

Residuals:
Min     1Q Median  3Q Max
-79.348 -5.207-1.553 3.234 91.730

Coefficients:
    Estimate Std. Error t value Pr(>|t|) (Intercept) 79.34808       0.76655
103.513 < 2e-16 ***
x1    44.33054    2.98082    14.872       < 2e-16 ***
x2    -0.25695    0.03052    -8.418       < 2e-16 ***
x3    -0.71926    0.20700    -3.475       0.000517 ***
x4    1.91673    0.21483     8.922       < 2e-16 ***
---Significance stars
Signif. codes: 0 '***' 0.001 '**' 0.01 '*' 0.05 '.' 0.1 ' ' 1
Residual standard error: 10.75 on 3795 degrees of freedom
Multiple R-squared: 0.295
Adjusted R-squared: 0.2943
F-statistic:    397 on 4 and 3795 DF, p-value: < 2.2e-16
***Extra Diagnostic information
Residual sum of squares: 438220.4
Sigma(hat): 10.7416
AIC: 28837.27
AICc: 28837.29
***********************************************************************
*  Results of Geographically and Temporally Weighted Regression  *
***********************************************************************
```

```
**********************Model calibration information*********************
Kernel function for geographically and temporally weighting: bisquare
Fixed bandwidth for geographically and temporally weighting: 1588 Regres-
sion points: the same locations as observations are used.
Distance metric for geographically and temporally weighting: A distance ma-
trix is specified for this model calibration.

****************Summary of GTWR coefficient estimates:**************** Min.
1st Qu.    Median 3rd Qu.      Max.
Intercept 77.51808 78.04941 78.90483 80.81359 81.6088
x1        38.92445    41.54890    45.24507    46.50348    48.8012
x2        -0.30704    -0.27220    -0.24765    -0.23827    -0.2258
x3        -0.89982    -0.83107    -0.69748    -0.66529    -0.6514
x4         1.60402     1.63729     1.88564     2.32509     2.4791
```

The GWmodel:: package has already accustomed its users to very extensive results. At the beginning, basic sample statistics and results of the global pooled regression model is returned. Then, at the end GTWR regression results are printed. The results of the parameters differ from those obtained for 2012 – in this way, it was possible to model time-space heterogeity in the dataset and capture changes in parameters not only in space but also in time. Unfortunately, the model at this stage does not return diagnostic information that could allow it to be compared with the linear model estimated in the full sample. Perhaps in the next version of the package this functionality will be available.

Notes

1 In LOOCV cross-validation, the model is estimated with N iterations (where N is the number of observations). In each iteration, another group of N points is excluded from the training set treated as a test set. GWR in this cross-validation variant generates N times out-of-sample prediction on each of the points. LOOCV can be approximated in linear models using statistics called general cross-validation (GCV), which significantly reduces its calculation time.

2 Patterns of these statistics are given in the book by Fotheringham, Brunsdon, and Charlton (2003).

3 The test was proposed in the publication of Leung et al. (2000a) and then simulated in the work of Mei, He, and Fang (2004).

4 The use of GWR metrics other than Euclidean has been studied by, among others, Lu, Charlton, Harris, and Fotheringham (2014) and Lu, Charlton, Brunsdon, and Harris (2016).

5 The main modification concerns the method of computing the weight matrix. The matrix needed to estimate the ERR, SMA and SLA models should be symmetrical (if A is the nearest neighbour of B, which is not the default result of the knn2nb() function for the dataset used in this chapter; however, this can be forced by setting the value of the argument *sym* = *TRUE*). The second, minor change is to define the bandwidth as a function parameter instead of computing it inside the procedure.

6 At the same time, one should not forget about the requirement for a causal relationship between the explanatory variable and the explained variable – the explanatory variable can improve the model purely by accident – the phenomenon is known in machine learning as overfitting.

7 In the GWmodel:: package, GWR model significance tests are built into functions, for example, gwr.basic() and **gwr.robust**() (provided that the functions are run with the argument *F123.test* = *TRUE*).

8 Studentised residuals are calculated for each observation separately. It is the difference between the real value of observation and its value prognosed in a model, which was calculated without this particular observation. Studentised residuals behave better than standard ones in a case of very influential observations (outliers), which in standard estimation have very small residuals (well fitted) even though they are significantly different than the dataset. If the studentised residual is high and the standard residual is small, this is a sign that this particular observation is influencing the estimates of at least part of the parameters.

9 The *cut1* and *cut2* values are then multiplied by the standard deviation of the residues – by default, large residues that are more than two standard deviations from the mean are considered, and the next threshold is set by default above three deviations.

10 The modification consisted of interference in the code of the get.ts() function, which did not allow for the use of *Date* format.

## Chapter 7

Spatial unsupervised learning

Katarzyna Kopczewska

Unsupervised learning (USL) methods do not assume any structural or hierarchical ordering of input data. Dependent and explanatory variables are not distinguished. The purpose of using USL methods is to determine the structure, distribution or common characteristics of the data, which provides for a better insight. The USL algorithms give as a result the clustering (grouping) of data.

In spatial problems, when one uses point data geolocalised in geographic space (x, y) and possibly assigned values of a characteristic (z), the default classification of variables into coordinates and characteristics appears. By nature, the data is partially structured – profiled using coordinates or values of features. There may be several approaches to analysing such data. First, it is possible to group with regard to the coordinates (x, y) only, where the spatial clusters of neighbouring localised points are detected – and further within the separated spatial clusters, the (z) values are examined. One can use the k-means, PAM and CLARA algorithms (7.1) based on the distance matrix or the DBSCAN statistics (7.2) based on the location density. Second, it is possible to group the (z) features in search of similarly multidimensional observation groups – and further to map them to determine if this similarity of features translates into spatial proximity. Spatial analysis of the principal components (7.3) or spatial drift (7.4) can be used here. Third, an approach combining data grouping and spatial clustering is possible – based on hierarchical clustering with spatial restrictions (7.5) or spatial oblique decision trees (7.6). This chapter presents these algorithms.

This chapter omits the detection of spatial disease clusters, which is based on GAM Openshaw, Besag-Newell and Kulldorff-Nagarwall statistics. This approach examines the number of cases (i.e. diseases, deaths, etc.) in a given area referred to the population (and the expected number of events) in this area. The testing of their significance uses permutation as well as the multinomial, Poisson and Poisson-gamma distributions. In R, there are DCluster::, smerc:: and SpatialEpi:: packages that contain commands for these methods. This chapter does not address spatial segregation of software in R in the seg:: and OasisR:: packages.

A good supplement to this chapter is the book titled *The Elements of Studying: Data Mining, Inference, and Prediction* (Hastie, Tibshirani & Friedman, 2017), which discusses in detail supervised and unsupervised machine learning methods, linear regression methods, regularisation, kernel smoothing, model selection and cross-validation, trees, neural networks, support vector machine, clustering based on the nearest neighbourhood criterion, random forests and graphic models – but mainly with respect to a-spatial data.

7.1 Clustering of spatial points with k-means, PAM (partitioning around medoids) and CLARA (clustering large applications) algorithms

Unsupervised (machine) learning can be used for spatial clustering of points located in specific geographic coordinates. In the traditional a-spatial approach, observations are clustered on the basis of the values of several variables, and one looks for the groups of observations with similar values of these variables. In the spatial approach, one clusters the geographical coordinates (latitude and longitude), which allows creating spatial clusters, regardless of the values of variables observed in these locations. The effect of clustering algorithms is the optimal division of points (x, y) into the indicated number of clusters (groups) based on the distance between points, where these points are concentrated around the centres of clusters. There are three aspects of these algorithms that should be discussed are the choice of the cluster's centre, the metric of the distance between the points and the optimal division into groups.

The selection of cluster centres might be twofold: 1) one can indicate a point on the surface that does not belong to the tested dataset, or 2) one can indicate the point from those existing in the dataset. In the first scenario, this is the k-means algorithm, while in the second, it is the PAM (partitioning around medoids) algorithm (also k-medoids). In the case of large datasets, the CLARA (clustering large applications) algorithm is used, which is a variant of the PAM algorithm, based on sampling to increase computational efficiency (while PAM operates iteratively throughout the entire dataset).

In both the k-means and PAM cases, the goal is to minimise intracluster variance and maximise intercluster variance. In other words, they aim to select centres so that the observations within the cluster are similar (homogenous), while the clusters are as diverse as possible (heterogeneous) between the clusters. The basic difference between k-means and PAM is the method of selecting points: optimising the selection of any k points on the plane in k-means and iteratively choosing the best combination of existing k points in PAM.

In the k-means method, the cluster locations are optimised by allocation on the plane. For the original random location of centres of k clusters, distance between points and cluster centres is determined, and successively the centres of clusters are allocated to optimal locations (x^*, y^*), regardless of whether at such points (x^*, y^*) on the examined surface there are empirical locations of points from the dataset.

In CLARA and PAM in the first phase (BUILD phase), k cluster centres are randomly selected (for k clusters) from among the available, existing and analysed points. The cost function is determined, examining the dispersion inside and outside of the separated groups. The cost function sums up the distances for each point from the nearest medoid to those points. In the second phase (SWAP), one iteratively changes the cluster points (medoids) and for these new points recalculates the cost function. Finally, the selected best cluster points minimise the cost function. The algorithm allows all points to be assigned to the assumed number of clusters around the cluster points. The use of k-means and PAM depends on the analytical needs and limitations of the solution. In the unrestricted approach, one can use k-means, while when the solutions (centroids) are to come from the tested sample, PAM is applied.

The distance between points can be measured in several ways. The most frequently used is the Euclidean distance measure, which for two points $X = (x_1, x_2, x_3, \ldots, x_n)$ and $Y = (y_1, y_2, y_3, \ldots, y_n)$ is expressed by the formula:

$$\sqrt{\sum_{i=1}^{n}(x_i - y_i)^2} \tag{7.1}$$

and is the root square of the sum of the squares of the differences of subsequent values. This distance can be calculated in a traditional way – as planar on a flat plane and in geographical terms – in a spherical manner, taking into account the curvature of the Earth, referred to as great-circle distance. It should be emphasised that the Euclidean measure is the most flexible and most commonly used, while R has implementations for both spherical and planar approaches.

Apart from the Euclidean distance, in data clustering, the urban distance (also called Manhattan distance) can be used, which is given by:

$$\sum_{i=1}^{n}|x_i - y_i| \tag{7.2}$$

and is the sum of the absolute value of the differences of subsequent dimensions. This distance is usually computed in the Cartesian (planar) system, and there is no simple spherical equivalent. In the planar system, Minkowski's distance metrics are also available, given the formula:

$$\sum_{i=1}^{n}\left(|x_i - y_i|^p\right)^{1/p} \tag{7.3}$$

which is the generalisation of Euclidean and Manhattan metrics, depending on the selected p, which can be natural or decimal number as well as positive or negative.

There is also the Canberra distance, which is given by the formula:

$$\sum_{i=1}^{n} \frac{|x_i - y_i|}{|x_i| + |y_i|} \tag{7.4}$$

It should be emphasised that the interpretation of distances other than Euclidean is not always obvious. Most clustering algorithms are sensitive to the selection of distance metrics, and changing the distance determination method can significantly affect the outcome of grouping.[1]

The *quality of clustering* can be examined with different measures. The most popular and commonly used is the silhouette statistic, which is used in the interpretation and checking of consistency within data clusters. It exists in two versions: as an individual statistic S_i and as an average of individual statistics S. The individual statistic S_i is given by the formula:

$$S_i = \frac{(b_i - a_i)}{\max(a_i, b_i)} \tag{7.5}$$

where a_i is the average distance from the point to all other objects in the cluster, and b_i is the minimum average distance from the point to other clusters (tested for each cluster separately). The total statistics of the S silhouette are given as:

$$S = \frac{\sum_{i=1}^{n} S_i}{n} \tag{7.6}$$

S_i and S statistics are limited: $s \in [-1,1]$. Negative values of the silhouette statistics are undesirable, because it means that $a_i > b_i$, so the remaining clusters are closer than the point's own cluster. On the contrary, positive values of the silhouette statistic are desirable. The optimal value of the S_i and S statistics is close to 1 ($s \sim 1$), which occurs when the distance between the observation and the middle point in the point's own cluster is minimal. In the interpretation, one looks for the highest values of the silhouette statistic for a different number of clusters k.

Another measure is the gap statistic, which compares the variance within the established clusters with the expected variance for the random distribution generated by sampling as part of the Monte Carlo simulation. The statistic was proposed by Tibshirani, Walther, and Hastie (2001). As a reference distribution for variable x, a vector from the uniform distribution is generated within the limits of [*min (x), max (x)*]. The gap statistic is given as:

$$Gap_n(k) = E_n^* \{\log(W)_k\} - \log(W)_k \tag{7.7}$$

where W_k, the combined intracluster sum of squares around the cluster average, is defined as:

$$W_k = \sum_{r=1}^{k} \frac{1}{2n_r} D_r \tag{7.8}$$

for the C_r clusters, n observations and squared Euclidean d distances:

$$D_r = \sum_{i,i' \in C_r} d_{i,i'} \tag{7.9}$$

W_k is determined for both the tested variable x and the reference variable x', which is a vector from the uniform distribution within the limits of [min (x), max (x)]. The optimal number of k clusters is in case of the biggest discrepancy (gap) between the expected values $E_n^* \{\log(W_k)\}$ (from the theoretical values) and the observed $\log(W_k)$ (from the empirical values) of the squares of the distance. The extension of this statistic is the weighted gap (Yan & Ye, 2007).

Another measure of the quality of clustering is the Hopkins H statistic, which checks the tendency to group data in more clusters. This is a statistic that checks whether a given cluster can be divided into subgroups. The null hypothesis (H_0) assumes that the dataset is evenly distributed and there are no significant clusters. The alternative hypothesis (H_1) assumes that the dataset is not evenly distributed and contains significant clusters

(H_1). The $h \sim 0.5$ statistic means that the data is from a random distribution. The separation of statistic h from the value of 0.5 means that data is ordered, either uniformly or, by contrast, clustered. The most common implementations assume that $h \sim 1$ values are for the grouped points and $h \sim 0$ for evenly distributed points. The implementation in R is inverse ($H = 1 - h$), which means that values of h below 0.5 are used for the grouped data and above 0.5 for the even distribution. The further h is from 0.5, the stronger the phenomenon. Statistics compares the total distances between the nearest actual neighbours – a pair of real points (w_j^d) and distances between hypothetical nearest neighbours – pairs of real points and uniformly, randomly distributed points u_j^d:

$$h = \frac{\sum_{j=1}^{n} u_j^d}{\sum_{j=1}^{n} u_j^d + \sum_{j=1}^{n} w_j^d} \tag{7.10}$$

where $\sum_{j=1}^{n} u_j^d$ is the average distance to the hypothetical nearest neighbour – the distance between the real point and the evenly generated random point (with the same variance as the actual data), and $\sum_{j=1}^{n} w_j^d$ is the average distance to the actual nearest neighbour – the distance between the actual data. As the statistic is based on randomly generated data, its values in iterations may vary.

The effect of the clustering procedure is the clustering vector, in which the elements are natural numbers from 1 to k, where k is the number of clusters. Each observation – a geolocation point – obtains the assigned cluster number. Membership in this cluster optimises the given measure of the quality of division into clusters, for example, silhouette.

In R software, k-means clustering can be executed with the command **kmeans**() from the stats::, where the first argument gives the data and the second is the number of clusters. PAM can be performed with the **clara**() and **pam**() commands from the cluster:: package, respectively.[2]

Alternatively, clustering can be performed with the command **eclust**() from the factoextra:: package (Kassambara & Mundt, 2017), specifying the type of algorithm used in the

FUNcluster option (*"kmeans"*, *"pam"*, *"clara"*, *"fanny"*, *"hclust"*, *"agnes"*, *"diana"*) and the distance measure in the *hc_metric* option (*"euclidean"*, *"manhattan"*, *"maximum"*, *"canberra"*, *"binary"*, *"minkowski"*, *"pearson"*, *"spearman"*, *"kendall"*). The *"hclust"*, *"agnes"* and *"diana"* algorithms belong to the hierarchical group, while *"fanny"* generates fuzzy clustering, in which each observation is assigned to several clusters, and is also available in the **fanny**() command from the cluster:: package. Distances *"pearson"*, *"spearman"* and *"kendall"* are based on appropriate correlations between variables and are used as input information in hierarchical clustering.

The optimal number of clusters can be determined using silhouette statistics. *Ex ante*, the number of clusters can be examined with the **fviz_nbclust**() command from the factoextra:: package, which shows global silhouette statistics for different numbers of clusters. *Ex post*, for a given subdivision, local silhouette statistics can be examined by the **fviz_silhouette**() command whose argument is an *eclust* class object. It happens that the differences between the n and $n + 1$ divisions of clusters are negligible – it is worth using the theoretical justification for a given number of clusters, which will be chosen for further analysis.

Gap statistics are available as the **clusGap**() command from cluster::, as well as the **fviz_gap_stat**() command from the factoextra:: package.

Example

The example subsequently uses point data (x, y) regarding the location of firms in the Lubelskie voivodeship (*firms*). Without preliminary assumptions, it was attempted to group geographical coordinates into five clusters, assuming the optimisation (maximisation) of the silhouette statistics. Due to the large size of the set (about 37,000 observations), a CLARA algorithm based on sampling and dedicated to large datasets was used. It is available in the **clara**() command from the cluster:: package and allows two different distance metrics (*metric* = *"euclidean"* or *"manhattan"*) to be used. Two parameters are determined in sampling: sample size (*sampsize* option), which should be larger than the number of clusters (k), and the number of samples (*samples* option), default 5, although it is recommended to use more samples. Of course, increasing both parameters increases the quality of calculations but at the cost of extending the operation time. The **clara**() command results are in a *clara* class object, which can be an argument for the commands from the factoextra:: package: **fviz_cluster**() which draws the clustering results and **fviz_silhouette**() which displays the individual silhouette statistics (see Figure 7.1). The graphics from the factoextra:: package are based on the functionality of the ggplot2:: package.

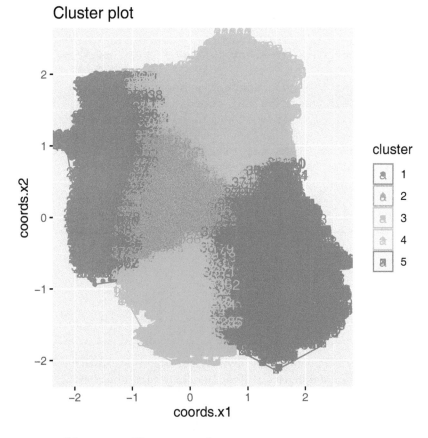

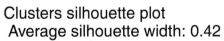

Figure 7.1 Spatial visualisation of clusters (*k* = 5) and silhouette statistics for this division

Source: Own study using R and cluster:: and factoextra:: packages

```
library(cluster)
library(factoextra)

firms<-read.csv("geoloc_data_firms.csv", header=TRUE, dec=",", sep=";")

# clustering points in the geographical space
# columns 12:13 contain geographical coordinates
c1<-clara(firms[,12:13], 5, metric="euclidean", sampsize=1000)

# alternative command of CLARA algorithm
c2<-eclust(firms[,12:13], k=5, FUNcluster="clara") # factoextra::
c1$clustering # clustering vector
fviz_cluster(c1) # graphics of division into clusters, factoextra::
fviz_silhouette(c1) # silhouette statistics, factoextra::
```

```
     cluster    size    ave.sil.width
1       1        188         0.30
2       2        282         0.54
3       3        120         0.44
4       4        159         0.33
5       5        251         0.41
```

From the previous result, it can be seen that the division into arbitrarily chosen $k = 5$ clusters results, according to individual statistics, is the not optimal classification of only a few observations – negative values of individual silhouette statistics (Figure 7.1b). The horizontal dotted line in Figure 7.1b is the average of individual S_i, the global silhouette S statistic, and is 0.44 (as described at the top of Figure 7.1b), which means that the quality of clustering is acceptable. The separated clusters are spatially coherent but not equally sized. The least number of observations is in cluster 3, while the highest number is in cluster 2. This results from the spatial distribution of the analysed points: relatively less frequently located in the southern part of the area and showing a strong spatial agglomeration in the central part of the studied area. The algorithm examines the sum of the distances from the centre; hence in an equilibrium view, a larger number of points close to each other is equivalent to a smaller number of spatially distributed points. It is worth noting that the result of the **fviz_silhouette**() command for the CLARA algorithm shows the cluster silhouette and size only for the selected subsample $n_i = 1000$ elements, while the **fviz_cluster**() command shows the extrapolation of clustering to the full dataset.

For a more complete picture of the quality of division into clusters, a graph of global silhouette statistics for a different number of clusters should be drawn up, and the clusters' divisibility should be assessed by Hopkins statistics, as shown subsequently (see Figure 7.2). This is a pre-analysis, independent of the clustering result obtained previously. The **fviz_nbclust**() command works for PAM, CLARA, k-means and other algorithms.

```
# global silhouette statistics chart, command from factoextra::
fviz_nbclust(firms[1:5000,12:13], clara, method="silhouette")

# Hopkins statistics chart, factoextra:: package
get_clust_tendency(firms[1:1000,12:13], 2, graph=TRUE,
gradient=list(low="red", mid="white", high="blue"), seed=123)
```

```
# $hopkins_stat
[1] 0.6039652
```

The chart of silhouette statistics for a different number of clusters (Figure 7.2a) shows that the optimal number of clusters is $k = 2$, the worst is $k = 4$, and $k = 8,9,10$ are also acceptable. Despite the dedicated CLARA algorithm for large datasets, the **fviz_nbclust**() command, which displays the silhouette statistic for a different number of k clusters, does not work well for large samples, and it is necessary to use the subsample. The same is true for the command **get_clust_tendency**(), which examines the divisibility of clusters and determines the Hopkins statistic – it was necessary to reduce the set to the subsample $n_i = 1000$ observations. The *seed* option (here *seed = 123*) allows for choosing a quasi-random distribution used in calculating Hopkins statistics, which

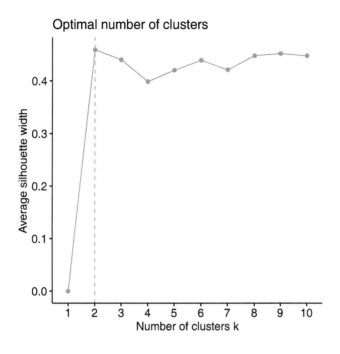

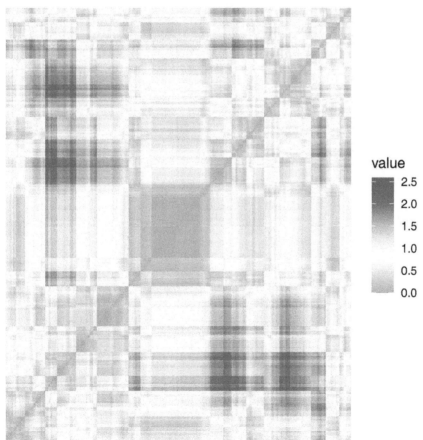

Figure 7.2 Silhouette and Hopkins statistics in the diagnostics of clustering quality

Source: Own study using R and cluster:: and factoextra:: packages

guarantees the replicability of the result. A Hopkins statistic equal to $H = 0.60$ is quite high, which means that the data is clustered and non-random. This is confirmed by the distance chart (Figure 7.2b), which has two colours: red, assigned to low values on the scale, and blue, assigned to high values on the scale. Low distance values mean a low value of impossibility, which is interpreted as high similarity. High distance values (on the scale) mean a high value of dissimilarity or low similarity. The chart is ordered according to the similarity of objects. When colour blocks are visible in the chart, the data shows a tendency to cluster what is happening in this situation.

In addition to silhouette and Hopkins statistics, the *gap statistic* can be determined, which also assesses the quality of clustering and allows setting the optimal number of clusters. Calculations and graphics codes are presented subsequently.

```
# gap statistic
# subset of 5000 obs., max. 10 clusters and 5 iterations
gap<-clusGap(firms[1:5000,12:13], FUN=kmeans, K.max=10, B=5)
gap
```

```
Clustering Gap statistic ["clusGap"] from call:
clusGap(x = firm[1:5000, 12:13], FUNcluster = kmeans, K.max = 10,    B = 5)
B=5 simulated reference sets, k = 1..10; spaceH0="scaledPCA"
--> Number of clusters (method 'firstSEmax', SE.factor=1): 2
            logW      E.logW        gap         SE.sim
 [1,]     6.972769   7.335051   0.3622819    0.002736380
 [2,]     6.626303   7.012408   0.3861042    0.004675658
 [3,]     6.438766   6.828529   0.3897631    0.003873580
 [4,]     6.289185   6.638857   0.3496720    0.003163805
 [5,]     6.190010   6.529624   0.3396138    0.004503275
 [6,]     6.073661   6.429847   0.3561857    0.005569402
 [7,]     5.967851   6.364239   0.3963880    0.009388580
 [8,]     5.928411   6.297622   0.3692110    0.005159287
 [9,]     5.818223   6.232120   0.4138975    0.009702025
[10,]     5.715451   6.175538   0.4600862    0.009318130
```

```
# gap plot with the default selection criterion: firstSEmax
fviz_gap_stat(gap)
```

```
# gap chart when changing the selection criterion to globalmax
fviz_gap_stat(gap, linecolor="red", maxSE=list(method="globalmax"))
```

When analysing the gap statistics, one looks for the largest difference between the expected values $E_n^*\{\log(W_k)\}$ and the observed $\log(W_k)$ of the squares of the distance. The **clusGap()** command from the cluster:: package presents both component statistics as well as the gap itself and its standard error (remember that the expected value is determined based on the bootstrap). In the previous example (see Figure 7.3), the maximum gap value appears for the maximum assumed number of clusters ($k = 10$) and can be higher for $k > 10$, while $k = 5$ clusters are the least optimal. Using the gap graphics, generated with the **fviz_gap_stat()** command from the factoextra:: package, one must specify the criterion of finding the maximum gap in the options. It should be emphasised that algorithms that are based on distances counted in pairs are quite slow in large sets. In the previous case, for the feasibility of calculations, only a subsample of 5000 elements was used.

The graphics of points assigned to clusters in the factoextra:: package on the basis of the ggplot2:: package (see Figure 7.1) are automatic and colourful (e.g. the **fviz_cluster()** command); however, this is not a mapping in the geographical space. Using typical plot commands (**plot()**, **points()**), one can easily draw the location of points within the region – both before and after the division into clusters (see Figure 7.4). Figure 7.4b uses the palette *GrandBudapest1* from the wesanderson:: package to mark the clusters with colours. It runs in the **wes_palette()** command and slot *name*.

The following graphic (see Figure 7.3a) uses a subset of the map (**see** Section 2.4 – **map section**). The colour layer in Figure 7.3a reflects the clustering vector, obtained as a slot *clustering* from a *clara* class object. In the transformation from the value of the clustering vector to the colour assignment to the point (*x, y*), the

Clustering of geolocation
with the CLARA algorithm

Clusters by k–means algorithm

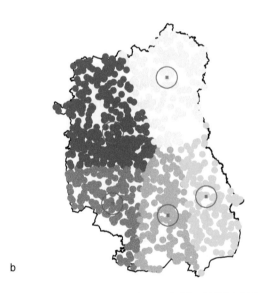

Figure 7.3 Locations of points as clusters: a) CLARA clustering, b) *k*-means clustering

Source: Own study with the use of R and wesanderson:: and sp:: packages

findInterval() command from the base:: package was used, which checks into which interval the observation of the tested variable falls. Next, the code assigns the indicated element (e.g. colour) from that interval. In other words, if the observation value falls into the third interval, the third colour from the colour vector is selected for drawing the colour layer. In the legend (command **legend**()) plotted in the bottom-left corner (the *"bottomleft"* argument) for colours (option *pt.bg = cols*), the lower intervals (option *legend = brks*) were specified. The figure was automatically saved in the *Working Directory* (**savePlot**() command).

```
voi<-readOGR(".", "wojewodztwa") # 16 units
voi<-spTransform(voi, CRS("+proj=longlat +datum=NAD83"))
```

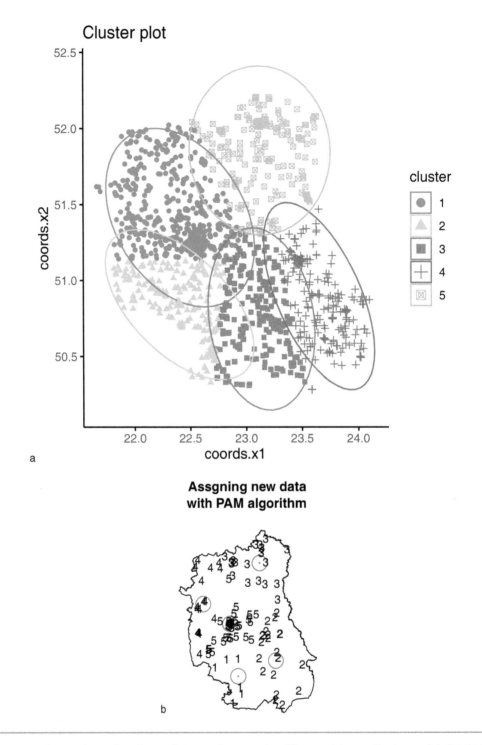

a

b

Figure 7.4 Clustering of points: a) PAM clusters in ellipses, b) prediction with PAM for new points

Source: Own study using R and factoextra:: package

```
# creation of a map subset
voi.df<-as.data.frame(voi)
lubelskie.voi<-voi[voi.df$jpt_nazwa_=="lubelskie",]

# drawing empirical data without division into
clusters plot(lubelskie.voi, main="Lubelskie NTS2")
points(firms$coords.x1, firms$coords.x2, pch=".")
```

Spatial unsupervised learning

```
# drawing empirical data divided into
clusters library(wesanderson) # colour palette
cols<-wes_palette(n=6, name="GrandBudapest1", type="continuous")
cols                          # displays the colours available in the palette
variable <-c1$clustering          # variable for colouring
summary(variable)                 # variable summary
brks<-c(0, 1, 2, 3, 4, 5)         # intervals


plot(lubelskie.voi)               # drawing a subset map
points(firms[,12:13], col=cols[findInterval(variable, brks)],
      pch=21, bg=cols[findInterval(variable, brks)], cex=0.2)
legend("bottomleft", legend=brks, pt.bg=cols, bty="n", pch=21)
title(main=" Clustering of geolocation \with the CLARA algorithm")
savePlot(filename="CLARA partitioning", type="jpg")
```

The previous algorithm for clustering the geographical coordinates can be successfully used in logistic tasks. For example, when managing sales representatives' contracts in the sales network, areas of employee activity should be determined. The PAM, CLARA and k-means algorithms can successfully propose such assignments. One can then examine which sales representative obtained more locations to serve and what the sum of the distances between these locations is, as well as who has the most dispersed points in space. Based on Figure 7.3, one can draw centre points on the map with a division into representatives.

Example

Unlike in the previous example, a random subset of *firms* was created, which contains point data (*x, y*) regarding the location of firms in the Lubelskie region. The first 2000 observations were selected from a randomly sorted full dataset. The aim of the analysis is to compare the effects of the k-means and PAM algorithms on a relatively small sample. The **kmeans()** command from the package stats:: and **pam()** from cluster:: were used. The vector of random numbers from the uniform distribution on the interval [0,1] was generated with the command **runif()**, and the number of elements was indirectly defined as the first element – number of rows – from the result of the command **dim()** returning the number of rows and columns of the *data.frame* class.

```
# creating a random identifier to sort the data
firms$los<-runif(n=dim(firms)[1],min=0,max=1) # vector of random numbers

library(doBy)
firms<-orderBy(~los, data=firms) # sorting the data set firms.sub<-
firms[1:2000,] # randomly selected 2000 observations

c3<-kmeans(firms.sub[,12:13], 5)
c3
```

```
K-means clustering with 5 clusters of sizes 445, 205, 375, 269, 706
```

```
Cluster means:
      coords.x1        coords.x2
1      23.44241         50.84765
2      22.73440         50.62611
3      22.11558         51.56237
4      23.08036         51.84815
5      22.53070         51.21261
```

```
Clustering vector:
15395 12063 2142 24573 593 32250 25693 32796 11807 33938 15838 18208 24230

. . . . .
Within cluster sum of squares by cluster:
[1] 54.73344 15.06266 42.37240 34.17406 45.33003
 (between_SS / total_SS = 78.8 %)

Available components:

[1] "cluster"      "centers"      "totss"       "withinss"    "tot.withinss"
[6] "betweenss"    "size"         "iter"        "ifault"
```

The previous result, the printout of the *c3* object, presents in turn: 1) the coordinates of centres of all created clusters (*coords.x1, coords.x2*), 2) a clustering vector listing the observation number in the original full set and the cluster number to which the observation has been assigned, 3) intracluster sums of squares of distance from the centre and 4) the available slots in this object.

```
library(cluster)
c4<-pam(firms.sub[,12:13], 5)
```

```
           ID   coords.x1   coords.x2
10019     257   22.56928    51.24039
18726     100   23.29519    50.77028
3024      484   22.43534    51.83888
15265     717   22.06887    51.26268
2615     1967   23.21300    51.92997

15395 12063 2142 24573  593 32250 25693 32796 11807 33938 15838 18208 24230
    1     1    2     3    2     2     2     1     2     3     1     4     2
 4805 32103 5277 22365 5696 10709 27948  7141 35953 24406  7730 31124 18739
    1     1    4     1    2     3     2     1     3     2     1     2     1

Clustering vector:

. . . .

Objective function:
    build        swap
0.2747580 0.2701297

Available components:
[1] "medoids"      "id.med"       "clustering"   "objective"   "isolation"
[6] "clusinfo"     "silinfo"      "diss"         "call"        "data"
```

A very similar result is obtained from the **pam()** command. In principle, there are always differences in coordinate values of cluster centres, because in PAM they are coordinates of the real points, and in *k*-means, this lies in the middle of the cluster. The quality of division into clusters is also different. The objective function assumes randomly chosen cluster centres in the *build* phase and optimised centres in the *swap* phase. The *swap* value should always be lower, because it shows the improvement in the selection of cluster centres after optimisation.

The scatterplot of cluster membership (see Figure 7.3b) shows that the random selection of start values and various criteria for the selection of centres may lead to different division of points into groups. The **viridis()** palette from the viridis:: package gives colours whose intensity and diversity are the same in the colour and black-and-white versions. In addition, they are distinguishable for most people who do not recognise colours (*colour blindness*).

```
library(viridis)
cols<-viridis(6)              # color palette
brks<-c(0, 1, 2, 3, 4, 5)     # intervals

variable<-c3$cluster          # clustering vector
plot(lubelskie.voi)           # drawing a map section
points(firms.sub[,12:13], col=cols[findInterval(variable, brks)],
    pch=21, bg=cols[findInterval(variable, brks)], cex=0.8)
points(c3$centers, col="red", pch=".", cex=3)
points(c3$centers, col="red", cex=3)
title(main="Clusters by k-means algorithm")

variable<-c4$clustering          # clustering vector
plot(lubelskie.voi)           # drawing a map section
points(firms.sub[,12:13], col=cols[findInterval(variable, brks)],
    pch=21, bg=cols[findInterval(variable, brks)], cex=0.8)
points(c4$medoids, col="red", pch=".", cex=3)
points(c4$medoids, col="red", cex=3)
title(main="Clusters by PAM algorithm")
```

For the grouping obtained previously, one can perform the analysis as in the previous example. The **fviz_cluster()** and **fviz_silhouette()** commands from the factoextra:: package will work for both *c3* and *c4* objects, despite the different classes of these objects (see Figure 7.4a). The *c3* object is in the *kmeans* class, while the *c4* object is in the *pam* and *partition* classes. Pre-analysis of the optimal number of clusters generated with the **fviz_nbclust()** command works on the base data, and one can get the result for *k*-means and PAM algorithms.

```
fviz_cluster(list(data=firms.sub[,12:13],              cluster=c3$cluster),
ellipse.type="norm",      geom="point",      stand=FALSE,      palette="jco",
ggtheme=theme_classic()) #factoextra::

fviz_cluster(list(data=firms.sub[,12:13],              cluster=c4$clustering),
ellipse.type="norm",      geom="point",      stand=FALSE,      palette="jco",
ggtheme=theme_classic()) #factoextra::
```

The essence of machine learning is the ability to automatically classify new data points on the basis of the predefined model. In the case of dividing geographical coordinates into clusters, the aim is to obtain the allocation of new data to the established clusters. Such possibilities are offered by the flexclust:: package, which contains **predict()** function for most groupings from different packages. The syntax of the **predict()** command requires a grouping object in the *kcca* class and new data. Conversion to the *kcca* class is possible with the **as.kcca()** command, which specifies the grouping result and the original data.

The following example presents the classification of new points for *k*-means and PAM grouping, written previously in objects *c3* and *c4*, respectively, in the *kmeans* and *pam/partition* classes. The example shows a conversion to the *kcca* class, preparation of new 100 points, and visualisation of the result of the automatic classification. The map of clusters shows the numbers of clusters in circles, and the new points are marked with the cluster number to which the point has been assigned. New points were marked with the **text()** command. The colours of the **viridis()** palette have become more transparent thanks to the *alpha=0.2* option. As one can see, grouping and prediction are in fully consistent (see Figure 7.4b).

```
firms.out<-firms[2001:2100,]

library(flexclust)
c3.kcca<-as.kcca(c3, firms.sub[,12:13]) # conversion to kcca
c3p<-predict(c3.kcca, firms.out[,12:13]) # prediction for k-means

c4.kcca<-as.kcca(c4, firms.sub[,12:13]) # conversion to kcca
c4p<-predict(c4.kcca, firms.out[,12:13]) # prediction for PAM

cols<-viridis(6, alpha=0.2)    # viridis::, and also plasma()
```

```
brks<-c(0, 1, 2, 3, 4, 5)
lubelskie.voi<-voi[voi$jpt_nazwa_=="lubelskie",]
plot(lubelskie.voi, main="Assigning new data
with k-means algorithm")

points(firms.sub[,12:13], pch=21, bg=cols[findInterval(c3$cluster, brks)],
col=cols[findInterval(c3$cluster, brks)], cex=0.8)
points(c3$centers, col="red", pch=".", cex=2)
points(c3$centers, col="red", cex=3)
text(c3$centers, labels=rownames(c3$centers), font=2)
text(firms.out[,12:13], as.character(c3p)) # new points
```

7.2 Clustering with the density-based spatial clustering of applications with noise algorithm

An alternative approach (discussed subsequently) to the clustering presented previously is carried out by looking for similarities between points based on distance metrics: density-based spatial clustering of applications with noise (DBSCAN) statistics. It does not use distance metrics to search for the nearest objects, like most traditional clustering algorithms, but rather examines the spatial density of points and extracts the areas with a high density of points, separated by areas with a low (or lower) density of points – so-called noise.[3] DBSCAN looks for spatial structures of points to reflect the density of these points to the best possible extent. Unlike other methods, DBSCAN does not assume any parametric distributions, cluster shapes or number of clusters. It is a method resistant to weak connections and outliers.

In the classification of points belonging to a cluster (core and border) and those outside the cluster (noise), one uses the approach based on a circle with a radius of epsilon ε and minimum number of points in this radius *MinPts*. For each point, the number of points in radius ε is examined and whether the points fall into the radius of other points. Core points are those that have at least the minimum number of *MinPts* points within a radius of ε. All these points in the radius ε from the core point (ε-neighbourhood) are referred to as directly reachable (or *directly density-reachable*). In turn, boundary points are those that are in the radius ε from the core point but do not themselves contain the minimum number of *MinPts* points in their radius ε. Points are classified as noise when they are outside the radius of core and boundary points. Points can also be (indirectly) available (*reachable, density-reachable*) when there is a sequence of points that are directly accessible to each other, which in practice means going from point to point through other points, always within a radius of ε. In the DBSCAN approach, not all points belong to clusters – clusters are formed from densely located (core and boundary) points, while points lying in low-density (noise) areas are outside the cluster. Clusters fulfil the condition that all of their points are interconnected by density (*density-connected*), which means that the points are available directly or indirectly (*directly density-reachable* and *density- reachable*).

The algorithm sets clusters one by one in a complete way. Starting from a randomly chosen point, it examines the neighbourhood in a given radius ε and marks the points belonging to the cluster and constituting noise. All points belonging to the cluster are iteratively tested until the capabilities are exhausted and the full cluster is formed. Subsequently, in the same procedure, points are examined, which constitute noise against the previously formed cluster.

The DBSCAN algorithm defines two parameters: the epsilon ε radius and the minimum number of points *MinPts* in this radius. They are crucial for the detection of clusters. They are not independent – for points with different density and assuming a high ε, it is harder to find a cluster for high *MinPts* than for low *MinPts*. Setting those parameters often requires expert knowledge. *MinPts* should be treated as the desired minimum cluster size. In order to obtain any reasonable grouping, *MinPts* should be 3 or more (with *MinPts* = 1, each point creates its own cluster, while with *MinPts* = 2, the basic hierarchical grouping appears). When setting ε, one may use the approach of the *k*-nearest neighbours, showing the distance to the *k*th nearest neighbour. It is usually recommended to use ε given at the knee/elbow of the chart. Too large a value of ε causes most observations to be placed in one cluster, while too small a radius does not allow clustering at all.

Despite the many undoubted advantages, the weak point of the algorithm is that it classifies border points into clusters that, with a weakly variable density, can belong to different clusters, and cross-referencing can combine two clusters into one spatial structure. In addition, the choice of distance metrics is important for the result, and, as advanced works show, with multivariate data and the use of Euclidean distance, finding ε is difficult or even impossible. There may also be a problem with the selection of radius ε at highly heterogeneous local point densities.

The dbscan:: package is available in R for calculating the scan statistics. It contains, among others, the **dbscan()** function for the discussed DBSCAN[4] algorithm and accompanying functions used to diagnose the environment of points discussed subsequently in the example.

Example

The example uses point data (x, y) regarding the location of firms in the Lubelskie voivodeship (*firms*). The analysis of the possibility of distinguishing location clusters was carried out on a limited collection of 5000 observations. In contrast to previously presented algorithms that assigned all observations to clusters, DBSCAN statistics leave some points as noise, while clusters only include points located in dense clusters.

To perform clustering with the **dbscan()** command from the dbscan:: package, one has to define parameters ε and *minPts* in its options. They can be derived from the neighbourhood analysis of points, which will be presented subsequently.

The basic method of analysing the neighbourhood of points is to assess the distance to nearest neighbours on the graph. The chart can be created with the command **kNNdistplot()**, also from the dbscan:: package. To use the **kNNdistplot()** command, the value of the MinPts parameter should be determined – the minimum number of points in a given radius ε. This is the number of *knns* of neighbouring points (k nearest neighbours), which will create clusters. The **kNNdistplot()** function calculates for each point in the dataset the distance to the *knn*th neighbour. In this graph, the y-axis gives the distance between neighbours (it is ε), and the points from the dataset are indexed on the x-axis. They have been ordered in ascending order according to the value of ε. The number of index points on the x-axis is equal to the number of points in the dataset multiplied by the number of *knn* neighbours[5] indicated. A practical recommendation often used assumes setting the ε value for the **dbscan()** command at the level at which the *knn* graph breaks down (the so-called knee/elbow) (see Figure 7.8). The values that the **kNNdistplot()** function draws can be specified by the command **kNNdist()**.

```
library(dbscan)
firms<-read.csv("geoloc_data_firms.csv", header=TRUE, dec=",", sep=";")
sub<-firms[1:5000,12:13] # 5000 obs have been selected for analysis
head(kNNdist(sub, k=5)) # knn = 5 neighbours established
```

	1	2	3	4	5
1	0.002053393	0.008291464	0.018161356	0.018161356	0.018161356
2	0.000000000	0.000000000	0.000000000	0.003361357	0.003361357
3	0.000000000	0.000000000	0.000000000	0.000000000	0.000000000
4	0.000000000	0.015644609	0.024134099	0.024134099	0.048522228
5	0.003204727	0.009461100	0.016644173	0.023602205	0.027245459
6	0.001598467	0.003779231	0.005449997	0.008990256	0.009163621

```
kNNdistplot(sub, k=5)   # distance chart for knn = 5
abline(h=0.01, col="red", lty=2)     # dashed red line

kNNdistplot(sub, k=20) # distance chart for knn=20
abline(h=0.01, col="red", lty=2)     # dashed red line
```

The increase in the number of analysed neighbours, of course, extended the index on the x-axis, as well as increasing the distance recorded on the y-axis. However, the knee/elbow of the graph remained approximately similar (see Figure 7.5a). This analysis shows that for the purposes of DBSCAN statistics, the reasonable value of parameter ε (*eps*) is $\varepsilon = 0.01$, although other scenarios are worth examining. The value of the second parameter, *minPts*, was adopted as 5 and 20.

In the diagnosis of k nearest neighbours[6] of a given point, one can also use the command **kNN()** from the dbscan:: package. The points to be analysed and the number of k nearest neighbours should be given as the input, and in the result, the matrix of distances to the next neighbours is obtained in the $dist$ slot (first – nearest, second, etc.) and in the id slot the index matrix whose observations are the nearest neighbours for a given point. Neighbours can be sought not only as the nearest but also within a given radius of ε. To do this, one can use the **frNN()** command from the dbscan:: package (*fr* is the abbreviation for fixed radius). Similarly to **kNN()**,

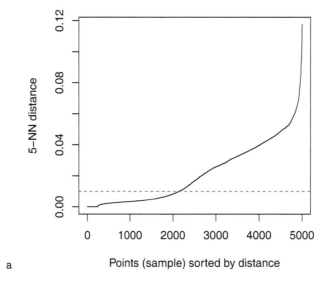

a

Points (sample) sorted by distance

Convex Cluster Hulls

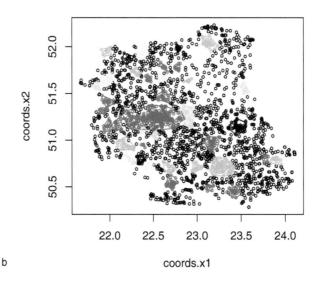

b

coords.x1

Figure 7.5 DBSCAN clustering: a) distance chart for *k* nearest neighbours *knn* = 5, b) clusters for *eps* = 0.05 and *knn* = 20

Source: Own study using R and dbscan:: packages

in **frNN**() output, one gets in the *$dist* slot a matrix of distances to all neighbours indicated in a given radius and in the *$id* slot of the index matrix, whose observations are the nearest neighbours for a given point. Neighbour connections found can be drawn with the **plot**() command, where the first argument is the result of the **kNN**() or **frNN**() commands and the second argument is the set of analysed points.

The command **frNN**() is supplemented by the **pointdensity**() command from the dbscan:: package, which counts neighbours in a given radius ε – the local density for each point. As arguments to this function, one gives the points for analysis, radius ε and the method of counting (usually counting as *type* = *"frequency"*). The result is a vector of the number of neighbours of each point under examination.

One can also conduct a comparative analysis of the local density of points. The **lof**() function (*local outlier factor*) from the dbscan:: package compares the local densities of the tested point and its *k* neighbours. If the local densities are similar (similar numbers of neighbours in a given radius), the *lof* is 1, and for outlier noise, *lof* is significantly higher than 1. As arguments of this function, one gives the dataset of the points and neighbourhood with the number of *k*-nearest neighbours used for comparison.

```
a<-kNN(sub, k=3) # searching for the nearest 3 neighbours of each point
head(a$dist) # distance to the k-nearest neighbours
```

```
            1             2             3
1 0.002053393   0.008291464   0.018161356
2 0.000000000   0.000000000   0.000000000
3 0.000000000   0.000000000   0.000000000
4 0.000000000   0.015644609   0.024134099
5 0.003204727   0.009461100   0.016644173
6 0.001598467   0.003779231   0.005449997
```

```
head(a$id) # id k nearest neighbours
```

```
       1      2      3
1  3287   4099   1801
2  1638   4635   4957
3    84   2708   4498
4  2924   3058   3634
5  3500   3972   4572
6  1022   3147    788
```

```
a2<-frNN(sub, eps=0.01) # searching for neighbours within a radius of 0.01
head(a2$dist) # distances to neighbours within a given radius
```

```
$'1'
[1] 0.002053393 0.008291464
$'2'
[1] 0.000000000 0.000000000 0.000000000 0.003361357 0.003361357
0.003361357
0.003361357 0.003361357
$'3'
 [1] 0 0 0 0 0 0 0 0 0 0
$'4'
[1] 0
$'5'
[1] 0.003204727 0.009461100
$'6'
[1] 0.001598467 0.003779231 0.005449997 0.008990256 0.009163621
```

```
head(a2$id) # id of neighbours in a given radius
```

```
$'1'
[1] 3287 4099
$'2'
[1]    1638 4635 4957    63 1552 2076 2721 3491
$'3'
 [1]    84    290  315   1611 2708 3377 3622 3940 4272    4498
$'4'
[1]    2924
```

```
$'5'
[1] 3500 3972
$'6'
[1] 1022 3147    788 1472 4999
```

```
a3<-pointdensity(sub, eps=0.01, type="frequency") # local density
head(a3) # the first point has 4 neighbors within a radius of ε=0.01
```

```
[1]  4 10 12    3    4    7
```

```
a4<-lof(sub, k=3) # comparison of the density of local points
head(a4)
```

```
[1]  Inf 1.0000000 1.0000000 1.0287263 0.9503234 1.0345568
```

In the clustering algorithm, in the **dbscan**() command, in addition to the ε and *minPts* parameters, one can specify whether the boundary points should be included in the cluster (the default option *borderPoints = TRUE*) or should be noise (*FALSE*). The result of the **dbscan**() command is the statistics of the number of created clusters. For *knn = 5* neighbours, 126 clusters and 2487 points classified as noise were created. For *knn = 20* neighbours, 17 clusters and 3602 points were classified as noise, while at *eps = 0.05* there are 38 clusters and 1922 noise points. The **hullplot**() command generates a scatterplot with clusters marked in colour. Its *solid = TRUE* option allows for drawing polygons based on the envelope of points belonging to the cluster, while the *alpha* option allows for controlling the degree of polygon transparency (*alpha=0* fully transparent, *alpha=1* not transparent) (see Figure 7.5b).

```
dbs1<-dbscan(sub, eps=0.01, minPts=5) # clustering
hullplot(sub, dbs1) # points chart with marked clusters

dbs2<-dbscan(sub, eps=0.05, minPts=20)
hullplot(sub, dbs2, solid=TRUE, alpha=0.7)
```

```
DBSCAN clustering for 5000 objects.
Parameters: eps = 0.05, minPts = 20
The clustering contains 38 cluster(s) and 1922 noise points.

    0    1    2      3    4    5    6      7      8    9     10    11    12   13   14    15
 1922  114   29   1044   36   90   95    101    150   38    109    94    91   67   41   262
   16   17   18     19   20   21   22     23     24   25     26    27    28   29   30    31
   30   22   58     67   31   40   49     26     48   23     23    42    29   26   23    38
   32   33   34     35   36   37   38
   27   20   21     20   20   20   14

Available fields: cluster, eps, minPts
```

The result of the clustering algorithm is separated dense clusters of points, spatially wider or narrower, depending on the assumed parameter values. A significant percentage of points was classified as noise, that is, points located sparsely. In Figure 7.5b, the largest clusters coincide with the location of the largest cities in the region: the city of Lublin is clearly visible, to which corresponds the large navy elliptic cluster located in the middle of the area and the adjacent light blue cluster for the city of Świdnik; the gray southern round cluster coincides with the location of the city of Zamość, the northeast green polygon with the location of the city of Biała Podlaska and the central-eastern black polygon with the location of the city of Chełm.

The previous clustering allows for the prediction for new points, within which it is checked whether they lie within the created cluster or belong to the noise. In spatial terms, the algorithm compares the coordinates (x,y) of a new point with the boundaries of geometry formed on the envelope of points belonging to the cluster. Subsequently, one can examine whether 10 random points belong to clusters created earlier.

```
sub1<-firms[5001:5010,12:13] # 10 new points were selected
#dbs2<-dbscan(sub, eps=0.05, minPts=20) # previous clustering
predict(dbs2, newdata=sub1, data=sub) # prediction
```

```
5001 5002 5003 5004 5005 5006 5007 5008 5009 5010
   0    0    3   11    0    8    0    3   15    0
```

The result of the prediction shows that half (5) of the observations were qualified for previously created clusters with numbers as in the results, while the others belong to noise; that is, they are located in areas of low density.

From the previous analysis, from the technical perspective, it can be seen that the number of clusters and points classified as noise differs significantly depending on the parameters adopted. In order to know the distribution of results for the analysed data, simulation of the result parameters was performed subsequently depending on the input parameters (*eps* and *minPts*). As a result, the number of created clusters (the *result.i* saved as the object) and the percentage of unassigned observations – noise (the *result.j* saved as the object) – were examined. In the resulting object from the **dbscan()** command in the *$cluster* slot, there is a vector with the numbers of clusters to which the given observation has been assigned, with the value 0 meaning noise. By determining the *max* from this vector, one can determine how many clusters were created, and counting the value of 0 and referring them to the number of observations gives the percentage of noise.

```
# simulation of clustering scenarios due to eps and minPts
vec.i<-(1:20)*0.005     # eps parameter, in rows
vec.j<-(1:20)*5         # minPts (knn) parameter, in columns
result.i<-matrix(0, nrow=20, ncol=20)
rownames(result.i)<-vec.i colnames(result.i)<-vec.j
result.j<-matrix(0, nrow=20, ncol=20)
rownames(result.j)<-vec.i colnames(result.j)<-vec.j

for(i in 1:20){
for(j in 1:20){
dbs.temp<-dbscan(sub, eps=vec.i[i], minPts=vec.j[j])
result.i[i,j]<-max(dbs.temp$cluster)
result.j[i,j]<-length(which(dbs.temp$cluster==0))/dim(sub)[1]
}}

result.i # the number of clusters created
```

	5	10	15	20	25	30	35	40	45	50	55	60	65	70	75	80	85	90	95	100
0.005	98	42	26	5	3	1	1	0	0	0	0	0	0	0	0	0	0	0	0	0
0.01	126	41	24	17	13	11	11	7	6	3	2	1	1	1	1	0	0	0	0	
0.015	150	46	27	21	16	15	12	8	8	7	7	4	3	3	2	1	1	1	1	1
0.02	162	53	27	21	19	15	14	11	10	8	8	8	6	4	4	3	2	2	2	2
0.025	185	59	32	22	19	15	14	14	10	9	9	8	7	6	5	5	2	2	2	2
0.03	202	71	36	25	20	17	14	14	10	10	9	9	9	6	5	5	4	3	3	2
0.035	169	82	43	24	18	16	15	14	12	10	9	9	9	9	7	5	5	4	4	4
0.04	123	82	43	28	19	18	14	13	13	10	9	9	9	9	8	6	5	4	4	4

	5	10	15	20	25	30	35	40	45	50	55	60	65	70	75	80	85	90	95	100
0.045	71	73	49	35	22	18	15	14	13	10	9	8	8	8	8	8	6	4	4	4
0.05	38	59	54	38	24	19	17	14	14	12	9	9	8	8	8	7	6	5	4	4
0.055	21	34	41	40	28	21	19	17	14	12	10	10	9	9	8	8	7	6	5	4
0.06	12	23	35	35	30	21	22	18	15	13	12	10	9	9	9	8	7	6	5	4
0.065	11	14	24	34	29	24	23	22	17	14	12	11	10	9	9	9	7	6	5	5
0.07	7	8	17	27	27	20	21	19	19	17	13	11	10	10	9	9	8	7	6	5
0.075	5	6	13	12	17	17	17	20	18	19	15	13	12	10	10	9	9	7	6	5
0.08	1	3	7	11	12	15	16	17	17	18	18	13	12	10	10	10	9	7	7	6
0.085	1	2	5	8	9	10	13	14	17	17	18	15	13	12	10	9	10	9	7	6
0.09	1	2	4	6	6	7	13	15	13	14	16	16	13	11	9	9	9	9	8	8
0.095	1	1	5	5	7	4	6	11	12	11	13	14	16	13	11	10	9	9	9	8
0.1	1	1	2	5	5	5	6	9	12	11	11	12	15	15	13	11	10	9	10	9

```
print(result.j, digits=2) # noise percentage
```

	5	10	15	20	25	30	35	40	45	50	55	60	65	70	75	80	85	90	95	100
0.005	0.6122	0.7418	0.8502	0.955	0.975	0.989	0.990	1.00	1.00	1.00	1.00	1.00	1.00	1.00	1.00	1.00	1.00	1.00	1.00	1.00
0.01	0.4974	0.6220	0.6814	0.720	0.746	0.779	0.811	0.87	0.89	0.93	0.95	0.97	0.97	0.97	0.98	0.98	1.00	1.00	1.00	1.00
0.015	0.4302	0.5746	0.6310	0.656	0.686	0.700	0.728	0.76	0.77	0.78	0.79	0.83	0.85	0.87	0.89	0.91	0.91	0.91	0.93	0.94
0.02	0.3640	0.5276	0.5984	0.628	0.640	0.665	0.676	0.71	0.72	0.74	0.75	0.75	0.78	0.81	0.81	0.83	0.84	0.85	0.85	0.85
0.025	0.2766	0.4862	0.5644	0.602	0.621	0.648	0.658	0.66	0.70	0.71	0.71	0.73	0.74	0.76	0.78	0.78	0.83	0.83	0.84	0.84
0.03	0.1804	0.4220	0.5244	0.571	0.600	0.619	0.641	0.64	0.68	0.68	0.70	0.70	0.70	0.74	0.76	0.76	0.79	0.81	0.81	0.83
0.035	0.1164	0.3428	0.4760	0.548	0.582	0.600	0.623	0.63	0.65	0.67	0.68	0.68	0.69	0.69	0.72	0.75	0.76	0.78	0.78	0.78
0.04	0.0688	0.2624	0.4264	0.513	0.560	0.572	0.602	0.62	0.62	0.66	0.67	0.67	0.68	0.68	0.70	0.73	0.75	0.77	0.77	0.77
0.045	0.0418	0.1852	0.3520	0.459	0.528	0.555	0.579	0.59	0.60	0.64	0.65	0.66	0.66	0.67	0.69	0.69	0.72	0.76	0.76	0.76
0.05	0.0252	0.1250	0.2760	0.384	0.485	0.518	0.542	0.58	0.58	0.61	0.64	0.64	0.66	0.66	0.66	0.68	0.70	0.72	0.75	0.75
0.055	0.0144	0.0708	0.2044	0.305	0.413	0.481	0.506	0.53	0.56	0.58	0.61	0.62	0.63	0.63	0.65	0.65	0.67	0.69	0.71	0.74
0.06	0.0092	0.0372	0.1402	0.252	0.352	0.433	0.464	0.50	0.53	0.55	0.57	0.60	0.62	0.62	0.62	0.64	0.66	0.68	0.70	0.73
0.065	0.0050	0.0236	0.0868	0.184	0.285	0.372	0.409	0.45	0.49	0.53	0.56	0.57	0.59	0.61	0.61	0.62	0.65	0.67	0.70	0.70
0.07	0.0036	0.0162	0.0504	0.116	0.228	0.319	0.361	0.41	0.43	0.47	0.52	0.55	0.57	0.58	0.60	0.60	0.62	0.65	0.67	0.69
0.075	0.0026	0.0136	0.0324	0.081	0.154	0.241	0.297	0.36	0.40	0.41	0.46	0.50	0.52	0.55	0.56	0.58	0.59	0.63	0.65	0.68
0.08	0.0010	0.0112	0.0186	0.054	0.101	0.185	0.249	0.30	0.36	0.38	0.40	0.47	0.49	0.53	0.55	0.55	0.57	0.62	0.62	0.65
0.085	0.0004	0.0080	0.0140	0.031	0.070	0.133	0.194	0.24	0.31	0.33	0.35	0.40	0.45	0.48	0.52	0.54	0.54	0.57	0.61	0.64
0.09	0.0000	0.0030	0.0100	0.022	0.052	0.091	0.137	0.21	0.26	0.30	0.32	0.35	0.40	0.44	0.48	0.52	0.53	0.54	0.58	0.58
0.095	0.0000	0.0020	0.0056	0.015	0.031	0.068	0.097	0.15	0.22	0.25	0.27	0.29	0.31	0.38	0.43	0.45	0.50	0.51	0.52	0.54
0.1	0.0000	0.0014	0.0038	0.010	0.020	0.046	0.077	0.10	0.16	0.21	0.24	0.26	0.27	0.29	0.34	0.39	0.44	0.48	0.49	0.50

As was visible in the two previous table printouts, the number of clusters created for a given dataset can be very different and range from a few to a few hundred. The largest number of clusters is for the combination of a small ray of *eps* and a small number of points around (*minPts*). Similarly, the percentage of noise can be significantly different, from promile to nearly all observations. The relation is visible: a large number of clusters reduces the noise.

The distributions of the number of clusters and the noise figure were determined by the **density**() command (see Figure 7.6). The surface and three-dimensional charts of these results are also presented. Various visuali-sation possibilities were used. The **image**() command from the basic built-in graphics:: package automatically assigns colours from the **heat.colours**() palette and allows for applying a contour with the **contour**() command but automatically converts the *x* and *y* axis scale to a range (0,1) regardless of the actual labels, which makes interpretation more complicated.

```
# density distributions # Figure 7.6
plot(density(result.i), main="distribution of the number of clusters
depending on the eps and minPts parameters")
plot(density(result.j), main="distribution of the noise percentage
depending on the eps and minPts parameters")
```

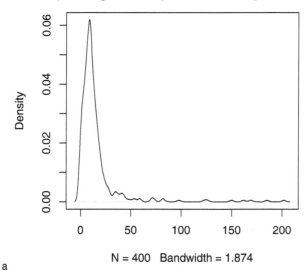

a

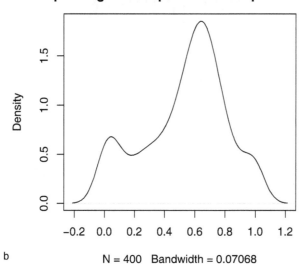

b

Figure 7.6 Distributions of density of simulation results: a) number of clusters, b) percentage of observations classified as noise

Source: Own study with the use of R

```
# surface charts with contours
image(result.i, axes=TRUE)
contour(result.i, add=TRUE, drawlabels=FALSE)
image(result.j)
contour(result.j, add=TRUE, drawlabels=TRUE) # with contour lines
```

The previous frequency charts (see Figure 7.6) confirm there exists a large range of results, as visible in the matrix.

Constantly developing 3D graphics give better possibilities for visualising the matrix results. The basic function **persp()** allows for drawing the planes, also with different graphic effects, but these are static images, without the possibility of animation (see Figure 7.7). An interesting dynamic solution is the **plot_ly()** command from

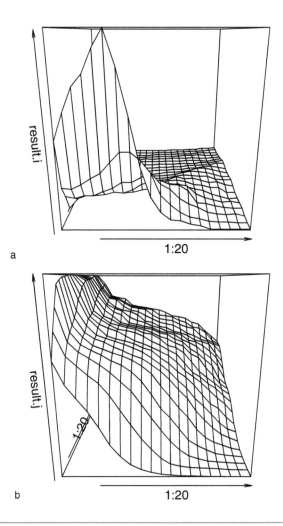

a

b

Figure 7.7 3D surface charts for the results of sensitivity analysis: a) number of clusters, b) percentage of observations classified as noise

Source: Own study with the use of R

the plotly:: package, which generates 3D charts with the possibility of animation and displays them automatically in the default web browser as a web page. Dynamic point inspection is possible on these charts, thanks to the displayed label with *x*- and *y*-axis values (row and matrix column labels) and the value of the *z* matrix. The command syntax uses pipe operators from the magrittr:: package. This passes the result to the next action by the operator %>%, which means that instead of notation f(*x*), the notation *x* %>% f() appears. The pipe syntax does not require the objects to be stored separately and allows for converting the results "on the air". In practice, after defining **plot_ly**() input data for *xyz* dimensions in the command, the chart type is determined – for surface charts, the %>% **add_surface**() code is added. Without this assignment, the command only displays the empty three-dimensional layout of the axis. In the case of line charts, the chart type is defined by %>% **add_lines**().

```
# surface 3D static charts # Figure 7.7
persp(1:20, 1:20, result.i)
persp(1:20, 1:20, result.j)

# dynamic 3D surface charts with inspection
library(plotly)
plot_ly(x=vec.i, y=vec.j, z=result.i) %>% add_surface()
plot_ly(x=vec.i, y=vec.j, z=result.j) %>% add_surface()
```

The analysis of the results shows that a high number of clusters, above 100, arises at the lowest assumed values of the *eps* radius (on the *x*-axis, from 0.005 to 0.1), while after exceeding the value of about 0.04, it drops to a few. The number of clusters is also dependent on the number of nearest *minPts* neighbours (on the *y*-axis, from 5 to 100). The percentage of non-assigned observations decreases with the increase of both parameters, for the lowest values of *x* and *y*, the unassigned rate is 100, which means that no clusters have been created.

7.3 Spatial principal component analysis

Standard a-spatial principal component analysis (PCA) is mainly used to reduce dimensions where there are many variables. It is based on eigenvalues and eigenvectors from the covariance matrix for standardised data and scaling of the result by rotation. As a result, the original dataset is projected onto a new *k*-dimensional space, usually two-dimensional. The principal components, being the product of the eigenvector and the dataset, are the axes of the largest variance of the dataset. Their interpretation is aimed at obtaining the greatest possible variance explained by a certain number of dimensions of the new space. An important element is the standardisation of variables. When it is missing, it may result in a change in the result, mainly in the direction of the influence of the variable with the highest absolute values. The results of a typical PCA can be treated as synthetic variables carrying the information of the basic dataset but using a significantly smaller number of variables. Rotated PCA is interpreted as a mechanism for grouping factors (variables) that exhibit the highest similarity.

There are several packages and commands for PCA in R. These include **prcomp**(), **princomp**() and **loadings**() from the stats:: package; **PCA**() and **dimdesc**() from the FactoMineR:: package; **get_pca_ind**() from the factoextra:: package and **principal**() from the psych:: package. Data normalisation is possible, among others with the **data.Normalization**() command from the clusterSim:: package. The most important differences between functions are the decomposition method – stats::**princomp**() uses spectral decomposition, while stats::**prcomp**() and FactoMineR::**PCA**() use singular value decomposition. Using the **principal**() command from the psych:: package, one can rotate factors (*loadings*).

The basic difference between spatial and a-spatial PCA concerns the desired dimension of variation. In a-spatial PCA, this is the axis of the maximum variance of data for different information variables in specific units. In spatial PCA, this is the dominant pattern of variability over space for specific locations and repeated measurements in them. This is a study of the so-called *eigenimage*.

In analysis of spatial panel data (*v* variables for *n* units – locations at time *t*), double variability is indicated: spatial structure and internal (genetic) variability. Static analysis of spatial autocorrelation, for example, using the Moran's *I* statistic, allows for studying only the spatial structure. The reference to both variations requires the use of more advanced methods, such as spatial PCA.

Wang and Huang (2017) point to problems related to space dimensions and making discrete results continuous in the full space. One of the basic problems of spatial panel data is the greater number of *n* locations than time units *t*. Traditional PCA may in this situation give results that are too sensitive, from which it is difficult to obtain stable estimates for new points. For this reason, Wang and Huang (2017) propose the use of regularisation, consisting of imposing restrictions (penalty function) on coefficients in linear variable combinations (similarly to LASSO) to obtain the right degree of smoothness and sparseness, controlling orthogonality.

In the SpatPCA:: package (Wang & Huang, 2018), one can find the regularised principal component method for spatial data. It assumes the inclusion of geographical coordinates for the analysed units in the algorithm and allows for regularly and irregularly spaced spatial units. In practice, this means that regularly distributed data comes from the grid, while units arranged irregularly are measuring points in space or coordinates of centres of centroids. In this approach, the spatial dimension is modelled on the coefficient of sparseness. The algorithm assumes a dual function of punishment: smoothing for internal variability and promoting spatially sparse patterns based on LASSO. In modelling, one uses the distance between the tested locations. This approach does not involve the use of a spatial weights matrix. The **spatpca**() command from the SpatPCA:: package uses kriging to prepare predictions of the main variation dimensions for new points within the bounded box enveloped by the initial points used in the estimation.

Spatial PCA is used in the study of natural phenomena (e.g. temperature in the ocean, atmospheric data, etc.), but it can be used in the study of economic phenomena (e.g. unemployment rate). The results of the spatial analysis of the principal components, except for giving the spatial patterns for controlling spatial variability and internal variability, can be used as information about the spatial process in panel models with random effects.

The **spatpca**() command from the SpatPCA:: as an argument sets the location matrix *x* (e.g. one-, two- or three-dimensional) sized $p \times d$ for *p* location and *d* location dimensions (coordinates), data matrix Y ($n \times p$) for *p* location and *n* variables. Additionally, the user can define such parameters as number of eigenfunctions (number of target dimensions after reduction) (option *K*) and non-negative smoothness parameters *tau1*, sparseness *tau2* and tuning *gamma*. These are selected automatically (options equal to *NULL*) for the estimation and defined when extrapolating the trend. The *center=TRUE* option allows for by-column standardisation. The result object contains the estimation parameters, prediction of *Y* in new locations and eigenfunctions.

There are also other packages for spatial analysis of the principal components. The adegenet:: package (Jombart, 2015) based on spatial classes with sp:: is dedicated to genetic data for allele[7] analysis. More about spatial PCA can be found in Demšar, Harris, Brunsdon, Fotheringham, and McLoone(2013).

Example

The following example uses monthly data on the unemployment rate in Poland at the level of poviats (NTS4) in 2017, which gives a dataset of 380 territorial units in 12 periods each year (2011–2018) (*unempl*). From the data, the trend was removed with the command **detrend**() from the pracma:: package (Borchers, 2018), which in practice means subtracting the average from each matrix column. It is worth noting that in the input data in the commands **detrend**() and **spapca**(), the data column is a time series for a given spatial unit.

The resulting graphics were made with the command **quilt.plot**() from the fields:: package (Nychka, Furrer, Paige, & Sain, 2017), which allows for drawing points (*x*, *y*) and their values (*z*) easily by setting the default colour scale. The functional equivalent for territories (polygons) is the **choropleth**() command from the GISTools:: package (Brunsdon & Chen, 2014), which similarly requires only the spatial dimension and the variable to be visualised.

```
unempl<-read.csv("unemp2018.csv", header=TRUE, dec=",", sep=";")
library(SpatPCA)
library(pracma)

dane<-unempl[,85:96]                    # data selection for one year-2017
dane.t<-t(dane)                         # data transposition
dane.td<-detrend(dane.t, "linear")      # time series without trend, pracma ::

pov<-readOGR(".", "powiaty") # 380 units
pov<-spTransform(pov, CRS("+proj=longlat +datum=NAD83"))
crds<-coordinates(pov)                  # a reminder of centroids of units

# 3D PCA with automatically set parameters
spca<-spatpca(x=crds, Y=dane.td)
attributes(spca)                        # available PCA spatial slots
```

```
$names
[1] "call"   "eigenfn" "Yhat"   "Khat"   "stau1"  "stau2"  "sgamma"
[8] "cv1"    "cv2"     "cv3"    "tau1"   "tau2"   "gamma"  "Yc"
$class
[1] "spatpca"
```

```
head(spca$eigenfn)
```

```
                  [,1]            [,2]
[1,]    -0.0565815454    2.604785e-02
[2,]    -0.0661522584   -2.640910e-02
[3,]    -0.0822298166    9.305130e-03
[4,]    -0.0374833082    3.086048e-02
[5,]    -0.0388219269    3.173066e-02
[6,]    -0.0404151327    1.709994e-02
```

The result of spatial PCA in the analysis uses several elements. The principal components are determined by an algorithm that projects by default onto two dimensions. Both principal components serve to maximise variance around their axes. The values of the first and second principal components in their locations are examined in order to determine the spatial patterns of the phenomenon. A two-dimensional graph (*scatterplot*) of the first component is also examined in order to determine outliers (see Figure 7.8a). These values most often form the centres of changes in the spatial trend. The principal component multiplied by observations allows the component trend over time to be studied. An interesting feature of this approach is the ability to extrapolate

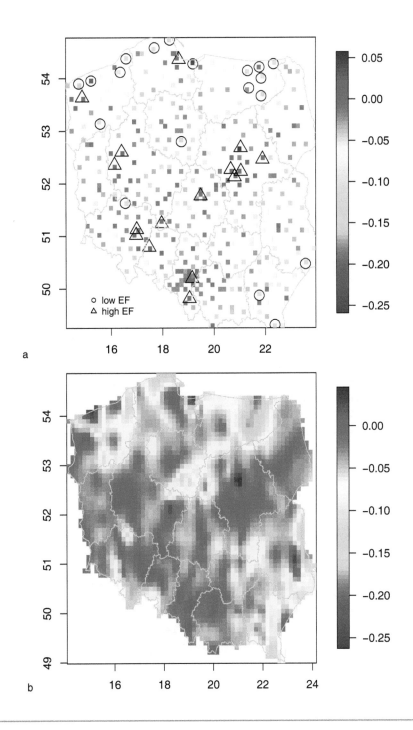

a

b

Figure 7.8 Spatial PCA: a) first principal component with outliers, b) extrapolation of spatial process

Source: Own study using R and SpatPCA:: and fields:: packages

the spatial trend to other points, determined arbitrarily in the space under examination (within a bounding box). Extrapolation of the trend is carried out using the kriging method. The PCA parameters are then selected from the result, estimated in the real data modelling and used further in the kriging (options *K=spca$Khat, tau1=spca$stau1, tau2= spca$stau2*).

```
# scatterplot with a contour map for the first principal component
# in search of spatial trends
library(fields)
quilt.plot(crds, spca$eigenfn[,1]) # from the fields::
plot(voi, add = TRUE, border="grey80")

# the first principal component over time - line chart plot(dane.
td%*%spca$eigenfn[,1], type="l", ylab="Thefirst principal component")

# scatterplot of both components
# in the search for outliers
plot(spca$eigenfn[,1], spca$eigenfn[,2]) # Figure 7.8a
abline(h=-0.1, lty=3, col="grey80")
abline(v=-0.1, lty=3, col="grey80")

# selection of negative and positive outliers from the eigenvector
a = -0.1 # minimum of 1.st quadrant - border of outliers
# the first component smaller than -0.1, negative outlier
outs.n<-which(spca$eigenfn[,1]<a)
unempl[outs.n, c(3,6,85)]
```

	poviat	region	X2017.01
8	Powiat strzyżowski	Podkarpackie	18.3
10	Powiat leski	Podkarpackie	20.6
67	Powiat tomaszowski	Lubelskie	10.3
220	Powiat lęborski	Pomorskie	11.4
223	Powiat pucki	Pomorskie	10.5

```
# selecting outliers from the dataset a=0
# first component greater than a=0, positive outlier
a=0
outs.p<-which(spca$eigenfn[,1]>a)
unempl[outs.p, c(3,6,85)]     # selecting outliers from the dataset
```

	poviat	region	X2017.01
23	Powiat pruszkowski	Mazowieckie	5.4
28	Powiat pułtuski	Mazowieckie	18.1
35	Powiat warszawski zachodni	Mazowieckie	3.2
36	Powiat węgrowski	Mazowieckie	10.3

```
# marking outliers on the map
quilt.plot(crds, spca$eigenfn[,1])

points(crds[outs.n,1], crds[outs.n,2], pch=21, col="black", cex=2)
points(crds[outs.p,1], crds[outs.p,2], pch=24, col="black", cex=2)
plot(voi, add = TRUE, border="grey80")
legend(15,50, c("low EF", "high EF"), pch=c(21,24), bty="n", cex=0.8)
```

A few observations can be made from the previous analysis. First of all, for the data without the trend, the time dimension was reduced, and a stable spatial pattern for one moment in time was presented (see Figure 7.8a). Second, it was possible to determine which observations were outliers and break the spatial trend. On the one hand, these are negative outliers, with the lowest values of the first principal component (*eigenfunction, EF*), located peripherally. On the other hand, they are positive outliers – centrally located locations, most often in big cities.

The extrapolation of the spatial trend (process) to new spatial points is presented subsequently (see Figure 7.8b). This allows for a more complete coverage of the area being examined. The **spatpca()** command assumes that new spatial points will be given in matrix form. The **spsample()** command from the sp:: package (Pebesma & Bivand, 2005) generates points within a given area (here national contour), creating an object in the *sp* (and *SpatialPoints*) class. It is possible to convert it to the *data.frame* class, with the command **as.data.frame()**, and then to the *matrix* class with the command **as.matrix()**. In the object names, the extensions have been added, just to indicate the object's class. Trend interpolation can also be performed by the **spatpca()** command, but there one gives the parameter values (obtained in the base estimation) and new points.

```
# kriging for new points – Figure 7.8b
# in search of extrapolation of the spatial trend

# drawing new points in the map area and their conversion to the matrix class
# draw options: random | regular | stratified
library(sp)
pl<-readOGR(".", "Panstwo")
pl<-spTransform(pl, CRS("+proj=longlat +datum=NAD83"))

newpoints<-spsample(pl, 20000, type="stratified") # from the sp:: package
newpoints.df<-as.data.frame(newpoints)
newpoints.m<-as.matrix(newpoints.df)

# interpolation by kriging on new points
new.output<-spatpca(x=crds, Y=dane.td, K=spca$Khat, tau1=spca$stau1,
          tau2=spca$stau2, x_new=newpoints.m)
quilt.plot(newpoints.m, new.output$eigenfn[,1])
plot(voi, add = TRUE, border="grey80")
```

The obtained result is an estimate of 380 points (x, y) on 20,000 points (x', y') drawn. One can see where there are centres of high unemployment (the lowest values on the map) and areas without the risk of unemployment (the highest values on the map). It is possible to assign selected points to administrative areas and map them (see Chapter 4). A vector of eigenvalues can be used as a spatial process structure in other models.

7.4 Spatial drift

Modern spatial research distinguishes two spatial processes that are competitive with each other: spatial autocorrelation, which is the similarity of observation in the closest neighbourhood, and spatial drift, which is the directionality and clustering of specific effects in specific locations. Spatial autocorrelation is modelled by typical spatial models with three, two or one spatial components (see Chapter 5). The methodological answer to the local variation of effects and spatial drift is the GWR method, which allows local coefficients for each observation to be obtained but does not examine the similarity in the space of these coefficients (see Chapter 6).

Müller, Wilhelm, and Haase (2013) proposed modelling spatial drift based on GWR regression coefficients, which are successively clustered using the *k*-means method, and cluster membership modelled with a dummy variable in a global spatially heterogeneous model. This approach allows for taking into account both spatial processes: spatial autocorrelation and spatial drift.

Example

In the following example, one looks for a spatial drift in the distribution of the unemployment rate at the NTS4 poviat level. For 380 observations (poviats) of the unemployment rate in a given month (May 2018) B_{t2}, i, the dynamic-spatial model was estimated. As explanatory variables, the unemployment rate was tested in previous months (time lags April 2018 B_{t1},i and March 2018 B_{t0}, i), local spatio-temporal lags (April 2018 lagged with contiguity matrix $W_{conti}B_{t1}$,i, March 2018 lagged with contiguity matrix $W_{conti}B_{t0}$,i), global spatio-temporal lags (April 2018 lagged with inverse distance matrix $W_{inv.dist} B_{t1}$,i, March 2018 lagged with inverse distance matrix $W_{inv.dist} B_{t0}$,i) and relative *dist*$_i$ location – the Euclidean distance of the poviat from the core city. The form of the final model is given by equation 7.11. Before the estimation, it is necessary to prepare the data (including spatial lags) and a spatial weights matrix.

```
# preparation of data for estimation
# selecting variables from the data set
data<-unempl[,c(1:10,99:101)]

# contiguity spatial weights matrix
crds<-coordinates(pov)
cont.nb<-poly2nb(as(pov, "SpatialPolygons"))
cont.listw<-nb2listw(cont.nb, style="W")

# spatial weights matrix by inverse distance
pov.knn<-knearneigh(crds, k=379)
pov.nb<-knn2nb(pov.knn)
dist<-nbdists(pov.nb, crds)
dist1<-lapply(dist, function(x) 1/x) # list class object
# list class object - weight according to the distance criterion
invdist.listw<-nb2listw(pov.nb, glist=dist1)

# time-space lags
data$X2018.03.Wconti<-lag.listw(cont.listw, data$X2018.03)
data$X2018.04.Wconti<-lag.listw(cont.listw, data$X2018.04)
data$X2018.03.Winvdist<-lag.listw(invdist.listw, data$X2018.03)
data$X2018.04.Winvdist<-lag.listw(invdist.listw, data$X2018.04)
```

The geographically weighted regression was estimated:

$$B_{t2,i} = \beta_0 + \beta_1..B_{t1,i} + \beta_2..B_{t0,i} + \beta_3..W_{Conti}B_{t1,i} + \beta_4..W_{Conti}B_{t0,i} + \\ +\beta_5..W_{inv.dist}B_{t1,i} + \beta_6..W_{inv.dist}B_{t0,i} + +\beta_5..dist_i + \varepsilon_i$$

(7.11)

The commands from the spgwr:: (Bivand & Yu, 2017) were used: **ggwr.sel**() to estimate the bandwidth and **ggwr**() to estimate the GWR coefficients. The bandwidth is estimated by optimisation – minimisation of RMSE. The **ggwr.sel**() command assumes two approaches to the spatial dimension by selecting the bandwidth: global, taking into account all observations as neighbours (*adapt = FALSE*), and adaptive, with variable bandwidth, based on *k* nearest neighbours (*adapt = TRUE*). Using the option *longlat = FALSE*, the bandwidth is expressed in degrees.

```
# model
library(spgwr)

eq<-X2018.05~X2018.04+X2018.03+X2018.04.Wconti+ X2018.03.Wconti+
+X2018.04.Winvdist+X2018.03.Winvdist+dist

# bandwidth
bw<-ggwr.sel(eq, data=data, coords=crds, family=poisson(), longlat=TRUE)

# GWR model # generalized geographically weighted regression
model.ggwr<-ggwr(eq, data=data, coords=crds, family=poisson(), longlat=TRUE,
bandwidth=bw)
model.ggwr
```

```
Call:
ggwr(formula = eq, data = dane, coords = crds, bandwidth = bw,
    family = poisson(), longlat = TRUE)
Kernel function: gwr.Gauss
Fixed bandwidth: 41.90926
Summary of GWR coefficient estimates at data points:
                      Min.      1st Qu.     Median    3rd Qu.      Max.    Global
X.Intercept.    -6.3969e+00  -7.5663e-03  7.5061e-01  1.3562e+00  6.9024e+00   0.7551
X2018.04        -2.5039e-01   8.1900e-02  1.6418e-01  2.4754e-01  7.3616e-01   0.1396
X2018.03        -5.7055e-01  -1.2738e-01 -4.7595e-02  3.3176e-02  3.5443e-01  -0.0377
X2018.04.Wconti -1.8713e+00  -2.5929e-01 -6.5021e-02  1.1806e-01  8.4582e-01  -0.0266

X2018.03.Wconti -8.7277e-01  -1.1168e-01  5.7311e-02  2.5010e-01  1.7783e+00   0.0222
X2018.04.Winvdist -1.9168e+01 -2.3712e+00 -1.5837e-01  1.8050e+00  2.3057e+01   0.7935
X2018.03.Winvdist -2.2300e+01 -1.5583e+00  2.3967e-01  2.2689e+00  1.8654e+01  -0.7121
dist            -3.4306e-03  -5.9592e-04  1.9023e-04  8.5557e-04  4.8293e-03   0.0005
```

The result of the GWR estimation is the variability intervals of the regression coefficients. Each coefficient can have a different value for each observation. The last column of the result, Global, are coefficients from a typical regression model with a global parameter for each variable. It is noticeable that in the GWR model, all coefficients achieve a significant level of volatility; that is, they take both positive and negative values. At the band estimation stage, the cross-validation optimisation in the **ggwr.sel**() command, subsequent optimisation steps were displayed. The band converged in 15 iterations from the initial value of 340 (CV = 729) to 41.9 (CV = 541).

Spatial patterns of regression coefficients can be illustrated in diagrams. The **choropleth**() command from the GISTools:: package was used. Each of the presented spatial distributions of GWR coefficients (see Figure 7.9) is characterised by quite strong spatial autocorrelation, although the patterns themselves are different. The distributions for subsequent time lags (b) and c) are their own mirror images.

```
library(GISTools) # Figure 7.9
choropleth(pov, model.ggwr$SDF$X2018.04)
choropleth(pov, model.ggwr$SDF$X2018.04.Wconti)
choropleth(pov, model.ggwr$SDF$X2018.04.Winvdist)
choropleth(pov, model.ggwr$SDF$dist)
```

In many studies, one can find an approach to analyse individual spatial distributions of coefficients. The spatial drift model proposes a combined analysis approach. All GWR coefficients are treated as a combined dataset, and the existence of clusters is examined due to all of these coefficients. Müller et al. (2013) proposed using the k-means method to search for spatial clusters in a set of GWR coefficients (see Section 7.1) (see Figure 7.10).

```
# clustering of GWR coefficients
# ex ante survey of the optimal number of clusters
library(factoextra)
fviz_nbclust(as.data.frame(model.ggwr$SDF[,2:9]), FUNcluster=kmeans)

# clustering - approach 1 - stats:: package
# kmeans() from the stats:: package gives the result in the kmeans class
clusters1<-kmeans(as.data.frame(model.ggwr$SDF[,3:5]), 2) # 2 clusters
choropleth(pov, clusters1$cluster) # the 2nd argument is a clustering vector
title(main="2 clusters, results from kmeans()")

# clustering - approach 2 - factoextra:: package
# eclust() from the factoextra:: - outputs in kmeans and eclust classes
```

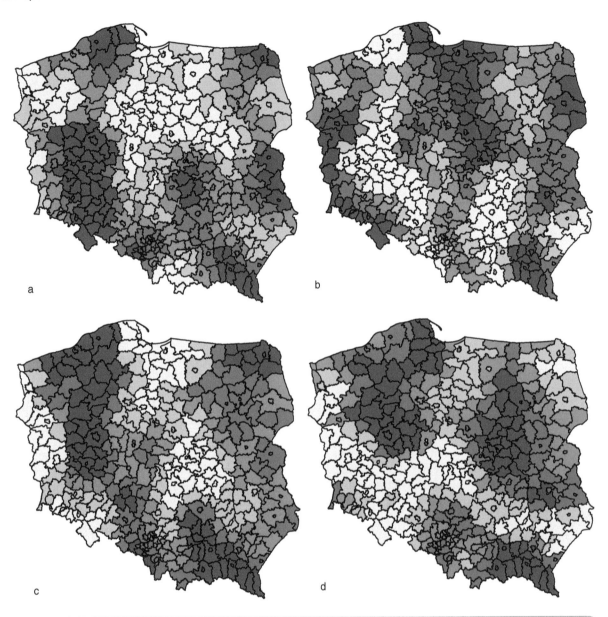

Figure 7.9 Spatial distribution of coefficients from GWR: a) time lag t_1 (B_{t1}, i), b) space-time local lag (t_1) with contiguity matrix ($W_{conti}\,B_{t1}$, i), c) local spatio-temporal lag (t_0) with contiguity matrix ($W_{conti}\,B_{t0}$, i), d) pov distance from the voivodeship city – relative location

Source: Own study using R and spgwr:: and GISTools:: packages

```
clusters2<-eclust(as.data.frame(model.ggwr$SDF[,2:5]), "kmeans", k=8)
choropleth(pov, clusters2$cluster) # Figure 7.10a
title(main="8 clusters, result from eclust()")
text(clusters2$centers[,5:6], labels=1:8, cex=2, col="green")
fviz_silhouette(clusters2)    # Figure 7.10b
```

	cluster	size	ave.sil.width
1	1	48	0.31
2	2	40	0.44

8 clusters, result from eclust()

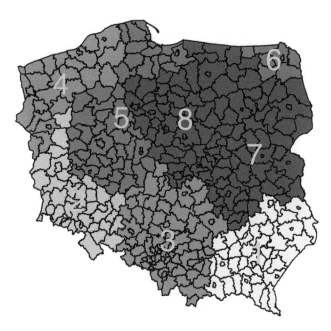

Clusters silhouette plot
Average silhouette width: 0.36

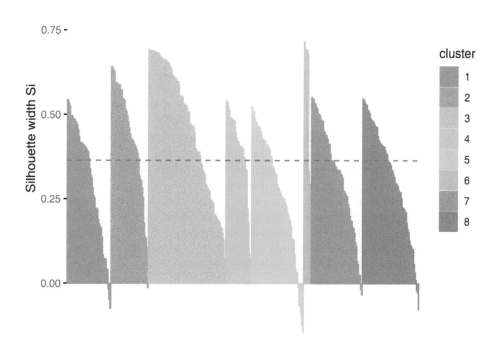

Figure 7.10 Spatial quality and visualisation of division of poviats into clusters based on all GWR coefficients

Source: Own study using R and factoextra:: stats:: and GISTools:: packages

3	3	83	0.47
4	4	28	0.38
5	5	56	0.27
6	6	8	0.51
7	7	55	0.32
8	8	62	0.32

```
fviz_cluster(clusters2, geom="point", ellipse.type="norm") # 2D projection
```

```
clusters2$silinfo # quality of clustering
```

```
$widths
cluster   neighbor      sil_width
5     1           7   0.5433873641
12    1           7   0.5412631026
142   1           7   0.5321054613
2     1           7   0.5253153912
140   1           7   0.4965884398
```

```
clusters2$size   # the size of clusters
```

```
[1]  48 40 83 28 56     8 55 62
```

The obtained *ex ante* result of searching for the optimal number of clusters suggests the creation of two clusters, because in all other cases, global silhouette statistics will reach a lower value. The spatial distribution of clusters is similar to distributions a) and c) from Figure 7.9, in other words, the first time lags and time-space (global) lags. This proves the strong dependence of the phenomenon on historical values (so-called path dependence). The result from the *silinfo* slot shows the affiliation of the observation (poviat) to the cluster and the nearest next cluster, as well as individual (local) statistics. Slot *size* shows that the sizes of clusters are not equal, which means that the algorithm divides the observations into groups of more and less typical ones. When assuming the division into eight clusters (see Figure 7.10), only a few marginal suboptimal assignments appear (negative values of the silhouette statistics), while clustering emphasises spatial correlations and developmental durability. The overlapping ellipses (from **fviz_cluster**()) show the hierarchical character of the phenomenon and the existence of subgroups within the broader classes. The previous analysis highlights the directions of spatial processes – spatial drift. This proves that the development process does not spread evenly in space, although, on the other hand, it is spatially continuous. Belonging to clusters is not random over space; on the contrary, it exhibits high spatial autocorrelation.

The last stage of spatial drift modelling together with spatial autocorrelation is the creation of dummy variables for each cluster and the estimation of a typical spatial model that takes these variables into account. Because the spatial lags of explained variables have been included in the GWR model, the estimated model of global coefficients should contain at most the spatial error model (SEM) or be an a-spatial model, as the entire spatial autocorrelation has been absorbed by the explanatory variables.

```
# creating dummy variables for clusters (divided into 8 clusters)
data$clust1<-rep(0, times=dim(data)[1])
data$clust1[clusters2$cluster==1]<-1
data$clust2<-rep(0, times=dim(data)[1])
data$clust2[clusters2$cluster==2]<-1
data$clust3<-rep(0, times=dim(data)[1])
data$clust3[clusters2$cluster==3]<-1
data$clust4<-rep(0, times=dim(data)[1])
data$clust4[clusters2$cluster==4]<-1
```

```
data$clust5<-rep(0, times=dim(data)[1])
data$clust5[clusters2$cluster==5]<-1
data$clust6<-rep(0, times=dim(data)[1])
data$clust6[clusters2$cluster==6]<-1
data$clust7<-rep(0, times=dim(data)[1])
data$clust7[clusters2$cluster==7]<-1

# new equation taking into account clusters of GWR coefficients
eq1<-X2018.05~ X2018.04 + X2018.03 + X2018.04.Wconti + X2018.03.Wconti +
dist+ clust1 + clust2 + clust3 + clust4 + clust5 + clust6 + clust7

cont.listw<-nb2listw(cont.nb, style="W") # reminder of the matrix W
invdist.listw<-nb2listw(pov.nb, glist=dist1) # reminder of the matrix W

# spatial error model
model.sem<-errorsarlm(eq1, data=data, cont.listw)
summary(model.sem)
```

```
Call:errorsarlm(formula = eq1, data = data, listw = cont.listw)

Residuals:
       Min         1Q      Median        3Q        Max
-0.9918458  -0.0938523   0.0071398  0.1025068  0.8359089

Type: error
Coefficients: (asymptotic standard errors)

                   Estimate    Std. Error   z value    Pr(>|z|)
(Intercept)       0.08967714   0.04803761    1.8668    0.06193
X2018.04          1.42854946   0.04260849   33.5273    < 2.2e-16
X2018.03         -0.44213576   0.04131734  -10.7010    < 2.2e-16
X2018.04.Wconti   0.34974536   0.08422245    4.1526    3.287e-05
X2018.03.Wconti  -0.34149787   0.08110111   -4.2108    2.545e-05
dist              0.00002808   0.00042135    0.0666    0.94687
clust1           -0.06362389   0.03713133   -1.7135    0.08662
clust2           -0.10079049   0.04005926   -2.5160    0.01187
clust3           -0.06181959   0.03589537   -1.7222    0.08503
clust4           -0.08623527   0.04536431   -1.9009    0.05731
clust5           -0.05207556   0.03886501   -1.3399    0.18028
clust6           -0.12967355   0.07487852   -1.7318    0.08331
clust7           -0.03875068   0.03604064   -1.0752    0.28229

Lambda: 0.082389, LR test value: 1.0048, p-value: 0.31615
Asymptotic standard error: 0.075017
    z-value: 1.0983, p-value: 0.27209
Wald statistic: 1.2062, p-value: 0.27209

Log likelihood: 117.3227 for error model
ML residual variance (sigma squared): 0.031536, (sigma: 0.17758) Number of
observations: 380
Number of parameters estimated: 15
AIC: -204.65, (AIC for lm: -205.64)
```

```
# a-spatial linear model
model.ols<-lm(eq1, data=data)
summary(model.ols)
```

```
Call:
lm(formula = eq1, data = data)

Residuals:
     Min      1Q  Median      3Q     Max
-0.98402 -0.09590 0.00577 0.10018 0.83537

Coefficients:
                   Estimate  Std. Error  t value  Pr(>|t|)
(Intercept)       8.708e-02   4.624e-02    1.883    0.0605 .
X2018.04          1.429e+00   4.360e-02   32.774   < 2e-16 ***
X2018.03         -4.426e-01   4.229e-02  -10.467   < 2e-16 ***
X2018.04.Wconti   3.489e-01   8.476e-02    4.117  4.75e-05 ***
X2018.03.Wconti  -3.406e-01   8.160e-02   -4.174  3.74e-05 ***
dist              6.603e-05   4.071e-04    0.162    0.8712
clust1           -6.317e-02   3.510e-02   -1.800    0.0727 .
clust2           -9.598e-02   3.805e-02   -2.522    0.0121 *
clust3           -6.057e-02   3.428e-02   -1.767    0.0781 .
clust4           -8.605e-02   4.306e-02   -1.999    0.0464 *
clust5           -5.233e-02   3.728e-02   -1.404    0.1613
clust6           -1.367e-01   7.197e-02   -1.899    0.0583 .
clust7           -3.655e-02   3.441e-02   -1.062    0.2888
---
Signif. codes:  0 '***' 0.001 '**' 0.01 '*' 0.05 '.' 0.1 ' ' 1

Residual standard error: 0.1811 on 367 degrees of freedom
Multiple R-squared: 0.998,     Adjusted R-squared: 0.9979
F-statistic: 1.533e+04 on 12 and 367 DF, p-value: < 2.2e-16
```

The previous results show that the separated clusters of GWR coefficients are mostly significant explanatory variables in the model. This shows the heterogeneity of the phenomenon in space. The unemployment rate, apart from a clear dependence on time (*path dependence*) and on the nearest neighbourhood, shows spatial supra-regional and directional patterns. In the spatial error model, the lambda coefficient (for the spatially correlated error term) is insignificant, which means that all spatial effects have been filtered out with explanatory variables, and the final cross-sectional model can be estimated by standard regression methods. The goodness of fit of the model to the data is very good ($R^2 = 0.998$), and all the coefficients are jointly significant (F test). In clusters 1–7, the unemployment rate is lower than in cluster 8 (baseline), which had to be omitted in the model.

7.5 Spatial hierarchical clustering

Yet another method of extracting clusters is based on a matrix of dissimilarities, which is equivalent to a matrix of distances between points. The resulting classification tree is used here, then cut at the selected height. The hierarchical clustering algorithm works iteratively, starting from the state in which each observation is its own cluster. In the next steps, the two most similar clusters are combined into one, until the state when a single cluster is created. In this method, there are different criteria of similarity, and it is also possible to introduce restrictions.

The basic methods of creating clusters include the Ward method, assuming the minimisation of variance between clusters and methods selecting differently located units for cluster connections: extremely separated (*complete linkage*), closest neighbours (*single linkage*), separated with average distance (*median linkage*) or centroids.[8]

In *traditional hierarchical clustering*, inertia is used as a measure of clustering quality. One can determine the intracluster inertia, between-cluster inertia and total inertia.

Within-cluster (intracluster) inertia W, assuming the existence of a P_K partition, is the sum of $I(C_K)$ inertia in all available K ($k = 1, \ldots, K$) clusters and is expressed by:

$$W = \sum_{K=1}^{k} I(C_k) \tag{7.12}$$

where the individual intracluster inertia $I(C_k)$ are determined as:

$$I(C_K) = \sum_{i \in C_k} w_i d_i^2(x_i, g_k) \tag{7.13}$$

where d_i is the distance between observation x_i and the centre of the cluster g_k, while w_i is the weight assigned to the observation – which in particular may be $1/n$ for n observations. Intracluster inertia for a single cluster is the sum of the weighted squared distances between the observations and the centre of the cluster; this measures the heterogeneity within clusters – the lower the inertia and thus the heterogeneity, the more coherent the clusters.

Between-cluster inertia B measures the separation between clusters and is expressed as the sum of the weighted squared distances d_k between the centres of g_k clusters and the centre g of all observations considered together. Hence, the intercluster inertia is given as:

$$B = \sum_{k=1}^{K} \mu_k d_k^2(g_k, g) \tag{7.14}$$

where μ_k is the sum of the weights assigned to the observations inside the given cluster k:

$$\mu_k = \sum_{i \in C_k} w_i \tag{7.15}$$

Total inertia T is the sum of the weighted squared distances d_g between individual observations x_i and the centre g of all observations taken together:

$$T = \sum_{i=1}^{n} w_i d_g^2(x_i, g) \tag{7.16}$$

The total inertia does not depend on the division into clusters. The total T inertia can also be expressed as the sum of intracluster inertia W and intercluster inertia B:

$$T = W + B \tag{7.17}$$

This implies that for a given total inertia, independent of the division into clusters, the reduction of inertia (diversity, heterogeneity) inside the cluster translates into an increase in intercluster inertia (moves the clusters away from each other).[9]

Good division into clusters is characterised by high intercluster inertia (diversity) and low intracluster inertia (heterogeneity).[10] The measure of the division quality is the percentage of total inertia explained by the division of P_k, which is equal to $Q = (1 - W/T)$. This percentage is 100% when each observation is its own cluster (so-called *singletons*) and is 0% when all observations are in one cluster.[11]

In spatial terms, there are several modifications to the classical approach (Chavent, Kuentz-Simonet, Labenne, & Saracco, 2018).[12] First, in the traditional a-spatial approach, clusters for observations are created based on a set of attributes assigned to these observations, while their diversity reflects the dissimilarity matrix D_0. The location, both absolute in the coordinate system and relative to other observations, is not considered. In spatial terms, Chavent et al. (2018) introduced spatial restrictions regarding the location of points. The clue is to take into account the second D_1 dissimilarity matrix reflecting the location distances between points. They introduce *mixing parameter α*, which weights both dissimilarity matrices D_0 and D_1. The goal is to find such an α parameter to increase the spatial coherence of clusters as much as possible without compromising the coherence resulting from the values of the features.

Second, non-Euclidean distances are considered, which requires the use of a generalised form of intraclustered inertia:

$$I_\alpha(C_K) = \sum_{i \in C_K} \sum_{j \in C_K} \frac{w_i w_j}{2\mu_k} d_{ij}^2 \tag{7.18}$$

where $\mu_k = \Sigma i \in C_K\ w_i$ is the weight of the C_K cluster.

Third, as a consequence, the full measure of the intracluster pseudo-inertia for cluster K, depending on the linking parameter α, is given as:

$$I\left(C_K^\alpha\right) = \left(1-\alpha\right)\sum_{i\in C_K}\sum_{j\in C_K}\frac{w_i w_j}{2\mu_k^\alpha}d_{0,ij}^2 + \alpha\sum_{i\in C_K}\sum_{j\in C_K}\frac{w_i w_j}{2\mu_k^\alpha}d_{1,ij}^2 \tag{7.19}$$

where the total intracluster inertia is, as in the classical approach, the sum of inertia for $W = \sum_{k=1}^{K}I_\alpha\left(C_K^\alpha\right)$ clusters.

In R, a-spatial hierarchical clustering can be done with the **hclust()** command from the stats:: package, which contains various clustering methods in the options and a command to trim the tree **cutree()**, also from the stats:: package. The **hclust()** command allows a full grouping structure of objects to be obtained, and the final result, depending on the desired data aggregation, is obtained by the **cuttree()** command, which reflects the selected cross-section of the whole structure. One can indicate, for example, the desired number of clusters or neighbourhood within a specified distance (on the scale of values from the distance matrix). The results of the **hclust()** command are in the *hclust* class. The interactive **identify.hclust()** command from the stats:: package allows the levels of cluster cuts and their structure to be tested by mouse clicks. There is also a *dendrogram* class, within which it is possible to draw and transform various objects. The **rect.hclust()** command from the stats:: package allows rectangles (boundaries) to be applied to the indicated observation groups/clusters.

Clustering with spatial constraints is available in the ClustGeo:: package, where the **hclustgeo()** command is available for clustering, a command to diagnose the linking parameter α – **choicealpha()** allowing selection of the empirical parameter and **plot.choicealpha()** to draw it. There are also measures of quality: inertia of individual clusters given by the command **inert()**, total pseudo-inertia given by the command **inertdiss()**, intracluster pseudo-inertia based on a dissimilarity matrix for a specific division **withindiss()** and the measure of Ward aggregation (when each point is a single cluster, so-called singletons) in the **wardinit()** command. The results of the **hclustgeo()** command are in the *hclust* class. The **cutree()** tree truncation command from the stats:: package also works for the **hclustgeo()** command, thanks to the compatibility of the result classes.

Example

Approximately 2000 random observations were selected from the dataset for company locations (*firms*). In the presented simple unconditional hierarchical analysis, only the geographical location (x, y) is considered, and the features of points (z) are not taken into account. In this example, the *de facto* typical problem of clustering has been reversed: while in typical a-spatial tasks, the features relate to the properties of non-location points, here it was assumed that the location coordinates are the only analysed points' features. In such a situation, strictly spatial analysis is carried out based on the a-spatial algorithm, which is possible only because non-localisation features of points are not taken into account. In the implementation of the example, there is one non-location variable (z), in the subset, but its role is mainly illustrative.

To determine the distance between the points, the **distm()** command from the geosphere:: package was used, which requires spherical coordinates as input data. The typical hierarchical clustering command **hclust()** will be used along with the **cutree()** tree truncation command, both from the stats:: package. The mapping of the result and the determination of cluster resources with the command **gCentroid()** from the sp:: package will also be shown.

In spatial analyses, typical coordinates recorded as *data.frame* are mostly planar. Hence, it is necessary to transform the projection of the coordinate into a spherical form. The most convenient data format is *SpatialPointsDataFrame*, within a wide *sp* class, defined by the sp:: package. The transition from *data.frame* to *SpatialPointsDataFrame* is possible, among others by defining a coordinate within a set with the **coordinates()** command from the sp:: package – this operation changes the class of the object and furthermore gives the projection to the dataset with the **proj4string()** command from the sp:: package. One sets the coordinates' type (spherical or planar) with the **spTransform()** command from the sp:: package by assigning different types of projections.

```
# creating a subset and formatting the coordinate
firms.sub<-firms[20000:22000, c(12:13,20)] # variables x,y,z
class(firms.sub)
```

```
[1] "data.frame"
```

```
# headings: x and y is the location, with the observation feature
colnames(firms.sub)<-c("x","y","z")
coordinates(firms.sub)<-c("x","y") # defining the coordinate
class(firms.sub) # visible change of the object class
```

```
[1] "SpatialPointsDataFrame"
attr(,"package")
[1] "sp"
```

```
# giving the projection to the data collection proj4string(firms.sub)<-
"+proj=longlat +datum=WGS84 +ellps=WGS84"

# projection formatting: defining the coordinate as planar firms.sub<-
spTransform(firms.sub, CRS("+proj=merc +datum=WGS84
+ellps=WGS84"))

# projection formatting: defining the coordinate as spherical firms.sub<-
spTransform(firms.sub, CRS("+proj=longlat +datum=WGS84
+ellps=WGS84"))
```

For a formatted dataset, one can specify a matrix of distances between points using the **distm()** command from the geosphere:: package. A *matrix* class object is obtained. The **distm()** command returns the distance in meters, and to obtain distances in km, the entire matrix should be divided by 1000 (m).

```
library(geosphere)
# requires spherical coordinates, result in meters
mdist<-distm(firms.sub)
mdist[1:5, 1:5]
```

```
           [,1]        [,2]       [,3]        [,4]        [,5]
[1,]       0.00    24632.78   139192.7    51380.40   113165.32
[2,]   24632.78        0.00   123467.1    56445.04    98274.53
[3,]  139192.75   123467.09        0.0   106951.67   185353.04
[4,]   51380.40    56445.04   106951.7        0.00   154438.57
[5,]  113165.32    98274.53   185353.0   154438.57        0.00
```

```
mdist.km<-mdist/1000 # result in km mdist.km[1:5, 1:5]
```

```
           [,1]        [,2]       [,3]        [,4]        [,5]
[1,]    0.00000    24.63278   139.1927    51.38040   113.16532
[2,]   24.63278     0.00000   123.4671    56.44504    98.27453
[3,]  139.19275   123.46709     0.0000   106.95167   185.35304
[4,]   51.38040    56.44504   106.9517     0.00000   154.43857
[5,]  113.16532    98.27453   185.3530   154.43857     0.00000
```

There are several algorithms for extracting clusters as part of the **hclust()** command, followed by one of the most popular (*method="complete"*). The clustering **hclust()** command requires an object in the *dist* class as the input, which can be obtained with the conversion by **as.dist()**. The clustering result is in the *hclust* class and specifies the full tree (dendrogram). Trimming a tree at a specified height allows selecting from this full tree the cluster structure at a satisfactory level of aggregation. In option k, the desired number of clusters is defined (e.g. $k = 3$). In the alternative option, h sets the neighbourhood cutoff point (h is on the distance scale from the

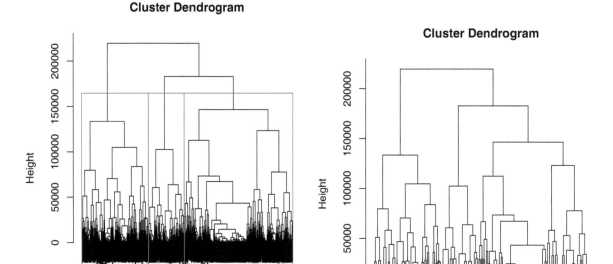

Figure 7.11 Hierarchical structure of the dendrogram

Source: Own study using R and the stats:: package

distance matrix) – the larger the *h* radius, the fewer the clusters. The values of *h* can be read on the *y*-axis of the dendrogram. The clustering vector from the result was saved in the data object (see Figure 7.11).

```
hc<-hclust(as.dist(mdist), method="complete")
firms.sub$clust<-cutree(hc, h=60000)# neighbors within a radius of 60km
firms.sub$clust<-cutree(hc, k=5) # division into 5 clusters
plot(hc)
plot(hc, hang=-1) # lower label management
```

An insight into the hierarchical structure of the dendrogram can be obtained by interactive examination of the separated groups. The **identify.hclust()** command from the stats:: package allows the level of cluster cuts to be selected with the mouse/cursor and their structure to be displayed, that is, the observations included in the cluster. To save the selection result as an object, the result of the **identify()** command should be assigned the object name.

```
plot(hc, hang=-1) # typical dendrogram
# interactive command that activates the cursor selection mode
x<-identify(hc)
x # displaying observations belonging to selected branches
```

```
[[1]]
 [1]      38   69  117  129  163  247  281  291  303  310  314  317  369  371  392
[16]     396  442  479  512  539  557  562  617  630  636  640  641  646  662  670
[31]     680  687  690  694  725  782  792  809  822  865  904  952  954  955 1005
[46]    1045 1076 1092 1110 1127 1139 1180 1208 1256 1267 1299 1374 1412 1416 1420
```

```
[61]   1455 1460 1499 1569 1576 1580 1584 1596 1635 1644 1656 1667 1686 1699 1716
[76]   1723 1744 1774 1788 1814 1911 1931 1943 1961
[[2]]
[1]      44  151  153  192  221  227  231  283  290  296  307  322  327  335  341
[16]    351  374  377  419  427  428  432  446  453  468  486  521  525  544  550
[31]    634  654  673  701  704  719  734  784  787  810  873  883  899  937  941
[46]    946  980 1052 1058 1064 1077 1082 1120 1146 1150 1164 1178 1198 1225 1250
[61]   1257 1261 1273 1303 1309 1383 1424 1426 1431 1497 1559 1612 1625 1690 1748
[76]   1775 1783 1785 1804 1820 1822 1849 1858 1888 1899 1923 1939 1947 1983
```

Similarly, for a dendrogram, one can indicate (non-interactively) where the cluster boundaries should be drawn. In the **rect.hclust()** command from the stats:: package, one can specify the option *k* if clusters are to be extracted or option *h* for the height at which the indicated clusters are to be cut off – they are indicated in the *which* option, with ordinal numbers given at a specified height counted from left to right. In the example subsequently, three clusters are marked, and in the next line of code, clusters 2 and 4 (*which=c (2,4)*) at a height of *h=100,000*. The *border* option (from 1 to 8) defines the *border* colours: 1 = black, 2 = red, 3 = green, 4 = dark blue, 5 = blue, 6 = pink, 7 = yellow, 8 = gray (see Figure 7.11b).

```
# a rectangle that divides the result into three clusters plot(hc)
rect.hclust(hc, k=3, border="red")
# two rectangles for the second and fourth cluster
plot(hc)
x<-rect.hclust(hc, h=100000, which = c(2,4), border=5:6)
```

The result obtained previously can be enriched by determining the means (centroids) of the separated clusters. The **gCentroid()** command from the rgeos:: package specifies the centre of gravity of the geometry, whereby the point group (belonging to a specific cluster) in the *SpatialPoints* object can be treated as the input geometry. The following code sets in the loop for subsequent clusters the coordinates of the geometric centres of such a cluster and stores them in the prepared matrix. The result of the **gCentroid()** command is in the *SpatialPoints* class and contains the cluster constraints (*bbox*) and coordinates of the measure (*coords*).

```
library(rgeos)
how.many.clust<- max(firms.sub$clust)
centr<-matrix(0, ncol=2, nrow=how.many.clust)
for (i in 1:how.many.clust)
centr[i,]<-gCentroid(subset(firms.sub, clust == i))@coords
centr
```

```
         [,1]      [,2]
[1,]   22.46760  51.18914
[2,]   23.29756  50.66582
[3,]   23.15874  51.91049
[4,]   22.33370  51.81257
[5,]   23.47162  51.24080
```

One can illustrate the previously mentioned division into clusters and derived centres of these clusters. A smart way to conditionally match the colours of points with the cluster number is to use the **factor()** command as part of the *col* option in the **plot()** command. The colours were chosen from the basic **rainbow()** palette. The centroids were marked separately as black points.

In centroids, one can also anchor the circles with specific radii to assess the spatial extent of clusters. The **circles()** command from the dismo:: package creates such circles for a given radius[13] as polygons. In the *dissolve = FALSE* option, the overlapping of circles can be switched on, and by default, the overlapping areas of these objects are treated as a union.

```
plot(firms.sub, col=rainbow(5)[factor(firms.sub$clust)], pch=".", cex=3)
points(centr, pch=".", cex=10, col="black")
plot(lubelskie.voi, add=TRUE)
library(dismo)
rings<-circles(centr, d=30000, lonlat=T, dissolve=TRUE) # r=30km
plot(rings@polygons, axes=TRUE, add=TRUE)
```

Example

Simple cluster hierarchical analysis has a version with spatial limitations. The example subsequently uses panel regional data on the monthly unemployment rate in 2017 (*unempl*). Centroids of regions – poviats (NTS4) – were used as a proxy of locations.

The basic clustering command from the ClustGeo:: package, **hclustgeo()**, assumes as input one or two dissimilarity matrices, which are objects in the *dist* class, as well as the α parameter and options for matrix scaling and data weighting. Objects in the *dist* class can be obtained from the **dist()** command or as a converted result from another class to the *dist* class using the **as.dist()** command.

Two matrices, the dissimilarity matrix of data (features) D_0 and the (geographic) distance matrix D_1 between locations were created. D_0 for regional features was obtained with the **dist()** command and **distm()** was used for distances between geographic coordinates. The key parameter of this analysis is the α parameter. It weighs the shares of D_0 and D_1. The choice of α may be arbitrary or based on the Q criterion function (or normalised Q – Qnorm) for both matrices. The intersection of these curves determines the optimal value of α. The Q criterion is defined as the percentage of intercluster pseudo-inertia to the total inertia ($Q = 1 - W/T = B/T$) (see Figure 7.12).

```
dane<-unempl[,85:96]
crds<-coordinates(pov)
D0<-dist(dane, method="euclidean")

library(geosphere)
D1<-as.dist(distm(crds))

library(ClustGeo)
range.alpha<-seq(0,1,0.1)
chosen<-choicealpha(D0,D1,range.alpha,K=5,graph=FALSE)
plot(chosen, cex=0.8,norm=FALSE) # Q - Figure 7.12a
#plot(chosen, cex=0.8,norm=TRUE)    # standardized Q

alpha<-0.35
hcg<-hclustgeo(D0,D1,alpha=alpha, wt=NULL)
plot(hcg,labels=FALSE) # typical dendrogram for full classification
hcg.cut<-cutree(hcg, k=5)
hcg.cut
```

```
  [1] 1 1 1 1 2 2 2 1 2 1 2 2 1 2 3 3 3 3 3 3 3 3 3 3 1 3 3 4 1 3 4 3 3 1 3 3 1
 [38] 3 1 4 3 3 1 1 3 1 3 3 4 1 2 1 1 2 1 1 1 3 3 3 1 3 3 3 1 3 1 1 1 1 3 1 1 2
 [75] 2 2 1 2 2 2 2 2 2 2 2 2 2 2 2 2 2 2 2 2 3 3 3 3 3 3 3 3 3 2 3 3 3 3 5
[112] 3 3 2 3 5 3 3 3 5 5 5 5 5 5 5 5 5 5 5 5 5 1 1 2 1 2 1 1 1 1 1 2 4 5 5 5 1
[149] 3 3 3 5 5 5 2 5 5 2 2 5 2 5 3 2 5 5 5 2 5 2 2 3 3 5 5 3 2 5 5 5 5 5 2 5 3 5
[186] 5 3 3 3 3 2 1 3 3 4 5 3 3 5 5 3 3 3 3 3 5 5 5 5 5 5 5 5 5 5 5 3 3 3 5 5 5 3
[223] 5 5 2 1 2 1 5 5 4 1 1 5 5 1 5 4 4 2 2 5 4 4 5 5 5 5 4 2 4 3 3 5 4 4 3 3 5
[260] 3 4 4 4 3 5 5 4 5 5 5 5 5 4 5 2 2 2 2 2 2 2 2 2 3 4 4 5 4 4 4 4 4 5 4 5 4
[297] 5 5 5 5 5 5 4 5 5 5 2 2 2 2 2 2 2 2 2 2 2 2 2 2 2 2 5 2 2 2 2 2 2 2 2 5
[334] 5 5 5 5 2 5 5 4 5 2 5 4 4 5 4 5 5 5 5 2 5 5 2 5 5 5 5 5 4 4 4 5 4 4 5 4
[371] 4 5 4 4 4 4 4 4 5 5
```

```
plot(pov, border="grey", col=hcg.cut) # map of Figure 7.12b
```

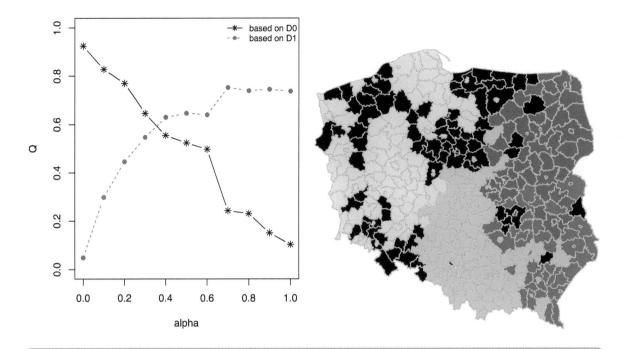

Figure 7.12 Q criterion dependent on parameter α and spatially conditioned clustering
Source: Own study using R and ClustGeo:: and sp:: packages

For both matrices, D_0 and D_1, one can determine the inertia, which is available in the ClustGeo:: package. The **inertdiss()** determines the total inertia for a given dissimilarity matrix, the **inert()** command determines the inertia of the indicated cluster (the indicated subset of data) and the **withindiss()** command determines the intra-cluster inertia for the specified clustering. The **inderdiss()** and **withindiss()** commands operate on dissimilarity matrices and **inert()** on the underlying data. In the **withindiss()** command, one gives as input data the clustering vector, which is the result of the **cutree()** command. In the **inert()** command, it is the same, but indicating the ID of the cluster, for which an individual intracluster inertia is to be calculated. The **intert()** command requests a vector that specifies which rows of the subset to use for the calculations. The **which()** command gives the observation numbers of the *hch.cut* vector that meet the condition (e.g. *hcg.cut == 2* -cluster No. 2), which gives equivalent indexing as in the full dataset.

```
inertion<-matrix(0, nrow=3, ncol=2) #object to store results
colnames(inertion)<-c("D0- features ","D1-localization")
rownames(inertion)<-c("intra-cluster", "total", "percentage")

inertion[1,1]<-withindiss(D0, part=hcg.cut)        # intra-cluster
inertion[1,2]<-withindiss(D1, part=hcg.cut)
inertion[2,1]<-inertdiss(D0)                       # overall
inertion[2,2]<-inertdiss(D1)
inertion[3,1]<-inertion[1,1]/ inertion[2,1]        #percentage
inertion[3,2]<-inertion[1,2]/ inertion[2,2]
inertion
```

```
                D0-features  D1-localization
intra-cluster   113.8413743  1.749762e+10
total           238.6487391  4.955921e+10
percentage      0.4770248    3.530649e-01
```

```
1-inertion[3,] # Q measure
```

```
D0-features D1-localization
0.5229752  0.6469351
```

```
# intra-cluster inertia for subsequent clusters - D0
i1<-inert(dane, indices=which(hcg.cut==1))
i2<-inert(dane, indices=which(hcg.cut==2))
i3<-inert(dane, indices=which(hcg.cut==3))
i4<-inert(dane, indices=which(hcg.cut==4))
i5<-inert(dane, indices=which(hcg.cut==5))

# values of individual inertia and their sum
cbind(i1, i2, i3, i4, i5, sum(i1, i2, i3, i4, i5))
```

```
          i1       i2       i3       i4       i5
[1,]  16.2077 11.26378 22.59072 13.86947 49.9097 113.8414
```

As one can see, the internal diversity of features in clusters is higher (0.477) than the distances in clusters (0.353); thus clusters based on geographical distances are more homogeneous than the features linked to them. The sum of individual intracluster inertia determined successively with the **inert**() command is equal to the total intracluster inertia determined by the **withindiss**() command.

7.6 Spatial oblique decision tree

An interesting algorithm for splitting the surface (*Spatial Partitioning*) into spatially irregular regions, possibly uniform in respect to the observed (*z*) values in points (*x*, *y*), is the spatial oblique decision tree (SpODT). The SpODT algorithm, using regression curves, determines the areas in which the values of the variable (*z*) are similar – for example, low. It compares neighbour values looking for a boundary (straight line) that will divide different values. The SpODT algorithm is an extension of the *classification and regression tree* (CART) algorithm, which also aims to extract areas of similar values but only uses rectangles instead of straight lines.

The SpODT algorithm is available in the SPODT:: package (Gaudart et al., 2015).[14] It works on data in the *SpatialPointsDataFrame* class.

Example

For monthly data on unemployment in poviats (*unempl*), only one month was selected, and together with the coordinates of poviat centres obtained from the **coordinates**() command, it was combined into a common object of the *SpatialPointsDataFrame* class. Coordinates were converted to planar using **proj4string**() and **spTransform**() because of the requirements of the **spodt**() command.

In order to illustrate the analysed phenomenon, a typical regional map of the observed values was presented, made with the **choropleth**() command from the GISTools:: package (see Figure 7.13a). Next, using the **spodt**() command from the SPODT:: package, one derives a spatial division structure, while using the command **spodt. tree**() from the SPODT:: package, a decision tree was drawn (see Figure 7.13b). In the command **spodt**(), one can determine optionally in the *rtwo.min* slot the minimum value of R^2 in regression – in this example it was defined as $R^2 = 0.01$, while the maximum value for which the allocation was possible was $R^2 = 0.1$. The basic classification of data based on the decision tree shows the division nodes, as well as partition parameters (i.e. cluster id, average, variance, and local R^2) and regression function to cutoff areas (otherwise it is at the extremes where cluster id, number of locations, average and variance) are given. The components of the tree graph can be

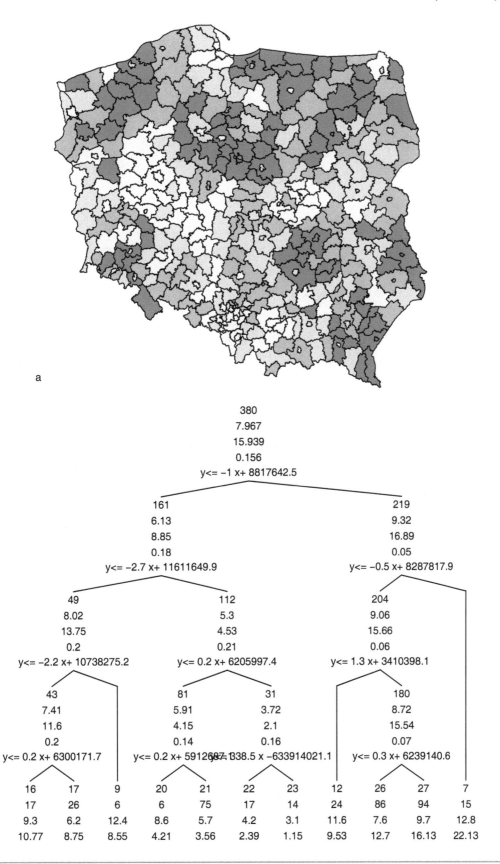

a

b

Figure 7.13 The use of spatial oblique decision trees: a) a typical regional map, b) an oblique decision tree, c) an illustration of oblique separating areas, d) abstract graphics with mis-specified parameters

Source: Own study using R and SPODT::, GISTools:: and RColorBrewer:: packages

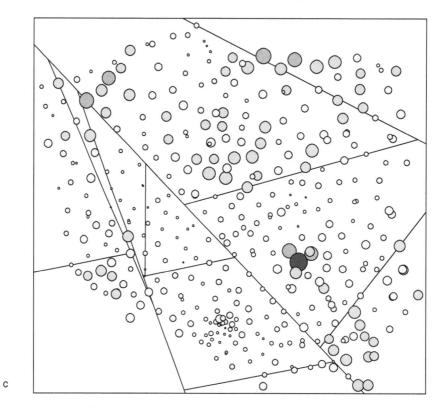

c

d

Figure 7.13 Continued

displayed with the **attributes**() command. The affiliation of each subsequent point to the clusters from the lowest division within the tree can be displayed from the slot *partition* (*division@partition*).

```
# data transformation
crds<-coordinates(pov)
# ID variable [,4], unemployment variable [,101] and population data [,12]
unempl1<-SpatialPointsDataFrame(crds, unempl[,c(4, 101, 12)])
proj4string(unempl1)<-"+proj=longlat +datum=WGS84 +ellps=WGS84"
unempl1<-spTransform(unempl1, CRS("+proj=merc +datum=WGS84 +ellps=WGS84"))
# map of the analyzed data # Figure 7.13a

library(GISTools)
library(RColorBrewer)
shades<-auto.shading(unempl1@data[,2], n=4, cols=brewer.pal(4, "Blues"))
choropleth(pov, unempl1@data[,2], shading=shades) # variable map

# decision tree library(SPODT)
division<-spodt(unempl1@data[,2]~1, unempl1, rtwo.min=0.01)
spodt.tree(division) # decision tree Figure 7.13b #attributes(division)
division@partition
```

```
  [1] 12 12 12 12 12 20 26 12 26 12 21 12 12 26 26 26 27 26 26 26 27 27 26 27 26
 [26] 26 27 27 26 26 27 26 26 26 26 26 26 27 26 27 26 26 26 26 26 26 26 26 27 26
 [51] 12 12 12 26 12 26 26 26 26 26 26 26 26 26 26 26 12 26 12 12 26 12 12 21 21
 [76] 21 26 20 21 21 21 21 20 21 21 21 21 21 20 21 21 21 21 21 26 26 26 26 26
[101] 26 26 26 26 26 21 26 26 21 26 22 26 26 21 26 22 26 26 26 17 17 17 17 17 17
[126] 17 17 17  9 17 17 17 12 12 26 12 26 20 20 12 12 12 26 27 21 27 22 26 26 27
[151] 26 23 23 21 21 23 23 21 21 21 21 27 26 21 23 22 21 21 21 21 27  7 27 22 26
[176] 26 22 22 22 22 27 21 22 27 23 22 26  7 27 27 21 26  7  7 27 23 27 27 26 27
[201] 26 27  7  7 23 23 17 22 23 23 27 23 27 22 22 26 26 27 27 27 27 21 27 27 26
[226] 26 26 26 27 27 27 26 26 27 27 26 27 27 27 21 26 27 27 27 16 16  9 23 27 21
[251] 27  7  7 27  7  7 27 27 27  7 27 27 27 27 16 27 17 16 17 17 27 27 17 21
[276] 21 21 21 21 21 21 21 21  7  7 27 27  7  7 27 27  9 17 27 27  9 17 27 27 17
[301]  9 27 27 27 27 27 21 21 21 21 21 21 21 21 21 21 21 21 21 21 21 21 21 21 21
[326] 21 21 21 21 21 21 21 17 17 16 16 16 22 17 16 27 16 22 16 27 27 16 27 23 16
[351] 27 27 21 17 16 22  9 16 16 17 16 27 27 27 27 27 27 27 27 27 27 27 27 27 27
[376] 27 27 27 16 17
```

The visualisation of the division with straight lines of points (*x, y, z*) in the geographical space requires the creation of a line object – in the *SpatialLines* class, which is possible with the command **spodtSpatialLines**() from the SPODT:: package. The extreme coordinates of the bounding box (*bbox*) can be recovered from the @ *bbox* slot. The object created with the command **spodtSpatialLines**() can be drawn with the **plot**() command and used as contour for further layers – in particular the points examined. It is also possible to add a colour layer for points (option *bg*) and manage their size (*cex* option) (see Figure 7.13c).

```
# division of surfaces with straight oblique lines
cols<-c("white", "cadetblue1", "deepskyblue", "deepskyblue3", "darkblue")
brks<-(0:5)*6
division.line<-spodtSpatialLines(division, unempl1)
plot(division.line) # 7.23c
points(unempl1, cex=unempl1@data$X2018.05/5, bg=cols[findInterval(unempl1@
data$X2018.05, brks)], pch=21)

division.line@bbox #bounding box - extreme coordinates
```

```
        min       max
x 1573523  2679203
y 6273550  7294148
```

The scaling of points on the graph is crucial for the creation of nice graphics. Randomly selected values lead to graphs that are not easily interpreted by analysts (see Figure 7.13d). The following code uses another variable from the dataset – the number of people in poviats. Due to missing observations, an automatic imputation was made using the mean value. The circles in the chart are too large – the radii defined as variable/500 are too wide.

```
# poorly chosen model parameters lead to abstract graphics
# supplementing missing data with a mean
unempl1@data[which(is.na(unempl1@data[,1])==TRUE),1]<-
mean(unempl1@data[,1], na.rm=TRUE)

sp<-spodt(unempl1@data[,1]~1, unempl1)
ssp<-spodtSpatialLines(sp, unempl1)
plot(ssp) #7.13d
points(unempl1, cex=unempl1@data$population_2013/500, pch=21)
```

The previous example shows the division into 11 clusters (the lowest classification presented by the tree). The tree specifies the cluster number (id), number of cases, average and variance of values within the cluster. For example, cluster 27 includes the highest number of observations – 94, with the average value of the unemployment rate being 9.7% and its variance equal to 16.13. One can easily verify the calculations displayed in the tree graph by selecting relevant rows of the dataset for observations from a given cluster and counting the statistics of the variable.

```
# checking which observations are in the selected cluster
a<-which(division@partition==27)

# determination of statistics as on the tree chart
length(unempl1[a,2])
#[1] 94

# double-check of the average and std.dev in cluster no.27
mean(unempl1[a,]$X2018.05)
#[1] 9.732979

var(unempl1[a,]$X2018.05)
#[1] 16.12697
```

Spatial oblique decision trees are a concept similar to hierarchical trees; however, they include a spatial dimension. The division in Figure 7.13c splits the space into possibly homogenous areas with regard to the value of analysed variable. Tree (Figure 7.13b) simplifies the overview of division into clusters using detailed information on the cluster parameters.

Notes

1 In the clusterSim:: package (Walesiak & Dudek, 2017), there are other distances: the Bray-Curtis distance for data expressed as a quotient – dist.BC() command; a generalised distance measure (GDM) for data measured on a metric scale (quotient and interval) or ordinal scale – **GDM()**, **GDM1()**, **GDM2()**; the Sokal-Michener distance for nominal data – **dist.SM()** command and distance for symbolic data (Ichino-Yaguchi or Hausdorff) – **dist.Symbolic()** command. However, these distances are poorly applicable to typical spatial data

2 The cluster:: package contains many interesting clustering commands (for specific data types) and accompanying commands.

3 The DBSCAN algorithm was proposed by Ester, Kriegel, Sander, and Xu (1996) and has been modified (e.g., DENCLUE, OPTICS or HDBSCAN) as well as employed in applications detecting financial crimes, streaming data, studying signals, gene expression, multimedia data and traffic. A good introduction is in *Vignettes to the dbscan:: Package* (Hahsler, Piekenbrock, & Doran, 2018).

4 The extension of the DBSCAN algorithm is HDBSCAN, based on a hierarchical tree of clusters. This allows for a more stable solution. In the dbscan:: package, the hdbscan() function is available, as well as **extractFOSC()**, which optimally selects clusters based on the output of the **hdbscan()** and **glosh()** commands comparing local densities with global ones. Yet another extension is the OPTICS algorithm, available via the **optics()** command from the dbscan:: package, which is based on data sorting, with the given radius treated as the upper limit. Parallel to the **optics()** command, the **reachability()** command specifying the availability structure is used.

5 In other words, the kNNdist() function creates a distance matrix, where the rows are successive points from the dataset and the columns are the kth neighbours, from first to kth. The kNNdistplot() function creates a vector from the matrix value created by the **kNNdist()** function, arranges it in ascending order and draws the distance function. The index reported on the x-axis does not have an interpretation.

6 One can also use the sNN() command from the dbscan:: package to study the number of shared neighbours. This implementation finds application in a clustering algorithm that takes into account the (fuzzy) sharing of neighbours, available in the **sNNclust()** command in the dbscan:: package.

7 Allele – one of two or more alternative forms of a gene that arise as a result of a mutation and are in the same place on the chromosome.

8
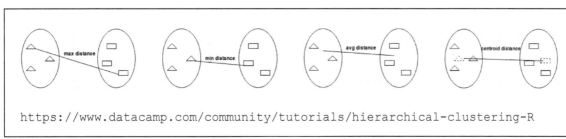

https://www.datacamp.com/community/tutorials/hierarchical-clustering-R

9 In a situation where $wi = 1/n$, the inertia is equivalent to the variance.

10 More in Murtagh and Legendre (2011), https://arxiv.org/pdf/1111.6285.pdf

11 In order to select the number of clusters, the literature uses Q, defined as $Q = (In +1 - In)/In +1$, where I is a measure of intraclustering inertia.

12 More on https://arxiv.org/pdf/1707.03897.pdf and in Vignettes on https://cran.r-project.org/web/packages/ClustGeo/vignettes/intro_ClustGeo.html. Hierarchical methods based on kernel estimation are developed by Fouedjio (2016).

13 An alternative is to use the gBuffer() command from the rgeos:: package.

14 https://www.jstatsoft.org/article/view/v063i16/v63i16.pdf

## Chapter 8

Spatial point pattern analysis and spatial interpolation

Kateryna Zabarina

This chapter presents the methods of spatial point pattern analysis and spatial interpolation. Studies on spatial point patterns have a long history since John Snow (1855) published his research on the spatial patterns of cholera in London. But modern methodologies started to appear in the mid-1950s, mostly in epidemiology. Gatrell, Bailey, Diggle, and Rowlingson (1996, p. 256) state that point pattern analysis became popular among geographers in the late 1950s–early 1960s, when "a spatial analysis paradigm began to take firm hold within the discipline". Then, the two main branches of point pattern analysis were developed: distance-based methods (*K* function, *G* and *F* functions) and area-based methods (*intensity* issue, *kernel density* estimation, *quadrat count* analysis). These two wide classes of techniques are still used in point pattern studies.

Methods developed from the mid-20th century (Ripley, 1976, 1977, 1981; Besag, 1974, 1977; Besag & Diggle, 1977; Diggle, 1975; Diggle, Besag, & Gleaves, 1976; Diggle, 1978, 1983; Cox & Isham, 1980; Cliff & Ord, 1981) were first applied in biology and ecology, and then later appeared in epidemiology, criminology, astronomy and, in recent years, in economic studies, which is becoming more popular. *Ripley's K function* was first used in economics by Arbia and Espa (1996)[1] and became a popular tool in investigation of point pattern clustering (Feser & Sweeney, 2000; Sweeney & Feser, 1998; Maoh & Kanaroglou, 2007; Kang, 2010; Arbia, Espa, Giuliani, & Mazzitelli, 2010; Arbia, Espa, Giuliani, & Mazzitelli, 2012; Espa, Arbia, & Giuliani, 2013; Arbia, Espa, & Quah, 2008; Marcon & Puech, 2017; Eckey, Kosfeld, & Werner, 2012). *Point process models* are not so widely used in economics; however, they started to appear, such as in the study of Bocci and Rocco (2016), who examine determinants of firms' locations using an inhomogeneous Poisson model.

The fundamental idea of point pattern studies is a comparison of theoretical assumptions about their behaviour and empirical evidence from the data. A classical example of a point pattern is a Poisson point process (also known as *complete spatial randomness* [CSR] or *complete spatial randomness and independence* [CSRI] for multitype point patterns), which reveals random behaviour, stationarity and isotropy (which means it preserves statistical properties under any shifts or rotations). In practice, such patterns rarely occur, though they are a useful benchmark, especially when initial analysis does not suggest CSR as a starting point. Performing point pattern analysis helps not only to answer the question of whether an underlying data generating process is random or non-random –but also to investigate possible relationships between points of one or more different types. Verification of a CSR/CSRI hypothesis may result either in non-rejection, and then one states that an examined pattern is Poisson, or in rejection, and then three scenarios are possible. All of them are shown in Figure 8.1.

Spatial interpolation uses available neighbourhood data to estimate an unknown value in a given location. Several interpolation methods are known: *inverse distance weighted* (IDW), *kriging, triangulated irregular network* (TIN), *thin plate splines* and so on. The range of use for spatial interpolation is quite wide: from biology (Ferrier, 2002; Moore, Grayson, & Ladson, 1991; Kozak, Graham, & Wiens, 2008; Tremblay

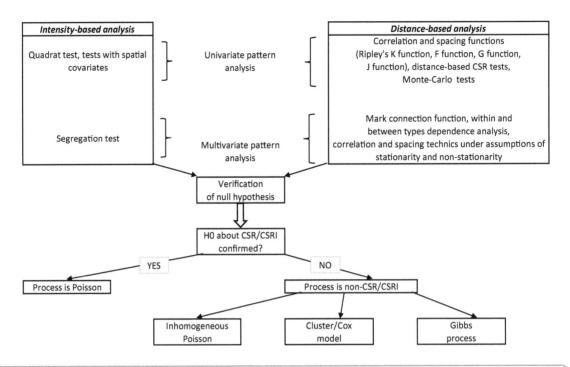

Figure 8.1 Algorithm of point pattern analysis performed in this chapter

Note: Only methods applied in the current study are mentioned

Source: Own analysis based on existing literature

et al., 2006) to environmental studies (Li, Heap, Potter, & Daniell, 2011; Kumar et al., 2016; Phillips, Henry Lee, Herstrom, Hogsett, & Tingey, 1997; Montero, Chasco, & Larraz, 2010; Li & Heap, 2008). Recently, the usage of spatial interpolation methods in economics and economic-related studies has increased: among socioeconomic studies, it is worth mentioning Goodchild, Anselin, and Deichmann (1993) and studies of homicide rates caused by socioeconomic factors by Poveda (2011). Environmental economics studies include interesting examples of economic costs caused by PM10 (Sun, An, Tao, & Hou, 2013) or the implication of forest cleaning on economics (D. Wheeler, Hammer, Kraft, Dasgupta, & Blankespoor, 2013). Pure economic studies cover a wide range of issues: economic growth (Amiri & Gerdtham, 2011; Amiri & Zibaei, 2012), insurance (Hejazi, Jackson, & Gan, 2017), real estate (L. Anselin, 1998), electricity price (Bello, Reneses, Muñoz, & Delgadillo, 2016), land price distribution (Hu, Cheng, Wang, & Xu, 2013), economic activity (Triantakonstantis & Stathakis, 2014), resource economics (Anselin, 2001), single region multipliers (Ooster-haven, 2005), economic hotspots (Veen & Logtmeijer, 2005) and choice experiments in economics (Campbell, Hutchinson, & Scarpa, 2009).

The structure of this chapter is as follows. First, an introduction and main definitions and methods for point patterns will be given (Section 8.1). Sections 8.2–8.4 describe methods of analysis and provide results for unmarked point patterns and Sections 8.5–8.7 for marked point patterns (with the main emphasis on second-order properties and Gibbs models in both cases). Section 8.8 provides an introduction to spatial interpolation methods with an emphasis on kriging. All point pattern analysis (Sections 8.1–8.7) is based on the R spatstat:: package, since it offers a wide range of instruments for spatial point patterns. Kriging is performed using the automap:: (Hiemstra, 2013) package. Other possible packages for point pattern analysis, point process models and kriging methods available in R are given in Tables 8.1 and 8.2 at the end of this chapter.

The full chapter is based on the spatstat:: package, as it is the most commonly used package, since it contains many functions and methods to work with point patterns (creation, visualisation, statistical inference etc.). Other packages offered by R (Bivand, 2020) are based on spatstat:: but may contain other interesting functions, methods or datasets:[2]

- For graphs, statistics and point pattern characteristics (mostly distance-based measures), one can use packages ads:: (Goreau, Pelissier, & Verley, 2018), aspace:: (Bui, Buliung, & Remmel, 2012), dbmss:: (Lang, Marcon, Pucch, & Traissae, 2019), ecespa:: (De La Cruz, 2018), mmpp:: (Hino, Murata, Takano, & Yoshikawa, 2017), SGCS::

(Rajala, 2019), spatgraphs:: (Rajala, 2017a), spatial:: (Bivand, Ripley, & Venables, 2015), spatstat::, splancs:: (Bivand, Diggle, Eglen, Petris, & Rowlingsson, 2017), stpp:: (Diggle, Gabriel, Rodriguez, & Rowlingsson, 2018);

- For model selection: selectspm:: (De La Cruz, 2015);
- For advanced cases, one can use replicatedpp2w:: (De La Cruz, 2017) (analysis of replicated point patterns) or spatialsegregation:: (Rajala, 2017b) (segregation measures for multitype point patterns).

8.1 Introduction and main definitions

The formal definition given by Baddeley, Rubak, and Turner (2015) states that one can call a set $\mathbf{x} = \{x_1, x_2, \ldots, x_n\}$ in two-dimensional space R^2 a point pattern. Points can overlap; that is, points x_i and x_j can have the same location ($x_i = x_j$ for $i \neq j$). One can distinguish between *finite* point patterns (i.e. where the number of points is finite) and *locally finite* point patterns (in this case, a finite number of points is selected from an infinite or bigger finite point pattern).

In the simplest version, information about each point includes only its longitude and latitude. However, very often, points inside a pattern belong to different values, defined by marks. A *mark* is an additional characteristic[3] of the point; it can be either numerical or categorical.[4] Depending on number of mark types, one can distinguish between uni-, bi- or multivariate point patterns.

A dataset may also contain **covariates.** This is explanatory data, which should be observable in each location, and it preserves its meaning across the study window (Baddeley et al., 2015, Section 5.6.4). Covariates can be expressed as a spatial function $Z(u)$, where u is a spatial location or other spatial pattern. In order not to confuse marks and covariates, one small example can be presented. Suppose one has a dataset showing locations of housing. As a mark, one can identify price, flat area, number of rooms, district and so on. Each of these features is meaningful for one particular location. Assume one wants to investigate how distance to airport or railway station affects the price of real estate. In this example, distance will be the covariate, as it is expressed as a spatial function.

The number of points within a point pattern is finite;[5] therefore, each point pattern also has boundaries (*window*), which can be regularly (represented by circle, square or other regular polygon) or irregularly (for example, ellipse, rectangle, irregular polygon, geographic/administrative border) shaped.

This section presents the first steps in dealing with point patterns – creation and visualisation and description of a dataset.

8.1.1 Dataset

The dataset used in this chapter contains 37,087 records about enterprises of Lubelskie voivodeship in Poland from the REGON database. Among others, there is information about the number of employees for each firm, industry sector and legal form. According to Polish National Statistics, an enterprise is classified as micro if it has <10 employees, small if it has <50 employees, medium if it has <250 employees and big otherwise. One can summarise the firms in a dataset according to number of employees; the majority of firms to be analysed are microenterprises.

```
firms<-read.csv("geoloc_data_firms.csv", header=TRUE, dec=",", sep=";")
names(firms)
```

```
[1]    "ID"          "ADDRESS"     "STREET"      "STREET.NO"   "ZIP"
[6]    "CITY_POST"   "CITY"        "region2"     "poviat"      "region3"
[11]   "subreg"      "coords.x1"   "coords.x2"   "LEGAL_FORM1" "LEGAL_FORM2"
[16]   "OWNERSHIP"   "PKD7"        "SEC_PKD7"    "GR_EMPL"     "empl"
```

```
# size of employment - variable GR_EMPL
# GR_EMPL = 1 -> between 1 and 9 persons
# GR_EMPL = 2 -> between 10 and 49 persons
# GR_EMPL = 3 -> between 50 and 249 persons
# GR_EMPL = 4 -> between 250 and 999 persons
# GR_EMPL = 5 -> more than 1500 persons
```

```
table(firms$GR_EMPL)
```

1	2	3	4	5
36392	568	116	6	5

```
firms$empl<-ifelse(firms$GR_EMPL==1, 5, ifelse(firms$GR_EMPL==2, 30,
ifelse(firms$GR_EMPL==3,150, ifelse(firms$GR_EMPL==4, 600, 1500))))
table(firms$empl)
```

5	30	150	600	1500
36392	568	116	6	5

Polish National Statistics distinguishes between three basic[6] legal forms: legal person (code 1), organisational entity without legal personality (code 2) and natural persons conducting economic activity (code 9). According to the statistics, most of the firms in a study dataset are provided by people conducting economic activity.

```
table(firms$LEGAL_FORM1)
```

1	2	9
2063	2129	32895

8.1.2 Creation of window and point pattern

A window is a particular border for a point pattern, and its shape can be different. Package spatstat:: allows one to create a custom window (with **owin()** or functions for geometric shapes: **square()**, **ellipse()**, **disc()**, **hexagon()**, **regularpolygon()**) or to convert to *owin* class object (**as.owin()** or alternatively **as(object, "owin")**).

Window-related studies usually raise a question about the window's shape or location (known as a "moving window"); work by Wiegand and Moloney (2004) analyses how a moving window impacts point pattern analysis. Turner (2009) notices that a properly specified window determines a point pattern, since it delineates space where data is known and available and where there is no data.

In the following example, for a given study point dataset, the administrative border of Lubelskie voivodeship is required. A joint map of Polish NUTS2 regions can be taken from the Polish GeoInformation Systems website,[7] and later one can extract the required region. The following code shows how to properly read data, transform to an *owin* class object, and then plot. Remember, as in the case of tessellation (see Chapter 4), to change the projection to planar for the **as.owin()** command.

```
# load packages
library(sp)        # for spTransform()
library(spatstat)  # for as.owin()
library(rgdal)     # for readOGR()
library(maptools)  # necessary to run as.owin()
voi<-readOGR(".", "wojewodztwa") # read data
region<-voi[voi$jpt_nazwa_=="lubelskie",] # still sp class
region<-spTransform(region, CRS("+proj=merc +datum=WGS84")) # planar coords

W<-as.owin(region) # conversion to owin class object
W<-as(region, "owin") # conversion to owin class object

class(W)
#[1] "owin"
```

A point pattern in a spatstat:: package is an object of class *ppp*. The creation of a point pattern can be done by calling function **ppp()**, for which the arguments are point coordinates (vectors *x* and *y*) in the same (planar) projection as the *owin* class object, study region (observation window, object of class **owin**) and (optionally) point characteristics (marks – vector[8] or data frame). The following code shows the creation of a point pattern.

Additional arguments which can be provided for **ppp()** are logical indicators checking whether all points lie inside the specified window (*check*) or there are duplicated[9] points (*checkdup*).

```
library(spatstat)

# change of projection of points
cord<-as.matrix(cbind(firms$coords.x1, firms$coords.x2))
cord.sp<-SpatialPoints(cord) # points in sp class - spherical
proj4string(cord.sp)<-CRS("+proj=longlat +datum=NAD83") # spherical
cord.sp<-spTransform(cord.sp, CRS("+proj=merc +datum=NAD83")) #planar

# unmarked point pattern
ppp_um<-ppp(x=cord.sp@coords[,1], y=cord.sp@coords[,2], window=W)

# marked point pattern
ppp_m<-ppp(x=cord.sp@coords[,1],    y=cord.sp@coords[,2],    window=W,
marks=firms$GR_EMPL)

# plotting unmarked point pattern from example above
par(mar=c(1,1,1,1))
plot(ppp_um, main= "Study point pattern, 37087 obs.") # with spatstat::
```

To create a custom window, several options are possible. A custom window shape is not so frequently used by researchers; however, it can be helpful in studies of phenomenon within some radius (e.g. in case of disease studies) or polygon (e.g. for animal or plant species–related studies). The most frequently used shapes are ellipse/circle, square/rectangle or another polygon. The following code shows function calls.[10]

```
# creating a circle
# center of the circle approximately in the center of region
# a and b - ellipse radii, centre - center of circle,
# phi - anticlockwise rotation angle
circ<-ellipse(a=200000, b=200000, centre=c(2547063.75417,6635810.07789),
phi=pi)
X.c<-ppp(x=cord.sp@coords[,1], y=cord.sp@coords[,2], window=circ)
plot(X.c) # plotting

# creating a square
# in first bracket - x coordinate of bottom left and top right vertices
# in second bracket - the same for y coordinates
boundingbox(W) # getting coordinates of bounding box
# window: rectangle = [2406231.2, 2687896.3] x [6457131, 6818514] units
sq<-owin(c(2406232, 2687897), c(6457131, 6818514)) # create a rectangle
X.sq<-ppp(x=cord.sp@coords[,1], y=cord.sp@coords[,2], window=sq)
plot(X.sq) #plotting
```

8.1.3 Marks

Marks are features of points. A marked point pattern can be either *multitype* or *multivariate*.[11] Baddeley et al. (2015, Section 14.2.5) explained the difference as follows. A multitype pattern is a pattern where each location x_i has a characteristic m_i, and points of all types are in the same point pattern. Multivariate patterns consist of m patterns, and each of them is a point pattern with corresponding mark value m_i.

The following code shows the creation of two marked point patterns – one with numerical marks representing the number of employees in each firm and the second with categorical marks representing legal form (since legal form is given as a code number to represent a categorical mark, it should be converted to a class factor).

```
# numerical marks
ppp_empl<-ppp(x=cord.sp@coords[,1],   y=cord.sp@coords[,2],   window=W,
marks=firms$empl) # marks representing employment level

# categorical marks
ppp_firms<-ppp(x=cord.sp@coords[,1],   y=cord.sp@coords[,2],   window=W,
  marks=as.factor(firms$LEGAL_FORM1)) # marks representing legal form
```

The plotting of a point pattern is possible with the function **plot.ppp()** (or equivalently **plot()**). Plots of unmarked patterns were shown in the previous section, and in general, they are easy to obtain: the user specifies only one obligatory argument – point pattern – and with other options, one can define points' size (*cex*) and shape (*pch*), axes labels (*xlab, ylab*) and scale (*xlim, ylim*), plot legend (*legend*) and title (*main*) and so on. In cases of marked point patterns, the user can provide the points' colour (*cols=c()*), shape (*chars=c()* or *shape=c()*) and size according to mark type value (*maxsize, meansize* or *markscale*) or also choose marks which will be plotted (*which.marks*, in the case of two or more marks per point). The following code shows the application of the **plot()** function for the two point patterns mentioned previously.

```
# marked point pattern with numerical marks - ppp_empl
# control for shape of points according to number of employees
# option markscale is a multiplier for all mark values
plot(ppp_empl, markscale=0.002, pch=1)

# marked point pattern with categorical marks - ppp_firms
# control for colors of points according to legal form
plot(ppp_firms, pch=1, cex=0.8, cols= c("red", "green", "blue"))
```

Both patterns contain several values of marks. Due to the large number of data points, R does not display all possible values correctly. Coming back to the dataset description, one can remember that the most prevalent value of employees number was 5, and the majority of firms are provided by natural persons. To plot data in a way that the layer of most frequent value lies on the bottom, the function **split()** can be used to divide point patterns to several different patterns and plot them one by one in study window. The function **split()** takes as arguments a point pattern, splitting factor(by default, it is factor mark value, but it can also be a logical vector, pixel image, tessellation etc.) and three logical values – deleting the empty sub-pattern (**drop()**, FALSE by default), deleting the column of marks used as splitting factor (**reduce()**, FALSE by default) and determining whether new patterns will be marked or unmarked (**un()**, NULL by default). In both cases, splitting is applied, since the point patterns are too big, and **plot()** cannot display values correctly. Plots of both patterns are presented in Figure 8.2.

```
# pattern contains 37087 points,
# some mark values cannot be displayed correctly with plot() function
# numerical values will be converted to class 'factor',
# pattern will be split and then plotted layer by layer

marks(ppp_empl)<-as.factor(marks(ppp_empl))
split_ppp_empl<-split(ppp_empl)

# window of plot
plot(W, main="Marked point pattern \n representing number of employees")
# consecutive layers of point pattern
# first layer - with the most observed value
plot(split_ppp_empl[[1]], add=TRUE, cex=0.75, pch=16, col="green")
plot(split_ppp_empl[[2]], add=TRUE, cex=1, pch=16, col="blue")
plot(split_ppp_empl[[3]], add=TRUE, cex=1.25, pch=16, col="red")
```

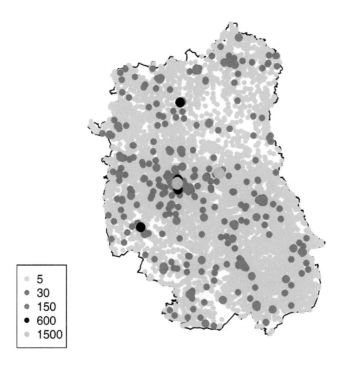

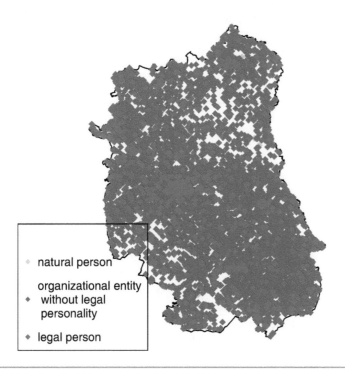

Figure 8.2 Marked and unmarked point patterns: a) with numerical marks, b) with categorical marks

Source: Own calculations using spatstat::

```
plot(split_ppp_empl[[4]], add=TRUE, cex=1.5, pch=16, col="black")
plot(split_ppp_empl[[5]], add=TRUE, cex=1.75, pch=16, col="orange")
legend("bottomleft", legend=c("5","30","150","600","1500"),
       col=c("green", "blue", "red", "black", "orange"), pch=16, cex=0.8)


# split
split_ppp_firms<-split(ppp_firms, reduce=TRUE)
cols <- c("red", "green", "blue")
# step by step
plot(W, main="Marked point pattern \n representing legal form")
plot(split_ppp_firms[[2]], add=TRUE, col=cols[2], pch=18)
plot(split_ppp_firms[[3]], add=TRUE, col=cols[3], pch=18)
plot(split_ppp_firms[[1]], add=TRUE, col=cols[1], pch=18)
legend("bottomleft", legend=c("natural person", "organizational entity \n
without legal \n personality", "legal person"),
       col=c("green", "blue", "red"), pch=18, cex=0.8)
```

Marks can be changed by simply assigning new values. One should be careful, as the operation: **marks(point_pattern) <- new_value** rewrites the dataset, and no backup can be made. However, it is possible to preserve original values with the command **setmarks(point_pattern, new_value)**[12] – after the function call, the environment returns a point pattern with replaced values, but initial marks remain unchanged. The example subsequently shows how to change mark values from legal form to industry sector.

In the example, due to the fact that there are 20 different industries and plotting them all together will result in an illegible graph, it was decided to divide industries into four groups.

```
firms$sector[firms$SEC_PKD7=="A"] <- "agriculture"
firms$sector[firms$SEC_PKD7=="F"] <- "construction"
firms$sector[firms$SEC_PKD7 %in% c("B", "C", "D", "E")] <- "production"
firms$sector[firms$SEC_PKD7 %in%
c("G","H","I","J","K","L","M","N","O","Q","P","R","S")] <- "service"
```

First, values of marks will be changes with replacement:

```
summary(marks(ppp_firms)) # old mark values
```

```
    1     2     9
 2063  2129 32895
```

```
marks(ppp_firms)<-as.factor(firms$sector) # change values
summary(marks(ppp_firms)) # new mark values
```

```
  agriculture construction    production   service    NA's
        20659         1946          1428     13050       4
```

```
plot(ppp_firms, cols=c("red", "blue", "green", "black"), pch=1, cex=0.75)
```

It can be observed that mark values have changed. Figure 8.3a shows a point pattern with new mark values. The code subsequently shows results of the **setmarks()** function call:

```
## using setmarks()
summary(setmarks(ppp_firms, as.factor(firms$sector)))
```

```
  Marked planar point pattern: 37087 points
  Average intensity 5.813575e-07 points per square unit
```

ppp_firms

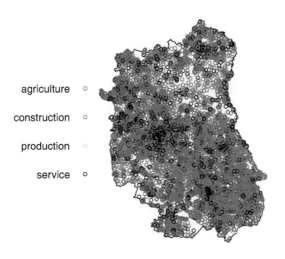

- agriculture   ○
- construction ○
- production   ○
- service   ○

mat_pix

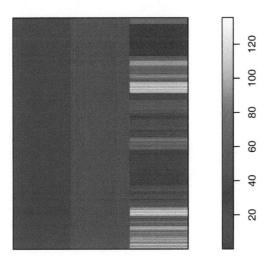

Figure 8.3 Visualisation of location: a) point pattern with new assigned mark values, b) pixel image of distances to Lublin

Source: Own calculations using spatstat::

```
*Pattern contains duplicated points*
Coordinates are given to 1 decimal place
i.e. rounded to the nearest multiple of 0.1 units


Multitype:

                frequency         proportion         intensity
Agriculture     20659             0.55710160         3.238403e-07
Construction     1946             0.05247688         3.050453e-08
Production        1428            0.03850821         2.238462e-08
service         13050             0.35191330         2.045654e-07
```

```
Window: polygonal boundary
single connected closed polygon with 32439 vertices
enclosing rectangle: [2406231.2, 2687896.3] x [6457131, 6818514] units Window
area = 63793800000 square units
Fraction of frame area: 0.627
```

```
summary(marks(ppp_firms)) # no changes can be observed
```

agriculture	construction	production	service	NA's
20659	1946	1428	13050	4

```
## alternative - using pipeline operator from package magrittr::
library(magrittr)
ppp_empl %mark% as.factor(firms$sector)
```

```
Marked planar point pattern: 37087 points
Multitype, with levels = agriculture, construction, production, service
window: polygonal boundary
enclosing rectangle: [2406231.2, 2687896.3] x [6457131, 6818514] units
```

To delete all mark values, one can either set them to NULL, or use **unmark()**.

```
unmark(ppp_firms)
ppp_firms
```

```
Planar point pattern: 37087 points
window: polygonal boundary
enclosing rectangle: [2406231.2, 2687896.3] x [6457131, 6818514] units
```

```
# or equivalently
marks(ppp_firms) <- NULL
ppp_firms
```

```
Planar point pattern: 37087 points
window: polygonal boundary
enclosing rectangle: [2406231.2, 2687896.3] x [6457131, 6818514] units
```

For further analysis, to examine interactions between firms of different industries, marks of legal form will be replaced with marks representing industry sector.

```
marks(ppp_firms) <- as.factor(firms$SEC_PKD7)
summary(marks(ppp_firms))
```

-	A	B	C	D	E	F	G	H	I	J	K	L	M
4	20659	24	1287	55	62	1946	4884	1145	496	336	506	499	1424

N	O	P	Q	R	S
382	226	703	941	315	1193

8.1.4 Covariates

As was stated in the introduction, covariates are treated as explanatory data, which preserve their meaning in the whole study region and are usually represented as spatial function. In spatstat::, covariates can be presented as a **pixel image** (object of class *im*, either one or list of images), **function** (*x,y*, . . .) calculated in location (*x,y*), **window** (object of class *owin*, logical indicator: TRUE when inside window and FALSE otherwise), **tessellation** (object of class *tess*, determining which tile contains each separate spatial location) or **single number** (some constant). Sometimes, when any covariates are absent, Cartesian coordinates *x* and/or *y* can be used as a spatial covariate. Three principles should be held when creating a spatial covariate:

1 Values of the *covariate* should be observed (in theory) in each location across study window;
2 Values of the *covariate* should be available for all points in study pattern;
3 Values of the *covariate* should be presented in locations not belonging to the observed pattern inside the study window.

In the case of the study dataset, covariates are not presented. However, this does not mean covariates cannot exist in economic data. When analysing business location, one can mention several reasons for firm establishment: localisation patterns (Pablo-Martí, Muñoz-Yebra, & Santos, 2014), local demand- and supply-based linkages (Karahasan, 2014), economic parameters (income per capita, market size, economic freedom etc. [Wei, Yuan, & Liao, 2013]), FDI distribution (Blanc-Brude, Cookson, Piesse, & Strange, 2014) and so on. The main disadvantage of all these indicators could be the fact that they are available either as aggregated data (for example, economic freedom index) or are not available for some locations. In the case of the study dataset, there is no data which can be treated as a covariate, but it is possible to create it.

Baddeley et al. (2015, p. 51) argue that a pixel image is the "most useful and effective format" of a covariate. To create a pixel image, the function **as.im()** can be used; it converts an object of class *im/owin/matrix/tess/function/data.frame* and so on to class *im*. The most appropriate option for the study dataset is to convert a matrix containing locations of all points in the study region and the value of the covariate to a pixel image, which can be used in further analysis.

Example

This example shows how to produce a pixel image from a whole dataset (in further analysis, locations for study point pattern will be randomly selected from a study dataset and new point patterns will be created, so a pixel image also will be re-created). It was suggested that a factor determining firms' location could be their distance to Lublin – the capital city of the voivodeship. To create it, the **spDistsN1()** function from the sp:: package was used. This function calculates either Euclidian or Great Circle distance (e.g. WGS84 ellipsoid projection) between two objects – the first one could be either a matrix of 2D points or an object of class *SpatialPoints/SpatialPointsDataFrame*, and the second one is either a single 2D point or an object of class *SpatialPoints/SpatialPointsDataFrame* with one point only. Argument *longlat* equals TRUE, which allows the calculation of the Great Circle distance and returns the result in kilometres. The following code shows the creation of a new variable.

```
# load libraries
library(rgdal)   # reading shapefiles
library(sp)      # for spDistsN1()
library(spatstat) # for window, pixel image

# specify coordinates for Lublin
# first value x/longitude, second value y/latitude
lublin<-c(22.568445, 51.246452) # in case of spherical

# create vector of distances
# object 'firms' should be of class SpatialPointsDataFrame
# so it should be transformed as follows:
coordinates(firms)<-~coords.x1+coords.x2

# calculation of distances, option longlat=TRUE gives results in km
firms$dist<-spDistsN1(firms, lublin, longlat=TRUE)
```

In the next step, three columns from the *firms* object will be selected in order to create a matrix:[13]

```
# creation of a matrix
mat_im<-cbind(firms@coords[,1], firms@coords[,2], firms$dist)
```

The last step is the creation of a pixel image. Function **as.im**() takes two arguments – the object to be transformed and study window as a determinant of pixel array geometry (colour scale shows distance in kilometres to Lublin) (see Figure 8.3b).

```
mat_pix <- as.im(mat_im, W)
plot(mat_pix) # Figure 8.3b
```

In further analysis, two covariate options will be used – a simple example of a spatial covariate (*x* or *y* coordinate) and a newly created pixel image covariate.

8.1.5 Duplicated points

Duplication of points can occur either from the fact of error while recording data or (as in the case of the study dataset) or from the fact that several points merely have the same location in space. Currently existing methods of point pattern analysis assume simplicity of point process, so a preliminary check for coincident points seems logical (Baddeley et al., 2015, Section 3.4.4). A decision should be made by the researcher, and this decision should be careful and balanced.

In case of unmarked points patterns, duplication is easy to detect – points simply have the same *x* and *y* coordinates. For multitype patterns, points are identical if their Cartesian coordinates coincide and their marks have the same value. To check for duplicated points, one can use **duplicated**(), which takes a point pattern as an only argument. The output will be the vector of logical values indicating the presence/absence of duplicated points; the length of this vector is equal to the number of points in the point pattern. In order to know how many coinciding points are present, one can use **duplicated**() embedded in the function **sum**() or **summary**(). To delete duplications, the function **unique**() can be used.

In the following example, the two point patterns analysed before (unmarked and marked, illustrating firms of different industries) will be considered. First, the unmarked pattern is examined.

```
sum(duplicated(ppp_um))
#[1] 12530

ppp_um <- unique(ppp_um)
ppp_um
```

```
Planar point pattern: 24557 points
window: polygonal boundary
enclosing rectangle: [2406231.2, 2687896.3] x [6457131, 6818514] units
```

```
summary(duplicated(ppp_um))
```

```
   Mode    FALSE
logical    24557
```

```
## alternative way
ppp_um<-ppp_um[!duplicated(ppp_um)] # result is the same as above
```

For marked point patterns, as stated previously, duplication includes not only the same coordinates but also the same mark values.

```
d2 <- summary(duplicated(ppp_firms))
d2
```

```
Mode      FALSE    TRUE
logical   27039    10048
```

Another option in helping to deal with duplicated points is the **rjitter**() function. This function takes two arguments – the vector of initial values and the radius of the circle within which duplicated points will be "shuffled". The following code shows the results of **rjitter**() call; no duplications can be found.

```
xr1 <- rjitter(ppp_um, 0.02)
summary(duplicated(xr1))
#    Mode    FALSE
#logical   37087

plot(xr1)

xr2 <- rjitter(ppp_firms, 0.02)
summary(duplicated(xr2))
#    Mode    FALSE
#logical  37087
```

As stated before, the decision on what to do with duplicated points should be made by the researcher. Both deleting coincided points and shuffling them in the neighbourhood are equally good options. The aim of the current study will be widely described in the next section, but a strategy in brief is as follows. Studies of marked point patterns of firms could be interesting from a perspective of firm locations studies, policy making and so on. When one examines the location of firms from particular industries, such analysis can shed light on agglomeration issues – do firms locate closer to competitors from the same industry or to another industry enterprise, do they interact despite information about their sector membership and so on? – and this analysis will be performed in the following sections, and often shuffling is recommended. For unmarked point patterns, the strategy of analysis is almost the same – to investigate the relationship of points independently of their industry membership. However, in this case, it seems logical to delete duplicated observations.

8.1.6 Projection and rescaling

Sometimes, when analysing a point pattern, it turns out that a measurement unit was not specified, and results become unclear. It can happen even if the coordinate system was provided. For a study region, which is Lubelskie voivodeship, the area is equal to 25,122.46 km². However, the area of study window W, calculated with the **area.owin**() function, which calculates the area of the *owin* class object, is much different – 63,793,793,864 square units:

```
area.owin(W)
#[1] 63793793864
```

Knowing the ratio between the "old" (or system default) and "new" (a real one) unit, the function **rescale**(), which takes two arguments – object *point_pattern* and ratio – rescales the whole point pattern[14] to a proper form. After rescaling, both the new and old pattern are equivalent. The pros of this method are proper visual and summary analysis of a point pattern (examination of intensity, definition of interaction distance etc.).

After simple calculations, 1 unit is equal to approximately $6.2754 * 10^{-4}$ km (equivalent to 0.00062754 km). The following code shows how to rescale the study point pattern (here, the unmarked point pattern after duplication removal was chosen).

```
summary(ppp_um)
```

```
Planar point pattern:  24557 points
Average intensity 3.849434e-07 points per square unit
Coordinates are given to 1 decimal place
i.e. rounded to the nearest multiple of 0.1 units
Window: polygonal boundary
single connected closed polygon with 32439 vertices
enclosing rectangle: [2406231.2, 2687896.3] x [6457131, 6818514] units
Window area = 63793800000 square units
Fraction of frame area: 0.627
```

```
ppp_um<-rescale(ppp_um, 1593,52, "km")
summary(ppp_um)
```

```
Planar point pattern:  24557 points
Average intensity 0.9774891 points per square km
Coordinates are given to 4 decimal places
Window: polygonal boundary
single connected closed polygon with 32439 vertices
enclosing rectangle: [1510.01, 1686.7666] x [4052.118, 4278.901] km
Window area = 25122.5 square km
Unit of length: 1 km
Fraction of frame area: 0.627
```

As can be seen from the output, the area of the study region has the value of the real Lubelskie voivodeship area (with a small error caused by rescaling). However, values of spatial coordinates for both points and border also are rescaled, which may be confusing.

Another way to switch from unknown units to known (for example, meters) is changing the coordinate reference system (CRS). The most widely used is WGS84 (*World Geodetic System'84*). However, shapefiles for Polish administrative borders have another CRS – the coordinate system from 1992, used mostly for medium- and small-scale mapping (1:10,000 and smaller). It is possible to set the same CRS for a study dataset.

```
library(sp)
library(rgdal)

# check for CRS of voivodeship borders
voi<-readOGR(".", "wojewodztwa") # 16 units
proj4string(voi)
#[1] "+proj=tmerc +lat_0=0 +lon_0=19 +k=0.9993 +x_0=500000 +y_0=-5300000
#+ellps=GRS80 +units=m +no_defs"

# first transform to sp object
coordinates(firms)<-~coords.x1+coords.x2
proj4string(firms)<-CRS("+proj=tmerc +lat_0=0    +lon_0=19    +k=0.9993
+x_0=500000 +y_0=-5300000 +ellps=GRS80 +units=m +no_defs")

summary(firms)
```

```
Object of class SpatialPointsDataFrame
Coordinates:
             min       max
coords.x1 21.64175 24.11703
coords.x2 50.26383 52.27605
Is projected: TRUE
```

```
proj4string :
[+proj=tmerc +lat_0=0 +lon_0=19 +k=0.9993 +x_0=500000 +y_0=-5300000
+ellps=GRS80
+units=m +no_defs]
```

However, leaving the projection for the study window unchanged, coordinates will also be unchanged and given in that projection; moreover, the creation of a point pattern will result in an error.

```
# choose proper voivodeship
region<-voi[voi$jpt_nazwa_=="lubelskie",]

ids<-sample(dim(firms)[1], size=1500)
firms.sub<-firms[ids,]

# check whether CRS are the same for region and SpatialPoints of firms
proj4string(region)
#[1] "+proj=tmerc +lat_0=0 +lon_0=19 +k=0.9993 +x_0=500000 +y_0=-5300000
#+ellps=GRS80 +units=m +no_defs"

# convert to class 'owin'
W<-as(region, "owin")
p<-ppp(firms.sub$coords.x1, firms.sub$coords.x2, W)
# Warning message:
# 1500 points were rejected as lying outside the specified window

summary(p)
```

```
Planar point pattern:   0 points
Average intensity 0 points per square unit
Window: polygonal boundary
single connected closed polygon with 32439 vertices
enclosing rectangle: [681015.8, 861895.7] x [274858.4, 499342.6] units
Window area = 2.5134e+10 square units
Fraction of frame area: 0.619
*** 1500 illegal points stored in attr(,"rejects") ***
```

Even if projections for both firms and the region are the same, the bounding boxes are different, and this is the reason for failure while creating a point pattern:

```
proj4string(firms)
# "+proj=tmerc +lat_0=0 +lon_0=19 +k=0.9993 +x_0=500000 +y_0=-5300000
# +ellps=GRS80 +units=m +no_defs"
proj4string(region)
# "+proj=tmerc +lat_0=0 +lon_0=19 +k=0.9993 +x_0=500000 +y_0=-5300000
# +ellps=GRS80 +units=m +no_defs"

bbox(firms)
#                   min       max
# coords.x1 21.64175 24.11703
# coords.x2 50.26383 52.27605
bbox(region)
#          min       max
# x 681015.8 861895.7
# y 274858.4 499342.6
```

For further analysis, the first option will be used, since rescaling does not change the basic properties of the point pattern (so rescaled patterns can be used in models' estimation etc.) As stated previously, the unmarked pattern contains information only about points' locations, so deleting duplicates should not change the "behaviour" type of the pattern. On the other hand, for the marked point pattern, it makes sense to jitter duplicated observations, as points of different types are supposed to affect each other. For further analysis, all duplicated points from the unmarked point pattern were deleted, and a sample of ~1500 points was chosen as a study unmarked pattern. For the marked point pattern, observations were jittered, and then a sample of ~1500 points was also chosen as a study marked point pattern.

The following code shows the creation of study patterns for further analysis.

```
# unmarked
ppp_um<-unique(ppp_um)
ppp_um<-rescale(ppp_um, 1593.52, "km")
set.seed(ppp_um$n) # to guarantee the same data in new drawing
s1<-(runif(ppp_um$n)<0.06)
um_s<-ppp_um[s1]
summary(duplicated(um_s)) # 1502 points, no duplicates found

# marked, marks are industry sectors
ppp_firms<-rjitter(ppp_firms, 0.02)
ppp_firms<-rescale(ppp_firms, 1593.52, "km")
set.seed(ppp_firms$n)
s2<-(runif(ppp_firms$n)<0.045)
mm_s<-ppp_firms[s2]
summary(duplicated(mm_s)) #1589 points, with no duplicates found
```

Analysis of both pattern types is provided in the next sections. For the intensity-based analysis of unmarked point patterns, spherical coordinate projection was used[15] in order to apply a pixel image as a possible covariate factor. The pixel image of covariate for CSR tests (see 8.2) was created on the basis of 24,557 points left after deleting duplicates from the unmarked point pattern: ~1500 points from this pattern were chosen for further analysis (as shown in the code previously), and the other points are considered locations across the study window, which do not belong to the point pattern, but where the value of the spatial covariate can be observed.

8.2 Intensity-based analysis of unmarked point pattern

In the literature, *complete spatial randomness* is also known as a *homogeneous Poisson process*. Points are homogeneous (there is no preferable location) and located independently, and their number has a Poisson distribution. CSR is a particular benchmark with which other patterns can be compared. Also, a lot of models have CSR as a starting point (for example, the formation of cluster processes such as Matérn or Thomas begins from the random generation of "parent" points, which has a Poisson distribution), so this process is important in point process studies.

The inhomogeneous Poisson process is a modification of CSR. In contradiction to a homogeneous version, the number $N(X \cap B)$ of points falling in a region B has the expected value, given as $\mathbb{E}[N(X \cap B)] = \int \lambda(u)du$. Moreover, given that $N(X \cap B) = n$, n points are independent and identically distributed, with joint probability density $f(u) = \lambda(u)/I$, where $I = \int_B \lambda(u)du$ (Baddeley et al., 2015, Sections 5.3 and 5.4).

The homogeneous Poisson process can be estimated in R with the **ppm()** function; the user should specify a study point pattern and trend as "~1" – in other words, there are no suggested covariates, and the trend is stationary. Estimating inhomogeneous process, one should express trend as function of Cartesian coordinates or spatial covariate(s). In general, the code structure for homogeneous and inhomogeneous Poisson process model estimation is as follows:

```
homo_model<-ppm(pattern ~1)
inhom_model_1<-ppm(pattern ~x)
inhom_model_2<-ppm(pattern ~covariate)
```

An intensity-based approach is usually used in epidemiology and criminology studies to determine whether the intensity of an observed phenomena is homogeneous or varies in space. Intensity may be a function of Cartesian coordinates or another spatial covariate. In firm-related studies, this analysis seems to be rather inappropriate, since firms interact with each other in terms of better location, resources and so on, and their intensity does not vary directly because of some underlying characteristics – there are more firms in bigger cities. However, if a dataset contained any covariate data, it would be interesting to examine the intensity of firms, for example, taking into account resource distribution, closeness to main transport routes and so on.

To perform intensity analysis, one can use a quadrat count test, a test of the presence of a spatial covariate or a kernel density estimation. There are not too many possible weaknesses of these methods, but they exist. For example, when performing a quadrat test, one can choose either too small quadrats, and then some of them will be empty, or quadrats may be too big, and the distribution of points will be unequal. Seeking a proper interaction function could be a problem when several covariates are available.

The intensity-based methods discussed in this section include only tests for inhomogeneity – a quadrat test and tests of dependence of intensity on a covariate. In counting quadrats, one can assume that regions of equal area contain more or less the same number of points (as in the homogeneous process). A second group of tests checks for the presence of a spatial covariate which can imply intensity. The null hypothesis of both tests is that a point process is completely random (homogeneous, stationary). The following subsections describe tests' performance with examples.

8.2.1 Quadrat test

A quadrat test is easy to perform – the window is divided into quadrats (number is specified manually). Then, an algorithm counts the number of points in each quadrat (n_k) and states CSR if all n_k are i.i.d. Poisson random variables and their expected values are equal (Baddeley et al., 2015, Section 6.4). However, ease is balanced by imperfection – rejecting H_0 about homogeneity, it is hard to say exactly which process this is.

A quadrat test can be performed with the function **quadrat.test()**. Inside the function call, the number of quadrats is specified manually by setting the number of tiles on the X axis as *nx* and on the Y axis as *ny*; the main object of quadrat test analysis can be either a point pattern (object of class *ppp*) or an estimated point process model (object of class *ppm*; in this case, a goodness-of-fit test is performed). The user also can choose a test method: by default, *method="Chisq"*, but it can be changed to *method="MonteCarlo"*. An alternative argument specifies a form of alternative hypothesis: by default, *alternative="two-sided"*.

The result of the test performance is an object of both classes "*quadrattest*" and "*htest*". Calling for object properties, one can find information about null and alternative hypotheses, test statistic, *p*-value and so on.

```
q1<-quadrat.test(um_s, nx=5, ny=5)
q1
```

```
            Chi-squared test of CSR using quadrat counts
            Pearson X2 statistic
data:       um_s
X2 = 879.39, df = 23, p-value < 2.2e-16
alternative hypothesis: two.sided
Quadrats: 24 tiles (irregular windows)
```

```
names(q1)
```

```
 [1] "statistic"    "parameter"    "p.value"    "method"     "data.name"
 [6] "alternative"  "observed"     "expected"   "residuals"    "CR"
[11] "method.key"
```

```
q1$method
```

```
[1] "Chi-squared test of CSR using quadrat counts"
[2] "Pearson X2 statistic"
```

The quadrat test performed previously shows that one should reject the null hypothesis, so the point process is not CSR, but one cannot say more. Probably 25 is too small a number of tiles, so the next four quadrat tests will be performed on 100, 625, 2500 and 10,000 tiles. Results of the quadrat test with other numbers of tiles confirm the rejection of the null hypothesis:

```
q2<-quadrat.test(um_s, nx=10, ny=10)
q2$p.value
#[1] 0

q3<-quadrat.test(um_s, nx=25, ny=25)
q3$p.value
#[1] 0
q4<-quadrat.test(um_s, nx=50, ny=50)
q4$p.value
#[1] 0
q5<-quadrat.test(um_s, nx=100, ny=100)
q5$p.value
#[1] 0
```

8.2.2 Tests with spatial covariates

The next cohort of tests (Kolmogorov-Smirnov test [first described by Berman (1986) and later among others by Baddeley et al. (2005)], Anderson-Darling test, Cramer-von Mises test, Berman test [Berman, 1986]) checks the same null hypothesis as previously – complete spatial randomness of a point pattern – but the method of analysis is slightly different. Under a test function call, one specifies a covariate (it can be the x or y coordinate or any other covariate), and the test checks whether it is significant (i.e. the pattern has a spatial trend).

To perform the first three tests, the function spatstat::**cdf.test**() can be used. It includes three main arguments – point pattern, covariate and test type (the Kolmogorov-Smirnov test is the default option). After performing a test, it is possible to plot test results – a difference can be detected between observed and expected distributions of the covariate.

The Berman test, performed by spatstat::**berman.test**(), besides the covariate, allows one to choose between two Z-statistics – $Z1$, known as the Waller-Lawson statistic, which uses the sum of covariate values of a pattern, and $Z2$, which considers the cumulative distribution function (a detailed description can be found in Baddeley et al., 2015, Section 6.7.2). The results of both **cdf.test**() and **berman.test**() are objects of class *htest*. Additionally, the results of **cdf.test**() can be plotted by calling **plot**().

The first possible covariate to use is the x coordinate. The following code shows the results of **cdf.test**() with three different options and two **berman.test**() calls (with statistics $Z1$ and $Z2$).

```
ks.rand<-cdf.test(um_s, "x")
ks.rand
```

```
        Spatial Kolmogorov-Smirnov test of CSR in two dimensions
data:  covariate 'x' evaluated at points of 'um_s'
    and transformed to uniform distribution under CSR
D = 0.1138, p-value < 2.2e-16
alternative hypothesis: two-sided
```

```
cvm.rand<-cdf.test(um_s, "x", "cvm")
cvm.rand
```

```
        Spatial Cramer-Von Mises test of CSR in two dimensions
data:  covariate 'x' evaluated at points of 'um_s'
    and transformed to uniform distribution under CSR
omega2 = 4.3666, p-value = 7.281e-11
```

```
ad.rand<-cdf.test(um_s, "x", "ad")
ad.rand
```

```
        Spatial Anderson-Darling test of CSR in two dimensions
data:  covariate 'x' evaluated at points of 'um_s'
    and transformed to uniform distribution under CSR
An = 19.656, p-value = 3.995e-07
```

```
bt1<-berman.test(um_s, "x")
bt1
```

```
        Berman Z1 test of CSR in two dimensions
data:  covariate 'x' evaluated at points of 'um_s'
Z1 = -0.12838, p-value = 0.8978
alternative hypothesis: two-sided
```

```
bt2<-berman.test(um_s, "x", "Z2")
bt2
```

```
        Berman Z2 test of CSR in two dimensions
data:  covariate 'x' evaluated at points of 'um_s'
        and transformed to uniform distribution under CSR
Z2 = -4.8442, p-value = 1.271e-06
alternative hypothesis: two-sided
```

As can be observed, Berman test's statistic $Z1$ does not reject H_0, whereas $Z2$ does, as well as the Kolmogorov-Smirnov, Cramer-von Mises and Anderson-Darling tests. Baddeley et al. (2015, Section 6.7.2) explain this as follows: since $Z1$ compares observed and expected mean values, it cannot capture a difference between the distributions. This discrepancy is clearly visible on the Kolmogorov-Smirnov test plot. The conclusion to be made is that an observable point pattern is rather not completely random (see Figure 8.4a).

The second step of analysis is to check whether the pixel image of the spatial covariate should be accounted for in further point pattern analysis. The syntax is as before; instead of "*x*", one should specify the name of the pixel image object (in our case – *mat_pix*). One should be careful – a Cartesian coordinate covariate should be given in quotes, whereas the name of other type of covariate should be given without quotes. The code subsequently provides function calls and results:

```
mat_im<-cbind(firms.sub$coords.x1, firms.sub$coords.x2, firms.sub$dist)
mat_pix<-as.im(mat_im, W)
ks.rand.im<-cdf.test(um_s, mat_pix)
ks.rand.im
```

```
        Spatial Kolmogorov-Smirnov test of CSR in two dimensions
data:  covariate 'mat_pix' evaluated at points of 'um_s'
    and transformed to uniform distribution under CSR
D = 0.32974, p-value < 2.2e-16
alternative hypothesis: two-sided
```

```
cvm.rand.im<-cdf.test(um_s, mat_pix, "cvm")
cvm.rand.im
```

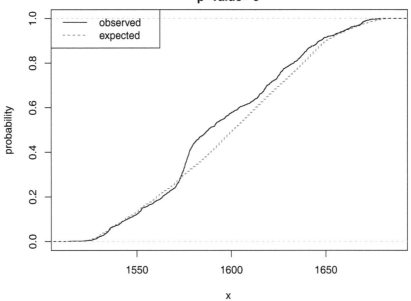

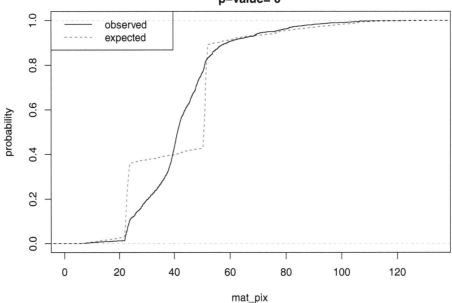

Figure 8.4 Plot of Kolmogorov-Smirnov test under the presence of a covariate: a) *x* covariate, b) pixel image covariate

Source: Own calculations using spatstat::

```
        Spatial Cramer-Von Mises test of CSR in two dimensions
data:  covariate 'mat_pix' evaluated at points of 'um_s'
      and transformed to uniform distribution under CSR
omega2 = 35.239, p-value = 3.883e-05
```

```
ad.rand.im<- cdf.test(um_s, mat_pix, "ad")
ad.rand.im
```

```
        Spatial Anderson-Darling test of CSR in two dimensions
data:  covariate 'mat_pix' evaluated at points of 'um_s'
      and transformed to uniform distribution under CSR
An = 161.97, p-value = 3.995e-07
```

```
bt1.im<-berman.test(um_s, mat_pix)
bt1.im
```

```
        Berman Z1 test of CSR in two dimensions
data:  covariate 'mat_pix' evaluated at points of 'um_s'
Z1 = -0.042189, p-value = 0.9663
alternative hypothesis: two-sided
```

```
bt2.im <- berman.test(um_s, mat_pix, "Z2")
bt2.im
```

```
        Berman Z2 test of CSR in two dimensions
data:  covariate 'mat_pix' evaluated at points of 'um_s'
      and transformed to uniform distribution under CSR
Z2 = -1.8306, p-value = 0.06716
alternative hypothesis: two-sided
```

The conclusion is the same as before – only in the case of the Berman test with the $Z1$ statistic, H_0 was not rejected, but as in the previous case, the Berman test did not capture any discrepancy between curves of the observed and theoretical distribution of covariate (see Figure 8.4b)

```
plot(ks.rand)
plot(ks.rand.im)
```

Both groups of tests suggested the rejection of homogeneity. Tests for the presence of a spatial covariate argue that both the x coordinate and the pixel image of distances to Lublin can be treated as a factor affecting point pattern intensity. However, the search for a proper intensity function can be troublesome, as the form of dependence can take a different shape. Also, the presence of other spatial covariates could have a significant impact on the result.

8.3 Distance-based analysis of the unmarked point pattern

A distance-based approach allows one to determine the presence and type of dependence (positive/negative, tendency to clustering or regularity) between points or also define its interval. Available instruments include functions for correlation analysis (K, L, *pair correlation,* function g), distance functions (G, F, J), distance-based tests (Clark-Evans test and Hopkins-Skellam test) and Monte Carlo tests (Diggle-Cressie-Loosmore-Ford (DCLF) test, MAD test and envelopes). It should be noticed that the K function cannot be treated as a universal mechanism of correlation detection. A possible problem is confusing inhomogeneous and stationary clustered patterns, which can happen when analysing K or a related function (L or g, which are transformations of K function).

Also, if the *K* function suggests a cluster pattern, it is not necessarily clustered – points can lie closer to each other in comparison to a random pattern but may interact. One more possible problem could be the presence of patterns, which for the empirical *K* function is πr^2 (as for the theoretical *K* function), but the process is not CSR.

The common idea of summary functions is to compare theoretical values with empirical, taking into account distances d_{ij} between points x_i and x_j ($i \neq j$) of a point pattern and assuming stationarity and homogeneity of the point pattern. Baddeley et al. (2015, Section 7.3) suggest that distances can tell us a lot about a point pattern. Distance-based tests and Monte Carlo tests check the null hypothesis about CSR (a detailed description is provided subsequently).

In the previous chapter, it was shown that the null hypothesis about homogeneity should be rejected. So, the analysis of distance-based measures will be done using an inhomogeneous version of each function.

8.3.1 Distance-based measures

8.3.1.1 Ripley's K function

Trying to derive a function of pairwise distances, Baddeley et al. (2015, Section 7.3) introduce a distance $r > 0$, which can be treated as a radius of circle around point. So, each point can have neighbours, lying inside this circle (for which $d_{ij} \leq r$), and some other points lying far away from it. Dividing this cumulative sum ($\sum_{i=1}^{n} \sum_{j=1, j \neq i}^{n} 1\{d_{ij} \leq r\}$) by the number of all points in the point pattern and taking into account the area of the study region ($|W|$), an empirical *K* function can be defined as:

$$\hat{K}(r) = \frac{|W|}{n(n-1)} \sum_{i=1}^{n} \sum_{j=1, j \neq i}^{n} 1\{d_{ij} \leq r\} e_{ij}(r)^{[16]} \text{ (Baddeley et al., 2015, p. 204)} \tag{8.1}$$

The *K* function for a completely random point pattern takes the form πr^2 (this is true for all versions of the *K* function). The empirical curve lying above the theoretical suggests clustering below – regularity. It is recommended to keep in mind that if the *K* function (or any other function) shows cluster behaviour, one can confuse clustering either with inhomogeneity or interaction.

The **Kinhom**() command estimates an inhomogeneous version of the *K* function. The main argument is a point pattern; others are optional. The argument "*correction*" is to specify the type of edge correction: isotropic, border, Ripley and so on. By default, all possible types are selected. If the function call is assigned to an object, the result will be of class *fv*, which is easy to plot. It should be noticed that for patterns with a number of points exceeding 1000, R computes only a border correction estimate. The following code shows the call and results:

```
Ki<-Kinhom(um_s)
Ki
Function value object (class 'fv')
```

```
for the function r -> K[inhom](r)
. . . . . . . . . . . . . . . . . . . . . . . . . . . . . . . . . . . . .
            Math.label
r           r
theo        K[pois](r)
border      {hat(K)[inhom]^{bord}}(r)
bord.modif  {hat(K)[inhom]^{bordm}}(r)
            Description
r           distance argument r
theo        theoretical Poisson K[inhom](r)
border      border-corrected estimate of K[inhom](r)
bord.modif modified border-corrected estimate of K[inhom](r)

. . . . . . . . . . . . . . . . . . . . . . . . . . . . . . . . . . . . .
Default plot formula:  .~r
where "." stands for 'bord.modif', 'border', 'theo'
Recommended range of argument r: [0, 44.189]
Available range of argument r: [0, 44.189]
Unit of length: 1 km
```

```
plot(Ki) #Figure 8.5a
```

From the output, one can see that values of the inhomogeneous K function were calculated for an interval [0, 44.189] km. Figure 8.5a presents the inhomogeneous K function of the unmarked pattern.

As can be seen, estimation for the inhomogeneous Poisson function lies below the observed curves (which are the border-corrected estimate $\hat{K}_{inhom}^{bord}(r)$ and the modified border-corrected estimate $\hat{K}_{inhom}^{bordm}(r)$. $\hat{K}_{inhom}^{bord}(r)$ lies close enough to $K_{pois}(r)$ in the middle of the graph, so the conclusion is either that the pattern reveals cluster behaviour or it is close to random. This can happen when some points simply have more neighbours than others. But Baddeley et al. (2015, Section 7.3.5.3) argue that a lack of any correlation observed on the plot does not mean that one is dealing with independence of points inside a point pattern. It seems that there is a need to apply other methods of analysis.

8.3.1.2 F function

The F function is called the empty spaces function. According to Baddeley et al. (2015), if one defines $d(u, \boldsymbol{X})$ = $min\{\|u - x_i\| : x_i \in \boldsymbol{X}\}$ as a distance from a fixed point u to the nearest point of process $\boldsymbol{X}$, then $d(u, \boldsymbol{X})$ is called the "empty-space distance". So, for a point process, $F(r)$ is simply a cumulative distribution function of the empty-space distance. The value of $F(r)$ for a completely spatial random pattern is given as as $F(r) = 1 - exp(-\lambda\pi r^2)$. The inhomogeneous process $F(r)$ takes another form; however, interpretation of results is the same as for the homogeneous function – the empirical curve lying below theoretical suggests clustered behaviour.

Finhom() estimates the inhomogeneous F function for a point pattern, taking one obligatory argument – point pattern. The result of the function call shows a tendency to cluster in a point pattern:

```
Fi <- Finhom(um_s)
plot(Fi) # Figure 8.5b
```

8.3.1.3 G function

For the derivation of the G function (also known as the nearest-neighbour function), one needs to introduce a definition of nearest neighbour distance d_i. The minimum distance between two distinct points x_i and x_j $(i \neq j)$ can be also treated as "the minimum distance from point x_i to all points of pattern $\boldsymbol{X}$ excluding x_i" (Baddeley et al., 2015, Section 8.3.1). Thus, one can express the G function as $G(r)=P\{d(u, X\backslash u)\leq r | X$ has a point at $u\}$ (Baddeley et al., 2015, Section 8.3.1). It is worth mentioning that $F_{pois} \equiv G_{pois}$, but for other point patterns, these functions differ. For the inhomogeneous pattern, the $G(r)$ formula looks different, but the interpretation of results for it is the same as for the homogeneous function – the theoretical curve lying below empirical means that pattern is clustered.

Ginhom() estimates the inhomogeneous G function for a point pattern; similarly to **Finhom**(), it takes one obligatory argument – point pattern, whereas others are optional. The results of the function call show that a pattern tends to reveal clustering behaviour:

```
Gi <- Ginhom(um_s)
plot(Gi) # Figure 8.5c
```

8.3.1.4 J function

One more distance-based measure is the J function, which combines the F and G functions: $J(r) = \dfrac{1-G(r)}{1-F(r)}$.

As stated previously, the values of the F and G functions for CSR are equivalent, so $J(r)$ for CSR takes a value of 1. If the numerator (or probability that "nearest neighbor distance will be greater than r" [Baddeley et al., 2015, Section 8.6.2]) is greater than the denominator (or the "corresponding probability of empty-space distances" [as before]), then $J(r)>1$ and the pattern is considered regular. For $J(r)<1$, a clustered pattern is suggested.

The results of the **Jinhom**() call subsequently confirm the conclusions from the F and G function plots, suggesting the pattern is clustered:

```
Ji<-Jinhom(um_s)
plot(Ji) # Figure 8.5d
```

8.3.1.5 Distance-based complete spatial randomness tests

Nearest-neighbour and empty-space distance concepts are also a theoretical background for the CSR test based on distances. There are two well-known tests checking for pattern behaviour – Clark-Evans (P. J. Clark & Evans, 1954)

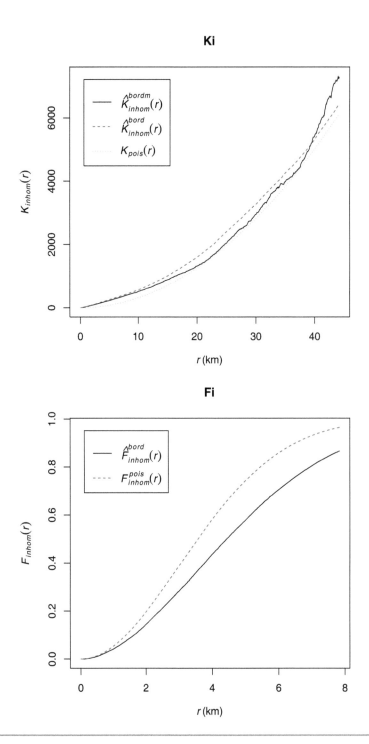

Figure 8.5 Inhomogeneous functions for unmarked point pattern: a) *K* function, b) *F* function, c) *G* function, d) *J* function

Source: Own calculations using spatstat::Kinhom(), spatstat::Ginhom(), spatstat::Finhom(), spatstat::Jinhom()

and Hopkins-Skellam (Hopkins & Skellam, 1954), which can be called by the functions **clarkevans.test**() and **hopskel.test**(), respectively. Both functions take one obligatory argument, which is point pattern; optional arguments are the test-performing *method* ("*asymptotic*" (default) or "*MonteCarlo*"), the alternative hypothesis (several options are possible; "*two-sided*" is default) and the number of simulations if the Monte Carlo test was chosen. The **clarkevans.test**() function also allows one to specify the type of edge correction. Results of both tests are of the *htest* class.

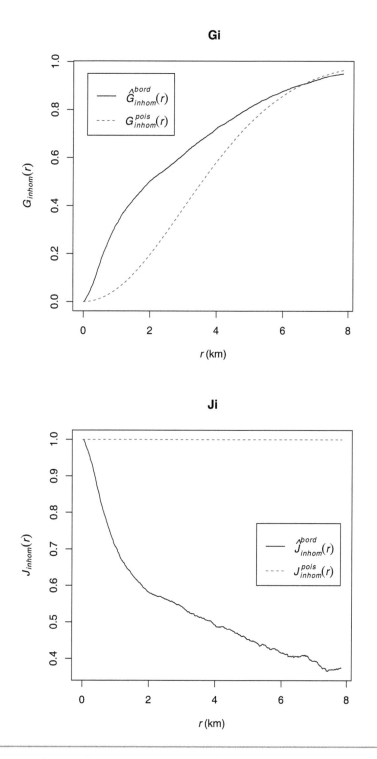

Figure 8.5 (Continued)

The Clark-Evans R index can be computed as a ratio between the "average of the nearest-neighbour distances d_i for m randomly sampled points in a point pattern (or for all data points) and expected value $\mathbf{E}[\mathbf{D}]$ for a completely random process with the same intensity" (Baddeley et al., 2015, Section 8.2.2). The value R determines the point pattern type: in the case of $R < 1$, there is evidence of a cluster pattern; if $R > 1$, the pattern is supposed to be regular.

In the Hopkins-Skellam test, test statistic H is also an indicator of point pattern type. It is calculated as the sum of squared distances between point and its nearest neighbour of pattern $\mathbf{X}$ over the sum of distances between the point of simulated point pattern U and nearest neighbour in pattern $\mathbf{X}$. $H = 1$ means complete spatial randomness, $H > 1$ suggests a regular pattern and $H < 1$ suggests clustering.

The following code shows function calls and the results of tests; both tests suggest rejection of the null hypothesis about complete spatial randomness:

```
ce.test <- clarkevans.test(um_s)
ce.test
```

```
        Clark-Evans test
        No edge correction
        Z-test
data:  um_s
R = 0.64388, p-value < 2.2e-16
alternative hypothesis: two-sided
```

```
ht<-hopskel.test(um_s)
ht
```

```
        Hopkins-Skellam test of CSR
        using F distribution
data:  um_s
A = 0.27318, p-value < 2.2e-16
alternative hypothesis: two-sided
```

8.3.2 Monte Carlo tests

The Monte Carlo test is based on replicated simulations. One first chooses a proper test statistic and then performs several simulations under the true null hypothesis (Baddeley et al. 2014). Two type of Monte Carlo tests are possible – graphical, based on simulation envelopes, and non-graphical – the DCLF test and MAD test.

8.3.3 Envelopes

Using an envelope, one can test the null hypothesis by constructing the non-rejection interval around the hypothesised value and see whether the empirical curve lies in that interval or whether there are some points where null should be rejected. Envelopes are possible to obtain calling the **envelope**() function. This function has a wide spectrum of uses – from the CSR assumption check to modelling goodness-of-fit. The main arguments of this function are either point pattern (*ppp* class object) or estimated model (*ppm* or *kppm*[17] class object), summary function and number of simulations. Other arguments are logical indicators, optional for the function call.

Envelopes can be estimated for each of the previously mentioned functions. This example will use the inhomogeneous L function to check whether it lies inside the envelope of simulated patterns. Any values of the empirical L function lying outside the envelope suggest that the point pattern is somehow different from the inhomogeneous L function but is not random, of course. The L function is a simple transformation of the K function proposed by Besag (1977): $L(r) = \sqrt{\dfrac{K(r)}{\pi}}$. Since the K function for a completely random point pattern takes the form πr^2, the L function for CSR is just r, which makes a graphical analysis easier. The inhomogeneous L function takes the same form. The following code shows the function call with results, and Figure 8.6a presents the estimated inhomogeneous function for a study pattern – it is visible that curves do not coincide, meaning that the pattern is non-random:

```
E <- envelope(um_s, Linhom, nsim = 5, verbose=FALSE) # Figure 8.6a
plot(E)
```

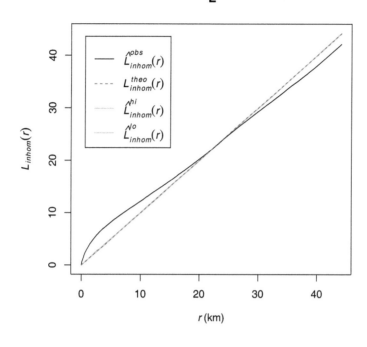

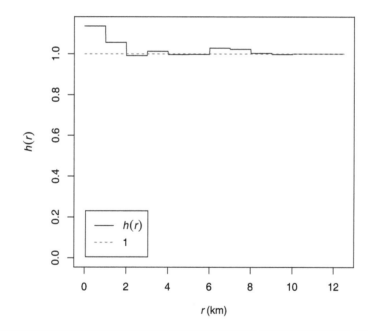

Figure 8.6 Estimation of point pattern: a) envelope of inhomogeneous *L* function for study point pattern, b) results of step-function pairwise interaction estimation

Source: Own calculations using spatstat::**envelope()**, **ppm()** and **fitin()** commands

8.3.4 Non-graphical tests

The DCLF test (named after Diggle, 1986; Cressie, 1991; Loosmore and Ford, 2006, and described in detail by Baddeley et al., 2014), and the MAD test (by Ripley, 1977, 1981) perform simulations to check the randomness of point patterns. Both tests have the same null hypothesis – the observed point pattern is CSR,[18] and both compare the discrepancy between the observed and theoretical summary function.[19] The number of simulations *m* is chosen by the researcher and is connected to test significance. According to Baddeley et al. (2014), *m* typically is 19, 39 or 99; however, many other options are also possible. To calculate the test significance α, one can compare the observed value with the simulated value. For one-sided tests, α can be calculated either using the minimum or maximum of simulated values as $^1/_{m+1}$ or $k/m+1$, where k is a rank of the largest/smallest value chosen for the test. Similarly, for the two-sided test, one can choose both the minimum and maximum of simulated values ($2/m+1$) or both kth largest and k th smallest simulated values ($2k/m+1$).

In the function call, it is sufficient to provide only a point pattern (or estimated model, if it is an object of analysis), summary function and number of simulations. The following code shows the results of nine simulations for each test. Since the alternative is a two-sided test (default option for **dclf.test()** and **mad.test()**), both types of deviations are significant. So, the desired p-value equals $2/(9+1) = 0.2$.

The results provided subsequently suggest that the null hypotheses of complete spatial randomness should be rejected:

```
dclf.xr <- dclf.test(um_s, Lest, nsim=19)
dclf.xr
```

```
Diggle-Cressie-Loosmore-Ford test of CSR
Monte Carlo test based on 19 simulations Summary function: L[inhom](r)
Reference function: theoretical Alternative: two.sided
Interval of distance values:
[0, 44.1891641673773]
Test statistic: Integral of squared absolute deviation Deviation = observed
minus theoretical
data:      um_s
u = 154.55, rank = 1, p-value = 0.1
```

```
mad.xr <- mad.test(X.rand, Lest, nsim=19)
mad.xr
```

```
Maximum absolute deviation test of CSR Monte Carlo test based on 19
simulations Summary function: L[inhom](r)
Reference function: theoretical Alternative: two.sided
Interval of distance values:
[0, 44.1891641673773]
Test statistic: Maximum absolute deviation Deviation = observed minus
theoretical

data:      um_s
mad = 3.8389, rank = 1, p-value = 0.1
```

All kinds of analysis showed that the studied pattern is non-random and tends to be clustered. Clustering in this case just means that distances between points are shorter than for CSR. Since the K function, for example, checks for correlation, one cannot say whether points interact.

8.4 Selection and estimation of a proper model for unmarked point pattern

After analysis performed in the previous sections, the conclusion is the following – the studied pattern reveals non-random behaviour. An intensity-based approach suggested that the pattern is at least inhomogeneous – several tests confirmed the presence of a spatial covariate. On the other hand, distance-based analysis argues that the pattern tends to reveal clustering behaviour. It can be modelled either with Cox/cluster processes or Gibbs processes. The Cox and cluster processes' common feature is that these processes are estimated in two steps – in the first, the Poisson process (as a process for characterising unseen "parent" points) is estimated; in the second, dependent on cluster type, the algorithm replaces each "parent" with a cluster of "descendant" points. The location of the "daughter" point depends on the parent's location and also those of other "daughter" points, and clusters are independent from each other. There are several cluster processes from which estimation is easy; however, one has to choose the right one and fit appropriate parameters. Plotting the realisation of such processes shows clearly visible clusters, which in the case of real objects' locations in space seems not so sensible.

Investigating firms' location, one can assume that the pattern of firms reveals cluster behaviour, but this could mean interaction instead. Firms can be considered units which compete for resources, better location and so on. The interaction concept is well investigated by the Gibbs models. Points either tend to locate far from each other (in other words, "inhibit") or can locate closer (this type of interaction can be named "attraction"; Baddeley et al., 2015, Section 13.1). Interactions are unobservable and cannot be captured by Ripley's K function, since it seeks correlation. But the K function (or L function) can be used in determination of the interaction radius, which is an important parameter for some processes from the Gibbs family. Under the analysis performed in the current study, putting more emphasis on Gibbs processes will be done. The inhomogeneity of firms' locations in space was proven by some studies (e.g. Li and Zhu, 2017). There are no studies investigating whether firms locate according to cluster process, but papers of Sweeney and Gómez-Antonio (Sweeney & Gómez-Antonio, 2016; Gómez-Antonio & Sweeney, 2018) apply Gibbs processes to analyse whether unobservable interaction implies on firm location.

The next section will describe, step by step, the estimation of the Gibbs process model for the study point pattern. Theoretical notes also will be provided.

8.4.1 Theoretical note

The main field of the Gibbs process models' use is spatial ecology, where relationships between species are examined. Economics seems to be a very untypical field for the Gibbs models; however, as suggested before, firms "fight" for resources, better location or performance and so interact with each other. Interaction can be pairwise or higher order.

Gibbs processes possess some key concepts (which can be found in Baddeley et al., 2015, Section 13.3), but the most interesting is the so-called "hard-core" process. Under it, there is some distance to which points are restricted and cannot come closer. This property is very useful in biology-related studies (for example, in modelling tree locations) or geography (when points are cities). The hard-core concept in firm location studies seems not to be logical, since there is not a clear restriction of firms from one or different industries where to locate; these restrictions (if they exist) are imposed rather by physical characteristics of the place and are not given in advance.

Several pairwise interactions are possible to investigate. Depending on the aim of research, one can choose an appropriate process. The previous paragraph suggests the rejection of hard-core processes (*Fiksel* interaction, *hard core* process, hybrid *Strauss-hard core* process). Hierarchical interactions also seem to be senseless – firms do not appear according to this principle. Also, some possible interaction types could be excluded due to the fact that they are suggested for regular patterns (*Diggle-Gates-Stibbard* potential model,[20] *Diggle-Gratton* potential model) or specific study fields (for example, the *Lennard-Jones* potential process is applied in physics). The *PairPiece* process can be used as a mechanism which helps to choose the appropriate kind of interaction, as it takes a range of possible radia and shows an interaction behaviour. The *softcore* process requires two arguments to determine the scale of interaction and its shape and strength, which have to be chosen arbitrarily. Without knowing the nature of a process, this choice seems tough. Playing with hybrids of several pairwise interaction processes may be considered an interesting but tedious and challenging task.

The *Strauss* process does not totally restrict a pair of points to lie closer but penalises for this. Among all pairwise processes, it seems to be the most appropriate to describe interaction between points. Baddeley et al. (2015, Section 13.3.7) argue that this process is "intermediate between hard core process and CSR".

Among higher-order interaction processes, there are several options as well; however, taking into account the choice of the Strauss process as a possible interaction model for the spatial pattern of firms, the *Geyer* saturation model, which was proposed by Geyer (1999) as an extension of the Strauss model, seems to be the most appropriate choice. The saturation process is "a modification of the Strauss process in which the overall contribution from each point is trimmed to never exceed a maximum value" (Baddeley et al., 2015, Section 13.7.2). For the saturation parameter $s > 0$, the interaction parameter γ can take values greater than 1, and it would be consistent with a clustered pattern; for values lower than 1, the pattern is considered ordered.

In terms of current research, it seems interesting to investigate and compare the performance of both interaction models. Though even without such assumptions, a couple of possible high-order interaction processes can be rejected, either because they are usually used in other study fields (*area interaction* process in physics, *Ord* process for forestry data) or too complicated a structure (*Baddeley-Geyer* process as a combination of several Geyer processes). The application of a hybrid of several high-order interaction processes (as in the case of pairwise interactions hybrids) requires justification of choice not only for the proper process but also the proper parameters.

8.4.2 Choice of parameters

The choice of interaction radius is crucial for the Strauss model – a wrongly chosen value causes the interaction parameter γ to exceed 1 (or the threshold for a process to be a CSR). If γ < 1, one deals with the Strauss process. A value of zero suggests the hard-core process. A proper radius can be obtained in two ways:

- By finding an interval of the biggest discrepancy between the theoretical and empirical K (or more often L) function using the **profilepl**() function, which by maximising the profile likelihood (other options are profile pseudolikelihood, profile composite likelihood and AIC) finds the best model for data given the assumed interaction process and possible radia interval (*R course on spatial point patterns*);
- By estimation of a model with the PairPiece interaction process, which also takes possible radia intervals and checks for the presence of higher-order interactions (see Baddeley et al., 2015, Section 13.6.3). The application of the K (or L) function to determine an appropriate interaction radius is risky to some extent, since it assumes from the beginning that one deals with the homogeneous Strauss process. Nevertheless, in the examples subsequently, both options are used.

Figure 8.6a represents the envelope test for the inhomogeneous L function, where a black solid line represents an empirical and a red dashed line a theoretical curve. The biggest discrepancy can be observed for small distances in the interval approximately (1, 10) km. The function **profilepl**() takes the following arguments: data frame with possible interaction distances (*s*), interaction process (*f*) and some logical values (for example, using AIC instead of profile likelihood [*aic=TRUE*], printing function progress [*verbose*]). The output subsequently shows the results of the function call:

```
r1 <- seq(1,10,by=0.05) # declaring the vector of possible radii
D1 <- data.frame(r=r1) # conversion to the class data.frame
fp1 <- profilepl(D1, Strauss, um_s ~ 1, aic=TRUE, fast=TRUE, verbose=FALSE)
fp1
```

```
profile log pseudolikelihood
for model: ppm(um_s~1,  aic=TRUE,  interaction=Strauss,  fast=TRUE)
fitted with rbord = 10
interaction: Strauss process
irregular parameter: r in [1, 10]
optimum value of irregular parameter:  r = 2.2
```

The results of running **profilepl**() suggest that the optimal interaction radius is 2.2 km. As stated previously, this kind of optimisation assumes the Strauss homogeneous model for a point process. The code and figure subsequently show the estimation of the step-function pairwise interaction; from a graph, it can be observed that interaction parameter is higher than 1, suggesting that the interaction is not pairwise but higher order:

```
m1 <- ppm(um_s ~ 1, PairPiece(1:10)) # check for all possible radii
f1 <- fitin(m1) # fitting step function
plot(f1, main="") # Figure 8.6b
```

The comparison of results suggests that the usage of the first optimisation method was inappropriate. Having in mind that the assumed interaction interval is from 1 to 10 km, it is also possible to find parameters of the stationary Geyer saturation model. For this purpose, again, the **profilepl()** function will be applied (see the following code):

```
# stationary Geyer saturation model
# setting r from 1 to 10 and s from 1 to 3
df <- expand.grid(r=seq(1,10, by=0.05), sat=c(1,3))
pG <- profilepl(df, Geyer, um_s ~ 1, aic=TRUE, verbose=FALSE)
pG
```

```
profile log pseudolikelihood
for model: ppm(um_s ~ 1,  aic = TRUE,  interaction = Geyer)
fitted with rbord = 20
interaction: Geyer saturation process
irregular parameters:
 r in [1, 10]
sat in [1, 3]
optimum values of irregular parameters:
 r = 1.05 and sat = 3
```

The optimisation results show the optimal radius as 1.05 km and the optimal saturation parameter as 3. In the next step, the same will be done to estimate the non-homogeneous Geyer saturation model with the presence of a spatial covariate (x coordinate). The results are totally the same, but the latter model is adjusted for possible inhomogeneity:

```
df1 <- expand.grid(r=seq(1,10, by=0.05), sat=c(1,3))
pG1 <- profilepl(df1, Geyer, um_s ~ x, aic=TRUE, verbose=FALSE)
pG1
```

```
profile log pseudolikelihood
for model: ppm(um_s ~ x,      aic = TRUE,  interaction = Geyer)
fitted with rbord = 20
interaction: Geyer saturation process
 irregular parameters:
r in [1, 10]
sat in [1, 3]
optimum values of irregular parameters:
 r = 1.05 and sat = 3
```

8.4.3 Estimation and results

In the previous step, two different Geyer saturation models were estimated – stationary and non-stationary. The function **as.ppm()** transforms the object of class *profilepl* and allows one to see the estimated coefficients. Both models have an interaction parameter greater than 1, suggesting that the pattern is clustered. The parameters (which are interaction radius and saturation value) are the same; however, the second model assumes the presence of a spatial covariate:

```
as.ppm(pG)
```

```
Stationary Geyer saturation process
First order term: beta = 0.03844399
Interaction distance:  1.05
Saturation parameter:  3
Fitted interaction parameter gamma:  2.0541475
Relevant coefficients:
Interaction
  0.7198609
```

```
as.ppm(pG1)
```

```
Nonstationary Geyer saturation process Log
trend:    ~x
Fitted trend coefficients:
 (Intercept)      x
 2.538169154-0.002877295
Interaction distance:  1.05
Saturation parameter:  3
Fitted interaction parameter gamma:    2.0493618
Relevant coefficients:
Interaction
  0.7175284
```

Goodness-of-fit analysis for the estimated Gibbs processes includes (among others) formal tests (**cdf.test**(), **berman.test**(), **quadrat.test**() and simulation **envelope**()), residual analysis (**residuals**()) and the predicted K or pair-correlation function for the estimated model (**Kmodel**(), **pcfmodel**()). Graphical analysis of the estimated model can be done using **qqplot**(), **diagnose.ppm**()[21] or one of the summary functions (K, F or G).

To choose between a pair of estimated models, one can apply the **anova**() test. Inside the **anova**() function call, such arguments as two models (model_1, model_2) and test ("Chi" for likelihood ratio test) are used. The null hypothesis assumes that the first model is true. The returned p-value allows one to make a proper conclusion.

The code subsequently shows the ANOVA test for the two models estimated previously. The p-value close to zero suggests that a proper model is a non-homogeneous Geyer:

```
anova(as.ppm(pG), as.ppm(pG1), test="Chi")
```

```
Analysis of Deviance Table
Model 1: ~1      Geyer
Model 2: ~x      Geyer
Npar Df AdjDeviance Pr(>Chi)
1    2
2    3 1    7.1773 0.007383 **
Signif. codes:   0 '***' 0.001 '**' 0.01 '*' 0.05 '.' 0.1 ' ' 1
```

Several graphical analysis methods are available for the estimated interaction model. The function **diagnose.ppm**() shows four different plots, as mentioned previously; lurking plots[22] are good in detecting trend misspecification in the estimated model. If the assumed trend is wrong, the discrepancy between the assumed and the real trend will be observed.

For the fitted non-homogeneous Geyer saturation model, a four-panel residual plot can be called using the function **diagnose.ppm**() with the following arguments: estimated model (**as.ppm(pG1)**) and type of residuals (several options are possible, but in this example, *Pearson* residuals will be used). Also, it is possible to add simulation pattern envelopes for the lurking variable plot analysis (*envelope=TRUE*) and provide the number of simulations (*nsim*). Figure 8.7a shows four-panel residual plot.

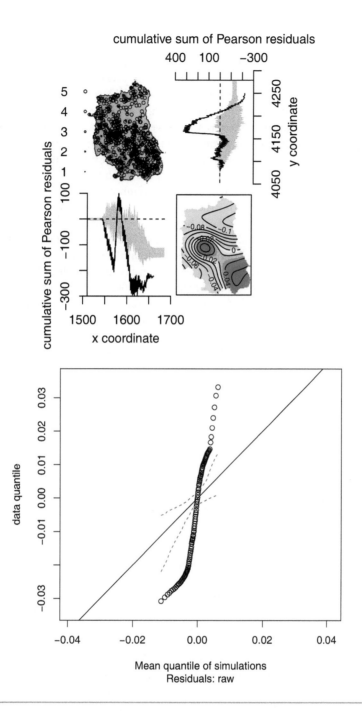

Figure 8.7 Diagnostics of estimation results: a) four-panel residual plot for the estimated model, b) qqplot for inhomogeneous Geyer saturation model

Source: Own calculations using spatstat::diagnose.ppm() and qqplot.ppm()

```
diagnose.ppm(as.ppm(pG1), type="Pearson", envelope=TRUE, nsim=19) # Figure 8.7a
```

Lurking plot variables (top right for y coordinate and bottom left for x coordinate) suggest that trend formula that was chosen is inappropriate. The plots' shapes do not suggest any clearly visible trend. In comparison with lurking plots, **qqplot.ppm**() checks for "misspecification of the **interaction** in a fitted model" (Baddeley et al., 2015, Section 13.10.5.1). Figure 8.7b shows the qqplot for estimated model; it is clearly visible that the interaction process is not Geyer:

```
qqplot.ppm(as.ppm(pG1), nsim=19) # Figure 8.7b
```

8.4.4 Conclusions

This section provided analysis of an unmarked point pattern based on currently known point pattern analysis techniques. Thanks to intensity-based and density-based analysis, it was found that the study of unmarked point patterns of firm locations in Lubelskie voivodeship reports clustered behaviour. It was suggested that the behaviour could be rather interactive, since the K function is typically used to capture correlation and not interaction. Among the available interaction for point patterns, two were examined – the Strauss process and the Geyer saturation process (which is an extension of Strauss). The step-function pairwise interaction plot rejected the assumption about pairwise interaction, so the Geyer saturation model was assumed to be the correct one. During its examination, it was found that both the interaction process and the specified trend were wrong.

What can be a possible solution? Probably, the estimation of inhomogeneous Geyer saturation process failed, because originally the Gibbs processes were constructed for other study fields, such as biology, epidemiology and so on. Moreover, the true trend is unknown and can be either a function of Cartesian coordinates, other covariates (which unfortunately were not provided in a study dataset) or a combination of both. Nevertheless, the true model describing the behaviour of firms in Lubelskie voivodeship can be still expressed as the Gibbs process. Since the hypothesis about pairwise interaction was rejected in favour of high-order interaction, it is suggested that firms truly interact with each other in groups of three and more. Unfortunately, for the unmarked process, it is not possible to detect the presence of industry agglomeration, but this is examined in the next section. It seems that either other interactions (such as the multiscale saturation process *SatPiece*, the connected component interaction *Concom* or the Geyer triplet interaction process *Triplet*), other trend formulae, a hybrid of several interaction processes or the application of all these factors can be possible tools to find a firm's true behaviour.

8.5 Intensity-based analysis of marked point pattern

CSRI is an extension of the CSR concept for multitype point patterns. According to Baddeley et al. (2015, Section 14.5.1), there are four properties which should be satisfied for CSRI:

1   After unmarking, the point pattern is CSR;
2   Labels (mark values) are i.i.d.;
3   Each sub-pattern of type m_i is CSR;
4   Sub-patterns $X^{(1)}, \ldots, X^{(m)}$ are independent.

In practice, 2 and 4 are sufficient to call a point process CSRI. In the literature, they are known as "random labelling" and "component independence", respectively.

For the inhomogeneous multitype Poisson process, 2 is slightly changed to 2' – for a given location x_i, "the type of each point is random, independent of the types of other points, and depends only on the location of the point" (Baddeley et al., 2015, Section 14.5.2). To be an inhomogeneous multitype Poisson process, the point process should satisfy either 1 and 2' or 3 and 4.

There are three different choices of null hypothesis to perform multitype pattern analysis, assuming: 1) the point pattern is CSRI, 2) random labelling and 3) independence of components. In practice, the second and third are sufficient to state that the point pattern is CSRI. It worth mentioning that random labelling does not entail component independence, as points may still reveal interaction. From another perspective, it is still possible to perform analysis based either on intensity measures (*segregation* test) or apply distance-based techniques (variations of multitype K function, *nearest neighbour* distances and correlation, *mark equality* function, I function, Monte Carlo tests). They, as in the case of an unmarked pattern, require one to choose a proper starting assumption about the pattern's homo- or heterogeneity.

This section describes intensity-based methods of marked pattern analysis.

8.5.1 Segregation test

The segregation test proposed by Kelsall and Diggle (1998) and later generalised by Diggle, Zheng, and Durr (2005) tests a null hypothesis which can be treated as "more specific assumption of **random labelling**: given the locations x_i, the type labels m_i are random variables, independent of each other, with common probability distribution $P\{m_i = m\} = p_m$" (Baddeley et al., 2015, Section 14.4.4). This is a Monte Carlo test, which randomly shuffles type labels with locations fixed. To perform this test, one can use the function **segregation.test**(), specifying the point pattern and number of simulations as arguments. The number of simulations can be chosen arbitrarily; 19 is the default option. The results of the test are presented subsequently:

```
s<-segregation.test(mm_s, nsim=19)
s
```

```
     Monte Carlo test of spatial segregation of types
data:       mm_s
T = 0.022802, p-value = 0.8
```

The test *p*-value is quite high, so the null hypothesis is not rejected. It means that firms from different industries are not located in clusters; they are spread randomly across the study window. Though the null hypothesis is not rejected, it does not mean that dependence between firms does not exist. This will be checked in the next section.

8.6 Correlation and spacing analysis of the marked point pattern

As in the case of unmarked patterns, before applying correlation and spacing techniques, one should assume either the stationarity or non-stationarity of a point pattern. The choice can be crucial, since the wrong assumption can lead to misinterpretation. The subsections subsequently present two methods of distance-based analysis for multitype patterns – with the assumption of stationarity and non-stationarity.

8.6.1 Analysis under assumption of stationarity

8.6.1.1 K function variations for multitype pattern

The *K* function for a stationary multitype point pattern can be estimated with **Kcross()**. The main idea of the function is to compare the distance not only between points of type *i* in point process *X* but also with neighbour points of type *j*. In the case of independence between *i* and *j* points, empirical $K_{ij}(r) = \pi r^2$.

To perform a test, firms of sectors F (construction) and M (professional, scientific and technical activities) were chosen. The arguments provided for **Kcross()** are the point pattern and the two mark labels for the chosen sectors:

```
Kfm<-Kcross(mm_s,"F", "M")
plot(Kfm)
```

From a plot, it can be concluded that the previously mentioned industries are rather dependent; that is, the location of one is connected with the location of other, since the discrepancy between the theoretical and empirical curves is clearly observed (see Figure 8.8a).

For wider analysis – checking how points of type *i* are located relative to other types (i.e. examining spatially inhomogeneous intensity (Baddeley, Rubak, & Turner, 2019) – the function **Kdot()** can be applied. For complete independence between points of type *i* and others, $K_{ij}(r) = \pi r^2$. Any discrepancy between curves suggests dependence between points of type *i* and *j* for $j \neq i$. The function **Kdot()** takes a couple of arguments – the marked point pattern, mark type to which distances are measured and edge correction (as a default option, all correction types are used). Suppose one wants to examine whether firms from sector G (wholesale and retail trade; repair of motor vehicles and motorcycles) are related to firms from other sectors. The **Kdot()** function is as follows:

```
Kdi<-Kdot(mm_s, "G")
plot(Kdi)
```

Figure 8.8b shows the estimated inhomogeneous *Kdot* function examining dependence between points of type G and others. It can be observed that, in general, differences between theoretical and empirical curves exist, so one can suggest locations of firms in wholesale and retail and repair of motor vehicles and firms from other industries to be dependent.

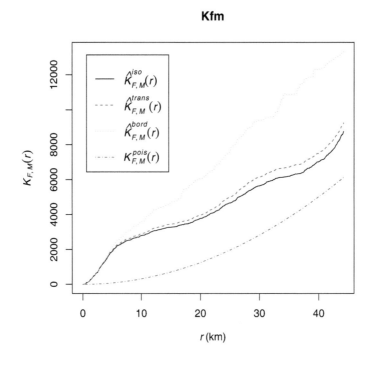

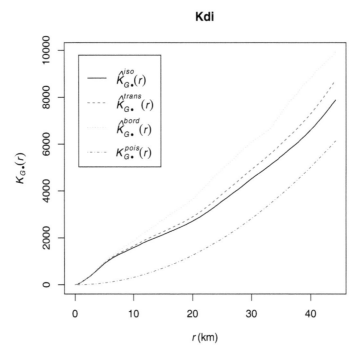

Figure 8.8 Graphic analysis of a multitype point pattern under assumption of stationarity: a) homogeneous **Kcross** function, b) homogeneous **Kdot** function, c) mark connection function, d) mark equality function, e) *I* function, f) randomisation of test components' independence

Source: Own calculations using spatstat::

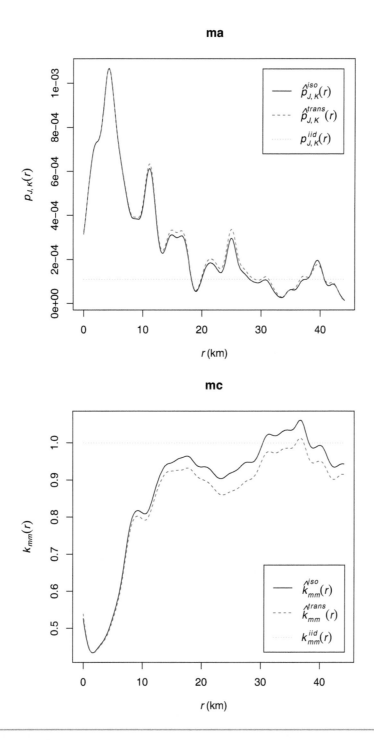

Figure 8.8: **(Continued)**

8.6.1.2 Mark connection function

One more function measuring dependence between points of different types is the mark connection function (as in Baddeley et al., 2015; Section 14.6.4.2):

$$p_{ij}(r) = \frac{\lambda_i \lambda_j g_{ij}(r)}{\lambda_\bullet^2 g(r)} \tag{8.2}$$

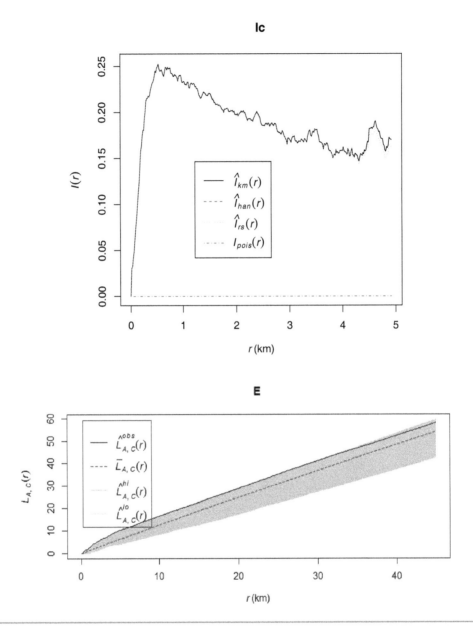

Figure 8.8: (Continued)

where $g_{ij}(\cdot)$ is the pair correlation function (see more in Section 8.2.3.3) between types i and j, and $g(\cdot)$ is the pair correlation function for the unmarked process. In spatstat::, it can be computed using **markconnect()**, which takes as arguments a point pattern and a pair of two different marks. The mark connection function allows one to check dependence between a point of different types for two and more types (the **alltypes()** function allows one to examine all pairwise connections). This function is time consuming in cases of big patterns (with number of points > 1000), so for this example, two industries will be chosen – J (information and communication) and K (financial and insurance activities). The plot (see Figure 8.8c) suggests that random labelling of these types can be observed only in short distances (up to ~20 km). So, a general assumption about random labelling in the case of J and K industries is not confirmed, though it tells nothing about the dependence between these types.

```
ma<-markconnect(mm_s, "J", "K")
plot(ma)
```

8.6.1.3 Analysis of within- and between-type dependence

Baddeley et al. (2015, Section 14.6.7) suggest that in the case of marked patterns with numerous possible mark values, estimations of summary functions (such as *K, G, F*) can be biased. To deal with this, one can either reduce the number of mark types by grouping or apply other kinds of summary functions, allowing one to investigate within- and between-type dependence.

One can check the random labelling property using the mark equality function **markcorr()**. This function is defined as a "sum of the mark connection functions $p_{ii}(r)$ between all pairs of points of the same type" (Baddeley et al., 2015, Section 14.6.7.1). Values close to 1 suggest a lack of correlation between marks.

The syntax of the function is simple: it takes only one argument – point pattern:

```
mc<-markcorr(mm_s)
plot(mc)
```

From the graph (see Figure 8.8d) it can be observed that for short distances (up to ~13–14 km), points rather have different types, which is consistent with random labelling. This coincides with the conclusion from the segregation test. However, there still is no evidence of firm independence.

The second kind of dependence can be checked with the *I* function, defined in Van Lieshout and Baddeley (1999) as the difference between the *J* function for the processes of type *i* only, multiplied by the proportion of type *i* points, and the *J* function for the whole point process without marks. In other words, it is supposed to measure "dependence between points of different types" (Baddeley et al., 2015, Section 14.6.7.2). Deviations from the independence threshold (which is zero) can suggest either positive (*I(r) > 1*) or negative (*I(r) < 1*) association.

Results of the components' independence analysis are presented in Figure 8.8e. The plot shows a positive association between types, as the curve lies much higher than the zero threshold. These results seem to be interesting, as it turns out that firms from different industry sectors tend to locate in the neighbourhood of each other, which from an economic point of view can suggest a Jacobian agglomeration type.

```
Ic <- Iest(mm_s)
plot(Ic)
```

8.6.1.4 Randomisation test of components' independence

This test is an envelope-based test allowing one check the presence/absence of interaction between points of different types. It is applicable only under the assumption of stationarity, and possible summary functions to use are the **Kcross()**, **Fcross()** and **Jcross()** functions. Under a test, "the simulated patterns **X** are generated from the dataset by splitting the data into the sub-patterns of points of each type, and randomly shifting each of these sub-patterns, independently of each other" (Baddeley et al., 2015, Section 14.7.1). The null hypothesis states that points of two different types are independent.

The following code shows an example of this test. Two different industries were selected for analysis – A (agriculture, forestry, fishing) and C (manufacturing). The argument *rshift*, specifying the radius of random shifting, was set to 35 km. It should be noted that for proper function performance, the study window should be rectangular:

```
mm_s$window # initial window
```

```
window: polygonal boundary
enclosing rectangle: [1510.01, 1686.7666] x [4052.118, 4278.901] km
```

```
# setting new window
window(mm_s) <- owin(c(1510.01,1686.7666), c(4052.118,4278.901))

mm_s$window # changed window
#window: rectangle = [1510.01, 1686.7666] x [4052.118, 4278.901] km
```

After 99 simulations of *Lcross* function, it can be clearly observed (see Figure 8.8f) that the chosen industries are independent of each other (which is consistent with the fact that these industries usually do not reveal any interaction):

```
E<-envelope(mm_s, Lcross, nsim=99, i="A", j="C",
            simulate=expression(rshift(mm_s, radius=35)))
plot(E)
```

Under the assumption of pattern stationarity, CSRI was not confirmed. Despite the fact that firms of different industries are located rather randomly (i.e. there is no evident "cluster" inside any particular industry, which was proved by the mark equality function and segregation test), several techniques suggested that points of certain types reveal positive association.

8.6.2 Analysis under assumption of non-stationarity

8.6.2.1. Inhomogeneous K function variations for multitype pattern

Under the assumption of non-stationarity, inhomogeneous versions of the previously mentioned *K* functions can be applied – **Kcross.inhom**() and **Kdot.inhom**(). Contrary to homogeneous versions, they require a specification of sub-processes' intensities – namely *lambdaI* and *lambdaJ*. The package spatstat:: offers several functions for intensity estimation, but one should be careful with them. According to Baddeley et al. (2015, Section 7.10.2), the so-called "leave-one-out" kernel smoother is the best option, since it provides the least bias. This smoother is a default choice for **Kcross.inhom**() and **Kdot.inhom**() (as well as for other inhomogeneous functions), so it will be used in the following analysis.

The *K* function for non-stationary multitype point pattern can be estimated with **Kcross.inhom**(). The main idea and syntax of the function are the same as for stationary patterns. As before, industry sectors F (construction) and M (professional, scientific and technical activities) were chosen for analysis:

```
Kfmi <- Kcross.inhom(mm_s, "F", "M")
plot(Kfmi)
```

From Figure 8.9a, it can be seen that the previously mentioned industries definitely interact with each other, since the discrepancy between the theoretical and empirical curves is clearly visible.

The syntax and main idea of **Kdot.inhom**() is the same as for its homogeneous version. For complete independence between points of type *i* and others, $K_{i\cdot}^{inhom}(r) = \pi r^2$. Any discrepancy between curves suggests dependence between points of type *i* and *j* for *j* ≠ *i*:

```
Kdii<-Kdot.inhom(mm_s, "G")
plot(Kdii)
```

From Figure 8.9b, it can be observed that, in general, differences between the theoretical and empirical curves exist, so one can suggest locations of firms in wholesale and retail and repair of motor vehicles and firms from other industries are dependent.

To name the marked point pattern CSRI, at least two of four assumptions should be satisfied – random labelling and independence between components. The first one does not imply the second, which is exactly what happened in the case of study dataset. Despite the fact that firms of different industries are located rather randomly (i.e. there is no evident "cluster" inside a particular industry, which was proved by the mark equality function and segregation test), several techniques suggested that points of different types reveal positive association.

8.7 Selection and estimation of a proper model for unmarked point pattern

The results of the performed tests and analysis suggest that the studied pattern cannot be classified as random but rather clustered. Possible models to be estimated for non-homogeneous patterns are the non-homogeneous multitype Poisson process, multitype Cox/cluster process or multitype Gibbs process. As in the case of unmarked

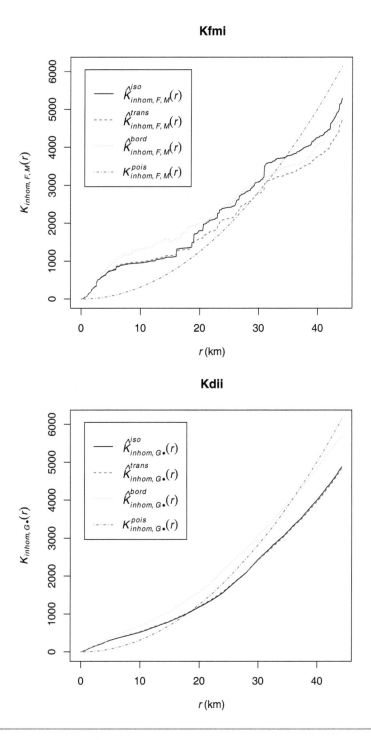

Figure 8.9 Graphic analysis of a multitype point pattern under assumption of non-stationarity: a) inhomogeneous **Kcross** function, b) inhomogeneous **Kdot** function

Source: Own calculations using spatstat::Kcross.inhom(), spatstat::Kdot.inhom

patterns, a search of the proper intensity function form could be tough, and there is a risk of fitting an improper model to the data. The multitype Cox/cluster processes unfortunately are not implemented in R yet, so the only type of model which can fit the data is the multitype Gibbs process. Firms reveal competitive behaviour on the one hand, but, on the other, can agglomerate either within one or between several industries. From a research point of view, it is interesting to examine the presence of any interaction in a given point pattern of enterprises.

Following the agglomeration issue concept, these processes are suggested to distinguish between Marshallian and Jacobian externalities. The only possible weakness can be the fact that under multitype Gibbs processes, only pairwise interactions are assumed.

8.7.1 Theoretical note

Multitype Gibbs processes can be easily estimated the same way as Poisson or cluster processes. However, the interesting fact is that inside the multitype Gibbs processes, one can explore interaction of three types: not depending on types m, m' (so interaction depends on locations of points x_m and $x_{m'}$), interaction within one type (so, it is assumed that points of two different types are independent) and the mark-dependent pairwise interaction. The latter can be divided into three models: multitype *hard core* (when the points of types m, m' can never be located closer than some distance $r_{m,m'} > 0$), multitype *Strauss* and multitype *hard core Strauss*. As for the unmarked pattern, it seems logical not to apply hard-core processes, as there are no distance restrictions for firms. Similarly, the hierarchical process also will be not taken into account. The following subsections describe the estimation of the multitype Strauss process for a study multitype pattern.

8.7.2 Choice of optimal radius

The preliminarily choice of possible interaction radius can be done using the **Kcross()** function, and then the proper radius will be estimated by **profilepl()**. As for the Strauss pairwise interaction for the unmarked pattern, there is the same validity criterion – the value of the interaction parameter γ cannot exceed 1.

The number of mark types in the study point pattern is too big (20), so the examination of possible between-interactions can be long. Due to the fact that previously, relationships between some industries (for example, between F and M with the **Kcross()** function or G vs the other using the **Kdot()** function) were examined, these three industries were chosen to create a new multitype point pattern with three mark types. Also, for study purposes, three smaller multitype patterns will be created – the combination of industries F and M, M and G and F and G. Figure 8.10 shows the estimated **Kcross** function for a new pattern. The blue dotted line represents the theoretical Poisson curve, and the other curves represents border-corrected versions. From all plots, it can be concluded that the possible interaction radius is interval (30, 40) km, either within or between industries.

Splitting the whole pattern can be done using the function **split()** (described briefly in Section 8.1), and the combination of two or more patterns can be done using the **superimpose()** function, which takes from two or more possible patterns to combine:

```
s<-split(mm_s) # splitting into 20 different patterns
s$F$marks<-as.factor(rep("F", 82)) # attaching mark values
s$G$marks<-as.factor(rep("G", 204))
s$M$marks<-as.factor(rep("M", 46))
X.r2<-superimpose(s$F, s$G, s$M) # big pattern of 3 industries
fm<-superimpose(s$F, s$M) # pattern of F and M industries
fg<-superimpose(s$F, s$G) # pattern of F and G industries
mg<-superimpose(s$G, s$M) # pattern of M and G industries
```

8.7.3 Within-industry interaction radius

As for the unmarked patterns, **profilepl()** chooses an optimal value of interaction radius. The code subsequently shows the results of the function performance in defining within-industry interaction radius; it was observed that the results of optimisation for within-industry interaction using the whole and separate patterns differ slightly:

```
# big marked pattern with 3 mark types
RR<-data.frame(R=seq(30,40,by=0.05)) # specifying interval
MS<-function(R) {MultiStrauss(radii=diag(c(R,R,R)))} # with interaction
pm<-profilepl(RR, MS, X.r2~marks, verbose=FALSE) # optimization
pm # suggested radius of within industry interaction is 30.4km
```

array of Kcross functions for big_pattern.

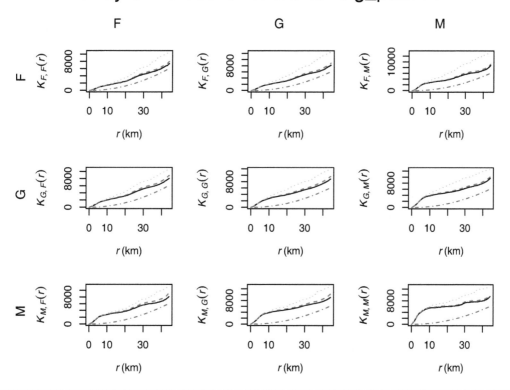

Figure 8.10 Estimation of the **Kcross** function for a new pattern

Source: own calculations using spatstat::Kcross()

```
profile log pseudolikelihood
for model: ppm(X.r2 ~ marks,  interaction = MS)
fitted with rbord = 40
interaction: Multitype Strauss process
irregular parameter: R in [30, 40]
optimum value of irregular parameter:  R = 30.4
```

```
# checking for each industry separately
r1 <- seq(30,40,by=0.05)
D1 <- data.frame(r=r1)
fp1 <- profilepl(D1, Strauss, s$F ~ 1)
fp1 # 30.55 km for F industry
```

```
profile log pseudolikelihood
for model: ppm(s$F ~ 1, interaction = Strauss)
fitted with rbord = 40
interaction: Strauss process
irregular parameter: r in [30, 40]
optimum value of irregular parameter:  r = 30.55
```

```
r2 <- seq(30,40,by=0.05)
D2 <- data.frame(r=r2)
fp2 <- profilepl(D2, Strauss, s$G ~ 1, verbose=FALSE)
fp2 # 30.45 km for G industry
```

```
profile log pseudolikelihood
for model: ppm(s$G ~ 1, interaction = Strauss)
fitted with rbord = 40
interaction: Strauss process irregular parameter: r in [30, 40]
optimum value of irregular parameter:  r = 30.45
```

```
r3 <- seq(30,40,by=0.05)
D3 <- data.frame(r=r3)
fp3 <- profilepl(D3, Strauss, s$M ~ 1, verbose=FALSE)
fp3 # 30.9 km for M industry
```

```
profile log pseudolikelihood
for model: ppm(s$M ~ 1, interaction = Strauss)
fitted with rbord = 40
interaction: Strauss process
irregular parameter: r in [30, 40]
optimum value of irregular parameter:  r = 30.9
```

The results suggest that there are two solutions for the within-industry radius: either 30.4 km for each separate industry or 30.55 km, 30.45 km and 30.9 km for industries F, G and M, respectively.

8.7.4 Between-industry interaction radius

To find the between-industry estimation radius for three new patterns, the function **profilepl()** will again be applied:

```
r4<-seq(30,40,by=0.01)
D4<-data.frame(r=r4)
fp4<-profilepl(D4, Strauss, fm~marks, aic=TRUE, fast=TRUE, verbose=FALSE)
fp4 # 30 km
```

```
profile log pseudolikelihood
for model: ppm(fm~marks,      aic=TRUE,    interaction=Strauss,    fast=TRUE)
fitted with rbord = 40
interaction: Strauss process irregular parameter: r in [30, 40]
optimum value of irregular parameter:    r = 30
```

```
r5<-seq(30,40,by=0.01)
D5<-data.frame(r=r5)
fp5<-profilepl(D5, Strauss, fg~marks, aic=TRUE, fast=TRUE, verbose=FALSE)
fp5 # 30.54 km
```

```
profile log pseudolikelihood
for model: ppm(fg~marks,      aic=TRUE,    interaction=Strauss,    fast=TRUE)
fitted with rbord = 40
```

```
interaction: Strauss process
irregular parameter: r in [30, 40]
optimum value of irregular parameter:  r = 30.54
```

```
r6<-seq(30,40,by=0.01)
D6<-data.frame(r=r6)
fp6<-profilepl(D3, Strauss, mg~marks, aic=TRUE, fast=TRUE, verbose=FALSE)
fp6 # 30.05 km
```

```
profile log pseudolikelihood
for model: ppm(mg~marks,        aic=TRUE,    interaction=Strauss,      fast=TRUE)
fitted with rbord = 40
interaction: Strauss process
irregular parameter: r in [30, 40]
optimum value of irregular parameter:  r = 30.05
```

The results of the optimal radius search in the patterns of two industries suggest that for industries F and M, the interaction radius equals 30 km, for industries F and G-30.54 km and for industries M and G-30.05 km.

8.7.5 Estimation and results

Estimation will be done using the MultiStrauss interaction process. After finding values of the proper interaction radius, it is suggested that two types of model will be estimated – first, one which assumes no interaction within industry, and the second allowing for all possible interactions. Estimations of both models are presented in the following subsections.

8.7.6 Model with no between-industry interaction

Due to the results of the separate optimisation for a big marked pattern and for each industry pattern, it should be assumed that the interaction radius within an industry does not have to be the same for all industries; however, both options will be checked.

To estimate the MultiStrauss interaction process, the function **ppm()** is called; the arguments are the model and matrix of interaction radia; in the case of only within-type interaction, the matrix is diagonal. The following code shows the results of the two models' estimation – 1) 30.4 km for each separate industry, and 2) 30.55 km, 30.45 km and 30.9 km for industries F, G and M, respectively:

```
rr1<-diag(c(30.4, 30.4, 30.4))
fit1<-ppm(X.r2~marks, MultiStrauss(rr1))
fit1
```

```
Stationary Multitype Strauss process Possible
marks: 'F', 'G' and 'M'
Log trend: ~marks
First order terms:
     beta_F        beta_G        beta_M
0.0004343718 0.0014539061 0.0001623045

3 types of points Possible types:
   [1] F G M
Interaction radii:
```

```
     F   G   M
F 30.4 NA
NA G NA 30.4
NA M NA NA 30.4
Fitted interaction parameters gamma_ij


          F         G         M
F   1.125336        NA        NA
G         NA  1.040544        NA
M         NA        NA  1.166552


Relevant coefficients:
   markFxF    markGxG     markMxM
0.11808169 0.03974385 0.15405239


For standard errors, type coef(summary(x))

*** Model is not valid ***
*** Interaction parameters are outside valid range ***
```

```
rr2 <- diag(c(30.55, 30.45, 30.9))
fit2 <- ppm(X.r2~marks, MultiStrauss(rr2))
fit2
```

```
Stationary Multitype Strauss process
Possible marks: 'F', 'G' and 'M'
Log trend: ~marks

First order terms:
      beta_F     beta_G     beta_M
2.395928e-04 1.324084e-03 4.850602e-05

3 types of points
Possible types:
[1] F G M
Interaction radii:
      F     G     M
F   30.55 NA    NA
G   NA 30.45    NA
M   NA    NA 30.9
Fitted interaction parameters gamma_ij
          F         G         M
F   1.149236        NA        NA
G         NA  1.041845        NA
M         NA        NA  1.219277


Relevant coefficients:
markFxF    markGxG     markMxM 0.13909747 0.04099277 0.19825817

For standard errors, type coef(summary(x))

*** Model is not valid ***
*** Interaction parameters are outside valid range ***
```

As can be seen, unfortunately, the models are not valid, since the interaction parameter exceeds 1. A possible explanation could be the fact that the assumption about the stationary MultiStrauss model failed, and the real model can contain a trend component or evidence of a higher-order interactions (which is not yet implemented in spatstat::).

In order to find the appropriate non-stationary trend, the same operations can be repeated for the two trend-covariate interactions: additive spatial interaction (*~marks* + *x*) or multiplicative spatial interaction (*~marks* × *x*):

```
# additive trend
rr3 <- diag(c(30.55, 30.45, 30.9))
fit3 <- ppm(X.r2~marks+x, MultiStrauss(rr3))
fit3
```

```
Nonstationary Multitype Strauss process
Possible marks: 'F', 'G' and 'M'
Log trend:  ~marks + x
```

```
Fitted trend coefficients:

  (Intercept)            marksG              marksM                   x

-13.273912384       1.715674530        -1.678310334         0.002416002


3 types of points
Possible types:
[1] F G M
Interaction radii:
F    G    M    NA
NA
F    30.55    NA
Fitted interaction parameters gamma_ij
          F         G          M
F 1.15443      NA         NA
G      NA 1.04348         NA
M      NA       NA 1.229226

Relevant coefficients:
   markFxF     markGxG     markMxM
0.14360659 0.04256274 0.20638437

For standard errors, type coef(summary(x))

*** Model is not valid ***
*** Interaction parameters are outside valid range ***
```

```
# multiplicative trend
rr4 <- diag(c(30.55, 30.45, 30.9))
fit4 <- ppm(X.r2~marks*x, MultiStrauss(rr4))
fit4
```

```
Nonstationary Multitype Strauss process
Possible marks: 'F', 'G' and 'M'
Log trend: ~marks * x

Fitted trend coefficients:
      (Intercept)          marksG        marksM                   marksG:x
marksM:x
-27.540006762     17.443842994 30.734882513 0.009379794  -0.007679857 -
0.015728176

3 types of points
Possible types:
[1] F G M
Interaction radii:
        F     G      M
F 30.55    NA     NA
G    NA 30.45     NA
M    NA    NA   30.9
Fitted interaction parameters gamma_ij
          F       G        M
F 1.170696      NA      NA
G         NA 1.04299     NA
M         NA      NA 1.195889

Relevant coefficients:
   markFxF     markGxG     markMxM
0.15759855 0.04209209 0.17888954

For standard errors, type coef(summary(x))
*** Model is not valid ***
*** Interaction parameters are outside valid range ***
```

Unfortunately, both non-stationary multiStrauss process models are non-valid. Probably, there is a need to take into account between-industry interaction radia, which is done in the next section.

8.7.7 Model with all possible interactions

This subsection presents the model estimation for the marked pattern, where all interactions – between or within types – are possible. The following code shows the results of the **ppm()** function call:

```
#        F        G        M
# F    30.55    30.54    30.
# G    30.54    30.45    30.05
# M    30       30.05    30.9
```

```
tt1<-matrix(c(30.55, 30.54, 30, 30.54, 30.45, 30.05, 30, 30.05, 30.9),
nrow=3, ncol=3)
fit5 <- ppm(X.r2~marks, MultiStrauss(tt1))
fit5
```

```
Stationary Multitype Strauss process
Possible marks: 'F', 'G' and 'M'
Log trend: ~marks
```

```
First order terms:
      beta_F        beta_G        beta_M
1.813110e-04 8.005018e-04 1.635988e-05


3 types of points
Possible types:
[1] F G M
Interaction radii:
       F     G     M
F 30.55 30.54 30.00
G 30.54 30.45 30.05
M 30.00 30.05 30.90
Fitted interaction parameters gamma_ij
          F         G         M
F 0.7295519 1.1803302 0.9988869
G 1.1803302 0.9571421 1.0980550
M 0.9988869 1.0980550 0.9698932


Relevant coefficients:
     markFxF       markFxG       markGxG       markFxM       markGxM       markMxM
-0.315324818 0.165794212 -0.043803396 -0.001113716 0.093540474 -0.030569302


For standard errors, type coef(summary(x))
*** Model is not valid ***
*** Interaction parameters are outside valid range ***
```

It can be observed that interaction parameters for within-industry relationships are chosen correctly, but for between-industries, unfortunately, they are not valid. The explanation is as in a previous case – perhaps another trend should be applied or interactions are of a higher order. The examination of non-stationary trends (additive and multiplicative, as previously) for the MultiStrauss model unfortunately also resulted in an invalid model:

```
### additive trend
tt1<-matrix(c(30.55, 30.54, 30, 30.54, 30.45, 30.05, 30, 30.05, 30.9),
nrow=3, ncol=3)
fit6<-ppm(X.r2~marks+x, MultiStrauss(tt1))
fit6
```

```
Nonstationary Multitype Strauss process

Possible marks: 'F', 'G' and    'M'
Log trend: ~marks + x
Fitted trend coefficients:
(Intercept)          marksG          marksM                 x
-20.337553379    1.364115326    -2.632379361    0.005774477

3 types of points
Possible types:
[1] F G M
Interaction radii:
        F       G       M
F   30.55   30.54   30.00
G   30.54   30.45   30.05
M   30.00   30.05   30.90
```

```
Fitted interaction parameters gamma_ij
            F          G          M
F   0.7229956  1.185277   1.0046892
G   1.1852766  0.959905   1.0986113
M   1.0046892  1.098611   0.9791146

Relevant coefficients:

     markFxF        markFxG        markGxG        markFxM        markGxM
markMxM
-0.324352156   0.169976202   -0.040920933   0.004678283   0.094046970
-0.021106568


For standard errors, type coef(summary(x))
*** Model is not valid ***
*** Interaction parameters are outside valid range ***
```

```
### multiplicative trend
tt1<-matrix(c(30.55, 30.54, 30, 30.54, 30.45, 30.05, 30, 30.05, 30.9),
nrow=3, ncol=3)
fit7<-ppm(X.r2~marks*x, MultiStrauss(tt1))
fit7
```

```
Nonstationary Multitype Strauss process
Possible marks: 'F', 'G' and 'M'
Log trend: ~marks * x


Fitted trend   coefficients:
 (Intercept)        marksG         marksM            x         marksG:x
marksM:x
-70.11917046   53.13764023  137.03899765   0.03009084   -0.02532008    -
0.06870391


3 types of points
Possible types:
[1] F G M
Interaction radii:

        F      G      M
F   30.55  30.54  30.00
G   30.54  30.45  30.05
M   30.00  30.05  30.90

Fitted interaction parameters gamma_ij
            F          G          M
F   0.6947789  1.198711   1.0662057
G   1.1987107  0.951892   1.1128304
M   1.0662057  1.112830   0.8329118

Relevant coefficients:
     markFxF      markFxG      markGxG      markFxM      markGxM      markMxM
-0.36416163  0.18124657  -0.04930371  0.06410630  0.10690665-  0.18282747


For standard errors, type coef(summary(x))
*** Model is not valid ***
*** Interaction parameters are outside valid range ***
```

The examination of multitype patterns was challenging but promising at the same time. Despite the fact that primary analysis detected stationarity, both intensity-based and distance-based analyses suggested that the point pattern rather reveals some kind of non-random behaviour. Summary functions, such as K, suggested a clustered pattern, but as in the case of unmarked pattern analysis, clustering can be confused with interaction. Choosing a proper instrument for model estimation, it was discovered that since firms do compete in space for location and resources, and since there is still the unsolved agglomeration issue, the best solution could be the use of the Gibbs processes, which are constructed exactly to capture interactions both within and between types. Unfortunately, the suggested interaction radius resulted in an invalid model both for the stationary and non-stationary MultiStrauss models. Possible further steps in R could be the examination of more advanced spatial trends, since higher-order interactions are not implemented for multitype patterns yet.

8.8 Spatial interpolation methods – kriging

For a finite sample of size n, spatial interpolation allows one to find unknown values of some measure for a set of m locations using available data from $(n-m)$ locations. Li and Heap (2008) overview plenty of interpolation methods. Among them, one can find local neighbourhood methods (inverse distance weighted, natural neighbourhood, triangulated irregular network), geostatistical methods (*kriging*), variational methods (*thin plate spline*) and some specific methods (*Voronoi* polygons, *area-to-surface* interpolation, interpolation on sphere) (Li & Heap, 2008, vol. 23, Chapter 2; Mitas & Mitasova, 1999, Section 3.2).

The term *kriging* was introduced by Matheron (1960) inspired by work by Krige (1951); however, Cressie (1990) also mentions researchers who contributed to kriging from 1938.[23] There are several definitions of this technique, though the most accurate was given by Krige (1981): "multiple regression procedure for arriving at the best linear unbiased [predictor] or best linear weighted moving average [predictor] of the ore grade of an ore block (of any size) by assigning an optimum set of weights to all the available and relevant data inside and outside the ore block". As in the case with point pattern analysis techniques, kriging is more widely used in non-economic-related studies (mostly in geology, mining, ecology); however, some studies in economics can be mentioned (Martínez, Lorenzo, & Rubio, 2000; Barrosa, Salles, & Ribeiro, 2016; Cousin, Maatouk, & Rullière, 2016; Fleming, 2000; Chaveesuk & Smith, 2005; Magri & Ortiz, 2000). Several kriging techniques are available currently, and the choice depends mainly on the aim of the research. The most widely used is ordinary kriging (Wackernagel, 1995); however, simple kriging, universal kriging and co-kriging are also popular (Geostatystyka w R).

Kriging accounts for spatial autocorrelation, which may help in explaining variation in data. This method seems to be useful in the interpolation of employment level, since this data may vary spatially from one location to another. The following subsections present estimation of two kriging methods (ordinary and universal) with the automap::[24] package and the comparison of their performance in finding the unknown value of employment level in chosen locations. The study dataset will be cleaned from outliers and divided into two parts[25] – in-sample data, containing 80% of observations, and out-of-sample, containing 20% of observations. The subsections subsequently will provide the reader with theoretical notes about kriging, the chosen interpolation methods and an example.

8.8.1 Basic definitions

The most important function for kriging is semivariance, defined as follow (from Li & Heap, 2008, vol. 23, Section 2.2.2):

$$\gamma(x_i, x_0) = \gamma(h) = \frac{1}{2} var\left[Z(x_i) - Z(x_0)\right] \tag{8.3}$$

where h is the distance between x_i and x_0. The function $\gamma(h)$ is then the semivariogram (or theoretical variogram). Kriging-related literature allows one to choose among several variations of functions for the theoretical variogram; four commonly used are spherical, exponential, gaussian and linear.

The plot of $\gamma(h)$ with respect to h is called an experimental variogram (see Figure 8.11d), and shows three important values – sill, nugget and range. *Nugget* can be explained either as an intercept or as a measurement error. According to Diggle and Ribeiro (2007, p. 57), "when the sampling design specifies a single measurement at each of n distinct locations, the nugget effect has a dual interpretation as either measurement error or spatial variation on a scale smaller than the smallest distance between any two points in the sample design, or any

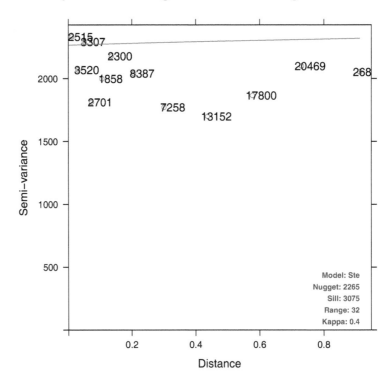

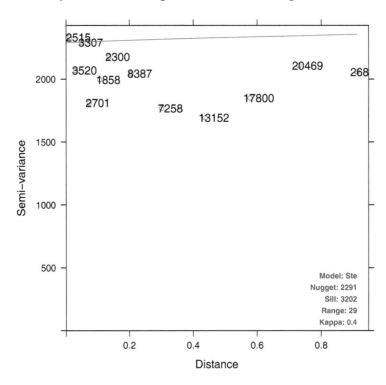

Figure 8.11 Variograms for three kriging models: a) ordinary kriging, b) universal kriging based on function of Cartesian coordinates, c) universal kriging based on distance to Lublin, d) experimental variogram

Source: a)–c) Own calculations using automap::autofitVariogram(), d) Li and Heap (2008)

Experimental variogram and fitted variogram model

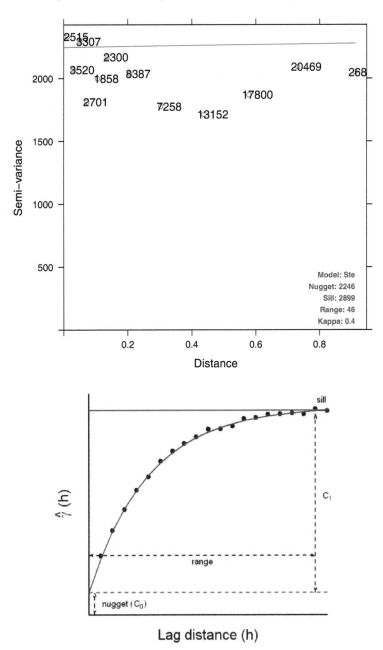

Figure 8.11 (Continued)

combination of these two effects". Hartkamp, De Beurs, Stein, and White (1999, p. 6) define nugget as "non-spatial variability of the variable".

Range is the distance where spatial autocorrelation exists. The last value of range determines *sill* – the maximum value of $\gamma(h)$. So, basically, maximum range shows where spatial correlation disappears (curves start to be flatter). In theory, for zero range, the value of semi-variance should also be zero; however, in practice, the nugget effect appears. According to Clark (2010, p. 312), there is still a dispute about this issue, but results of his experiment argue that "allowing the semivariogram model to intercept the axis rather than go to zero produces estimates which are apparently more reliable than assuming the data are accurate". The last important indicator of a variogram is the sill/nugget ratio – Hartkamp et al. (1999, p. 6) state that if the value of this ratio is close to 1, it means that "most of the variability is non-spatial".

For all kriging methods, weighting coefficients are used to calculate a value of some measure for the point where it is not known. Coefficients take into account the overall relationship between points, as well as the distance between measured and unmeasured locations. The general form of the spatial interpolation formula $\hat{Z}(x_0) = \sum_{i=1}^{n} \lambda_i Z(x_i)$ (Li & Heap, 2008, vol. 23, p. 21), where λ_i is a weighting coefficient for each sampling point, Z is the observed value measured in the location x_i (so a set of $z(x_i)$ is in-sample data) and $\hat{Z}$ is the interpolated value in the out-of-sample location x_0 that can be extended to a form:

$$\hat{Z}(x_0) - \mu = \sum_{i=1}^{n} \lambda_i [Z(x_i) - \mu(x_0)]$$

(8.4)

where μ "is a known stationary mean, assumed to be constant over the whole domain and calculated as the average of the data" and $\mu(x_0)$ is "mean of samples within the search window" (Li & Heap, 2008, vol. 23, Section 2.2.3).

8.8.2 Description of chosen kriging methods

As stated before, kriging was developed for other research purposes, meaning environmental studies, where unknown values of soil acidity, elevation or presence of mineral deposits were the aim of the computation. Among the previously mentioned economic related studies, kriging was used, for example, for portfolio optimisation, financial term structures, estimation of economic loss or housing price. Employment level as a value to interpolate seems to be an interesting issue, since it varies from region to region and depends not only on population characteristics (age, skills, density), job specifics and so on but also on closeness to capital city.

The choice of kriging method to perform under the current study was based not only on how well kriging techniques are implemented in R but also on their prevalence in currently existing studies and theoretical assumptions. Simple kriging assumes constant a mean across the study region, which is not true in the case of the study dataset – dispersion of values from 5 to 1500 and right-skewedness reject this assumption (see output subsequently):

```
summary(firms$empl)
```

Min.	1st Qu.	Median	Mean	3rd Qu.	Max.
5.000	5.000	5.000	6.134	5.000	1500.000

For the current study, ordinary and universal kriging were chosen as the most widely used and easy-to-estimate methods. Following Pebesma (2004), this choice is sensible, since no simulations are required and no trend coefficients are provided.

Ordinary kriging assumes that the mean value is not known across the study region, which is reasonable in the case of the study dataset – the distribution is right skewed, with the majority of firms with five employees, so the mean differs across voivodeship. Universal kriging takes into account the presence of a spatial trend, in the simplest case expressed as a function of Cartesian coordinates, but can also use any additional data (known as a spatial covariates). There is no information about the presence of a trend in the study dataset; however, it can be assumed that the number of employees varies spatially across the region. In order to verify whether distance to capital city has any impact on kriging results, universal kriging will be performed in two versions – with trend specified as a sum of Cartesian coordinates and as a distance to Lublin.

8.8.3 Data preparation for the study

The dataset contains 37,087 observations, representing enterprises of Lubelskie voivodeship. For the current study, only location and information about employment level are required.

The dataset contains some outliers (600 and 1500) as well as the dominating value of 5, which may result in interpolation bias. Before kriging interpolation, data should be cleaned; for this purpose, subsetting was used.

```
table(firms$empl)
```

```
      5    30   150   600  1500
  36392   568   116     6     5
```

```
# cleaning data
firms <- firms [which(firms$empl!=5),]
firms <- firms [which(firms$empl!=600),]
firms <- firms [which(firms$empl!=1500),]
```

To investigate whether distance to regional main city has any impact on kriging results, a new variable was created – distance from each point to Lublin.

8.8.4 Estimation and discussion

After cleaning, the dataset contains 684 observations. This sample was divided in two – the input with 80% of data and the output with 20% of data.

```
firms$r<-runif(dim(firms)[1]) #random variable - uniform distribution [0,1]
# new objects are also of class SpatialPointsDataFrame:
input<-firms[firms$r<0.8,] # choosing 80% of data as input data
output<-firms[firms$r>0.8,] # choosing 20% of data as output data
```

Before kriging interpolation, variograms will be examined, mostly in order to check the sill/nugget ration. The function **autofitVariogram()** was used for variogram calculation. To obtain a variogram, one needs first to specify a formula which shows the form of linear dependence between the predicted variable and regressor – for ordinary kriging, the formula is '*predicted_variable~1*', for universal – either '*predicted_variable~x+y*' (or any other function of Cartesian coordinates) or '*predicted_variable~regressor*'. The second important argument is input dataset – an object of class *SpatialPointsDataFrame*. Other arguments (for example, start values of sill, nugget and sill (*start_vals*), list of variogram models to test (*model*)) are optional.

The following code shows the **autofitVariogram()** call and fitted variograms – for ordinary kriging and both variants of universal kriging (see Figure 8.11a–c):

```
library(automap)
ok.var<-autofitVariogram(empl~1, input) # ordinary kriging
plot(ok.var)

# universal kriging based on function of Cartesian coordinates
uk.var.1<-autofitVariogram(empl~coords.x1+coords.x2, input)
plot(uk.var.1)

# universal kriging based on distance to Lublin
uk.var.2<-autofitVariogram(empl~dist, input)
plot(uk.var.2)
```

Analysis of these plots show that variability in data seems to be mostly non-spatial, since the sill/nugget ratio in all cases is very close to 1: for ordinary kriging, $\frac{nugget}{sill} = \frac{2265}{3075} = 0.7365$ for universal kriging with function of Cartesian coordinates $\frac{nugget}{sill} = \frac{2291}{3202} = 0.7154$ and for universal kriging based on distance to capital city $\frac{nugget}{sill} = \frac{2246}{2899} = 0.7747$. The range of spatial dependence is 32 km, 29 km and 46 km, respectively.

The fitted variogram model for all cases is Stein's parametrisation of the *Matérn* covariance function[26] with the shape parameter $\kappa = 0.4$. Sills of all variograms are quite flat, which also shows low spatial variation in data.

To perform kriging interpolation, the function **autoKrige()** was used. As the main arguments,[27] one should specify the kriging formula (the same as in an **autofitVariogram()** call), the input data (object of class *SpatialPointsDataFrame*) and output data (any *sp* object – polygon, grid or points set). The result is an object of the *autoKrige* class. To extract kriging results, one can call for estimated model attributes: *estimated_model$krige_output* returns interpolation results (predicted values, variance and standard deviation), *estimated_model$exp_var* returns the sample variogram and *estimated_model$sserr* returns the sum of squares between the sample and the fitted variogram models. The following code shows the call of the **autoKrige()** function for three kriging methods:

```
## ordinary
ok.xy <- autoKrige(empl~1, input, output, verbose=FALSE)

## universal, x+y
uk.xy <- autoKrige(empl~x+y, input, output, verbose=FALSE)

## universal, dist
uk.dist <- autoKrige(empl~dist, input, output, verbose=FALSE)
```

To plot estimated values in unknown locations, the function **automapPlot()** was used. The main arguments to be specified are the spatial object to plot and the column/list of this object to plot; other arguments are the same as for the **spplot()** function.

For each of the **automapPlot()** calls shown subsequently, the first argument is an object of the *SpatialPoints-DataFrame* object containing interpolated values of employment level, variance and standard deviation. The variable of interest is predicted employment level, which was chosen for a plot as *var1.pred*. The plot of original values was also provided for visual analysis:

```
# Figure 8.12
class(ok.xy$krige_output)
#[1] "SpatialPointsDataFrame"
attr(,"package")
#[1] "sp"

names(ok.xy$krige_output)
#[1] "var1.pred" "var1.var"  "var1.stdev"

## original values
result0<-automapPlot(output,"empl", main="Original \n data")
result0

## ordinary
result1<-automapPlot(ok.xy$krige_output,"var1.pred", main="Ordinary
kriging")
result1

## universal, x+y
result2<-automapPlot(ok.xy$krige_output,"var1.pred", main="Universal
kriging, \n Cartesian coordinates function")
result2

## universal, dist
result3<-automapPlot(ok.xy$krige_output,"var1.pred", main="Universal
kriging, \n distance to Lublin")
result3
```

A brief visual analysis suggests that all methods have more or less the same results. In the initial dataset, only two values were present, 30 and 150; however, all the kriging methods produced a set of interpolated values between these limits. Measure of goodness of fit can be RMSE (8.5):

$$RMSE = \sqrt{\frac{1}{n}\sum_{i=1}^{n}(initial\ value - interpolated\ value)^2} \tag{8.5}$$

**Original
data**

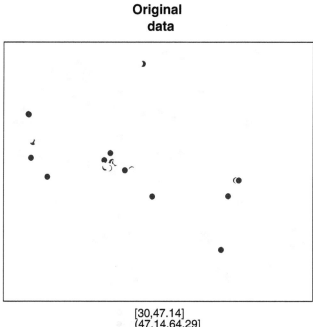

```
[30,47.14]
(47.14,64.29]
(64.29,81.43]
(81.43,98.57]
(98.57,115.7]
(115.7,132.9]
(132.9,150]
```

**Ordinary
kriging**

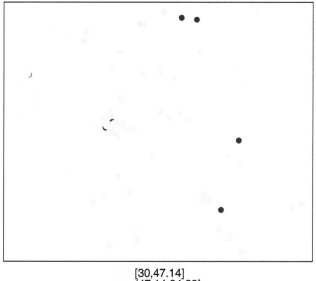

```
[30,47.14]
(47.14,64.29]
(64.29,81.43]
(81.43,98.57]
(98.57,115.7]
(115.7,132.9]
(132.9,150]
```

Figure 8.12 Plotted results of kriging interpolation: a) original data, b) ordinary kriging, c) universal kriging with Cartesian coordinates function, d) universal kriging based on distance to Lublin

Source: Own calculations using **automapPlot()**

**Universal kriging,
Cartesian coordinates function**

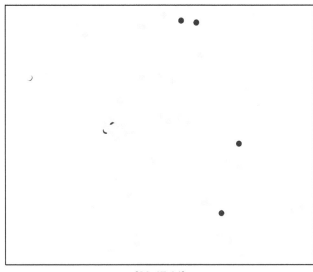

[30,47.14]
(47.14,64.29]
(64.29,81.43]
(81.43,98.57]
(98.57,115.7]
(115.7,132.9]
(132.9,150]

**Universal kriging,
distance to Lublin**

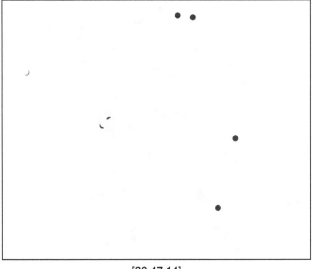

[30,47.14]
(47.14,64.29]
(64.29,81.43]
(81.43,98.57]
(98.57,115.7]
(115.7,132.9]
(132.9,150]

Figure 8.12 (Continued)

The following code shows the calculation of RMSE for all the interpolation methods with the function **rmse()** from the Metrics:: package, which takes only two arguments – original data and predicted data:

```
library(Metrics)
rmse_ord<-rmse(output$empl, ok.xy$krige_output@data$var1.pred)
rmse_un_xy<-rmse(output$empl, uk.xy$krige_output@data$var1.pred)
rmse_un_dist<-rmse(output$empl, uk.dist$krige_output@data$var1.pred)
rmse_ord #[1] 42.64517
rmse_un_xy #[1] 42.80971
rmse_un_dist #[1] 42.62353
```

The value of RMSE is the smallest for universal kriging based on distance to Lublin, which may suggest that spatial variation of employee numbers across Lubelskie voivodeship is caused by closeness to the capital city. The RMSE value for ordinary kriging differs for only ~0.02, which is consistent with the fact that usually, results of ordinary and universal kriging with a planar trend give very close results. When suggesting trends in the data (which is possibly distance to the capital city), one would do better to choose the universal kriging interpolation. However, one should be careful – primary analysis of the study data showed that a variation in the data is mostly non-spatial, so probably, kriging is not the best tool for such a prediction.

Table 8.1 R packages for point process models

Point process model/R package	DSpat	ecespa	fractal[28]	inlabru	lgcp	NScluster	ppmlasso	ptproc	PtProcess	SDALGCP	spatstat	sppmix	stpp
Homogeneous/inhomogeneous Poisson point process	+[29]/-			+/+			+/+	+?			+/+	+/+	+/+
Moran cascade process			+										
Poisson cluster process		+[30]											+
Gibbs/Markov point process											+		
Strauss process											+		
Neyman-Scott process						+					+		
Matern[31] point process											+		
Thomas[32] process						+?					+		
Cauchy[33] process											+		
VarGamma[34] process											+		
Multidimensional point process								+					
Cox process/log-Gaussian Cox process										+?/+	+/+		-/+
Determinant point process											+		
Time-dependent point process models									+				
Spatio-temporal point process[35]												+[36]	+
Spatio-temporal log-Gaussian Cox process				+									

Source: Own analysis using SSLib manuals (http://www.statsresearch.co.nz/dsh/sslib/), inlabru web page (https://sites.google.com/inlabru.org/inlabru), Baddeley (2008), Møller and Díaz-Avalos (2010), Ogata and Stoyan (2008), Peng (2003), Illian, Sørbye, Rue, and Hendrichsen (2010), Illian et al. (2013) and R package documentation (Johnson, Laake, & VerHoef, 2014; De La Cruz, 2018; Harte, 2017a, 2017b; Bachl, Lindgren, Borchers, & Illian, 2018; Davies, Diggle, Rowlingsson, & Taylor, 2018; Nakano, Saga, & Tanaka, 2019; Renner, 2015; Peng, 2002; Diggle, Giorgi, & Johnson, 2019; Chen & Micheas, 2017)

Table 8.2 R packages for spatial interpolation

Kriging method / R package		*automap::*	*autoFRK::*	*constrainedKriging::*	*DiceKriging::*	*DiceOptim*	*fields::*	*geoR::*	*geoRglm::*	*gstat::*	*KriSp::*	*KRIG::*	*KrigInv::*	*Kriging::*	*LatticeKrig::*	*ltsk::*	*MuFiCokriging::*	*OmicKriging::*	*sgeostat::*	*RandomFields::*	*SK::*	*Spatfd::*	*spatial::*
con-strained-Kriging	Simple kriging	+			+	+		+	+	+	+	+	+						+	+			
	Ordinary kriging	+			+	+	+		+	+		+		+						+			
	Universal kriging	+		+	+	+	+	+		+		+	+										+
	Global/local kriging									+						−/+ [37]							
	Point/block kriging	−/+		−/+						−/+													
	Constrained kriging			+																			
	Univariate kriging										+												
Kriging	Functional kriging																					+	
	Regression kriging											+											
	Bayesian kriging							+															
	External trend kriging							+		+													
	Conventional kriging							+							+								
	Fixed rank kriging		+												+								
	Spatio-temporal kriging									+						+ [38]							
	Segment-based kriging																			+ [39]			
	Omic kriging																	+					
Co-kriging	Simple co-kriging								+	+							?			+			
	Ordinary co-kriging									+													
	Functional co-kriging																					+	
	Universal co-kriging									+							?						
	Multifidelity co-Kriging																+						

Note: Names in italics are for packages which are not in the CRAN repository but have individual webpages
Source: Own analysis using Li and Heap (2008), Furrer (2006), Hofer and Papritz (2011), Villamil, Bohorquez, Giraldo, and Mateu (2018), Wheeler et al. (2014) and R packages documentation (Hiemstra, 2013; Gillespie, Huang, Nychka, Tzeng, & Wang, 2018; Hofer, 2015; Deville, Ginsbourger, & Roustant, 2018; Ginsbourger, Roustant, & Picheny, 2016; Furrer, Nychka, Paige, & Sain, 2018; Diggle & Ribeiro, 2016; Christensen & Ribeiro, 2017; Graeler & Pebesma, 2019; Guarderas, Lagos, & Lopez, 2018; Azzimonti, Chevalier, Ginsbourger, & Picheny, 2018; Olmedo, 2014; Hammerling, Lenssen, Nychka, Sain, & Smirniotis, 2018; Chen, Chen, Kumar, & Liang, 2019; Le Gratiet, 2012; Gebhardt, 2016; Schlather et al., 2019; Song, 2018; Bivand, Ripley, & Venables, 2015; Im, Michaels, Trubetskoy, & Wheeler, 2016)

This section provided the reader with an example of using the automap:: package for the kriging of economic data (interpolation of unknown employment level value for out-of-sample dataset). The main benefit of automap:: is the automatic estimation of both the variogram as well as the kriged model based on the variogram. For illustration, ordinary and universal kriging were chosen as the most often used and most obvious kriging methods. Postestimation visual analysis showed that due to the fact that the study dataset is highly unbalanced (and there is no other available data for enterprises' locations), performance of all methods seems to be unreliable. The RMSE value was the smallest for universal kriging with the trend expressed as distance to capital city, which could suggest the importance of this factor in interpolation. However, due to the fact that the initial variogram analysis detected a big proportion of non-spatial variability in the study dataset (values of sill/nugget ratio are above 0.73), kriging is probably is not the best tool for interpolation. Variation in employment level can be explained by other, non-spatial characteristics.

Notes

1 The first mention of Ripley's K function can be found in Barff (1987); however, for analysis, the L function was used.
2 Except packages SGCS::, spatgraphs::, spatial::.
3 Sometimes, more than one mark is provided for each point.
4 Spatial analysis in this chapter will be performed on multitype point patterns with categorical mark values. However, numerical marks can be used for visualisation.
5 Hereafter, we use finite point patterns.
6 A wider list can be found at https://stat.gov.pl/en/metainformations/glossary/terms-used-in-official-statistics/97, term.html or checked in Dziennik Ustaw Rzeczypospolitej Polskiej. Rozporządzenie Rady Ministrów z dnia 30 listopada 2015 r. w sprawie sposobu i metodologii prowadzenia i aktualizacji krajowego rejestru urzędowego podmiotów gospodarki narodowej, wzorów wniosków, ankiet i zaświadczeń. http://prawo.sejm.gov.pl/isap.nsf/download.xsp/WDU20150002009/O/D20152009.pdf
7 *Spatial datasets for Poland.* https://gis-support.com/spatial-datasets-for-poland/
8 Vectors x, y and marks should have the same length.
9 Duplication and solution of this problem are provided in one of the following sections.
10 In the case of spherical coordinates projected with "+proj=longlat +datum=NAD83", the *owin* objects would be created with different coordinate values, as following. Transformation of projections can be done with, for example, https://mygeodata.cloud/cs2cs/

```
circ<-ellipse(a=1.5, b=1.5, centre=c(22.88066,51.26983), phi=pi) sq<-owin(c(21,25),
c(50,53))
```

11 Currently available spatstat:: tools allow only investigation of multitype patterns.
12 Or also: **point_pattern %mark% new_value**.
13 Three is the minimum number of columns for the matrix – the first two represent coordinates, and the last is for the value of interest.
14 Marks remain unchangeable; only spatial coordinates are changed.
15 Results and conclusions of tests will be the same as in the case of Mercator projection and (further rescaling), since such transformations do not change behaviour of points and their locations.
16 $e_{ij}(r)$ is called an "edge correction", and it is introduced because of the existence of irregularly shaped windows.
17 Function which estimates Cluster or Cox model.
18 Similarly to **envelope**(), these functions also can be used in modelling goodness-of-fit analysis.
19 As a summary function, Ripley's K is a default option. But Baddeley et al. (2014) recommend replacing it with the L function, since "MAD test statistic and the DCLF test statistic tend to be more influenced by fluctuations of $K(r)$ occurring at larger distances r" (Baddeley et al., 2014, p. 485).
20 There is no clear information about suggested usage provided.

21 Residual plots: Poisson residual measure, smoothed Poisson residual field, two lurking variable plots for x and y coordinates, respectively.

22 Lurking plot draws spatial point process residuals against a covariate.

23 See Table 1 in Cressie, N. (1990). The Origins of Kriging. *Mathematical Geology, 22*(3), 239–252.

24 The main advantage of the automap:: package is that it does all the "heavy" work and fits the model to the variogram without the need of an initial estimation of parameters. Though this package is not the best option in all cases, it is useful when manual estimation is tough and requires a lot of iterations. The main disadvantage could be the fact that in some cases, the automatically fitted variogram has flat sills.

25 Kriging-related studies suggest an optimal size for in-sample data as 50–100 observations or more (Brus & Heuvelink, 2007), which does not mean one should follow the rule "the more points the better prediction". But a more important question is to choose locations which are reliable in terms of prediction (Heuvelink, Griffith, Hengl, & Melles, 2012). Since studies about optimal sample size were made mostly for environmental data, the choice of in-sample and out-of-sample size in this chapter was done arbitrarily.

26 Matern correlation functions form a big family of functions used in geospatial studies (Hoeting, Davis, Merton, & Thompson, 2006). Diversity of Matern functions is determined by the smoothing parameter kappa – for example, $\kappa = 0{,}5$ transforms a Matern model to an exponential model. Application of this class of functions varies from soil sciences to geophysics (Pardo-Iguzquiza & Chica-Olmo, 2008).

27 Other arguments are optional and are the same as for **autofitVariogram**().

28 As a part of the Statistical Seismology Library (Harte, 2007, SSLib, http://www.statsresearch.co.nz/dsh/sslib/).

29 Function spatstat::rpoispp() is imported.

30 It is possible to simulate homogeneous or inhomogeneous Poisson cluster processes.

31 Belongs to the Neyman-Scott processes family.

32 As previously.

33 Neyman-Scott process with Cauchy kernel.

34 Neyman-Scott process with variance gamma kernel.

35 Including spatio-temporal cluster point processes.

36 Including marked space-time Poisson process.

37 Here, implemented as local ordinary kriging.

38 Here are methods for local/global spatio-temporal and block spatio-temporal kriging.

39 Including segment-based ordinary and segment-based regression kriging.

## Chapter 9

Spatial sampling and bootstrapping

Katarzyna Kopczewska and Piotr Ćwiakowski

The problems of spatial sampling relate to predominantly geolocalised point data, which, in addition to geographical coordinates, contains information on the value of phenomena. This is an *xyz* system, where *(x,y)* is a two-dimensional location with the observed value of a specific phenomenon.

The data can be independent – that is, the point *(x,y)* is assigned one value for one phenomenon *(z)*, and this is the simplest analytical situation. Much more theoretical complications arise when a given point *(x,y)* is assigned more than one value for one or more phenomena and there is a correlation between *z* values – they are then dependent data. Even more analytical problems are generated by the situation when data are dependent in space; that is, the correlation between values from phenomena concerns not only the observations for which they are examined but also neighbouring observations.

The method of testing point data depends on many factors. One of the most important, though little discussed in the literature, is the *scale of the phenomenon*. In a small sample situation, when too few points are observed, statistical methods can be unreliable, and the calculated statistics are disturbed. *Bootstrapping* is a data quality improvement procedure that has been applied successfully for a long time. It consists of drawing subsamples with a replacement (often the same length as the original sample) and testing the properties of many bootstrapped samples rather than the original one. For each subsample, a statistic is calculated, for example, an average of one variable, correlation of two variables, directional coefficients from an econometric model and so on. From the estimates in the subsamples, a statistical distribution is created that serves to make inferences about the value of the analysed statistics. Most often, as an estimate of bootstrap statistics, the average or median from the distribution of statistics from samples is taken. Despite the same information being available to the researcher, the method allows for more precise estimations of standard deviations, confidence intervals and significance tests – bootstrapped statistical variances are usually lower than those determined classically from the original sample (review of methods e.g. DiCiccio & Efron, 1996; MacKinnon, 2002, 2006; Davison, Hinkley, & Young, 2003; Fox, 2015). The problem of small data is still present in many industries (e.g. medicine); however, technological advances have caused more and more researchers to face the opposite problem – reducing the dataset so as not to lose the information contained in the original set. The bootstrap method or sampling can help in a statistical way, not involving software or hardware, to solve big data problems.

For methods based on sampling over space (e.g. in a region), there are several issues to be aware of:

1   When, in the literature, we refer to sampling, two different approaches may be behind it: a subsample from random data is selected (randomly or targeted) or random points are generated on the surface. In this chapter, both approaches will be shown.
2   The drawing may be single or multiple. When one-time sampling is carried out, it is necessary to ensure stratification of the sampling (multicriterial representativeness, drawing from pre-defined categories).[1] When multiple drawing (resampling) is carried out, in the form of bootstrapping or cross-validation[2] (*k*-fold

cross-validation), the appropriate number of repetitions and sample size ensure that the multicriteria representation will be maintained.

3   It can be assumed that the points have a specific distribution (e.g. Poisson) or that one deals with complete randomness – CSR. Studies on spatial patterns of point distributions are very well developed; thus, these issues are not discussed in this chapter.

4   Randomisation of spatial data (generation and sampling) very often requires the introduction of spatial constraints, that is, independent drawings within regions or defined shapes. This chapter presents the following problems of constraints.

5   Consideration should be given to the length of the subsample (if shorter than the original set), called the coverage level, and the number of repeat draws. In the literature, one can find considerations on bootstrapping efficiency for example, due to assumptions of randomness (e.g. Andrews and Buchinsky, 2000) or due to the size of the subsample (Griffith, 2005, 2008a, 2013; Martino, Elvira, & Louzada, 2017).

Yet another issue is the *aggregation* of point data to separate areas. The *modified areal unit problem* (MAUP), which has been known for years, requires a balanced research approach. Aggregation, by reducing data dimensions, promotes the synthesis of information and *de facto* its simplification. However, badly performed aggregation can change the result or hide existing patterns and dependencies. A lot of articles and research papers have been published on this subject and are worth studying. One of the best articles is the text "Dots to Boxes: Do the Size and Shape of Spatial Units Jeopardize Economic Geography Estimations?" (Briant, Combes & Lafourcade, 2010).

Finally, analyses on points located in space are used in various fields of science, with use of methodology specific for the topic. This book is written from an economic perspective, where the location of firms or relationships between regions are examined, but these models are also excellent for real estate valuation. In the literature, there are also many applications of these methods applied to such areas as astronomy and the distribution of stars (e.g. Loh, 2008; Kemball & Martinsek, 2005), medicine (Austin, 2008) or natural sciences – for example, the location of trees, climate measurements, earth and so on. These methods are also of great importance in epidemiology and many other fields.

The distribution structure is as follows: first, the classes of point data and the method of analysing them to be employed are presented (9.1); second, ways of generating new points on the plane are shown, which can be used as test data or reference distributions for comparisons (9.2); third, there are ways of random drawing from existing points – a simple drawing (9.3.1), a more advanced block bootstrap (9.3.3) and moving block bootstrap (9.3.4), also showing how the sperrorest:: package works: (9.3.2). The use of spatial sampling and bootstrapping in cross-validation models (9.4) is also presented.

The spsann:: package is not discussed in this chapter – it allows for optimising the spatial configuration of the sample using the spatial simulated annealing (SSA) method. The spcosa:: package is also not discussed: it allows for sampling from a stratified division based on *k*-means (*compact geographical strata*). The gridsample:: package has been omitted – it allows for sampling for grid data. Econometric modelling was also omitted using the random forest method, which is based on a simple bootstrap (case resampling) of both observations and explanatory variables.

An interesting addition to this chapter is the book *An Introduction to Bootstrap Methods with Applications to R* (Chernick & LaBudde, 2014), which generally discusses the a-spatial bootstrap (also including spatial elements), and an open access book titled *Handbook of Spatial Analysis. Theory and Application with R* (INSEE/EUROSTAT, 2018), which contains in Chapter 10 a very good description of the sampling methods of a balanced sample.

9.1 Spatial point data – object classes and spatial aggregation

The analysis of spatial point data is most often used in two classes of objects: 1) based on the sp:: – classes *sp*, *SpatialPolygons* and *SpatialPoints* and 2) based on of the spatstat:: – package *owin* and *ppp* classes.

In the sp:: package, the point data are in the *SpatialPoints* class, while the contours of the regions are in the *SpatialPolygons* class. In both cases, objects can have defined datasets and extended classes for *SpatialPointsDataFrame* and *SpatialPolygonsDataFrame*, respectively. Subsequently, in the first stage, visualisation of points on the map was made – provincial and poviat contours. With the command **over**() from the sp:: package, the affiliation of points to individual poviats was checked. The **over**() command requires an agreed spherical coordinate format between objects; hence, there was a conversion of point coordinates with the **SpatialPoints**() command. After assigning points to poviats, it was possible to count points in regions. The **aggregate**() command was used for this purpose, for which the **length**() function was used to measure the length of a vector in a subgroup conditioned by elements indicated in the *by* argument. The bar graph made with the **barplot**() command is arranged horizontally with option *horiz=TRUE*. The headers in the *names.arg* argument come from the summary of the **aggregate**() command. The wider left margin of the chart, which holds the full headers, was obtained by the *par* option of the command **par**() – the margins in the vector **c**() are defined as the bottom, left, top and right. The *las* option defines the direction of the *x*-axis labels: *las=1* means labels parallel to the axis and *las=2* perpendicular labels. The *xlim* option specifies the range of values on the *x*-axis. The **abline**() command allows for placing the auxiliary lines – vertical with the option *v=* or horizontal *h=*, dashed with the option *lty=3*.

In further analysis, there are three spatial object subsets used (see Chapter 2): *SpatialPolygons* class objects – *lub.voi* (contour map for single voivodeship), *lub.pov* (contour map for selected poviats) and *SpatialPoints* class object: *company.lim.sp* (only point locations).

```
# point data set - 5000 observations
firms<-read.csv("geoloc_data_firms.csv", header=TRUE, dec=",", sep=";")
firms.lim<-firms[1:5000,]

#reminder - reading spatial objects
voi<-readOGR(".", "wojewodztwa") # 16 units
voi<-spTransform(voi, CRS("+proj=longlat +datum=NAD83"))

pov<-readOGR(".", "powiaty") # 380 units
pov<-spTransform(pov, CRS("+proj=longlat +datum=NAD83"))

#reminder - creating subsets for lubelskie region
voi.df<-as.data.frame(voi)
lub.voi<-voi[voi.df$jpt_nazwa_=="lubelskie",]
data<-read.csv("data_nts4_2019.csv", header=TRUE, dec=",", sep=";")
data15<-data[data$year==2015,]
lub.pov<-pov[data15$region_name=="Lubelskie",]

# Figure 9.1a - subset of map (see Chapter 2) and point data
plot(lub.voi, lwd=2) # subset of the provincial map created earlier
plot(lub.pov, add=TRUE) # subset of the poviat map points(firms.lim[,12:13],
pch=".") # locations xy points

# change the point data class from data.frame to SpatialPoints
firms.lim.sp<-SpatialPoints(firms.lim[,12:13],
proj4string=CRS(proj4string(lub.pov)))

pts<-over(firms.lim.sp, lub.pov) # affiliation of points to areas
pts.ag<-aggregate(pts$jpt_nazwa_, by=list(pts$jpt_nazwa_), length)
head(pts.ag)
```

```
                Group.1    x
1          powiat bialski  254
2   powiat Biała Podlaska   89
3      powiat biłgorajski  269
4           powiat Chełm  112
5        powiat chełmski  195
6   powiat hrubieszowski  182
```

```
# horizontal bar chart of aggregated data
par(mar=c(2,10,2,2)) # order of margins: bottom, left, top, right
barplot(pts.ag$x, horiz=TRUE, names.arg=pts.ag$Group.1, las=1,
xlim=c(0,700))
abline(v=(1:7)*100, lty=3)
par(mar=c(2,2,2,2))
```

firms.lim.ppp

firms.lim.ppp.m

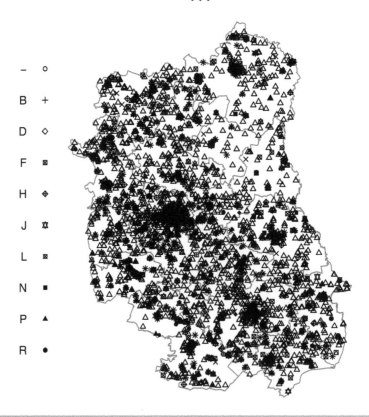

Figure 9.1 Visualisation of *ppp* class objects: a) locations only, b) locations and values in these locations

Source: Own study using R and spatstat:: package

An alternative is the spatstat:: package, dedicated to the analysis of spatial points, which requires the creation of its own object classes. First, one needs to convert the contour layer – the *SpatialPolygons* class object – to *owin* class. This is possible with the command **as()** with the option "*owin*". Second, the coordinates of points should be converted to the *ppp* class, taking into account the shape of the region – given in the **ppp()** command as an object of the *owin* class. In the **ppp()** command, one can also assign values observed in points – the *marks* option is used for this.

The following subjects will be used for further analysis: objects of the class of *owin:lub.pov.owin* and objects of the class *ppp*: *firms.lim.ppp* (only the location of points on the poviat map) and *firms.lim.ppp.m* (locations and characteristics of points on the poviat map).

```
library(spatstat)
library(rgdal)
library(maptools) # necessary to run as.owin() or as(*, "owin") # change of
contour projection
lub.pov<-spTransform(lub.pov, CRS("+proj=merc +datum=NAD83")) # planar
lub.pov.owin<-as(lub.pov, "owin") # conversion of SpatialPolygon to owin

# change of points projection
proj4string(firms.lim.sp)<-CRS("+proj=longlat +datum=NAD83") # spherical
firms.lim.sp<-spTransform(firms.lim.sp, CRS("+proj=merc +datum=NAD83"))

# ppp object - without marks and with marks
firms.lim.ppp<-ppp(firms.lim.sp@coords[,1], firms.lim.sp@coords[,2],
window=lub.pov.owin)
firms.lim.ppp.m<-ppp(firms.lim.sp@coords[,1],        firms.lim.sp@coords[,2],
window=lub.pov.owin, marks=firms.lim[,18])

# points diagram (ppp class) - location and value in point
plot(firms.lim.ppp) # points diagram (ppp class) - location only, Figure 9.1a
plot(firms.lim.ppp.m, cex=0.8, border="red") #locations & values, Figure 9.1b
```

In the example in Figure 9.1b, the values in the point are discrete – these are business categories from B to R. The symbols shown in the figure have been assigned to the categories automatically.

9.2 Spatial sampling – randomisation/generation of new points on the surface

Quantitative research often uses artificially generated point locations. Such theoretical distributions most often constitute reference points for comparisons with empirical distributions. They are also a great starting point for testing statistical models under controlled conditions. Pragmatically, they can be a kind of "rescue" when there is a lack of access to empirical data.

Generating points inside a specific area is possible with the **spsample()** command from the sp:: package.[3] In the case of drawing from a spatially regular distribution, the points on the surface are arranged in grid order as centres of square grids dividing the irregular area. Random distribution does not assume any regularities in the location of points. In stratified draws, points are selected from "invisible" shapes (e.g. tiles) dividing the examined area. Non-aligned and hexagonal schemes assume regularity of location but are not as simple as a regular distribution. In the clustered distribution, all observations are located at one point. Drawn points are in a planar (non-spherical) coordinate system.

```
# creation of a map section - voivodship (NTS2) Lubelskie, Figure 9.2
voi.df<-as.data.frame(voi)
lub.voi<-voi[voi.df$jpt_nazwa_=="lubelskie",]

# randomly drawn points - regular distribution plot(lub.voi, main="Regular
dots in Lubelskie NTS2")
points(spsample(lub.voi, n=1000, "regular"), pch=".", cex=2)
```

```
# randomly drawn points - random distribution
plot(lub.voi, main="Random dots in Lubelskie NTS2")
points(spsample(lub.voi, n=1000, "random"), pch=".", cex=2)

# randomly drawn points - stratified drawing
plot(lub.voi, main="Stratified dots in Lubelskie NTS2")
points(spsample(lub.voi, n=1000, "stratified"), pch=".", cex=2)

# randomly drawn points - non-aligned drawing
plot(lub.voi, main="Nonaligned dots in Lubelskie NTS2")
points(spsample(lub.voi, n=1000, "nonaligned"), pch=".", cex=2)

# randomly drawn points - hexagonal distribution plot(lub.voi,
main="Hexagonal dots in Lubelskie NTS2")
points(spsample(lub.voi, n=1000, "hexagonal"), pch=".", cex=2)

# randomly drawn points - clustered distribution plot(lub.voi,
main="Clustered dots in Lubelskie NTS2")
points(spsample(lub.voi, n=1000, "clustered"), pch=".", cex=2)
```

The area for which points are drawn can be defined differently. It may be a single shape, as previously, with the separated region (see Figure 9.2). It can be a multiregional map, as well as a fragment of a multiregional map. In the code subsequently, this is the entire map of Poland, consisting of 16 regions (*voi*) and one indicated region from this map (*voi@polygons [[1]]*).

Regular dots in Lubelskie NTS2   **Random dots in Lubelskie NTS2**

Figure 9.2 Drawing points in space – managing the distribution of points

Source: Own study with the use of R and the sp:: package

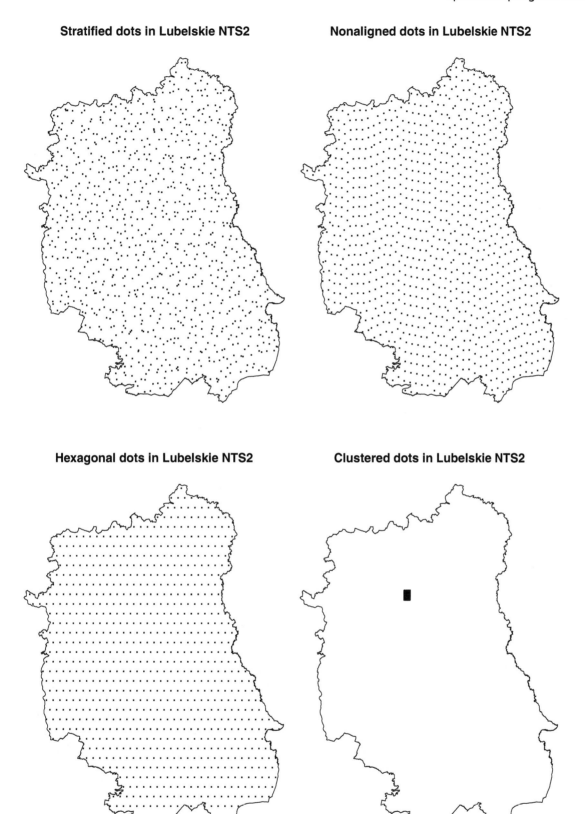

Figure 9.2 (Continued)

```
# Figure - random points - the whole multi-regional map
plot(voi, main="Random dots in Poland (16 regions)")
points(spsample(voi, n=2000, "random"), pch=".", cex=2)

# Figure - random points - one region from a multi-regional map
plot(voi, main="Random dots in selected region (1 out of 16)")
points(spsample(voi@polygons[[1]], n=200, "random"), pch=".", cex=2)
```

In the environment of the spatstat:: package, one can also carry out different types of sampling. In the background, the **rstrat**() command performs the division of the *owin* window into regular tiles as indicated in the options (*nx* and *ny* – number of tiles in each column and row) and successively draws *k* observations in each tile. The **rsyst**() command generates regular points. Parameters *nx* and *ny* in **rsyst**() mean that the distance between the extreme vertical and horizontal coordinates is divided into the indicated number of *nx* and *ny* points, and, finally, points are determined only inside the contour. Thus, the maximum number of points (with a regular shape of the region – for example, a rectangle) is $xn \cdot ny$, and with an irregular shape, it is always lower. Both commands sample within the defined area/region. In both commands, one can easily extract the *xy* coordinates of generated points, referring to the *x* and *y* slots, for example, *r1$x* and *r1$y*.

```
lub.voi<-spTransform(lub.voi, CRS("+proj=merc +datum=NAD83")) # planar
lub.voi.owin<-as(lub.voi, "owin") # conversion SpatialPolygon to owin

# stratified sampling, with background divided into tiles r1<-
rstrat(win=lub.voi.owin, nx=10, ny=10, k=5)
plot(r1, main="66 tiles, k = 5 observations in the tile ")
r2<-rstrat(win=lub.pov.owin, nx=15, ny=15, k=3)
plot(r2, main="431 tiles, k = 3 observations in the tile ")

# generating points evenly distributed in space
r3<-rsyst(win=lub.voi.owin, nx=10, ny=10)
plot(r3, main="63 regular points") # regular points and regional contour

r4<-rsyst(win=lub.pov.owin, nx=15, ny=15)
plot(r4, main="140 regular points") # regular points and poviat contours
```

9.3 Spatial sampling – sampling of sub-samples from existing points

The simplest variant of the nonparametric bootstrap – random picking from existing points – is a simple drawing, technically carried out by drawing the ID of the observation, which will be selected from the set (so-called *case resampling*). A more advanced drawing procedure is *block bootstrapping*. In the basic version, it consists of the division of the dataset into non-overlapping groups, and instead of drawing individual observations, whole groups are drawn with replacements. For *n* observations, blocks of length *b* are created, which gives a total of *n/b* blocks for the dataset. Replacements are drawn with *n/b* blocks to obtain a total of *n* observations.[4] A modification of this procedure is a *moving block bootstrap* in which blocks are assumed to overlap. Assuming a block shift of 1, *n* blocks are created for *n* observations, which gives a total of $n - b + 1$ blocks for the dataset. It draws with replacement from *n/b* blocks to get a total of *n* observations.[5] For spatial data, it is also possible to bootstrap the values (*z*) assigned to the location (*xy*). In the regression bootstrap, a residual drawing (e.g., Wu, 1986) can be used. For the estimated regression model, the residual (as the difference between the theoretical and empirical value of the variable explained *y*, $e_i = y_i - \hat{y}_i$) is determined. The bootstrapped sample contains the same values of variables x_i and modified values of y_i – these are matches of explanatory values derived from the model corrected by bootstrapped residuals ($y_i = \hat{y}_i + e_i$) – residuals e_i are drawn with replacement from the original residual vector. It is also possible to locate, for example, local coefficients from GWR (e.g. Harris et al., 2017), a curve corridor in kriging (Franco-Villoria & Ignaccolo, 2017) or test statistics (de Lima, dos Santos, Duczmal, & da Silva Souza, 2016), although the simplest approach was originally used (Freedman, 1981).

In addition to the previous approaches, other modifications of the bootstrap procedure can be found in the literature. At the same time, there is a debate about the advantages and disadvantages of various drawing methods and their applications. A good overview of the basic issues in the bootstrap method can be found in Hesterberg (2015), where a-spatial applications in the determination of t statistics, consequences for variance, confidence interval construction and regression are shown. A good review of spatial and a-spatial approaches to the board is presented by Loh (2008). The literature can be summarised in the following way.

First of all, the *number of variables and the degree of correlation* between them. One can distinguish situations in which one variable (univariate dataset) and many correlated variables (multivariate dataset) are examined. The early research on bootstrapping concerned a single variable. The relationship between variables is a challenge that is still being studied. A summary of the earlier theoretical achievements for dependent variables can be found in a book by Lahiri (2003), but new solutions are systematically appearing in this area.

Second, the *number of observations* drawn from the original sample. Initially, starting from the work of Efron and Tibshirani (1993), the goal was to improve the statistical properties of the small sample and/or better infer the population based on the sample. This was a methodological answer to the problems of those times. The beginnings of the digital era were related to the modesty of the available data. In many research areas, this problem has disappeared, and in principle, has been supplanted by the opposite problem – big data. In many other areas, such as medicine, the problem remains valid. While the use of bootstrapping in solving small sample problems is very well researched and described, the opposite problem – its use in a large sample and increasing the efficiency of calculations without losing data quality – is still a rare research approach.

Third, the *applications of the bootstrap* method. For the most part, bootstrap is used to assess the quality of test stats, variances and confidence[6] intervals, when moving away from theoretical assumptions about distributions. It may also be a correction of errors in theoretical models (e.g. for point pattern – Loh, 2008). Many researchers point to the added value of the improvement of the properties of estimators (e.g. Hesterberg, 2015). However, doubts can be found (e.g. Loh, 2008), which shows that even if the approximation accuracy by bootstrapping is slightly higher than using the normal distribution (but not significant practically), the high required theoretical assumptions reduce the attractiveness of this method.

Fourth, the *design of the drawing procedure*. In many works, it is written that a simple random sampling (*case resampling*) does not work for spatial dependent data (Lahiri, 2003; Loh, 2008), because it does not retain the dependency structure well and may unrealistically multiply observations at one point. The popular bootstrapping block (Hall, 1985; Kunsch, 1989; Liu & Singh, 1992) is used, transplanted directly from the time series analysis. Block bootstrapping allows randomly selected blocks that contain reasonably large subsamples to be drawn. However, its disadvantage is low computational efficiency, especially for large datasets. The correlation structure may also be disturbed, resulting in an unnatural narrowing of confidence intervals (Loh & Stein, 2004). As an alternative to sampling points or blocks of points, Loh (2008) suggested bootstrapping values (z) assigned to the location (xy). It is argued that this allows for maintaining a greater degree of data dependency (as in the original data), high consistency in the case of cloned data and high efficiency in implementation. Another option for dependent data is subsampling, proposed by Kemball and Martinsek (2005), in which the random part of the observation is removed (this is the equivalent of a generalised jackknife method or just random drawing without replacement). The determined standard deviation approximates the error, but a correction is necessary due to the smaller sample. A comparison of the previous methods (Radovanov & Marcikić, 2014) shows that the efficiency of the method varies due to the size of the subsample and the length of the block.

Fifth, the *value added from using the bootstrap* procedure. Bootstrapped residuals from the regression model replicate the situation in which x values are constant and only y are random. This is a technical procedure for preventing the accidental or statistical omission of observations with a rare value/level of the factor. As Hesterberg (2015) argues, a complete omission of the variable level will make it impossible to estimate this variable, and omitting several observations from the group and rare values will increase the variance of estimates. There may also be a problem in the case of regression with interactions. However, in the case of continuous variables, the indicated issues are not a problem. An advantage of this approach is the technical possibility of limiting heteroscedasticity.

9.3.1 Simple sampling

In R, a simple drawing is made using the **sample**() function from the base:: package, which randomises the subsample. The first argument (x) of this command is the vector (or other object) from which the subset will be selected, and the second argument (*size*) is the number of elements to be selected. The *replace=FALSE* option allows for drawing without replacement and *replace=TRUE* to draw with replacement. Usually, the sampling is applied uniformly – with equal probabilities for all elements of the vector, from which subset observations are selected. It is also possible to assign different probabilities – then the *prob* option is used.

The results of multiple sampling (drawing *i* times in loop) are examined subsequently. The results of all drawing iterations are summarised with the command **table()**, which counts for how many times the given observation was drawn. This information has been added to the basic dataset with the **merge()** command – as the arguments, the base and attached objects are given, the *by.x* and *by.y* options are given the variable names from both objects that contain the same ID allowing connection and the *all.x=TRUE* option preserves all observations of the base file – its omission causes the base set to be trimmed only for merged observations. In the last stage, a column with colours depending on the value was added to the set – this is one of the options for colouring points on the chart.

```
#firms.lim<-firms[1:5000,] # reminder of the data set
result<-matrix(0, nrow=50, ncol=30)
for(i in 1:30){
result[,i]<-sample(1:5000, size=50, replace=TRUE)}
result[1:5, 1:5] # the result of drawing random ID
```

```
        [,1]    [,2]    [,3]    [,4]    [,5]
[1,]     159     240    3943     650    3561
[2,]    2410    3707    2142    3989    4369
[3,]    3543     506    1698    1740    1814
[4,]    3184    1543    1832    4595    2469
[5,]    1889    3064    3105    1284    4079
```

```
a1<-as.data.frame(table(result))
head(a1)
```

```
    result  Freq
1        3     2
2        6     1
3        8     2
4       18     1
5       19     1
6       28     2
```

```
firms.lim$ID<-1:5000
firms.m<-merge(firms.lim, a1, by.x="ID", by.y="result", all.x=TRUE)

firms.m$color<-"grey70"
firms.m$color[firms.m$Freq==1]="red"
firms.m$color[firms.m$Freq==2]="blue"
firms.m$color[firms.m$Freq==3]="green"
plot(firms.m[,12:13], bg= firms.m$color, pch=21) # Figure 9.3
legend("bottomleft", pch=21, pt.bg=c("grey70", "red", "blue", "green"),
c("NA", "drawn 1 time", "drawn 2 times", "drawn 3 times"), bty="n",
cex=0.8)
```

In the previous example, in the random drawing (*replace=FALSE*), a total of 1500 observations were randomly drawn (1 × 1058 + 2 × 188 + 3 × 22 = 1500), of which about 70% were unique. In the drawing with replacement (*replace=TRUE*) of 1500 selected observations (1 × 1115 + 2 × 157 + 3 × 21 + 4 × 2 = 1500), approximately 74% were unique.

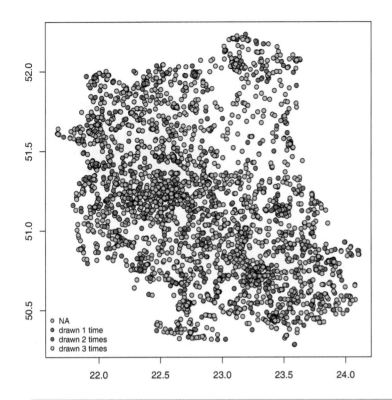

```
Option replace=FALSE

table (al$Freq)

#    1      2      3
#1058    188     22

Option replace=TRUE

table (al$Freq)

#    1      2      3      4
#1115    157     21      2
```

Figure 9.3 **Results of multiple sampling: a) location of drawn observations, b) frequency of drawing the same observation in the draw without replacement and with replacement**

Source: Own study with the use of R

9.3.2 The options of the sperrorest:: package

One of the packages dedicated to solving bootstrap problems is sperrorest:: (Brenning, 2012). The package has several functionalities:

1   It allows the spatial dataset to be divided into spatial fragments (among others in a simple manner, according to *k*-means, according to rectangular and round tiles) in order to carry out cross-validation of the model. After the division, one part of the dataset is treated as a test, and the remaining part as training in simple linear regression models. The functions used are: **partition_cv()**, **partition_cv_strat()**, **partition_disc()**, **partition_factor()**, **partition_factor_cv()**, **partition_kmeans()**, **partition_tiles()**.
2   It allows bootstrapping based on the division of space as for cross-validation to be performed, that is, drawing observations from separated areas. As a result, a division is obtained as for cross-validation (test and training part of the set), with the difference that not all observations (only the drawn part) are included in these subsets. The functions used are: **represampling_bootstrap()**, **represampling_disc_bootstrap()**, **resample_factor ()**, **represampling_tile_bootstrap()**, **resample_uniform()**, **resample_strat_uniform ()**, **represampling_kmeans_bootstrap()**, **represampling_factor_bootstrap()**.
3   It allows for testing the econometric model with the **sperrorest()** command in terms of cross-validation and bootstrapping. The current implementation assumes only simple functional forms of linear models (without interaction, logarithms, spatial components). The construction of objects containing data for validation and bootstrapping limits their use outside the **sperrorest()** function. Model quality measures can be determined with the **err_default()** command. Although the function in the name assumes spatial error, these are not the spatial models known in which the spatial lag is created using the spatial-dimensional matrix and the estimated coefficient for the autoregressive error structure. Spatial error refers to the method of data sampling, taking into account the mutual location determined on the basis of given coordinates (*x*, *y*) and distance between points calculated by the **add.distance()** command (the average distance in the test set) and **dataset_distance()** (the Euclidean distance between two collections).

4   The package has technical functions, including for managing tiles, including **as.tilename()**, **get_small_tiles()**, **tile_neighbors()** and so on.

The following are examples of how the sperrorest:: package divides and prepares data for validation and bootstrapping. The mode of operation of commands for the area (poviat) and point datasets are presented in the following.

To the set of *unempl* concerning the poviat unemployment rate measured every month (380 regions), geographical coordinates of centroid centres (centres of gravity) of poviats have been added, which were designated by the **coordinates()** command from the sp:: package. The **partition_cv()** command randomly divides the set into a training and test part. An average distance between test points has also been added. The point pattern (Figure 9.4a) (unfortunately, an uncluttered graph is generated) has geographic coordinates of the data on the axes – red points are test data, while black points are a training subset. Quite an unusual reference within the object (*unemployment.parti.cv.dist [[1]] [[1]]*) allows for viewing the result for the first iteration. By default, there are 10 iterations of cross-validation.

```
# preparation of area data
crds<-coordinates(pov) # geographical coordinates of centroids
unempl<-read.csv("unemp2018.csv", header=TRUE, dec=",", sep=";")
unempl$crds<-crds # adding xy coordinates to the data set

# preparation of a map section for point data
voi.df<-as.data.frame(voi)
lub.voi<-voi[voi.df$jpt_nazwa_=="lubelskie",]

# a-spatial division of data for cross-validation
# data and geographical coordinates are given
# coordinates are not included in the calculations, only in the graphics
library(sperrorest)
unempl.parti.cv<-partition_cv(unempl, coords=crds)
plot(unempl.parti.cv, unempl, coords=crds) # Figure 9.4a - selected points
unempl.parti.cv.dist<-add.distance(unempl.parti.cv, unempl, coords=crds)
unempl.parti.cv.dist[[1]][[1]] # structure of result - first iteration
```

```
$train
  [1]    2   3   4   5   6   7   8   9  10  12  13  14  15  16  17  18  19  21  22  25  26  27  28  29  30  31  32
 [28]   33  34  35  36  37  38  39  40  41  42  43  44  46  47  48  49  50  51  52  53  54  55  56  57  58  60  61
 [55]   62  63  65  66  67  68  69  70  71  72  73  74  75  76  77  78  79  80  82  83  84  85  86  87  88  89  90
 [82]   91  92  93  94  95  96  99 100 101 102 103 105 106 108 109 110 111 112 113 114 115 117 118 119 120 121 122
[109]  123 124 125 126 127 128 129 130 132 133 135 136 137 138 140 141 142 143 144 146 147 148 149 150 151 152 153
[136]  154 155 156 157 158 159 160 161 162 163 164 165 167 168 169 170 171 172 173 174 175 176 177 178 179 181 182
[163]  183 185 186 187 188 189 190 191 192 193 194 195 196 198 199 200 201 202 203 204 205 206 207 208 209 210 211
[190]  212 213 214 215 216 217 218 219 220 221 223 224 225 226 227 228 229 230 231 232 233 234 235 236 238 239 241
[217]  243 244 245 246 248 249 250 251 254 255 256 257 258 259 260 261 262 263 264 265 266 267 269 270 271 272 273
[244]  275 276 277 278 280 281 283 284 285 286 287 288 289 290 291 292 293 294 295 296 297 298 299 300 301 302 303
[271]  304 306 307 308 309 310 311 312 313 314 315 316 317 319 320 321 322 323 326 327 328 329 330 331 332 333 334
[298]  335 336 337 338 339 340 341 342 343 344 345 346 347 348 349 350 351 352 353 354 355 356 357 358 359 360 361
[325]  362 363 364 365 366 367 368 369 371 372 373 374 375 376 377 378 379 380

$test
 [1]   1  11  20  23  24  45  59  64  81  97  98 104 107 116 131 134 139 145 166 180 184 197 222 237 240 242 247
[28]  252 253 268 274 279 282 305 318 324 325 370

$distance
[1] 0.04093776
```

The result is not very user friendly. To draw selected points more clearly, as areas on the map, it is necessary to create a new variable that takes values equal to 0 for training data and values equal to 1 for test data.

```
# drawing selected observations on the map
resampl<-rep(0, times=dim(unempl)[1])
resampl[unempl.parti.cv.dist[[1]][[1]]$test]<-1
variable <-as.data.frame(resampl)
```

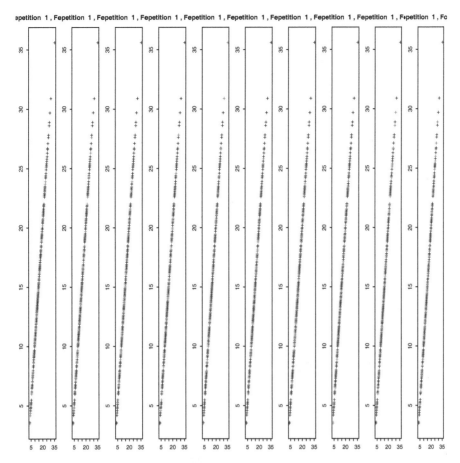

A–spatial division for cross–validation

□ training
□ test

Figure 9.4 A-spatial partitions for cross-validation of area data: a) all iterations in dot plots, b) mapping of iteration results No. 1

Source: Own study using the sperrorest:: package

```
# chart of selected test areas - Figure 9.4b
library(RColorBrewer) brks<-c(0, 0.8, 1)
cols=brewer.pal(3,'Blues')
plot(pov, col=cols[findInterval(variable$resampl, brks)])
legend("bottomleft", legend=c("training", "test"), fill=cols, cex=0.8,
bty="n")
title(main="A-spatial division for cross-validation")
```

Similarly, one can split point data. From the *firms*, the first 500 observations were selected. The procedure as in the previous example for area data was repeated. The chart was slightly different. While in the previous example, the **findInterval()** function was used for area data and intervals [0; 0.8) and [0.8; 1) were defined, in the example subsequently, the **factor()** function is used for point data, which divides the variables into levels (0 and 1) and assigns them the appropriate colour and shape of the symbol. Otherwise, the colours were also specified, previously from the blue palette (blues) using the **brewer.pal()** command from the RColorBrewer:: package and subsequently using the colour names (*coral4, cornflowerblue*). The following result shows that the test data was randomly selected in space.

```
# preparation of point data
firms<-read.csv("geoloc_data_firms.csv", header=TRUE, dec=",", sep=";")
firms.lim<-firms[1:500,]

# a-spatial partitioning of points Figure 9.5a
firms.lim.parti.cv<-partition_cv(firms.lim, coords=c("coords.x1",
"coords.x2"))
plot(firms.lim.parti.cv, firms.lim, coords=c("coords.x1", "coords.x2"))

firms.lim.parti.cv.dist<-add.distance(firms.lim.parti.cv, firms.lim, coords
=c("coords.x1", "coords.x2"))
firms.lim.parti.cv.dist[[1]][[1]] # structure of result - first iteration

# creation of a dummy variable
resampl<-rep(0, times=dim(firms.lim)[1]) resampl[firms.lim.parti.cv.dist[[1]]
[[1]]$test]<-1
variable<-as.data.frame(resampl)

# chart of randomly selected test points - Figure 9.5b
par(mfrow=c(1,1)) cols<-c("coral4","cornflowerblue")
pchset<-c(21,22,23) # vector symbols for subsequent levels of a variable

plot(firms.lim$coords.x1, firms.lim$coords.x2, col=cols[factor(variable$resam
pl)], pch=pchset[factor(variable$resampl)], cex=1.1, bg=cols[factor(variabl
e$resampl)])

plot(lub.voi, add=TRUE)
```

Similar to the previous examples, it is possible to divide the area and point data according to spatial algorithms. The *k*-means algorithm is shown subsequently, in which one should indicate the number of clusters to be created (*nfold* option). Clusters are created only on the basis of location data – geographical coordinates indicated by the *coords* option. As one can see, the algorithm selects several observations from different clusters.

```
# division according to the k-means algorithm
unempl.parti.km<-partition_kmeans(unempl, coords=crds, nfold=10, order_
clusters=FALSE)
plot(unempl.parti.km, unempl, coords=crds)

# creation of a dummy variable
resampl<-rep(0, times=dim(unempl)[1])
resampl[unempl.parti.km[[1]][[1]]$test]<-1
variable<-as.data.frame(resampl)
```

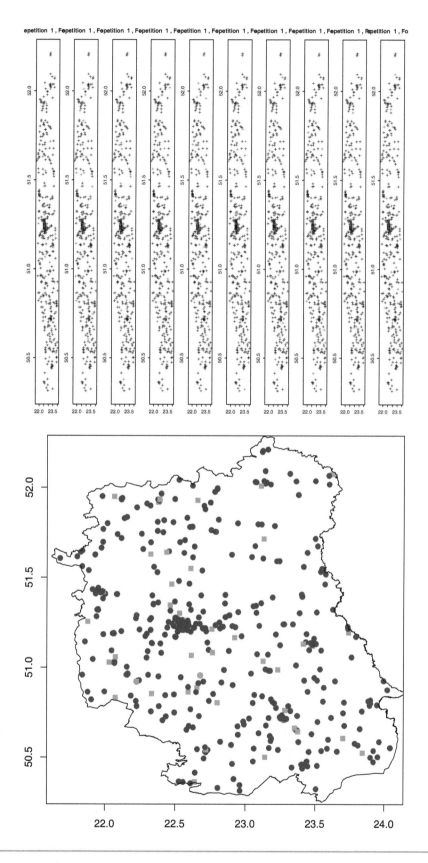

Figure 9.5 A-spatial partitions for cross-validation of point data: a) all iterations in dot plots, b) mapping of iteration results no. 1

Source: Own study using the sperrorest:: package

```
brks<-c(0, 0.8, 1)
cols=brewer.pal(3,'Blues')
plot(pov, col=cols[findInterval(variable$resampl, brks)], main="Selection of
areas according to the k-means algorithm")
legend("bottomleft", legend=c("training","test"), fill=cols, cex=0.8,
bty="n")
```

The result for point data is slightly different. Points were selected from one spatial cluster. Moreover, the axes of collective charts are on a geographical scale.

```
# k-means algorithm for point data
firms.lim.parti.km<-partition_kmeans(firms.lim, coords=c("coords.x1",
"coords.x2"), nfold=10)
plot(firms.lim.parti.km, firms.lim, coords=c("coords.x1", "coords.x2"))

# creation of a dummy variable
resampl<-rep(0, times=dim(firms.lim)[1]) resampl[firms.lim.parti.km.dist[[1]]
[[1]]$test]<-1
variable<-as.data.frame(resampl)

# map of points locations with colours reflecting partitions
cols<-c("coral4","cornflowerblue")
pchset<-c(21,22,23)
plot(firms.lim$coords.x1, firms.lim$coords.x2, col=cols[factor(variable$resam
pl)], pch=pchset[factor(variable$resampl)], cex=1.1, bg=cols[factor(variabl
e$resampl)])
plot(lub.voi, add=TRUE)
```

The result is very similar when the **es()** command is used instead of the **partition_kmeans()** command. Partitioning with tiles creates rectangles in the background that cover the points being examined. The difference is in the command options – instead of the *nfold* option (the number of clusters, e.g. *nfold=10*), the number of rows and columns of tiles is specified in the *nsplit* option (e.g. *nsplit=c(4,3)* – four rows and three tile columns). The methods of using the sperrorest:: package in cross-validation are presented in Section 9.4.

9.3.3 Sampling points from areas determined by the *k*-means algorithm – block bootstrap

A block bootstrap is the easiest way to move away from the usual random drawing. This method assumes the creation of observation blocks whose elements do not overlap. In a simple case, for example, a time series, assuming block length b, n/b blocks are drawn for n observations to cover the dataset 100%. For example, for 100 observations sorted ascending, one can create 10 blocks of 10 observations each. Observations 1:10 form block1, and observations 61:70 block7. Then 10 blocks are drawn and "glued" to them, for example, block3, block7, block3, block8. . . Block length has been the subject of several studies (including Nordman, Lahiri, & Fridley, 2007) – in the literature, one can find, among others, hints that the block length should be, among others $n^{1/3}$, $n^{1/4}$ or $n^{1/5}$, depending on whether the variance or bias is estimated as a one-sided or double-sided density function (see Hall, Horowitz, & Jing, 1995).

The division of points on the plane into n/b blocks can be accomplished by the *k*-means algorithm, which optimises cluster assignments based on intracluster distances (striving for the minimum), distances between clusters (reaching the maximum) and distance to the next nearest cluster (striving for the maximum). In the case of regular distribution points, clusters can be parallel. For points located irregularly on the plane, the *k*-means algorithm allocates points to n/b cluster means but without a guarantee that in each cluster there will be b observations.

The following is code that allows for running a block bootstrap for spatial data. Data from *unempl* (poviat monthly unemployment rates) were used, where $n = 380$ locations, and data on the unemployment rate in the following months are strongly correlated in time and space. Geographical coordinates of centroids of poviats on which clustering was carried out were determined. The correlation between the observed unemployment rates in the bootstrapped and the full sample was examined. The result was compared with a simple drawing, so-called case resampling.

Assuming an average block length of $b = 10$, $n = 380$ observations, $n/b = 38$ blocks can be created. This means that in the *k*-means procedure (k-means) $k = n/b = 38$ clusters was assumed. The block bootstrap in the simplest view consists of drawing instead of observing whole blocks. In the previous example, one should draw $k = 38$ blocks, which on average should give *n = 380* observations. In the case of very unevenly distributed data on the plane (e.g. strong agglomeration, outliers, areas without coverage), the clusters will be of different sizes, and the sample will be centred around *n = 380*. It is also possible to draw a sample spatially stratified based on *k*-means. Then, *b = 10* observations from each cluster are drawn. With regular spatial distribution, it allows for drawing points from each part of the space. Moreover, the size of the cluster when drawing with replacement does not matter.[7]

In the division of points into *k* clusters using the *k*-means algorithm, the command **kmeans()** from the base:: package was used. Its result from the cluster slot is a clustering vector. A clustering vector was added to the dataset, the values of which indicate affiliation to specific clusters.

The choice of colours is seemingly a trivial matter. However, most palettes define at most a few or a dozen or so colours, which does not allow for the colouration of, for example, 38 clusters. The viridisLite:: package allows for specifying up to 256 colours selected from several available palettes: **viridis()**, **magma()**, **inferno()**, **plasma()**, **cividis()**. Display of colours is well guided in help to the **viridis()** command.

The following code divides the points into clusters. It is a random division – in subsequent iterations, other divisions for *k* given groups can be obtained (see Figure 9.6).

```
# data preparation
unempl<-read.csv("unemp2018.csv", header=TRUE, dec=",", sep=";")
crds<-coordinates(pov) # geographical coordinates of centroids
unempl$crds<-crds # adding xy coordinates to the data set

n<-dim(unempl)[1]      # number of observations
b<-10                  # average cluster length
k<-n/b                 # the number of clusters

c1<-kmeans(unempl$crds, k)    # k-means algorithm
unempl$clust<-c1$cluster      # clustering vector

library(viridisLite)
#cols<-plasma(k)   #  color palette
cols<-inferno(k)   #  color palette
brks<-1:k          # intervals

pl<-readOGR(".", "Panstwo")
pl<-spTransform(pl, CRS("+proj=longlat +datum=NAD83"))
plot(pl)  # drawing a map, Figure 9.6a
```

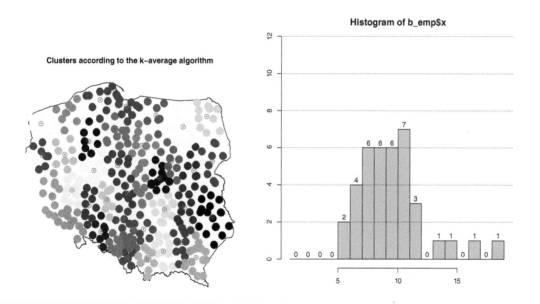

Figure 9.6 Clusters generated with *k*-means: a) centroids in *k* clusters, b) histogram of cluster sizes

Source: Own study using the viridisLite:: package

```
points(unempl$crds, col=cols[findInterval(unempl$clust, brks)],
     pch=21, bg=cols[findInterval(unempl$clust, brks)], cex=2.5)
points(c1$centers, col="red", pch=".", cex=3) # plot of cluster centres
points(c1$centers, col="red", cex=1.5) # plot of cluster centers
title(main="Clusters according to the k-average algorithm")
```

Before making the sampling, it is worth having an overview of the sizes of the created clusters. In the analysed example, the spatial distribution of poviat centres is not fully uniform, although certainly more even than in the *firms* collection regarding the location of firms. The degree of agglomeration of point data translates into the evenness of the allocation of observations to clusters: the greater the concentration of points in space, the more diverse the clusters. Cluster size statistics were examined with the command **aggregate**() and illustrated with a histogram from the **hist**() command and the equivalent of ggplot2::. The dominant size of clusters is the length of $b = 8$, and the largest cluster, for the Silesian agglomeration, has 24 observations, due to the high spatial agglomeration of points.

```
b_emp<-aggregate(unempl$X2018.05, by=list(unempl$clust), length)
head(b_emp)
```

```
  Group.1        x
1        1       10
2        2        9
3        3        6
4        4       11
5        5       10
6        6       11
```

```
mmax<-max(b_emp$x) # Figure 9.6b
hist(b_emp$x, breaks=1:mmax, ylim=c(0,12), labels=TRUE, col="orange")
abline(h=(1:6)*2, lty=3)

library(ggplot2) # alternative graphics
ggplot(b_emp,    aes(x=x))    +    geom_histogram(binwidth=1,    fill="grey50",
color="white")
```

A simple block bootstrap involves drawing whole blocks of data. Technically, it involves drawing the numbers of clusters and building a bootstrap sample from observations belonging to those clusters. Combining fragments of the original sample starts with creating an empty set of *data.frame* class (*boot1* object) to which the blocks of observations are attached from below with the **rbind**() function. In this situation, it is not worth creating a blank matrix for a sample with a predetermined number of rows (e.g. $n = 380$), because the number of cases in the bootstrapped sample depends on the blocks drawn. Adding more than the defined number is a problem, while leaving empty lines requires deleting them so that they are not included in the calculations. The following block bootstrap code has been inserted into the **for**() loop to repeat the sampling and to examine the sampling parameters – the percentage of unique cluster IDs, the percentage of unique observations and the length of the sample. The history of drawings is recorded in the *selected* object.

```
iter<-100
result<-matrix(0, nrow=iter, ncol=4) # to write the results of analysis
colnames(result)<-c("unique   clusters",   "sample       length",   "unique
observations", "correlation")
selected<-matrix(0, nrow=k, ncol=iter) # empty object to save results

for(i in 1:iter){
ss<-sample(1:k, size=k, replace=TRUE)
selected[,i]<-ss # record of randomly drawn clusters
result[i,1]<-length(unique(ss)) # number of unique IDs

# empty data.frame
boot1<-data.frame(cluster=numeric(0), var1=numeric(0), var2=numeric(0),
crds1=numeric(0), crds2=numeric(0))

# appending from the bottom of the blocks of observation
for(j in 1:k){
boot1<-rbind(boot1, unempl[unempl$clust==ss[j],c(103, 100, 101, 102)])}}
```

```
result[i,2]<-dim(boot1)[1] # length of bootstrapped sample
result[i,3]<-dim(unique(boot1[,4]))[1] # number of unique observations
result[i,4]<-cor(boot1[,2], boot1[,3]) # correlation
}
head(result)
```

	unique clusters sample	length unique	observations correlation
[1,]	23	383	239 0.9988164
[2,]	22	380	228 0.9984175
[3,]	25	394	256 0.9979915
[4,]	25	380	255 0.9988699
[5,]	23	394	237 0.9990510
[6,]	23	381	241 0.9989378

```
selected[1:6, 1:6]
```

	[,1]	[,2]	[,3]	[,4]	[,5]	[,6]
[1,]	2	19	7	19	16	16
[2,]	18	13	14	24	1	5
[3,]	17	23	21	23	12	20
[4,]	7	6	3	2	10	31
[5,]	35	30	26	33	22	2
[6,]	20	32	14	34	20	37

The analysis of the loop results was made based on density functions. All results are presented as a percentage – relative to the full basic sample. The value of the correlation coefficient ranges ±0.02% around the average of the full sample. The percentage of unique observations that were drawn randomly is 60%–67%, while the sample size most often differs up to 4% from the baseline.

```
summary(result[,1]/38) # the percentage of unique clusters
```

Min.	1st Qu.	Median	Mean	3rd Qu.	Max.
0.5263	0.5987	0.6316	0.6268	0.6579	0.7368

```
summary(result[,2]/380) # length of the trial relative to the full sample
```

Min.	1st Qu.	Median	Mean	3rd Qu.	Max.
0.9053	0.9651	1.0013	1.0039	1.0421	1.1711

```
summary(result[,3]/380) # the percentage of unique observations
```

Min.	1st Qu.	Median	Mean	3rd Qu.	Max.
0.4921	0.5914	0.6276	0.6285	0.6711	0.7421

```
# correlation coefficient
summary(result[,4]/cor(unempl[,100], unempl[,101]))
```

Min.	1st Qu.	Median	Mean	3rd Qu.	Max.
0.9991	0.9997	1.0000	1.0000	1.0002	1.0008

```
plot(density(result[,4]), main="Distribution of the correlation coefficient")
#Figure 9.7a
abline(v=cor(unempl[,100], unempl[,101]), lty=3, col="coral", lwd=2)
```

```
plot(density(result[,1]/38), xlim=c(0,2), main="Structure
of the bootstrapped sample") # unique clusters, Figure 9.7b
lines(density(result[,2]/380), lwd=2) # the length of the trial
lines(density(result[,3]/380), lty=3) # unique observations
legend(1.10,6,c("unique clusters%", "sample length%", "unique
observations%"), lty=c(1,1,3), lwd=c(1,2,1), cex=0.8, bty="n")
```

One can also make a **stratified bootstrap**. This involves drawing with the replacement of *b* observations from each cluster *k*. Thanks to this, the spatial distribution of the sample is preserved. In the analysed situation, *k = 38* clusters were separated with the expected length of *b = 10*, because the number of observations is *n = 380*. Technically, a conditional sampling is carried out; that is, from each of *k = 38* subsets (clusters), *b = 10* observations are drawn with the replacement. The observations are successively added to the result set. The following example uses a triple loop. The first loop, by *i* from 1 to *iter*, is for subsequent iterations – repeats of the entire bootstrap. The second loop, by *m* from 1 to *k*, is for each cluster within a given division. *k* subsets are sequentially created here, and for each subset, one draws with the replacement from the pool of observations in the subset/cluster. The third loop, by *j* from 1 to *b*, replaces the **merge**() command. It attaches from the bottom, row by row, the observations from the base file to the result set. The **for**() loop protects against the reduction of repeating lines (there are duplicates due to drawings with replacement). The range up to *b* results from the assumed block length. For each iteration, a new division into clusters is generated – a new assignment of individual observations to *k = 38* clusters. The history of drawings is recorded in the *selected2* object. Distributions of iteration results are presented as densities using **ggplot**() with the **geom_density**() option. Vertical lines are added with the **geom_vline**() command with the *xintercept* option. The alpha parameter determines the degree of transparency.

```
iter<-100         # the number of iterations of the entire model
n<-dim (unempl)[1] # number of observations in the basic set
b<-10             # average cluster length - block length
k<-n/b            # the number of clusters
unempl$ID<-1:380   # adding ID to the basic file

result<-matrix(0, nrow=iter, ncol=2)
colnames(result)<-c("unique observations", "correlation")
selected2<-matrix(0, nrow=380, ncol=iter) # object of history of draws

for(i in 1:iter) {# new iterations - repeating the test iter times
c1<-kmeans(unempl$crds, k)    # k-means algorithm
unempl$clust<-c1$cluster       # clustering vector

# empty data.frame - new for each iteration
boot1<-data.frame(cluster=numeric(0), var1=numeric(0), var2=numeric(0),
crds1=numeric(0), crds2=numeric(0), ID=numeric(0))

for(m in 1:k){   # loop after all k clusters
sub<-unempl[unempl$clust==m,]
ile.obs<-dim(sub)[1] # checking the subset length - of cluster
ss<-sample(1:ile.obs, size=b, replace=TRUE) # drawing b obs. from a subset

# appending from the bottom of the blocks of observation
for(j in 1:b) {# a loop for each observation from a given cluster
boot1<-rbind(boot1, sub[ss[j],c(103, 100, 101, 102, 104)])}}
selected2[,i]<-boot1[,5]

result[i,1]<-dim(unique(boot1[,4]))[1] # number of unique observations
result[i,2]<-cor(boot1[,2], boot1[,3])} # correlation

# Figure 9.8a
ggplot(as.data.frame(result), aes(x=result[,2])) + geom_density
(fill = "lightblue") + geom_vline(xintercept = cor(unempl[,100],
```

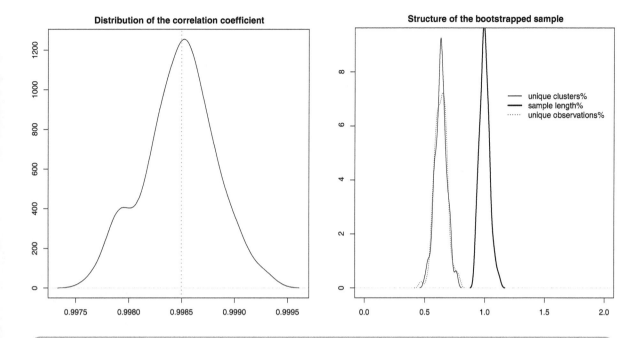

Figure 9.7 Block bootstrap results: a) correlation of variables, b) structure of a bootstrapped sample

Source: Own study using the base:: package

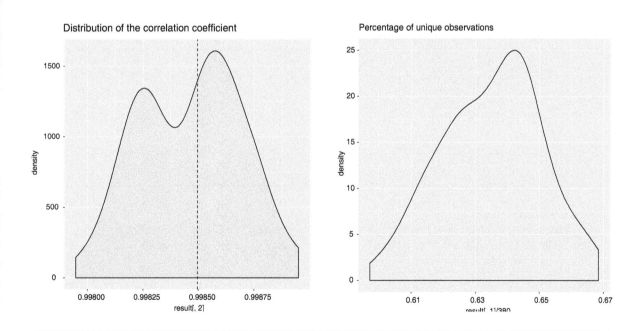

Figure 9.8 Results of stratified block bootstrap: a) density of correlation coefficient between variables, b) percentage of unique observations

Source: Own study using the ggplot2:: package

```
unempl[,101]), linetype = "dashed") + ggtitle("Distribution of the
correlation coefficient")

# Figure 9.8b
ggplot(as.data.frame(result), aes(x=result[,1]/380)) + geom_density
(fill = "lightcoral", alpha=0.8) + ggtitle("Percentage of unique
observations")

summary(result[,2]/cor(unempl[,100], unempl[,101])) # correlation coefficient
```

```
   Min. 1st Qu.  Median    Mean 3rd Qu.    Max.
 0.9994  0.9998  0.9999  0.9999  1.0001  1.0005
```

From the simulation previously, it can be seen that the dominant value of the correlation coefficient calculated from the bootstrapped sample coincides with the determined correlation coefficient from the full original sample. For a fairly even spatial distribution, as in the previous example, the effective coverage of the sample is about 64%. The fluctuation of the bootstrapped correlation coefficient is small – about 0.01%.

For the previously mentioned sampling, the number of randomly drawn points was counted. In the randomisation model according to a simple block bootstrap, a selected matrix was created that contains the ID of the randomly drawn clusters (with replacement) in all drawings. In each iteration, the same division into clusters according to k-means was used. One of the counting options is to assign the number of drawings to each frequency point in all iterations. The points were coloured due to the frequency of occurrence in subsamples using the **choropleth**() command from the GISTools:: package. The contour map was created by combining poviats by assigning to clusters – the **unionSpatialPolygons**() command from the maptools:: package was used. The frequency of clusters was determined by the **table**() command.

```
f1<-as.data.frame(table(selected))
head(f1)
```

```
  selected    Freq
1        1     115
2        2     103
3        3     111
4        4     105
5        5      84
6        6      99
```

```
library(maptools)
reg.kmeans<-unionSpatialPolygons(pov, IDs= unempl$clust) #maptools
plot(reg.kmeans, lwd=2) # poviats combined by grouping k-means
plot(pov, add=TRUE)

library(GISTools) # Figure 9.9a
choropleth(reg.kmeans, f1$Freq, main="Frequency of drawing given cluster in
100 iterations")
shades<-auto.shading(f1$Freq)
choro.legend(14, 50.25, shades, cex=0.65, bty="n")
```

Another option for counting of sampling frequency is the raster approach, which counts the observations in fixed windows and creates a colour map for the grids. Using the raster:: package, a dividing grid with the **raster**() command was created, defining the number of boxes in rows and columns (*nrow* and *ncols* options). Coordinates limiting the outline of the map can be created with the command **bbox**() from the sp:: package.

Frequency of drawing given cluster in 100 iterations

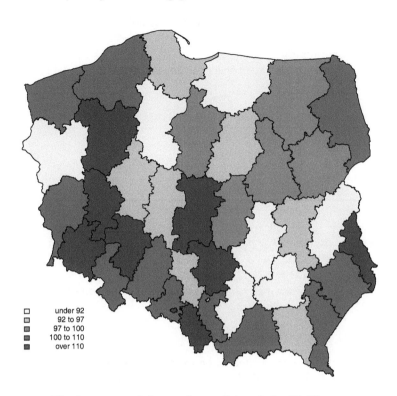

under 92
92 to 97
97 to 100
100 to 110
over 110

The frequency of drawn observations, raster 30x30

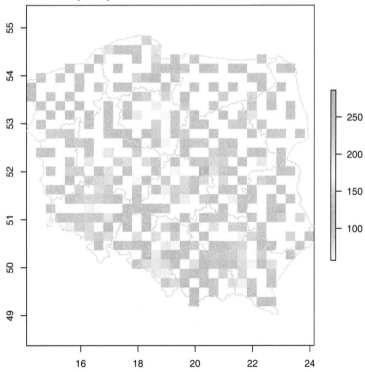

Figure 9.9 Statistics of randomly selected clusters in stratified block bootstrap: a) the frequency of randomly drawn clusters in 100 iterations, b) raster plot of frequency of drawn observations

Source: Own study using the maptools:: and GISTools:: packages

In Figure 9.9b, the overrepresentation in the south (19°, 50.5°) results from a very dense settlement network and very small territorial units. It can be seen that the less defined the grates, the stronger the aggregation of the result and the higher the possible values (different scale of the legend).

```
# coordinates and frequency
p1<-data.frame(crds, z=as.data.frame(table(selected2))[,2])

bb<-bbox(p1)
```

```
        min      max
x  14.12289  24.14578
y  49.00205  54.83579
```

```
library(raster)
r<-raster(nrows=30, ncols=30, ymn=bb[2,1], ymx=bb[2,2], xmn=bb[1,1],
xmx=bb[1,2]) # Figure 9.9b
r1<-rasterize(crds, r, field=p1$z, fun=sum)
plot(r1, main="The frequency of drawn observations, raster 30x30")
plot(voi, add=TRUE, border="grey80")
```

9.3.4 Sampling points from moving blocks (moving block bootstrap)

Moving block bootstrapping (MBB) was created to be used with time series. This is a method of extracting blocks whose elements overlap. Assuming a block shift of 1, n blocks are created for n observations, which gives a total of $n - b + 1$ blocks for the dataset. This draws with replacement n/b blocks to obtain a total of n observations. In the time series, for $n = 100$ observations sorted ascending by time, one can create $100 - 10 + 1 = 91$ blocks, 10 observations each. Observations 1:10 form block1, observations 2:11 block2, observations 3:13 block3 and so on. Within a bootstrap, 10 blocks are randomly drawn and merged together, for example, block 23, block 47, block 35, block 8 and so on.

Due to the directionality of the time series, it is hard to point out some neighbours for extreme observations: "backwards" for the oldest observations and "forward" for the latest observations – hence, the habitual "forward" neighbours are counted from the oldest observation, and the last block is for observation number $n - b + 1$. In specific solutions, one can take typical (full) blocks and atypical blocks – incomplete – for extreme observations. It is then only contractual whether the "lead" observation for which the block is formed is the first, middle or last in a given block. Neighbourhood can be defined as "forward" (with chronologically ordered data) or "forward" and "backward" or only "backwards".

In the case of spatial data arranged irregularly, ordering the space dimension (as in the example previously after the time dimension) is not unambiguous. Due to the multidirectional spatial processes (in contrast to the one-directional time series – from the oldest to the newest), one cannot directly indicate the "next" observation. One can indicate a block of the next observations. The method of division of data does not change: for n observations, blocks of length b are created, which gives a total of maximum n blocks for the dataset – due to the asymmetry of the neighbourhood, there may be a situation where each point has a different set of neighbours. It draws with replacement n/b blocks to get a total of n observations. With a problem defined in this way, this means that the nearest neighbours are qualified for the bootstrap block, where $k = b - 1$. The knn algorithm is well known and implemented in spatial research. In this context, the moving block bootstrap is identical to the drawing of n/b points along with their knn surroundings. For example, for $n = 100$ observations and blocks of length $b = 10$, one draws $n/b = 100/10 = 10$ points with their surroundings. In the surroundings of each point, there are $k = b - 1 = 10 - 1 = 9$ neighbours. The total sample size (drawn and surrounding) is $10 + 10 \times 9 = 100$ observations. Such an algorithm will look for neighbours in the circles surrounding the point, which will translate into the shape of tiles bootstrap.[8]

The code subsequently presents a spatial implementation of a moving block bootstrap. It was based on data from *unempl* (poviat monthly unemployment rates), where there are $n = 380$ locations, and data on the unemployment rate in the following months are strongly correlated in time and space. The geographical coordinates of the centroids of the poviats were determined, and the nearest neighbours were selected based on the Euclidean distance. The correlation between the observed unemployment rates in the bootstrapped sample and the full sample was examined. The result was compared with a simple draw, so-called case resampling.

Regardless of the block length (if $b < n$), for $n = 380$ one can create a maximum of $n = 380$ blocks. For blocks with a length of $b = 10$, $n/b = 38$ blocks were drawn with replacement. For drawn $n/b = 38$ leading points, $knn = 9$ neighbours were determined using the **knearneigh()** function from the spdep:: package. Together, they are blocks with the assumed length of $b = 10$. The object containing the neighbour ID (380 rows × 9 columns) has been transformed into a vector (length 380 × 9) and connected in a longer vector with the ID of the leading points ($380 \times 9 + 38 = 380$). A new bootstrapped file has been created by copying subsequent rows from the base file in the **for()** loop. Due to the bootstrap drawings with replacement, some drawn IDs overlap. For this reason, the combining command (e.g. **merge()**) was not used, because it would take into account unique values, not duplicated ones. It should be remembered that statistics (e.g. correlation) calculated for unique values and for all, including duplicated observations, differ. Finally, the correlation of unemployment rates in two periods was calculated.

```
# data preparation
unempl<-read.csv("unemp2018.csv", header=TRUE, dec=",", sep=";")
crds<-coordinates(pov) # geographical coordinates of centroids
unempl$crds<-crds # adding xy coordinates to the data set

# determining k nearest neighbors based on coordinates
knn.set<-knearneigh(crds, k=9) # matrix k nearest neighbors
knn.set$'nn' # fragment of the knn matrix
```

```
       [,1]  [,2]  [,3]  [,4]  [,5]  [,6]  [,7]  [,8]  [,9]
[1,]    140     7    54   142     3    56     5    51   137
[2,]     13   136     3   134   142    10   140   133     6
[3,]    140   136   142     2    13     1    51     5    12
[4,]      8   137   135   143    12    11     5   138   139
[5,]     12   142   133     8   137     4   140    11     1
[6,]    133    10   139    11   134     8    12     5     2
. . . . . . . . . . . . . . . . . . . . . . . . . . . . . . . . . . . . . . . .
```

```
# bootstrapped trials
lead<-sample(1:380, size=38, replace=TRUE) # drawing of main points
boot<-as.vector(knn.set$'nn'[lead,]) # vector from randomly drawn neighbors
boot2<-c(boot, lead) # joining the main points and neighbors

length(boot2) # total length of the drawn IDs
#[1] 380
length(unique(boot2)) # number of unique IDs
#[1] 242

# rewriting the drawn values from the complete data set according to the
new ID
# studied variables in columns 100: 101
boot3<-matrix(0, nrow=380, ncol=2)
for(i in 1:380){
boot3[i,1:2]<-as.numeric(unempl[boot2[i], 100:101])}

cor(boot3[,1], boot3[,2]) # correlation in the bootstrapped sample
#[1] 0.9979167

a<-cor(unempl[,100], unempl[,101]) # correlation in the full sample a
#[1] 0.9984963
```

In the previous example, two parameters were accepted without a broad discussion: the degree of coverage of the sample and the length of the blocks as 100% and $b = 10$ observations, respectively. A question arises, as the change in the block length (for longer or shorter), as well as the reduction in the sample coverage, will affect the quality of the estimation. The simulations of these changes are presented subsequently.

When testing the coverage level, a step by 5 percentage points was taken. For each assumed coverage threshold (*100% = 380 obs, 95% = 361 obs, 90% = 342*), 30 bootstrap iterations were performed and the mean correlation coefficient and its standard deviation (based on the vector obtained correlation coefficients). In determining the number of leading points to be drawn (the *size* option in the *lead* object), the rounding up (**ceiling()** command was used. The result (with dimensions of 20 columns and 30 lines) contains successive correlation coefficients for subsequent iterations (rows) and various coverage indicators (columns). The **apply()** command has been assigned statistics (average and standard deviation) in the columns (option 2), and objects are simplified with the command **as.vector()**. The result was combined into one object with the command **c()**, and in order to give the names of the columns, the object was converted to the class *data.frame* with the command **as.data.frame()**. The following variables have been added to the set: the lower and upper "corridor" (determined based on the standard deviation), as well as the type of drawing and the coverage of the sample as the *x*-axis label.

The **ggplot()** command from the ggplot2:: package was used to create a transparent graph containing the average value and the corridor from the standard deviation. In the basic part of the command, the values of the *x*-axis (variable interval) and *y*-axis (variable av) and the type of the drawing – with the line (*geom_line*) and points (*geom_point*) – were defined in *aes*. Subsequently a "ribbon" (*geom_ribbon*) was added, that is, a corridor defined on the basis of the defined lower and upper limits (as one standard deviation downwards and upwards from the average), with a set degree of alpha transparency. The names of the *x* (*xlab*) and *y* (*ylab*) axes have also been specified. An auxiliary line (correlation value in the full sample) was also added using the **geom_hline()** command, where the line level and line type were defined. The title of the graph was marked with the command **ggtitle()**. The graphics components have been written to the object (classes *gg* and *ggplot*) and are called with the **plot()** command.

```
vec<-(20:1)*5/100
#[1] 1.00 0.95 0.90 0.85 0.80 0.75 0.70 0.65 0.60 0.55 0.50 0.45 0.40
#[14] 0.35 0.30 0.25 0.20 0.15 0.10 0.05
b=20# block length (possibly b = 20)
knn.set<-knearneigh(crds, k=b-1) # matrix k nearest neighbors

result<-matrix(0, nrow=30, ncol=20)
colnames(result)<-paste(rep("cov", times=20), vec)
rownames(result)<-paste(rep("iter", times=30),1:30)

for(i in 1:20) {# coverage levels - result columns
samp.size<-380*vec[i] # the total number of observations to be drawn

for(j in 1:30) {# iterations for each coverage level - result rows
lead<-sample(1:380, size=ceiling(samp.size/b), replace=TRUE)
boot<-as.vector(knn.set$'nn'[lead,]) # vector from randomly drawn neighbors
boot2<-c(boot, lead) # joining the main points and neighbors
boot3<-matrix(0, nrow=samp.size, ncol=2)
for(k in 1:samp.size){
boot3[k,1:2]<-as.numeric(unempl[boot2[k], 100:101])}
result[j,i]<-cor(boot3[,1], boot3[,2])}} # correlation
result[1:6, 1:6]
```

		cov 1	cov 0.95	cov 0.9	cov 0.85	cov 0.8	cov 0.75
iter	1	0.9985300	0.9986096	0.9984907	0.9989637	0.9988412	0.9984415
iter	2	0.9984141	0.9990997	0.9986531	0.9991253	0.9987931	0.9987609
iter	3	0.9982426	0.9990557	0.9985787	0.9986018	0.9982330	0.9984496
iter	4	0.9985746	0.9981365	0.9984090	0.9979848	0.9985347	0.9986322
iter	5	0.9981815	0.9984626	0.9984531	0.9986378	0.9992491	0.9983372
Iter	6	0.9986263	0.9987568	0.9986992	0.9987861	0.9984838	0.9988003

```
mea<-as.vector(apply(result, 2, mean))
sd<-as.vector(apply(result, 2, sd))
data<-as.data.frame(cbind(mea, sd))
colnames(data)<-c("mea", "sd")
data$lower<-data$mea-data$sd
```

```
data$upper<-data$mea+data$sd
data$model<-rep("block20", times=20) #also: block20
data$interval<-(20:1)*5/100

data1<-data
data11<-rbind(data, data1)

library(ggplot2)

# Figure 9.10a
p<-ggplot(data=data11, aes(x=interval, y=mea, colour=model)) + geom_point()
+ geom_line()
p<-p + geom_ribbon(aes(ymin=data11$lower, ymax=data11$upper), linetype=2,
alpha=0.1) + xlab("Coverage of the sample") + ylab("Average correlation")
p<-p+ geom_hline(yintercept=a, linetype="dashed") + ggtitle("MBB for b=10")
plot(p)

# Figure 9.10b
p1<-ggplot(data=data11, aes(x=interval, y=mea, colour=model)) + geom_point()
+ geom_line()
p1<-p1 + geom_ribbon(aes(ymin=data11$lower, ymax=data11$upper), linetype=2,
alpha=0.1) + xlab("Coverage of the sample") + ylab("Average correlation")
p1<-p + p1+ geom_hline(yintercept=a, linetype="dashed") + ggtitle
("MBB for b=10")
plot(p1)
```

The previous graph shows that the correlation coefficient variance (corridor) is relatively stable for the coverage ranging from 50%–100%, which may mean that using half of the sample will give similar results as an estimate on the entire sample. The extension of the block (wider neighbourhood) generally slightly decreases the variance of the coefficient.

To compare the result, the simulation was repeated using simple random sampling (so-called *case resampling*). The results from the moving block bootstrap model and from the simple case resampling (CR) were connected by lines (sticking up from the bottom) and drawn as two lines with two corridors. This is possible thanks

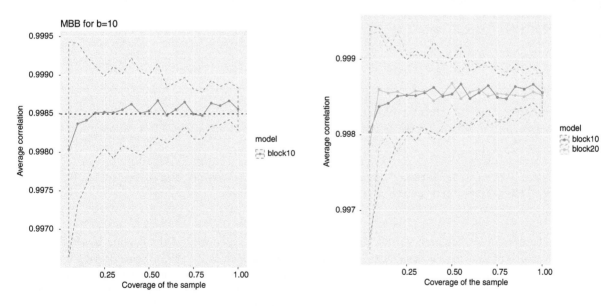

Figure 9.10 Average correlation and fluctuation corridor (by std.dev) in the moving block bootstrap model for: a) block length *b* = 10, b) block length *b* = 20

Source: Own study, graphics using the ggplot2:: package

to the colour option in *aes*. As a last comparison, the result for the drawing with the replacement was checked, changing the **sample**() command options from *replace=TRUE* to *replace=FALSE*.

```
# simple sampling
result2<-matrix(0, nrow=30, ncol=20)
colnames(result2)<-paste(rep("cov", times=20), vec)
rownames(result2)<-paste(rep("iter", times=30),1:30)

for(i in 1:20) {# coverage levels - result columns
ile<-380*vec[i] # the total number of observations to be drawn

for(j in 1:30) {# iterations for each coverage level - result rows
lead<-sample(1:380, size=ile, replace=TRUE) # possible replace=FALSE
boot3<-matrix(0, nrow=ile, ncol=2)
for(k in 1:ile){
boot3[k,1:2]<-as.numeric(unempl[lead[k], 100:101])}
result2[j,i]<-cor(boot3[,1], boot3[,2])}} # correlation

mea2<-as.vector(apply(result2, 2, mean))
sd2<-as.vector(apply(result2, 2, sd))
data2<-as.data.frame(cbind(mea2, sd2))

colnames(data2)<-c("mea", "sd")
data2$lower<-data2$mea-data2$sd
data2$upper<-data2$mea+data2$sd
data2$model<-rep("simple", times=20)
data2$interval<-(20:1)*5/100

data3<-rbind(data, data2) # combining MBB and CR results
```

	mea	sd	lower	upper	model	interval
1	0.9984860	0.0002494889	0.9982366	0.9987355	block	1.00
2	0.9985708	0.0003016183	0.9982692	0.9988724	block	0.95
3	0.9986270	0.0003267457	0.9983002	0.9989537	block	0.90
4	0.9985777	0.0002737754	0.9983039	0.9988515	block	0.85
5	0.9985038	0.0003659026	0.9981379	0.9988697	block	0.80
6	0.9985405	0.0003209114	0.9982196	0.9988614	block	0.75

```
# joint MMB and CR chart - thanks to the color option in aes
p<-ggplot(data=data3, aes(x=interval, y=mea, colour=model)) + geom_point()
+ geom_line()
p<-p+geom_ribbon(aes(ymin=data3$lower, ymax=data3$upper), linetype=2,
alpha=0.1) + xlab ("sample coverage") + ylab ("average correlation") +
ggtitle("drawing with replacement")
plot(p) # Figure 9.11a,b
```

In the literature, a lot has been written theoretically about the superiority of the moving block bootstrap over other models of drawing the dependent spatial data (including Hall et al., 1995). Remembering the genesis of bootstrapping as the method of multiplying the sample, and not its reduction, it has become established to assume drawing with replacement. Previously is a comparison of four sampling scenarios: in Figure 9.11a with replacement and in Figure 9.11b without replacement. Additionally, for both models, the results of a moving block bootstrap with straight case resampling were compared. The conclusions from the following empirical analysis are as follows. In the case of drawings with replacement, the variance of the correlation coefficient is similar in both models at high coverage and lower in the simple drawing with low coverage. In the case of no-replacement draws, the moving block bootstrap model has a relatively constant variance, for coverage from 40% to 100%. In the same interval, the variance in the case resampling is significantly lower with the same precision

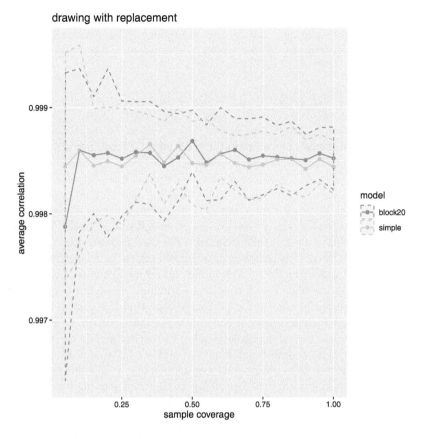

drawing with replacement

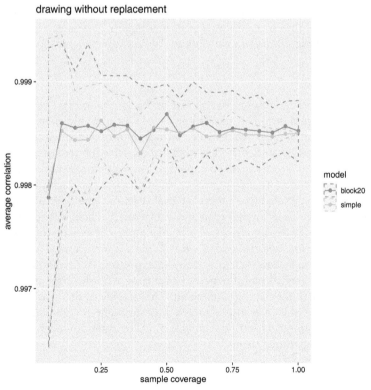

drawing without replacement

Figure 9.11 Comparison of movable block bootstrap and simple drawing typical for: a) bootstrap drawing with replacement, b) modification – drawing without replacement

Source: Own study, graphics using the ggplot2:: package

of the correlation coefficient estimation. The lowest variance and the highest convergence of the correlation coefficient have the least popular solution – simple case resampling without replacement. The moving block bootstrap generates similar (moderately good) results with 50% coverage or more, regardless of how one draws.

9.4 Use of spatial sampling and bootstrapping in cross-validation of models

In this section, the use of spatial sampling methods in the verification of econometric models using cross-validation will be presented. Quantitative models, both econometric, based on theoretical premises, as well as, above all, a-theoretical machine learning models, require validation on a test sample. A typical way to check the quality of models is so-called cross-validation. In general, the cross-validation algorithm relies on multiple division of the set into training and validation – the first is used to estimate the model, while the second determines the forecast and the selected forecast error metric for each observation (most often RMSE or MAE). The procedure is repeated several times on different sets of training and validation data. The forecast errors obtained in this way are averaged later between all validation tests. The average error obtained in this way is an estimate of the forecast error.

There are various methods for dividing data into training and testing in cross-validation. The most popular of these is the so-called k-fold cross-validation. It consists of randomly dividing the sample into k subsamples (most often $k = 5$ in the small sample situation and $k = 10$ in the large sample situation) and executing k times the model on $k – 1$ samples, leaving in each iteration one subsample as a validation set (different in each iteration). In this way, each observation is once in the validation set and $k – 1$ times in the training set. Often, for more stable results, the procedure can be repeated several dozen times so that the average forecast error is calculated on the basis of a larger number of iterations. The estimator of forecast error obtained in this procedure is usually biased (inflates the forecast error), because it is not used for full information training, but it has a smaller variance.

The role of cross-validation in addition to 1) estimating the error of the forecast also consists of 2) detecting overfitting of the model, and 3) tuning hyper-parameters (mainly in machine learning models). The phenomenon of overfitting the model is related to the placement in the equation of seemingly significant variables, which significantly increase the fit of the model to the training data. If the model actually contains too many parameters (too many degrees of freedom) – the error on the validation sample will be much higher than on the training sample. Similarly, based on this algorithm, one can choose the best value for the model's hyper-parameter.[9]

Example

The example uses monthly data on the unemployment rate in the county section (380 units) in the period from January 2011 to May 2018 (89 months). The aim will be to forecast unemployment in May 2018. In addition to the historical unemployment rates, the model will also include information about the population in the poviat and geographical coordinates of poviat centroids. The obtained model will be completely a-theoretical. Through the geolocation of the centres of gravity, the spatial component is included in the model – the model will be able to build decision rules explaining the level of unemployment based on these two variables (latitude and longitude).

From the technical perspective, to slightly shorten the codes and increase their transparency, the object *formula_ml* of class *formula* was created using the command **as.formula()**, which will be an argument of the model function, expressing the structure of the matrix X and Y (explanatory and explanatory variables). After creating the formulas, a variable voivodeship was also added – it will be used for stratified and factored sampling. The right side of the equation (RHS, right-hand side), typically an expression of the sum of variables (e.g. variable1 + variable2 + variable3. . .) was created with the **paste()** command with the argument collapse = '+' based on the variable headers obtained with the **names()** command. The left side of the equation (LHS, left-hand side), expressing a single dependent variable, was also obtained with the **names()** command. The right and left sides of the equation are connected with the **paste0()** command, which by definition does not contain separators but only "glues" elements without separating them.

```
# Data preparation - selection of variables
unempl<-read.csv("unemp2018.csv", header=TRUE, dec=",", sep=";")
unempl_ml<-unempl[, c(13:101)] # a subset of data dedicated to analysis
crds_ml<-unempl$crds

# imputation of missing data
unempl_ml[c(248, 286, 344, 359), 1]<-c(631188, 123659, 119171, 102422)
```

```
# attaching geographic coordinates to a data set
unempl_ml$x<-crds[,1]
unempl_ml$y<-crds[,2]

# formula of the equation
RHS<-paste(names(unempl_ml)[-90], collapse='+')
LHS<-names(unempl_ml)[90]

formula_ml<-as.formula(paste0(LHS, '~', RHS))
formula_ml

# Adding a voivodship variable to the data set
unempl_ml$voi<-unempl$region
```

The analysed set does not have many observations (380 territorial units), but there are many variables for this – for this reason, an attempt was made to match a random forest. A random forest is an often-used machine learning model that uses the concept of combining forecasts – so-called ensembling. The random forest generates M independent regression or decision trees (depending on whether a discrete or continuous variable is predicted). For each tree, the learning set (bootstrap) is retrieved with replacement, which is the sub-sample used in the estimation. In addition, only a random subset of variables (columns) is involved in the estimation of each random tree (in each iteration, other variables are drawn – at least one column, maximally all). Thanks to these two treatments (drawing of selected variables and observations), different models are created in the random forest in terms of the distribution of traits of the training sample (63.2% of unique observations are randomly counted in each iteration) and in terms of the number of variables, on the basis of which are generated rules. The tree is called a single model estimated on a fragment of the base set, the most important variables being the widest part of the trunk (at the ground). The random forest forecast is the average of the forecasts randomly drawn from each tree.

In the estimation of random forests, several parameters should be specified. The number of columns randomly selected to build a single tree is the most important hyperparameter of the model. Next to it, the number of trees, their depth,[10] and the size of the bootstrapped training sample should also be mentioned (it may be smaller than the original data). Because the tuning of hyper-random forest parameters is not the goal in this chapter, only the optimal number of random variables (columns) and the total number of trees have been selected here.

The estimation was made using the randomForest:: package by starting the random forest on the full row of columns (91) and for a large number of trees (500). The selected random seed, set with the command **set.seed**(), allows for obtaining of the reproducible (though not necessarily representative!) results.

```
library(randomForest) # loading the package
set.seed(1234)
model<-randomForest(formula_ml, data=unempl_ml, ntree=500)# estimation
model # results printout
```

```
Call:
randomForest(formula = formula_ml, data = unempl_ml, ntree = 500, mtry = 90)
            Type of random forest: regression
                  Number of trees: 500
No. of variables tried at each split: 91

    Mean of squared residuals: 0.14321
              % Var explained: 99.1
```

```
plot(model) #fit of the model, Figure 9.12
```

The *print* method of the *randomForest* object returns the basic model parameters and the error obtained on the *out-of-bootstrap* (OOB) trials for *ntree = 500* trees. After many simulations and further analysis of the results, it can be concluded that 500 trees is far too much, and after 50–100 trees, the OOB error is stabilising. This result is repeated in other drawings. It is not surprising – the set is simple and the variables are highly correlated – hence, the forest does not require too many iterations to obtain the optimal result. A higher number of trees causes it to overfit and deteriorate the error of the forecast. For further examples, a random forest with 50 trees was used.

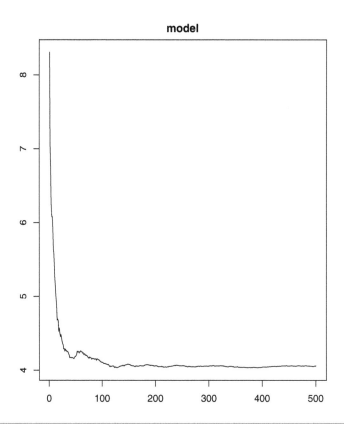

Figure 9.12 Out-of-bootstrap error graph for *i*-first trees

Source: Own study, graphics using the graphics:: package in R

Next, the sperrorest:: package will be used to assess the forecast error. The results of the cross-validation procedure using spatial sampling can easily be obtained from the **sperrorest**() function. The form of the model and the database are traditionally established in the formula and date arguments. The names of columns containing spatial coordinates needed for spatial sampling are defined using the text vector in the *coords* argument. Thanks to the arguments *model_fun* and *model_args*, one can validate almost any model available in packages in R; one can also manually define its hyperparameters (the latter using the list – **list**()). If the model does not have a predict function defined in the package, one can write the algorithm with its own function[11] thanks to the *pred_fun* and *pred_args* arguments. The function for spatial sampling and its arguments are set in the *smp_fun* and *smp_args* arguments – obviously, it concerns the sperrorest:: procedure, which returns a *sperrorest* structure that is understandable: the family of the **partition_***() function and the resamples function. The size of the training and test sample as well as the distribution of the explained variable can be corrected (also with the use of the sperrorest package function::) in the *train_fun, train_param, test_fun* and *test_param* arguments. In turn, by the *error_fun* argument, the algorithm is powered by a function for counting errors on the training and validation set – by default, this is an **err_default**() function, the results of which will be discussed further. By using the *error_rep* and *error_fold* arguments, one can decide whether to store the error values for each cross-validation replication in the results or for each replication folder as well. The function also offers control over the process of counting the importance ranking of variables (*importance, imp_variables, imp_permutations*). Because calculations can be time consuming and memory intensive, and individual iterations of cross-validation are independent of each other, the user also has control over memory management (*do_gc* – when a garbage collector is to be cleaned), parallel calculations that are started by default (additional options: *par_args* allow configuration parallel calculations, including the **makeCluster**() function) and the way to report progress in calculations (*progress* and *progress_out*).

Various functions from the sperrorest:: package will be used for sampling. The **partition_cv**() function was used first – to run a-spatial sampling to *k = 5-step* cross validation. The following code presents sampling and mapping. Each of the five graphs produced with the simple **plot**() command shows selected random coordinates of centroid poviats divided into training data (black) and validation data – test (red). Each point is used four times as a training and once as a validation, which is easy to check for selected (e.g. extreme) points.

```
# five maps with locations of sampled points
library(sperrorest)
resamp<-partition_cv(unempl_ml, nfold=5, repetition=1, seed1=1)
plot(resamp, unempl_ml, coords=c("x","y"))
```

Further, one can go to the cross-validation estimation. The following is an example of the configuration of the **sperrortest()** function, which allows for a partial (due to randomisation) replicability of results. In the random forest, the number of 100 repetitions in five-step validation was determined, and the random seed was set. It should be remembered that in bootstrap-based methods, the results are slightly different from each other.

```
res_rf_cv<-sperrorest(formula_ml, # function formula
                data=unempl_ml, # database
                model_fun=randomForest, # function name
                model_args=list(ntree=50, # number of trees
                    mtry = 90), # number of variables
                smp_fun=partition_cv, # sampling function
                smp_args=list(repetition=100, # number of repetitions
                nfold=5, # number of folds
                seed1=1234), # seed for sampling
                err_fun=err_default, # goodness-of-fit statistics
                error_rep=TRUE, # saving errors in each repetition
                progress='all') # printout of progress in the calculation

round(summary(res_rf_cv$error_rep), 3) # Result printout
```

	mean	sd	median	IQR
train_bias	-0.001	0.002	-0.001	0.002
train_stddev	0.186	0.006	0.186	0.009
train_rmse	0.186	0.006	0.186	0.009
train_mad	0.088	0.002	0.088	0.003
train_median	0.002	0.001	0.002	0.002
train_iqr	0.119	0.003	0.119	0.003
train_count	1520.000	0.000	1520.000	0.000
test_bias	-0.003	0.007	-0.002	0.010
test_stddev	0.417	0.022	0.415	0.024
test_rmse	0.417	0.022	0.415	0.024
test_mad	0.220	0.010	0.222	0.013
test_median	0.006	0.007	0.006	0.010
test_iqr	0.297	0.013	0.297	0.016
test_count	380.000	0.000	380.000	0.000

Goodness-of-fit statistics are returned by the **err_default()** function, which is the default value of *err_fun*. As one can see from the printout, this function returns the characteristics of the distributions of the following goodness-of-fit statistics separately for the training and test set:

- **train_bias** and **test_bias** – the average difference between the forecast and the observed value (rest – here called bias),
- **train_stddev** and **test_stddev** – standard deviation of residuals,
- **train_rmse** and **test_rmse** – RMSE forecast errors,
- **train_mad** and **test_mad** – MAD, the median of absolute deviations from the median,
- **train_median** and **test_median** – average absolute deviation from the median,
- **train_iqr** and **test_iqr** – interquartile range of residuals and
- **train_ count** and **test_count** – the number of samples taken together in the training and test sample.

For each of the seven statistics, the mean, median, standard deviation and interquartile range from all 100 iterations were reported. Thus, one can find out about both the distribution of in-sample/out-of-sample residuals and the distribution of individual characteristics between successive iterations of cross-validation. Focusing on the interpretation of the first column of the table: the average error allows determining if the model is biased – if the expected value of residuals would be much higher/lower than zero, it would mean that the model systematically overestimates/underestimates the modelled phenomenon. Both on the teaching and the test sample, there are no significant differences. The median, like the average, provides information about the central tendency but not so much about the expected value as about the central observation. If the median differs from the average, that is, the distribution is asymmetrical (different from normal), it may be useful information – for example, one can check whether the model overestimates or underestimates the results more often. The interquartile range (the difference between the third and the first quartile), the standard deviation and the mean absolute deviation make it possible to estimate how much the rest are dispersed around the measures of the central tendency. Finally, the RMSE (average square error of residuals) is a generally accepted measure of the forecast error in the validation of forecast quality. In the presented example, it can be seen for all sizes that both the scale of errors and their scattering is small.

Big irregularities in the distribution of errors are also not visible. In addition, as expected, the error on the test sample (estimation of the forecast error) is greater than on the training sample and has a higher standard deviation.

Simple sampling in small samples may result in unrepresentative results (they may then bias a forecast estimator error). In particular, it may happen that no observation will be drawn for certain areas. In order to check the representativeness of the random cross-check sampling, the distribution of selected poviats (NTS4) broken down by voivodeships (NTS) was examined and compared with the actual values appearing in this territorial division. In the case of the sample, lack of representativeness did not take place – each province is well represented, although the proportions are not accurate.

```
# A set with an example sample of data
df<-data.frame(prop=c(round(prop.table(table(unempl_ml$voi)), 2),
round(prop.table(table(unempl_ml$voi[resamp$'1'$'3'$train])), 2)),
labels=c(rep('whole sample (380 poviats)', 16), rep('Sample training set',
16)), voi=c(names(table(unempl_ml$voi)), names(table(unempl_ml$voi))))
head(df)
tail(df)
```

```
prop                          labels                    voi
1 0.08    whole    sample    (380  poviats)          Dolnośląskie
2 0.06    whole    sample    (380  poviats)   Kujawsko-Pomorskie
3 0.06    whole    sample    (380  poviats)            Lubelskie
4 0.04    whole    sample    (380  poviats)            Lubuskie
5 0.06    whole    sample    (380  poviats)              Łódzkie
6 0.06    whole    sample    (380  poviats)          Małopolskie
.........
   prop              labels                    voi
27 0.06 Sample training set            Pomorskie
28 0.09 Sample training set              Śląskie
29 0.04 Sample training set        Świętokrzyskie
30 0.05 Sample training set   Warmińsko-Mazurskie
31 0.10 Sample training set         Wielkopolskie
32 0.05 Sample training set    Zachodniopomorskie
```

```
# Figure - empirical vs. sampled frequency of poviats.
ggplot(data=df, aes(x=voi, y=prop, fill=labels)) +
  geom_col(position='dodge') +
  theme(axis.text.x=element_text(angle=45, hjust=1, size=7),
      legend.position='bottom',
      legend.title=element_blank()) +
    scale_y_continuous(labels=scales::percent, name='Frequency')
```

In order to ensure the same distribution of voivodeships (and poviats) in each subsample (fold), stratified sampling should be performed, involving the drawing of a certain number of observations from defined subgroups (here voivodeships). An exemplary result of the spatial distribution of points for a five-step validation is presented in Figure 9.13. The difference with the previous simple sampling is small – but the distribution of features by provinces in training trials is already identical with the entire sample. Small differences are simply

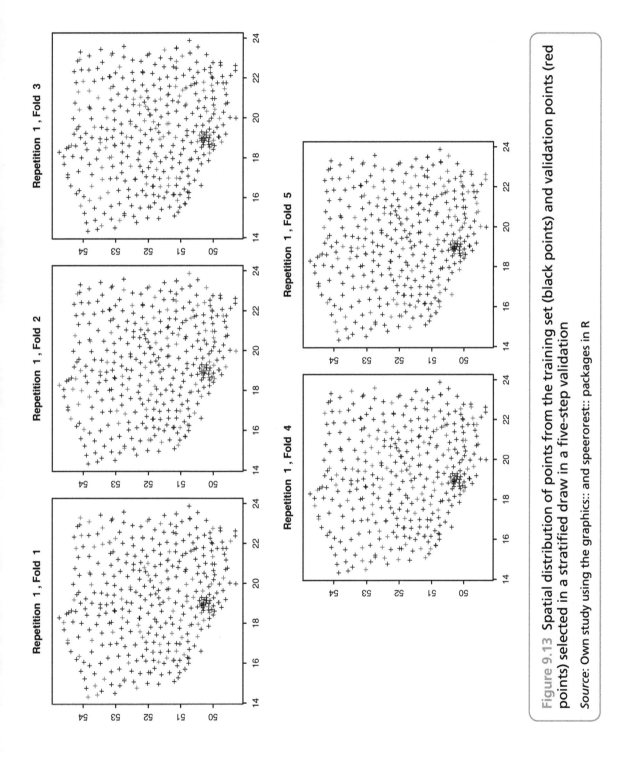

Figure 9.13 Spatial distribution of points from the training set (black points) and validation points (red points) selected in a stratified draw in a five-step validation

Source: Own study using the graphics:: and speerorest:: packages in R

due to the fact that 80% of the observations from each province and the whole sample cannot be drawn to the training set at the same time.

```
# Generation of sample folders for stratified sampling
resamp<-partition_cv_strat(unempl_ml,      nfold=5,      repetition=1,
seed1=1234, strat='voi')

plot(resamp, unempl_ml, coords=c("x","y")) # Figure 9.13

# Preparation of data for sampling and histogram
df<-data.frame(prop=c(round(prop.table(table(unempl_ml$voi)), 2),
round(prop.table(table(unempl_ml$voi[resamp$'1'$'3'$train])),2)),
labels=c(rep('whole sample (380 poviats)', 16), rep('Sample training set',
16)), voi=c(names(table(unempl_ml$voi)), names(table(unempl_ml$voi))))

# barplot comparing sampled and empirical frequencies ggplot(data=df,
aes(x=voi, y=prop, fill=labels)) +
  geom_col(position='dodge') +
  theme(axis.text.x=element_text(angle=45, hjust=1, size=10),
      legend.position='bottom', legend.title=element_blank()) +
      scale_y_continuous(labels=scales::percent, name='Frequency')
```

Subsequently, cross-validation was calculated for the sampling scheme in the accepted standard of a 100-fold five-step validation. The code is analogous to the previous one, except that this time, the function **partition_cv_strat()** was passed to the argument *smp_fun* and the arguments of the function (*smp_args*) were supplemented with the name of the start variable (*strat= 'voi'*).

```
set.seed(1234)
res_rf_strat<-sperrorest(formula_ml, # formula function
                data=unempl_ml, # database
                model_fun=randomForest, # function name
                model_args=list(ntree=50, # number of trees
                        mtry=90), # number of variables
                smp_fun = partition_cv_strat, # sampling function
                smp_args = list(repetition=100,
                        nfold=5,
                  seed1=1234,
                  strat='voi'),
                error_rep=TRUE, # saving errors in each repetition
                Progres='all') # printout of progress in the calculation

round(summary(res_rf_strat$error_rep), 3) # result printout
```

	mean	sd	median	IQR
train_bias	-0.001	0.002	-0.001	0.002
train_stddev	0.186	0.006	0.186	0.008
train_rmse	0.186	0.006	0.186	0.008
train_mad	0.088	0.002	0.088	0.002
train_median	0.002	0.001	0.002	0.002
train_iqr	0.119	0.003	0.119	0.003
train_count	1520.000	0.000	1520.000	0.000
test_bias	-0.001	0.008	-0.002	0.010
test_stddev	0.419	0.022	0.416	0.024
test_rmse	0.419	0.022	0.415	0.024
test_mad	0.220	0.009	0.220	0.009
test_median	0.007	0.007	0.006	0.011
test_iqr	0.297	0.012	0.296	0.014
test_count	380.000	0.000	380.000	0.000

In the previous results, the most important is the RMSE error on the training and test set. The estimation of the standard deviation of the forecast (RMSE error) has not changed significantly. The distribution of poviats in relation to voivodeships in training trials may not have been identical to the distribution in the whole sample, but it was similar enough to bias the estimator of the forecast error. It is worth checking the properties of other sampling methods.

The estimates for block-based sampling based on *k*-means are presented subsequently. In this sampling scheme, on the basis of the *k-means* algorithm, *k* spatial blocks of observations with an average number *n/k* are determined (where *k* is the number of folds and *n* is the number of observations). Their characteristic feature is the spatial blocking of observations. Then, *k – 1* folds create a training set, while *k*th, a spatially homogeneous cluster, is a test set. Figure 9.14 shows an example of the result of the sampling algorithm for *k = 5* folds. To carry out *k-means* clustering, the column names should be given, which contain the geographic coordinates of poviat centroids.

```
# resampling kmeans
resamp<-partition_kmeans(unempl_ml, nfold=5, coords=c('x', 'y'),
                        repetition=1, seed1=1234)
plot(resamp, unempl_ml, coords=c("x","y")) # Figure 9.14
```

The spatial pattern is very different from the previous ones. The algorithm has found five clusters, starting from the left: 1) the western wall, 2) Pomerania and Kujawy (central-northern areas), 3) Małopolska and Świętokrzyskie (south-central areas), 4) widened Podlasie (northeastern areas) and 5) extended Podkarpacie (southeastern areas). From a geometric perspective, Poland has been divided into three north-south divisions, two of which are further divided around half of the country's area.

Cross-validation using the **partition_kmeans()** function has been performed as shown subsequently. Interestingly, the average forecast errors (*train_RMSE* and *test_RMSE*) are at a similar level as in previous validations. This means that the a-spatial validation gives the correct results. The lower variance of the error (compared to previous validation types) may be due to the similar structure of the folds in subsequent iterations (*k*-means similarly allocates observations to clusters, so in practice, each of 100 repetitions draws similar cluster structures – therefore, trained models and the generated forecasts will not differ significantly between replicates of the folded cross-validation).

	mean	sd	median	IQR
train_bias	-0.001	0.002	-0.001	0.002
train_stddev	0.189	0.006	0.189	0.008
train_rmse	0.189	0.006	0.189	0.008
train_mad	0.088	0.002	0.088	0.003
train_median	0.002	0.001	0.002	0.002
train_iqr	0.120	0.003	0.120	0.003
train_count	1520.000	0.000	1520.000	0.000
test_bias	0.000	0.008	-0.002	0.010
test_stddev	0.416	0.011	0.417	0.016

	mean	sd	median	IQR
test_rmse	0.415	0.011	0.416	0.016
test_mad	0.229	0.009	0.227	0.013
test_median	0.008	0.007	0.008	0.010
test_iqr	0.308	0.013	0.306	0.018
test_count	380.000	0.000	380.000	0.000

Yet another option for sampling is factor sampling – based on drawing into a training set and validation at the level of a nominal variable – factor (e.g. province) – all observations with a given value of a factor variable are sent to a given set. For example, if the factor variable on which one draws is the voivodeship, poviats from other provinces will fall into the training and validation set (see an exemplary assignment to the folds in Figure 9.14b).

The results in Figure 9.14b show that in each iteration, three (and four) voivodeships were selected as validators, while the remaining 13 voivodeships were selected as learning ones. In the following case, 1) Lubuskie, Podlasie and Łódź; 2) Zachodniopomorskie, Pomorskie, Mazowieckie and Lubelskie; 3) Warmian-Masurian, Lower Silesian and Świętokrzyskie; 4) Opolskie, Śląskie and Podkarpackie and 5) Kujawsko-Pomorskie, Wielkopolskie and Malopolskie, Poland.

```
# factor sampling
resamp<-partition_factor_cv(unempl_ml, nfold=5, fac='voi', coords=c('x',
'y'), repetition=1, seed1=1234)
```

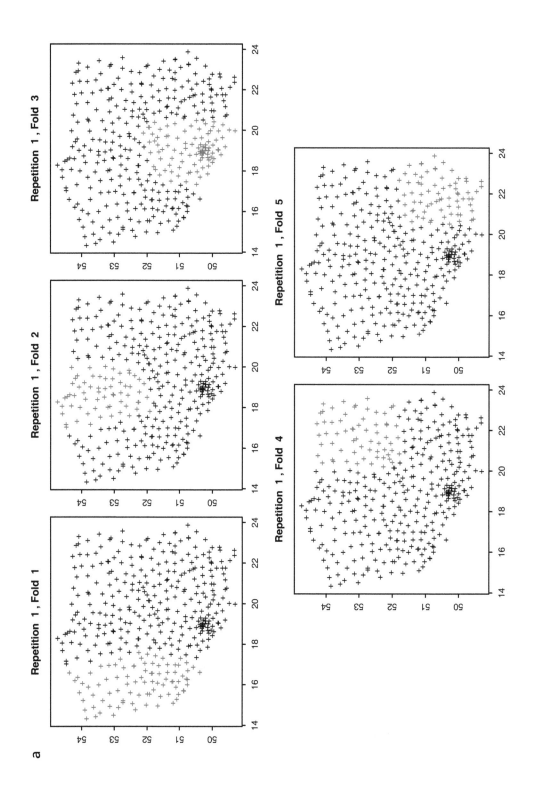

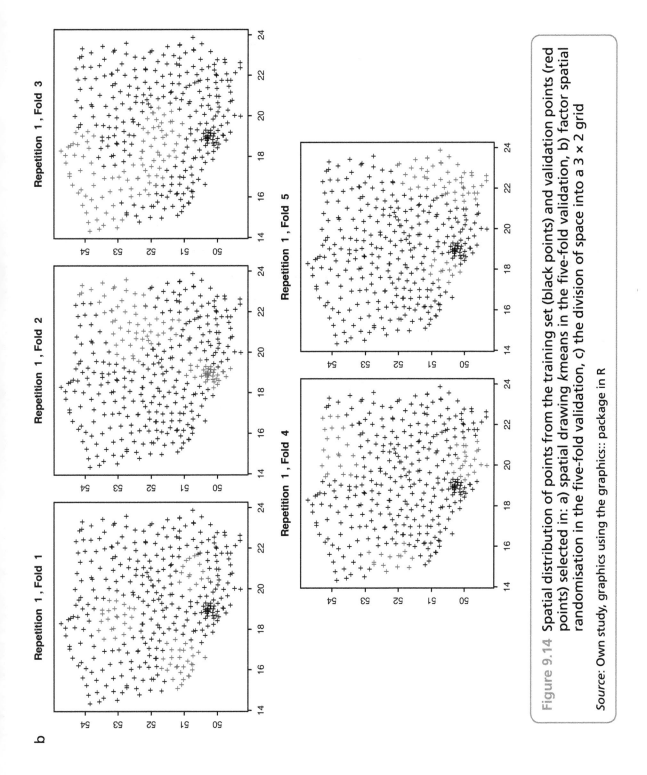

Figure 9.14 Spatial distribution of points from the training set (black points) and validation points (red points) selected in: a) spatial drawing *kmeans* in the five-fold validation, b) factor spatial randomisation in the five-fold validation, c) the division of space into a 3 × 2 grid

Source: Own study, graphics using the graphics:: package in R

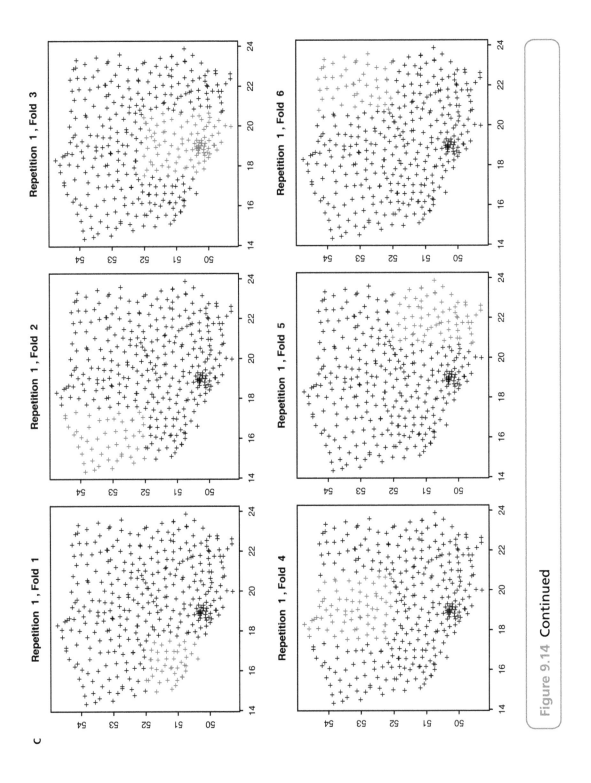

Figure 9.14 Continued

In analysing the results, it is again worth focusing on RMSE. The error obtained as well as its variance on the test set are larger than in previous experiments. What could be the reason for this? It is true that the algorithm draws compact groups of points but from different parts of the territory. The test group is therefore more difficult to forecast than in the types of sampling discussed previously. The closest neighbours in the training group have only border points from the test group. This may be a source of slightly worse results in the test (0.428 in sampling, versus 0.416, 0.417 and 0.419 in earlier examples).

	mean	sd	median	IQR
train_bias	-0.001	0.002	-0.001	0.002
train_stddev	0.186	0.007	0.186	0.010
train_rmse	0.186	0.007	0.186	0.010
train_mad	0.088	0.002	0.088	0.003
train_median	0.002	0.001	0.002	0.002
train_iqr	0.120	0.003	0.120	0.004
train_count	1520.000	0.000	1520.000	0.000
test_bias	0.001	0.011	0.001	0.012
test_stddev	0.429	0.032	0.425	0.026
test_rmse	0.428	0.032	0.424	0.026
test_mad	0.224	0.009	0.225	0.012
test_median	0.010	0.007	0.010	0.010
test_iqr	0.304	0.013	0.302	0.015
test_count	380.000	0.000	380.000	0.000

As in the last model, sampling from tiles will be checked using the **partition_tiles**() function. Because it is not possible to design a grid with an odd number of windows (to maintain the rule $k = 5$) – the closest division was chosen for a 3 × 2 grid, giving a total of $n = 6$ tiles. The division illustration is shown in Figure 9.14c. Similar to the division of k-means, the territory was divided into three north-south strips and each of them into two parts – northern and southern.

```
# resampling tiles
resamp<-partition_tiles(unempl_ml, nsplit=c(3,2),       coords=c('x', 'y'),
repetition=1)
plot(resamp, bezrob_ml, coords = c("x","y"), mfrow =c(3, 2)) # Figure 9.14c
```

It is worth noting that this is not a drawing but a deterministic division. Moreover, it is not useful in small collections. It is most often used in large datasets, with no clear clusters or clear administrative districts. The results of the estimation are presented subsequently. The RMSE error on the test set in this approach is the lowest compared with the previous results and has the lowest variance. However, it is difficult to compare the results with other types of sampling. Due to the small sample, only six samples (six folds) were generated – which look identical in each of the 100 iterations. Therefore, the error (RMSE) has a very low variance (0.008), and the average error RMSE (0.396) cannot be considered a representative estimator of the forecast error.

	mean	sd	median	IQR
train_bias	-0.001	0.001	-0.001	0.001
train_stddev	0.179	0.004	0.179	0.005
train_rmse	0.179	0.004	0.179	0.005
train_mad	0.086	0.001	0.086	0.001
train_median	0.002	0.001	0.002	0.001
train_iqr	0.117	0.001	0.117	0.002
train_count	3040.000	0.000	3040.000	0.000
test_bias	-0.004	0.004	-0.004	0.005
test_stddev	0.397	0.008	0.396	0.011
test_rmse	0.396	0.008	0.396	0.011
test_mad	0.212	0.007	0.212	0.009
test_median	0.007	0.006	0.008	0.009
test_iqr	0.286	0.009	0.285	0.013
test_count	380.000	0.000	380.000	0.000

It is time to compare the results of existing models. In the previous examples, five drawing models were implemented: 1) classic *k*-fold – simple drawing, 2) stratified *k*-fold, 3) division by *k*-means, 4) factor division and 5) division by tiles. The results obtained from these methods will be compared subsequently. Figure 9.15 presents the forecast error density functions for each model. Vertical dashed lines indicate distribution dominants. The lowest values come from the tiles model and the highest from the factor model. Figure 9.15 presents the functions of the density of RMSE error distributions with 100 iterations of all five types of data sampling described previously. On the basis of the graphical analysis of the distributions, it turns out that the a-spatial sampling is similar to the *k*-means and stratified sampling. Slightly different, a shifted higher error is noted for factorial sampling (the reasons for this result were described previously), while a much lower result was obtained for tiled sampling (tiles).

```
# Random forest with kmeans resampling
resamp<-partition_kmeans(unempl_ml, nfold=5, coords=c('x', 'y'),
                         repetition=1, seed1=1234)
set.seed(1234)
res_rf_km<-sperrorest(formula_ml, # formula function
                data=unempl_ml, # database
                model_fun=randomForest, # function name
                model_args=list(ntree = 50, # number of trees
                            mtry=90), # number of variables
                smp_fun=partition_kmeans,, # sampling function
                smp_args=list(repetition=100, # number of reps
                        nfold=5, # number of folds
                        seed1=1234), # seed for sampling
                    error_rep=TRUE, # saving errors in each rep
                    error_fold=TRUE, # saving errors in each fold
                    progress='all') # printout of progress
# Random forest with resampling factor
resamp<-partition_factor_cv(unempl_ml, nfold=5, fac= 'voi', coords=
c('x', 'y'), repetition=1, seed1=1234)
```

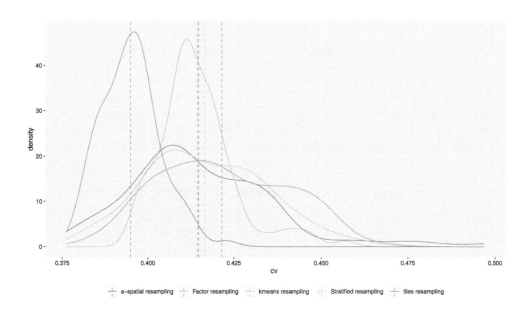

Figure 9.15 Graphs of forecast error density in validation *k* = 5 (100 repetitions) in different sampling schemes

Source: Own study, graphics using the ggplot2:: package in R

```
set.seed(1234)
res_rf_fa<-sperrorest(formula_ml, # formula function
                 data=unempl_ml, # database
                 model_fun=randomForest, # function name
                 model_args=list(ntree=50, # number of trees
                               mtry=90), # number of variables
                 smp_fun=partition_factor_cv, # sampling function
                 smp_args=list(repetition=100, # number of reps
                               fac='voi', # name of factor var
                               nfold=5, # number of folds
                               seed1=1234), # seed for sampling
                     error_rep=TRUE, # saving errors in each rep
                     error_fold=TRUE, # saving errors in each fold
                     progress='all') # printout of progress
# Random forest with resampling tiles
resamp<-partition_tiles(unempl_ml, nsplit=c(3, 2), coords=c('x', 'y'),
                 repetition=1)
set.seed(1234)
res_rf_ti<-sperrorest(formula_ml, # formula function
                 data=unempl_ml, # database
                 coords=c("x","y"), #coordinations of obs'
                 model_fun=randomForest, # function name
                 model_args=list(ntree=50, # number of trees
                               mtry=90), # number of variables
                 smp_fun=partition_tiles, # sampling function
                 smp_args=list(repetition=100, # number of reps
                               nsplit=c(3, 3),# number of splits
                               seed1=1234), # seed for sampling
                     error_rep=TRUE, # saving errors in each rep
                     error_fold=TRUE, # saving errors in each fold
                     progress='all') # printout of progress

# Preparing data with aggregate results.
df<-data.frame(cv=res_rf_cv$error_rep$test_rmse,
                 strat=res_rf_strat$error_rep$test_rmse, factor= res_rf_
                 fa$error_rep$test_rmse, kmeans= res_rf_km$error_rep$test_
                 rmse, tile= res_rf_ti$error_rep$test_rmse)
# Figure 9.15
ggplot(data=df) +
  stat_density(aes(x=cv, color='Classic CV'), geom='line') +
  geom_vline(aes(xintercept=mean(cv), color=' Classic CV'), linetype=
'dashed') +
  stat_density(aes(x=strat, color='stratified CV'), geom='line') +
  geom_vline(aes(xintercept=mean(strat), color=' stratified CV'), linetype
='dashed') +
  stat_density(aes(x=factor, color='factor CV'), geom='line') +
geom_vline(aes(xintercept=mean(factor), color=' factor CV'), linetype
='dashed') +
  stat_density(aes(x=kmeans, color='kmeans CV'), geom='line') +
  geom_vline(aes(xintercept=mean(kmeans), color='kmeans CV'), linetype =
'dashed') +
  stat_density(aes(x=tile, color='tiles CV'), geom='line') +
  geom_vline(aes(xintercept=mean(tile), color='tiles CV'), linetype =
'dashed') +
  theme(legend.position='bottom', legend.title=element_blank())
```

To sum up, spatial validation using the grid is the least useful method – it gives slightly different combinations of sets – in the presented example, it was difficult to obtain a reliable estimate of the forecast error. Perhaps if the sample were bigger and points instead of regular distributions would accumulate into irregular clusters – then cross-validation on a more accurate grid would be more accurate. Interestingly, a-spatial validation (classic k-fold, stratified k-fold) and spatial (in the block variant by k-means) give very similar estimators of errors, which is a premise that in the discussed example, a-spatial sampling allowed to obtain representative results. However, this is known only after comparing the results of the a-spatial and spatial procedures.

Notes

1 In simplified terms, stratified spatial randomising means that randomly drawn points will be systematically distributed over the whole area (this is a *de facto* drawing from the separated sub-areas), not completely random about the location. It is also possible to impose further conditions, for example, which, apart from spatial representativeness, the same number of observations for women and men should be selected.

2 Cross-validation is a way of estimating the quality of estimation from the out-of-sample class. It consists of the division of the whole sample into k subsamples. In each of the iterations, one part becomes the test data and the $k - 1$ parts become the training data. The estimation of the model takes place on the training data and its testing (validation) on the test data. More about bootstrapping in cross-validation can be found in, among others, Efron and Tibshirani (1997).

3 There are also other commands, such as in the spatialEco:: sample.annulus() or **sample.poly**() package, dedicated to drawing points from the ring or drawing polygons. However, their results are sometimes unreliable, hence the lack of presentation in the text.

4 For example, for 100 observations sorted ascending, one can create 10 blocks of 10 observations each. Observations 1:10 form block1, and observations 61:70 block7. Then 10 blocks are drawn and merged into them, for example, block 3, block7, block3, block8 . . .

5 For example, for 100 observations sorted ascending, one can create $100 - 10 + 1 = 91$ blocks, 10 for each observation. Observations 1:10 form block1, observations 2:11 block2, observations 3:13 block3 and so on. Then 10 blocks are drawn and "glued" to them, for example, block 23, block 47, block 35, block8 . . .

6 The simple confidence interval is the cutting off of the centiles of ordered values of the measure being measured. For example, the 90% confidence interval for the average lies in finding the 5th and 95th percentile among the average values determined in subsequent iterations. In the simplest terms, the standard error of the measure being measured is equal to the standard deviation from the bootstrapped values of the measure in subsequent iterations. There are studies expanding this approach, which include influence of spatial autocorrelation (Strand, 2017).

7 A spatially stratified drawing with replacement may give a misrepresentation of the representation in the case of an extremely uneven spatial allocation of points and clusters with very few observations. One can imagine a lonely point far away from all other points. It is likely to be a one-element cluster in the k-means procedure. Randomised stratification, assuming the drawing of, for example, $b = 10$ observations from each cluster, will make the outlier and singular value unjustifiably multiplied.

8 A similar alternative formula is the subsemble estimator, which also depends on the bootstrap of leading points and their surroundings and based on this estimation and prediction (Barbian & Assunção, 2017).

9 A hyperparameter is a parameter of the model whose value must be determined before the algorithm is enabled. Usually, there is no one best value – different values can perform best in each set, and one should discover it, among others, using cross validation. There are of course some good practices and "rules of thumb", though they should be treated as a starting point rather than for solving the problem. Since this book focuses on the specifics of spatial data modelling and not on the mechanics of machine learning, the aspects of hyper-parameter tuning are omitted.

10 The depth of the tree is understood as the maximum number of levels of the decision tree. The decision tree algorithm is referred to ad recursive partitioning – the dataset is divided into smaller and smaller parts according to the values of explanatory variables. Each division divides the set into two parts. There is one set at zero level, at level 1 – maximum 2 subsets (tree branches), on level 2 – maximum 4 subsets (branches), and so on. Limiting the depth of the tree simplifies the model (fewer divisions), which limits its adjustment to the training data and reduces the risk of overtraining.

11 Instructions on how to create a user predict function can be found here: https://cran.r- project.org/web/packages/ sperrorest/vignettes/custom-pred-and-model-functions.html

## Chapter 10

Spatial big data

Piotr Wójcik

The term "big data" refers to large files containing different data, which can also be in unstructured form (e.g. text data). Their scale causes statistical processing and analysis using traditional computers and statistical tools to be difficult and sometimes impossible due to the limited size of disk space, operational memory or the lack of appropriate statistical tools. The added value of analysing this type of data is the ability to discover surprising, previously unknown relationships that allow one to gain new, often profitable knowledge.

The concept of a big dataset is relative and depends on the industry. Most often it refers to the data size measured in terabytes (1 TB = 1024 GB) or even petabytes (1 PB = 1024 TB), although it can also refer to data measured in megabytes (Bennett, 2017).

Big data is a relatively new term – it appeared at the beginning of the 21st century with the development of the Internet, with the increase in computing power and disk spaces, especially those available online in the so-called "clouds" (*cloud computing*). The cost of storing huge resources in digital format has dropped significantly, and computers have become ubiquitous. It resulted in all kinds of information being saved in digital form (digitised), stored and analysed using appropriate software.

Usually, the appearance of big data is associated with the META Group (2001) report prepared by analyst Doug Laney (2001), describing the so-called "3V model" (from: *volume, velocity, variety*). *Volume* refers to huge data sizes. However, the large volume of data alone is not enough to define it as big data. One can easily imagine, for example, a digital archive of text, audio or video documents that stores petabytes of data, but it is not used to carry out any advanced operations on them, except for collection, cataloguing and accessing resources. *Velocity* refers to high variability and speed of producing new data, which is then delivered to analytical tools. Often, this data should be handled and processed equally quickly (in real time) (e.g. detection of fraud related to the use of payment cards by bank customers, safe control of an autonomous car, and so on). *Variety*, on the other hand, refers to a wide variety of data that can be delivered in an array of formats – from classic structured data in numerical form to unstructured text documents or data in the form of audio or video files. This data is often characterised by an excess of information – it must be pre-filtered and processed – adapted for analysis in typical analytical solutions.

A few years later, the model was completed with a fourth component, whose name also begins with the letter V – *veracity* – and henceforth it is called the "4V model". Veracity or value refers to a potentially large value of knowledge that a company can obtain through big data analysis. Extracting this knowledge from the collected data requires processing of large data volumes, searching for non-obvious relations, often going beyond standard statistical analyses.

Currently, many large datasets have a spatial character – they contain geolocation data of individual objects. The purpose of this chapter is to discuss spatial big data, as well as the associated challenges and tools available in the R environment. The chapter structure is as follows: first, examples of using big data in data analysis (10.1) are discussed; second, spatial big data is described, briefly recalling the spatial data types and indicating the challenges associated with the analysis of large datasets of this type (10.2). Next, extensive applications of the selected functions of the sf:: package for spatial and point data analysis (10.3) and the use of the dplyr:: (10.4) package functions for effective analysis of large datasets, including spatial data frames of the *sf* class, are discussed. Finally, an exemplary analysis of a large raster dataset is presented (10.5).

10.1 Examples of big data applications

An obvious example of big data analysis is the collection and processing of information by companies providing Internet technologies. A list of websites visited on the Internet, phrases typed in a web browser or in a free e-mail service (even those finally deleted before sending), time spent on individual pages, clicked links, products bought online, a list of friends or content published on social networks are a mine of knowledge about every internet user. Currently, based on the history of the websites visited by the user and information made available on social networks, it is possible to predict gender, marital status, education, job position, religion, sexual orientation or various preferences and tendencies with a high degree of precision – for example, a tendency to change jobs or terminate the current relationship in the near future can be predicted. One can also predict many behaviours of Internet users and anticipate needs (for example, shopping) – even those one is not yet aware of. Companies that collect this type of information often know more about Internet users than their relatives do. An often-cited, anecdotal example is the Target Corporation, whose analysts built an algorithm that predicts the probability of pregnancy for Internet users based on the history of online shopping – they identified products purchased by pregnant women. The task of the model was to identify pregnant women early enough to hit a marketing message to women in the third trimester of pregnancy. The message went to a teenager whose nervous father complained to the company for demoralising his daughter. Some time later, he called with apologies, because it turned out that his daughter was in fact pregnant, and the algorithm used detected this fact faster than her own father (Fry, 2018).

Using analytics of huge datasets makes life of today's Internet users easier. Thanks to the extensive recommendation systems, one can often receive information about new books, movies or songs, perfectly matched to individual interests and needs. However, be aware that this is done at the expense of a large limitation of privacy. Many internet users are willing to incur this cost. The data analysed by big data algorithms is collected not only online but also by means of GPS receivers located in smartphones, smartwatches or other devices that measure the daily activity of their users. These devices are called "wearables" due to the fact that they are usually worn by a consumer. Never before has data on various human activities been generated at such a high rate. Within a minute, hundreds of thousands of emails are sent around the world, Facebook users create millions of instances of new content, 500 hours of video material appears on Youtube, and Dropbox is enriched by almost a million new files (Wassén, 2017). Every month, petabytes of data are sent via smartphones (source: Ericsson Mobility Report 2013, quoting for Dapp & Heine, 2014).

Therefore, typical applications of big data analytics include predicting shopping preferences, creating personalised ads, identifying potential fraud attempts among billions of transactions made around the world using credit cards.

In an interesting way, big data is used in sport. Each player's movement is recorded and analysed for optimal performance. Athletes swallow special sensors that constantly monitor the state of their organism and the biological processes taking place in it. The pioneer of this type of research was Formula 1 racing, where not only the staff of mechanics but also IT specialists and statisticians watch over the car drivers. The situation is similar in many football teams, where the coaching staff use professional support from data collection and analytics. Technological progress in the world means a further increase in the scale of big data. There are already talks about smart apartments, household appliances (such as the smart fridge) and even smart clothes, equipped with sensors that collect a variety of information and communicate with other devices without the participation of their owner.

10.2 Spatial big data

The increasing amount of collected information is de facto spatial data – containing geolocation information. It includes, for example, data on the route of return from work or walk (collected using a GPS transmitter on a smartphone or smartwatch) or the location of the most visited stores or service outlets (card payments). But also information posted on Twitter, queries entered in the search box, entries or photos posted on social networks very often contain geolocation data. Hence the appearance of the concept of spatial big data (Eldawy & Mokbel, 2016). In addition to data collected by electronic devices used by individuals, the concept of spatial big data also includes information collected by satellites orbiting the Earth, including climatic and weather data.

10.2.1 Spatial data types

Spatial big data can have a variety of forms – vector (points, lines, polygons or any combinations thereof) or raster images (Cugler, Oliver, Evans, Shekhar, & Medeiros, 2013) – and also can be stored in different formats – shapefile, GeoJSON or KML files.[1]

An example of large spatial data in the raster form are satellite images of the Earth. About 35% of the satellites currently orbiting the Earth are used for commercial purposes. Raster data are used in meteorology to analyse the location and composition of clouds; study transparency, humidity and air density; detect and analyse the scale and composition of pollutants in the atmosphere, detect areas of different temperature or measure air movements over long distances (Wikipedia, 2019a).

Numerous economic applications of data and analyses of this kind concern agriculture. Data from satellite images can be used to better understand how crop yields are influenced by factors observable from space, such as weather, sun exposure, air quality or pest activity. This allows one to determine the optimal conditions for running an agricultural business (Marr, 2017). Raster data from satellites can also be used to classify soil types and register and assess damage caused by natural disasters, which is used, for example, in insurance to verify farmers' declarations of crop losses. Another application is collecting data on areas that are difficult to access; obtaining vegetation parameters, such as tree height, crown diameter and afforestation density; biomass determination; forest boundaries; measurement of snow-covered and ice-covered areas; monitoring of glaciers, wetlands, waste heaps and dumps; hydrographic measurements up to a depth of 70 m or detecting the remains of archaeological objects below the earth's surface.[2]

In large cities, this type of data can in turn be used to analyse the intensity of vehicle traffic, create a numerical model of land cover or create three-dimensional models of cities used for spatial planning or analysis of the spread of noise and pollution. Another application of raster data from satellite images is the design of routes of roads, railways and pipelines; the recording of high voltage lines and detection of their collisions with tree crowns and creating a numerical model of land cover for forest areas used for road planning, drainage systems and so on.

Big spatial vector data (BSVD) can be represented in various forms – points, lines or areas. It covers different categories (see Yao & Li, 2018). One of them is location-based data; this is geographic data that contains spatial location and time identification. This may be data on the location of companies or individuals, which, in addition, does not have to be constant over time. Location data is most often collected by positioning systems (e.g. GPS) used by smartphones, smartwatches or other devices carried by users or installed in vehicles. A separate category is data from social media – Internet data with an additional attribute of spatial location. This includes data posted by users on social media (e.g. Twitter, Facebook et al.) or typed in web browsers. Data of this kind plays a large role in monitoring public opinion, natural disasters or the spread of disease and epidemics.

10.2.2 Challenges related to the use of spatial big data

10.2.2.1 Processing of large datasets

The challenges for large datasets are their storage and efficient processing. In this context, it is often referred to as distributed (parallel) processing and the storage of data in the cloud (*cloud computing*). The concept of cloud computing means a service provided by an external operator via the Internet – a scalable platform with adequate IT infrastructure (servers with software). It also means the possibility of spreading calculations on many computers and performing many partial calculations at the same time. The cloud computing user does not have to own the infrastructure (computing servers) or software, and the service charge is usually related to the intensity of its use.

The first idea to solve the problem of processing large datasets, named MapReduce, was presented in 2004 by two Google employees – Dean and Ghemawat (2004). Google collects an enormous amount of data in connection with the operation of the popular Internet search engine. Along with the dynamic development of the Internet, the complexity of the problem grew non-linearly.

The basic assumption of the MapReduce approach is to divide the computational problem into two main stages called *mapping* and *reducing*. They are responsible for performing calculations in parallel on many computers (servers) at the same time, handling possible machine failures and communication between them, in order to effectively use the entire available network. The computational problem is automatically divided into many smaller parts, which allows it to be performed in a distributed environment. Storing huge data volumes also requires a distributed file system. For this reason, the MapReduce system was very closely related to the Google file system (GFS) – a distributed file system offered by Google. Tasks written in the MapReduce model retrieved input data from GFS and stored the results there.

The original MapReduce system is a closely guarded Google secret. However, alternative open source implementations of the algorithm have been created that allow one to take advantage of the MapReduce approach. The most popular of them, the free implementation of the distributed model of MapReduce calculation, is the Hadoop system (created by Yahoo!) inspired by the original MapReduce. As in the Google solution, Hadoop uses the distributed Hadoop file system (HDFS).

An alternative to the Hadoop system, focused more on analytics than just efficient processing of huge data, is the Apache Spark system using Spark's resilient distributed datasets (RDDs). Thanks to the ability to process data in memory, Spark is faster than Hadoop MapReduce. There are also specialised solutions focused on the storage and processing of spatial data, both based on the Hadoop system (HadoopGIS, Spatial-Hadoop) as well as on the Spark environment (SpatialSpark, GeoSpark, Simba, LocationSpark, Magellan).

A detailed discussion of these tools goes beyond the scope of this book – for more information, the interested reader is suggested to read the article by Pandey, Kipf, Neumann, and Kemper (2018).

10.2.2.2 Mapping and reduction

The first step of the calculation (*map*) usually consists of two parts. In the first place, the computational problem (input dataset) is divided into many smaller problems (sets). Proper mapping involves downloading data from the source and projection – defining pairs consisting of *key* and *value* (key, value) on their basis. The mapping step is strictly parallel – each row of data is treated as a separate piece of information that does not affect the reading or processing of the others.

The next step is to organise the mapped data by sorting and shuffling it – a list of values is created for each unique value of the key. This is usually the most expensive phase of the MapReduce algorithm (Busłowska & Juźwiuk, 2014).

The last step is to reduce, which means to combine the data and prepare the final result for each of the partial sets. The results of reduction are saved to files, with each fragment of partial data generating one resulting file. Importantly, the reduction step cannot be started before the mapping process has finished on all parts of the set.

Before starting calculations, one has to specify the number of mapping objects (M) and the number of reducing objects (R). Processed data is divided into M parts. Each of them is processed on one of the computing servers.[3] The same procedure applies to the reduction task. Data on each server is divided into R parts and processed in a distributed form. Ultimately, M times R resulting files are created, which, after aggregation, gives the final result of the calculations.

A common illustration of the operation of a distributed computing system is the example of counting instances of individual words in text.[4] The starting point is a text file. It is divided into M parts, which are then transferred to the mapping function. The mapping process means generating a pair of <word, 1> for each word – each word (*key*) is assigned a value (*value*) of 1, its single occurrence in the text. At the shuffle stage, a list of 1 value is created for each unique word, which is then passed to the reduction task. The length of the list received is the number of instances of the word in the document.

The condition for applying a distributed approach to data analysis is the ability to perform independent calculations on fragmented, partial datasets. This often turns out to be impossible in the case of spatial data – for example, in the case of estimating the parameters of the spatial regression model, which is performed in an iterative manner, where the next iteration uses the previous result (Cugler et al., 2013).

10.2.2.3 Spatial data indexing

As in the case of non-spatial big datasets, the key element determining the effective search of a spatial object database is the use of indexing. The most common spatial indexing algorithm is called R-tree (see Leutenegger, Lopez, & Edgington, 1997). The essence of the R-tree indexing algorithm is to approximate each geometry through the smallest rectangle containing this geometry (minimum bounding rectangle, MBR), as well as grouping objects close to each other and building a hierarchical relationship model in the form of a tree. At the lowest level of the tree (leaves), each MBR rectangle describes a single spatial object. At higher levels of aggregation, MBR rectangles include more and more objects relatively close together in space (adjacent to each other). The indexing of spatial data allows one to significantly accelerate the operations of searching for the nearest neighbours. The R-tree algorithm has been used in the functions of the sf:: package, which will be discussed in the next part of this chapter.

10.3 The sd:: package – simple features

The sp:: package, with additional tools from the rgeos:: and rgdal:: packages, discussed in earlier chapters, have been standard tools for analysing spatial data in R. Recently there appeared an alternative, the sf:: package, which allows one to store spatial data in a form similar to datasets that do not contain spatial information. This facilitates the analysis of spatial data and allows the use of modern effective data processing and visualisation tools consistent with the tidyverse:: approach (dplyr::, ggplot2::), also for spatial data. The functions of the dplyr:: package allow for the efficient processing of large datasets. Their counterparts are also used in popular environments for the analysis of large datasets.[5] The sf:: package combines functionalities previously available in three different packages sp::, rgdal:: and rgeos:: in one place. Therefore, it quickly gained popularity among people dealing with spatial analysis in R, becoming a new standard for this type of analysis.

The package name sf:: is an abbreviation of simple features – the name of the standard for vector data created jointly by the Open Geospatial Consortium (OGC) and the International Organization for Standardization (ISO). This standard (ISO 19125–1: 2004) defines the method of representing real objects in computer databases – mainly two-dimensional vector data – with particular emphasis on their geometry. Objects can have descriptive attributes (numeric, textual or logical type – for example, the height of a point above sea level, the size of a region's population, city name, information or road represented by a highway etc.). In addition, each object has an attribute associated with its geometry. The geometry of objects is defined in a two-dimensional coordinate system and uses linear interpolation of the distance between coordinates. Numerous types of geometry of spatial objects are possible: point, line, polygon, many points, many lines and so on.

The simple features standard is commonly used in open spatial databases (such as PostGIS) or commercial GIS tools (e.g. ESRI ArcGIS). In turn, the GeoJSON standard is a simplified version of the simple features standard.

10.3.1 *sf* class – a special data frame

In the sf:: package, a new class of spatial objects is defined, with the same name as the package, that is, *sf*. The *sf* class allows for storing spatial objects of all types. SFs are *de facto* data frames (*data.frame, tibble*) containing spatial data. This allows, as mentioned earlier, one to easily and efficiently process this type of data using the functionalities of the dplyr:: package. It is also possible to visualise spatial data stored in this way using the functions of the ggplot2:: package.

Rows of this special data frame contain consecutive spatial objects (*features*), while columns describe their attributes – as in the standard data frame. However, the *sf* class objects contain a special column called *geometry*, which defines the geometry of individual objects. This column is an object of the *sfc* class, which in turn is a list of *sfg* class objects representing the geometry (shapes) of individual objects (data rows). The *sfg* class contains the same basic information as the individual fields (slots) of the *Spatial* class objects from the sp:: package: CRS, geographical coordinates and geometry type (object shape). Of the 17 defined types of geometry, the most frequently used ones are:[6]

- POINT,
- MULTIPOINT – many points,
- LINESTRING – line: a sequence of two or more points connected by straight lines,
- MULTILINESTRING – many lines,
- POLYGON – a closed figure consisting of sections that can have empty areas in the middle,
- MULTIPOLYGON – many polygons,
- GEOMETRYCOLLECTION – any combination of the previous types.

Theoretically, *sf* class objects can contain more than one column defining the geometry of objects, but there is usually only one such column. More details about the sf:: package can be found in Pebesma (2018). Subsequently one can find examples of applications of functions from the sf:: package on the data representing POLYGON and POINT geometries.

First of all, it is necessary to load the sf:: package with **library**() command. To simplify the identification of functions from this package, their names start with the prefix st_, which makes it easier to autocomplete their names in RStudio.

```
library(sf)
```

10.3.2 Data with POLYGON geometry

The first step is to import data from the shapefile file covering the poviats of Poland. The **st_read**() function is used to load data in the sf:: package, which automatically recognises the type of the file (the format of the data being read) – one can display a list of recognised formats by running the **st_drivers**() command.

```
poviats<-st_read("data/poviats.shp")
```

```
driver 'ESRI Shapefile'
## Simple feature collection with 380 features and 29 fields
## geometry type:  MULTIPOLYGON
## dimension:       XY
## bbox:            xmin: 171677.6  ymin: 133223.7  xmax: 861895.7
ymax: 775019.1
## epsg (SRID):     NA
## proj4string:     +proj=tmerc +lat_0=0 +lon_0=19 +k=0.9993 +x_0=500000
+y_0=-5300000 +ellps=GRS80 +units=m +no_defs
```

When loading the data, basic information about it is displayed. The **st_read**() function recognises the data type as an ESRI Shapefile and uses the appropriate engine to import it. The data concerns 380 objects (features) and contains 29 attributes (fields). The type of geometry is MULTIPOLYGON, the data is two-dimensional (*dimension=XY*) and the boundary box coordinates and the cartographic projection (*proj4string*) of the geographical coordinates are printed. Analogously, it is possible to save the *sf* object by using the **st_write**() function. Checking the class and structure of the loaded object:

```
class(poviats)
## [1] "sf"            "data.frame

str(poviats)
## Classes 'sf' and 'data.frame':  380 obs. of  30 variables:
##  $ iip_spat: Factor w/ 1 level "PL.PZGIK.200": 1 1 1 1 1 1 1 1 1 1...
##  $ iip_identy: Factor w/ 380 levels "00d02105-cbd3-45ef-837b-
ea0fa86bf60c",..: 12 306 25 169 56 62 370 151 29 99...
##  $ jpt_sjr_ko: Factor w/ 1 level "POW": 1 1 1 1 1 1 1 1 1 1...
##  $ jpt_kod_je: Factor w/ 380 levels "0201","0202",..: 26 28 6 4 14 1 27
..
##  $ jpt_nazwa_: Factor w/ 370 levels "aleksandrowski",..: 361 130 90 70
193 18 89 106 142 324...
## $ jpt_nazw01: Factor w/ 0 levels: NA NA NA NA NA NA NA NA NA NA...
## $ jpt_organ_: Factor w/ 0 levels: NA NA NA NA NA NA NA NA NA NA...
## $ jpt_orga01: Factor w/ 1 level "NZN": 1 1 1 1 1 1 1 1 1 1...
##  $ jpt_jor_id: num  0 0 0 0 0 0 0 0 0 0...
##  $ wazny_do  : Date, format: NA NA...
##  $ jpt_wazna_: Factor w/ 2 levels "1   NZN",..: 2 2 2 2 2 2 2 2 2 2
. . .
## $ wersja_do : Date, format: NA NA...
## $ jpt_powier: num  57581 5629 62714 73827 104931...
##  $ jpt_kj_iip: Factor w/ 1 level "0000000000EGIB": 1 1 1 1 1 1 1 1 1 1
. . .
##  $ jpt_kj_i01: Factor w/ 380 levels "0201","0202",..: 26 28 6 4 14 1 27
. . .
##  $ jpt_kj_i02: Factor w/ 0 levels: NA NA NA NA NA NA NA NA NA NA...
##  $ jpt_code_01: Factor w/ 0 levels: NA NA NA NA NA NA NA NA NA NA...
##  $ id_bufora_: num  0 0 0 0 0 0 0 0 0 0...
##  $ id_bufor01: num  13888 13888 13888 13888 13888...
```

```
##   $ id_technic: num  0 0 0 0 0 0 0 0 0 0. . .
##   $ jpt_opis  : Factor w/ 380 levels "1296647","828986",..: 330 331 337
336 347 357 332 340 342 1. . .
##   $ jpt_sps_ko: Factor w/ 1 level "UZG": 1 1 1 1 1 1 1 1 1 1. . .
##   $ gra_ids   : Factor w/ 0 levels: NA NA NA NA NA NA NA NA NA NA. . .
##   $ status_obi: Factor w/ 1 level "VALID": 1 1 1 1 1 1 1 1 1 1. . .
##   $ opis_bledu: Factor w/ 0 levels: NA NA NA NA NA NA NA NA NA NA. . .
##   $ typ_bledu : Factor w/ 0 levels: NA NA NA NA NA NA NA NA NA NA. . .
##   $ wersja_od : Factor w/ 12 levels "2012-09-26","2013-07-09",..: 1 1 10
1 1 1 10 10 10 10. . .
##   $ wazny_od  : Factor w/ 3 levels "2012-09-26","2013-01-01",..: 1 1 1 1
1 1 1 1 1. . .
##   $ iip_wersja: Factor w/ 31 levels "2012-09-27T07:36:28+02:00",..: 1 1
29 1 1 1 29 29 29 29. . .
##   $ geometry  :sfc_MULTIPOLYGON of length 380; first list element: List of 1
##   ..$ :List of 1
##   .... $ : num [1:3361, 1:2] 269084 269164 269176 269160 269157. . .
##   ..- attr(*, "class")= chr"XY" "MULTIPOLYGON" "sfg"
##   - attr(*, "sf_column")= chr "geometry"
##   - attr(*, "agr")= Factor w/ 3 levels "constant","aggregate",..: NA NA
NA NA NA NA NA NA NA NA. . .
##   ..- attr(*, "names")= chr"iip_spat" "iip_identy" "jpt_sjr_ko"
"jpt_kod_je" . . .
```

The whole object has a *sf* class, while the last geometry column has the *sfc_MULTIPOLYGON* class, which is a subclass of the *sfc* class and consists of 380 objects, each of which has the following classes: *XY, MULTIPOLYGON* and *sfg*.

In the case of spatial data, it is important to determine whether and which map projection of geographical coordinates has been applied. It is a mathematically determined way of rescaling the surface of the globe in order to present it on a plane. In general, it is worth using scaled data in the analyses, which allows measuring distances in units easier to understand – meters or kilometres, not angular seconds.

The function **st_is_longlat**() allows checking whether the geographical coordinates have been projected (the result TRUE means not), and the function **st_crs**() allows checking which mapping was applied.

```
st_is_longlat(poviats)
## [1] FALSE

st_crs(poviats)
## Coordinate Reference System:
##   No EPSG code
##   proj4string: "+proj=tmerc +lat_0=0 +lon_0=19 +k=0.9993 +x_0=500000
+y_0=-5300000 +ellps=GRS80 +units=m +no_defs"
```

The change of the cartographic mapping used is done by the **st_transform**() function, which will be used in the examples later in the chapter. The projection method is defined using the *proj4string* object or the EPSG code.

An important concept associated with spatial geometries is their correctness. This refers generally to two-dimensional geometry ([MULTI]POLYGON). Intuitively, the geometry of a polygon (POLYGON) is valid if its boundaries consist of consecutively connecting but not overlapping and non-crossing sections. In turn, the MULTIPOLYGON object is correct if it consists of correct polygons. Checking the correctness of the geometry of the analysed objects can be time consuming (especially for very complex geometries), but it saves problems when they are later visualised or analysed. In the sf:: package, validation is enabled by the **st_is_valid**() function.

```
table(st_is_valid(poviats))
## FALSE   TRUE
##     1    379
which_incorrect<-which(!st_is_valid(poviats))
```

```
poviats[which_incorrect,]
## Simple feature collection with 1 feature and 29 fields
## geometry type:      MULTIPOLYGON
## dimension:          XY
##  bbox: xmin: 467876.4    ymin: 728503.4    xmax: 473502.9    ymax:
733423.5
##  epsg (SRID): NA
##  proj4string: +proj=tmerc +lat_0=0    +lon_0=19    +k=0.9993
+x_0=500000
+y_0=-5300000 +ellps=GRS80 +units=m +no_defs
##        iip_spat iip_identy jpt_sjr_ko
## 378  PL.PZGIK.200 60ce8df5-82b9-47e8-9193-92051c398480 POW
##      jpt_kod_je jpt_nazwa_ jpt_nazw01 jpt_organ_ jpt_orga01 jpt_jor_id
## 378      2264    Sopot     <NA>      <NA>      NZN          0
##      valid_to  jpt_valid_ version to jpt surface     jpt_kj_iip
jpt_kj_i01
## 378    <NA> 6     NZN     <NA>        0 0000000000EGIB    2264
##      jpt_kj_i02 jpt_code_01 id_bufor_ id_bufor01 id_technic jpt_opis
## 378    <NA>     <NA>       0      13884        0    829215
##      jpt_sps_ko  gra_ids status_obi description of error type of error
version_from
## 378    UZG    <NA>    VALID       <NA>      <NA> 2012-09-26
##     valid from                  iip_version
geometry
## 378 2012-09-26 2012-09-27T09:33:51+02:00 MULTIPOLYGON (((467876.4 73. . .
```

One of the poviats (city of Sopot) has an incorrect geometry. Its automatic correction can be done with the **st_make_valid**() function from the lwgeom:: package, which is an extension of the sf:: package – its description can be found in (Pebesma, 2019b). Why was the geometry for Sopot incorrect? After correction, the geometry of this poviat is of the GEOMETRYCOLLECTION type. This means that it consists of more than one geometry, and they have different types.

```
poviats2<-lwgeom::st_make_valid(poviats)

class(st_geometry(poviats2[which_incorrect,]))
## [1] "sfc_GEOMETRYCOLLECTION" "sfc"
```

Checking the classes of elements included in the GEOMETRYCOLLECTION geometry shows the source of the problem.

```
lapply(st_geometry(poviats2[which_incorrect,])[[1]], class)
## [[1]]
## [1] "XY"      "POLYGON" "sfg"
##
## [[2]]
## [1] "XY"         "LINESTRING" "sfg"
st_geometry(poviats2[which_incorrect,])[[1]][[2]]
## LINESTRING (470402.8 732660.7, 470402.8 732660.8)
```

In addition to the POLYGON, the original geometry of Sopot included a separate line (LINESTRING) which was not connected to the rest of the geometry or coincided with another section defining the geometry. Therefore, it is a good idea to remove this unnecessary line from the geometry for Sopot and leave only a POLYGON object in it. To select objects of the selected type(s) from a complex geometry (collection of objects), one can use the function **st_collection_extract**().

```
poviats2[which_incorrect, "geometry"]<-st_collection_extract(st_geometry
(poviats2[which_incorrect,]), type= c("POLYGON"), warn = FALSE) table(st_
is_valid(poviats2))
## TRUE
##  380
```

Despite the growing popularity of the sf:: package, many analyses using spatial data can be carried out only on *Spatial* objects defined in the sp:: package. Conversion between *Spatial* and *sf* types is very easy. Conversion of an *sf* object into a *Spatial* object can be done with the **as_Spatial()** or **as()** function from the sf:: package with the second argument equal to "Spatial".

```
poviats_sp<-as(poviats2, "Spatial") class(poviats_sp)
## [1] "SpatialPolygonsDataFrame"
## attr(,"package")
## [1] "sp"
```

The reverse conversion can be done using the **st_as_sf()** function.

```
poviats_sf<-st_as_sf(poviats_sp)
class(poviats_sf)
## [1] "sf"        "data.frame"
```

Displaying the *sf* class object in the R console will show its first 10 rows and all columns by default. The display of more lines is possible using the **print()** function with the argument *n=*. The output can be also restricted to selected columns from the data frame.

```
print(poviats2[, c("jpt_kod_je", "jpt_nazwa_")], n=5)
## Simple feature collection with 380 features and 2 fields
## geometry type:  GEOMETRY
## dimension:      XY
## bbox:           xmin: 171677.6 ymin: 133223.7 xmax: 861895.7 ymax:
775019.1
## epsg (SRID):    NA
## proj4string:    +proj=tmerc +lat_0=0 +lon_0=19 +k=0.9993 +x_0=500000
+y_0=-5300000 +ellps=GRS80 +units=m +no_defs
## First 5 features:
##   jpt_kod_je jpt_nazwa_    geometry
## 1       0226     złotoryjski    MULTIPOLYGON (((269083.6    37...
## 2       0262        Legnica    MULTIPOLYGON (((296226.6    37...
## 3       0206   jeleniogórski    MULTIPOLYGON (((244044.1    33...
## 4       0204       górowski    MULTIPOLYGON (((311163.1    42...
## 5       0214      oleśnicki    MULTIPOLYGON (((377122.1    37...
```

ATTENTION! Limiting the output to one column using the $ operator results in omitting the geometry column and returning the vector, not the *sf* class object.

```
head(poviats2$jpt_kod_je)
## [1] 0226 0262 0206 0204 0214 0201
## 380 Levels: 0201 0202 0203 0204 0205 0206 0207 0208 0209 0210 0211. . .
3263
```

Generally, when limiting *sf* data to selected columns, a geometry column is always included. To omit it, one should use the **st_set_geometry(x, value)** function on the *sf* class object and assign the geometry a NULL value, similar to deleting columns from data, or use the **st_drop_geometry()** function.

```
head(st_set_geometry(poviats2[, c("jpt_kod_je", "jpt_
nazwa_")],value = NULL))
##     jpt_kod_je      jpt_nazwa_
## 1        0226      złotoryjski
## 2        0262          Legnica
## 3        0206     jeleniogórski
## 4        0204         górowski
## 5        0214         oleśnicki
## 6        0201     bolesławiecki

head(st_drop_geometry(poviats2[, c("jpt_kod_je", "jpt_nazwa_")]))
##     jpt_kod_je      jpt_nazwa_
## 1        0226      złotoryjski
## 2        0262          Legnica
## 3        0206     jeleniogórski
## 4        0204         górowski
## 5        0214         oleśnicki
## 6        0201     bolesławiecki
```

This operation will omit the geometry column and convert the *sf* object to *data.frame*. Referring to the geometry column is possible with standard $ operator or **st_geometry**() function.

```
class(poviats2$geometry)
## [1] "sfc_GEOMETRY" "sfc"
class(st_geometry(poviats2)[[1]])
## [1] "XY"              "MULTIPOLYGON" "sfg"
```

The column containing the geometry has the *sfc* class, while its first element is a collection of polygons and has the class *sfg* and MULTIPOLYGON. For simplicity, only the columns denoting the poviat code (*jpt_kod_je*) and its name (*jpt_nazwa_*) will be kept in the dataset *poviats2* (with corrected geometries).

```
poviats2<-poviats2[, c("jpt_kod_je", "jpt_nazwa_")]

head(poviats2)
## Simple feature collection with 6 features and 2 fields
## geometry type:  MULTIPOLYGON
## dimension:      XY
## bbox:           xmin: 235157.1 ymin: 323842.8 xmax: 415844.4 ymax:
438816.5
## epsg (SRID):    NA
## proj4string:    +proj=tmerc +lat_0=0 +lon_0=19 +k=0.9993 +x_0=500000
+y_0=-5300000 +ellps=GRS80 +units=m +no_defs
##     jpt_kod_je      jpt_nazwa_                            geometry
## 1        0226      złotoryjski   MULTIPOLYGON (((269083.6 37. . .
## 2        0262          Legnica   MULTIPOLYGON (((296226.6 37. . .
## 3        0206     jeleniogórski  MULTIPOLYGON (((244044.1 33. . .
## 4        0204         górowski   MULTIPOLYGON (((311163.1 42. . .
## 5        0214         oleśnicki  MULTIPOLYGON (((377122.1 37. . .
## 6        0201     bolesławiecki  MULTIPOLYGON (((235157.1 37. . .
```

One can also add new attributes to the *sf* class object (fields, columns). In this case, the region (voivodeship) code will be added, which can be created based on the first two characters from the poviat code.

```
poviats2$code_voi<-substr(poviats2$jpt_kod_je, 1, 2)
```

The frequency table for the new column will show how many poviats (counties) are located in each of the voivodeships (regions).

```
table(poviats2$code_voi)
## 02 04 06 08 10 12 14 16 18 20 22 24 26 28 30 32
## 30 23 24 14 24 22 42 12 25 17 20 36 14 21 35 21
```

For objects of the *sf* class, the **plot()** method is defined. Specifying the whole *sf* object as an argument will generate a separate map for each characteristic stored in the set (which is usually not desirable). To draw a map of the selected characteristic (see Figure 10.1), the appropriate column should be provided, but, as mentioned earlier, NOT using the $ operator, because then the geometry column will not be included in the data, which is necessary to draw the map.

```
plot(poviats2[, "code_voi"])
```

However, for a dataset containing many rows, the map can take a long time to draw due to complicated shapes (geometries) of individual units (here poviats). To check how complex these objects are, the total number of all vertices in all geometries for individual poviats will be calculated. The **st_cast()** command allows one to change the geometry for individual cases – here it will be changed from POLYGON to MULTIPOINT. The resulting object is a simplified geometry of the *sfc_MULTIPOINT* class.

```
poviats_points<-st_cast(poviats2$geometry, "MULTIPOINT")

class(poviats_points)
## [1] "sfc_MULTIPOINT" "sfc"
```

For each county, it is a list of points, which are its vertices, whose connection allows one to draw a map of this poviat. One can check how many vertices the first poviat in the data has using the functions **length()**. The sum of the number of vertices (points) for all counties will be calculated using the **length()** command applied to the geometry of each of them using **sapply()** and summing them up with **sum()**. The sum of vertices is nearly 4.5 million! That is why drawing the map for poviats is taking a while.

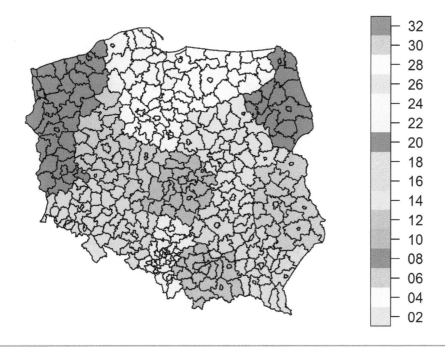

Figure 10.1 Assignment of poviats to voivodeships

Source: Own study using R and the sf:: package

487

```
length(poviats_points[[1]]) # for first poviat
## [1] 6720

poviats_points_count<-sapply(poviats_points, length)
sum(poviats_points_count) # Checking how many vertices are in all counties
## [1] 4484190
```

By using the **object_size()** function from the pryr:: package, one can check the size of the object containing spatial data for poviats. The result shows it is over 36 megabytes.

```
pryr::object_size(poviats2)
## 36.1 MB
```

The plotted map does not show all the details of the shape of individual poviats. A good idea would be to simplify it – change the shape of individual polygons to be more schematic, created using fewer vertices and connecting sections. To transform (simplify) an object, the **st_simplify()** function is used,[7] whose first argument is the object to be transformed and second is the tolerance (*dToterance*), which is the distance. All resulting segments will have a length not less than the given tolerance value (in this case, it will be the distance measured in meters).

The boundaries of the poviats will be simplified so that they are crossed out with sections of not less than 50 meters. The correct size of the tolerance parameter should be chosen by trial and error – it cannot be too big.

```
poviats_simple<-st_simplify(poviats2, dTolerance = 50)
```

One can check the size and number of vertices for simplified data. The size of data dropped from 36 MB to 2.8 MB, and the number of vertices from almost 4.5 million to 314,000.

```
pryr::object_size(poviats_simple)
## 2.8 MB

poviats_simple_points<-st_cast(poviats_simple$geometry, "MULTIPOINT")
sum(sapply(poviats_simple_points, length))

## [1] 314428
```

10.3.3 Data with POINT geometry

Sample point spatial data will be created by finding the geographical centroids of poviats. In the sf:: package, the **st_centroid()** function is used for this.

```
poviats_central<-st_centroid(poviats2)

## Warning in st_centroid.sf(poviats2): st_centroid assumes attributes are
## constant over geometries of x

poviats_central
## Simple feature collection with 380 features and 3 fields
## geometry type:   POINT
## dimension:       XY
## bbox:            xmin: 192697.2  ymin: 159415.1  xmax: 843942.1  ymax:
763121.8
## epsg (SRID):     NA
## proj4string:     +proj=tmerc +lat_0=0 +lon_0=19 +k=0.9993 +x_0=500000
+y_0=-5300000 +ellps=GRS80 +units=m +no_defs
## First 10 features:
##   jpt_kod_je   jpt_nazwa              geometry code_voi
## 1       0226   złotoryjski   POINT (281548 364257.5) 02
```

```
##   2          0262              Legnica    POINT (302183.2 374365.8)     02
##   3          0206          jeleniogórski  POINT (267389.5 337558.1)     02
##   4          0204             górowski       POINT (330797.1 421145)    02
##   5          0214            oleśnicki    POINT (395990.7 378529.8)     02
##   6          0201          bolesławiecki  POINT (258866.1 391276.9)     02
##   7          0261          Jelenia Góra   POINT (268458.5 338942.1)     02
##   8          0208             kłodzki        POINT (330392 281113.9)    02
##   9          0210             lubański     POINT (239479.3 363197.3)    02
##   10         0221            wałbrzyski   POINT (306671.4 323318.4)     02
```

```
class(poviats_central$geometry) # Checking the column class with geometry
## [1] "sfc_POINT" "sfc"
```

As expected, this is an *sfc* class object of the *sfc_POINT* subclass. It can be displayed on the map. Of course, one can superimpose the visualisation of data from different spatial sets (provided that they have consistent projections of geographic coordinates). In the next **plot()** function, use the *add=TRUE* argument.

```
plot(st_geometry(poviats2))
plot(poviats_central$geometry, add=TRUE, pch=20, col="red")
```

10.3.4 Visualisation using the ggplot2:: package

Much more effective maps can be obtained by using the functions of the ggplot2:: package. The following examples have been limited to the simplest ones to show the possibility of an easy application of the ggplot2:: package to the spatial visualisation of *sf* data.

The ggplot2:: package is currently the leading tool in R for effective visualisations of data, including spatial data. In includes functionalities (geometries) appropriate for spatial *sf* objects – **geom_sf()**. Data visualisation with the use of **geom_sf()** differs slightly from other geometries used in ggplot2::, for example, **geom_point()**. This is due to the specificity of spatial data objects – their visualisation may consist of drawing points, lines or polygons, depending on the data content.

Subsequently one can find a simple example – if the goal is only to draw the geometry contained in the data, no arguments are used in the **geom_sf()** function. Adding the **theme_bw()** command changes the default gray background of the chart – see the result in Figure 10.2a,b.

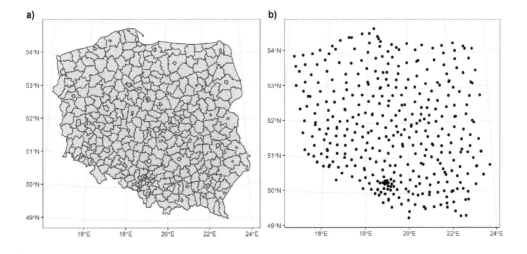

Figure 10.2 Map of poviats drawn with the use of the ggplot2:: package: a) border contours, b) centroids

Source: Own study using R and sf:: and ggplot2::.

```
ggplot(poviats_simple) + geom_sf() + theme_bw()  # 10.2a
ggplot(poviats_central) + geom_sf() + theme_bw() # 10.2.b
```

Presentation of the value of a selected variable on the map requires giving a name of this variable and linking it to the selected aesthetics with the **aes()** function – here it will be the *fill=. . .* aesthetic, that is, the fill colour of the areas on the map (Figure 10.3a).

```
ggplot(poviats_simple) + geom_sf(aes(fill = code_voi)) + theme_bw()
```

Also here, data from two different objects can be combined into one visualisation (see Figure 10.3b). One should then use the **geom_sf()** function as many times as many different datasets are used and give the name of the appropriate dataset as an argument of this function.

```
ggplot() + geom_sf(data=poviats_simple, aes(fill=code_voi)) +
geom_sf(data=poviats_central, col="black") + theme_bw()
```

10.3.5 Selected functions for spatial analysis

The **st_area()** function allows one to calculate the area of territorial units represented in subsequent data rows. Its use makes sense when the geometry of individual objects has the POLYGON or MULTIPOLYGON type – if the geometry is point-like or the linear, the area will be equal to 0.

```
head(st_area(poviats_simple), 10)
## Units: [m^2]
##   [1]    575771506    56235123  627168248  737724533 1048039067 1303343157
##   [7]    109273268  1642279556  428444393  429815417

head(st_area(poviats_central), 10)
## Units: [m^2]
##   [1] 0 0 0 0 0 0 0 0 0 0
```

The **st_distance()** function allows one to calculate distances between groups of spatial objects. It requires two arguments, *x* and *y*, and a matrix of distances between each element of the object *x* and each element of the object *y* is returned. Arguments *x* and *y* can have any type of geometry. If the *y* argument is not given, distances are calculated for each pair of elements in the *x* object.

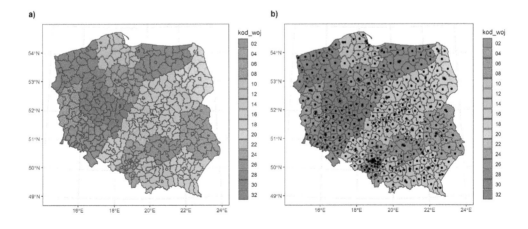

Figure 10.3 Map of poviats' affiliation to voivodeships using the ggplot2:: package: a) simple data, b) combined data

Source: Own study using R and the sf:: package.

In the first example, the distances from the borders and centroids of three cities (Warsaw, Bydgoszcz, Katowice) to the geographical centroids of five selected poviats will be calculated (one of them is Warsaw). The indexes of the mentioned cities in the poviat dataset will be stored as separate objects. The resulting distances are given in meters.

```
(which_warszawa<-which(poviats$jpt_nazwa_ == "powiat Warszawa"))
## [1] 151
(which_bydgoszcz<-which(poviats$jpt_nazwa_ == "powiat Bydgoszcz"))
## [1] 369
(which_katowice<-which(poviats$jpt_nazwa_ == "powiat Katowice"))
## [1] 329


# distances from the city borders
st_distance(poviats[c(which_warszawa, which_bydgoszcz, which_katowice),],
poviats[141:145,])

## Units: [m]
##              [,1]      [,2]      [,3]      [,4]      [,5]
## [1,] 233644.9 221106.8 180248.9  44118.0 258673.04
## [2,] 446662.6 420668.4 364924.2 181442.3 230518.94
## [3,] 262013.0 211493.8 143846.5 305688.9  97895.02


# distances from city centers st_distance(poviats_central[c(which_warszawa,
which_bydgoszcz, which_katowice),],
poviats[141:145,])

## Units: [m]
##              [,1]      [,2]      [,3]      [,4]      [,5]
## [1,] 248647.7 235519.0 195031.7  59538.84 272347.5
## [2,] 456708.8 430933.6 375283.8 190686.91 239574.1
## [3,] 269292.6 218624.2 151121.4 315094.02 106753.4


# lets omit the second argument
st_distance(poviats_central[c(which_warszawa, which_bydgoszcz,
which_katowice),])

## Units: [m]
##             [,1]      [,2]      [,3]
## [1,]         0.0 225908.2 264528.7
## [2,] 225908.2       0.0 330780.6
## [3,] 264528.7 330780.6       0.0
```

However, for the POLYGON type geometry, the function works quite slowly. A simple way to speed it up is to change the geometry type into MULTIPOINT.

```
system.time(distances_polygon<-st_distance(poviats[c(which_warszawa, which_
bydgoszcz, which_katowice),], poviats[141:145,]))

##    user  system elapsed
##   42.34    0.03   43.94

system.time(distances_multipoint<-st_distance(st_cast(poviats[c
(which_warszawa, which_bydgoszcz, which_katowice),], "MULTIPOINT"),
poviats[141:145,]))

##    user  system elapsed
##   10.15    0.00   10.75
```

In the previous example, calculations on MULTIPOINT geometry are performed several times faster. Calculated distances should not (significantly) differ. In this example, they do not differ at all.

```
distances_polygon - distances_multipoint
## Units: [m]
##           [,1]   [,2]   [,3]   [,4]   [,5]
##   [1,]       0      0      0      0      0
##   [2,]       0      0      0      0      0
##   [3,]       0      0      0      0      0
```

Unlike most of the functions of the sf:: package, examples of the use of which are listed subsequently, the **st_distance**() function does not use spatial indexing. The other functions also operate on two groups of spatial objects given x and y as arguments. Indexing is based on the elements of the first group of spatial objects (x) and significantly speeds up data processing (details in Pebesma & Bivand, 2017). As mentioned earlier, spatial indexes allow very fast spatial analyses to be made, for example, by determining whether a pair of geometries intersect, touch or are contained within another geometry. In the functions of the sf:: package, the R-tree indexing algorithm has been implemented (Pebesma, 2019a).

The **st_intersects**() function for each element of the x object returns a list of element indexes from the y object that intersect with it (have a common part). By adding the argument *sparse=FALSE*, one gets the result stored in the form of a matrix of logical values. In the following example, the city of Warsaw intersects with the fourth element of the second object (also Warsaw or its geographical centre).

```
st_intersects(x=poviats[c(which_warszawa, which_bydgoszcz, which_
katowice),], y=poviats[148:152,])

## Sparse geometry binary predicate list of length 3, where the predicate
was 'intersects'
##   1: 4
##   2: (empty)
##   3: (empty)

st_intersects(poviats[c(which_warszawa, which_bydgoszcz, which_katowice),],
poviats_central[148:152,], sparse=FALSE)
##          [,1]    [,2]    [,3]    [,4]    [,5]
## [1,]    FALSE   FALSE   FALSE   TRUE    FALSE
## [2,]    FALSE   FALSE   FALSE   FALSE   FALSE
## [3,]    FALSE   FALSE   FALSE   FALSE   FALSE
```

Using the **st_intersects**() function on polygon data, giving both arguments of a complete dataset, will result in a list of neighbours for specific areas.

```
poviats_neighbou<-st_intersects(poviats, poviats) poviats_neighbou
## Sparse geometry binary predicate list of length 380, where the predicate
was 'intersects'
## first 10 elements:
##   1: 1, 5, 7, 51, 54, 137, 140
##   2: 2, 3, 5, 6, 13, 133, 134, 136
##   3: 2, 3, 5, 51, 136, 140, 141, 142
##   4: 4, 5, 8, 135, 137, 143
```

Subsequently there are few examples of selected other functions with analogous syntax, which use spatial indexing.

```
# Checking if the geometries of individual objects are identical
st_equals(poviats[c(which_warszawa, which_bydgoszcz, which_katowice),],
poviats[148:152,], sparse = FALSE)
```

```
##         [,1]    [,2]    [,3]    [,4]    [,5]
## [1,]   FALSE   FALSE   FALSE    TRUE   FALSE
## [2,]   FALSE   FALSE   FALSE   FALSE   FALSE
## [3,]   FALSE   FALSE   FALSE   FALSE   FALSE

# Checking if geometries of individual objects are disjoint
st_disjoint(poviats[c(which_warszawa, which_bydgoszcz, which_katowice),],
poviats_central[148:152,], sparse = FALSE)

##         [,1]    [,2]    [,3]    [,4]    [,5]
## [1,]    TRUE    TRUE    TRUE   FALSE    TRUE
## [2,]    TRUE    TRUE    TRUE    TRUE    TRUE
## [3,]    TRUE    TRUE    TRUE    TRUE    TRUE

# Checking if the geometries are touching (are tangent)
st_touches(poviats[c(which_warszawa, which_bydgoszcz, which_katowice),],
poviats_central[148:152,],    sparse = FALSE)

##         [,1]    [,2]    [,3]    [,4]    [,5]
## [1,]   FALSE   FALSE   FALSE   FALSE   FALSE
## [2,]   FALSE   FALSE   FALSE   FALSE   FALSE
## [3,]   FALSE   FALSE   FALSE   FALSE   FALSE

# Checking if the geometry is located within a distance
# no greater than the value given as the argument dist (in meters)
st_is_within_distance(poviats[c(which_warszawa, which_bydgoszcz, which_
katowice),], poviats_central[148:152,], dist = 1000, sparse = TRUE)

## Sparse geometry binary predicate list of length 3, where the predicate
was 'is_within_distance'
##   1: 4
##   2: (empty)
##   3: (empty)

# Checking if the geometries of individual elements from the first object
# contain elements from the second object
st_contains(poviats[c(which_warszawa, which_bydgoszcz, which_katowice),],
poviats_central[148:152,],    sparse = FALSE)

##         [,1]    [,2]    [,3]    [,4]    [,5]
## [1,]   FALSE   FALSE   FALSE    TRUE   FALSE
## [2,]   FALSE   FALSE   FALSE   FALSE   FALSE
## [3,]   FALSE   FALSE   FALSE   FALSE   FALSE

# Checking if the geometries of individual elements from the first object
# cover the elements from the second object
st_covers(poviats[c(which_warszawa, which_bydgoszcz, which_katowice),],
poviats_central [148:152,],    sparse = FALSE)

##         [,1]    [,2]    [,3]    [,4]    [,5]
## [1,]   FALSE   FALSE   FALSE    TRUE   FALSE
## [2,]   FALSE   FALSE   FALSE   FALSE   FALSE
## [3,]   FALSE   FALSE   FALSE   FALSE   FALSE

# Checking if the geometries of individual elements from the first object
# are covered with elements from the second object
```

```
st_covered_by(poviats_central[148:152,], poviats[c(which_warszawa, which_
bydgoszcz, which_katowice),], sparse = FALSE)
```

```
##           [,1]    [,2]    [,3]
## [1,]    FALSE   FALSE   FALSE
## [2,]    FALSE   FALSE   FALSE
## [3,]    FALSE   FALSE   FALSE
## [4,]     TRUE   FALSE   FALSE
## [5,]    FALSE   FALSE   FALSE
```

Some other functions from the sf:: package will be also discussed in the following examples in the next sections.

10.4 Use the dplyr:: package functions

This part will discuss the use of selected functions of the dplyr:: package. This package contains functions that allow for efficient data processing in R – they work very quickly even on large datasets. Outstanding performance (speed) on data frames is one of the most important advantages of the dplyr:: package. The key elements of the package are implemented in C++ using the Rcpp:: package (allowing integration of R with C++). A very important advantage is also the consistency of the syntax of individual functions, which definitely facilitates their use. Therefore, currently the way of processing data from this package is standard in R. The package dplyr:: is a part of tidyverse:: – a collection of useful packages developed by RStudio (for more information, see RStudio, 2019b).

Subsequently one can find a brief overview of the functions of this package, which does not exhaust all of its capabilities. The purpose of this chapter is to show efficient data processing using the tidyverse:: approach on sample big spatial data, stored in the form of a spatial data frame using the previously mentioned sf:: package. All functions from the tidyverse:: packages have methods for *sf* class objects defined (see Pebesma, 2019e).

Let's get to know the following basic functions of the dplyr:: package:

- **filter**() – selection of observations that meet the indicated conditions;
- **select**() – selection of columns; inside the **select**() function, one can use:

  - **contains**() – selection of columns whose name contains a given string of characters,
  - **starts_with**() – selection of columns whose name starts with a given string,
  - **ends_with**() – selection of columns whose name ends with a given string of characters,
  - **matches**() – selection of columns whose name contains a given pattern (one can use regular expressions here),

- **rename**() – change of column names,
- **arrange**() – sorting data by indicated columns (by default ascending, use **desc**() for descending order),
- **mutate**() – creating new variables,
- **glimpse**() – showing the data structure, alternative to **str**(),
- **group_by**() – division of data into subgroups with the use of selected columns, allows processing data in subgroups,
- **summarize**() – summary of data; under **summarize**() one can use:

  - **n**() – counting observations
  - **n_distinct**() – counting unique values
  - **sample_n**() – selection of random sample of *n* elements (by default without replacement),
  - **sample_frac**() – selection of random sample – a given fraction of all observations (by default without replacement).

Each of these functions expects a data frame as the first argument. Subsequent arguments often require the names of columns, which in these functions from dplyr:: are provided *without* quotes (more details in Wickham, François, Henry, & Müller, 2019a, 2019b).

The necessary packages are loaded first. In the examples, some functions from the stringr:: package will also be used, which is also part of the tidyverse:: collection and is used to process text data.

```
library(dplyr)
library(stringr)
library(sf)
```

Then, data from the shapefile covering the cadastral regions of Poland is imported. These are the country division units used for the purposes of land and building register (real estate cadastre). The data was downloaded from GUGiK (2019). The loaded object has, of course, the *sf* class. The data structure can be checked using the **glimpse()** function.

```
areas<-st_read("data/registration_areas.shp")
## Reading layer 'registration_areas' from data source
'data/registration_areas.shp' using driver 'ESRI Shapefile'
## Simple feature collection with 54010 features and 29 fields (with 22
geometries empty)
## geometry type:    MULTIPOLYGON
## dimension:        XY
## bbox:             xmin: 171677.6  ymin: 133223.7  xmax: 861895.7  ymax:
775019.1
## epsg (SRID):      NA
## proj4string: +proj=tmerc +years_0=0 +lon_0=19 +k=0.9993 +x_0=500000
+y_0=-5300000 +ellps=GRS80 +units=m +no_defs


class(areas)
## [1] "sf"        "data.frame"
glimpse(areas)
## Observations: 54,010
## Variables: 30
## $ iip_spat  <fct> PL.PZGIK.200, PL.PZGIK.200, PL.PZGIK.200,
PL.PZGIK. . . .
## $ iip_identy <fct> a3acab38-fd08-4396-8d5d-c1fae1ee7ee8, 90de059f-
ef12. . .
## $ iip_wersja <fct> 2012-09-21T16:26:46+02:00, 2012-09-
21T16:26:46+02:0. . .
## $ jpt_sjr_ko <fct> OBR, OBR, OBR, OBR, OBR, OBR, OBR, OBR, OBR, OBR, O. . .
## $ jpt_kod_je <fct> 022308_5.0003, 022308_5.0012, 022308_5.0009, 022309. . .
## $ jpt_nazwa_ <fct> Bogusławice, Radwanice, Mokry Dwór, Wojkowice, PIEŃ . . .
## $ jpt_nazw01 <fct> NA, NA, NA, NA, NA, NA, NA, NA, NA, NA, NA, NA, NA, . . .
## $ jpt_organ_ <fct> NA, NA, NA, NA, NA, NA, NA, NA, NA, NA, NA, NA, NA, . . .
## $ jpt_orga01 <fct> NZN, NZN, NZN, NZN, NZN, NZN, NZN, NZN, NZN, NZN, N. . .
## $ jpt_jor_id <dbl> 0, 0, 0, 0, 0, 0, 0, 0, 0, 0, 0, 0, 0, 0, 0, 0, 0, . . .
## $ wazny_od   <date> NA, NA, NA, NA, NA, NA, NA, NA, NA, NA, NA, NA, NA. . .
## $ wazny_do   <date> NA, NA, NA, NA, NA, NA, NA, NA, NA, NA, NA, NA, NA. . .
## $ jpt_wazna_ <fct> 7NZN, 7NZN, 7NZN, 7 . . .
## $ wersja_od  <date> NA, NA, NA, NA, NA, NA, NA, NA, NA, NA, NA, NA, NA. . .
## $ wersja_do  <date> NA, NA, NA, NA, NA, NA, NA, NA, NA, NA, NA, NA, NA. . .
## $ jpt_powier <dbl> 21, 21, 21, 21, 21, 21, 21, 21, 21, 21, 21, 21, 21, . . .
## $ jpt_kj_iip <fct> 0000000000EGIB, 0000000000EGIB, 0000000000EGIB, 000. . .
## $ jpt_kj_i01 <fct> 022308_5.0003, 022308_5.0012, 022308_5.0009, 022309. . .
## $ jpt_kj_i02 <fct> NA, NA, NA, NA, NA, NA, NA, NA, NA, NA, NA, NA, NA, . . .
## $ jpt_code_01 <fct> NA, NA, NA, NA, NA, NA, NA, NA, NA, NA, NA, NA, NA, . . .
## $ id_bufora_ <dbl> 0, 0, 0, 0, 0, 0, 0, 0, 0, 0, 0, 0, 0, 0, 0, 0, 0, . . .
## $ id_bufor01 <dbl> 9088, 9088, 9088, 9089, 9105, 9086, 9088, 9089, 908. . .
## $ id_technic <dbl> 0, 0, 0, 0, 0, 0, 0, 0, 0, 0, 0, 0, 0, 0, 0, 0, 0, . . .
## $ jpt_opis   <fct> 542714, 542715, 542716, 542738, 542913, 542688, 542. . .
## $ jpt_sps_ko <fct> NA, NA, NA, NA, NA, NA, NA, NA, NA, NA, NA, NA, NA, . . .
## $ gra_ids    <fct> NA, NA, NA, NA, NA, NA, NA, NA, NA, NA, NA, NA, NA, . . .
## $ status_obi <fct> VALID, VALID, VALID, VALID, VALID, A. . .
## $ opis_bledu <fct> NA, NA, NA, NA, NA, NA, NA, NA, NA, NA, NA, NA, NA, . . .
## $ typ_bledu  <fct> NA, NA, NA, NA, NA, NA, NA, NA, NA, NA, NA, NA, NA, . . .
## $ geometry   <MULTIPOLYGON [m]> MULTIPOLYGON (((366069.6 34. . . ., MULTIP. . .
```

The dimensions of the data frame were also checked, and a few of its first rows were displayed.

```
dim(areas)
## [1] 54010    30


head(areas) # first three only
## Simple feature collection with 6 features and 29 fields
## geometry type:   MULTIPOLYGON
## dimension:       XY
## bbox:            xmin: 222863.5  ymin: 338284.7  xmax: 369885.1  ymax:
383750.3
## epsg (SRID):     NA
## proj4string:     +proj=tmerc +lat_0=0 +lon_0=19 +k=0.9993 +x_0=500000
+y_0=-5300000 +ellps=GRS80 +units=m +no_defs
##      iip_spat                           iip_identif
## 1 PL.PZGIK.200 a3acab38-fd08-4396-8d5d-c1fae1ee7ee8
## 2 PL.PZGIK.200 90de059f-ef12-449a-aa41-c5f2042bde74
## 3 PL.PZGIK.200 c984f31d-4fc4-4bac-b5b2-b330602b5bc9


          iip_version jpt_sjr_ko          jpt_kod_je    jpt_nazwa
## 1 2012-09-21T16:26:46+02:00     OBR 022308_5.0003  Bogusławice
## 2 2012-09-21T16:26:46+02:00     OBR 022308_5.0012    Radwanice
## 3 2012-09-21T16:26:46+02:00     OBR 022308_5.0009   Mokry Dwór


##     jpt_name01 jpt_body_  jpt_bod01 jpt_jor_id  valid_from    valid_to
## 1        <NA>      <NA>        NZN          0        <NA>        <NA>
## 2        <NA>      <NA>        NZN          0        <NA>        <NA>
## 3        <NA>      <NA>        NZN          0        <NA>        <NA>


##        jpt_valid_ version_from  version_to jpt_powier jpt_kj_iip jpt_kj_i01
## 1 7         NZN        <NA>        <NA>         21 0000000000EGIB 022308_5.0003
## 2 7         NZN        <NA>        <NA>         21 0000000000EGIB 022308_5.0012
## 3 7         NZN        <NA>        <NA>         21 0000000000EGIB 022308_5.0009


##     jpt_kj_i02 jpt_code_01  id_bufor_   id_bufor01  id_technic jpt_descr
## 1        <NA>      <NA>          0        9088          0       542714
## 2        <NA>      <NA>          0        9088          0       542715
## 3        <NA>      <NA>          0        9088          0       542716


##     jpt_sps_ko   gra_ids   status_obi  descr_error  type_error
## 1        <NA>      <NA>      VALID        <NA>        <NA>
## 2        <NA>      <NA>      VALID        <NA>        <NA>
## 3        <NA>      <NA>      VALID        <NA>        <NA>


##                           geometry
## 1 MULTIPOLYGON (((366069.6 34. . .
## 2 MULTIPOLYGON (((366596.1 35. . .
## 3 MULTIPOLYGON (((367154.5 35. . .
```

The following is an example of using the **select**() function with the dplyr:: package – the first argument is the *data frame* on which the operation will be performed, and the next arguments are the selected variables, where the variable names are given without the quotes.

```
select(areas, jpt_kod_je, jpt_nazwa_, geometry) # selected columns

## Simple feature collection with 54010 features and 2 fields (with 22
geometries empty)
```

```
## geometry type:    MULTIPOLYGON
## dimension:        XY
## bbox:             xmin: 171677.6  ymin: 133223.7  xmax: 861895.7  ymax:
775019.1
## epsg (SRID):      NA
## proj4string:      +proj=tmerc +years_0=0 +lon_0=19 +k=0.9993 +x_0=500000
+y_0=-5300000 +ellps=GRS80 +units=m +no_defs ## First 10 features:
##      jpt_kod_je         jpt_nazwa_                              geometry
## 1  022308_5.0003     Bogusławice     MULTIPOLYGON (((366069.6 34. . .
## 2  022308_5.0012      Radwanice      MULTIPOLYGON (((366596.1 35. . .
## 3  022308_5.0009     Mokry Dwór      MULTIPOLYGON (((367154.5 35. . .
## 4  022309_2.0028      Wojkowice      MULTIPOLYGON (((363396.1 34. . .
```

As mentioned earlier, in the case of *sf* objects, the geometry column is always selected. When selecting adjacent columns, one can use a colon and indicate only the first and the last column.

```
select(areas, jpt_kod_je:jpt_orga01)
## Simple feature collection with 54010 features and 5 fields (with 22
geometries empty)
## geometry type:    MULTIPOLYGON
## dimension:        XY
## bbox:             xmin: 171677.6  ymin: 133223.7  xmax: 861895.7  ymax:
775019.1
## epsg (SRID):      NA
## proj4string: +proj=tmerc +lat_0=0 +lon_0=19 +k=0.9993 +x_0=500000
+y_0=-5300000 +ellps=GRS80 +units=m +no_defs
## First 10 features:
##         jpt_kod_je     jpt_nazwa_    jpt_name01   jpt_organ_   jpt_orga01
## 1    022308_5.0003     Bogusławice      <NA>         <NA>          NZN
## 2    022308_5.0012      Radwanice       <NA>         <NA>          NZN
## 3    022308_5.0009     Mokry Dwór       <NA>         <NA>          NZN
## 4    022309_2.0028      Wojkowice       <NA>         <NA>          NZN
##                              geometry
## 1 MULTIPOLYGON (((366069.6 34. . .
## 2 MULTIPOLYGON (((366596.1 35. . .
## 3 MULTIPOLYGON (((367154.5 35. . .
## 4 MULTIPOLYGON (((363396.1 34. . .
```

Within the **select()** method, special functions are defined, for example, **starts_with()**, **ends_with()**, **contains()**, **matches()**:

```
select(areas, starts_with("jpt"))
## Simple feature collection with 54010 features and 15 fields (with 22
geometries empty)
## geometry type:    MULTIPOLYGON
## dimension:        XY
## bbox:             xmin: 171677.6  ymin: 133223.7  xmax: 861895.7  ymax:
775019.1
## epsg (SRID):      NA
## proj4string: +proj=tmerc +years_0=0 +lon_0=19 +k=0.9993 +x_0=500000
+y_0=-5300000 +ellps=GRS80 +units=m +no_defs
## First 10 features:
##     jpt_sjr_ko        jpt_kod_je       jpt_nazwa_   jpt_nazw01   jpt_organ_
## 1        OBR       022308_5.0003     Bogusławice      <NA>         <NA>
## 2        OBR       022308_5.0012      Radwanice       <NA>         <NA>
## 3        OBR       022308_5.0009     Mokry Dwór       <NA>         <NA>
## 4        OBR       022309_2.0028      Wojkowice       <NA>         <NA>
```

```
## First 10 features:
##     jpt_orga01      jpt_jor_id    jpt_wazna_   jpt_powier       jpt_kj_iip
## 1        NZN             0 7           NZN           21   0000000000EGIB
## 2        NZN             0 7           NZN           21   0000000000EGIB
## 3        NZN             0 7           NZN           21   0000000000EGIB
## 4        NZN             0 7           NZN           21   0000000000EGIB
##       jpt_kj_i01     jpt_kj_i02    jpt_kod_01    jpt_opis    jpt_sps_ko
## 1   022308_5.0003          <NA>          <NA>      542714          <NA>
## 2   022308_5.0012          <NA>          <NA>      542715          <NA>
## 3   022308_5.0009          <NA>          <NA>      542716          <NA>
## 4   022309_2.0028          <NA>          <NA>      542738          <NA>
##                                      geometry
## 1 MULTIPOLYGON (((366069.6 34. . .
## 2 MULTIPOLYGON (((366596.1 35. . .
## 3 MULTIPOLYGON (((367154.5 35. . .
## 4 MULTIPOLYGON (((363396.1 34. . .

select(areas, ends_with("01"))

## Simple feature collection with 54010 features and 5 fields (with 22
geometries empty)
## geometry type:   MULTIPOLYGON
## dimension:       XY
## bbox:            xmin: 171677.6  ymin: 133223.7  xmax: 861895.7  ymax:
775019.1
## epsg (SRID):     NA
## proj4string:     +proj=tmerc +years_0=0 +lon_0=19 +k=0.9993 +x_0=500000
+y_0=-5300000 +ellps=GRS80 +units=m +no_defs
## First 10 features:
##     jpt_nazw01    jpt_orga01       jpt_kj_i01    jpt_kod_01    id_bufor01
## 1         <NA>          NZN    022308_5.0003          <NA>          9088
## 2         <NA>          NZN    022308_5.0012          <NA>          9088
## 3         <NA>          NZN    022308_5.0009          <NA>          9088
## 4         <NA>          NZN    022309_2.0028          <NA>          9089
##                                      geometry
## 1 MULTIPOLYGON (((366069.6 34. . .
## 2 MULTIPOLYGON (((366596.1 35. . .
## 3 MULTIPOLYGON (((367154.5 35. . .
## 4 MULTIPOLYGON (((363396.1 34. . .

select(areas, contains("wersja")) # "wersja" means version

## Simple feature collection with 54010 features and 3 fields (with 22
geometries empty)
## geometry type:   MULTIPOLYGON
## dimension:       XY
## bbox:            xmin: 171677.6  ymin: 133223.7  xmax: 861895.7  ymax:
775019.1
## epsg (SRID):     NA
## proj4string:     +proj=tmerc +lat_0=0 +lon_0=19 +k=0.9993 +x_0=500000
+y_0=-5300000 +ellps=GRS80 +units=m +no_defs
## First 10 features:
##                        iip_wersja      wersja_od      wersja_do
## 1      2012-09-21T16:26:46+02:00           <NA>           <NA>
## 2      2012-09-21T16:26:46+02:00           <NA>           <NA>
## 3      2012-09-21T16:26:46+02:00           <NA>           <NA>
## 4      2012-09-21T16:27:31+02:00           <NA>           <NA>
```

```
##                        geometry
## 1 MULTIPOLYGON (((366069.6 34. . .
## 2 MULTIPOLYGON (((366596.1 35. . .
## 3 MULTIPOLYGON (((367154.5 35. . .
## 4 MULTIPOLYGON (((363396.1 34. . .
```

```
select(areas, matches("_.{3}_"))
```

```
## Simple feature collection with 54010 features and 5 fields (with 22
geometries empty)
## geometry type:   MULTIPOLYGON
## dimension:       XY
## bbox:            xmin: 171677.6  ymin: 133223.7  xmax: 861895.7  ymax:
775019.1
## epsg (SRID):     NA
## proj4string:     +proj=tmerc +lat_0=0 +lon_0=19 +k=0.9993 +x_0=500000
+y_0=-5300000 +ellps=GRS80 +units=m +no_defs
## First 10 features:
##      jpt_sjr_ko      jpt_kod_je   jpt_jor_id   jpt_code_01   jpt_sps_ko
## 1           OBR   022308_5.0003            0         <NA>         <NA>
## 2           OBR   022308_5.0012            0         <NA>         <NA>
## 3           OBR   022308_5.0009            0         <NA>         <NA>
## 4           OBR   022309_2.0028            0         <NA>         <NA>
##                        geometry
## 1 MULTIPOLYGON (((366069.6 34. . .
## 2 MULTIPOLYGON (((366596.1 35. . .
## 3 MULTIPOLYGON (((367154.5 35. . .
## 4 MULTIPOLYGON (((363396.1 34. . .
```

In the last case, with the **matches**() function, a regular expression is used, which means a pattern consisting of an underscore, then three arbitrary characters and then another underscore.

In the **select**() function, one can also specify several conditions. In this case, all columns that meet at least one of the given conditions are returned.

```
select(areas, jpt_kod_je:jpt_orga01, ends_with("01"), matches("_.{3}_"))
```

```
## Simple feature collection with 54010 features and 11 fields (with 22
geometries empty)
## geometry type:   MULTIPOLYGON
## dimension:       XY
## bbox:            xmin: 171677.6  ymin: 133223.7  xmax: 861895.7  ymax:
775019.1
## epsg (SRID):     NA
## proj4string:     +proj=tmerc +lat_0=0 +lon_0=19 +k=0.9993 +x_0=500000
+y_0=-5300000 +ellps=GRS80 +units=m +no_defs
## First 10 features:
##          jpt_kod_je    jpt_nazwa_   jpt_nazw01   jpt_organ_   jpt_orga01
## 1      022308_5.0003   Bogusławice        <NA>         <NA>          NZN
## 2      022308_5.0012    Radwanice         <NA>         <NA>          NZN
## 3      022308_5.0009   Mokry Dwór         <NA>         <NA>          NZN
## 4      022309_2.0028    Wojkowice         <NA>         <NA>          NZN
##          jpt_kj_i01 jpt_kod_01 id_bufor01 jpt_sjr_ko jpt_jor_id jpt_sps_ko
## 1      022308_5.0003       <NA>       9088        OBR          0       <NA>
## 2      022308_5.0012       <NA>       9088        OBR          0       <NA>
## 3      022308_5.0009       <NA>       9088        OBR          0       <NA>
## 4      022309_2.0028       <NA>       9089        OBR          0       <NA>
```

```
##                            geometry
## 1 MULTIPOLYGON (((366069.6 34. . .
## 2 MULTIPOLYGON (((366596.1 35. . .
## 3 MULTIPOLYGON (((367154.5 35. . .
## 4 MULTIPOLYGON (((363396.1 34. . .
```

Similarly, one can indicate which columns should be excluded from the data using the additional operator – (minus). In the following example, all columns that meet at least one of the given conditions are omitted.

```
select(areas, -ends_with("01"), -matches("_.{3}_"))
```

```
## Simple feature collection with 54010 features and 20 fields (with 22
geometries empty)
## geometry type:   MULTIPOLYGON
## dimension:       XY
## bbox:            xmin: 171677.6  ymin: 133223.7  xmax: 861895.7  ymax:
775019.1
## epsg (SRID):     NA
## proj4string:     +proj=tmerc +lat_0=0 +lon_0=19 +k=0.9993 +x_0=500000
+y_0=-5300000 +ellps=GRS80 +units=m +no_defs
## First 10 features:
##       iip_przest                        iip_identy
## 1 PL.PZGIK.200 a3acab38-fd08-4396-8d5d-c1fae1ee7ee8
## 2 PL.PZGIK.200 90de059f-ef12-449a-aa41-c5f2042bde74
## 3 PL.PZGIK.200 c984f31d-4fc4-4bac-b5b2-b330602b5bc9
## 4 PL.PZGIK.200 e57061af-034a-4562-8916-a0b57473ada2
##                      iip_wersja    jpt_nazwa_ jpt_organ_  wazny_od
## 1     2012-09-21T16:26:46+02:00  Bogusławice       <NA>      <NA>
## 2     2012-09-21T16:26:46+02:00  Radwanice        <NA>      <NA>
## 3     2012-09-21T16:26:46+02:00  Mokry Dwór       <NA>      <NA>
## 4     2012-09-21T16:27:31+02:00  Wojkowice        <NA>      <NA>
##      wazny_do jpt_wazna_ wersja_od    wersja_do jpt_powier   jpt_kj_iip
## 1     <NA> 7       NZN      <NA>         <NA>        21 0000000000EGIB
## 2     <NA> 7       NZN      <NA>         <NA>        21 0000000000EGIB
## 3     <NA> 7       NZN      <NA>         <NA>        21 0000000000EGIB
## 4     <NA> 7       NZN      <NA>         <NA>        21 0000000000EGIB
##      jpt_kj_i02 id_bufora_ id_technic  jpt_opis  gra_ids   status_obi
opis_bledu
## 1        <NA>          0          0    542714      <NA>     AKTUALNY
<NA>
## 2        <NA>          0          0    542715      <NA>     AKTUALNY
<NA>
## 3        <NA>          0          0    542716      <NA>     AKTUALNY
<NA>
## 4        <NA>          0          0    542738      <NA>     AKTUALNY
<NA>
##       typ_bledu                        geometry
## 1        <NA>     MULTIPOLYGON (((366069.6 34. . .
## 2        <NA>     MULTIPOLYGON (((366596.1 35. . .
## 3        <NA>     MULTIPOLYGON (((367154.5 35. . .
## 4        <NA>     MULTIPOLYGON (((363396.1 34. . .
```

In the following example, only columns containing the code (*jpt_kod_je*) and name (*jpt_nazwa_*) of the territorial unit will be kept in the data. The dataset will be overwritten in a limited form.

```
select(areas, jpt_kod_je, jpt_nazwa_)->areas
```

The following example uses the **filter()** function from the dplyr:: package. Only rows with the name of the area including the string "WARSZA" will be selected. Here, also, the names of variables are listed without the quotes.

```
filter(areas, str_detect(jpt_nazwa_, "WARSZA"))

## Simple feature collection with 6 features and 2 fields
## geometry type:   MULTIPOLYGON
## dimension:       XY
## bbox:            xmin: 627776.5  ymin: 363273.3  xmax: 733350.6  ymax:
## 473778.2
## epsg (SRID):     NA
## proj4string:     +proj=tmerc +years_0=0 +lon_0=19 +k=0.9993 +x_0=500000
## +y_0=-5300000 +ellps=GRS80 +units=m +no_defs
##           jpt_kod_je          jpt_nazwa_                          geometry
## 1     060910_2.0021          WARSZAWIAKI    MULTIPOLYGON (((731601.4  36. . .
## 2     141803_2.0027         WARSZAWIANKA    MULTIPOLYGON (((627776.5  46. . .
## 3     141803_2.0009  KOLONIA WARSZAWSKA    MULTIPOLYGON (((627936 4666. . .
## 4     141707_2.0022          WARSZAWICE    MULTIPOLYGON (((657090.9  45. . .
```

The query can be expanded by an additional condition. Let's select only areas from the Mazowieckie voivodeship. There is no voivodeship code in the table – it can be obtained by extracting the first two digits from the poviat code (*jpt_kod_je*). The code *str_sub(jpt_kod_je, 1, 2) == "14")* means selecting rows for which the first two digits in the variable *jpt_kod_je* are equal to 14 (the code of the Mazowieckie voivodeship). In general, one can enter several conditions separated by commas – subsequent conditions are then interpreted as subsequent arguments to the **filter()** function. The result includes records that meet all given conditions together.

```
filter(areas, str_detect(jpt_nazwa_, 'WARSZ'),
str_sub(jpt_kod_je,1,2)=="14")

## Simple feature collection with 7 features and 2 fields
## geometry type:   MULTIPOLYGON
## dimension:       XY
## bbox:            xmin: 568469.7  ymin: 380410.7  xmax: 661516.4  ymax:
## 538571.3
## epsg (SRID):     NA
## proj4string:     +proj=tmerc +years_0=0 +lon_0=19 +k=0.9993 +x_0=500000
## +y_0=-5300000 +ellps=GRS80 +units=m +no_defs
##           jpt_kod_je          jpt_nazwa_                          geometry
## 1     141803_2.0027         WARSZAWIANKA   MULTIPOLYGON   (((627776.5 46. . .
## 2     141803_2.0009  KOLONIA WARSZAWSKA   MULTIPOLYGON   (((627936 4666. . .
## 3     141707_2.0022          WARSZAWICE   MULTIPOLYGON   (((657090.9 45. . .
## 4     141707_2.0023          WARSZÓWKA   MULTIPOLYGON   (((658621.9 46. . .
```

The resulting units are not, however, the areas within Warsaw. These will later be identified in another way, by the TERYT code.

If there is no provincial code in the data, one can add it. For this purpose, the **mutate()** function will be used to create a new variable. The resulting object will be stored under the same name as the source dataset. The **str_sub()** function used subsequently has the same syntax as the **substr()** used earlier but is more effective on large datasets.

```
areas<-mutate(areas, code_voi=str_sub(jpt_kod_je, 1, 2))
```

The name of the *jpt_kod_je* variable will be changed to *code_area*, while the *jpt_nazwa_* variable will be changed to *name_area*. For this purpose, the function **rename(new_name = old_name)** will be used. In addition, a list of column names is displayed both before and after the change.

```
names(areas)
## [1] "jpt_kod_je" "jpt_nazwa_" "geometry"       "kod_woj"
areas<-rename(areas, code_area = jpt_kod_je, name_area = jpt_nazwa_,
code_voi= kod_woj)

names(areas)
## [1] "code_area"       "name_area" "code_voi"     "geometry"
```

There are a few commands to summarise the dataset. One can check how many observations (areas) are in the data frame by using the **nrow**() command. Alternatively, one can also use the **summarize**() function of dplyr::. ATTENTION! In order for the code operating on columns containing features other than geometry to be executed quickly, the geometry column will be omitted during data processing (it will be removed by changing its value to NULL). For this purpose, the aforementioned **st_set_geometry**() function will be used. In this case, the result will be a data frame consisting of all other columns.

```
nrow(areas)
## [1] 54010

summarize(st_set_geometry(areas, NULL), n())
##      n()
## 1 54010
```

Another advantage of the function from the dplyr:: package is returning a result that always has the same class as the input object – in this case, it is a *data.frame*. The column with the number of poviats can be given a name.

```
summarize(st_set_geometry(areas, NULL), number=n())
##   number
## 1 54010
```

At first glance, using this function seems more complicated than using the usual **nrow**() function. Advantages of the dplyr:: approach can be seen better when analysing data in subgroups. In the next example, the number of poviats was calculated separately for each voivodeship. This can easily be done using the **group_by**() function, which converts the data frame into a form containing subgroups (created on the basis of unique combinations of variable values given as successive function arguments) – the resulting object has one more class – *grouped_df*. Using the **summarize**() function on an object of this type will result in the calculation of the summary for each subgroup separately. Because the aggregation refers to attributes not related to the geometry of the analysed territorial units, the geometry column was omitted for the acceleration of the calculations. The **summarize**() function is then applied to the grouped object.

```
areas_voi<-group_by(st_set_geometry(areas, NULL), code_voi)
class(areas_voi)
## [1] "grouped_df" "tbl_df" "tbl" "data.frame"

summarize(areas_voi, number=n()) # first 6 only
## # A tibble: 16 x 2
##       code_voi number
##       <chr>    <int>
## 1     02       2897
## 2     04       3609
## 3     06       4040
## 4     08       1210
## 5     10       4733
## 6     12       2642
```

The indicated statistics (number of observations, here the areas) were counted in subgroups for each of 16 regions separately. The result will be additionally sorted with respect to the resulting variable *number*. The **arrange**() function is used here, in which the previously used **summarized**() function is nested.

```
arrange(summarize(areas_voi, number=n()), number) # first 6 only
## # A tibble: 16 x 2
##       code_voi  number
##       <chr>     <int>
## 1     16        1086
## 2     08        1210
## 3     18        1777
## 4     22        2152
## 5     24        2184
## 6     26        2280
```

Additional use of **desc**() allows for sorting the data in descending order of the given variable.

```
arrange(summarize(areas_voi, number=n()), desc(number)) # first 6 only

## # A tibble: 16 x 2
##       code_voi  number
##       <chr>     <int>
## 1     14        11292
## 2     10        4733
## 3     30        4574
## 4     06        4040
## 5     04        3609
## 6     20        3503
```

The previous code is not very transparent. Its readability can be improved by rewriting it with the pipeline operator %>% from the magnittr:: package, which is also automatically loaded by the dplyr:: package. This operator is used when performing many nested operations, where the next function uses the result of the previous step as its first argument (usually input data). When using the pipeline operator, in the function entered to its right, its first argument is omitted, which by default is the result of the expression entered on the left of the operator. So the sequence *a %>% b(. . .) %>% c(. . .)* is a more readable alternative to the series of nested references c(b(a(), . . .), . . .). Let's see an alternative formulation of the previous code using a pipeline operator.

```
areas_voi %>% summarize(number=n())%>%arrange(desc(number)) # first 6 only
## # A tibble: 16 x 2
##       code_voi  number
##       <chr>     <int>
## 1     14        11292
## 2     10        4733
## 3     30        4574
## 4     06        4040
## 5     04        3609
## 6     20        3503
```

One can read this as follows: take a dataset, then count the number of observations (in subgroups, because it is a *grouped_df* object) and write it to the column *number*, and finally sort the resulting summary in descending order with respect to the column *number*. If the result of the previous function should be used as an argument other than the first one, must be used in an unusual way (e.g. not as a function argument) or needs to be used in several places, its place of insertion is indicated by a dot. – for example, the representation of the number of preareas in voivodeships by means of a histogram can be obtained with the code:

```
areas_voi %>% summarize(number=n()) %>% arrange(desc(number)) %>%
.[["number"]] %>% hist()
```

This set of operations can be read as follows: take a dataset, then count the number of observations (in subgroups, because it is a *grouped_df* object) and write it to the column *number*, and then sort the resulting summary in descending order with respect to the column *number*. From the sorted dataset, select the column named *number* and display its histogram.

All previous operations can also be formulated in this way. The following example starts with the loading of data, so as not to change the setup created and processed in the previous examples. Data transformations and all calculations will be performed "on the fly" – without storing the imported data in the workspace.

```
st_read("data/areas_registration.shp") %>% select(jpt_kod_je, jpt_nazwa_)
%>%
mutate(code_voi=str_sub(jpt_kod_je, 1, 2)) %>% st_set_geometry(NULL) %>%
group_by(code_voi) %>% summarize(number=n()) %>% arrange(desc(number))

## Reading layer 'areas_ registration' from data source 'data/areas_
registration.shp' using driver 'ESRI Shapefile'
## Simple feature collection with 54010 features and 29 fields (with 22
geometries empty)
## geometry type:   MULTIPOLYGON
## dimension:       XY
## bbox:            xmin: 171677.6  ymin: 133223.7  xmax: 861895.7  ymax:
775019.1
## epsg (SRID):     NA
## proj4string:     +proj=tmerc +years_0=0 +lon_0=19 +k=0.9993 +x_0=500000
+y_0=-5300000 +ellps=GRS80 +units=m +no_defs ## # A tibble: 16 x 2
##         code_voi   number
##         <chr>      <int>
## 1       14         11292
## 2       10         4733
## 3       30         4574
## 4       06         4040
## 5       04         3609
## 6       20         3503
## 7       32         3074
## 8       28         2957
## 9       02         2897
## 10      12         2642
## 11      26         2280
## 12      24         2184
## 13      22         2152
## 14      18         1777
## 15      08         1210
## 16      16         1086
```

In the same way, the total area of all areas will be calculated separately for each province – with the use of the aforementioned **st_area**() function. A new variable *area* was added to the dataset; then the geometry was omitted and the observations were grouped into voivodeships so that for each of them, one can calculate the total area separately for the constituents included in it. Disabling geometry (**st_set_geometry**(**NULL**)) accelerates calculations.

```
areas %>% mutate(area=st_area(.)) %>% st_set_geometry(NULL) %>%
group_by(code_voi) %>% summarise(area=sum(as.numeric(area))/1e6) # 1-6 only

## # A tibble: 16 x 2
```

```
##         code_voi area
##         <chr>          <dbl>
## 1       02             19936.
## 2       04             17948.
## 3       06             25134.
## 4       08             13990.
## 5       10             18194.
## 6       12             15166.
```

It is worth mentioning that, alternatively to the combination of the **group_by**() and **summarize**() functions, one can use the **aggregate**() command from the stats:: package, which also has the appropriate method defined for the *sf* class objects.

```
areas %>% mutate(area=st_area(.)) %>% aggregate(area~code_voi, data=.,
FUN=sum, on.rm=TRUE) # First 6 only
```

```
##         code_voi    area
## 1       02          19936140275 []
## 2       04          17947525674 []
## 3       06          25134000860 []
## 4       08          13989815872 []
## 5       10          18194435193 []
## 6       12          15166028482 []
```

Thus, it can be seen that the functions of the *sf* package are adapted to process data using the tidyverse:: approach and the pipeline operator. This method of data processing is very convenient and effective, which is why storing spatial data in the form of a data frame is very useful.

At the end, a map of areas within Warsaw will be displayed after their identification with the use of the TERYT number (1465) – see Figure 10.4.

```
areas %>% filter(str_detect(code_area, "1465")) %>% ggplot() + geom_sf() +
theme_bw()
```

A potential alternative to the tidyverse:: approach for processing large datasets in R is the data.table:: package. This tool is considered even more efficient than dplyr:: when performing complex data operations. The use of the dplyr:: package function by many R users is, however, perceived as easier. What's more, the data.table:: package does not support (for now) the analysis of data containing spatial attributes. There is a spatialdatat-able:: (SymbolixAU, 2019) package, but its functionality is limited to the effective calculation of the distance between spatial objects. In addition, it has only a development version and has not been developed since 2017. An interesting solution combining the speed of data.table:: and readability dplyr:: is the recently developed dtplyr:: package, which allows one to write dplyr-like code, which is automatically translated into the equivalent, but usually much faster, data.table code.

10.5 Sample analysis of large raster data

10.5.1 Measurement of economic inequalities from space

The purpose of the analysis in the following example is to check whether the intensity of night-time light intensity (NTLI) is correlated with the size of gross domestic product (GDP) in Poland at the level of voivodeships.

GDP *per capita* or per employee is a commonly used measure of the level of economic development on a national and regional level. There are also attempts to create a synthetic multidimensional measure that would take into account other than income-like aspects of economic or social activity of the population. One of them is the *Human Development Index* (HDI), which includes education and health in addition to income. Synthetic

Figure 10.4 Map of areas within Warsaw

Source: Own study using R and sf:: and ggplot2::.

indicators, using data from official public statistics, often cannot be applied at the local level due to the unavailability of necessary data.

In recent years, the intensity of night lights, assessed on the basis of satellite images of the Earth, is used as a potential alternative to indicators of the level of economic development in national and regional terms. The use of this data allows an approximate measurement of the level of socioeconomic development for areas other than administrative or for countries with less reliable public statistics (e.g. sub-Saharan Africa). Analyses comparing this measure with the values of various indicators from official public statistics for countries where public statistics are reliable show high correlation between the intensity of night lights and GDP at the national level but weaker on a regional level. At the regional level, the intensity of night lights strongly correlates, for example, with the size of the population.

Data on the intensity of night lights is based on satellite images collected and processed by the *National Oceanic and Atmospheric Administration* (NOAA). This institution provides averaged annual data for the period 1992–2013 in unprocessed and processed form (so-called stable lights). The intensity of the lights is measured for pixels of size 30 × 30 arc seconds. This corresponds to less than one square kilometre near the equator and covers an area of about half a square kilometre for the latitude of Poland. For each pixel, the intensity of night lights is given in units called digital numbers (DNs), taking integer values on the 0–63 scale – standardised separately for data from each year or satellite. The scale limitation has serious drawbacks, widely discussed in the scientific literature. For example, it is impossible to distinguish between the intensity of light in city centres and their periphery. Furthermore, low light intensity can be zeroed in the filtering process, so there is no certainty that the zero value means no lighting. DN values of 1 and 2 are underrepresented in the data. Due to the standardisation of data for each year or satellite independently, measurements from different years and/or satellites are not directly comparable.

However, these drawbacks do not limit the frequent use of night-time data in empirical analyses, whose values can be compared over time for different areas in relation to the average value for all analysed units. The huge advantage of this type of data is the possibility of aggregating them to the level of any territorial units, not necessarily administrative ones.

Subsequently one can find an analysis of sample data from the F18 satellite with averaged light intensity measurements for 2013 taken from the NOAA website (2019). The analysis includes stable lights. A single file in. tif format has a size of approximately 720 MB; therefore, for the purposes of the following example, its fragment was prepared limited to the area of Europe.

The standard package for analysing raster data in R is raster::. It stores spatial data in the form of a stack of layers, in which further characteristics are stored – for example, various meteorological indicators and/or data for subsequent moments in time.

The following analysis will be carried out using the raster:: tools. Before performing the analysis, it is necessary to load the needed packages.

```
library(rgdal)
library(rgeos)
library(dplyr)
library(readr)
library(raster)
library(pryr)
library(tabularaster)
```

10.5.2 Analysis using the raster:: package functions

To import data in a raster package, the **raster()** function is used if the data has only one layer (that is, only one value for each cell/pixel) or the **brick()** function if there are more layers and thus more data stored in them. At the same time, only one value can be stored for a given cell in one layer. In the case of night-time lights, the data contains only one layer.

```
data_raster<-raster("data/F182013.v4c_web.cf_cvg_PART.tif")
## [1] "RasterLayer"
## attr(,"package")
## [1] "raster"
## 11.4 kB
```

The loaded data is an object of the *RasterLayer* class. Multilayer data, loaded with the **brick()** function, would be an object of the *RasterBrick* class. Importantly, the **raster()** and **brick()** functions do not automatically load data into the computer's memory. This is due to the fact that raster objects can be very large; they are not always processed completely. To save memory, data values are loaded only when they are processed, and the memory is immediately released after processing. Thanks to this, it is possible to process and display even large raster data on a regular computer. The **inMemory()** function is used to check if data has been loaded into memory. When the answer is FALSE, data has not been loaded. Nevertheless, it can be presented graphically (see Figure 10.5).

```
inMemory(data_raster)
## [1] FALSE
plot(data_raster)
```

The **plot()** method can be used for single-layer data or for single layers of complex objects. Here one can see the outline of the European continent. Displaying the raster object in the console will show basic data, including its dimensions, number of cells, range of geographical coordinates, information about the coordinate projections used, the file name and the range of values stored in it.

```
print(data_raster)

## class   : RasterLayer
## dimensions : 7201, 14400, 103694400    (nrow, ncol, ncell)
## resolution : 0.008333333, 0.008333333  (x, y)
## extent ymax): -29.99583, 90.00417, 14.99583, 75.00417 (xmin, xmax, ymin,
## crs        : +proj=longlat +datum=WGS84 +no_defs +ellps=WGS84
```

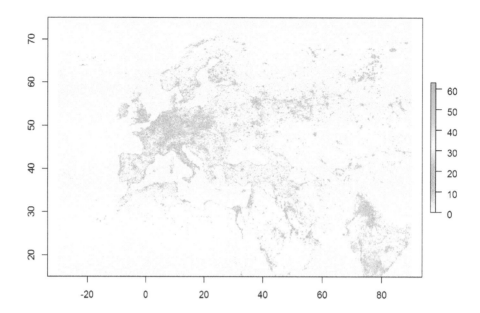

> **Figure 10.5** Intensity of night lights in 2013 – full map of Europe
>
> *Source*: Own study using R and raster:: package.

```
+towgs84=0,0,0
## source  : data/F182013.v4c_web.stable_lights.avg_vis_PART.tif
## names   : F182013.v4c_web.stable_lights.avg_vis_PART
## values  : 0, 63      (min, max)
```

Raster objects can contain numeric data (numeric or integer type) and logical or qualitative (factor) but not textual data. The dimensions of the raster object are stored in the ncols and nrows elements. They can also be checked using the appropriate **ncol**() and **nrow**() commands, and the total number of cells using the **ncell**() command. The number of pixels covering the entire world map is checked once again. The *raster* class is created in the S4 class system.

```
data_raster@ncols
## [1] 14400
data_raster@nrows
## [1] 7201
ncol(data_raster)
## [1] 14400
nrow(data_raster)
## [1] 7201
ncell(data_raster)
## [1] 103694400
dim(data_raster)
## [1] 7201 14400      1
```

In total, there are over 100 million cells. The last (third) dimension displayed by the **dim**() function means the number of layers of the object being analysed, which can also be checked using the **nlayers**() function. The value for a particular cell (pixel) can be read by referring to it using the operator [, as in the case of a matrix or a data frame. In the case of raster objects, the operator [, except for indexing by line number and column, also allows the cell identifier to be provided. Raster objects can be expanded by adding subsequent layers – the **stack**() or **brick**() function is used for this.

```
nlayers(data_raster)
## [1] 1
```

```
data_raster[1111, 9999]
## 0
data_raster[1e6]
## 0
```

Before joining successive layers, it is worth checking if they have compatible characteristics – coordinate range, number of rows and columns, projection, rotation used and resolution. The **compareRaster()** function is used for this. Arguments in **compareRaster()** are the following compared layers. In order not to load additional large data into memory, the following example creates an object with three identical layers containing data about the intensity of night lights in 2013. The results of TRUE mean that all characteristics of all compared objects are compatible. So one can combine them and assign names to individual layers. The output of stacking the raster layers is in class *RasterStack*.

```
compareRaster(data_raster, data_raster, data_raster)
## [1] TRUE

data_raster_3layers<-stack(data_raster, data_raster, data_raster)
names(data_raster_3layers)=c("layer1", "layer2", "layer3")
class(data_raster_3layers) ## [1] "RasterStack"
## attr(,"package")
## [1] "raster"

data_raster_3layers
## class      : RasterStack
## dimensions : 7201, 14400, 103694400, 3 (nrow, ncol, ncell, nlayers)
## resolution : 0.008333333, 0.008333333  (x, y)
## extent     : -29.99583, 90.00417, 14.99583, 75.00417 (xmin, xmax,
ymin, ymax)
## coord. ref.: +proj=longlat +datum=WGS84 +no_defs +ellps=WGS84
+towgs84=0,0,0
## names      :  layer1,     layer2,     layer3
## min values :       0,          0,          0
## max values :      63,         63,         63
```

One can refer to a specific layer in three ways, by using **subset()**, with [[]] operators and with $ operators, as shown subsequently.

```
raster::subset(data_raster_3layers, "layer1")
## class      : RasterLayer
## dimensions : 7201, 14400, 103694400, 3  (nrow, ncol, ncell, nlayers)
## resolution : 0.008333333, 0.008333333   (x, y)
## extent     : -29.99583, 90.00417, 14.99583, 75.00417   (xmin, xmax,
## coord. ref.: +proj=longlat +datum=WGS84 +no_defs +ellps=WGS84
+towgs84=0,0,0
## data source: dane/F182013.v4c_web.stable_lights.avg_vis_PART.tif
## names      : layer1
## values     : 0,     63    (min, max)

data_raster_3layers[["layer2"]]
## class      : RasterLayer
## dimensions : 7201, 14400, 103694400, 3  (nrow, ncol, ncell, nlayers)
## resolution : 0.008333333, 0.008333333 (x, y)
## extent     : -29.99583, 90.00417, 14.99583, 75.00417   (xmin, xmax,
## coord. ref. : +proj=longlat +datum=WGS84 +no_defs +ellps=WGS84
+towgs84=0,0,0
## data source : dane/F182013.v4c_web.stable_lights.avg_vis_PART.tif
## names       : layer 2
## values      : 0,     63    (min, max)
```

509

```
data_raster_3layers$layer3 ## class : RasterLayer
## dimensions : 7201, 14400, 103694400, 3 (nrow, ncol, ncell, nlayers) ##
resolution     : 0.008333333, 0.008333333 (x, y)
## extent     : -29.99583, 90.00417, 14.99583, 75.00417  (xmin, xmax,
## coord. ref. : +proj=longlat +datum=WGS84 +no_defs +ellps=WGS84
+towgs84=0,0,0
## data source : dane/F182013.v4c_web.stable_lights.avg_vis_PART.tif
## names      : layer 3
## values     : 0,   63   (min, max)
```

Before performing further analyses, the data will be limited only to the area of Poland. For this purpose, a shapefile file with a map of Polish voivodeships will be used.

```
map_voi<-st_read(dsn="data/voivodeships.shp")
## Reading layer 'voivodeships' from data source 'data/voivodeships.shp'
using driver 'ESRI Shapefile'
## Simple feature collection with 16 features and 29 fields
## geometry type:  MULTIPOLYGON
## dimension:      XY
## bbox:           xmin: 171677.6  ymin: 133223.7  xmax: 861895.7  ymax:
775019.1
## epsg (SRID):    NA
## proj4string:    +proj=tmerc +lat_0=0 +lon_0=19 +k=0.9993 +x_0=500000
+y_0=-5300000 +ellps=GRS80 +units=m +no_defs
## [1] "sf"         "data.frame"
```

In order to be able to impose a map on raster data, it must be ensured that an identical projection of geographic coordinates is used in both datasets. The function **crs()** enables checking the current values of these parameters for both *sf* and *raster* objects.

```
crs(data_raster)
## CRS arguments:
##   +proj=longlat +datum=WGS84 +no_defs +ellps=WGS84 +towgs84=0,0,0

crs(map_voi)
## [1] "+proj=tmerc +lat_0=0 +lon_0=19 +k=0.9993 +x_0=500000 +y_0=-5300000
+ellps=GRS80 +units=m +no_defs"
```

To standardise the projection used, one can simply copy the one used in the raster file to the set with the map. The **st_transform()** function is used for this:

```
map_voi<-st_transform(map_voi, crs(data_raster)@projargs)
```

For the purpose of further analysis, the dataset with the map will be limited to columns with the name (*jpt_kod_je, jpt_nazwa_*) and the area (*jpt_area*) of territorial units.

```
map_voi<-map_voi %>% dplyr::select(jpt_kod_je, jpt_nazwa_, jpt_area)
```

The **crop()** function is used to spatially limit the raster dataset, in which the target spatial range is given as the second argument. This must be an *Extent* class object resulting from the **extent()** function. It returns the minimum and maximum values of the *x* and *y* coordinates of the analysed raster data. In the case of data on the intensity of night lights, these are the extreme values of geographical coordinates – length (*x*) and latitude (*y*), respectively.

```
extent(data_raster)   # Checking the spatial extent of the analysed raster data

## class   :    Extent
## xmin    :    -29.99583
```

```
## xmax    :    90.00417
## ymin    :    14.99583
## ymax    :    75.00417

extent(map_voi) # checking a set with a map of Polish regions:

## class   :    Extent
## xmin    :    14.12289
## xmax    :    24.14578
## ymin    :    49.00205
## ymax    :    54.83642
```

The raster dataset will be cut to the rectangle in which Poland is located. In the **crop()** function, the first argument is the source data, while the second is a crop range. If necessary, the transformed raster data can be saved to a file using the **writeRaster()** function. The resulting object with the superimposition of the border maps of Polish voivodeships on the data on the intensity of night lights is shown in Figure 10.6.

```
data_raster_PL<-crop(data_raster, extent(map_voi)) # crop range
writeRaster(data_raster_PL, filename="data_raster_PL.tif", overwrite=TRUE)
plot(data_raster_PL)
plot(st_geometry(map_voi), add=TRUE)
```

The highest illuminance values in the largest cities and metropolitan areas are clearly visible. However, this dataset still contains areas located outside of Poland. The **mask()** function is used to identify pixels of a raster object that lie inside particular polygons defined by a spatial object.

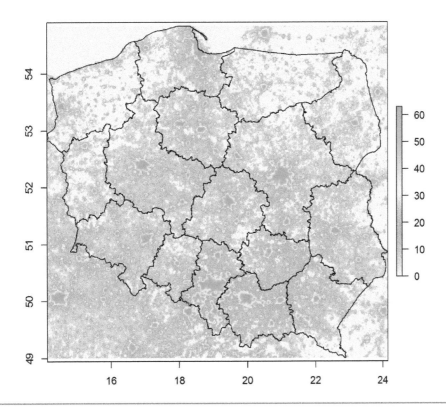

Figure 10.6 Intensity of night lights in 2013 – limited to the area of Poland

Source: Own study using R and raster package ::.

It creates a new raster object that has the same values as the source object. However, if a spatial object provided as the mask argument is a map (*SpatialPolygon*, *SpatialPolygonDataFrame*, *sf*), the resulting raster object will be limited to pixels inside the map area (values for the pixels outside the polygons are converted to missing). In the following example, the resulting object will be stored under the same name as the source object. Once again, the resulting object (data truncated to the area of Poland) will be displayed with the imposition of a map of regional borders. The result is shown in Figure 10.7. Now the data was correctly limited to the area of Poland.

```
data_raster_PL<-mask(data_raster_PL, mask = map_voi)
plot(data_raster_PL)
plot(st_geometry(map_voi), add = TRUE)
```

The next step in the analysis will be the aggregation of the values of the night lights intensity to the level of the voivodeships (regions). Before this, one can check the number of pixels with **ncell**() command, omitting those with missing values with the **!is.na**() command. It is also worth checking in which order the voivodeships appear in the dataset with the map, because the order of aggregation results will be analogous. Voivodeship codes in the Polish Central Statistical Office are even numbers from 2 to 32 assigned in alphabetical order of voivodeship names. For simplicity, the map dataset will be sorted by a column with the region code using the **arrange**() command.

```
ncell(data_raster_PL[!is.na(data_raster_PL)])
## [1] 591102

map_voi$jpt_kod_je
##  [1] 16 26 04 14 22 24 28 32 02 30 10 20 12 08 18 06
## Levels: 02 04 06 08 10 12 14 16 18 20 22 24 26 28 30 32

map_voi<-map_voi %>% arrange(jpt_kod_je)
```

In order to aggregate data, it is necessary to identify in which unit (voivodeship) each pixel lies. The **extract**() function from the raster:: package will be used. It allows one to identify pixel values of a raster object lying in the range of selector, which can be of various types of spatial objects – points, lines or polygons. In the following case, the voivodeship map, which consists of 16 polygons, will be used as a selector. Because the **extract**() function is present in many R packages, to avoid unexpected errors, use it with the package name. By the way, the time of executing the following code will be measured (in seconds) with the **system.time**() command. The resulting object is a list of 16 vectors containing values from cells of a raster object located in subsequent voivodeships – it can be checked with the **class**() and **str**() commands.

```
system.time(data_raster_PL_voi<-raster::extract(data_raster_PL, map_voi))
##    user    system   elapsed
##   48.90    0.02      49.12

class(data_raster_PL_voi)
## [1] "list"

str(data_raster_PL_voi) ## List of 16
##  $ : num [1:36851] 5 6 6 6 5 6 6 6 6 6. . .
##  $ : num [1:34689] 7 7 7 7 7 7 7 8 8 8. . .
##  $ : num [1:46549] 6 6 7 0 6 6 6 7 7 7. . .
##  $ : num [1:26472] 0 0 0 0 0 0 0 0 0 0. . .
##  $ : num [1:34058] 7 8 8 7 8 8 8 7 8 8. . .
##  $ : num [1:27351] 10 10 11 10 11 11 11 11 12 12. . .
##  $ : num [1:67536] 0 0 0 0 0 0 0 0 0 0. . .
##  $ : num [1:17219] 0 0 0 6 5 0 0 0 0 0. . .
##  $ : num [1:32186] 7 7 7 7 7 8 8 9 10 10. . .
##  $ : num [1:39141] 0 0 0 0 0 0 0 0 0 0. . .
##  $ : num [1:36295] 0 0 0 0 0 7 8 10 10 10. . .
```

```
##   $ : num [1:22418] 13 15 14 10 11 11 10 10 9 9. . .
##   $ : num [1:21488] 8 8 8 8 8 7 8 8 8 8. . .
##   $ : num [1:47515] 0 0 0 0 0 0 0 0 0 0. . .
##   $ : num [1:56606] 6 6 0 0 7 7 6 0 0 0. . .
##   $ : num [1:44728] 0 0 0 0 7 0 0 0 0 0. . .
```

The final step will be adding up values for individual areas with the **sapply()** command. Much more time-efficient identification and aggregation of data for selected areas can be obtained with the use of the additional function of **cellnumbers()** from the tabularaster:: package by Michael D. Sumner (2018). Another advantage of this approach is the compatibility of the resulting objects with the tidyverse:: environment. The result is a *data frame (tibble)* that assigns each cell the object number (region, voivodeship). Checking the time efficiency of this solution is possible with the **system.time()** function. This method is 500 times faster than using simply the **extract()** function. It is worth checking if the results of both codes are identical.

```
lights_voi<-sapply(data_raster_PL_voi, sum, na.rm = TRUE)
print(lights_voi)

##   [1]  435658 368045 427221 186357 373306 514639 733677 171916 374102
221116
## [11] 352334 514245 268449 226330 605407 225656

## Classes 'tbl_df', 'tbl' and 'data.frame':       591102 obs. of       2
variables:
##   $ object_: num  11 11 11 11 11 11 11 11 11 11. . .
##   $ cell_  : int  467 468 469 470 471 472 473 474 475 476. . .
##      User   system   elapsed
##      0.0    80.00    0.08
identical(lights_voi, lights_voi2$lights)
## [1] TRUE


system.time ({cell<-cellnumbers(data_raster_PL, map_voi) str(cell)
cell %>% mutate(light=raster::extract(data_raster_PL, cell$cell_)) %>%
group_by(object_) %>% summarise(lights=sum(light, na.rm = TRUE)) ->
lights_voi2})
```

Finally, the data on the intensity of night lights aggregated to the level of voivodeships will be compared with the value of gross domestic product and population in 2013. Let's load relevant data downloaded from the Local Data Bank of the Polish Central Statistical Office.

```
dataCSO<-read_csv("lights_data/dataCSO.csv")
## Parsed with column specification: cols(voiid = col_double(), name =
col_character(),
## population2013 = col_double(), GDP2013 = col_double())


head(dataCSO)
## # A tibble: 6 x 4
##      voiid   name                   population2013    GDP2013
##      <dbl>   <chr>                           <dbl>      <dbl>
## 1        2   "DOLNO\x8cL\xa5SKIE"          2911045     140251
## 2        4   KUJAWSKOPOMORSKIE             2094090      73880
## 3        6   LUBELSKIE                     2160513      65786
## 4        8   LUBUSKIE                      1022253      36582
## 5       10   "\xa3\xd3DZKIE"               2517787     101077
## 6       12   "MA\xa3OPOLSKIE"              3356805     128119
```

The column *voiid* contains the region code analogous to the unit code in the set with the map *map_voi*. In the next step, the aggregate values of night lights are added to CSO data on population and GDP, and the correlation between these variables is checked. From the output, one can see that the correlation of light intensity with GDP and population is very high (0.89 and 0.94, respectively).

```
map_voi$jpt_kod_je %>% as.character() %>% as.numeric() %>%
identical(dataCSO$voiid)
## [1] TRUE

dataCSO$lights2013 <- lights_voi2$lights
dataCSO %>% dplyr::select(ends_with("2013")) %>% cor()

##                 population2013    GDP2013    lights2013
##  population2013     1.0000000    0.9490750    0.9404741
##  GDP2013            0.9490750    1.0000000    0.8905807
##  lights2013         0.9404741    0.8905807    1.0000000
```

10.5.3 Other functions of the raster:: package

In addition, the raster:: package contains functions that allow for calculating descriptive statistics for raster objects. The **summary**() method is defined for them, displaying the minimum, maximum, median, first and third quartiles and the number of missing observations. The calculation of other descriptive statistics, for example, average, standard deviation or other summaries defined with the use of own functions, can be performed using the **cellStats**() function. If the function **summary**() or **cellStats**() is applied to an object with multiple layers, the summary will be performed for each layer separately.

```
summary(data_raster_PL) # summary of object created before
##         F182013.v4c_web.stable_lights.avg_vis_PART
## Min.                                              0
## 1st Qu.                                           5
## Median                                            8
## 3rd Qu.                                          12
## Max.                                             63
## NA's                                         250998

cellStats(stack(data_raster_PL), mean, na.rm = TRUE)
## [1] 10.14792

cellStats(stack(data_raster_PL), sd, na.rm = TRUE)
## [1] 11.2282

summary(data_raster_3layers) # applying function on more than one layer

## Warning in. local(object, . . .) : summary is an estimate based on a
sample of 1e+05 cells (0.01% of all cells)
##          layer1     layer2     layer3
## Min.          0          0          0
## 1st Qu.       0          0          0
## Median        0          0          0
## 3rd Qu.       0          0          0
## Max.         63         63         63
## NA's          0          0          0
```

The use of atypical (also own) summarising functions may not work for very large raster datasets. In this case, only basic functions are supported, such as: **sum**(), **mean**(), **min**(), **max**(), **sd**(), 'skew'() and 'rms'() (see the **cellStats**() documentation).

```
cv <- function(x, na.rm) 100 * sd(x, na.rm=na.rm)/mean(x, na.rm=na.rm)
cellStats(data_raster_PL, cv, na.rm = TRUE)
## [1] 110.6453
```

For raster objects, a number of methods are also defined that allow for easy visualisation of their values – **boxplot**(), **density**(), **hist**(). If the function summarising or visualising the data does not have the appropriate method defined for raster files, one can first extract the values for all cells and only in the next step apply the appropriate summary function. To get all values stored in a given layer, the **values**() or **getValues**() function should be used. As an example, a frequency table for the intensity of night lights in Poland was generated. Values of 0 make up almost 25% of them, but there are no values of 1, 2 and 3 at all, which is related to the filtering of *stable lights*, mentioned in the introduction to this chapter.

```
table(getValues(data_raster_PL))
##      0       4       5       6       7       8       9      10      11      12
## 141244      31    7026   63459   74326   57058   41148   30287   23037   18352
##     13      14      15      16      17      18      19      20      21      22
##  14715   12095   10202    8357    7143    6315    5576    4949    4466    3934
##     23      24      25      26      27      28      29      30      31      32
##   3548    3164    2962    2690    2567    2286    2191    1974    1843    1771
##     33      34      35      36      37      38      39      40      41      42
##   1650    1535    1473    1311    1369    1208    1185    1147    1102    1032
##     43      44      45      46      47      48      49      50      51      52
##    971     983     922     912     926     838     842     773     805     811
##     53      54      55      56      57      58      59      60      61      62
##    797     794     784     806     822     773     844     920    1063    1307
##     63
##   1681
```

10.5.4 Potential alternative – stars:: package

Recently, an alternative package to analyse raster data appeared, called stars:: (abbreviation for *scalable, spatiotemporal tidy arrays*) by Edzer Pebesma and Roger Bivand (2019). The package allows storing more complex raster data – defining dynamic stacks, supporting rotated, cut, rectilinear and curved rasters – and ensures tight integration with the sf:: package and the data processing approach used in tidyverse:: packages. According to the authors' declaration, the stars:: package is intended to be more scalable than the raster package, enabling computational processing of raster data of sizes far exceeding the RAM volume and computing capabilities of individual computers (which is not difficult – real-time climatic data has volume measured in terabytes and even in petabytes). The package stars:: is supposed to be in line with the tidyverse:: approach, but is still under development, and its functionality is limited. Therefore, it will not be discussed here. Progress in the development of the package can be tracked in its repository on the github portal (Pebesma, 2019d).

Notes

1 Keyhole markup language (KML) – a markup language that uses XML. It is an open standard approved by the Open Geospatial Consortium allowing visualisation of three-dimensional spatial data. Is used, among others, in Google Earth, Google Maps, Bing Maps, Flickr and NASA World Wind applications.
2 Using laser scanning from satellites or planes using a device called LIDAR (from the English acronym LIDAR, derived from the expression light detection and ranging or laser imaging detection and ranging), which is a combination of a laser and a telescope – see Wikipedia (2019a).
3 If the number of mapping objects M is greater than the number of servers, some servers will perform several tasks.
4 A detailed description of the example along with the visualisation can be found, for example, in Posa (2019).
5 See, for example, the sparklyr:: package, which is an R interface to the Spark system, allowing the tidyverse:: RStudio (2019a) approach to data processing.
6 A detailed description of all types of geometries can be found for example, in Pebesma, (2019c).
7 The function uses the Douglas-Peucker algorithm, also used, for example, in the *ST_simplify* function in PostGIS (Wikipedia, 2019b).

## Chapter 11

Spatial unsupervised learning – applications of market basket analysis in geomarketing

Alessandro Festi

Market basket analysis is a widely employed technique in marketing which provides product suggestions for purchases to a customer. These suggestions are based on past purchases made by customers through the statistical methodology of association rules. This technique was initially conceived to analyse transactions of products where the dataset is composed of sets of items purchased in different periods of time by N individuals, for instance, formed from a large number of receipts collected at the point of sale of a large-scale retail outlet. This chapter extends the application of market basket analysis to the geolocation points of N people in a specific space in order to identify associations among places that individuals have visited. Typical market basket analysis deals with products, which are linked into groups depending on frequency and composition of purchases. This analysis is to link locations. Similarly, one needs sets of geolocation points for each individual considered on many different occasions.

This chapter presents first the type of data needed for the analysis and the current limitations, namely the sensitivity and availability of this data. Second, the model used is presented: the simulation of the data, the assumptions under the validity of the results and the essential role of spatial open data to generate useful insights. Finally, the outputs will be discussed, presenting some areas of interest in which the model could be implemented. The text contains selected code in order to make the analysis reproducible, while the whole script can be downloaded from my github profile: https://github.com/alessandrofesti/Spatial-Association- Rules-for-geomarketing.

11.1 Introduction to market basket analysis

The purpose of the market basket algorithm is to investigate the purchasing habits of customers searching for associations among items which they have put in their shopping carts. Three basic measures asses how good the discovered rules are: support, confidence and lift. *Support* is an indication of how frequently the itemset (set of items purchased) appears in the dataset. This is the proportion of transactions in the dataset that contain the itemset. *Confidence* is an indication of how often the rule has been found to be true. For instance, the rule *productA => productB* represents the proportion of transactions containing *productA* that also contain *productB*. *Lift* compares the support of the rule with the expected support under the hypothesis of independence among items. The higher the value for support, confidence and lift, the better the rule. On the other hand, if one of the indices is high, that does not imply that the others are high as well; rather, this means that the indices need to be jointly evaluated. This means also that the best and worst rules depend on the index being considered.

An R output showing the associations of products is usually as follows.

```
   lhs       =>    rhs               support     confidence   lift        count
 {herbs}    => {root vegetables}    0.0032982   0.1341330    234.34574   52
 {berries}  => {sour cream}         0.0030857   0.6438059    122.68394   51
```

The first rule *{herbs} => {root vegetables}* links herbs with root vegetables. Thus, on the basis of people's preferences, it suggests that people who are interested in buying herbs could also be interested in buying root vegetables. Considering this rule, a supermarket could adopt a cross-selling strategy towards its customers and, for instance, offer them a certain discount for the combined purchase of *herbs & root vegetables*. It is important to remark that rules composed of one item on the left-hand side of the association and one on the right-hand side are always bi-directional: *product A => product B* implies that *product B => product A* with the same values for support and lift.

When one deals with geolocation data, the reasoning does not change: one just needs to consider that the items, instead of products, will be geolocated locations with specific coordinates. These locations can be any kind of public or private activity for which one can get and use any spatial identifier. In the example, private companies are linked. The following table shows the output of the market basket analysis model applied to geolocation data. The numbers one the LHS and RHS (instead of products) represent the index of the linked companies.

```
 lhs     =>  rhs      support       confidence    lift          count
 {525}   => {716}     0.003241783   0.7912088     156.89953     72
 {716}   => {525}     0.003241783   0.6428571     156.89953     72
 {926}   => {3162}    0.003151733   0.8974359     142.37179     70
 {3162}  => {926}     0.003151733   0.5000000     142.37179     70
 {3415}  => {2764}    0.003241783   0.7826087     140.17532     72
 {2764}  => {3415}    0.003241783   0.5806452     140.17532     72
 {1158}  => {1184}    0.003061684   0.6296296     78.12332      68
 {1184}  => {1158}    0.003061684   0.3798883     78.12332      68
 {665}   => {2848}    0.003196758   0.8452381     145.52510     71
 {2848}  => {665}     0.003196758   0.5503876     145.52510     71
```

One can read the first association {525} => {716} in this way: people who on average get close to the indexed activity 525 tend to get close to the indexed activity 716 with a support of 0.0032, confidence of 0.7912 and lift of 156.899, and vice versa. Thus, 0.0032 represents the proportion of individual paths which contain the itemset {525,716}, 0.7912 represents the proportion of individual paths containing the indexed company {525}, which also contains the indexed company {716}, while 156.899 represents the ratio between support and expected support of the rule under the hypothesis of independence among the two indexed companies. Going further, one can state that given that a person has been to the place represented by the first indexed company (525), which, for instance, could be a cinema or a flower shop, it is likely, with respect to the associated indices of support, confidence and lift, that that person will go (or has already gone) to the second company (716). In this way, locations are linked, allowing general patterns to be found with respect to people's movement in a specific geographic area.

11.2 Data needed in spatial market basket analysis

Spatial market basket analysis for tracking the path of individual customers requires the data of N individuals in a particular area for multiple paths. A path includes the set of locations that a customer has visited on a particular occasion. Intuitively, think about yourself: every day, you may go to work, then go to some shops or for a walk, and maybe after that you meet some friends or go dining outside with your family. Imagine tracking your position, geocoding it and writing down the latitude and longitude at every interval of time t (t can be 1 second,

3 seconds or 30 seconds . . . it is an arbitrary value). The smaller the interval, the more precise the data and the heavier and more difficult to manage the dataset is. In this way, you would be able to get a set of dots, connect them with lines and draw your daily movements.

Obviously, individual paths can be of any kind and any shape. In reality, especially within a city, linear paths are very unlikely because streets are curved and one can encounter many physical obstacles. Technically, paths can be simply simulated using simple functions.

```
x <- runif(50, -1,1) # Inputs uniformly distributed [-1,1]
y   <-   3*x   +   5plot(x,   y,   xlim=range(x),   ylim=range(y),   xlab="x",
ylab="y",pch=16)
lines(x[order(x)], y[order(x)], xlim=range(x), ylim=range(y), pch=16)

y2<-12*x^4-10*x^2+x-4
plot(x, y2, xlim=range(x), ylim=range(y2), xlab="x", ylab="y2", pch=16)
lines(x[order(x)], y2[order(x)], xlim=range(x), ylim=range(y2), pch=16)

y3<-2*x^7+7
plot(x, y3, xlim=range(x), ylim=range(y3), xlab="x", ylab="y3", pch=16)
lines(x[order(x)], y3[order(x)], xlim=range(x), ylim=range(y3), pch=16)
```

Figure 11.1 shows just three examples out of an infinite set of possibilities: consider, for instance, the second curved figure, and imagine it on a map where the points are the locations visited by an individual (in a day, for instance), while the lines represent the connections between them. Imagine writing down the latitude and longitude of your movements every day for many days, and imagine many other people doing the same as you. This data generated by many people in a geographic/spatial area is exactly the data needed for this analysis. Dealing with many paths for many individuals means that lines intersect and points fall close or overlap, creating connections and showing associations among people's movements. This is basically what one is interested in discovering, analysing and making use of.

An important issue is whether this real-path data exists. According to a survey by Carto, *The State of Location Intelligence* (2018), location data has become more and more appealing on the market to identify new customers and market strategies and improve customer service. It is mainly collected through websites, web-based applications and mobile devices.[1] Seventy-eight percent of executives interviewed were planning on investing in location data within a year and 84% within three years. Spatial data is expected to increase and spatial analysis to become more and more important. Despite this, this data is not easy to get, mainly because of two reasons: 1) costs and 2) privacy. Intuitively, the *cost* of gathering and processing the data of the positions of many individuals for many different intervals of time is an important reason for business managers to evaluate if it is worthwhile and profitable. What is evident is that blockchains, cloud architectures and peer-to-peer technologies are becoming more widely used; in the same way, data storage costs are destined to decrease proportionally. So one should expect that the best is yet to come.

Concerning the second reason, privacy, there are a lot of geolocation-based applications such as Google Maps, Uber, Lime, Mobike and Fitbit, just to mention a few, which keep track of our positions. There are also many other apps that one uses daily which do not directly offer geobased services but which track our position

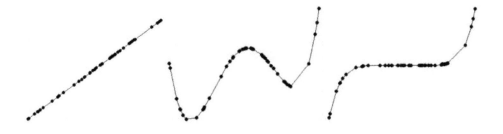

Figure 11.1 Three different possible individual paths

Source: Own code in R

as well. When one installs them on our devices, they ask us for permissions to gather, manage, integrate and perform analysis on this data. For instance, by having Google Maps installed on a device and giving it permission to track our position, every time one goes to a restaurant or a museum, the application asks you to evaluate your experience at that precise place. That could seem particularly scary to some people, but not to the majority. According to a McKinsey & Co. study (2016) on car use, 55% of more than 3000 interviewees would be comfortable with sharing their location data.[2]

When giving an application access to a user's position, the engineers behind the platform can have access to any spatial data concerning customer movements: it is up to the company, because of the costs and its ethics, to limit the use of the data to the authorisations required. Given that the issue is particularly delicate, and remembering that this is not a book about IT laws and regulations, just a few pieces of information will be added. Nevertheless it is important to remark that this data already exists, and there are some companies, especially in the United States, which have emerged specifically to analyse it.[3]

The issue of privacy has already been noted by national governments and international organisations, which is why it is important that recent regulations have been approved. The *General Data Protection Regulation* (GDPR) is an EU regulation that was adopted in April 2016 and came into force in May 2018. This responds to the necessity to unify all the EU member state data regulations into one programme. The GDPR describes requirements for data processing for companies and organisations and establishes that processors are required to offer customers explicit and transparent notification about their data practices. In other words, from May 2018 on, companies which operate in Europe have been obliged to ask customers transparently, through non-pre-checked checkboxes, to accept the way their data would be collected, used and managed. Furthermore, under the GDPR, companies, besides asking for consent, need to have "legitimate reasons" to process employees' personal data. Those which also process sensitive data must appoint an official data protection officer to inform and advise the organisation about any privacy issue to be considered.[4] Google was accused in November 2018 of GDPR privacy violation by seven European countries for tracking phone users regardless of their privacy settings.[5] Therefore, this data exists but requires specific procedures to be processed and employed.

11.3 Simulation of data

This analysis will be conducted on the simulated spatial data. The fundamental assumption is that people's movements tend to show some general "correlated" trends. The idea here is intuitive and informal. It is to stress that people, on average, tend to show some recurrent tendency in their movements both at the individual and aggregate levels. At the *individual level*, for instance, one can expect that a person who has gone to work today will go to work tomorrow, too. At the *aggregate level*, one may expect that people who have a common interest are more likely to go to the same places. For instance, people who have a dog, on average, tend to go to the veterinary surgeon and to the park more often than others. So, one can expect a general shared trend because of that common trait. It is also evident that if all the people went every day to the same places, the market basket algorithm would identify rules associating all these locations. On the contrary, if all the people changed paths every day, going every day to new and different places, one would expect the algorithm to find random associations. Thus, realistically, one should expect some degree of correlation with previous movements both for each individual and among sets of individuals. For simplicity, the simulated data includes just the correlation among the same individual's paths. The concept is summarised in Figure 11.2.

Figure 11.2 shows some correlated paths. As one can see, the patterns look similar in the three cases, but there are some locations (points) which change among the paths. So, to strengthen the concept, it is important for this spatial analysis, which uses individual spatial data, to take into account the correlation among people's shifts. Here the correlation among the points of the individual paths is guaranteed by using the multivariate normal (MVN) distribution.

The area which is considered in this example is the metropolitan city of Bologna (Italy). This is because of the availability of open data sources concerning spatial coordinates of private activities. Details of data are described in Section 11.4.

First, some "starting points" are created, and then one generates the paths using the MVN. This distribution allows for simulating the data with an arbitrary level of correlation among the univariate normal distributions which compose it. This means that for this analysis, in order to simulate many people's paths, one simulates many multivariate normal distributions, one for each individual. Then, given a multivariate normal distribution for an individual, the sets of individual locations composing each of his paths will correspond to the set of points generated by each normal distribution.

The first step is to choose a set of random points in the area of Bologna: they represent the "central" location visited by the simulated individuals. The idea is to consider a central location of the city; then, in order to

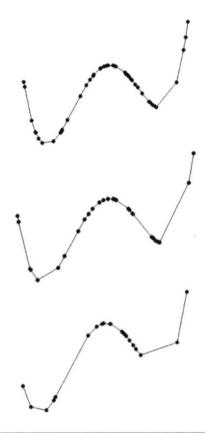

Figure 11.2 Correlated set of individual spatial locations

Source: Own code in R

simulate points with a certain variability from the city centre, one generates 20 locations through a normal distribution using the **rnorm**() function starting from this central point. In the case of the city of Bologna, Palazzo d'Accursio, which has, respectively, latitude and longitude coordinates (44.493674, 11.342220) is chosen as the central location. Thus, the set of initial points has a latitude and longitude equal to the latitude and longitude of Palazzo d'Accursio plus some variability determined by the normal distribution. The variability of the normal distribution then defines the distance of the simulated points from Palazzo d'Accursio. Formally, it is assumed that the latitude and longitude of the central position c of the individual i are normally distributed. As one can see from the following code, the normal variability is divided by 250, an arbitrary value, which allows for making the new points lie not too far from the city centre. When changing from 250 to a greater value, one gets smaller variability and points in a smaller area. In contrast, when choosing a value greater than 250, one gets points in a bigger area because of the bigger variability from the centred point.

$latitude_{ic} \sim (0,1)/250 + 44.493674$
$longitude_{ic} \sim (0,1)/250 + 11.342220$

```
# A set of 20 random points representing the central location
points<-as.data.frame(cbind(rnorm(20)/250+11.342220,
rnorm(20)/250+44.493674))
```

After obtaining the first 20 initial points, one can visualise them (as in Figure 11.3 for Palazzo d'Accursio) in Figure 11.4. For this purpose, one can use the packages shiny:: and leaflet::. shiny:: is particularly useful in creating interactive web applications for R, while leaflet:: is an interactive open-source JavaScript library to make interactive maps. shiny:: operates in a reactive programming framework. This means that any time any UI element that affects the result changes, so does the result. In the following code, the shinyApp is called through two arguments: UI, which is the user interface of the map, and server, which builds a list-like object named

Figure 11.3 Location of Palazzo d'Accursio (Bologna, Italy) – central location of the area under the study

Source: Own visualisation generated in R with shiny:: and leaflet:: packages

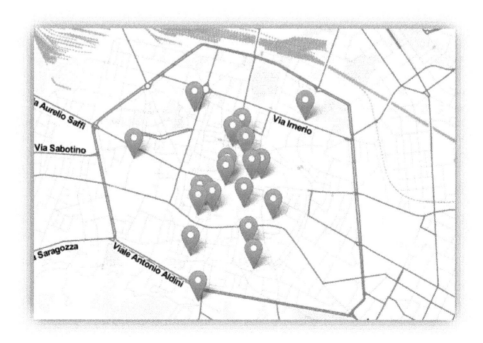

Figure 11.4 Location of the 20 initial points

Source: Own visualisation generated in R with shiny:: and leaflet:: packages

output that contains all of the code needed to update the R objects in the app. One first specifies the colours for the map through the function **rgbt()**, where "250" defines the intensity of the tone; then the function **fluidPage()** constructs a HTML file with R functions. The **leafletOutput()** function creates a UI element as a map, and the **actionButton()** function creates an action button which allows recalculating the value every time the button "New points" is triggered. On the server side, the function **eventReactive()**, from the input points, updates the output through the action button. Then, inside the **renderLeaflet()** expression, a leaflet map object is returned, and with the **addProviderTitles()** function, a title layer from a known map provider is added. One can use names(providers) in the R console to view all the possible options. Dealing with a map, the attribute *noWrap* is necessary to specify that the content should not wrap. Then, through the **addMarkers()** function, one lets the function know which columns contain the latitude/longitude data, and finally, a Shiny app object is generated through the function **shinyApp()** from the UI/server pair.

When one wants to plot spatial locations on a map with the same settings, latitudes and longitudes need to be defined, in matrix format, in an object. That object, called *points*, stemming from the **eventReactive()** command, is in the following code. It is a matrix which contains latitude and longitude of the previously generated random locations. Then, from these starting locations, the individual paths will be simulated.

```
r_colors<-rgb(t(col2rgb(colors()) / 255))
names(r_colors)<-colors()
UI<-fluidPage(
  leafletOutput("mymap"),
  p(),
  actionButton("recalc", "New points")
)
server<-function(input, output, session) {
  points<-eventReactive(input$recalc, {
  points
  }, ignoreNULL = FALSE)
  output$mymap <- renderLeaflet({
    leaflet() %>%
      addProviderTiles(providers$Stamen.TonerLite,
                       options = providerTileOptions(noWrap = TRUE)
      ) %>%
      addMarkers(data = points())
  })
}
shinyApp(UI, server)
```

The number of initial points will be equal to the number of individuals who are considered in the simulation – here it is 20. Then, starting from each of these points, one generates the sets of correlated paths for each individual using the multivariate normal distribution. In this way, one gets a dataset composed of the data of N individuals. For each of them, there is a set of correlated paths in the same spatial area. The degree of correlation in this example is set to 0.7. A multivariate normal distribution was generated with the **mvrnorm()** function from the MASS:: package. The following code allows, for each of the 20 simulated individuals, one to generate 10 paths composed of 10 locations each. *Sigma* and *mu* specify, respectively, the covariance matrix with a level of correlation of 0.7 and the vector of the means of the MVN, which in this case is composed of all zeros.

```
# Setting parameters
n_individuals<-20
n_paths<-10
n_locations<-10
points<-data.matrix(cbind(rnorm(n_individuals)/200 + 11.342220, rnorm(n_
individuals)/200 + 44.493674))
correlation<-0.7

# Defining path-generator function
paths_gen<- function(points) {
  paths_one<-c()
```

```
      for(j in 1:nrow(points)) {
        for (i in 1:n_paths) {
          mu <- rep(0,n_locations)
          Sigma<-matrix(correlation, nrow=n_locations, ncol=n_locations) +
    diag(n_locations)*.3
          rawvars<-mvrnorm(n=n_locations, mu=mu, Sigma=Sigma)
          geo_points<-as.data.frame(cbind(rawvars[,i]/250 +
    as.double(points[j]), rawvars[,i]/250 + as.double(points[j+nrow(points)]))))
          gg<-cbind(geo_points, i, j)
          paths_one<-rbind(paths_one, gg)
      return(as.data.frame(paths_one))
    }

    # Applying path-generator function to generated points paths<-paths_
    gen(points)
```

Running the code, one generates for each individual (*j*) 10 locations (*i*) for each path, each of them showing latitude and longitude. The head of the resulting table is shown in Figure 11.5, where line 10 shows the latitude and longitude of the last position generated for the first individual for the first path, while line 11 shows the latitude and longitude of the first position generated for the first individual for the second path.

With this approach, 20 individuals with 10 paths for each showing 10 locations each have been simulated. Figures 11.6a and b show two paths for the same individual (sets of correlated locations). One can interpret them as two different sets of spots in which the same individual has been geolocated in two different occasions. In contrast, Figure 11.6c shows two paths of two different individuals. Here, the two starting points are different, and there is no correlation among the generated locations. One should note that correlation refers to the locations visited and not to the shape of the paths. In fact, as one can see, the individual paths were considered linear directions in space because of their simplicity for this simulated example. Of course, one could apply non-linear curved transformations in order to make them curved and more realistic.

	long	lat	i	j
1	11.34309	44.49177	1	1
2	11.34585	44.49453	1	1
3	11.34611	44.49479	1	1
4	11.34451	44.49319	1	1
5	11.35183	44.50051	1	1
6	11.34846	44.49714	1	1
7	11.34618	44.49486	1	1
8	11.35200	44.50068	1	1
9	11.34513	44.49381	1	1
10	11.34303	44.49171	1	1
11	11.35127	44.49995	2	1

Figure 11.5 **Visualisation of the dataset (paths) – generated locations and correlated paths**

Source: Own visualisation generated in R

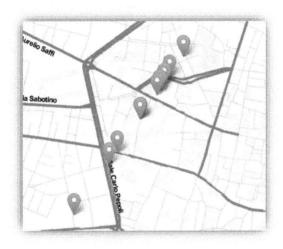

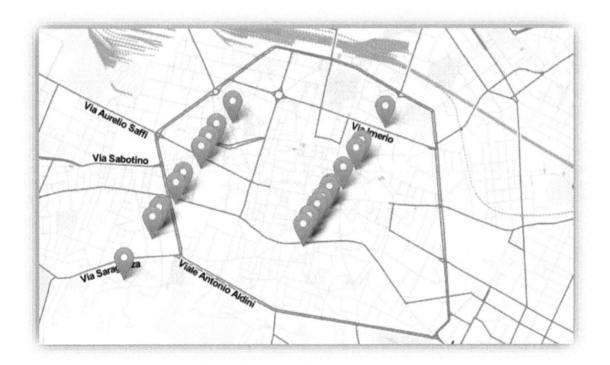

Figure 11.6 Simulation of different paths: a) first possible path for the given individual, b) other possible path for given individual, c) two paths for two different individuals

Source: Own visualisation generated in R with shiny:: and leaflet:: packages

11.4 The market basket analysis technique applied to geolocation data

The solutions presented in Section 11.3 worked with a small amount of data in order to have a better understanding of the operational arrangements behind the model and visualise the results in an understandable way. Following the same procedure as shown before, 50 paths for each of the 200 individuals were simulated with 50 locations for each path. This generated a big matrix of spatial points composed of latitude and longitude coordinates. For such a simulated dataset, there are a few analytical options. The market basket algorithm most likely would not find any association, because it is very unlikely for two randomly generated locations to have the same latitude and longitude. Moreover, that would not be useful at all for business purposes. Rather, it is far more useful to combine this data with open data sources representing the real private activities in the city of Bologna. The great open data portal of the city of Bologna is called *Iperbole* (http://dati.comune.bologna.it/node/640). In this analysis, the dataset called "*attività commerciali in sede fissa*" (commercial activities in a fixed place) will be employed. *Iperbole* gives the addresses of all the commercial companies of the metropolitan area of Bologna in 2018; there is also other data like the typology of the commercial activity and their size. The dataset contains many variables; among them, the street, street number and sector will be used for the analysis.

By running the analysis on open data, one can pose the following questions: How can open data be used in this analysis? How is it helpful? The idea behind this is the following: if a person has been to a location that is close enough to a public or private space that is geocoded in the analysed open dataset, one can assume that the person has also been to that specific location. In this way, by running the market basket algorithm, one can find associations among real activities rather than random spatial points.

This is a strong assumption, and there are a lot of arguments that can be used against it: the idea is that for this specific analysis and dealing with geomarketing purposes, stating that a person has been to a location or has been very close to that location does not make a big difference. However, it is worth highlighting that the analysis is not done on products but locations; thus, there is no interest in people's purchases but in their movements.

Unfortunately, until now, the *Iperbole* dataset has not provided the latitude and longitude of the activities, only the addresses. Thus, one has to geocode them using specific functions.[6] Good software to do that is Mapbox (https://docs.mapbox.com/), which unfortunately has not yet created an API for R users. Other chapters present ways of geocoding. In any case, through the geocoding function, one obtains the latitude and longitude starting with the addresses of the activities of the *Iperbole* dataset. The Mapbox API works well in Python. The following code allows obtaining the latitude and longitude for all locations.

```python
# Reading Iperbole dataset dataset=
pd.read_csv("elenco_esercizi_commercio_in_sede_fissa_anno_2018.csv",
sep = ';', header='infer', encoding='latin-1')

# Pasting, for each commercial activity, the variables
# "street", "address" and the city (Bologna)
# to create a new variable from which latitude and longitude can be
inferred
dataset['quartiere_settore']=dataset.ESERCIZIO_VIA+'
'+dataset.ESERCIZIO_CIVICO+' '+dataset.QUARTIERE+' Bologna'

dataset['lat'] = float
dataset['lon'] = float

# Importing Mapbox token
token=yourtoken
geocoder=Geocoder(access_token=token)

# Defining mapbox geocoding function
def mapbox_geocode(dataset):
    for i in range(len(dataset)):
        try:
            response=geocoder.forward(dataset.quartiere_settore[i])
            response = response.content
```

```
response=json.loads(response)
coordinates = response["features"][0]['geometry']['coordinates']
dataset.iat[i,29] = coordinates[1]
dataset.iat[i,30] = coordinates[0]except:
dataset.iat[i,29] = np.nan
dataset.iat[i,30] = np.nan
return(dataset

# Applying mapbox geocoding function to the Iperbole dataset geocoded_
dataset = mapbox_geocode(dataset)
```

It is worth noting that, despite its great efficiency, some of the addresses of the *Iperbole* dataset have been geocoded in countries like Nigeria and Saudi Arabia. By having latitude and longitude for all the activities in the *Iperbole* dataset, they can be plotted on a map. Figure 11.7 shows a Tableau visualisation of the geolocated commercial activities in a fixed place in the metropolitan city of Bologna.

Moreover, if, in addition to this analysis, one is able to combine the data of the simulated individual locations with a measure of length describing how long a person has been close to a specific company, the analysis could be run just on those observations which have a length greater than a specific value, thus avoiding the presence of noise. In fact, if one can record a person's movements every *t* seconds while they are going to work, the dataset ends up containing a lot of noise regarding all the locations that the customer has walked past. If, by contrast, one keeps just the locations in front of which the customer has stopped for some time, the analysis could be expanded and show more precise and useful results.

The next step in the analysis is to measure the distance and then to assign the simulated locations to the closest geocoded activity. Therefore, the algorithm computes the distance between each individual location and all the geocoded activities in the city of Bologna. Intuitively, the selected type of distance will affect the results. Dealing with spatial data, the best measure of distance to use would take into account the ellipsoidal shape of the earth. The function **distGeo**() from the geosphere:: package computes the shortest distance between two points on an ellipsoid, called the geodesic. The Great Circle distance, instead, aims to identify the shortest distance between two points, assuming that the earth is spherical rather than ellipsoidal. The function **distm**() in

Figure 11.7 Tableau visualisation of business units in Bologna

Source: Own visualisation generated with Tableau

527

the package geosphere:: computes the Great Circle distance based on three different methods: the **distCosine()**, according to the law of cosines; the **distHaversine()**, according to the Haversine method and the **distVincenty()**, according to the Vincenty method. It can accept as input a vector, a matrix or a *SpatialPoints* object. Moreover, the function **pointDistance()** from the raster:: package computes distances just between two points. The argument *longlat* can be set to *TRUE* or *FALSE*, respectively, to compute the ellipsoid distance or the planar Euclidean distance. More on calculating distances was presented in other chapters.

The problem with these measures is that they are computationally demanding, and when one deals with big data, as in this case, the algorithm could take a long time to complete the process. This is because the measures of distance have to make comparisons and find matches for each individual position with all the geocoded activities, which requires a significant workload for the software.

A relatively small spatial range of analysed data causes the Euclidean distance not to significantly alter the results. Obviously, the results are not as accurate as they would be if the ellipsoidal measures of distance were used, but it is a good compromise when dealing with average-sized cities like Bologna. With the Euclidean distance metric between two geolocated points, one obtains a measure of distance in the metric of the points; this means that given that in the analysis the spatial points are latitude and longitude coordinates, one obtains a measure of distance based on their magnitude. In further analysis, the empirical threshold of the Euclidean distance d was chosen as less than 0.0003, which is an arbitrary value. Intuitively, if one chooses a minimum distance d that is big enough, one would always find a closest neighbour which could be relatively far from the position. On the contrary, if one sets a measure of distance that is very small, the algorithm would hardly ever find a close neighbour.

In the analysis subsequently, calculations were based on the **spDistN1()** function from the sp:: package. This function computes both Great Circle and Euclidean distances between a matrix of spatial points and single spatial points. The algorithm considers the smallest distance (closest neighbour) for each of the simulated points from the geocoded activities. If the computed distance is less than an arbitrary value d, then the coordinates of the individual location are replaced with the coordinates of the geocoded activity. In other words, if a person is close enough to a certain activity, one can say that that person actually was at that precise activity. Thus, the algorithm replaces the person's coordinates with the coordinates of that activity. In this way, one can perform the analysis on real activities rather than on random points. The following code, for each of the individual positions, computes the distance with respect to all the geocoded companies. Among them, the minimum distance (nearest) is sorted, and if it is less than 0.0003, the latitude and longitude of the company are assigned to that individual position.

```
# Defining minimum distance
min_dist<-0.0003

# Defining function to assign individual positions
# to the geocoded commercial activities from the Iperbole dataset
assign<-function(geo_points) {
  geo_pointsTOplaces<-list()
    for(j in 1:nrow(path_v)) {
      x<-spDistsN1(places,path_v[j,], longlat = FALSE)
      x<-as.data.frame(x)
      dist<-which(x==min(x), arr.ind=TRUE)
      dist<-as.data.frame(dist)
      nearest<-sort(x[x>0],decreasing=F)[1]
      geo_pointsTOplaces$distance[j]<-nearest
      if(nearest < min_dist) {
        geo_pointsTOplaces$lon[j]<-places[dist[1,1],1]
        geo_pointsTOplaces$lat[j]<-places[dist[1,1],2]
      } else {
        geo_pointsTOplaces$lon[j]<-NA
        geo_pointsTOplaces$lat[j]<-NA
      }
    }
return(geo_pointsTOplaces)
```

Running the previous code allows modifying the previously generated table of simulated individual paths. For all those comparisons among individual positions and geocoded activities for which the distance d is greater than 0.0003 and there is no match, one changes the coordinates to NA. The number of observations one gets in the new data frame has length equal to the original one, showing latitude and longitude for places for which there is a match and NA for the other locations. In Figure 11.8a, one can see that for positions 1 to 10, when setting the minimum Euclidean distance between individual locations and private activities to 0.0003, there are five observations (obs. 1, 2, 4, 7, 10) which satisfy the condition. For those observations, the real coordinates of the location were replaced with the coordinates of the geocoded activity.

Given that the purpose of this phase is to understand if the individual positions are close enough to the geolocated activities, there is no interest in observations showing NA for which there is not a match, as it is useless in the process of finding association rules among companies. Those rows were excluded from the dataset with the function **na.omit()**, and the resulting table is shown in Figure 11.8b.

```
head(dataset, 10) # full dataset with Nas
dataset=na.omit(dataset)
head(dataset, 10) # limited dataset, without Nas
```

The previous step limited a dataset containing 500,000 observations to a new dataset containing only 86,725. This is a huge decrease, and, as stated previously, it would be very useful to have access to other geocoded public locations and obtain more matches: less data would be wasted and more general results would be obtained.

In the last step, when there is all the data to run the market basket algorithm, it is necessary to identify itemsets and items. The itemsets will be the sets of all positions (items) recorded for individual for one of the paths.

	long	lat	distance	i	j
1	11.34488	44.49355	7.773202e-05	1	1
2	11.34546	44.49448	2.500322e-04	1	1
3	NA	NA	3.479887e-04	1	1
4	11.34104	44.48999	2.352656e-04	1	1
5	NA	NA	5.028617e-04	1	1
6	NA	NA	6.126312e-04	1	1
7	11.34673	44.49548	1.909676e-04	1	1
8	NA	NA	3.398517e-04	1	1
9	NA	NA	3.705155e-04	1	1
10	11.34685	44.49577	2.147765e-04	1	1

	long	lat	distance	i	j
1	11.34488	44.49355	7.773202e-05	1	1
2	11.34546	44.49448	2.500322e-04	1	1
4	11.34104	44.48999	2.352656e-04	1	1
7	11.34673	44.49548	1.909676e-04	1	1
10	11.34685	44.49577	2.147765e-04	1	1
11	11.34309	44.49188	1.183682e-04	2	1
12	11.34343	44.49244	2.442104e-04	2	1
19	11.34343	44.49244	2.672490e-04	2	1
22	11.34685	44.49577	2.012285e-04	3	1
23	11.34461	44.49326	1.492796e-04	3	1

Figure 11.8 Dataset: a) after combining with the *Iperbole* dataset, b) resulting dataset after excluding the NA

Source: Own visualisation generated in R

So, in this analysis, for 50 generated latitude/longitude couples for each individual's path, there are itemsets composed of 50 locations each. Nevertheless, after applying the function **na.omit()** and deleting the NAs, some items were excluded and the itemsets were finally composed of a different (smaller) number of items.

Moreover, the items need to be uniquely identified. There are a few possibilities. The first one is to **paste()** latitude and longitude; in this way, one would obtain results showing associations among the latitude and longitude of the geocoded activities. The second way, which makes it far more understandable to read and interpret the results, is to **merge()** the variables latitude and longitude of the obtained dataset with the open dataset of geocoded private activities. In this way, as a final result, one gets readable associations among the activities. As all the real activities in the analysed dataset are indexed, the output will show associations among those indices rather than among the named companies. Figure 11.9 shows the obtained dataset, in which, apart from the variables already discussed, the variable *ID* represents an index associated to each geocoded activity, *customer* represents the individuals and *itemset* represents the customers' paths.

```
final<-merge(x=dataset, y=places, by.x = c(long,lat), by.y = c(long,lat))
head(final, 10)
```

11.5 Spatial association rules

The package arules:: is specifically designed to run market basket analysis. Additionally, the package arulesViz:: allows for interesting and interactive visualisations. The package arules:: requires data in class *transaction*; thus, a typical *data.frame* needs to be transformed into a *transaction* object. This can be done with the function **ddply()** from the package plyr::, which enables data in R to be easily manipulated. To do that, along with the items, the itemsets need to be uniquely identified for each customer. Thus a new variable *itemset_customer* is generated by pasting the variables *itemset* and *customer*. In the following code, all the items (ID) are aggregated with respect to the variable *itemset_customer*.

```
transactionData<-ddply(final, c(itemset_customer),
function(final)paste(final$ID, collapse=,))
```

For simplicity, after renaming the variable *itemset_customer* as *itemset*, the transaction object is presented in Figure 11.10.

The first line represents the first path of the first customer (first itemset), in which the activities (firms) visited are indexed as 1764, 3569, 1429, 1347 and 1316 (items). So, the column *items* represents the set of indexed

	long	lat	distance	itemset	customer	ID
1	11.33821	44.48682	2.196592e-04	4	1	1918
2	11.33821	44.48682	1.917560e-04	10	1	1918
3	11.34104	44.48999	2.352656e-04	1	1	1764
4	11.34104	44.48999	2.720240e-04	8	1	1764
5	11.34309	44.49188	2.634160e-04	10	1	1585
6	11.34309	44.49188	1.233773e-04	6	1	1585
7	11.34309	44.49188	2.285332e-04	7	1	1585
8	11.34309	44.49188	2.652281e-04	3	1	1585
9	11.34309	44.49188	1.183682e-04	2	1	1585
10	11.34343	44.49244	2.672490e-04	2	1	1580

Figure 11.9 Final dataset in which *ID* represents the matched indexed company

Source: Own visualisation generated in R

	itemset	items
1	1 1	1764,3596,1429,1347,1316
2	10 1	1918,1585,3628,3596,3596,1448
3	2 1	1585,1580,1580
4	3 1	1585,1481,1467,1438,1316
5	4 1	1918,1448,1429,1429,1347
6	5 1	3596,1429,1347,1347,1148
7	6 1	1585,1580,1347
8	7 1	1585,3628,1474,1429,1148
9	8 1	1764,3223
10	9 1	3628,1489,1489,3223,1330

Figure 11.10 **Transaction object**

Source: Own visualisation generated in R

companies that a customer has visited on a particular path and for which there is a match between the recorded individual position and the geocoded commercial activity. It is immediately evident that some itemsets contain the same firms multiple times. That means that during the same path, the same individual was geocoded more than once at the same activity (e.g. company indexed 1580 in Figure 11.10, obs. 3). Note that this could be the case when the people's recorded movements are very close or when there are rather empty areas of geocoded activities. In this case, this is the result of the simulation process. However, the duplicated items in each itemset will be deleted because they are not relevant for the market basket algorithm. In fact, when R reads the dataset as a transaction object through the function **read.transactions**(), a warning message on removing duplicated items appears.

```
tr<-read.transactions(transactionData, format='basket', sep=',')

#Warning message:
#In asMethod(object): removing duplicated items in transactions
```

The most frequent items are the companies which are mostly visited by the customers or, consequently, those which have better intercepted their movements. In fact, in terms of marketing, they need to be considered spatially aggregating points. On the other hand, however, being located in an area which is particularly covered by individuals does not necessary mean that a business can benefit from it. It may depend on a wide range of factors, starting from the very nature of the business. The first 20 most frequent companies are displayed in the absolute frequency plot in Figure 11.11. One can see that the three most visited companies are indexed as 1811, 3315 and 1683.

```
library(RcolorBrewer)
itemFrequencyPlot(tr,    topN=20,                   type='absolute', col=brewer.
pal(8,'Pastel2'), main='Absolute Item Frequency Plot')
```

As described before, the individual data was simulated through an MVN distribution starting from the latitude and longitude of Palazzo Re Enzo. Because of the bell-shaped distribution, the majority of the generated starting positions are also in the city centre, following the direction of the paths.

It is worth emphasising that the purpose of this specific analysis in not to discover real useful business insights, as it would be senseless dealing with simulated data. Rather, the purpose is to show a method of application which could be implemented using real data and offer valuable information from an uncommon perspective. Indeed, dealing with non-simulated data could generate very different patterns and could also lead

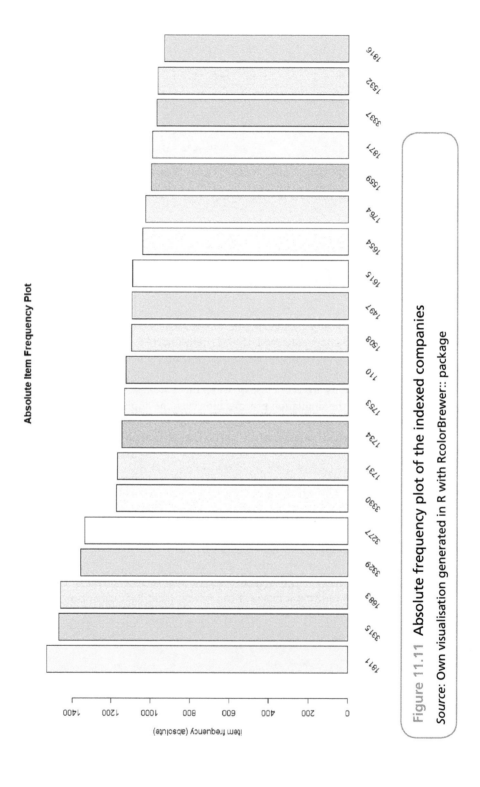

Figure 11.11 Absolute frequency plot of the indexed companies

Source: Own visualisation generated in R with RcolorBrewer:: package

to the identification of strategic points of aggregation which may not be immediately evident. These spots, with the noise filtered out, can reveal profitable business opportunities.

Now, having created and investigated the transaction object, one can run the market basket analysis. The association rules can be discovered using many different algorithms. Here the *Apriori* algorithm was applied, as it strongly decreases the computational weight of the analysis. The idea behind this algorithm is that a pair of items is frequent only if each of them is frequent. So, one considers the itemset {*productA, ProductB, ProductC*} if and only if the itemsets {*productA, ProductB*} {*productA, ProductC*} {*ProductB, ProductC*} are frequent with respect to a defined threshold. Thus, when running the **apriori()** function from the arules:: package, one specifies the thresholds for the minimum support, confidence and maximum length of the rule. In this example, these are: minimum support equal to 0.03, minimum confidence equal to 0.5 and maximum length equal to 10 items. Intuitively, the more geocoded individual positions and commercial activities, the higher the chance that the itemsets of locations are diversified and the indices of fit of the rules are small. The following code runs the *apriori* algorithm on the analysed transaction object.

```
association.rules<-apriori(tr,        parameter=list(supp=0.03,      conf=0.5,
maxlen=10))
summary(association.rules)
```

```
set of 198 rules

rule length distribution (lhs + rhs):sizes
2      3       4
104    66      28


Min. 1st Qu.    Median  Mean 3rd  Qu.      Max.
2.000      2.000  2.000   2.616     3.000   4.000


summary of quality measures:
support         confidence     lift    count
Min.    :0.03013    Min.    :0.5131    Min.    : 6.694    Min.    : 590.0
1st Qu.:0.03258    1st Qu.:0.7156    1st Qu.:12.332    1st Qu.: 638.0
Median :0.03575    Median :0.7959    Median :14.512    Median : 700.0
Mean    :0.03726    Mean    :0.7791    Mean    :13.920    Mean    : 729.5
3rd Qu.:0.04091    3rd Qu.:0.8612    3rd Qu.:15.338    3rd Qu.: 801.0
Max.    :0.05516    Max.    :0.9597    Max.    :21.772    Max.    :1080.0
```

For the parameters previously specified, the algorithm has identified a set of 198 rules of order 2, 3 and 4, which are rules respectively associating 2, 3 and 4 companies. Considering all the rules that the algorithm has found, which are a lot, one would get a huge network that would be messy and difficult to interpret. Thus, in the following, just a subset of them is taken into account to visualise and analyse the results. In general terms, the more the clusters of activities are separated, the more the aggregating points result is defined and the more useful the model is for business purposes. The following code inspects the first five rules generated, sorted by support.

```
rules<-head(association.rules, n=5, by=support)
inspect(rules)
```

lhs	=>	rhs	support	confidence	lift	count
{3632}	=>	{1798}	0.03524004	0.8801020	19.73929	690
{1798}	=>	{3632}	0.03524004	0.7903780	19.73929	690
{3632}	=>	{1781}	0.03469969	0.8915816	19.88288	699
{1781}	=>	{3632}	0.03469969	0.7961276	19.88288	699
{3308}	=>	{3337}	0.03299285	0.8873626	17.94893	646

The first rule suggests that people who on average tend to go to indexed activity 3632 also tend to go to the activity indexed 1798 with support equal to 0.03524004, confidence equal to 0.8801020 and lift equal to 19.73929690. Thus one knows that 3.52% of all the paths contain the companies {3632, 1798}, that the estimated conditional probability of visiting the company 1798 in a path under the condition that the path also contains the company 3632 is 0.88 and that the two companies visited together present at almost 20 times the rate one would expect under the assumption of visiting them independently.

The second rule is complementary with the same indices of fit for support and lift. In the same way, depending on the aims and needs in the analysis, one can sort the rules by confidence or lift.

The clusters of connected companies have been generated with respect to people's movements in Bologna and can be plotted. Figures 11.12a and 11.12b, respectively, show the networks for 100 and 20 rules, where the size of the circles defines the support of the rule, while its colour indicates the lift. Both figures are in fact important because, on the one hand, it is crucial to have an idea of the general structure of the network, and, on the other hand, it is certainly easier to work on a subset of them.

```
plot(association.rules[1:100], method=graph) plot(association.rules[1:20],
method=graph)
```

11.6 Applications to geomarketing

"Imagine a world in which data on companies, households and governments were widely available. Imagine, further, that researchers and decision-makers acting in the public interest had tools enabling them to test and model such data to explore different scenarios of the future. People would be able to make more informed decisions, based on the best available evidence" (Lovelace, Nowosad, & Muenchow, 2019).

In the world, people change, companies open and close and data is always messy. At the moment, the evidence just shows that this "technocratic dreamland" does not exist.

At least not yet. In any case, it is also very hard to represent the complexity of the reality: even if one had all the data desired, it could be rather challenging to make a good model of it. A model does not reproduce reality; it is just useful for understanding it and taking better decisions (Becker, 1978).

Having generated and plotted the association rules, how can they be useful to companies? How could they be advised? To answer these questions, three fields of application in marketing are presented.

Marketing is a management science which seeks to strengthen and enlarge the relationship between a company and its customers. Lovelace et al. (2019) defines the set of questions that geomarketing tries to answer: Where do target groups live, and which areas do they frequent? Where are competing stores or services located? How many people can easily reach specific stores? Do existing services over- or underexploit the market potential? What is the market share of a company in a specific area? On this path, the analysis will be presented as an innovative approach to marketing which helps decision-makers get useful information and plan business strategies thanks to location data. In particular, it will be discussed as a tool to help companies analytically plan the best location of a new branch, target customers, identify competitors or forge partnerships.

11.6.1 Finding the best location for a business

Location theory typically has microeconomic foundations. It generally investigates the location choices of firms and households along with territorial disequilibria and hierarchies.[7] Apart from the spatial information already used, the *Iperbole* dataset also contains, for each company, its commercial sector. The majority of the geocoded activities are clothing shops, followed by food, luxury objects and shoe shops. One can show their spatial distribution grouped by sector (Figure 11.13)

It appears that there are not any particularly clear patterns. More advanced techniques described in this book can provide more accurate procedures for their identification. Now, imagine running the market basket algorithm, not on companies as has been done in the past but on their sector. In other words, suppose that the itemsets, instead of being composed of companies with latitudes and longitudes, were composed of their respective sectors. One would obtain new association rules linking sectors rather than companies. The procedure does not change, apart from the transaction object. The network for the first 20 rules associating commercial sectors is presented in Figure 11.14.

The sector "LIBRI e FUMETTI" (books and graphic novels) is linked with the sector "Bevande-Vini-oli-birra" (beverages, wines, oils, beer). If the data was not simulated, one could easily try to interpret this rule by

Graph for 100 rules

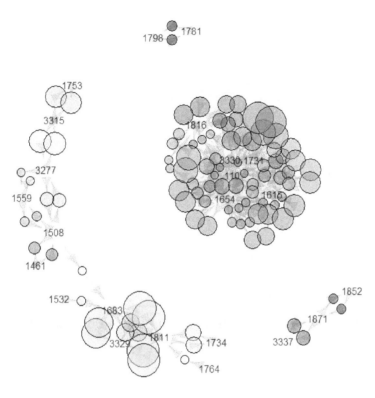

Graph for 20 rules

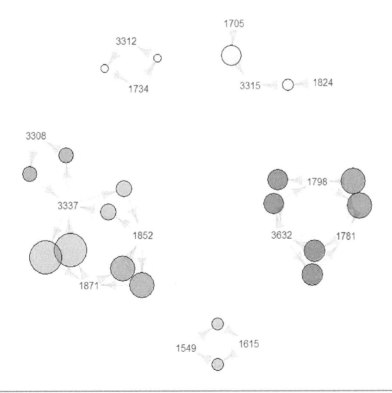

Figure 11.12 Network of rules: a) the first 100 rules, b) the first 20 rules

Source: Own visualisation generated in R with arulesViz:: package

Figure 11.13 **Tableau visualisation of business units in Bologna grouped by sector**
Source: Own visualisation generated with Tableau

thinking of all the students in Bologna who every day, at odd hours, enjoy reading books while drinking a glass of wine or a beer in a pub.

A business which wants to find the best place to open a new branch could model its strategy by taking into account this information. Obviously, here only businesses which have direct contact with the public or those which can take any advantage from being a spatial aggregating point are taken into account. Businesses which base their relationship with their customers only online act in a different space, where the measures of distance among nodes assume a different meaning.

Now, going beyond the previous information offered by the visualisations, it is interesting to explicitly advise a company where it should open a new branch. Among the others, in the figure previously there are four rules linking *tobacco shops* with *fabric stores* with a high index of lift. One can think of a company which is a fabric store and consider the rule previously *{tobacco shops} => {fabric stores}*. As usual, it can be read in this way: people who on average go to tobacco shops generally tend to go to fabric stores as well (with respect to the indices of fit of the rule). Knowing that, one can search for all the tobacco shops in the analysed area if there is a fabric store within a certain distance. If there is not, the area around these remaining tobacco shops could be a good choice to start searching for a place. Following this reasoning and setting the measure of distance (Euclidean distance is used), in the analysed dataset, there are just two tobacco shops for which there is not a match with any fabric store. The density plot in Figure 11.15 shows in green the areas around these two tobacco shops in which it would be convenient for a fabric store to open a new branch. Moreover, if one could integrate the data of the available properties from real estate agencies, real estate options to offer a business would be available.

The reasoning does not change considering rules of order greater than 2 or considering multiple rules at the same time.

11.6.2 Targeting

One of the biggest marketing mistakes for a company is to attempt to appeal to everyone at once. This is a waste of time and resources and can easily lead to a loss in competitiveness and market share. A successful company is one which constantly knows who its customers are. Nowadays, tools like Google Analytics and Facebook Insights help marketers have a better understanding both of their customers and potential customers. They can

Graph for 20 rules

Figure 11.14 **Network of the first 20 rules for companies' sectors**

Source: Own visualisation generated in R with arulesViz:: package

easily extract interaction data from the web and put it into statistical predictive models or simple descriptive dashboards to understand their target and constantly update it. Targeting is a procedure through which a company seeks to identify the customers who are most likely to buy its products or services. The target is thus the group of customers which is expected to be the most responsive to a campaign and consequently the most profitable regarding some specific products of the company. There are a great variety of ways to segment a market (Wedel & Kamakura, 2000). Behavioural segmentation, for instance, works on features like indices of usage and loyalty, while geographical segmentation mainly works on features like address, city, climate and region. Intuitively, if a company wants to sell sun cream, its geographical target would probably include those people who already live in or are likely to go to a sunny destination. Certainly, there are many other possibilities, usually complementary, which help marketers and decision-makers to better understand who their customers are.

The proposed model explains the relationship among companies with respect to people's movement. People who are likely to physically go to a company tend, on average, to go to other companies or geocoded spots as well. For instance, according to the rule {2064} => {151}, people who tend to go to the indexed activity (2064) are likely to go to indexed activity (151) as well. In this way, for company (151), individuals who go to activity (2064) are part of their target.

Knowing that, the target can be designed both on the basis of this geographical information and the characteristics of the associated company. There is no doubt that including in the model some specific information of a company would definitely improve it and make this analysis an intermediary for further more complicated models. One could, for instance, cluster those individuals or schedule specific on-site marketing campaigns.

Figure 11.15 **Density plot for tobacco shops not close to any fabric store**

Source: Own visualisation generated with Tableau

11.6.3 Discovery of competitors

Another intuitive approach to those results would be to look at them as tools for discovering local competitors. Imagine, for instance, managing a clothing store. Together with online purchases, customers can buy products on site. Imagine the usual rule *{2064} => {151}* showing good indices of fit. Thus, people who go to activity (151) also tend to go to activity (2064). Consequently, if the company (2064) is another clothing store, this implies that the two companies are competing for the same customers. Thus, given that the two companies are located on that path, they share their selling potential. The same reasoning, depending on the business's models and necessities, is valid for collaborations and partnerships.

11.7 Conclusions and further approaches

In this chapter, market basket analysis has been applied, through the *Apriori* algorithm, to the geolocation data of N individuals in the metropolitan city of Bologna. While the results of the analysis cannot be generalised because of the data simulation process, the idea behind them, on the other hand, can.

A strong assumption has been made by considering a company as a circular, spherical or ellipsoidal entity on the basis of the measure of distance employed. However, this is a condition hardly ever satisfied in reality. A rather interesting and innovative approach is represented by *geofencing*.[8] A geofence is a virtual perimeter which is built to define the limits of specific spatial objects. Nowadays many companies are working on that, especially thanks to Google API.[9] Using geofences rather than arbitrary measures of distance can definitely establish if a person was at a particular activity and not just very close to it, and that would bring this analysis a great step forward.

Furthermore, through the R package arulesSequences::, one can extract sequential rules from the dataset. Until now, we have read the rule {*ProductA*} => {*ProductB*} in this way: buying *productA*, a customer is expected to buy *ProductB* as well. The sequence of purchased items was not important. Dealing with sequential rules means adding a temporal component. The new rules have to be read as: buying *productA*, in the next purchase, a customer is expected to buy *productB*. Regarding this spatial example, the time variable could be represented by the specific time at which a location has been visited. In this way, given a present location of an individual, sequential destinations can be discovered.[10]

Finally, another interesting approach, though computationally more complicated, would be to run the algorithm just on the data of a single individual. Doing that, instead of discovering general rules that are valid on average, the component of correlation which characterises the individual movements would be stressed and a company could plan customised marketing campaigns towards each specific customer.

Notes

1. http://go.carto.com/hubfs/The%20State%20of%20Location%20Intelligence%202018%20v10042017.pdf?hsCtaTracking=cb25d151-ee5f-4593–9f6a-5388d9749ebcl7320ab2b-7f46–481a-9b8e-a47cc20f70a1
2. htttps://www.mckinsey.com/~/media/McKinsey/Industries/Automotive%20and%20Assembly/Our%20Insights/Monetizing%20car%20data/Monetizing-car-data.ashx, 2016
3. An interesting example is offered by the Thasos group: http://news.mit.edu/2018/startup-thasos-group- measuring-economy-smartphone-location-data-0328
4. https://www.geospatialworld.net/blogs/what-will-be-the-impact-of-gdpr-on-location-data/ https://www.didomi.io/en/blog/does-geolocation-require-consent
5. https://www.reuters.com/article/us-eu-google-privacy/european-consumer-groups-want-regulators-to-act-against-google-tracking-idUSKCN1NW0BS
6. The most well-known command, **geocode**(), from the **ggmap::** package uses the Google API through the R function. Google has recently decided to change its pricing policy, and its usage is limited.
7. For an exhaustive approach to location theories, see Roberta Capello (2011).
8. For interesting articles about geofencing, read: Garzon, Deva (2014), *Geofencing 2.0: Taking Location-Based Notifications to the Next Level* (conference paper) and Küpper, Bareth, and Freese (2011), *Geofencing and Background Tracking*, Informatik Schafft Communities.
9. https://developers.google.com/location-context/geofencing/
10. For an example with code, visit: https://rpubs.com/Mahsa_A/Part4_AnalyzeTransactionData

Appendix A
Datasets used in examples

The examples presented in the book were made on sample data for Poland. Data can be downloaded from https://github.com/kkopczewska/spatial_book.

To load the data, use a few basic R packages and set the working directory in which the files for reading are stored.

```
library(spdep)
library(rgdal)
library(maptools)
library(sp)

setwd("D:/SpatialBook")
```

A1. Dataset no. 1 / dataset1/ – poviat panel data with many variables

The dataset *data_nts4_2019.csv* has a panel structure: the data is collected successively for 380 units – poviats (NTS4) in 12 consecutive periods of time (years 2006–2017). The data is sorted first after years, then by unit ID (blocked in *year* groups). The data contains 4560 rows and 39 columns, which gives a total of 177,840 observations. Territorial units – poviats – are ordered as in a poviat map, which is defined by the *ID_MAP* variable.

Variable name	Description
ID_MAP	Order of spatial units as in shapefile
code_GUS	Code of poviat given by GUS (Polish Statistical Office)
poviat_name1	Name of poviat (with info on region for duplicated names)
poviat_name2	Name of poviat (no info on region for duplicated names)
subreg72_name	Name of subregion (division into 72 territorial units)
subreg72_nr	Number of subregion (division into 72 territorial units)
region_name	Name of region (voivodeship)
region_nr	Number of region (voivodeship)
year	Year
core_city	Core city of region (1 = yes, 0 = no)
dist	Distance from poviat to core regional city
XA01	Total own revenues
XA02	Personal income tax (PIT) revenues
XA03	Corporate income tax (CIT) revenues
XA04	Total revenues
XA05	Investment expenditures
XA06	Inhabitants total
XA07	Inhabitants in pre-working age < 17
XA08	Inhabitants in working age
XA09	Inhabitants in post-working age
XA10	Inhabitants per 1 km²
XA11	Population density of built-up and urbanised area (people/km2)
XA12	Registered unemployed labour force

(continued)

(*continued*)

Variable name	Description
XA13	Firms per 10,000 inhabitants in productive age
XA14	Salaries Poland = 100%
XA15	Employed in total
XA16	Employed in agriculture, forestry, hunting and fishing
XA17	Employed in industry and construction
XA18	Employed in trade, repair of motor vehicles; transport and storage; accommodation and gastronomy; information and communication
XA19	Employed in financial and insurance activities; real estate market service
XA20	Employed in other services
XA21	Unemployment rate
XA22	Unemployment rate Poland = 100%

```
data<-read.table("data_nts4_2019.csv", sep=";", dec=",", header=TRUE)
summary(data)
```

```
    ID_MAP              code_GUS                          poviat_name1

Min.   :  1.00    Min.   : 201000    Powiat    st. Warszawa   : 12
1st Qu.: 95.75    1st Qu.: 1004750   Powiat    aleksandrowski : 12
Median :190.50    Median : 1636000   Powiat    augustowski    : 12
Mean   :190.50    Mean   : 1720679   Powiat    bartoszycki    : 12
3rd Qu.:285.25    3rd Qu.: 2475250   Powiat    bełchatowski   : 12
Max.   :380.00    Max.   : 3263000   Powiat    będziński      : 12
                                     (Other)                   4488
          poviat_name2                           subreg72_name
Powiat bielski     :   24 Podregion 01 - jeleniogórski     : 108
Powiat brzeski     :   24 Podregion 59 - leszczyński       : 108
Powiat grodziski   :   24 Podregion 04 - wrocławski        :  96
Powiat krośnieński:   24 Podregion 10 - chełmsko-zamojski:  96
Powiat nowodworski:   24 Podregion 14 - zielonogórski     :  96
Powiat opolski     :   24 Podregion 27 - radomski          :  96
(Other)            : 4416 (Other)                           3960
  subreg72_nr            region_name         region_nr        year
Min.   : 1.00    Mazowieckie  : 504    Min.  : 2     Min.    :2006
1st Qu.:17.75    Śląskie      : 432    1st Qu.:10    1st Qu.: 2009
Median :35.50    Wielkopolskie: 420    Median :16    Median : 2012
Mean   :35.60    Dolnośląskie : 360    Mean   :17    Mean   : 2012
3rd Qu.:54.00    Podkarpackie : 300    3rd Qu.:24    3rd Qu.: 2014
Max.   :72.00    Lubelskie    : 288    Max.   :32    Max.   : 2017
                 (Other)       2256

   core_city          dist            XA01              XA02
Min.   :0.00000  Min.   :  0.00  Min.   :1.169e+07  Min.   :3.832e+06
1st Qu.:0.00000  1st Qu.: 37.61  1st Qu.:6.519e+07  1st Qu.:2.023e+07
Median :0.00000  Median : 53.91  Median :1.038e+08  Median :3.443e+07
Mean   :0.04211  Mean   : 57.80  Mean   :1.944e+08  Mean   :7.193e+07
3rd Qu.:0.00000  3rd Qu.: 77.33  3rd Qu.:1.783e+08  3rd Qu.:6.346e+07
Max.   :1.00000  Max.   :170.09  Max.   :1.163e+10  Max.   :5.041e+09
```

```
          XA03                    XA04                    XA05                    XA06
Min.    :  -1989467   Min.    :4.664e+07   Min.    :2.225e+06   Min.    :   20270
Median  :   1572082   Median  :2.337e+08   Median  :3.956e+07   Median  :   76412
Mean    :   6072587   Mean    :3.611e+08   Mean    :6.639e+07   Mean    :  101247
3rd Qu. :   3713525   3rd Qu. :3.568e+08   3rd Qu. :6.585e+07   3rd Qu. :  112482
Max.    : 811358089   Max.    :1.548e+10   Max.    :2.584e+09   Max.    : 1764615

          XA07                    XA08                    XA09                    XA10
Min.    :     3405   Min.    :    12546   Min.    :    2953   Min.    :   19.0
1st Qu. :    10561   1st Qu. :    34942   1st Qu. :    9054   1st Qu. :   61.0
Median  :    14974   Median  :    48498   Median  :   12857   Median  :   90.0
Mean    :    18870   Mean    :    64293   Mean    :   18085   Mean    :  379.7
3rd Qu. :    21216   3rd Qu. :    70920   3rd Qu. :   19402   3rd Qu. :  188.0
Max.    :   310319   Max.    :  1113637   Max.    :  424236   Max.    : 4128.0

          XA11                    XA12                    XA13                    XA14
Min.    :     782   Min.    :      337   Min.    :  593.5   Min.    :   60.10
1st Qu. :    1571   1st Qu. :     2665   1st Qu. : 1109.8   1st Qu. :   77.30
Median  :    2072   Median  :     3924   Median  : 1328.8   Median  :   81.70
Mean    :    2390   Mean    :     4725   Mean    : 1401.1   Mean    :   84.55
3rd Qu. :    2918   3rd Qu. :     5624   3rd Qu. : 1577.8   3rd Qu. :   88.10
Max.    :    7061   Max.    :    54842   Max.    : 4396.8   Max.    :  183.60
NA's    :    3420

          XA15                    XA16                    XA17                    XA18
Min.    :    4155   Min.    :       0   Min.    :     0   Min.    :       0
1st Qu. :   13262   1st Qu. :    2145   1st Qu. :  3067   1st Qu. :    1404
Median  :   20051   Median  :    4549   Median  :  5632   Median  :    2458
Mean    :   28896   Mean    :    6004   Mean    :  8103   Mean    :    5403

3rd Qu. :   29408   3rd Qu. :    8228   3rd Qu. :  9619   3rd Qu. :    4434
Max.    :  955785   Max.    :   31747   Max.    :135370   Max.    :  300528

          XA19                    XA20                    XA21                    XA22
Min.    :       0.0   Min.    :     1047   Min.    :  1.40   Min.    :   17.90
1st Qu. :     201.0   1st Qu. :     2965   1st Qu. :  9.20   1st Qu. :   85.67
Median  :     334.0   Median  :     4252   Median  : 13.20   Median  :  117.40
Mean    :    1110.2   Mean    :     8255   Mean    : 14.07   Mean    :  125.38
3rd Qu. :     659.2   3rd Qu. :     6769   3rd Qu. : 18.20   3rd Qu. :  159.72
Max.    :  116797.0   Max.    :   409657   Max.    : 38.70   Max.    :  389.40
```

names(data)

```
 [1] "ID_MAP"        "code_GUS"       "poviat_name1"   "poviat_name2"
 [5] "subreg72_name" "subreg72_nr"    "region_name"    "region_nr"
 [9] "year"          "core_city"      "dist"           "XA01"
[13] "XA02"          "XA03"           "XA04"           "XA05"
[17] "XA06"          "XA07"           "XA08"           "XA09"
[21] "XA10"          "XA11"           "XA12"           "XA13"
[25] "XA14"          "XA15"           "XA16"           "XA17"
[29] "XA18"          "XA19"           "XA20"           "XA21"
[33] "XA22"
```

dim(data)

```
[1] 4560   39
```

A2. Dataset no. 2 / dataset2/ – geolocated point data

The *geoloc_data.csv* dataset contains 37,378 observations regarding the location of companies in 2012 in the Lubelskie voivodeship. These are companies from the REGON register, and their headquarters were geolocated based on the address entered in the register. Geolocated points contain information about the unique ID (*ID*), full address (*ADDRESS*) with postal code (*ZIP*), street (*STREET*), property number (*STREET NO*), city address (*CITY*) and mail address (*CITY_POST*), belonging to poviat (*region2* and *poviat*) and to municipalities (*region3*) and the subregion number (*subreg*). Geolocation or geographical coordinates (*latitude* and *longitude*) of a point are given as *coords.x1* and *coords.x2*. There is also information on the basic legal form (*LEGAL_FORM1*) (takes the values as in the table subsequently), the detailed legal form (*LEGAL_FORM2*) (takes the values as in the table subsequently), the ownership form (*OWNERSHIP*) (takes the values as in the table subsequently), PKD code (*PKD7*), sector according to PKD code (*SEC_PKD7*) and employment size class (*GR_EMPL* and *empl*).

Basic legal form (*LEGAL_FORM1*)

Variable level	Description of code
1	Legal entity
2	Organisational unit without legal personality
9	Natural person conducting business activity

Detailed legal firm (*LEGAL_FORM2*)

Variable level	Description of code
01	Authorities, government administration
02	State control and law protection bodies
03	Local government communities
06	Courts and tribunals
09	Treasury
14	European economic interest groupings
15	Partner firms
16	Joint-stock firms
17	Limited liability firms
18	General partnerships
19	Civil partnerships operating on the basis of a contract concluded on the basis of the Civil Code
20	Limited partnerships
21	Limited joint-stock partnerships
22	European firms
23	Firms provided for in provisions of laws other than the Code of Commercial Firms and the Civil Code or legal forms to which the provisions on firms apply
24	State-owned enterprises
26	Mutual insurance firms
28	State organisational units
29	Municipal self-government organisational units
30	Poviat self-government organisational units
31	Voivodeship self-government organisational units
32	Budgetary economy institutions
34	Mutual reinsurance firms
35	Main branches of foreign mutual reinsurance undertakings
40	Cooperatives
42	European cooperatives
44	Universities
46	Independent public health care facilities
48	Foundations
49	Funds
50	The Catholic church

Variable level	Description of code
51	Other churches and religious associations
53	European groupings of territorial cooperation
55	Associations
60	Social organisations not specifically mentioned
65	Research institutes
70	Political parties
72	Unions
73	Employers' organisations
76	Economic and professional self-government
79	Branches of foreign entrepreneurs
80	Foreign representative offices
81	Public kindergartens
82	Private kindergartens
83	Primary public schools
84	Public schools, junior high schools
85	Housing associations
86	Upper secondary public schools
87	Public art schools
88	Private primary schools
89	Private junior high schools
90	Associations of agricultural producer groups
91	Private upper secondary schools
92	Private art schools
93	Public education system institutions
94	Non-public education system institutions
95	Other organisational units of the public education system
96	Other organisational units of the non-public education system
97	Schools and public education system groups
98	Schools and institutions of the non-public education system groups
99	Without any particular legal form

Ownership form

Variable level	Code description
	Public sector
111	State-owned property
112	Property of state legal persons
113	Local government property
	Mixed ownership in the public sector
121	Mixed ownership in the public sector with a predominance of state ownership
122	Mixed ownership in the public sector with a predominance of state ownership of legal entities
123	Mixed ownership in the public sector with a predominance of local government ownership
127	Mixed ownership in the public sector with no advantage of any type of public property
	Mixed ownership between sectors with a predominance of public sector ownership
131	Mixed ownership between sectors with a predominance of public sector ownership, including a predominance of state ownership
132	Mixed ownership between sectors with a predominance of public sector ownership, including a predominance of ownership by state-owned legal entities
133	Mixed ownership between sectors with a predominance of public sector ownership, including predominance of ownership of local government units or local government legal entities
137	Mixed ownership between sectors with a predominance of public sector ownership with no advantage of any kind of public property

(*continued*)

(continued)

Variable level	Code description
	Private sector
214	Property of domestic natural persons
215	Private domestic ownership remaining
216	Foreign ownership
	Mixed ownership in the private sector
224	Mixed ownership in the private sector with a predominance of ownership by domestic natural persons
225	Mixed ownership in the private sector with a predominance of other private domestic ownership
226	Mixed ownership in the private sector with a predominance of foreign ownership
227	Mixed ownership in the private sector with no advantage of any type of private property
338	Mixed ownership between sectors with or without private sector ownership predominance (50% share of public sector property and 50% share of private sector property)
231	mixed ownership between sectors with or without a predominance of private sector ownership and a predominance of state ownership in total capital
232	Mixed ownership between sectors with a predominance of private sector ownership or no sectoral advantage with a predominance of ownership of state-owned legal entities in total capital
233	Mixed ownership between sectors with a predominance of private sector ownership or no sectoral advantage with a predominance of local government ownership in total capital
234	Mixed ownership between sectors with a predominance of private sector ownership or no sectoral advantage with a predominance of ownership by domestic natural persons
235	Mixed ownership between sectors with a predominance of private sector ownership or no sectoral advantage with a predominance of other private domestic ownership
236	Mixed ownership between sectors with a predominance of private sector ownership or no sectoral advantage with a predominance of foreign ownership
237	Mixed ownership between sectors with or without private sector ownership predominance with no private ownership predominating

```
firms<-read.csv("geoloc_data_firms.csv", header=TRUE, dec=",", sep=";")
summary(firms)
```

```
      ID                                                          ADDRESS
Min.    :      4   23-200 KraÂśnik KraÂśnik ul. Lubelska 75A       : 16
1st Qu.:  92300   20-703 Lublin Lublin ul. Cisowa 9               : 12
Median : 185158   20-315 Lublin Lublin al. Wincentego Witosa 3  : 11
Mean   : 185857   20-346 Lublin Lublin ul. DÂługa 5              : 10
3rd Qu.: 279016   20-843 Lublin Lublin ul. Koncertowa 7          : 10
Max.   : 373094   22-100 CheÂłm CheÂłm al. Armii Krajowej NN     : 10
(Other)                                   37309

                    STREET           STREET.NO          ZIP
ul. Lubelska            : 372    2    :    926   22-400 :  1229
ul. PartyzantĂłw        : 205    1    :    925   21-500 :   968
ul. Polna               : 182    3    :    910   22-100 :   914
ul. Tadeusza KoÂściuszki: 176    5    :    876   21-400 :   772
ul. Warszawska          : 148    4    :    847   24-100 :   681
(Other)                  15884   6    :    822   22-600 :   602
NA's                     20411   (Other): 32072  (Other): 32212
```

```
              CITY_POST                   CITY
Lublin          :   5362   Lublin          :   5003   Lublin               :  5035
ZamoÂśÄ¦       :   1288   ZamoÂśÄ¦       :   1006   powiat lubelski      :  2755
BiaÂła Podlaska:   966    CheÂłm         :   804    powiat Âłukowski     :  2180
CheÂłm         :   949    BiaÂła Podlaska:   764    powiat zamojski      :  2132
ÂŁukÂłw        :   772    PuÂławy         :   636    powiat bialski       :  1985
PuÂławy         :   695    ÂŁukÂłw        :   474    powiat biÂłgorajski:  1969
(Other)         : 27346   (Other)         : 28691   (Other)              : 21322
              poviat                  region3           subreg
Powiat Lublin   :   5035   Lublin          :   5035   Min.   : 1.000
Powiat lubelski :   2755   ZamoÂśÄ¦       :   1006   1st Qu.: 2.000
Powiat łukowski :   2180   heÂłm          :   810    Median : 3.000

powiat zamojski :   2132   BiaÂła Podlaska:   765    Mean   : 2.635

powiat bialski  :   2022   PuÂławy         :   645    3rd Qu.: 3.000

powiat biłgorajski: 1969   ÂŁukÂłw        :   557    Max.   : 4.000

(Other)         :21285   (Other)         28560


coords.x1        coords.x2      LEGAL_FORM1      LEGAL_FORM2
Min.   :21.65 Min.   :50.26 Min.   :1.000 Min.   : 19.0
1st Qu.:22.42 1st Qu.:50.87 1st Qu.:9.000 1st Qu.: 99.0
Median :22.67 Median :51.23 Median :9.000 Median : 99.0
Mean   :22.77 Mean   :51.21 Mean   :8.157 Mean   :104.3
3rd Qu.:23.18 3rd Qu.:51.47 3rd Qu.:9.000 3rd Qu.: 99.0
Max.   :24.13 Max.   :52.27 Max.   :9.000 Max.   :999.0


OWNERSHIP               PKD7            SEC_PKD7   GR_EMPL
Min.   :   0.0   0150Z : 19251   A :  20864   Min.   : 1.000
1st Qu.: 214.0   4941Z :   667   G :   4772   1st Qu.: 1.000
Median : 214.0   4711Z :   537   F :   2031   Median : 1.000
Mean   : 212.6   4120Z :   526   C :   1374   Mean   : 1.024
3rd Qu.: 214.0   0143Z :   447   M :   1341   3rd Qu.: 1.000
Max.   : 338.0   4782Z :   430   S :   1268   Max.   : 5.000


(Other):15520   (Other): 5728
empl
Min.   :    5.000
1st Qu.:    5.000
```

```
Median :      5.000
Mean   :      6.336
3rd Qu.:      5.000
Max.   :   1500.000
```

names(firms)

```
 [1] "ID"          "ADDRESS"      "STREET"       "STREET.NO"    "ZIP"
 [6] "CITY_POST"   "CITY"         "region2"      "poviat"       "region3"
[11] "subreg"      "coords.x1"    "coords.x2"    "LEGAL_FORM1"  "LEGAL_FORM2"
[16] "OWNERSHIP"   "PKD7"         "SEC_PKD7"     "GR_EMPL"      "empl"
```

dim(firms)

```
[1] 37378  21
```

A3. Dataset no. 3 / dataset3/ – monthly unemployment rate in poviats (NTS4)

The *unemp2018.csv* dataset is cross-sectional data. The same variable – the registered unemployment rate – is observed in 380 poviats in the following months, from January 2011 to May 2018, which is 89 periods. Data comes from the Central Statistical Office. Data on the unemployment rate is supplemented with information on the statistical codes (*code1, code2*), name of the poviat (*poviat*), order of poviats in the shapefile map (*ID_MAP*), belonging to the subregion (*subregion*) and voivodeship (*region*), subregion number (*subregion_nr*), voivodeship number (*region_nr*), status of the capital of the region (*core_city*), distance of the poviat from the voivodeship city (dist) and the number of inhabitants in 2006 (population_2006) and in 2013 (population_2013). The structure of variable names for subsequent periods of unemployment rate is as follows: in the name X2011.01, part 2011 means the year and part .01 the month.

```
unemp<-read.csv("unemp2018.csv", header=TRUE, dec=",", sep=";")
summary(unemp)
```

```
    code1                   code2                   poviat
 Min.   :1.102e+09      Min.   :201000      Powiat bielski : 2
 1st Qu.:2.245e+09      1st Qu.:1004750       Powiat    brzeski    : 2
 Median :3.235e+09      Median :1636000       Powiat    grodziski  : 2
 Mean   :3.584e+09      Mean   :1720679       Powiat    krośnieński: 2
 3rd Qu.:5.020e+09      3rd Qu.:2475250       Powiat    nowodworski: 2
 Max.   :6.286e+09      Max.   :3263000       Powiat    opolski    : 2
                                              (Other)              :368
 ID_MAP                                             subregion  region
 Min.   :  1.00  Podregion 26 - ostrołęcko-siedlecki: 12 Mazowieckie : 42
 1st Qu.: 95.75  Podregion 08 - włocławski           : 10 Śląskie     : 36
 Median :190.50  Podregion 01 - jeleniogórski        :  9 Wielkopolskie: 35
 Mean   :190.50  Podregion 07 - grudziądzki          :  9 Dolnośląskie : 30
 3rd Qu.:285.25  Podregion 59 - leszczyński          :  9 Podkarpackie : 25
 Max.   :380.00  Podregion 63 - koszaliński          :  9 Lubelskie    : 24
                 (Other)                             :322 (Other)      188
   subregion_nr   region_nr core_city dist
 Min.   : 1.00   Min.  : 2    Min.   :0.00000    Min.   :  0.00
 1st Qu.:15.75   1st Qu.:10   1st Qu.:0.00000    1st Qu.: 35.89
 Median :32.00   Median :16   Median :0.00000    Median : 52.42
 Mean   :32.79   Mean :17     Mean   :0.04474    Mean   : 53.77
 3rd Qu.:50.25   3rd Qu.:24   3rd Qu.:0.00000    3rd Qu.: 71.77
```

```
names(unemp)
```

```
 Max.   : 66.00  Max.   :   32 Max.   : 1.00000  Max.  : 138.48
 population_2006  population_2013 X2011.01 X2011.02
 Min.   :  21235 Min.   :  20891 Min.   : 3.60  Min   : 3.70
 1st Qu.:  54920 1st Qu.:  55723 1st Qu.: 11.80 1st Qu. : 12.07
 Median :  75185 Median :  76369 Median : 15.25 Median  : 15.70
 Mean   :  99231 Mean   :  99915 Mean   : 16.33 Mean    : 16.59
 3rd Qu.: 109821 3rd Qu.: 111048 3rd Qu.: 20.93 3rd Qu. : 21.12
 Max.   :1702139 Max.   :1724404 Max.   : 36.80 Max.    : 37.50
 NA's   :4       NA's   :4
```

```
     X2011.03      X2011.04      X2011.05      X2011.06      X2011.07.
Min.    : 3.80 Min.    : 3.7 Min.    : 3.60 Min.    : 3.40 Min.    : 3.30
1st Qu.:12.00 1st Qu.:11.5 1st Qu.:10.90 1st Qu.:10.55 1st Qu.:10.47
Median :15.60 Median :14.9 Median :14.40 Median :13.90 Median :13.70
Mean    :16.43 Mean    :15.8 Mean    :15.27 Mean    :14.72 Mean    :14.59
3rd Qu.:20.73 3rd Qu.:19.9 3rd Qu.:19.20 3rd Qu.:18.70 3rd Qu.:18.52
Max.    :37.80 Max.    :36.4 Max.    :35.20 Max.    :34.80 Max.    :35.10
```

names(unemp)

```
  [1]  "code1"       "code2"       "poviat"          "ID_MAP"
  [5]  "subregion"   "region"      "subregion_nr"    "region_nr"
  [9]  "core_city"   "dist"        "population_2006" "population_2013"
 [13]  "X2011.01"    "X2011.02"    "X2011.03"        "X2011.04"
 [17]  "X2011.05"    "X2011.06"    "X2011.07"        "X2011.08"
 [21]  "X2011.09"    "X2011.10"    "X2011.11"        "X2011.12"
 [25]  "X2012.01"    "X2012.02"    "X2012.03"        "X2012.04"
 [29]  "X2012.05"    "X2012.06"    "X2012.07"        "X2012.08"
 [33]  "X2012.09"    "X2012.10"    "X2012.11"        "X2012.12"
 [37]  "X2013.01"    "X2013.02"    "X2013.03"        "X2013.04"
 [41]  "X2013.05"    "X2013.06"    "X2013.07"        "X2013.08"
 [45]  "X2013.09"    "X2013.10"    "X2013.11"        "X2013.12"
 [49]  "X2014.01"    "X2014.02"    "X2014.03"        "X2014.04"
 [53]  "X2014.05"    "X2014.06"    "X2014.07"        "X2014.08"
 [57]  "X2014.09"    "X2014.10"    "X2014.11"        "X2014.12"
 [61]  "X2015.01"    "X2015.02"    "X2015.03"        "X2015.04"
 [65]  "X2015.05"    "X2015.06"    "X2015.07"        "X2015.08"
 [69]  "X2015.09"    "X2015.10"    "X2015.11"        "X2015.12"
 [73]  "X2016.01"    "X2016.02"    "X2016.03"        "X2016.04"
 [77]  "X2016.05"    "X2016.06"    "X2016.07"        "X2016.08"
 [81]  "X2016.09"    "X2016.10"    "X2016.11"        "X2016.12"
 [85]  "X2017.01"    "X2017.02"    "X2017.03"        "X2017.04"
 [89]  "X2017.05"    "X2017.06"    "X2017.07"        "X2017.08"
 [93]  "X2017.09"    "X2017.10"    "X2017.11"        "X2017.12"
 [97]  "X2018.01"    "X2018.02"    "X2018.03"        "X2018.04"
[101]  "X2018.05"
```

dim(unemp)

```
#[1] 380 101
```

A4. Dataset no. 4 / dataset4/ – grid data for population

The dataset contains a grid in which the number of Polish residents was counted – women, men and total, in 315,857 cells, each with an area of 1 km². The data comes from the Central Statistical Office portal from the https://geo.stat.gov.pl/inspire platform. The data is in shapefile format – files named *PD_STAT_GRID_CELL_2011* with the extensions shx, shp, sbx, sbn, prj, dbf, cpg.

```
# loading grid for population and converting projections
pop<-readOGR(".", "PD_STAT_GRID_CELL_2011")
pop<-spTransform(pop, CRS("+proj=longlat +datum=NAD83"))

pop.df<-as.data.frame(pop) # extracting data to data.frame
head(pop.df)
```

TOT	TOT_0_14	TOT_15_64	TOT_65	TOT_MALE	TOT_FEM	MALE_0_14	MALE_15_64
0	0	0	0	0	0	0	0
97	15	71	11	51	46	9	38
0	0	0	0	0	0	0	0
0	0	0	0	0	0	0	0
0	0	0	0	0	0	0	0
0	0	0	0	0	0	0	0

	MALE_65	FEM_0_14	FEM_15_64	FEM_65	FEM_RATIO	SHAPE_Leng	SHAPE_Area
0	0	0	0	0	0.00000	2070.834	238411.4
1	4	6	33	7	90.19608	4002.039	1001019.4
2	0	0	0	0	0.00000	4002.403	1001201.8
3	0	0	0	0	0.00000	4002.396	1001198.0
4	0	0	0	0	0.00000	2800.241	310494.5
5	0	0	0	0	0.00000	2009.130	128915.9

```
# extracting grid
pop.grid<-as(pop, "SpatialPolygons")

# conversion to numerical data of subsequent columns of the data set
for(i in 1:15){
pop.df[,i]<-as.numeric(levels(pop.df[, i]))[pop.df[,1]]}

# cutting the grid to the contour of Warsaw, using contour from A5
pow.waw<-pov[pov@data$jpt_nazwa_=="powiat Warszawa",] # contour of Warsaw
lim<-over(pop.grid, pow.waw) # overlay of the grid and contour map
a<-which(lim$jpt_nazwa_=="Warsaw poviat") # lines matching the condition

# conditional grid and data.frame limitation to the selected area
pop.grid.waw<-pop.grid[lim$jpt_nazwa_=="powiat Warszawa",]
pop.df.waw<-pop.df[a,]

# figure - administrative borders and grid, Figure A1a
plot(pop.grid.waw)
plot(pow.waw, add=TRUE)

# figure - values of the examined variable, Figure A1b
library(GISTools)
choropleth(pop.grid.waw, pop.df.waw$TOT)
plot(pow.waw, add=TRUE)
```

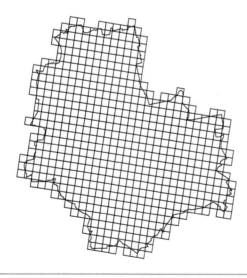
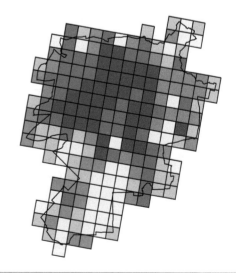

Figure A1 Grid data

Source: Own study

A5. Shapefiles of contour maps – for poviats (NTS4), regions (NTS2), country (NTS0) and registration areas

Shapefile contour maps for poviats (NTS4), voivodeships/regions (NTS2) and country (NTS0) are available for download from the Central Office of Geodesy and Cartography. The maps contain, respectively: 380 territorial units – poviats and cities with poviat rights, corresponding to the NTS4 classification; 16 regions – voivodeships, corresponding to the NTS2 classification; 1 national contour, corresponding to the NTS0 classification; 54,010 registration areas. The data is in shapefile format – files called *powiaty*, *wojewodztwa* and *Panstwo* with the extensions shx, shp, sbx, sbn, prj, dbf, cpg.

```
# map for poviats
pov<-readOGR(".", "powiaty")
pov<- spTransform(pov, CRS("+proj=longlat +datum=NAD83"))
plot(pov)

# map for regions
voi<-readOGR(".", "wojewodztwa")
voi<-spTransform(voi, CRS("+proj=longlat +datum=NAD83"))
plot(voi)

# map for country
pl<-readOGR(".", "Panstwo")
pl<-spTransform(pl, CRS("+proj=longlat +datum=NAD83"))
plot(pl)

# map for registration areas
obreby<-st_read("dane/areas_registration.shp")
obreby %>% filter(str_detect(kod_obrebu, "1465")) %>% ggplot() + geom_sf() +
theme_classic()
```

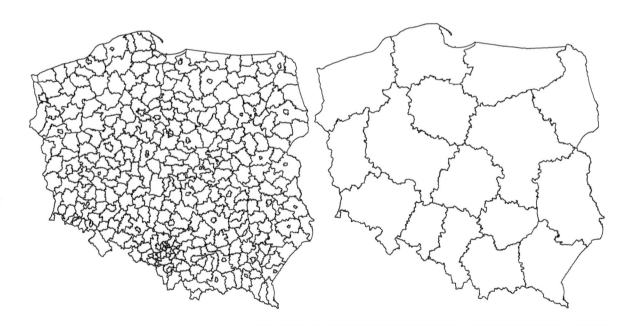

Figure A2 **Contour maps – for subregions (poviats), regions, country and registration areas**
Source: Own study

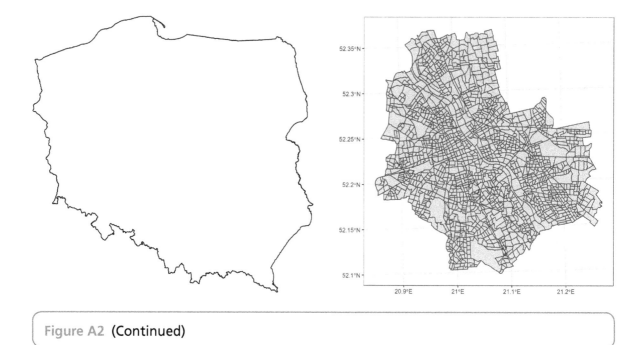

Figure A2 **(Continued)**

A6. Raster data on night light intensity on Earth in 2013

Raster files containing data on the intensity of night lights on Earth are provided by the US National Oceanic and Atmospheric Administration.[1] Light intensity is measured for pixels with dimensions of 30 × 30 arc second. This corresponds to less than 1 square kilometre around the equator and covers an area of about half a square kilometre for the latitude of Poland. For each pixel, the intensity of the night lights is given in units called digital numbers, taking integer values on a scale of 0–63 – standardised separately for each year's data or satellite. The book uses data from the F18 satellite with average measurements of the intensity of night lights for 2013 limited to the area of Europe. This is a raster file with the extension tif called *F182013.v4c_web.stable_lights.avg_vis_PART.tif*.

```
dane_raster<-raster("F182013.v4c_web.stable_lights.avg_vis_PART.tif")
```

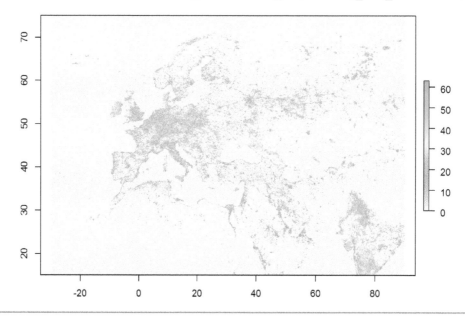

Figure A3 **Raster map of night light intensity in 2013**
Source: Own study

A7. Population in cities in Poland

The *populationxy.csv* dataset contains 30 rows – the largest cities in Poland for which data on population (*total, men, women*) is reported. The variable *voi.* indicates in which region (voivodeship) the city is located. The precise geolocation of cities is given with coordinates *xx* and *yy*.

```
# loading prepared data
populationxy<-read.table("populationxy.csv", sep=";", dec=".",header=TRUE)
populationxy
```

ID	City	Region	total	men	women	xx	yy
1	M.st.Warszawa	MAZOWIECKIE	1744351	800800	943551	52.25900	21.02000
2	Kraków	MAŁOPOLSKIE	761069	354954	406115	50.10220	19.96223
3	Łódź	ŁÓDZKIE	700982	318978	382004	51.77636	19.48418
4	Wrocław	DOLNOŚLĄSKIE	635759	296654	339105	51.11240	17.07775
5	Poznań	WIELKOPOLSKIE	542348	252870	289478	52.40637	16.92517
6	Gdańsk	POMORSKIE	462249	218901	243348	54.25925	18.66179

Note

1 https://ngdc.noaa.gov/eog/dmsp/downloadV4composites.html

Appendix B
Links between packages

The R software, like a typical network project, is created on the basis of available solutions. Authors of packages use other available, previously created packages in their solutions, which increases the efficiency of creating new algorithms. It is worth being aware what the links are between the most important packages discussed in this book. Subsequently we present the links for selected packages.

Analysis of links between packages is enabled by the miniCRAN:: package. The code syntax is as follows. One should remember that the graphic layout of links is generated randomly in each call, so replicating the code will, with high probability, give a slightly different graphic form of the tested links.

```
install.packages("miniCRAN")
library(miniCRAN)
tags<-"sp" # indication of packages for which links are to inspect
pkgDep(tags)
dg<-makeDepGraph(tags, depends=TRUE, suggests=TRUE, enhances=TRUE)
plot(dg, legendPosition=c(-1, 1), vertex.size=10, cex=1.4)
```

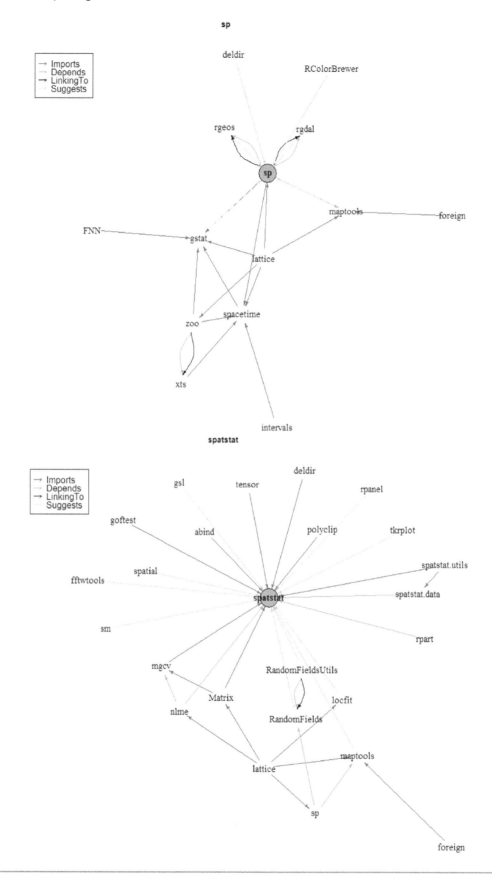

Figure B1 Links within sp:: and spatstat:: packages

Source: Own work with the use of miniCRAN:: package

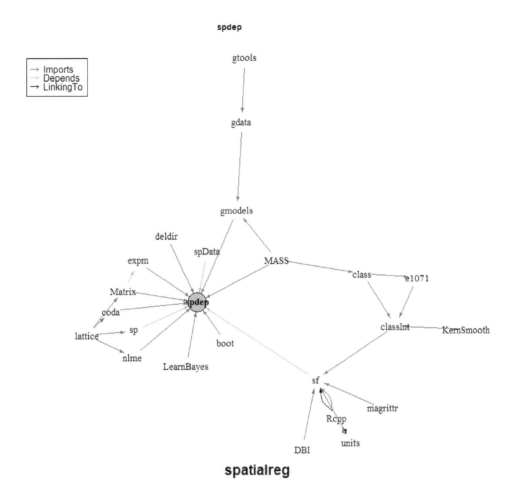

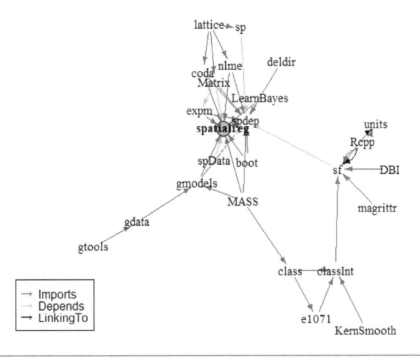

Figure B2 Links within spdep:: and spatialreg:: packages

Source: Own work with the use of miniCRAN:: package

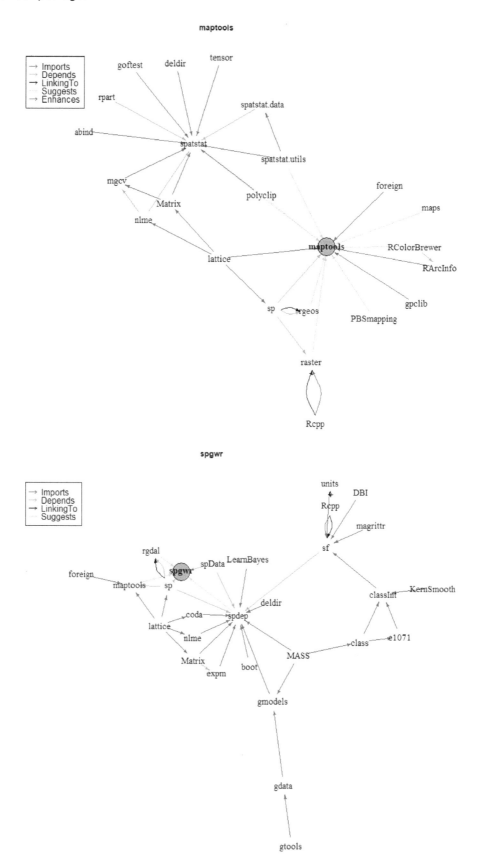

Figure B3 Links within maptools:: and spgwr:: packages

Source: Own work with the use of miniCRAN:: package

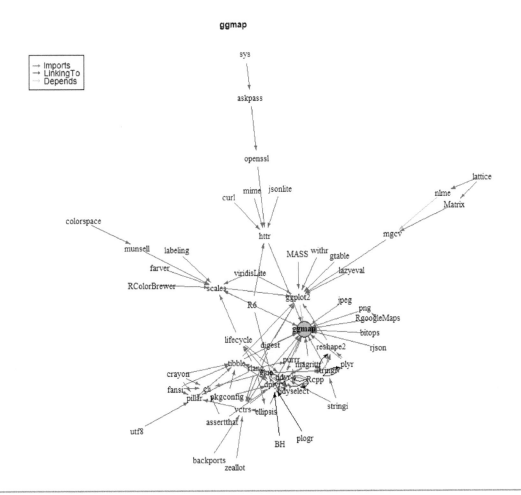

Figure B4 Links within ggmap:: package

Source: Own work with the use of miniCRAN:: package

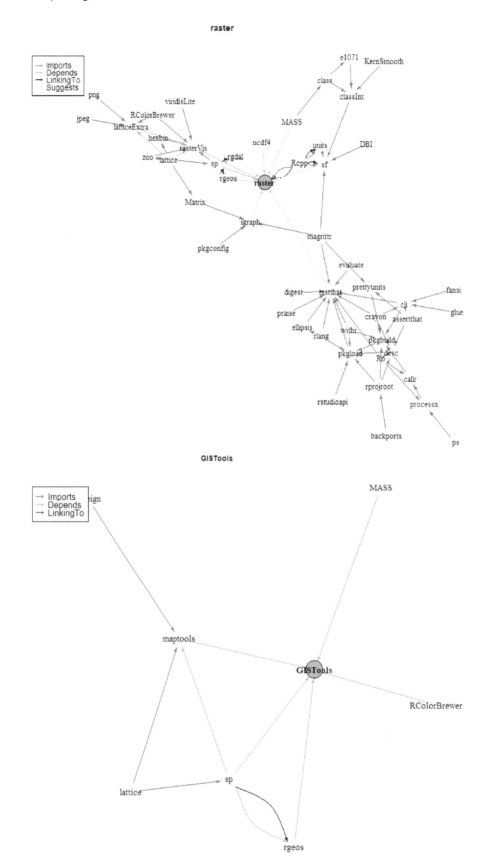

Figure B5 Links within raster:: and GISTools:: packages

Source: Own work with the use of miniCRAN:: package

References

R packages and software resources

Azzimonti, D., Chevalier, C., Ginsbourger, D., Picheny., V. (2018). *KrigInv: Kriging-based inversion for deterministic and noisy computer experiments*. R package version 1.4.1. Retrieved from https://cran.r-project.org/web/packages/KrigInv/KrigInv.pdf

Bachl, F. E., Lindgren, F., Borchers, D. L., & Illian, J. B. (2018). *inlabru: Spatial inference using integrated nested Laplace approximation*. R package version 2.1.9. Retrieved from https://cran.r-project.org/web/packages/inlabru/inlabru.pdf

Baddeley, A., Rubak, E., & Turner, R. (2019). *spatstat: Spatial point pattern analysis, model-fitting, simulation, tests*. R package version 1.58–2. Retrieved from https://cran.r-project.org/web/packages/spatstat/spatstat.pdf

Bivand, R. (2020). *CRAN task view: Analysis of spatial data*. Retrieved from https://cran.r-project.org/web/views/Spatial.html

Bivand, R., Diggle, P., Eglen, S., Petris, G., & Rowlingsson, B. (2017). *splancs: Spatial and space-time point pattern analysis*. R package version 2.01–40. Retrieved from https://cran.r-project.org/web/packages/splancs/splancs.pdf

Bivand, R., & Lewin-Koh, N. (2017). *maptools: Tools for reading and handling spatial objects*. R package version 0.9–2. Retrieved from https://CRAN.R-project.org/package=maptools

Bivand, R., Ripley, B., & Venables, W. (2015). *Spatial: Functions for kriging and point pattern analysis*. R package version 7.3–11. Retrieved from https://cran.r-project.org/web/packages/spatial/spatial.pdf

Bivand, R., & Yu, D. (2017). *spgwr: Geographically weighted regression*. R package version 0.6–32. Retrieved from https://CRAN.R-project.org/package=spgwr

Borchers, H. W. (2018). *pracma: Practical numerical math functions*. R package version 2.1.4. Retrieved from https://CRAN.R-project.org/package=pracma

Brunsdon, C., & Chen, H. (2014). *GISTools: Some further GIS capabilities for R*. R package version 0.7–2.

Bui, R., Buliung, R. N., & Remmel, T. K. (2012). *aspace: A collection of functions for estimating centrographic statistics and computational geometries for spatial point patterns*. R package version 3.2. Retrieved from https://cran.r-project.org/web/packages/aspace/aspace.pdf

Chavent, M., Kuentz-Simonet, V., Labenne, A., & Saracco, J. (2018, December). ClustGeo: An R package for hierarchical clustering with spatial constraints. *Computational Statistics, 33*(4), 1799–1822. https://doi.org/10.1007/s00180-018-0791-1

Chen, J., & Micheas, A. C. (2017). *sppmix: Modeling spatial poisson and related point processes*. R package version 1.0.2. Retrieved from https://cran.r-project.org/web/packages/sppmix/sppmix.pdf

Chen, Jin, Chen, Jun, Kumar, N., & Liang, D. (2019). *ltsk: Local time space kriging*. R package version 1.0.7. Retrieved from https://cran.r-project.org/web/packages/ltsk/ltsk.pdf

Christensen, O. F., & Ribeiro, P. J. (2017). *geoRglm: A package for generalised linear spatial models*. R package version 0.9–11. Retrieved from https://cran.r-project.org/web/packages/geoRglm/geoRglm.pdf

Davies, T. M., Diggle, P. J., Rowlingsson, B. S., & Taylor, B. M. (2018). *lgcp: Log-Gaussian cox process*. R package version 1.5. Retrieved from https://cran.r-project.org/web/packages/lgcp/lgcp.pdf

De La Cruz, M. (2015). *selectspm: Select point pattern models based on minimum contrast, AIC and goodness of fit*. R package version 0.2. Retrieved from https://cran.r-project.org/web/packages/selectspm/selectspm.pdf

De La Cruz, M. (2017). *replicatedpp2w: Two-way ANOVA-like method to analyze replicated point patterns*. R package version 0.1–2. Retrieved from https://cran.r-project.org/web/packages/replicatedpp2w/replicatedpp2w.pdf

De La Cruz, M. (2018). *ecespa: Functions for spatial point pattern analysis*. R package version 1.1–10. Retrieved from https://cran.r-project.org/web/packages/ecespa/ecespa.pdf

Deville, Y., Ginsbourger, D., & Roustant, O. (2018). *DiceKriging: Kriging methods for computer experiments.* R package version 1.5.6. Retrieved from https://cran.r-project.org/web/packages/DiceKriging/DiceKriging.pdf

Diggle, P. J., Gabriel, E., Rodriguez, F. J., & Rowlingsson, B. (2018). *stpp: Space-time point pattern simulation, visualisation and analysis.* R version package 2.0–3. Retrieved from https://cran.r-project.org/web/packages/stpp/stpp.pdf

Diggle, P. J., Giorgi, E., & Johnson, O. (2019). *SDALGCP: Spatially discrete approximation to log-Gaussian Cox processes for aggregated disease count data.* R package version 0.2.0. Retrieved from https://cran.r-project.org/web/packages/SDALGCP/SDALGCP.pdf

Diggle, P. J., & Ribeiro, P. J. (2016). *geoR: Analysis of geostatistical data.* R package version 1.7–5.2.1. Retrieved from https://cran.r-project.org/web/packages/geoR/geoR.pdf

Fox, J. (2019). *The R commander developers' page.* Retrieved from http://socserv.mcmaster.ca/jfox/Misc/Rcmdr/

Fox, J., & Bouchet-Valat, M. (2017). *Getting started with the R commander, version 2.4–0.* Retrieved from http://socserv.mcmaster.ca/jfox/Misc/Rcmdr/

Furrer, R., Nychka, D., Paige, J., & Sain, S. (2018). *fields: Tools for spatial data.* R package version 9.6. Retrieved from https://cran.r-project.org/web/packages/fields/fields.pdf

Gebhardt, A. (2016). *sgeostat: An object-oriented framework for geostatistical modeling in S+.* R package version 1.0–27. Retrieved from https://cran.r-project.org/web/packages/sgeostat/sgeostat.pdf

Gillespie, C., Huang, H.-C., Nychka, D., Tzeng, S., & Wang, W.-T. (2018). *autoFRK: Automatic fixed rank kriging.* R package version 1.0.0. Retrieved from https://cran.r-project.org/web/packages/autoFRK/autoFRK.pdf

Ginsbourger, D., Roustant., O., & Picheny, V. (2016). *DiceOptim: Kriging-based optimization for computer experiments.* R package version 2.0. Retrieved from https://cran.r-project.org/web/packages/DiceOptim/DiceOptim.pdf

Goodman, S. (2007). *R seek.* Custom Google search. Retrieved from http://www.rseek.org/

Goreau, F., Pelissier, R., & Verley, P. (2018). *ads: Spatial point pattern analysis. R package version 1.5–3.* Retrieved from https://cran.r-project.org/web/packages/ads/ads.pdf

Graeler, B., & Pebesma, E. (2019). *gstat: Spatial and spatio-temporal geostatistical modelling, prediction and simulation.* R package version 2.0–0. Retrieved from https://cran.r-project.org/web/packages/gstat/gstat.pdf

Guarderas, P., Lagos, D., & Lopez, A. (2018). *KRIG: Spatial statistic with kriging.* R package version 0.1.0. Retrieved from https://cran.r-project.org/web/packages/KRIG/KRIG.pdf

Hahsler, M., Piekenbrock, M., & Doran, D. (2018). dbscan: Fast density-based clustering with R. *Journal of Statistical Software, 25,* 409–416.

Hammerling, D., Lenssen, N., Nychka., D., Sain, S., & Smirniotis, C. (2018). *LatticeKrig: Multiresolution kriging based on Markov random fields.* R package version 7.0. Retrieved from https://cran.r-project.org/web/packages/LatticeKrig/LatticeKrig.pdf

Hamner, B., Frasco, M., LeDell, E. (2018). *Metrics: Evaluation metrics for machine learning.* R package version 0.1.4. Retrieved from https://cran.r-project.org/web/packages/Metrics/Metrics.pdf

Harte, D. (2007) *Statistical seismology library.* Retrieved from http://www.statsresearch.co.nz/dsh/sslib/

Harte, D. (2017a). *Fractal: Fractal analysis.* R package version 1.4–4. Retrieved from ftp://ftp.gns.cri.nz/pub/davidh/sslib/manuals/fractal.pdf

Harte, D. (2017b). *PtProcess: Time dependent point process modelling.* R package version 3.3–13. Retrieved from https://cran.r-project.org/web/packages/PtProcess/PtProcess.pdf

Hiemstra, P. (2013). *automap: Automatic interpolation package.* R package version: 1.0–14. Retrieved from https://cran.r-project.org/web/packages/automap/automap.pdf

Hino, H., Murata, N., Takano, K., & Yoshikawa, Y. (2017). *mmpp: Various similarity and distance metrics for marked point processes.* R package version 0.6. Retrieved from https://cran.r-project.org/web/packages/mmpp/mmpp.pdf

Hofer, C. (2015). *constrainedKriging: Constrained, covariance-matching constrained and universal point or block kriging.* R package version 0.2.4. Retrieved from https://cran.r-project.org/web/packages/constrainedKriging/constrainedKriging.pdf

Im, H. K., Michaels, K. A., Trubetskoy, V., & Wheeler, H. E. (2016). *OmicKriging: Poly-omic prediction of complex traits. R package version 1.4.0.* Retrieved from https://cran.r-project.org/web/packages/OmicKriging/OmicKriging.pdf

Jean Gaudart, N. G., Coulibaly, D., Barbet, G., Rebaudet, S., Dessay, N., Doumbo, O. K., & Giorgi, R. (2015). SPODT: An R package to perform spatial partitioning. *Journal of Statistical Software, 63*(16), 1–23. Retrieved from http://www.jstatsoft.org/v63/i16/

Johnson, D., Laake, J., VerHoef, J. (2014). *DSpat: Spatial modelling for distance sampling data.* R package version 0.1.6. Retrieved from https://cran.r-project.org/web/packages/DSpat/DSpat.pdf

Kassambara, A., & Mundt, F. (2017). *factoextra: Extract and visualize the results of multivariate data analyses*. R package version 1.0.5. Retrieved from https://CRAN.R-project.org/package=factoextra

Lang, G., Marcon, E., Pucch, F., & Traissae, S. (2019). *dbmss: Distance-based measures of spatial structures*. R package version 2.7–0. Retrieved from https://cran.r-project.org/web/packages/dbmss/dbmss.pdf

Le Gratiet, L. (2012). *MuFiCokriging: Multi-fidelity cokriging models*. R package version 1.2. Retrieved from https://cran.r-project.org/web/packages/MuFiCokriging/MuFiCokriging.pdf

Nakano, J., Saga, M., & Tanaka, U. (2019). *NScluster: Simulation and estimation of the Neyman-Scott type SpatialCluster models*. R package version 1.3.1. Retrieved from https://cran.r-project.org/web/packages/NScluster/NScluster.pdf

Nychka, D., Furrer, R., Paige, J., & Sain, S. (2017). *Fields: Tools for spatial data*. doi:10.5065/D6W957CT, R package version 9.6. Retrieved from www.image.ucar.edu/~nychka/Fields.

Olmedo, O. E. (2014). *Kriging: Ordinary kriging*. R package version 1.1. Retrieved from https://cran.r-project.org/web/packages/kriging/kriging.pdf

Pebesma, E. J., & Bivand, R. (2005). sp: Classes and methods for spatial data. *R. R News, 5*(2). Retrieved from https://cran.r-project.org/doc/Rnews/

Peng, R. D. (2002). *ptproc: Multi-dimensional point process models*. R package version 1.0. Retrieved from http://ftp.uni-bayreuth.de/math/statlib/R/CRAN/doc/packages/ptproc.pdf

Rajala, T. (2017a). *spatgraphs: Graph edge computations for spatial point patterns*. R package version 3.2–1. Retrieved from https://cran.r-project.org/web/packages/spatgraphs/spatgraphs.pdf

Rajala, T. (2017b). *spatialsegregation: Segregation measures for multitype spatial point patterns*. R package version 2.44. Retrieved from https://cran.r-project.org/web/packages/spatialsegregation/spatialsegregation.pdf

Rajala, T. (2019). *SGCS: Spatial graph based clustering summaries for spatial point patterns*. R package version 2.7. Retrieved from https://cran.r-project.org/web/packages/SGCS/SGCS.pdf

Renner, I. (2015). *ppmlasso: Point process models with LASSO penalties*. R package version 1.1. Retrieved from https://cran.r-project.org/web/packages/ppmlasso/ppmlasso.pdf

RStudio, Inc. (2019). *RStudio landing page*. Retrieved from http://www.rstudio.com/products/rstudio/

Schlather, M. et al. (2019). *RandomFields: Simulation and analysis of random fields*. R package version 3.3.6. Retrieved from https://cran.r-project.org/web/packages/RandomFields/RandomFields.pdf

Song, Y. (2018). *SK: Segment-based ordinary kriging and segment-based RegressionKriging for spatial prediction*. R package version 1.1. Retrieved from https://cran.r-project.org/web/packages/SK/SK.pdf

The R Foundation (2019a). *R (programming language)*. Retrieved from https://www.r-project.org/about.html

The R Foundation. (2019b). *R documentation. Class definitions*. Retrieved from https://stat.ethz.ch/R-manual/R-devel/library/methods/html/Classes_Details.html

The R Foundation. (2019c). *R documentation. Classes corresponding to basic data types*. Retrieved from https://stat.ethz.ch/R-manual/R-devel/library/methods/html/BasicClasses.html

The R Foundation. (2019d). *The comprehensive R archive network*. Retrieved from https://cloud.r-project.org/

Walesiak, M., & Dudek, A. (2017). *clusterSim: Searching for optimal clustering procedure for a data set*. R package version 0.47–1. Retrieved from https://CRAN.R-project.org/package=clusterSim

Wang, W. T., & Huang, H. C. (2018). *SpatPCA: Regularized principal component analysis for spatial data*. R package version 1.2.0.0. Retrieved from https://CRAN.R-project.org/package=SpatPCA

Scientific literature

Abramson, I. (1982). On bandwidth variation in kernel estimates – a square root law. *Annals of Statistics, 10*(4), 1217–1223.

Amiri, A., & Gerdtham, U-G. (2011, October 10). *Relationship between exports, imports, and economic growth in France: Evidence from cointegration analysis and Granger causality with using geostatistical models*. MPRA Paper. Retrieved from https://mpra.ub.uni-muenchen.de/34190/

Amiri, A., & Zibaei, M. (2012, February 2). *Granger causality between energy use and economic growth in France with using geostatistical models*. MPRA Paper. Retrieved from https://mpra.ub.uni-muenchen.de/36357/

Amrhein, V., Greenland, S., & McShane, B. (2019, March 20). Scientists rise up against statistical significance. *Nature*. https://www.nature.com/articles/d41586-019-00857-9?fbclid=IwAR3K6PysQ9FY4togs39BSciW3YsK-Pf6EE0Il9R8zxkW4GvrGBHFuz8yF5c

Andrews, D. W., & Buchinsky, M. (2000). A three-step method for choosing the number of bootstrap repetitions. *Econometrica, 68*(1), 23–51.

References

Anselin, L. (1988). Lagrange multiplier test diagnostics for spatial dependence and spatial heterogeneity. *Geographical Analysis, 20*(1), 1–17.

Anselin, L. (1995). Local indicators of spatial association – LISA. *Geographical Analysis, 27*(2), 93–115.

Anselin, L. (1998). GIS research infrastructure for spatial analysis of real estate markets. *Journal of Housing Research, 9*(1), 113–133.

Anselin, L. (1999). Interactive techniques and exploratory spatial data analysis. *Geographical Information Systems: Principles, Techniques, Management and Applications, 1*(1), 251–264.

Anselin, L. (2001). Spatial effects in econometric practice in environmental and resource economics. *American Journal of Agricultural Economics, 83*(3), 705–710.

Anselin, L. (2002). Under the hood issues in the specification and interpretation of spatial regression models. *Agricultural Economics, 27*(3), 247–267.

Anselin, L. (2019). A local indicator of multivariate spatial association: Extending Geary's C. *Geographical Analysis, 51*(2), 133–150.

Anselin, L., & Bao, S. (1997). Exploratory spatial data analysis linking SpaceStat and ArcView. In *Recent developments in spatial analysis* (pp. 35–59). Berlin, Heidelberg: Springer.

Anselin, L., & Bera, A. K. (1998). Introduction to spatial econometrics. In *Handbook of applied economic statistics* (p. 237). New York: Marcel Dekker.

Anselin, L., Bera, A. K., Florax, R., & Yoon, M. J. (1996). Simple diagnostic tests for spatial dependence. *Regional Science and Urban Economics, 26*(1), 77–104.

Anselin, L., & Florax, R. J. (1995). Small sample properties of tests for spatial dependence in regression models: Some further results. In *New directions in spatial econometrics* (pp. 21–74). Berlin, Heidelberg: Springer.

Anselin, L., & Kelejian, H. H. (1997). Testing for spatial error autocorrelation in the presence of endogenous regressors. *International Regional Science Review, 20*(1–2), 153–182.

Arbia, G. (2014). *A primer for spatial econometrics: With applications R.* Basingstoke: Palgrave Macmillan.

Arbia, G., & Espa, G. (1996). *Statistica Economica Territoriale.* Padua: Cedam.

Arbia, G., Espa, G., Giuliani, D., & Mazzitelli, A. (2010). Detecting the existence of space–time clustering of firms. *Regional Science and Urban Economics, Advances in Spatial Econometrics, 40*(5), 311–323. https://doi.org/10.1016/j.regsciurbeco.2009.10.004

Arbia, G., Espa, G., Giuliani, D., & Mazzitelli, A. (2012). Clusters of firms in an inhomogeneous space: The high-tech industries in Milan. *Economic Modelling, Frontiers in Spatial Econometrics Modelling, 29*(1), 3–11. https://doi.org/10.1016/j.econmod.2011.01.012

Arbia, G., Espa, G., & Quah, D. (2008). A class of spatial econometric methods in the empirical analysis of clusters of firms in the space. *Empirical Economics, 34*(1), 81–103. https://doi.org/10.1007/s00181-007-0154-1

Arraiz, I., Drukker, D. M., Kelejian, H. H., & Prucha, I. R. (2010). A spatial Cliff-Ord-type model with heteroscedastic innovations: Small and large sample results. *Journal of Regional Science, 50*, 592–614.

Assuncao, R. M., & Reis, E. A. (1999). A new proposal to adjust Moran's I for population density. *Statistics in Medicine, 18*(16), 2147–2162.

Austin, P. C. (2008). Using the bootstrap to improve estimation and confidence intervals for regression coefficients selected using backwards variable elimination. *Statistics in Medicine, 27*(17), 3286–3300.

Baddeley, A. (2008). Analysing spatial point patterns in R. 171.

Baddeley, A., Diggle, P. J., Hardegen, A., Lawrence, T., Milne, R. K., & Nair, G. (2014). On tests of spatial pattern based on simulation envelopes. *Ecological Monographs, 84*(3), 13.

Baddeley, A., Rubak, E., & Turner, R. (2015). *Spatial point patterns: Methodology and applications.* Boca Raton: R. Chapman and Hall/CRC.

Baddeley, A., Turner, R., Møller, J., & Hazelton, M. (2005). Residual analysis for spatial point processes (with discussion). *Journal of the Royal Statistical Society, B (Methodological), 67*(5), 617–666.

Baller, R. D., Anselin, L., Messner, S. F., Deane, G., & Hawkins, D. F. (2001). Structural covariates of US county homicide rates: Incorporating spatial effects. *Criminology, 39*(3), 561–588.

Baltagi, B. H., Song, S. H., & Koh, W. (2003). Testing panel data regression models with spatial error correlation. *Journal of Econometrics, 117*, 123–150.

Baltagi, B. H., & Yang, Z. (2013). *Non-normality and heteroscedasticity robust LM tests of spatial dependence.* Research Collection of School of Economics, Singapore Management University.

Bao, S. D. (2000). Determination of humic acid content in organic fertilizers. In *Soil and agricultural chemistry analysis* (pp. 438–440). China Agricultural Press Beijing.

Barbian, M. H., & Assunção, R. M. (2017). Spatial subsemble estimator for large geostatistical data. *Spatial Statistics, 22*, 68–88.

Bárcena, M. J., Ménendez, P., Palacios, M. B., & Tusell, F. T. (2014). Alleviating the effect of collinearity in geographically weighted regression. *Journal of Geographical Systems, 16*(4), 441–466.

Barff, R. A. (1987). Industrial clustering and the organization of production: A point pattern analysis of manufacturing in Cincinnati, Ohio. *Annals of the Association of American Geographers*, 77(1), 89–103. https://doi.org/10.1111/j.1467-8306.1987.tb00147.x

Becker, G. S. (1978). *The economic approach to human behavior*. Chicago, IL: University of Chicago Press.

Beenstock, M., & Felsenstein, D. (2012). Nonparametric estimation of the spatial connectivity matrix using spatial panel data. *Geographical Analysis, 44*(4), 386–397.

Bello, A., Reneses, J., Muñoz, A., & Delgadillo, A. (2016). Probabilistic forecasting of hourly electricity prices in the medium-term using spatial interpolation techniques. *International Journal of Forecasting, 32*(3), 966–980. https://doi.org/10.1016/j.ijforecast.2015.06.002

Belsley, D. A. (1991). A guide to using the collinearity diagnostics. *Computer Science in Economics and Management, 4*(1), 33–50.

Benjanuvatra, S., & Burridge, P. (2015). *QML estimation of the spatial weight matrix in the MR-SAR model*. DERS University of York Working Paper.

Berman, M. (1986). Testing for spatial association between a point process and another stochastic process. *Journal of the Royal Statistical Society: Series C (Applied Statistics), 35*(1), 54–62. https://doi.org/10.2307/2347865

Besag, J. (1974). Spatial interaction and the statistical analysis of lattice systems. *Journal of the Royal Statistical Society: Series B (Methodological), 36*(2), 192–225. https://doi.org/10.1111/j.2517-6161.1974.tb00999.x

Besag, J. (1977). Discussion of Dr Ripley's paper. *Journal of the Royal Statistical Society, B, 39*(2), 193–195.

Besag, J., & Diggle, P. J. (1977). Simple Monte Carlo tests for spatial pattern. *Journal of the Royal Statistical Society: Series C (Applied Statistics), 26*(3), 327–333. https://doi.org/10.2307/2346974

Betz, T., Cook, S. J., & Hollenbach, F. M. (2019). Spatial interdependence and instrumental variable models. *Political Science Research and Methods*, 1–16.

Bhattacharjee, A., & Jensen-Butler, C. (2013). Estimation of the spatial weights matrix under structural constraints. *Regional Science and Urban Economics, 43*(4), 617–634.

Bivand, R. (2002). Spatial econometrics functions in R: Classes and methods. *Journal of Geographical Systems, 4*(4), 405–421.

Bivand, R. (2020). CRAN task view: Analysis of spatial data. Retrieved from https://cran.r-project.org/web/views/Spatial.html

Bivand, R., Pebesma, E., & Gomez-Rubio, V. (2013). *Applied spatial data analysis with R* (2nd ed.). New York: Springer. Retrieved from http://www.asdar-book.org/

Blanc-Brude, F., Cookson, G., Piesse, J., & Strange, R. (2014). The FDI location decision: Distance and the effects of spatial dependence. *International Business Review, 23*(4), 797–810.

Bocci, C., & Rocco, E. (2016). Modelling the location decisions of manufacturing firms with a spatial point process approach. *Journal of Applied Statistics, 43*(7), 1226–1239. https://doi.org/10.1080/02664763.2015.1093612

Bottasso, A., Conti, M., Ferrari, C., & Tei, A. (2014). Ports and regional development: A spatial analysis on a panel of European regions. *Transportation Research Part A: Policy and Practice, 65*, 44–55.

Bowyer, A. (1981). Computing Dirichlet tessellations. *The Computer Journal, 24*(2), 162–166.

Brenning, A. (2012). Spatial cross-validation and bootstrap for the assessment of prediction rules in remote sensing: The R package 'sperrorest'. *IEEE International Symposium on Geoscience and Remote Sensing IGARSS*. doi:10.1109/igarss.2012.6352393, https://cran.r-project.org/web/packages/sperrorest/sperrorest.pdf

Briant, A., Combes, P. P., & Lafourcade, M. (2010). Dots to boxes: Do the size and shape of spatial units jeopardize economic geography estimations? *Journal of Urban Economics, 67*(3), 287–302.

Brunsdon, C., & Comber, L. (2015). *An introduction to R for spatial analysis and mapping*. Sage Publications Ltd. https://uk.sagepub.com/en-gb/eur/an-introduction-to-r-for-spatial-analysis-and-mapping/book258267#resources

Brunsdon, C., Charlton, M., & Harris, P. (2012). Living with collinearity in local regression models. In *Proceedings of the 10th international symposium on spatial accuracy assessment in natural resources and environmental sciences*. Brazil.

Brunsdon, C., Fotheringham, A. S., & Charlton, M. E. (1996). Geographically weighted regression: A method for exploring spatial nonstationarity. *Geographical Analysis*, 28(4), 281–298.

Brunsdon, C., Fotheringham, A. S., & Charlton, M. E. (1999). Some notes on parametric significance tests for geographically weighted regression. *Journal of Regional Science, 39*(3), 497–524.

Brus, D. J., & Heuvelink, G. B. M. (2007). Optimization of sample patterns for universal kriging of environmental variables. *Geoderma, 138*(1–2), 86–95. https://doi.org/10.1016/j.geoderma.2006.10.016

Busłowska, E., & Juźwiuk, Ł. (2014). Wprowadzenie do optymalnego wykorzystania MapReduce. *Logistyka*, (4), 3870–3875.

Byrne, G., Charlton, M., & Fotheringham, S. (2009). Multiple dependent hypothesis tests in geographically weighted regression. In *Proceedings of the 10th international conference on GeoComputation*. University of New South Wales.

Campbell, D., Hutchinson, W. G., & Scarpa, R. (2009). Using choice experiments to explore the spatial distribution of willingness to pay for rural landscape improvements. *Environment and Planning A: Economy and Space, 41*(1), 97–111. https://doi.org/10.1068/a4038

Capello, R. (2011). Location, regional growth and local development theories. *AESTIMUM, 58*. Firenze University Press.

Cardinali, S. (2010). *Geomarketing e nuove metriche per un differente approccio alla competizione. Prospettive di ricerca e strumenti di gestione*. Esculapio. https://www.editrice-esculapio.com/cardinali-geomarketing-e-nuove-metriche-per-un-differente-approccio-alla-competizione/

Case, A., Rosen, S. H., & Hines, J. R. (1993). Budget spillovers and fiscal policy interdependence: Evidence from the states. *Journal of Public Economics, 52*(3), 285–307.

Chagas, A. L. S., Toneto, R., & Azzoni, C. R. (2012). A spatial propensity score matching evaluation of the social impacts of sugarcane growing on municipalities in Brazil. *International Regional Science Review, 35*(1), 48–69.

Chaveesuk, R., & Smith, A. E. (2005). Dual kriging: An exploratory use in economic metamodeling. *The Engineering Economist, 50*(3), 247–271. https://doi.org/10.1080/00137910500227182

Chawla, S. D. (2017). 'One-size-fits-all' threshold for P values under fire. Scientists hit back at a proposal to make it tougher to call findings statistically significant. *Nature News*. https://www.nature.com/news/one-size-fits-all-threshold-for-p-values-under-fire-1.22625?WT.mc_id=TWT_NatureNews

Chen, Y. (2015). A new methodology of spatial cross-correlation analysis. *PLOS ONE, 10*(5), e0126158. doi:10.1371/journal.pone.0126158, https://arxiv.org/ftp/arxiv/papers/1503/1503.02908.pdf

Chernick, M. R., & LaBudde, R. A. (2014). *An introduction to bootstrap methods with applications to R*. Hoboken, NJ: John Wiley & Sons.

Chong, Z., Qin, C., & Chen, Z. (2019). Estimating the economic benefits of high-speed rail in China: A new perspective from the connectivity improvement. *Journal of Transport and Land Use, 12*(1).

Clark, I. (2010). Statistics or geostatistics? Sampling error or nugget effect? *Journal of the Southern African Institute of Mining and Metallurgy, 110*(6), 307–312.

Clark, P. J., & Evans, F. C. (1954). Distance to nearest neighbor as a measure of spatial relationships in populations. *Ecology, 35*(4), 445–453.

Cliff, A. D., & Ord, J. K. (1975a). Model building and the analysis of spatial pattern in human geography. *Journal of the Royal Statistical Society: Series B (Methodological), 37*(3), 297–328.

Cliff, A. D., & Ord, J. K. (1975b). Space-time modelling with an application to regional forecasting. *Transactions of the Institute of British Geographers*, 119–128.

Cliff, A. D., & Ord, J. K. (1975c). The comparison of means when samples consist of spatially autocorrelated observations. *Environment and Planning A, 7*(6), 725–734.

Cliff, A. D., & Ord, J. K. (1981). *Spatial processes: Models & applications*. Abingdon: Taylor & Francis.

Cliquet, G. (2006). *Geomarketing: Methods and strategies in spatial marketing*. London: iSTE.

Comber, A., & Harris, P. (2018). Geographically weighted elastic net logistic regression. *Journal of Geographical Systems, 20*(4), 317–341.

Conley, T. G. (1999). GMM estimation with cross sectional dependence. *Journal of Econometrics, 92*(1), 1–45.

Conley, T. G., & Tsiang, G. (1994). *Spatial patterns in labor markets: Malaysian development*. Working Paper, University of Chicago.

Corrado, L., & Fingleton, B. (2012). Where is the economics in spatial econometrics? *Journal of Regional Science, 52*(2), 210–239.

Cousin, A., Maatouk, H., & Rullière, D. (2016). Kriging of financial term-structures. *European Journal of Operational Research, 255*(2), 631–648. https://doi.org/10.1016/j.ejor.2016.05.057

Cox, D. R., & Isham, V. (1980). *Point processes*. Boca Raton: CRC Press.

Cressie, N. A. (1990). The origins of kriging. *Mathematical Geology, 22*(3), 239–252.

Cressie, N. A. (1991). *Statistics for spatial data*. John Wiley and Sons. Retrieved from https://onlinelibrary.wiley.com/doi/abs/10.1002/bimj.4710350210

Cressie, N. A. (1993). *Statistics for spatial data*. New York: John Willy and Sons, Inc.

Cugler, D. C., Oliver, D., Evans, M. R., Shekhar, S., & Medeiros, C. B. (2013). Spatial big data: Platforms, analytics and science. *GeoJournal, 14*, 1–12.

da Barrosa, M. R., Salles, A. V., & de Oliveira Ribeiro, C. (2016). Portfolio optimization through kriging methods. *Applied Economics, 48*(50), 4894–4905. https://doi.org/10.1080/00036846.2016.1167827

Dacey, M. F. (1968). An empirical study of the areal distribution of houses in Puerto Rico. *Transactions of the Institute of British Geographers*, 51–69.

Dapp, T. F., & Heine, V. (2014). *Big data. The untamed force*. Deutsche Bank Research, Frankfurt am Main.

Davison, A. C., Hinkley, D. V., & Young, G. A. (2003). Recent developments in bootstrap methodology. *Statistical Science*, 141–157.

De Castris, M., & Pellegrini, G. (2015). *Neighborhood effects on the propensity score matching*. Working Papers 0515, CREI Università degli Studi Roma Tre, revised 2015.

De Lara, E., LaMarca, A., & Satyanarayanan, M. (2008). *Location systems: An introduction to the technology behind location awareness* (p. 88). Morgan & Claypool Publishers. https://www3.cs.stonybrook.edu/~mdasari/courses/cse570/lamarca-location-awareness-tutorial.pdf

de Lima, M. S., dos Santos, V. S., Duczmal, L. H., & da Silva Souza, D. (2016). A spatial scan statistic for beta regression. *Spatial Statistics, 18*, 444–454.

Dean, J., & Ghemawat, S. (2004). MapReduce: Simplified data processing on large clusters. In *OSDI04: Sixth symposium on operating system design and implementation* (pp. 137–150). San Francisco, CA.

Debarsy, N., & Ertur, C. (2016). Interaction matrix selection in spatial econometrics with an application to growth theory. Retrieved from SSRN 2737402.

Delgado, M. S., & Florax, R. J. (2015). Difference-in-differences techniques for spatial data: Local autocorrelation and spatial interaction. *Economics Letters, 137*, 123–126.

Demšar, U., Harris, P., Brunsdon, C., Fotheringham, A. S., & McLoone, S. (2013). Principal component analysis on spatial data: An overview. *Annals of the Association of American Geographers, 103*(1), 106–128.

DiCiccio, T. J., & Efron, B. (1996). Bootstrap confidence intervals. *Statistical Science*, 189–212.

Diggle, P. J. (1975). Distance methods applied to a semi-deterministic clustering process. *Advances in Applied Probability, 7*(3), 450–451. https://doi.org/10.1017/S0001867800040519

Diggle, P. J. (1978). On parameter estimation for spatial point processes. *Journal of the Royal Statistical Society: Series B (Methodological), 40*(2), 178–181. https://doi.org/10.1111/j.2517-6161.1978.tb01660.x

Diggle, P. J. (1983). *Statistical analysis of spatial point patterns*. London: Academic Press.

Diggle, P. J. (1986). Displaced amacrine cells in the retina of a rabbit: Analysis of a bivariate spatial point pattern. *Journal of Neuroscience Methods, 18*(1), 115–125. https://doi.org/10.1016/0165-0270(86)90115-9

Diggle, P. J., Besag, J., & Timothy Gleaves, J. (1976). Statistical analysis of spatial point patterns by means of distance methods. *Biometrics, 32*(3), 659–667. https://doi.org/10.2307/2529754

Diggle, P. J., & Ribeiro, P. J. (2007). *Model-based geostatistics*. Springer Series in Statistics. New York: Springer-Verlag.

Diggle, P. J., Zheng, P., & Durr, P. (2005). Nonparametric estimation of spatial segregation in a multivariate point process: Bovine tuberculosis in Cornwall, UK. *Journal of the Royal Statistical Society. Series C (Applied Statistics), 54*(3), 645–658.

Doreian, P. (1980). Linear models with spatially distributed data: Spatial disturbances or spatial effects? *Sociological Methods & Research, 9*(1), 29–60.

Dormann, C. F., McPherson, J. M., Araújo, M. B., Bivand, R., Bolliger, J., Carl, G., . . Wilson, R. (2007). Methods to account for spatial autocorrelation in the analysis of species distribution data: A review. *Ecography, 30*, 609–628.

Dubé, J., Legros, D., Thériault, M., & Des Rosiers, F. (2014). A spatial difference-in-differences estimator to evaluate the effect of change in public mass transit systems on house prices. *Transportation Research Part B: Methodological, 64*, 24–40.

Duncan, E. W., White, N. M., & Mengersen, K. (2017). Spatial smoothing in Bayesian models: A comparison of weights matrix specifications and their impact on inference. *International Journal of Health Geographics, 16*(1), 47.

Earnest, A., Morgan, G., Mengersen, K., Ryan, L., Summerhayes, R., & Beard, J. (2007). Evaluating the effect of neighbourhood weight matrices on smoothing properties of conditional autoregressive (CAR) models. *International Journal of Health Geographics, 6*, 54.

Eckey, H.-F., Kosfeld, R., & Werner, A. (2012). Bivariate K functions as instruments to analyze inter-industrial concentration processes. *Jahrbuch Für Regionalwissenschaft, 32*(2), 133–157. https://doi.org/10.1007/s10037-012-0067-0

Efron, B., & Tibshirani, R. J. (1993). *An introduction to the bootstrap*. London: Chapman and Hall.

Efron, B., & Tibshirani, R. J. (1997). Improvements on cross-validation: The 632+ bootstrap method. *Journal of the American Statistical Association, 92*(438), 548–560.

Eldawy, A., & Mokbel, M. F. (2016). The era of big spatial data: A survey. *Foundations and Trends in Databases, 6*(3–4), 163–273.

Elhorst, J. P. (2010). Applied spatial econometrics: Raising the bar. *Spatial Economic Analysis, 5*(1), 9–28.

Elhorst, J. P. (2014a). Dynamic spatial panels: Models, methods and inferences. In *Spatial econometrics* (pp. 95–119). Berlin, Heidelberg: Springer.

Elhorst, J. P. (2014b). Spatial panel data models. In *Spatial econometrics* (pp. 37–93). Berlin, Heidelberg: Springer.

Espa, G., Arbia, G., & Giuliani, D. (2013). Conditional versus unconditional industrial agglomeration: Disentangling spatial dependence and spatial heterogeneity in the analysis of ICT firms' distribution in Milan. *Journal of Geographical Systems, 15*(1), 31–50. https://doi.org/10.1007/s10109-012-0163-2

Ester, M., Kriegel, H. P., Sander, J., & Xu, X. (1996). A density-based algorithm for discovering clusters in large spatial databases with noise. In *KDD* (Vol. 96, No. 34, pp. 226–231).

Fahrmeir, L., & Kneib, T. (2011). Spatial smoothing, interactions and geoadditive regression. In L. Fahrmeir & T. Kneib (Eds.), *Bayesian smoothing and regression for longitudinal, spatial and event history data*. New York: Oxford University Press.

Ferrier, S. (2002). Mapping spatial pattern in biodiversity for regional conservation planning: Where to from here? *Systematic Biology, 51*(2), 331–363. https://doi.org/10.1080/10635150252899806

Feser, E. J., & Sweeney, S. H. (2000). A test for the coincident economic and spatial clustering of business enterprises. *Journal of Geographical Systems, 2*(4), 349–373. https://doi.org/10.1007/PL00011462

Fleming, M. M. (2000). Spatial statistics and econometrics for models in fisheries economics: Discussion. *American Journal of Agricultural Economics, 82*(5), 1207–1209.

Florax, Raymond J. G. M., & Nijkamp, P. (2003). *Misspecification in linear spatial regression models*. Tinbergen Institute Discussion Papers, 2003–081: 3.

Florax, Raymond J. G. M., & Rey, S. (1995). The impacts of misspecified spatial interaction in linear regression models. In *New directions in spatial econometrics* (pp. 111–135). Berlin, Heidelberg: Springer.

Fortin, M. J., Dale, M. R. T., & Ver Hoef, J. (2002). Spatial analysis in ecology. *Encyclopedia of Environmetrics, 4*.

Fotheringham, A. S. (2001). Spatial interaction models. In N. J. Smelser & P. B. Baltes (Eds.), *International encyclopedia of the social & behavioral sciences* (Vol. 11). Amsterdam: Elsevier. https://doi.org/10.1016/B0-08-043076-7/02519-5

Fotheringham, A. S., Brunsdon, C., & Charlton, M. (2003). *Geographically weighted regression: The analysis of spatially varying relationships*. John Wiley & Sons.

Fotheringham, A. S., Crespo, R., & Yao, J. (2015). Geographical and temporal weighted regression (GTWR). *Geographical Analysis, 47*(4), 431–452.

Fotheringham, A. S., Kelly, M. H., & Charlton, M. (2013). The demographic impacts of the Irish famine: Towards a greater geographical understanding. *Transactions of the Institute of British Geographers, 38*(2), 221–237.

Fotheringham, A. S., & O'Kelly, M. E. (1989). *Spatial interaction models: Formulations and applications* (Vol. 1, p. 989). Dordrecht: Kluwer Academic Publishers.

Fotheringham, A. S., & Oshan, T. M. (2016). Geographically weighted regression and multicollinearity: Dispelling the myth. *Journal of Geographical Systems, 18*(4), 303–329.

Fox, J. (2015). *Applied regression analysis and generalized linear models*. Sage Publications.

Franco-Villoria, M., & Ignaccolo, R. (2017). Bootstrap based uncertainty bands for prediction in functional kriging. *Spatial Statistics, 21*, 130–148.

Franzese, R. J., & Hays, J. C. (2007). Spatial econometric models of cross-sectional interdependence in political science panel and time-series-cross-section data. *Political Analysis, 15*(2), 140–164.

Freedman, D. A. (1981). Bootstrapping regression models. *The Annals of Statistics, 9*(6), 1218–1228.

Fry, H. (2018). *Hello world: How to be human in the age of the machine*. Transworld. https://www.amazon.com/Hello-World-Being-Human-Algorithms-ebook/dp/B07BLHQMY9/ref=tmm_kin_swatch_0?_encoding=UTF8&qid=&sr=

Furrer, R. (2006). *KriSp*. Retrieved from https://user.math.uzh.ch/furrer/software/KriSp/

Gargallo, P., Miguel, J. A., & Salvador, M. J. (2018). Bayesian spatial filtering for hedonic models: An application for the real estate market. *Geographical Analysis, 50*(3), 247–279.

Garzon Sandro Rodriguez, B. D. (2014). *Geofencing 2.0: Taking location-based notifications to the next level, conference: 2014 ACM international joint conference on pervasive and ubiquitous computing*. Seattle, WA, USA.

Gatrell, A. C., Bailey, T. C., Diggle, P. J., & Rowlingson, B. S. (1996). Spatial point pattern analysis and its application in geographical epidemiology. *Transactions of the Institute of British Geographers, 21*(1), 256–274. https://doi.org/10.2307/622936

Geary, R. C. (1954). The contiguity ratio and statistical mapping. *The Incorporated Statistician, 5*(3), 115–146.

Getis, A., & Aldstadt, J. (2008). *Constructing the spatial weights matrix using a local statistic., Advances in spatial science: Perspectives on spatial data analysis* (pp. 147–163). Berlin: Spring.

Geyer, C. J. (1999). Likelihood inference for spatial point processes: Likelihood and computation. In *Stochastic geometry: Likelihood and computation* (pp. 141–172). London: Chapman and Hall/CRC.

Gollini, I., Lu, B., Charlton, M., Brunsdon, C., & Harris, P. (2015). GWmodel: An R package for exploring spatial heterogeneity using geographically weighted models. *Journal of Statistical Software, 63*(17), 1–50.

Gómez-Antonio, M., & Sweeney, S. (2018). Firm location, interaction, and local characteristics: A case study for Madrid's electronics sector. *Papers in Regional Science, 97*(3), 663–685.

Gonzales, R. A., Aranda, T. P., & Mendizabal, J. (2017). A Bayesian spatial propensity score matching evaluation of the regional impact of micro-finance. *Review of Economic Analysis, 9*, 127–153.

Goodchild, M. F. (1986). *Spatial autocorrelation*. Norwich: Geo Books, CATMOG, 47.

Goodchild, M. F., Anselin, L., & Deichmann, U. (1993). A framework for the areal interpolation of socioeconomic data. *Environment and Planning A: Economy and Space, 25*(3), 383–397. https://doi.org/10.1068/a250383

Goulard, M., Thibault, L. & Thomas-Agnan, Ch. (2017). About predictions in spatial autoregressive models: Optimal and almost optimal strategies. *Spatial Economic Analysis, 12*(2–3), 304—325. https://doi.org/10.1080/17421772.2017.1300679

Griffith, D. A. (2005). Effective geographic sample size in the presence of spatial autocorrelation. *Annals of the Association of American Geographers, 95*(4), 740–760.

Griffith, D. A. (2008a). Geographic sampling of urban soils for contaminant mapping: How many samples and from where. *Environmental Geochemistry and Health, 30*(6), 495–509.

Griffith, D. A. (2008b). Spatial-filtering-based contributions to a critique of geographically weighted regression (GWR). *Environment and Planning A, 40*(11), 2751–2769.

Griffith, D. A. (2013). Establishing qualitative geographic sample size in the presence of spatial autocorrelation. *Annals of the Association of American Geographers, 103*(5), 1107–1122.

Grolemund, G., & Wickham, H. (2017). *R for data science*. Canada: O'Reilly.

Haining, R. (1990). *Spatial data analysis in the social and environmental sciences* (p. 258). Cambridge: Cambridge University Press.

Haining, R. (1991). Bivariate correlation with spatial data. *Geographical Analysis, 23*(3), 210–227.

Hall, P. (1985). Resampling a coverage pattern. *Stochastic Processes and Their Applications, 20*(2), 231–246.

Hall, P., Horowitz, J. L., & Jing, B. Y. (1995). On blocking rules for the bootstrap with dependent data. *Biometrika, 82*(3), 561–574.

Halls, P. J., Bulling, M., White, P. C. L., Garland, L., & Harris, S. (2001). Dirichlet neighbours: Revisiting Dirichlet tessellation for neighbourhood analysis. *Computers, Environment and Urban Systems, 25*(1), 105–117.

Harris, P., Brunsdon, C., Lu, B., Nakaya, T., & Charlton, M. (2017). Introducing bootstrap methods to investigate coefficient non-stationarity in spatial regression models. *Spatial Statistics, 21*, 241–261.

Harris, P., Stewart Fotheringham, A., & Juggins, S. (2010). Robust geographically weighted regression: A technique for quantifying spatial relationships between freshwater acidification critical loads and catchment attributes. *Annals of the Association of American Geographers, 100*(2), 286–306.

Hartkamp, A. D., De Beurs, K., Stein, A., & White, J.W. (1999). *Interpolation techniques for climate variables*. Geographic Information Systems, Series 99–01, International Maize and Wheat Improvement Center (CIMMYT), Mexico.

Harvey, A. C., & Todd, P. H. J. (1983). Forecasting economic time series with structural and Box-Jenkins models: A case study. *Journal of Business & Economic Statistics, 1*(4), 299–307.

Hejazi, S. A., Jackson, K. R., & Gan, G. (2017, January). *A spatial interpolation framework for efficient valuation of large portfolios of variable annuities*. ArXiv:1701.04134 [q-Fin]. Retrieved from http://arxiv.org/abs/1701.04134

Herrera, M., Ruiz, M., & Mur, J. (2013). Detecting dependence between spatial processes. *Spatial Economic Analysis, 8*(4), 469–497.

Herrera, M., Mur, J., & Ruiz, M. (2019). A comparison study on criteria to select the most adequate weighting matrix. *Entropy, 21*(2), 160.

Hesterberg, T. C. (2015). What teachers should know about the bootstrap: Resampling in the undergraduate statistics curriculum. *The American Statistician, 69*(4), 371–386. Retrieved from https://www.ncbi.nlm.nih.gov/pmc/articles/PMC4784504/

Heuvelink, G. B., Griffith, D. A., Hengl, T., & Melles, S. J. (2012). *Sampling design optimization for space-time kriging* (pp. 207–230). John Wiley. https://doi.org/10(9781118441862)

Hoeting, J. A., Davis, R. A., Merton, A. A., & Thompson, S. E. (2006). Model selection for geostatistical models. *Ecological Applications: A Publication of the Ecological Society of America, 16*(1), 87–98.

Hofer, C., & Papritz, A. (2011). ConstrainedKriging: An R-package for customary, constrained and covariance-matching constrained point or block kriging. *Computers & Geosciences, 37*(10), 1562–1569. https://doi.org/10.1016/j.cageo.2011.02.009

Hopkins, B., & Skellam, J. G. (1954). A new method for determining the type of distribution of plant individuals. *Annals of Botany, 18*(70), 213–227.

Hu, S., Cheng, Q., Wang, L., & Xu, D. (2013, February). Modeling land price distribution using multifractal IDW interpolation and fractal filtering method. *Landscape and Urban Planning, 110*, 25–35. https://doi.org/10.1016/j.landurbplan.2012.09.008

Huang, B., Wu, B., & Barry, M. (2010). Geographically and temporally weighted regression for modelling spatio-temporal variation in house prices. *International Journal of Geographical Information Science, 24*(3), 383–401.

Illian, J. B., Martino, S., Sørbye, S. H., Gallego-Fernández, J. B., Zunzunegui, M., Paz Esquivias, M., & Travis, J. M. J. (2013). Fitting complex ecological point process models with integrated nested Laplace approximation. Edited by David Warton. *Methods in Ecology and Evolution, 4*(4), 305–315. https://doi.org/10.1111/2041-210x.12017.

Illian, J. B., Sørbye, S. H., Rue, H., & Hendrichsen, D. K. (2010). Fitting a log Gaussian Cox process with temporally varying effects – A case study. *Preprint Statistics, 17*. Retrieved from https://www.semanticscholar.org/paper/Fitting-a-log-Gaussian-Cox-process-with-temporally-Illian-S%C3%B8rbye/253a31d20eb19dbb82088ca002c52169d910393b.

INSEE/EUROSTAT. (2018). *Handbook of spatial analysis. Theory and application with R.* Retrieved from https://ec.europa.eu/eurostat/web/products-manuals-and-guidelines/-/INSEE-ESTAT-SPATIAL-ANA?inheritRedirect=true&redirect=%2Feurostat%2Fpublications%2Fmanuals-and-guidelines

Jombart, T. (2015). *A tutorial for the spatial analysis of principal components (sPCA) using adegenet 2.0. 0.* Retrieved from http://adegenet.r-forge.r-project.org/files/tutorial-spca.pdf

Kang, H. (2010). Detecting agglomeration processes using space–time clustering analyses. *The Annals of Regional Science, 45*(2), 291–311. https://doi.org/10.1007/s00168-009-0303-x

Kapoor, M., Kelejian, H. H., & Prucha, I. R. (2007). Panel data model with spatially correlated error components. *Journal of Econometrics, 140*(1), 97–130.

Karahasan, B. (2014). The spatial distribution of new firms: Can peripheral areas escape from the curse of remoteness? *Romanian Journal of Regional Science, 8*(2), 1–28.

Kelejian, H. H., & Piras, G. (2017). *Spatial econometrics.* Academic Press. https://www.elsevier.com/books/spatial-econometrics/kelejian/978-0-12-813387-3

Kelejian, H. H., & Prucha, I. R. (2007*b*). HAC estimation in a spatial framework. *Journal of Econometrics, 140*, 131–154.

Kelejian, H. H., & Prucha, I. R. (2010). Specification and estimation of spatial autoregressive models with autoregressive and heteroscedastic disturbances. *Journal of Econometrics, 157*(1), 53–67.

Kelsall, J. E., & Diggle, P. J. (1998). Spatial variation in risk of disease: A nonparametric binary regression approach. *Journal of the Royal Statistical Society. Series C: Applied Statistics, 47*(4), 559–573.

Kemball, A., & Martinsek, A. (2005). Bootstrap resampling as a tool for radio interferometric imaging fidelity assessment. *The Astronomical Journal, 129*(3), 1760.

Kolak, M., & Anselin, L. (2019). A spatial perspective on the econometrics of program evaluation. *International Regional Science Review*, http://dx.doi.org/10.1177/0160017619869781

Kooijman, S. A. L. M. (1976). Some remarks on the statistical analysis of grids especially with respect to ecology. In *Annals of systems research* (pp. 113–132). Boston, MA: Springer.

Kopczewska, K. (2016). Efficiency of regional public investment: NPV-based spatial econometric approach. *Spatial Economic Analysis, 11*(4), 413–431 http://dx.doi.org/10.1080/17421772.2016.1217346

Kopczewska, K. (2017). Distance-based measurement of agglomeration, concentration and specialisation. In K. Kopczewska, P. Churski, A. Ochojski, & A. Polko (Eds.), *Measuring regional specialisation* (pp. 173–216). Cham: Palgrave Macmillan.

Kopczewska, K. (2020). Spatial bootstrapped microeconometrics: Forecasting for out-of-sample geo-locations in big data. Manuscript submitted for publication.

Kopczewska, K., Churski, P., Ochojski, A., & Polko, A. (2017). *Measuring regional specialisation: A new approach.* Cham: Palgrave Macmillan/Springer.

Kopczewska, K., Churski, P., Ochojski, A., & Polko, A. (2019). *SPAG: Index of spatial agglomeration.* Papers in Regional Science. http://dx.doi.org/10.1111/pirs.12470

Kopczewska, K., Kudła, J., & Walczyk, K. (2017). Strategy of spatial panel estimation: Spatial spillovers between taxation and economic growth. *Applied Spatial Analysis and Policy, 10*(1), 77–102.

Kozak, K. H., Graham, C. H., & Wiens, J. J. (2008). Integrating GIS-based environmental data into evolutionary biology. *Trends in Ecology & Evolution, 23*(3), 141–148. https://doi.org/10.1016/j.tree.2008.02.001

Krige, D. G. (1951). A statistical approach to some basic mine valuation problems on the Witwatersrand. *Journal of the Southern African Institute of Mining and Metallurgy, 52*(6).

Krige, D. G. (1981). *Lognormal-de Wijsian geostatistics for ore evaluation*. Johannesburg: South African Institute of Mining and Metallurgy.

Kuhnert, P. M. (2003). *New methodology and comparisons for the analysis of binary data using Bayesian and tree based methods* (PhD Thesis). Queensland University of Technology.

Kumar, A., Gupta, I., Brandt, J., Kumar, R., Dikshit, A. K., & Patil, R. S. (2016). Air quality mapping using GIS and economic evaluation of health impact for Mumbai city, India. *Journal of the Air & Waste Management Association, 66*(5), 470–481. https://doi.org/10.1080/10962247.2016.1143887

Kunsch, H. R. (1989). The jackknife and the bootstrap for general stationary observations. *The Annals of Statistics*, 1217–1241.

Küpper, A., Bareth, U., & Freese, B. (2011). Geofencing and background tracking–The next features in LBSs. In *Proceedings of the 41th Annual Conference of the Gesellschaft für Informatik eV*.

Lahiri, S. N. (2003). *Resampling methods for dependent data*. New York: Springer.

Laney, D. (2001). *3D data management: Controlling data volume, velocity, and variety*. Retrieved from META Group website: http://blogs.gartner.com/doug-laney/files/2012/01/ad949-3D-Data-Management-Controlling-Data-Volume-Velocity-and-Variety.pdf

Le Gallo, J., López, F. A., & Chasco, C. (2019). Testing for spatial group-wise heteroskedasticity in spatial autocorrelation regression models: Lagrange multiplier scan tests. *The Annals of Regional Science*, 1–26. Retrieved from https://link.springer.com/epdf/10.1007/s00168-019-00919-w?author_access_token=rOCx_vSFXnqW_gCOjTw1c_e4RwlQNchNByi7wbcMAY6T8OW5mApbrPBfX6N5YSF6_Ha3v_4EKPadSd8QgISvdmJsg_uW1RAMf9DNY91CmfaCqQBor-3ytD1TIJ_K0mUPfTOLADmv0uEu2mOmMxQDhw%3D%3D

Le Gallo, J., & Páez, A. (2013). Using synthetic variables in instrumental variable estimation of spatial series models. *Environment and Planning A, 45*(9), 2227–2242.

Lee, L. F., & Yu, J. (2012). QML estimation of spatial dynamic panel data models with time varying spatial weights matrices. *Spatial Economic Analysis, 7*(1), 31–74.

LeSage, J. P. (2014). What regional scientists need to know about spatial econometrics. *The Review of Regional Studies 44*(1), 13–32.

LeSage, J. P., & Pace, R. K. (2009). *Introduction to spatial econometrics*. Boca Raton: Chapman and Hall/CRC.

LeSage, J. P., & Pace, R. K. (2014). The biggest myth in spatial econometrics. *Econometrics, 2*(4), 217–249.

Leung, Y., Mei, C-L., & Zhang, W-X. (2000a). Statistical tests for spatial nonstationarity based on the geographically weighted regression model. *Environment and Planning A, 32*(1), 9–32.

Leung, Y., Mei, C-L., & Zhang, W-X. (2000b). Testing for spatial autocorrelation among the residuals of the geographically weighted regression. *Environment and Planning A, 32*(5), 871–890.

Leutenegger, S. T., Lopez, M. A., & Edgington, J. (1997). STR: A simple and efficient algorithm for r-tree packing. *Proceedings 13th International Conference on Data Engineering*, 497–506.

Li, J., & Heap, A. D. (2008). A review of spatial interpolation methods for environmental scientists. *Geoscience Australia, 23*, 137.

Li, J., Heap, A. D., Potter, A., & Daniell, J. J. (2011). Application of machine learning methods to spatial interpolation of environmental variables. *Environmental Modelling & Software, 26*(12), 1647–1659. https://doi.org/10.1016/j.envsoft.2011.07.004

Li, K., & Lam, N. S. N. (2018). Geographically weighted elastic net: A variable-selection and modelling method under the spatially nonstationary condition. *Annals of the American Association of Geographers, 108*(6), 1582–1600.

Li, Y., & Zhu, K. (2017). Spatial dependence and heterogeneity in the location processes of new high-tech firms in Nanjing, China. *Papers in Regional Science, 96*(3), 519–535. https://doi.org/10.1111/pirs.12202

Liu, R. Y., & Singh, K. (1992). *Exploring the limits of bootstrap* (R. LePage & L. Billard, Eds., p. 225). New York: Wiley.

Loh, J. M. (2008). A valid and fast spatial bootstrap for correlation functions. *The Astrophysical Journal, 681*(1), 726. Retrieved from https://web.njit.edu/~loh/Papers/ApJ.Loh.2008a.pdf

Loh, J. M., & Stein, M. L. (2004). Bootstrapping a spatial point process. *Statistica Sinica*, 69–101.

Loosmore, N. B., & Ford, E. D. (2006). Statistical inference using the g or K point pattern spatial statistics. *Ecology, 87*(8), 1925–1931.

López, F. A., Mur, J., & Angulo, A. (2014). Spatial model selection strategies in a SUR framework. The case of regional productivity in EU. *The Annals of Regional Science, 53*(1), 197–220.

Lovelace, R., Nowosad, J., & Muenchow, J. (2019). *Geocomputation with R*. Boca Raton: CRC Press.

Lu, B., Charlton, M., Brunsdon, C., & Harris, P. (2016). The Minkowski approach for choosing the distance metric in geographically weighted regression. *International Journal of Geographical Information Science, 30*(2), 351–368.

Lu, B., Charlton, M., Harris, P., & Fotheringham, A. S. (2014). Geographically weighted regression with a non-Euclidean distance metric: A case study using hedonic house price data. *International Journal of Geographical Information Science, 28*(4), 660–681

MacKinnon, J. G. (2002). Bootstrap inference in econometrics. *Canadian Journal of Economics/Revue canadienne d'économique, 35*(4), 615–645.

MacKinnon, J. G. (2006). Bootstrap methods in econometrics. *Economic Record, 82*, S2–S18.

Magri, E. V., & Ortiz, J. C. (2000). Estimation of economic losses due to poor blast hole sampling in open pits. *Proceegings of Geostat, 12*.

Manski, C. F. (1993). Identification of endogenous social effects: The reflection problem. *The Review of Economic Studies, 60*(3), 531–542.

Maoh, H., & Kanaroglou, P. (2007). Geographic clustering of firms and urban form: A multivariate analysis. *Journal of Geographical Systems, 9*(1), 29–52. https://doi.org/10.1007/s10109-006-0029-6

Marcon, E., & Puech, F. (2017, January). A typology of distance-based measures of spatial concentration. *Regional Science and Urban Economics, 62*, 56–67. https://doi.org/10.1016/j.regsciurbeco.2016.10.004

Martínez, M. G., Lorenzo, J. M. M., & Rubio, N. G. (2000). Kriging methodology for regional economic analysis: Estimating the housing price in Albacete. *International Advances in Economic Research, 6*(3), 438–450. https://doi.org/10.1007/BF02294963

Martino, L., Elvira, V., & Louzada, F. (2017). Effective sample size for importance sampling based on discrepancy measures. *Signal Processing, 131*, 386–401.

Matheron, G. (1960). Krigeage d'un Panneau Rectangulaire Par Sa Périphérie. *Note Géostatistique, 28*.

McKinney, W. (2017). *Python for data analysis: Data wrangling with Pandas, Numpy, and IPython*. O'Reilly.

Mei, L-M., He, S-Y., & Fang, K.-T. (2004). A note on the mixed geographically weighted regression model. *Journal of Regional Science, 44*(1), 143–157.

Millo, G., & Piras, G. (2012). splm: Spatial panel data models in R. *Journal of Statistical Software, 47*(1), 1–38.

Ministeri, A. (2003). *Evolution of spatial distribution of innovate activity*. Working Paper, Universita Ca Foscari di Venezia.

Mitas, L., & Mitasova, H. (1999). Spatial interpolation. In P. Longley, M. F. Goodchild, D. J. Maguire, & D. W. Rhind (Eds.), *Geographical information systems: Principles, techniques, management and applications* (pp. 481–492). Wiley. Retrieved from https://www.geos.ed.ac.uk/~gisteac/gis_book_abridged/files/ch34.pdf

Møller, J., & Díaz-Avalos, C. (2010). Structured spatio-temporal shot-noise Cox point process models, with a view to modelling forest fires. *Scandinavian Journal of Statistics, 37*(1), 2–25. https://doi.org/10.1111/j.1467-9469.2009.00670.x

Montero, J-M., Chasco, C., & Larraz, B. (2010). Building an environmental quality index for a big city: A spatial interpolation approach combined with a distance indicator. *Journal of Geographical Systems, 12*(4), 435–459. https://doi.org/10.1007/s10109-010-0108-6

Moore, I. D., Grayson, R. B., & Ladson, A. R. (1991). Digital terrain modelling: A review of hydrological, geomorphological, and biological applications. *Hydrological Processes, 5*(1), 3–30. https://doi.org/10.1002/hyp.3360050103

Moran, P. A. P. (1950). Notes on continuous stochastic phenomena. *Biometrika, 37*(1), 17–23. doi:10.2307/2332142. JSTOR 2332142

Müller, S., Wilhelm, P., & Haase, K. (2013). Spatial dependencies and spatial drift in public transport seasonal ticket revenue data. *Journal of Retailing and Consumer Services, 20*(3), 334–348.

Mur, J., López, F., & Herrera, M. (2010). Testing for spatial effects in seemingly unrelated regressions. *Spatial Economic Analysis, 5*(4), 399–440.

Murtagh, F., & Legendre, P. (2011). *Ward's hierarchical clustering method: Clustering criterion and agglomerative algorithm*. arXiv preprint arXiv:1111.6285.

Mutl, J., & Pfaffermayr, M. (2008). *The spatial random effects and the spatial fixed effects model: The Hausman test in a Cliff and Ord panel model* (Economics Series No. 229). Vienna: Institute for Advanced Studies.

Nordman, D. J., Lahiri, S. N., & Fridley, B. L. (2007). Optimal block size for variance estimation by a spatial block bootstrap method. Sankhyā. *The Indian Journal of Statistics*, 468–493.

Oden, N. L. (1984). Assessing the significance of a spatial correlogram. *Geographical Analysis, 16*(1), 1–16.

Ogata, Y., & Stoyan, D. (2008, February). Parameter estimation and model selection for Neyman-Scott point processes. *Biometrical Journal. Biometrische Zeitschrift, 50*, 43–57. https://doi.org/10.1002/bimj.200610339

Oosterhaven, Jan. (2005). Spatial interpolation and disaggregation of multipliers. *Geographical Analysis, 37*(1), 69–84. https://doi.org/10.1111/j.1538-4632.2005.00522.x

Ord, J. K., & Getis, A. (1995). Local spatial autocorrelation statistics: Distributional issues and an application. *Geographical Analysis, 27*(4), 286–306.

Ord, J. K., & Getis, A. (2012). Local spatial heteroscedasticity (LOSH). *The Annals of Regional Science, 48*(2), 529–539.

Otto, P., Schmid, W., & Garthoff, R. (2018). Generalised spatial and spatiotemporal autoregressive conditional heteroscedasticity. *Spatial Statistics, 26*, 125–145.

Pablo-Martí, F., Muñoz-Yebra, C., & Santos, J. L. (2014). *An agent-based model of firm location with different regional policies*. Social Simulation Conference. Retrieved from https://ddd.uab.cat/record/128111

Pace, R. K., & LeSage, J. P. (2008). A spatial Hausman test. *Economics Letters, 101*(3), 282–284.

Pace, R. K., & LeSage, J. P. (2009). A sampling approach to estimate the log determinant used in spatial likelihood problems. *Journal of Geographical Systems, 11*(3), 209–225.

Pace, R. K., LeSage, J. P., & Zhu, S. (2012). Spatial dependence in regressors and its effect on performance of likelihood-based and instrumental variable estimators. *Advances in Econometrics, 30*, 257–295.

Paelinck, J. H. P., & Klaassen, L. H. (1979). *Spatial econometrics. Saxon House Farnborough*. Kiel: Kiel Institute for World Economics.

Páez, A., Farber, S., & Wheeler, D. (2011). A simulation-based study of geographically weighted regression as a method for investigating spatially varying relationships. *Environment and Planning A, 43*(12), 2992–3010.

Pardo-Iguzquiza, E., & Chica-Olmo, M. (2008). Geostatistics with the Matern semivariogram model: A library of computer programs for inference, kriging and simulation. *Computers & Geosciences, 34*(9), 1073–1079. https://doi.org/10.1016/j.cageo.2007.09.020

Pebesma, E. J. (2004). Multivariable geostatistics in S: The Gstat package. *Computers & Geosciences, 30*(7), 683–691.

Pebesma, E. J. (2018). Simple features for R: Standardized support for spatial vector data. *The R Journal, 10*(1), 439–446.

Peng, R. D. (2003). Multi-dimensional point process models in R. *Journal of Statistical Software, 8*(16). https://doi.org/10.18637/jss.v008.i16

Phillips, D. L., Henry Lee, E., Herstrom, A. A., Hogsett, W. E., & Tingey, D. T. (1997). Use of auxiliary data for spatial interpolation of ozone exposure in southeastern forests. *Environmetrics, 8*(1), 43–61. https://doi.org/10.1002/(SICI)1099-095X(199701)8:1<43::AID-ENV237>3.0.CO;2-G

Piras, G. (2010, June). Sphet: Spatial models with heteroskedastic innovations in R. *Journal of Statistical Software, 35*(1).

Piras, G., & Arbia, G. (2007). Convergence in per-capita GDP across EU-NUTS2 regions using panel data models extended to spatial. *Statistica, 67*(2), 157–172.

Poveda, A. C. (2011). Economic development, inequality and poverty: An analysis of urban violence in Colombia. *Oxford Development Studies, 39*(4), 453–468. https://doi.org/10.1080/13600818.2011.620085

Radovanov, B., & Marcikić, A. (2014). A comparison of four different block bootstrap methods. *Croatian Operational Research Review, 5*(2), 189–202.

Rey, S. J., & Montouri, B. D. (1999). US regional income convergence: A spatial econometric perspective. *Regional Studies, 33*(2), 143–156.

Ripley, B. D. (1976). The second-order analysis of stationary point processes. *Journal of Applied Probability, 13*(2), 255–266. https://doi.org/10.2307/3212829

Ripley, B. D. (1977). Modelling spatial patterns. *Journal of the Royal Statistical Society, B (Methodological), 39*(2), 172–192.

Ripley, B. D. (1981). *Spatial statistics*. New York: John Wiley & Sons, Inc. Retrieved from https://www.researchgate.net/publication/2905718_Ripley_B_D_1981_Spatial_statistics_John_Wiley_Sons_New_York

Rodríguez-Pose, A., & Villarreal Peralta, E. M. (2015). Innovation and regional growth in Mexico: 2000–2010. *Growth and Change, 46*(2), 172–195.

Romero, A. A., & Burkey, M. L. (2011). Debt overhang in the eurozone: A spatial panel analysis. *The Review of Regional Studies, 41*, 49–63.

Rubin, D. B. (1974). Estimating causal effects of treatments in randomized and nonrandomized studies. *Journal of Educational Psychology, 66*, 688–701.

References

Scarpa, B., & Azzalini, A. (2004). *Data mining e analisi dei dati*. Springer.

Shmueli, G. (2010). To explain or to predict? *Statistical Science, 25*(3), 289–310.

Sibson, R. (1980). A vector identity for the Dirichlet tessellation. In *Mathematical proceedings of the Cambridge philosophical society* (pp. 151–155). Cambridge: Cambridge University Press.

Snow, John. (1855). *On the mode of communication of cholera*. London: John Churchill.

Strand, G. H. (2017). A study of variance estimation methods for systematic spatial sampling. *Spatial Statistics, 21*, 226–240.

Sun, Z., An, X., Tao, Y., & Hou, Q. (2013). Assessment of population exposure to PM10 for respiratory disease in Lanzhou (China) and its health-related economic costs based on GIS. *BMC Public Health, 13*(1), 891. https://doi.org/10.1186/1471-2458-13-891.

Sweeney, S., & Gómez-Antonio, M. (2016). Localization and industry clustering econometrics: An assessment of Gibbs models for spatial point processes. *Journal of Regional Science, 56*(2), 257–287.

Sweeney, S. H., & Feser, E. J. (1998). Plant size and clustering of manufacturing activity. *Geographical Analysis, 30*(1), 45–64. https://doi.org/10.1111/j.1538-4632.1998.tb00388.x

Tibshirani, R., Walther, G., & Hastie, T. (2001). Estimating the number of clusters in a data set via the gap statistic. *Journal of the Royal Statistical Society: Series B (Statistical Methodology), 63*(2), 411–423.

Tobler, W. R. (1970). A computer movie simulating urban growth in the Detroit region. *Economic Geography, 46*(suppl 1), 234–240.

Tremblay, Y., Shaffer, S. A., Fowler, S. L., Kuhn, C. E., McDonald, B. I., Weise, M. J., . . Costa, D-P. (2006). Interpolation of animal tracking data in a fluid environment. *Journal of Experimental Biology, 209*(1), 128–140. https://doi.org/10.1242/jeb.01970

Triantakonstantis, D., & Stathakis, D. (2014). Cokriging areal interpolation for estimating economic activity using night-time light satellite data. In B. Murgante, S. Misra, Ana Maria A. C. Rocha, C. Torre, J. G. Rocha, M. I. Falcão, . . O. Gervasi (Eds.), *Computational science and its applications – ICCSA 2014* (pp. 243–252). Lecture Notes in Computer Science. Cham: Springer International Publishing.

Turner, R. (2009). Point patterns of forest fire locations. *Environmental and Ecological Statistics, 16*(2), 197–223.

Upton, G., & Fingleton, B. (1985). *Spatial data analysis by example. Volume 1: Point pattern and quantitative data*. Chichester/New York: John Wiley & Sons Ltd.

Van Lieshout, M. N. M., & Baddeley, A. J. (1999). Indices of dependence between types in multivariate point patterns. *Scandinavian Journal of Statistics, 26*(4), 511–532.

Veen, A. Van Der, & Logtmeijer, C. (2005). Economic hotspots: Visualizing vulnerability to flooding. *Natural Hazards, 36*(1), 65–80. https://doi.org/10.1007/s11069-004-4542-y

Vidaurre, D., Bielza, C., & Larrañaga, P. (2012). Lazy lasso for local regression. *Computational Statistics, 27*(3), 531–550.

Villamil, A. C., Bohorquez, M., Giraldo, R., & Mateu, J. (2018). Spatfd: An R package for functional kriging, functional cokriging and optimal sampling of functional data. *Latin American Conference on Statistical Computing (LACSC-2019)*. Guayaquil, Ecuador.

Wackernagel, H. (1995). Ordinary kriging. In *Multivariate geostatistics: An introduction with applications* (H. Wackernagel, Ed., pp. 74–81). Berlin, Heidelberg: Springer. https://doi.org/10.1007/978-3-662-03098-1_11

Wand, M. P., & Jones, C. M. (1995). *Kernel smoothing*. London: Chapman & Hall.

Wang, W. T., & Huang, H. C. (2017). Regularized principal component analysis for spatial data. *Journal of Computational and Graphical Statistics, 26*(1), 14–25.

Wasserstein, R. L., & Lazar, N. A. (2016). The ASA's statement on p-values: Context, process, and purpose. *The American Statistician, 70*(2), 129–133.

Wedel, M., & Kamakura, W. A. (2000). *Market segmentation*. Norwell, MA: Springer.

Wei, Y. D., Yuan, F., & Liao, H. (2013). Spatial mismatch and determinants of foreign and domestic information and communication technology firms in urban China. *The Professional Geographer, 65*(2), 247–264. https://doi.org/10.1080/00330124.2012.679443

Wheeler, D., & Tiefelsdorf, M. (2005). Multicollinearity and correlation among local regression coefficients in geographically weighted regression. *Journal of Geographical Systems, 7*(2), 161–187.

Wheeler, D. C. (2007). Diagnostic tools and a remedial method for collinearity in geographically weighted regression. *Environment and Planning A, 39*, 2464–2481.

Wheeler, D. C. (2009). Simultaneous coefficient penalization and model selection in geographically weighted regression: The geographically weighted lasso. *Environment and Planning A, 41*, 722–742.

Wheeler, D. C., Hammer, D., Kraft, R., Dasgupta, S., & Blankespoor, B. (2013, January). Economic dynamics and forest clearing: A spatial econometric analysis for Indonesia. *Ecological Economics, New Climate Economics, 85*, 85–96. https://doi.org/10.1016/j.ecolecon.2012.11.005.

Wheeler, D. C., Páez, A., Spinney, J., Waller, L. (2014). A Bayesian approach to hedonic price analysis. *Papers in Regional Science, 93*(3), 663–683.

Wheeler, D. C., & Waller, L. (2008). Comparing spatially varying coefficient models: A case study examining violent crime rates and their relationships to alcohol outlets and illegal drug arrests. *Journal of Geographical Systems, 11*(1), 1–22.

Wheeler, H. E., Aquino-Michaels, K., Gamazon, E. R., Trubetskoy, V. V., Dolan, M. E., Huang, R. S., Cox, N. J., & Im, H. K. (2014). Poly-omic prediction of complex traits: OmicKriging. *Genetic Epidemiology, 38*(5), 402–415. https://doi.org/10.1002/gepi.21808.

Wickham, H. (2014). *Advanced R*. New York: Chapman & Hall/CRC The R Series. (chapter: Functional programming).

Wiegand, T., & Moloney, K. A. (2004). Rings, circles, and null-models for point pattern analysis in ecology. *Oikos, 104*(2), 209–229. https://doi.org/10.1111/j.0030-1299.2004.12497.x.

Wilson, A. G. (1971). A family of spatial interaction models, and associated developments. *Environment and Planning A, 3*(1), 1–32.

Wu, B., Li, R., & Huang, B. (2014). A geographically and temporally weighted autoregressive model with application to housing prices. *International Journal of Geographical Information Science, 28*(5), 1186–1204.

Wu, C. F. J. (1986). Jackknife, bootstrap and other resampling methods in regression analysis. *The Annals of Statistics, 14*(4), 1261–1295.

Yan, M., & Ye, K. (2007). Determining the number of clusters using the weighted gap statistic. *Biometrics, 63*(4), 1031–1037.

Yao, X., & Li, G. (2018). Big spatial vector data management: A review. *Big Earth Data, 2*(1), 108–129.

Yoneoka, D., Saito, E., & Nakaoka, S. (2016). New algorithm for constructing area-based index with geographical heterogeneities and variable selection: An application to gastric cancer screening. *Scientific Reports, 6*, 26582.

Online resources

Bennett, K. (2017). *Understanding how big data really is*. Retrieved from https://apextinc.com/blog/defining-big-data/

Dziennik Ustaw Rzeczypospolitej Polskiej. *Rozporządzenie Rady Ministrów z dnia 30 listopada 2015 r. w sprawie sposobu i metodologii prowadzenia i aktualizacji krajowego rejestru urzędowego podmiotów gospodarki narodowej, wzorów wniosków, ankiet i zaświadczeń*. Retrieved from http://prawo.sejm.gov.pl/isap.nsf/download.xsp/WDU20150002009/O/D20152009.pdf

GADM Project (2019). *Mapa GADM*. Retrieved from https://gadm.org/index.html

Geostatystyka w R. (2016). Retrieved from http://www.wbc.poznan.pl/Content/382515/Nowosad_Jakub_Geostatystyka_w_R.pdf

Giraud, T. (2019). *R interface to the photon API*. Retrieved from https://github.com/rCarto/photon

GUGiK (2019). *Państwowy rejestr granic (PRG). główny urząd geodezji i kartografii*. Retrieved from https://gis-support.pl/granice-administracyjne/

Holtz, Y. (2018). *Bubble map*. Retrieved from https://www.r-graph-gallery.com/bubble-map/

inlabru. Retrieved from https://sites.google.com/inlabru.org/inlabru

Marr, B. (2017). *Why space data is the new big data*. Retrieved from https://www.forbes.com/sites/bernardmarr/2017/10/19/why-space-data-is-the-new-big-data/

NOAA (2019). *Version 4 DMSP-OLS Nighttime Lights Time series*. Retrieved from https://ngdc.noaa.gov/eog/dmsp/downloadV4composites.html

OpenStreetMap Contributors (2019a). *Mapa OpenStreetMap*. Retrieved from https://www.openstreetmap.org/

OpenStreetMap Contributors (2019b). *Taginfo. Nominatim*. OSM Search engine. Retrieved from https://taginfo.openstreetmap.org/projects/nominatim#tags

OpenStreetMap Contributors (2019c). *Welcome to nomination*. Retrieved from https://nominatim.openstreetmap.org/

OpenStreetMap Wiki (2019). *Map features*. Retrieved from https://wiki.openstreetmap.org/wiki/Map_Features

References

Overpass Project (2019). *About overpass API*. Retrieved from http://overpass-api.de/.

Overpass Turbo Project (2019). *Mapa overpass turbo*. Retrieved from http://overpass-turbo.eu/.

Pandey, V., Kipf, A., Neumann, T., & Kemper, A. (2018). How good are modern spatial analytics systems? *Proceedings of the VLDB Endowment, 11*(11), 1661–1673. Retrieved from http://www.vldb.org/pvldb/vol11/p1661-pandey.pdf

Pebesma, E. (2019a). *Geometric operations on pairs of simple feature geometry sets*. Retrieved from https://r-spatial.github.io/sf/reference/geos_binary_ops.html

Pebesma, E. (2019b). *Lwgeom. R bindings to the liblwgeom library*. Retrieved from https://github.com/r-spatial/lwgeom/

Pebesma, E. (2019c). *Simple features for R*. Retrieved from https://r-spatial.github.io/sf/articles/sf1.html

Pebesma, E. (2019d). *Spatiotemporal arrays, raster and vector datacubes*. Retrieved from https://github.com/r-spatial/stars

Pebesma, E. (2019e). *Tidyverse methods for sf objects*. Retrieved from https://r-spatial.github.io/sf/reference/tidyverse.html

Pebesma, E., & Bivand, R. (2017). *Spatial indexes coming to sf*. Retrieved from https://www.r-spatial.org/r/2017/06/22/spatial-index.html

Pebesma, E., & Bivand, R. (2019). *Spatial data science*. Retrieved from https://keen-swartz-3146c4.netlify.com/

Posa, R. (2019). *MapReduce algorithm example*. Retrieved from https://www.journaldev.com/8848/mapreduce-algorithm-example

R course on spatial point patterns. Statistical Society of Australia. Lab 10: Gibbs processes. Retrieved from http://spatstat.org/SSAI2017/solutions/solution10.html

REGON. Retrieved from https://bip.stat.gov.pl/en/regon/

RStudio (2019a). *Sparklyr: R interface for apache spark*. Retrieved from https://spark.rstudio.com/

RStudio (2019b). *Tidyverse: R packages for data science*. Retrieved from https://www.tidyverse.org/

Spatial datasets for Poland. Retrieved from https://gis-support.com/spatial-datasets-for-poland/

Sumner, M. D. (2018). *Tabularaster – package vignette*. Retrieved from https://cran.r-project.org/web/packages/tabularaster/vignettes/tabularaster-usage.html

SymbolixAU (2019). *Spatial data table*. Retrieved from https://github.com/SymbolixAU/spatialdatatable/

Terms used in official statistics. Legal form. Retrieved from https://stat.gov.pl/en/metainformations/glossary/terms-used-in-official-statistics/97,term.html

Urząd Geodezji i Kartografii (2019). *Dane udostępniane bez opłat na podstawie ustawy Prawo geodezyjne i kartograficzne*. Retrieved from http://www.gugik.gov.pl/pzgik/dane-udostepniane-bez-oplat

Wassén, O. (2017). *Big data facts – how much data is out there?* Retrieved from https://www.nodegraph.se/big-data-facts/

Wickham, H., François, R., Henry, L., & Müller, K. (2019a). *Dplyr. Overview*. Retrieved from http://dplyr.tidyverse.org/

Wickham, H., François, R., Henry, L., & Müller, K. (2019b). *Introduction to dplyr*. Retrieved from https://cran.r-project.org/web/packages/dplyr/vignettes/dplyr.html

Wikipedia (2019a). *Lidar*. Retrieved from https://en.wikipedia.org/wiki/Lidar/

Wikipedia (2019b). *Ramer–Douglas–Peucker algorithm*. Retrieved from https://en.wikipedia.org/wiki/Ramer%E2%80%93Douglas%E2%80%93Peucker_algorithm

Wikipedia (2019c). *R (programming language)*. Retrieved from https://en.wikipedia.org/wiki/R_(programming_language)

Index

Note: Page numbers in *italic* indicate a figure and page numbers in **bold** indicate a table on the corresponding page.

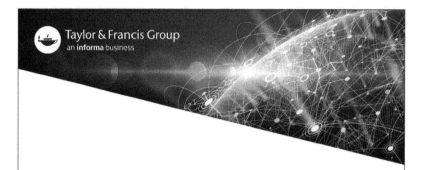